CORNISH

Metalliferous and Associated Minerals 1845–1913

Roger Burt

Peter Waite

Ray Burnley

The University of Exeter

in association with

The Northern Mine Research Society

First published 1987 by the University of Exeter in association with the Northern Mine Research Society

Other publications in this series

The Derbyshire Mineral Statistics 1845–1913
The Yorkshire Mineral Statistics 1845–1913
The Cumberland Mineral Statistics 1845–1913
The Lancashire and Westmorland Mineral Statistics
    with the Isle of Man 1845–1913
The Durham and Northumberland Mineral Statistics 1845–1913
Devon and Somerset Mines 1845–1913
The Mines of Cardiganshire 1845–1913

ISBN 0 85989 287 5

Printed in Great Britain by A. Wheaton & Co., Ltd., Exeter

CONTENTS

PREFACE

This is the seventh volume in our continuing series of county studies of metalliferous mining in the United Kingdom from the mid-nineteenth century to the First World War. It is the second to be published by the University of Exeter and completes the coverage of the South West. The series arises from a project funded by the Social Science Research Council in the Department of Economic History at Exeter to create a computer based data bank of material originally published in the annual volumes of *The Mineral Statistics of the United Kingdom*. We have used the facility of the computer to re-arrange this material in a mine-by-mine format and this book, like the earlier volumes for the northern counties, has been reproduced directly from the output of a high quality printer. The data is stored in a form permitting flexible programming for the internal and cross analysis of the production, ownership, management and employment series, and readers wishing to make use of the data bank, through terminals in Exeter or elsewhere on the network, should write to the Department of Economic History, University of Exeter.

The material presented here is in a very similar form to that which appeared in the original returns, with minimal editorial adjustments. The common prefix "Wheal" has has been dropped to save space but otherwise mine names have been adopted and spelt in a similar way to their most common occurrence in the returns made to the Mining Record Office by the actual owners of the mines; information under the different headings starts and stops just as it did in the original annual publications; and only limited attempts have been made to merge material where the same mine was worked under different names. Problems caused by multiple entries for mines of the same name *but with no clear and readily apparent connection*, have been resolved by recording them as separate numbered entries in these listings, eg. Mount and Indian Queens Nos.1 & 2. These matters are discussed in greater detail in the Introduction. The only important changes and new information that we have introduced relate to locations. Following the precedent first set in the Yorkshire volume, we have changed the locations given in the original returns to more appropriate current usage and have added approximate Ordnance Survey Grid References where known. For some sites, the conglomeration of mines and shafts, many of which were known by several different names, makes accurate location extremely difficult. In such cases several mines have sometimes been given the same Grid Reference. For many mines it has not been possible to be sure of a location and no references have been given. The authors would be very pleased to hear of any errors that may have crept in during these editorial adjustments or of any other major improvements that might be made so that the data bank can be amended.

INTRODUCTION

I

The Mineral Statistics

The official government collection and publication of details of the output of the British mineral industry was instituted in the 1840s under the auspices of the Geological Survey and Museum, directed by Sir Henry de la Beche. We have discussed the development of the ensuing series of annual publications, which have continued down to the present day, at some length elsewhere. It is useful, however, to outline the main stages of that development. The work was conducted by the Mining Record Office, which was initially part of the Geological Survey and Museum, and was effectively the responsibility of Robert Hunt, Keeper of the Mining Records. The first published series, relating to the output of copper and lead, principally in the years 1845 to 1847, appeared in *Memoirs of the Geological Survey of Great Britain and the Museum of Practical Geology in London* Vols I and II, (H.M.S.O. 1845 and 1847). This was regarded at the time as a limited exercise and no commitment was made for the publication of further annual series. In 1853, however, Hunt took the opportunity to update and extend these earlier series in a volume published by the Geological Survey, under his name, entitled *Records of the School of Mines and of Science Applied to the Arts,* Vol. I Pt IV. In that same year a Treasury Committee inquiring into the working of the Geological Survey and Museum reported favourably on the activities of the Mining Record Office and recommended that its activities should be placed, 'on a more regular footing'. This signalled the beginning of the regular collection and publication of a widening range of data relating to mineral extraction and related manufacturing and transportation.

The first of the new annual series, published under the title *Memoirs of the Geological Survey of Great Britain and of the Museum of Practical Geology: Mining Records: The Mineral Statistics of the United Kingdom of Great Britain and Ireland,* appeared in 1855 and was for the year 1853, so continuing an unbroken run for some production data, eg lead and copper, from 1845. Thereafter the volumes were published by the Geological Survey in unbroken series until 1882, usually appearing in the autumn of the year following the record year. Robert Hunt retired following the preparation of the 1881 volume and the opportunity was taken to reorganise and consolidate the now considerably expanded range of data collection. This was done by transferring the Mining Record Office from the Museum of Practical Geology to the Home

Office, where the Mine Inspectors, established by the *Coal Mine Inspection Act* of 1850, had for many years been publishing a similar range of output data in their annual reports. In particular, the *Coal and Metalliferous Mines Regulation Acts* of 1872 had required all active mines in the country to furnish the Mines Inspectorate with details of their output and employment and it was thought wasteful to continue the collection of voluntary returns and computations of the Mining Record Office alongside these.

The first of the new series, which was prepared by Hunt's old staff and closely followed the format of the earlier publications, appeared in 1884 under the title *The Mining and Mineral Statistics of the United Kingdom of Great Britain and Ireland for 1882* and was produced, unlike its earlier counterparts, as a Parliamentary Paper. The following year the title changed again to *Summaries of the Reports of the Inspectors of Mines to Her Majesty's Secretary of State, and the Mineral Statistics of the United Kingdom of Great Britain and Ireland, including Lists of Mines and Mineral Works* and for the years 1884 to 1887 inclusive they appeared annually as *The Mineral and Mining Statistics of The United Kingdom of Great Britain and Ireland, including Lists of Mines and Mineral Worked.* During these years the volumes included returns of a) the quantity and value of all minerals wrought b) the numbers of people employed in and about the mines and open works c) the number of fatal accidents in the mines d) a list of the mine owners, managers and agents e) a list of the recorded plans of abandoned mines that had been deposited at the Home Office f) an appendix showing the production of minerals in the British colonies and possessions.

From 1888 to 1896 the returns appeared as *The Mineral Statistics of the United Kingdom of Great Britain and Ireland with the Isle of Man* and this time the change in title was accompanied by important changes in the content. The details of accidents in the mines, previously held over from the early annual reports of the Inspectors of Mines, were dropped for separate publication, as was *The List of Mines,* a regular appendix of the names of the mine owners, agents, managers and numbers employed, which had been included at the back of *The Mineral Statistics* since 1853 for coal mines, 1859 for most metalliferous mines and 1863 for Iron mines. Finally in 1897 the title and content of the annual returns were changed to a format that they were to keep through to the First World War. For that year the volume appeared under the general title *Mines and Quarries: General Report and Statistics.* It was divided into four separate sections, Part III containing details of output and being subtitled *General Report and Statistics Relating to the Output and Value of the Mineral Raised in the United Kingdom, the Amount and Value of the Metals Produced and the Exports and Imports of Minerals.*

It is notable that until 1897 the 'Clerks of the Mineral Statistics', who had responsibility for preparing the publications, were still largely the same men that had worked with Robert Hunt from the earliest days of the Mining Record Office. This provided a strong element of continuity and an unparalleled wealth of experience in collecting and editing the material. Following Hunt's retirement in 1882, the new Home Office department had been jointly run by Richard Meade and James B. Jordan. Meade had originally been appointed by Hunt as early as 1841 and had been joined by Jordan in 1858. Meade retired from the Home Office in 1889 but Jordan continued until 1897, so consolidating a period of more than half a century's data gathering under three close associates.

II

Cornish Mineral Production

Cornwall was the most important metal mining county in the United Kingdom. It probably had the longest history of continuous production and a total value of output that dwarfed its nearest rivals. Together with associated districts just to the east of the Tamar, it produced nearly all of the country's tin and arsenic and most of its copper. Lead and silver deposits, which dominated the non-ferrous mining of the Pennines and Wales, were less plentiful but the county briefly claimed the country's largest single lead mine – East Wheal Rose – in the mid-1840s. Cornwall produced a wider range of minerals than any other district (it was the only source of some of the rarest minerals) and was the only one to see large-scale mining continuing down to recent times. The only important mineral that the county did not possess in commercial quantities was coal. Although this prevented it from maximising the benefits of its mineral wealth by developing processing and manufacturing industries, the county nevertheless became a leader in the early stages of British industrialisation. During the eighteenth and early nineteenth centuries Cornwall pioneered deep mining and steam pumping technology and its miners and managers were eagerly welcomed in mining districts throughout the world.

The position of Cornwall within the British non-ferrous mining industry in the late nineteenth and early twentieth centuries is shown in Table 1. Estimates of the average annual value of output of the principal non-ferrous ores and associated minerals by decade 1850-1909 suggest that the county was responsible for around half of U.K. production for most of that period. From around 53 per cent of national output in the 1850s, Cornwall's contribution dropped slightly to just over 40 per cent during the third quarter of the century as a rapid decline in its copper and lead output failed to be compensated by rising sales of tin. Thereafter, an increasing concentration on tin and arsenic production began to restore its earlier position. The markets for those metals held up better than for lead/silver, on which most other districts were primarily dependent, and by the beginning of the new century Cornwall was again producing more than half of the country's non-ferrous minerals. In 1900 tin revenues accounted for over 90 per cent of the county's income from metal mining, which was a marked reversal of the situation at the mid-century when copper was dominant and tin accounted for just over a third of the total. The *copper* county had become a *tin* county as mines responded to changing market conditions and exploited new tin deposits encountered at depth in old copper mines. The common progression from copper to tin at depth in the mineralisation of large parts of Cornwall and its commercial demonstration in the mid-nineteenth century, was the critical factor in the survival of Cornish mining during the next hundred years. Lead/silver output was of no real significance in the county after the third quarter of the nineteenth century and zinc, barytes, and fluorspar production, which had provided a brief salvation for lead mining districts in other parts of the U.K., never became of any real importance in Cornwall. Table 1 also clearly shows the very dominant position of Cornwall within the south western mining region as a whole. Somerset had ceased to be a major producer of non-ferrous ores by the end of the eighteenth century and although Devon still counted many important mines – including Devon Great Consols, for sometime the largest copper producer in the world – Cornwall contributed around 87 per cent of the region's total estimated value of output during the third quarter of the nineteenth century. By the 1900s this had increased to over 96 per cent. In quantitative terms, the story of the south west from the mid-nineteenth century is synonymous with the story of Cornwall. It is worth noticing, however, that the considerable additional output from Devon mines during the 1850s increased the region's overall value of output to almost two thirds of the national total, asserting its leading role in the industry even more convincingly.

From the published annual returns, it would appear that there were over fifteen hundred separate mining ventures producing ore of various types in Cornwall between 1845 and 1913. It is important to stress, however, that this does not necessarily mean that there were a similar number of mining sites. Many of the separate commercial ventures worked the same ground over again, as independent mines and as consolidated holdings with other areas of mineralised ground. Mining sets were constantly being given up and re-let. They were redefined, divided and amalgamated. A simple count of mine names or mining companies does not give a reliable figure of separate 'holes in the ground' or areas of independent working. A glance at a geological map of Cornwall shows large areas of intensively mineralised ground, complex vein systems and other deposits, which have been variously worked by different companies, in different combinations, at different times. In general, the changes and improvements in mining techniques which took place during the eighteenth and nineteenth centuries favoured activity on an increasingly large-scale. The multitudinous small and shallow workings of groups of independent miners which had still been very important in the early eighteenth century – each partnership working just one part of one lode – gave way to larger, deeper enterprises, organised by groups of capitalists from inside and outside of the region, to work several vein systems in combination. Increasing returns to scale at many sites, offered by improved pumping, mining and dressing machinery, encouraged further combinations of capital and enterprise in the late nineteenth century, so that whole areas of mineralised ground were worked as a single operation. At the same time, some "exhausted" sections of large workings were sold off and reverted to small independent ventures which might still scrape a living, conduct further exploration, and possibly discover new deposits. Together with other new prospects and discoveries, this had the effect of keeping up the total number of ventures in operation.

In the twentieth century the process of amalgamation has continued unabated, though the smaller marginal ventures have become less common. The history of South Crofty mine, which still operates between Camborne and Redruth, clearly illustrates the continuous process of consolidation and growing scale of operations. Today's mine is a very different enterprise from that which started operations under that name in 1854. It now works an area of ground many times larger than its original leases and includes the sets of many of its once famous and far more illustrious neighbours. Formed in 1854 from three previously independent mines (viz. Longclose, Dudnance and Penhellick), it had a troubled and chequered history down to the beginning of the twentieth century, expanding its operations only slightly. It then began a period of more rapid expansion, mainly at the expense of its neighbours, which gave up operations in the difficult inter-war years. By the late 1960s it controlled a set measuring nearly 6 square miles which included what had been the most productive mines in Cornwall – viz. all of the Carn Breas, Cooks Kitchins, Croftys, Dolcoaths, Pools, Roskears, Setons and Tincrofts, as well as Agar, Harriet, Tehidy and others. The total recorded output of all of the mines in its current set between 1845 and 1913 was well over 800,000 tons of copper concentrate and nearly a quarter of a million tons of tin. This amounts to over a fifth of Cornwall's total recorded output of copper and well over a third of its total tin production. The production of those metals was usefully supported by the sale of a wide range of associated minerals, particularly arsenic. In a similar way, today's Geevor mine has expanded far beyond its nineteenth century boundaries, to include the once highly productive sets of the Boscasswells and Levants, as well as Botallack, Carne, Carnyorth and Pendeen

xi

Consols. In both cases, the scale of the modern mining venture is hardly hinted at by the comparatively puny returns listed under South Crofty and Geevor in the ninteenth century. They inherited some of the most productive mineral ground in Cornwall and clearly showed how apparently exhausted lodes could be reworked to a profit given new, more efficient techniques and a buoyant metal market.

The process of steady amalgamation and consolidation of mines working the richer deposits gave the economic structure of the industry two important characteristics. Firstly, different metals tended to be produced in combination from the same mines, particularly the larger producers. The original process of mineralisation had created a mixture of minerals in many of the lodes and large-scale integrated production often meant raising and processing varying combinations of ores through the same shafts and dressing procedures. The economics of the Cornish mining industry can only be understood in terms of the production of changing "cocktails" of minerals which were sold in different and often diverging metal markets. The income of many mines was rather like that of the mixed farmer – many and various, with perhaps some specialisation within the mixture. Of around 1,530 mines producing metalliferous minerals in the county between the mid-nineteenth century and the First World War, about 380 derived income from the sale of more than one mineral, or approximately a quarter of the total. These multi-mineral producers included all of the major mines and accounted for more than four-fifths of the county's total output. Of the multiple producers, over 260 mines sold both copper and tin ores in varying combinations, frequently in association with arsenic. About 100 mines sold lead ores in association with other minerals, usually silver, zinc, copper and/or tin and occassionally iron or manganese. In general, the only mines that depended for the whole of their income on the production of just one mineral were small developing workings, those that had long-since passed their best, or those that held very small leases of ground. Almost everywhere, large-scale consolidated working of multiple lodes resulted in the production of several different ores, though usually the minerals were not produced in equal proportions and frequently the secondary minerals played only a minor supporting role.

The second major characteristic of the Cornish mining industry was the pronounced pyramidal structure of its production. The process of amalgamation and consolidation of the mines working the richer deposits meant that the bulk of the county's output was derived from a handful of big mines, supported by a larger number of medium-sized producers and a plethora of small workings which came and went on the margins of the industry. This can be demonstrated for the whole of the late nineteenth and twentieth centuries taken together or for shorter sub-periods or individual years. There was clearly a tendency for the number of medium and smaller mines to decline in absolute numbers over the years as metal prices fell and marginal producers found it more difficult to make a living, but the general structure of the industry remained unchanged. In 1855, for example, there were 133 mines producing tin in Cornwall, of which 17 produced over 200 tons of dressed ore, 30 produced 50-199 tons, and 84 produced less than 50 tons. The largest mines were therefore responsible for just over 52 per cent of the total, the medium mines nearly 36 per cent and the small mines just 12 per cent. By 1913 the total number of tin producers in Cornwall had fallen to 60. The effective measure of a large mine had now risen to a production of more than 500 tons of dressed ore and a medium mine to between 100 and 500 tons. The first category included six mines and was responsible for 67 per cent of total output. The medium category included eight mines and was responsible for 25 per cent of output, leaving 46 mines with a production of less than 100 tons, which were responsible for just eight per cent of output. The only noticeable trend was towards an even greater domination of a few large producers with a relatively larger number of small producers within the total. Similar patterns may be shown for the production of other metals but the more rapid decline of their total output tends to distort the general trend. For example, the 97 copper producing mines of 1855 had been reduced to just five by 1913. In 1855 there were six very large producers, with an output of over 5,000 tons of dressed ore and they accounted for nearly 29 per cent of total output. By 1913 there was just one "large" mine, with an output of 383 tons, and that accounted for 91 per cent of output. Similar structural patterns can be shown for the combined mineral output of mines – viz. the total value of the output of tin, copper and other minerals year-by-year – with the largest percentage being derived from a few major producers.

The structure of the industry can also be examined taking the period as a whole, from the mid-nineteenth century to the First World War. This enables the identification of the mines with the largest overall production record. It can be done either by looking at individual metals or by taking the total value of the mines' output of all ores. Returns of the value of tin production began in 1853 and by 1913 well over £39 million of dressed ore had been sold by nearly 700 mines in the county. Of this total, three quarters was derived from just 33 mines, all of which had a tin output in excess of £250,000. Just less than two thirds of the total was produced from 17 mines with an aggregate tin output of over £500,000. The value of copper output was returned from 1845 and provides the longest continuous series. By 1913 the total value of the output of dressed copper ore amounted to nearly £21 million. Of nearly 450 mines responsible for this total, just ten, with an output of over £500,000, accounted for more than a third and 24 mines, each with an aggregate value of production over £250,000 accounted for two thirds. Because some mines were large producers of both copper and tin – viz. Basset, West Basset, and Carn

Brea and Tincroft appeared in both lists – the number of producers dominating the two industries together was less than the aggregate number suggested above. Just 24 mines, with a joint output of copper and tin ore valued in excess of £500,000, were responsible for over half of the total output of those metals in Cornwall during the period.

A similar picture is presented if the total value of the mines' output of all minerals is taken. There are some problems here because the value of the output of some metals, eg. lead, were not regularly returned until the last quarter of the century. However, by simply taking those that are available for lead, arsenic, manganese and other associated minerals and adding them together with those for copper and tin, the number of large mines, producing over £500,000 of saleable ore during the period, is increased to 28. See Appendix A. Taken together, these large mines were responsible for nearly two thirds of the entire mineral output of Cornwall, which had a declared market value of just over £60 million. Just ten of these mines – classifiable as very large producers, with an aggregate output of over 1 million – contributed more than 40 per cent of the total. The medium-sized mines, with a total declared value of output between £100,000 and £500,000, numbered 73 and were responsible for just over a quarter of the county's output. Again, the largest contribution was made by the largest producers in this group. The small mines, with a value of output of all minerals that was less than £100,000, numbered over 1,420. However, they produced only 14 per cent of the total. Figure 1 illustrates the broadly based structure of the industry, both in terms of the number of mines within each production category and, inversely, the contribution of each group to Cornwall's total output.

It has been suggested that the emergence of a few large mines to dominate the Cornish industry was the result of a combination of mineralisation and production characteristics that offered important returns to scale. Richer, more easily worked lodes, economies in pumping, hauling and dressing operations, particularly as mining increased in depth, would suggest that the productivity of the largest mines was greater than that of the medium producers and their's was greater again than the smaller mines. Unfortunately, this is difficult to prove. It is not possible to assemble comprehensive figures of capital investment and for much of the early part of the period there are no reliable figures of employment. However, from 1878 the Mine Inspectors began to publish regular details of employment at all active mines and it is possible to estimate some crude figures of output per employee. This can be done on the basis of either the tonnage of ore produced or its value. Unfortunately, the tonnage figures are of little real value. As already noticed, most mines produced a mixture of different ores and the employment figures are not divided by metal. Dressed copper ore was easier to produce than black tin and considerably higher tonnages per employee could be achieved. Since different

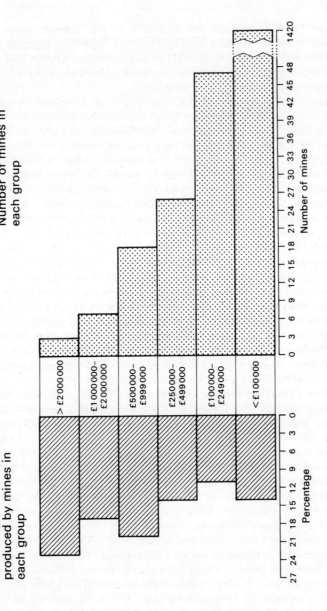

Percentage of total output produced by mines in each group

Number of mines in each group

Percentage

Number of mines

> £2 000 000

£1 000 000–£2 000 000

£500 000–£999 000

£250 000–£499 000

£100 000–£249 000

< £100 000

Fig. 1. The structure of Output from Cornish mines, taking the value of all metals produced 1845–1913.

mines produced different tonnages of copper in their total output, tonnage output could vary considerably without reflecting real differences in the efficiency of working. The only meaningful and comparable measure of output per employee is therefore derived by taking the total value of output, aggregated for all of the minerals produced.

Estimates of the value of output per employee for all large and medium-sized mines listed in Appendix A have been calculated for the years 1880, 1890, 1900, and 1910. They have been divided into three groups; Group 1 is very large producers, with a total value of output of more than £1 million; Group 2 is large producers, with a total value of output of between £500,000 and £1 million: and Group 3 is medium-sized producers, with a total value of output of between £250,000 and £500,000. See Table 2. Only returns for mines that were still working on a significant scale during the survey years (viz. employing more than 50 people) were included, since productivity varied wildly as they were coming into or going out of production. Within each group productivity varied considerably between mines, but some clear patterns do emerge. At the largest mines, the value of output per employee appears to have fallen off noticeably during the difficult years of low tin prices in the 1880s and early 1890s but to have recovered to new high levels by 1900, following reorganisation, rationalisation, and a revival of the market. The first decade of the new century saw a further strengthening of productivity, with the annual value of output per employee standing well above £100 at many mines by 1910. This represented an increase of nearly a third on the levels of 1880, at current prices. In the Group 2 mines, the average value of output per worker also seems to have declined slightly during the 1880s and possibly the early 1890s but to have been recovering by 1900. In 1910 it stood markedly above any earlier figures at about £90 per worker per annum but this may be a statistical aberration, produced by a very small sample. The medium-sized mines in Group 3 saw remarkably little change in the average value of output per worker during the period. The levels of 1880 still applied in 1890 and were hardly changed by 1910. Although these figures are at best only rough approximations, it does seem fairly clear that the value of output per employee was highest in the larger mines and diminished with the scale of operations. In this context it might be noted that it was the gradual concentration of the industry into the hands of larger producers, together with the general process of technological change and increased investment, that was largely responsible for the doubling of average productivity per worker during the period. Taking all types of Cornish metal mining together in 1854, around 36,500 men, women and children achieved a total output valued at around £1.5 million – or just over £40 per worker. In 1910, slightly more than 7,200 workers, now almost entirely adult males, produced metalliferous minerals to a value of about £636,000 – or around £88 per employee.

It is interesting to note, that while productivity per worker varied between mines of different size in Cornwall, there were at least some similarities with mines of the same size in other districts. An earlier volume in this series considered output per employee in Cardiganshire. It was noticed that the production of that county was not so heavily dominated by a few large producers as many other mining fields but was derived from a relatively large number of medium-sized ventures. They were all primarily lead/silver/zinc producers, unlike the copper/tin producers of Cornwall, but in most of the technical aspects of their operation they were very similar to the Cornish mines. Unfortunately, it is not possible to divide them into production categories that are comparable with those for Cornwall, because the value of lead output was not regularly returned until the mid-1870s, which was after their most productive years. However, comparing like with like in terms of the level of employment for spot years after that date produces some very similar levels of value of output per employee. In 1880, for example, productivity per person at Frongoch, the largest Cardiganshire producer, with a labour force of 294 working in a deep shaft and fairly highly capitalised operation, averaged £42.6. This compares closely with the £41.9 averaged at Cooks Kitchin in that same year and the £48.8 at Wheal Owles and £48.5 at Wheal Seton. All three of these medium-large Cornish mines employed similar numbers in similar workings. At Cwmystwyth, a shallow deposit worked through short shafts and adits without much expensive equipment, working was maintained in 1880 and 1890 with an average output per employee of around £30. This was not unlike the £29 average achieved at Pednandrea in 1880 or the £24 average at the Prince of Wales in 1890, both Cornish mines operating with a comparable labour force. Similarly, comparing Cornish mines with those in other parts of the U.K., the high £100 per employee achieved at the immensely productive Dolcoath in 1880, was closely matched by the £103 achieved at the highly capitalised Laxey mines in the Isle of Man in the same year. Equally, the £107 averaged at Greenside mine in Westmorland in 1900 was only a little less than the figure at East Pool and Agar at the turn of the century. The geographical flexibility of investment seems to have had some effect in evening out regional variations in the productivity of labour, but the degree of similarity achieved has yet to be estimated.

It is important, however, not to push these comparisons too far. They may simply be odd coincidence in a wide range of very divergent performances. Average productivity in the mines mentioned above often varied greatly in other years and there were clearly important differences in the productivity of many similar sized mines at any given point in time. It is also important to note that nothing that has been said should be taken to indicate differences in the profitability of different sized mines. Large workings employed more capital than small mines – in the form of underground equipment, steam engines, dressing machinery etc. as well as larger outlays on sinking deep shafts and driving long exploratory levels. The cost of all this investment needed to be paid for in addition to current labour charges and could sometimes be very high. It was therefore not unusual when some smaller mines, with a low value of output per employee, nevertheless returned higher profits on total outlays than some of the larger workings.

The gradual concentration of the Cornish mining industry into the hands of a diminishing number of large enterprises appears to have been achieved without a major change in the methods of ownership and management, at least until the very last years of the century. The cost book form of organisation, developed in the eighteenth century and earlier, proved very flexible and well equal to the task of financing and operating the largest enterprises through the heyday of the industry. As early as the mid-1830s, Sir Charles Lemon had listed five cost book mines that employed over 1,000 men, women and children, with Consolidated and United each employing well in excess of 3,000. They all involved capital investment of tens of thousands of pounds and were among the largest enterprises of the world's leading industrial nation. When a major change did finally take place, with a general shift to joint stock organisation following the tin crisis of the early 1890s, it was precipitated not by the inherent needs of large-scale finance and organisation but by a desire to limit liability and to increase security for an investing public that had become pessimistic about the future of the industry. The change resulted from the new requirements of investors rather than the operating conditions of the mines themselves.

The changing pattern of ownership is clearly documented in the mine returns. There is some problem for the third quarter of the nineteenth century as, unlike other counties, the *List of Mines* inexplicably furnished details only of mine managers and agents. However, from the mid-1870s details of the ownership of most mines in progress began to appear regularly. At that time many mines, particularly the minor producers, appear to have been in the hands of individuals or small partnerships, frequently referred to by their own names. They, like the adventurers adopting a more formal company name, may have been organised on the cost book principal, but this was not made clear. From 1891, however, the *List of Mines* began specifically to identify cost book companies and we have used the initials C.B. in the ownership sections to signify where this was done. Again it is not clear whether all cost book companies were properly identified from that date, but the great majority probably were. The ownership entries for Wheal Owles illustrate these points. The increasing importance of joint stock companies, identifiable by the abbreviation Ltd at the end of the company name, is noticeable from the 1870s. Until the 1890s this form of organisation was largely confined to the smaller, progressive mines and was almost entirely ignored by the larger established producers. Thereafter most of the surviving mines took advantage of the security of limited liability and registered as joint stock companies. There was no inevitable progression, however, with some ventures, such as Park of Mines, going in the other direction and other small workings remaining in private or partnership hands. It should be noticed that the original ownership returns also often included brief comments on the suspension, abandonment of workings etc. These have been included in the mine tables under the ownership comments section, as has necessary cross referencing material.

The management section of the Mine Tables lists in order the main officers of the mining companies, viz. manager, chief agent, and secretary. For Cornwall, the original *List of Mines* also sometimes listed the companies' pursers. In the returns individuals were frequently moved between categories and it would appear that for many mines, particularly the smaller ventures, one person might perform the functions of several offices. By contrast, for some larger ventures, multiple names were returned for the various offices. These have been allocated under the standard headings, with pursers and secretaries sometimes being identified with the abbreviations (P) and (S). Using this name information it is possible to follow the developing careers of managers and agents: to trace their progress from mine to mine: and to examine the wide range of interests that some individuals and families established at given points in time. The sorting facility provided with the computer data base is particularly useful for this work. The family management and promotional partnership of John Taylor and Sons, for example, can be shown to have had a direct involvement with almost 30 Cornish mines during the period, including such well known ventures as Balleswiddan, Polberro, Drakewalls and the Tolguses. From these mines they produced a wide range of different minerals, including copper, tin, lead, iron and arsenic, and they established an important position in the markets for most of them. Thus in the early 1860s, the combined output of mines in which the Taylors held a managerial interest placed them within the top five tin and copper producers in the county. In the 1870s their involvement in the great Herodsfoot mine made them the second largest Cornish lead producer and their control of Restormel Royal made them the county's main iron producers. However, there are limitations to this technique of analysis. The Taylors undoubtedly controlled other mines in the county through a network of agents but without careful information on their names, derived by other research, these mines cannot be readily identified. A similar problem presents itself for the analysis of family involvement in the industry. Were managers of the same surname all members of the same family? A search on the name Hosking, for example, identifies at least eight different individuals of that name involved in the management of 35 different mines. William Henry Hosking, widely known in several other mining districts up and down the country, was connected with ten of those mines. His possible family connection with others of that name must be established from other records as must be the possible degree of family cooperation and collaboration in the operation of their business interests. Nevertheless, the material presented here clearly has important uses in genealogical and business history.

Notwithstanding the growing level of concentration in the Cornish mining industry and the extended ownership and managerial control developed by some individuals and family partnerships, there is no evidence of any group of mines or investors developing an influence over market supply and prices. This was undoubtedly because of the rapid increase of imports from the third quarter of the nineteenth century and the sharply declining market share of domestic producers. Table 3 shows estimates of the total U.K. production, imports and exports of tin and lead at ten yearly intervals 1850–1913 and indicates the quantities available for domestic consumption. The diminishing role of home output is apparent in both sectors. Similar trends can be shown for copper, manganese, zinc and the other non-ferrous metals but it is difficult to calculate accurately the quantities of actual metal in trade from the available official statistics. As indicated above, differences in the timing and scale of the import invasion, as well as variations in the expansion of the industrial demand for the metals, and differences in the geological availability and costs of working the minerals, caused the structure of output to change significantly during the period. Table 1 showed how the relative balance of copper, tin and lead production in Cornwall at the mid-century had been turned into a near total domination of tin output by the beginning of the twentieth century as a result of the more rapid decline of copper and lead. These changes in the production of different minerals may now be taken in turn.

In the middle of the nineteenth century, as during the previous hundred years or more, *copper* was the most important mineral by quantity and value produced in Cornwall. See Table 1. The figures used here were those given in the ticketing returns 1845–1876, distinguishing the mines selling at the Swansea ticketings by (S) and ore sold by private contract by (P). From 1877 to 1881, the returns were compiled by the Mining Record Office from a combination of the Inspectors of Mines' returns and their own calculations of metal and value. From 1882, the figures are those produced by the Mine Inspectors only. In the 1840s and 1850s Cornwall was responsible for well over 80 per cent of British copper production and was probably the most important mining district in the world, accounting for nearly a quarter of total recorded output. Thereafter, however, the district went into dramatic, irreversible decline as its aging mines were unable to compete with rapidly increasing production from the rich, low cost workings of North America, Chile and Spain. Cornish copper production was halved between 1860 and 1870 and halved again during the next decade. Output fell more rapidly than in other parts of the U.K. and by the late 1880s its share of national production stood at only just over 40 per cent. See Table 4. Leadership in the small remaining rump of the industry was now shared almost equally with the mines on the other side of the Tamar, dominated by Devon Great Consols. By the 1890s and 1900s, the long decline had almost levelled off and Cornwall's share again began to climb gradually. This was mainly because its few remaining producers derived the bulk of their income from the production of other minerals, particularly tin, and were able to survive better than mines in other parts of the country. Nevertheless, in the years immediately preceding the First World War copper production in the county stopped almost entirely.

The changing level of copper production in Cornwall was accompanied by important changes in the geographical distribution of output within the county. During the second half of the nineteenth century this amounted to a dramatic decline in the output of the mines in the western districts of the county, with a rising, but not fully compensating, output from the eastern districts. The old traditional centre of the industry was declining while newly discovered eastern deposits were being seriously exploited for the first time. The 1840s and 1850s saw the peak of production from the western workings, which were mainly concentrated in a heavily mineralised area between Bissoe and Camborne. Mines like Consolidated and United, Buller and the Bassets, Carn Brea and Tincroft, that had dominated the county through the early part of the century, still played the leading role. They were powerfully supported by other neighbouring mines, such as East Crofty, North Pool, Seton and North Roskear. The only other important producers, such as Par and Fowey Consols, were to be found in the centre of the county, near St Austell. The mines in the eastern district, clustered mainly around Caradon Hill, were already beginning to make their presence felt but had not yet come fully into their own. Their star was soon to rise, however, paralleling that of Devon Great Consols just across the Devon border. By the end of the 1850s, the western mines still held the lead in output but South and West Caradon and neighbouring Phoenix were now well established among the top ten copper producers. While the output of United Mines of Gwennap, long Cornwall's most important copper producer, had fallen by nearly 20 per cent between 1848 and 1858, that of South Caradon achieved the highest value of copper output in the county in 1858 – a position which it held for the next three years. In the early 1860s the reorganisation of the United, Consolidated and Clifford workings into Clifford Amalgamated again gave west Cornwall the county's largest copper producer but the eastern mines continued to increase their relative importance, with East Caradon and Mark Valley also taking their place among the top ten producers. A revival of Seton and West Seton during the late 1860s also helped to keep up the traditional role of west Cornwall but with sharply falling prices the older mines were rapidly being forced to curtail production. East Cornwall's convenient topography for relatively low cost adit working enabled its mines to compete more effectively and by the early 1870s it had become the main source of output. In 1873, for example, Phoenix, the Caradons, Glasgow Caradon, Marke Valley, Hingston Downs, Prince of Wales, Gunnislake Clitters and Craddock Moor, out of a total of 68 working Cornish copper mines, were responsible for just over half of the county's surviving output. They retained something like this for the remainder of the decade, just ahead of the western district which increasingly relied on West Tolgus, West Seton, Crenver and Abraham, East Pool and particularly a revitalised Mellanear, located away from the main mining area, near Hayle on the north coast. With total output now down to less than a quarter of its mid-century level by volume, Mellanear took over from South Caradon as the largest copper producer in the county in 1879 and gradually consolidated and increased its lead during the 1880s as production elsewhere continued to fall off. In 1883 the east Cornwall mines still accounted for over 40 per cent of output but five years later they had all virtually disappeared. The new mining district that had given so much during the mid-century had ceased to exist as a force in the Cornish mining industry. To the west Mellanear gave way to Levant as the largest single producer in 1889. Sustained mainly by its revenues from tin and arsenic production, Levant continued as the largest copper producer down to the First World War and was responsible for more than two thirds of the county's total output 1890–1913. The far west had completely eclipsed the east as the last bastion of the county's copper production.

Wheras Cornish copper output declined rapidly from the mid-1850s, *tin* production increased until the early 1870s and held fairly steady at a high level until the early 1890s. See Table 5. The figures used here are the Stannary returns 1852–1872; a combination of Stannary and Mine Inspectors' returns collated by the Mining Record Office 1873–1881; and the Mine Inspectors' returns only after 1882. During this period Cornwall became, and remained, primarily a tin mining district. As already suggested this was mainly because many erstwhile copper mines encountered tin mineralisation at depth and switched to the production of the metal. The change was encouraged and facilitated by the relative firmness of tin prices. While both metals saw a major loss of overseas and domestic markets to foreign suppliers, the tin market generally remained stronger, sustained by the expanding tin plate industry. The price of tin actually increased by over 50 per cent during the third quarter of the century, from an average £84 per ton 1846–1850 to £130 per ton 1870–1874. During the same period copper prices remained relatively stable, dropping by a marginal five per cent from an average £87 per ton 1846–1850 to £82 per ton 1870–1874. The production switch which started as a simple response to the increasing profitability of tin became a rout for copper as sharply rising imports caused copper prices to collapse during the next two decades. By the end of the 1880s the price of copper had fallen by nearly a third to an average of £56 per ton 1886–1890, while tin prices held up above £100 and were still 25 per cent above their mid-century level. In 1854 the value of Cornish copper production stood around £1 annually and tin output was only a third of that, at just over £330,000. By 1890 the value of the county's copper production had slumped to just £10,000 while tin output had more than doubled to almost £750,000. Copper mining had become hopelessly uneconomic while tin production still sustained over 70 mines.

The fortunes of the tin industry changed dramatically in the mid-1890s and it too began a dramatic decline. Difficulties in the tin market caused prices to slump by nearly a third between 1890 and 1896 and output was suddenly halved. At one point, at the depths of the crisis, production was suspended in nearly all of the county's mines. Some were reorganised but many never recovered. In 1893 there were 67 mines in Cornwall producing tin but by 1897 only 31 were still making returns. Prices and production remained depressed for the next few years and when the market began to pick up again at the end of the decade, there was no perceptible surge in output. In 1900 tin prices were more than twice their level in 1895 and in the years immediately preceding the First World War they had rocketed to more than £200 per ton but production remained persistently low and unresponsive. See Table 5. By the late 1900s there was again a similar number of mines making returns annually but the output from the larger producers, and particularly the medium-sized mines, failed to pick up. For some reason that sector of the industry that was traditionally the most productive had suddenly lost its sensitivity to price changes. A comparison of tin production in 1892 and 1910 illustrates the change. In both years just over 60 mines made returns and output was dominated by a few large producers. However, in 1892 the seven largest mines, all producing over 500 tons of black tin, achieved a combined output of more than 8,000 tons. In 1910 the six mines in this category had a combined output of only just over 5,000 tons. The decline was even more pronounced in the medium production range. In 1892 there were 15 mines producing between 100 and 500 tons of black tin while in 1910 there were only three. In 1892 the combined output of mines in this production range was over 4,000 tons while in 1910 it was only just over 700 tons. The consequence of the tin crisis of the mid-1890s had been a permanent reduction in the output from Cornwall's largest mines and an almost complete disappearance of its medium-sized producers. This was partly because of the merger of some medium producers into new large consolidated ventures; partly because of the exhaustion of some tin deposits, particularly after their eyes had been picked during the crisis years of the mid-1890s; and partly because of capital starvation resulting from the increasing fascination of mine investors with more profitable foreign ventures. The relative importance of these and other factors has long been debated and undoubtedly varied considerably between different mines. However, the particularly poor performance of the medium-sized mines strongly points towards the paramountcy of the deficiency of capital and enterprise argument. Not all were swallowed up by larger neighbours and it is unlikely that so many deposits should have approached exhaustion at the same time. Rising prices should have given commercial viability to some previously underworked deposits. Here at least Cornwall apppears to have been clearly lacking in promotional drive, resources for development and expansion and, possibly, skilled management and labour. All were haemorrhaging to other mining districts worldwide and depriving the Cornish industry of its previous health and vigour.

Of the several leading Cornish tin producers, one mine, Dolcoath, consistently led the field as the standard bearer for the industry. It was the largest single producer in every year throughout the period. Dolcoath's output grew progressively until the late 1880s; levelled off and remained fairly steady during the difficult 1890s; and, though gradually contracting, continued at a high level through to the First World War. The trebling of its production during the 1850s and 1860s was significant in carrying Cornish tin output to its all-time peak in 1871. In that year Dolcoath produced just less than ten per cent of Cornwall's tin output. In the mid-1880s, with production again more than doubled and at its own peak of output, it was responsible for more than 20 per cent of the county's total. This share rose to well over a third in the mid-1890s as other mines faltered and closed but Dolcoath's production hardly hiccuped from peak levels. In the years before 1914 its tin output was still above that achieved in the prosperous 1870s and notwithstanding increasing depth and production problems, it was still accounting for around a quarter of the tin produced in the U.K.

Next to Dolcoath, the largest tin producers were Carn Brea and Tincroft, East Pool and Agar, Grenville, Basset and Levant. Most of these mines were found leading the production tables in every year during the period. However, they were joined and even surpassed in some years by a number of other large producers which also briefly appeared high in the reckoning. In the 1850s and 1860s, for example, Drakewalls, Great Wheal Vor and Providence mines consistently produced large quantities of tin. Similarly, in the 1870s and 1880s Botallack, South Condurrow, Eliza Consols and Phoenix made important contributions. South and West Francis also rose to prominence in the 1880s and 1890s and West Kitty (St Agnes), which saw a surge in its output from the mid-1880s, continued as one of the few very large producers well into the new century. The late 1890s also saw the emergence of South Crofty and it became well established as a future leader of the industry in the years preceeding the First World War. Nearly all of these mines were located in west Cornwall, particularly around the central Camborne – Redruth area and the neighbouring St Agnes and Lelant districts on the north coast. There was no significant eastward shift of production in the county on the pattern of copper mining. However, the occasional importance of Eliza Consols, Charlestown United, Par Consols and Great Polgooth in the St Austell area, as well as Phoenix, Drakewalls, Hingston Downs, Prince of Wales, Clitters United and others in the Calstock/Gunnislake area clearly indicates that the mineralisation of that part of the county was not unimportant.

The main by-product of Cornish copper and tin mining – rising to the status of primary material at some mines by the end of the century – was *arsenic*. In 1855 the production of refined white arsenic, which was used in the manufacture of pigments, dyes, glass, etc., was already well established. Total output was estimated by the Mining Record Office to be around 1,400 tons per annum. Recorded production was near that level in the late 1860s but then began to expand rapidly, mainly to supply a new export demand for use as a pesticide. The value of output peaked in the early 1880s, fell off, and then recovered to a second, lesser peak in the early 1890s. See Table 6. The increasing income that it produced usefully compensated for the sharp fall in copper revenues during these years and the levelling off in the growth of tin production from the early 1870s. The cushioning effect rapidly began to disappear from the mid-1890s, however, as production and the value of output plummeted. By 1903 production had fallen to less than a quarter of its 1890 level and only two mines received any significant benefit from its sale. Arsenic production had lived in a mutually supporting relationship with tin mining – the revenues from one helping to keep open mines that would also produce the other. When the markets for both products began to be severely undercut by low cost foreign competition at the end of the century, a mutual downward spiral was inevitable. By the early 1900s only a rump of the old industry was left. When both tin and arsenic prices revived in the middle of that decade there were not enough mines still operating to produce a significant increase in the output of either mineral. The downward decline was checked and arsenic production returned to just over 2,000 tons in 1907 but the great days of the industry were finished.

Over the period 1854–1913, nearly 140 mines marketed crude and refined arsenic and/or unprocessed arsenical pyrite. Most of the mines calcined the ore themselves but a few sold the ore to other mines and manufacturers. These were usually small producers or mines with an output that temporarily exceeded the capacity of their own processing plant. Thus around 110 mines marketed white arsenic during the period and just over 60 sold arsenical pyrites. Only a small proportion of these producers were in operation at any one time. In 1873 nearly 30 mines sold arsenic but through the peak years of production in the 1880s the number was always less than 20. The number of mines selling arsenical pyrite was usually less than ten until the 1890s. There was then a sudden increase in this section of the industry as mines everywhere desperately tried to maximise income from any source during the crisis years of the middle of the decade.

Most arsenic production in Cornwall took place in the central western district, between Camborne and Redruth, and in the east of the county, near Callington. These were the principal copper and tin districts and offered the widest facilities for by-product production. In the 1850s and early 1860s, when arsenic production was still at a low level, most output was derived from East Pool, Carn Brea and the Setons in the west and Okel Tor in the east. Prosper United, near Penzance, had a brief period of importance in the mid-1860s but it was the rise of New Great Consols, near Callington, that was primarily responsible for the sudden surge of output in the early 1870s. In the peak production year of 1873, New Great Consols was responsible for nearly a third of total output. East Pool resumed the leadership in production in the mid-1870s to be overtaken by the newcomer to the industry, Holmbush, and an expanded Okel Tor in the early 1880s. Agar and Levant also began to achieve some prominence in this period and by 1887, Agar and its neighbour East Pool were producing almost half of the total output. The last years of the 1880s also saw the rapid rise of arsenic production at Callington United and Drakewalls, and by 1890 the eastern half of the county was contributing well over half of the county total. This was short-lived, however, and by the latter part of that decade control of production was firmly back in the hands of the now amalgamated East Pool and Agar and Carn Brea and Tincroft, usefully supported by Levant. This situation continued down to the First World War, though South Crofty also became of importance in the early years of the new century.

The sale of arsenical pyrites was generally on a much smaller scale than refined arsenic and it was not separately recorded until 1875. The great majority of the pyrite producers were in the eastern district of the county and were led in the late 1870s and early 1880s by New Great Consols, Holmbush and Okel Tor, all three of which also sold refined arsenic. In the 1880s Calstock and Danescombe entered the market on a fairly large scale and were joined briefly at the end of the decade by Queen, located near Calstock. In the 1890s and 1900s Danescombe, Prince of Wales and Trelawney in the east and Great Busy in the west were the main producers, to be superseded by Tamar Consolidated in the years immediately preceding the War.

All-in-all a large number of Cornwall's copper and tin mines derived a useful additional income from the sale of arsenical by-products. It is important to emphasise, however, that the great majority of mines never found these products of lasting significance. They were consequential at some mines in some years and may even have kept a few going during crisis periods which would otherwise have seen their suspension or closure. However, it never succeeded in checking the general decline in the industry precipitated by the fall in tin prices. It must also be stressed that the majority of copper and tin producers never found arsenic in commercially viable quantities. Even some of the largest and otherwise highly diversified producers, such as the Bassets, Phoenix and Grenville, failed to find any commercially viable ore. The great Dolcoath rarely succeeded in producing arsenical products valued at more than a few hundred pounds per year.

The production of *lead* in Cornwall was taking place on a very large scale during the first half of the nineteenth century but collapsed during the third quarter and had dwindled to nothing by the late 1880s. See Table 7. The figures used here were those collected by the Mining Record Office, together with the Stannary returns and, from 1872, the Inspectors of Mines' returns. In 1850 the production of lead in Cornwall was equal to that of Cumberland and second in England only to that of Durham and Northumberland. By the late 1870s, however, it had fallen to the third smallest producer in England, ahead only of its southwestern neighbours, Devon and Somerset. As Table 1 shows, in the 1850s and 1860s, the value of lead and its associated silver, accounted for around one tenth of the total value of all metals sold in Cornwall and it was the third most valuable mineral in production. By the 1880s, with an output down to a few hundred tons per year, lead and silver contributed less than one per cent of the value of all metals and were in fifth position, behind tin, copper, arsenic and zinc. The demise of lead production in Cornwall and the south west appears to have been more the result of exhaustion than adverse price movements. Elsewhere in the U.K. relatively high and buoyant prices encouraged an expansion of production and produced record levels of output in the early 1870s. Nationally lead production only began to decline significantly during the last quarter of the century.

The problem for lead production in Cornwall was that output was heavily concentrated in the hands of a few large producers. When they went into decline there were no others to take their place. In the peak years of production in the county, just one mine reigned supreme. East Wheal Rose, perhaps the most productive mine in England at that time, accounted for 80 per cent of the county's recorded output in 1845. Although its production was already falling off rapidly and had nearly halved by 1850, it still accounted for over 40 per cent of the county's output in that year. East Wheal Rose was supported by a handful of other significant producers, notably Trelawney, Mary Anne, Callington and Herodsfoot. Unlike East Wheal Rose, located near Newlyn in the centre of the county, these other mines were all in the eastern district, close to Callington. In 1850, these four mines had an aggregate output of 4,131 tons of lead ore, which was similar to that of East Wheal Rose and gave the five mines together over 80 per cent of Cornwall's total lead production in that year. The output of Mary Anne and Trelawney held firm during the 1850s and even expanded slightly, whilst that from East Wheal Rose continued its steady decline. By the end of the decade, the reduction in the county's lead production was almost entirely accounted for by the fall-off in the output of East Wheal Rose and that mine had now slipped to third place in the list of major producers. The 1850s also saw the rise of another important producer, Swanpool, while the 1860s saw the emergence of West Chiverton, Cargol and Ludcot, all located in the west of the county. However, these mines could not compensate for the declining fortunes of Mary Anne and Trelawney in the mid-1860s and although there was some overall surge in output at the end of the decade, based on a burst of production from West Chiverton, serious decline was evident from the early 1870s. The fall in lead prices after 1873 greatly increased the industry's difficulties and a further fall in the 1880s virtually finished it off. Taking the period 1845–1913 as a whole, around 170 Cornish mines sold some quantity of lead ore. Just five of these mines – East Wheal Rose, West Chiverton, Mary Anne, Trelawney and Herodsfoot – accounted for over two thirds of total production. When their lodes began to run out, none were found to replace them.

One of the principal problems in sustaining Cornish lead mining was a lack of commercially viable secondary minerals. Old Cornish copper mines had been sustained from the 1870s by their ability to produce increasing quantities of tin; the income of many tin mines was bolstered by their ability to produce and market arsenic. Unfortunately, none of these minerals were found in substantial quantities by the major lead producers. Their only hope was provided by the possibilities of refining *silver* from some lead ores and occasionally by developing associated *zinc* deposits. Sometimes these minerals offered a substantial supplement to income from lead sales but they never succeeded in fully offsetting a decline in lead output. In the case of silver, this was because the metal was contained in the lead ore and the exhaustion of lead deposits meant an associated and equal decline in silver output. Of the lead producers, almost three quarters registered a *silver* content in their ores. They were responsible for almost all of the silver output in Cornwall. Over 40 lead mines also recorded some output of zinc, though their role in the total production of that metal was less complete. The assayed silver content of lead ores was reported in the Mineral Statistics from 1852 and its importance to the income of some mines can be seen throughout their productive life. In 1852, for example, the lead ores produced by East Wheal Rose were reported to contain 48,000 oz. of silver which, at around five shillings an oz., increased the total value of the lead by around £12,000. With total receipts from all sales in that year of around £20,000, silver had an almost equal role with lead in the economics of the enterprise. Similarly, at the peak of its output in 1870, West Chiverton produced 3,582 tons of lead which was estimated to contain over 160,000 oz. of silver, eg. 45 oz. to the ton. With refining becoming profitable at anything over 5 oz. per ton, this represented a very considerable extra income to the mine. Again the values of the lead and silver contents of the ores were very similar, at about £40,000 each. However, some caution must be exercised in using these calculations. Although the proportionate importance of the various minerals probably remained the same, it must be emphasised that the mines did not receive the full market values of the ores that they produced. Clearly allowance was made for transport, reduction, refining and other processing costs. For example, in 1873 West Chiverton sold 2,224 tons of lead ore with an estimated lead content of 1,668 tons of lead and 70,056 oz. of silver. The average price of lead in that year was £15-8-0 per ton and silver stood at about five shillings an ounce. This would suggest a potential market value for the metals of £43,201. However, the mine received only £29,929, with the other third of market value going to the merchants and manufacturers. On this estimate the mines received only just over two thirds of the market value of the metals they produced, the remainder going to the merchants and manufacturers.

Whatever the real return to the individual mining operations, there is no doubt that Cornwall was the most important silver mining county in the U.K. during the third quarter of the nineteenth century. See Table 8. For most of that period it averaged about a third of national production and rose to around 40 per cent in the mid-1850s. With a silver content averaging around 35 oz. per ton during the peak years of lead production in the early 1850s, the richness of its ores was second only to those of its Devonian neighbours, who were working effectively the same ores on the other side of the Tamar. Significantly, the silver content of the ores seems to have increased as lead production declined during the 1850s and 1860s. Silver production in Cornwall rose to a peak of well over 300,000 oz. in 1869, which represented an average of nearly 47 oz. per ton of lead. This was because (a) the silver content of some lodes increased as they were exploited at depth and (b) there was a shift in the centre of lead production towards the more argentiferous districts of the county. However, by the early 1870s the failure of the lead deposits began to precipitate a rapid decline in the output of both metals. Cornwall lost its lead as a silver producer to the Isle of Man in 1872 and within ten years production had fallen to less than 10,000 oz. annually. By 1887 the production of lead and silver in the county had stopped entirely.

It should be noticed that not all of Cornwall's silver came from lead mines. In a few places the lodes were so argentiferous that they supported operations which were primarily silver producing, with little or no associated lead output. At least 11 mines operated on this basis during the late 1870s and early 1880s, of which Newton, Great Crinnis, and the Prince of Wales were the most important. With the single exception of Talnotry in Scotland, they were the only silver mines in the U.K. Production levels were very low, however, and these ventures had no significant effect on the overall level of silver output. Similarly, three operations – Brothers, Duchy and Peru and New Consols Tin and Arsenic Works – returned an output of silver from copper and silver precipitate but it never became of any importance.

The production of *zinc* was less significant for Cornwall's lead mines than silver. Only around 45 mines produced the two metals in combination during the period and many of these had very low levels of lead output and made no important contribution to that industry. West Chiverton and Cargol were the only important lead producers to achieve a high level of zinc output and even here the main productive period was comparatively short. However, zinc sales undoubtedly made a very major contribution to Cargol's total income as its lead production fell off in the 1860s and it briefly kept the mine alive and profitable. Similarly, zinc production first grew in importance and then took over as the primary product of West Chiverton in the 1870s and significantly extended the life of Cornwall's second largest lead producer. In those years it became one of Britain's largest zinc mines. The supportive influence of zinc production was less diffuse than silver and concentrated in a shorter period but it undoubtedly played a crucial role in the lives of some mines.

With the exception of West Chiverton and Cargol, most of Cornwall's zinc came from mines other than lead mines, produced either alone or in association with copper, tin, iron and other minerals. Over 50 mines fell into this category. They included Great Retallack, Pencourse Consols and Budnick Consols, which were the leading producers of the 1850s and early 1860s. During that period they were responsible for up to a third of total U.K. zinc production. However, at the point when the introduction of new galvanising techniques began to expand demand and the producers of other districts began to enter the market, their production began to fall off. By the early 1870s Cornwall's diminished output accounted for only around five per cent of national production. See Table 9. The sudden surge of output from West Chiverton in the late 1870s, powerfully supported by that from Duchy and Peru in the early 1880s, revitalised the industry. It carried Cornish zinc output to an all-time peak during those years and it became the largest single county producer, contributing up to a fifth of total national output. However, new producers in Wales, Cumberland and the Isle of Man were also expanding their operations and soon resumed the lead. With the closure of West Chiverton and Duchy and Peru in the mid-1880s, Cornish zinc production collapsed and rarely again amounted to more than a few hundred tons per year.

Unlike some other metal mining districts, Cornwall produced very little *manganese* and *iron*. Across the county border, Devon mines increased their output of manganese ore to over 8,000 tons a year in the early 1870s but nothing was found in Cornwall with a potential of more than a few hundred tons per year. See Table 10. The only period of significant production in the county was during the 1870s and early 1880s, with a little sporadic output at the end of that decade. Most of the production came from Ruthers, near Newquay, which produced almost half of Cornwall's manganese, mainly in the early 1880s. It was briefly one of the largest producers in the U.K. in 1880, giving the county its only moment of real importance in the industry. The other half of Cornwall's output came from mines in the Launceston area, immediately adjacent to the large Devonian producers, which were located near Marystowe and Chillaton. New Phoenix, Ludcott, West End Down and Greystone Wood were the most important producers in this district, of which New Phoenix was clearly the most productive. In all there were only ten manganese mines in Cornwall, all of which were small operations by the standards of the main producing districts. They usually mined manganese in combination with a little lead, silver or tin but never in sufficient quantities to make a significant difference to the economics of the operations.

In terms of its contribution to national output, *iron* production in Cornwall was even less significant than manganese. The county never produced more than a fraction of one per cent of the U.K. total, even during its best years in the 1850s, early 1860s and early 1870s. See Table 11. It was the smallest producer in the south west, falling behind both Devon and Somerset. Compared with non ferrous mines, the tonnages of ore produced were impressively high but the value of iron output was relatively low. Few mines produced iron ore to a value of more than £2,000 a year. A comparison of the volume of iron and tin sold over the period 1855–1913 produces very similar results but the value of Cornwall's iron output was only 0.6 per cent of its tin production. Put another way, the value of all of the iron ore raised in the county during this period was roughly equal to the output of just one medium-sized copper and tin mine, such as North Basset.

Over 80 mines marketed iron ore in Cornwall during the period, either alone or in combination with many of the main non-ferrous minerals. Like other types of mining, output was dominated by a few large producers. The 11 mines with a total production in excess of 10,000 tons were responsible for almost three quarters of the county's total ouput. The four largest of these – Restormel Royal, Ruby, Pawton, and Trebisken – accounted for around half of the total. Most of the output came from the eastern part of the county, particularly the St Austell area, though the Perranzabuloe district was also important. With the exception of Restormel Royal, none of the mines produced on a significant scale for more than a few years. Once located, the ores were quickly ripped out and mining was soon abandoned. The most common type of ore produced was brown/red haematite but some of the smaller workings also had a limited output of spathose and magnetite. These are indicated in the mine table by the abbreviations (BH), (SP), and (MO).

Iron pyrites were also mined at some mines but like the sometimes associated minerals ochre, umber and redding, they have not been included in this study because of inconsistencies in the original data.

Finally mention must be made of the "rare" and minor, minerals that were sporadically produced by a few mines, usually as a by-product of other operations. The most important of these by volume and value was *wolfram/tungsten*. Like manganese, it was in increasing demand for the manufacture of new specialised hard steels. Between 1859 and 1913, Cornwall produced 4,857 tons of ore, virtually the entire national output, valued at over a quarter of a million pounds. It was more valuable than the output of iron in the county. Of the 20 or more mines recording sales of the ore during these years, four producers – East Pool, South Crofty, Carn Brea and Tincroft and Clitters United – dominated the trade. East Pool pioneered production during the 1860s and 1870s, joined briefly by Kit Hill United, and consistently remained the largest producer through to the early 1890s. At that point Carn Brea and Tincroft rapidly expanded production to take the lead but they were eclipsed themselves by Clitters United early in the new century. A revived East Pool was back in the lead by 1906 but was quickly overtaken by South Crofty in the years just before the War. With difficulties in the tin market, the production of this high value by-product could periodically play a strategic role in mine finances. In 1898, for example, the value of tungsten produced at Carn Brea and Tincroft accounted for 28 per cent of total receipts from the sale of all ores. Together with the value of arsenic sales, it produced 38 per cent of the mine's income in that year. Similarly, at South Crofty in 1907, tungsten accounted for over 20 per cent of the receipts from the sale of all ores and together with arsenic provided around 40 per cent of total income. Most of the mines that survived into the twentieth century did so on the production of a very mixed bag of minerals, exploiting all and every opportunity for profit.

The production of *antimony* was never as important as tungsten but three mines – Bodannon, Pengenna and Trebullet – tried to produce it as a single main product. Trevennick also had some production in association with argentiferous lead ore. Total output during the period only amounted to 35 tons, however, with a value of £400. *Bismuth* was produced in even smaller quantities. East Pool, Dolcoath and Wheal Owles were the only producers, with the largest output – 4 tons, valued at £120 – being derived from East Pool between 1872 and 1877. *Cobalt* was produced in the 1850s at Great Dowgas and from the 1870s at East Pool. Total recorded production from both mines amounted to only just over 8 tons for the whole period from both mines, with an estimated value of £320. *Uranium* production was also limited for most of the nineteenth century but expanded considerably from the 1890s. Three mines – Wheal Owles, East Pool and St Austell – produced very small quantities during the third quarter of the century but their combined output rarely rose much above one ton a year. They all stopped production in the 1880s. In 1890, however, Uranium Mines produced 22 tons in its first year of operation and it dominated production through to 1913 with sales of up to 103 tons per annum. It was supported in the first years of the new century by output from New Crow Hill and Trenwith, where uranium was produced in association with lead and copper and tin. Cornwall dominated the production of all of these rare minerals with one mine, East Pool, standing out as a major supplier of almost the whole range of them. The only exception to this rule was *nickel*. Mainly produced in Scotland, the only Cornish mines to market this mineral were Fowey Consols and St Austell Consols. During the 1850s and 1860s, they had a combined output of 17.6 tons of ore valued at £653. *Plumbago* was said to have been found at several mines in the county but production and sales were never recorded. Although Cornwall was the major domestic source of rare metallic minerals, it was particularly poor in the minor "earthy" products *barytes* and *fluorspar*. These did much to support metal mining in other parts of the U.K. but never made an important contribution to Cornish mines. East Basset was the only venture in the county to record an output of *barytes*, with a small production in the 1870s and early 1880s. It is likely that the "sundries" production of this mineral for 1876 related mainly to Devon. *Fluorspar* production was slightly more common, with nine mines recording some output during the second half of the century. The most important of these was Damsel mine in the St.Day district. The total value of fluorspar production in Cornwall, however, was less than £600.

The greatest *deficiency* in Cornwall's mineral wealth was without doubt its lack of *coal*. This proved to be a near fatal flaw in terms of the county's general industrialisation and long-term economic welfare. The smelting and manufacture of its metallic minerals was generally forced outside of the region to areas where fuel was plentiful and cheap. The secondary industries which developed within the county related only to primary mining and quarrying activity. With the long-term decline of these activities – which now amount only to the rump of a tin mining industry and large-scale quarrying confined to the St Austell area – there has been a progressive contraction of the industrial base on which to build a new future.

III

The Mine Tables

The information which appears in the following tables is the same as that published in the annual editions of the *Mineral Statistics,* now re-arranged in a mine-by-mine format. Some editorial judgement has been exercised in re-compiling the material, in order to link returns together and provide for necessary cross-referencing, but wherever possible the mine names, production record, ownership, management, employment and even comments are given just as they appeared in the original. If any category of information is missing under a mine name, or if the series cover only a few years, or indeed if an entire entry is not included, it is because there were no returns in the *Mineral Statistics.* To facilitate the alphabetical arrangement of the material, we have omitted the common prefix "Wheal" from mine names.

A problem that has occured in earlier volumes in this series and which has also been apparent here, is that of the random transfer of returns for some mines between neighbouring counties. The original *Mineral Statistics* may have incorrectly located some of the returns for 21 Cornish mines in Devon and 11 Devon mines in Cornwall. There is also the possibility that some Cornish material was also wrongly distributed under some other English and Welsh counties. The totals for all of the counties concerned were accordingly distorted by these errors. In the past we have sometimes attempted to correct these problems by transferring the relevant material between the counties. Here, however, we have left them unchanged in the interests of the faithful reproduction of the original and acknowledging the possibility of our own error. The probable misplaced entries are indicated in the Mine Tables and are as follows:

Cornish mines incorrectly located in the Devon returns

Arthur	Martha, Great
Atway	Martha, West
Calstock Consols	Mary Great Consols
Drakewalls	New Consols
Emmens United	New Great Consols
Friendship & Prosper	Pawton
Great Rodd	Polharman
Gunnislake	Prince of Wales
Gunnislake, East	Richards Friendship
Gunnislake, Old	Toy Tor
Laddock	

Devon mines incorrectly located in the Cornish returns

Adams	Friendship, North
Bedford, South	Girt Down
Bowden Down	Haytor
Collacombe	Langstone
Devon & Cornwall United	Whitleigh
Devon Consols, West	

Not all entries for all of these mines were wrongly located. Some of their returns were under the correct county. One particular problem arose for South Bedford and East Gunnislake which were worked for a long time as one mine, straddling the river Tamar border between Devon and Cornwall. The ownership returns appeared in one listing but production was given in different years under both counties. It is hoped that on completion of this series it will be possible to publish a full calendar of mislocated returns and to adjust the county totals tables accordingly.

The only new material that has been added to the *Mine Tables* is some additional cross-referencing information, the adjustment of some locations for current usage, and Ordnance Survey Grid Reference numbers. The Grid References have been derived from our own research, checked against those provided by A.K.Hamilton Jenkin in his seminal series *The Mines and Miners of West Cornwall*. In editing the material, close attention has been paid to J.H.Collins, *Observations on the West of England Mining Region* and H.G.Dines, *The Metalliferous Mining Region of South West England*, as well as other established reference works on Cornish mining. A number of the mines making returns were not referred to in any of these texts, however, and their precise location frequently remains unknown. We would like to thank Alasdair Neill and Justin Brooke for their invaluable advice in resolving many particularly difficult locational problems.

Four principal categories of information are included for every mine, where they are available. *Production* data are given by type of mineral produced and often include ore output, metal content, and value – the latter sometimes being calculated by assay and average market price. The absence of detailed production returns under a particular mine name should not necessarily be taken to imply that the mine was not working in those years, since its output might be disguised in the county aggregate figures, sundries or joint returns listed under other mines. Wherever possible we have tried to indicate such listings through cross references in the *Comment* section. For copper mines, the *comment* section has also sometimes been used to indicate the method of sale of the ore; the sales at the Cornish ticketings being indicated by (C), those at the Swansea ticketings by (S) and private contract agreements by (P). For iron mines, the *comment* is used to indicate the types of ore being produced; this was commonly red/brown haematite (RH), spathose (SP) and magnetite (M0).

The returns for the production of the different minerals on a mine-by-mine basis began to appear from different dates: lead from 1845; copper from 1845; tin and silver from 1852; zinc from 1854; arsenic and barytes from 1855; iron, manganese and tungsten from 1858; and fluorspar from 1874. In the early years there was undoubtedly some under recording and the number of the mines included in the annual listings usually increased noticeably as the Mining Record Office developed its procedures for collecting information. It should be noticed that this study does not systematically include material listed under sales of ore in stone, streamworks and foreshores.

For many counties, the *Ownership* and *Management* returns give a good indication of the years of activity at a mine, even when it was not producing and selling ore. These returns are derived from the *List of Mines*, which was appended to the back of the early *Mineral Statistics* and later published separately. They are normally available for non-ferrous mines from 1859 and iron mines from 1863. However, for some unknown reason, the Cornish section of the returns gave no regular details of the ownership until the mid-1870s and confined itself only to managers and agents. For many mines, ownership returns continued to be patchy well into the last quarter of the century. Careful examination of the *Lists of Mines* also reveals that major revisions were periodically conducted when unusually large numbers of mines suddenly disappeared from the returns. This suggests that they were not carefully edited for every edition and that they therefore continued to include some mines long after they had been suspended.

The *Employment* returns were also drawn from the *List of Mines* and provide a further check on the periods, level and type of activity at the mines. For example, a prospecting or developing working might show significant levels of employment even though there was little or no production or sales of ore. Similarly, a changing distribution of workers between underground and surface operations can give an idea of when ventures began to run down their mining operations and concentrate only on the redressing of spoil heaps. It should be remembered that many of the smaller mines at best provided only part-time employment for their labour force and that the same miners may have been counted several times over at different workings. The decision to publish detailed employment returns appears to have been left to the discretion of the local Mine Inspectors. For most Cornish mines this fortunately started at the relatively early date of 1878.

BIBLIOGRAPHY

D.B.Barton, *A History of Tin Mining and Smelting in Cornwall* (Truro, 1967)

D.B.Barton, *Essays in Cornish Mining History* (Truro, 1971) 2 vols

D.B.Barton, *A History of Copper Mining in Cornwall and Devon* (Truro, 1978)

G.Burke and P.Richardson, "The decline and fall of the cost book system in the Cornish tin industry 1895–1914" *Business History*, XXIII No.1 (1981) 4–18

J.A.Buckley, *A History of South Crofty Mine* (Redruth, 1982)

R.Burt, *John Taylor; Mining Entrepreneur and Engineer* (Buxton, 1977)

R.Burt, *The British Lead Mining Industry* (Redruth, 1984)

R.Burt, *Cornish Mining* (Newton Abbot, 1969)

R.Burt, "The Mineral Statistics of the United Kingdom: An Analysis of the Accuracy of the Copper and Tin Returns for Cornwall and Devon" *The Journal of the Trevithick Society*, No.8 (1981)

R.Burt and P.Waite, "An Introduction to the Mineral Statistics" *British Mining* No.21 (Spring, 1983)

R.Burt and I.Wilkie, "Manganese Mining in the South West of England" *Journal of the Trevithick Society*, No.11 (1984)

R.Burt, "The Development of the Uses of Some Minor Elements and Minerals since the Eighteenth Century" *Geology Teaching* Vol.9 No.4 (Dec. 1984)

J.H.Collins, *Observations on the West of England Mining Region* (Plymouth, 1912)

H.G.Dines, *The Metalliferous Mining Region of South West England* (Memoirs of the Geological Survey of Great Britain, H.M.S.O., 1956) 2 vols

H.L.Douch, *East Wheal Rose* (Truro, 1964)

A.K.Hamilton Jenkin, *The Mines and Miners of West Cornwall* (Truro, 1961–1969)

T.R.Harris, *Dolcoath, Queen of Cornish Mines* (Trevithick Society, 1974)

T.A.Morrison, *Cornwall's Central Mines* (Penzance, 1980) 2 vols

C.Noall, *Botallack* (Truro, 1972)

C.Noall, *The St.Just Mining District* (Truro, 1973)

C.Noall, *The St.Ives Mining District* (Redruth, 1982)

C.Noall, *Geevor* (Penzance, 1983)

C.J.Schmitz, *World Non-Ferrous Metal Production and Prices, 1700–1976* (1979)

H.R.Shambrooke, *The Caradon and Phoenix Mining Area* (Northern Mine Research Society, British Mining No.20, 1982)

Table 1
Value of U K and Cornwall Mineral Production
Decennial Averages 1850–1909

Mineral	1850–59 UK Ann.Av.	%Tot	1850–59 Cornwall Ann.Av.	%Tot	1860–69 UK Ann.Av.	%Tot	1860–69 Cornwall Ann.Av.	%Tot
Tin	645,847	18.6	638,820	34.8	775,151	22.9	761,305	50.7
Copper	1,229,997	35.5	1,020,804	55.5	988,393	29.3	580,831	38.7
Lead & Silver	1,547,219	44.6	172,557	9.4	1,560,701	46.3	152,406	10.2
Zinc	29,767	0.9	5,137	0.3	39,073	1.1	4,764	0.3
Arsenic	760	0.0	549	0.0	2,554	0.1	1,510	0.1
Manganese	3,947	0.1	0	0.0	2,512	0.1	0	0.0
Barytes	7,170	0.2	0	0.0	5,141	0.2	0	0.0
Fluorspar	904	0.1	246	0.0	10	0.0	19	0.0
Total	3,465,611	100.0	1,838,113	100.0	3,373,535	100.0	1,500,835	100.0
Corn.as % UK		53.0%				44.5%		

Mineral	1870–79 UK Ann.Av.	%Tot	1870–79 Cornwall Ann.Av.	%Tot	1880–89 UK Ann.Av.	%Tot	1880–89 Cornwall Ann.Av.	%Tot
Tin	785,047	30.7	779,726	72.9	753,184	45.2	749,847	88.9
Copper	320,928	12.6	190,214	17.8	106,946	6.4	61,748	7.3
Lead & Silver	1,313,781	51.4	81,886	7.7	636,260	38.2	4,861	0.6
Zinc	69,420	2.7	7,669	0.7	86,993	5.2	5,898	0.7
Arsenic	27,793	1.1	9,882	0.9	50,394	3.0	20,883	2.5
Manganese	20,659	0.8	129	0.0	5,319	0.3	270	0.0
Barytes	16,266	0.7	57	0.0	25,815	1.7	0	0.0
Fluorspar	139	0.0	6	0.0	371	0.0	14	0.0
Total	2,554,033	100.0	1,069,569	100.0	1,665,282	100.0	843,521	100.0
Corn.as % UK		41.9%				50.7%		

Mineral	1890–99 UK Ann.Av.	%Tot	1890–99 Cornwall Ann.Av.	%Tot	1900–09 UK Ann.Av.	%Tot	1900–09 Cornwall Ann.Av.	%Tot
Tin	499,038	46.3	497,090	93.5	573,436	52.3	568,806	95.1
Copper	20,869	2.0	13,620	2.6	20,829	1.9	13,446	2.2
Lead & Silver	376,707	34.9	12	0.0	324,972	29.6	943	0.2
Zinc	91,834	8.5	58	0.0	100,406	9.2	758	0.1
Arsenic	61,413	5.7	20,590	3.9	26,980	2.5	14,402	2.4
Manganese	2,098	0.2	8	0.0	6,551	0.6	0	0.0
Barytes	25,952	2.4	0	0.0	30,555	2.8	0	0.0
Fluorspar	282	0.0	19	0.0	12,398	1.1	0	0.0
Total	1,078,193	100.0	531,397	100.0	1,096,127	100.0	598,355	100.0
Corn.as % UK		49.3%				54.6%		

Table 2

Average Value of Output per Employee at Ten Year Intervals 1880–1910

1880	Average	Comment
Group 1 (9 observations)	£77.6	(5 mines between £69.6 and £73.2)
Group 2 (9 observations)	£59.3	(3 mines between £41.9 and £48.8)
		(3 mines between £68.2 and £87.5)
Group 3 (11 observations)	£46.5	(6 mines between £38.8 and £47.8)

1890	Average	Comment
Group 1 (9 observations)	£70	(4 mines between £62 and £73.3)
Group 2 (8 observations)	£52.9	(5 mines between £52.2 and £61.7)
Group 3 (7 observations)	£47.6	(4 mines between £34.2 and £49.2)

1900	Average	Comment
Group 1 (6 observations)	£98	(3 mines between £111 and £119)
		(2 mines between £88.9 and £90.9)
Group 2 (4 observations)	£53.8	(4 mines between £4.2 and £131.5)
Group 3 (2 observations)	£37.5	(1 mine £35.4 and 1 mine £39.5)

1910	Average	Comment
Group 1 (7 observations)	£104.4	(4 mines over £100)
Group 2 (4 observations)	£90.4	(2 mines over £100)
Group 3 (3 observations)	£48.7	(2 mines between £52.9 and £55.1)

An inexplicably high return for West Kitty in 1890 and an exceptionally low return for Phoenix in 1910 have been disregarded in producing these averages.

Table 3

Production, Trade and Estimated New Metal Available for Consumption:

Tin and Lead 1860–1910 (tons)

	1860 Tin	1860 Lead	1870 Tin	1870 Lead	1880 Tin	1880 Lead
UK Mine Output	6,695	63,323	10,200	73,421	8,918	56,949
Imports	2,900	23,150	4,700	72,738	19,500	110,005
Exports	2,700	29,729	5,100	61,810	4,400	33,821
Re-exports	500	66	1,100	2,372	8,700	5,600
Available for Consumption	6,395	56,678	8,700	82,577	15,318	127,533
% Consumption from UK Mines	100%	100%	100%	89%	58.2%	44.7%

	1890 Tin	1890 Lead	1900 Tin	1900 Lead	1910 Tin	1910 Lead
UK Mine Output	9,602	33,590	4,268	24,364	4,797	23,742
Imports	28,755	173,000	37,699	218,800	62,743	232,475
Exports	5,100	55,600	5,600	36,000	12,400	46,800
Re-exports	14,800	15,700	19,800	13,200	31,500	13,500
Available for Consumption	18,457	135,290	16,567	193,964	23,640	195,917
% Consumption from UK Mines	52%	24.8%	25.8%	12.6%	20.3%	12.1%

Metal content of imported ore estimated as the same as domestic ores.

Table 4

Cornwall copper ore production and its share of total U K output 1845 to 1913

Year	Corn. Ore(ton)	UK Ore(ton)	%UK Prod	Year	Corn. Ore(ton)	UK Ore(ton)	%UK Prod
1845	149,834	184,740	81.11	1880	26,737	52,128	51.29
1846	132,444	169,568	78.11	1881	24,510	52,556	46.64
1847	137,665	171,104	80.46	1882	25,641	52,237	49.09
1848	127,226	147,701	86.14	1883	23,250	46,820	49.66
1849	126,650	146,326	86.55	1884	21,539	42,149	51.10
1850	134,020	155,025	86.45	1885	19,734	36,379	54.25
1851	127,285	150,380	84.64	1886	7,541	18,617	40.51
1852	139,172	165,593	84.04	1887	3,415	9,359	36.49
1853	151,507	181,944	83.27	1888	6,838	15,550	43.97
1854	153,297	185,145	82.80	1889	4,959	9,310	53.27
1855	161,576	207,770	77.77	1890	5,271	12,481	42.23
1856	163,958	260,609	62.91	1891	4,039	9,158	44.10
1857	152,729	218,689	69.84	1892	2,813	6,265	44.90
1858	147,330	226,852	64.95	1893	2,673	5,576	47.94
1859	146,093	236,789	61.70	1894	3,362	5,994	56.09
1860	145,359	236,696	61.41	1895	5,504	7,791	70.65
1861	143,119	231,487	61.83	1896	5,616	9,168	61.26
1862	141,800	224,171	63.26	1897	4,140	7,352	56.31
1863	129,229	210,947	61.26	1898	5,293	9,131	57.97
1864	127,633	214,604	59.47	1899	5,172	8,319	62.17
1865	121,253	198,298	61.15	1900	5,926	9,488	62.46
1866	103,670	180,378	57.47	1901	4,251	6,792	62.59
1867	88,660	158,544	55.92	1902	4,547	6,112	74.39
1868	86,722	157,335	55.12	1903	5,351	6,867	77.92
1869	71,790	129,953	55.24	1904	4,433	5,465	81.12
1870	56,526	106,698	52.98	1905	4,651	7,153	65.02
1871	46,766	97,129	48.15	1906	3,053	7,758	39.35
1872	41,756	91,893	45.44	1907	2,802	6,792	41.25
1873	40,285	80,189	50.24	1908	1,556	5,441	28.60
1874	40,455	78,521	51.52	1909	1,580	3,717	42.51
1875	39,393	71,528	55.07	1910	870	4,178	20.82
1876	43,016	79,252	54.28	1911	988	3,262	30.29
1877	39,225	73,141	53.63	1912	590	1,933	30.52
1878	36,871	56,094	65.73	1913	421	2,732	15.41
1879	30,371	51,032	59.51				
				Total	3,973,751	6,030,155	65.90

Table 5

Cornwall tin ore production and its share of total U K output 1852 to 1913

Year	Corn. Ore(ton)	UK Ore(ton)	%UK Prod	Year	Corn. Ore(ton)	UK Ore(ton)	%UK Prod
1852	9,497	9,674	98.17	1883	13,999	14,469	96.75
1853	8,587	8,866	96.85	1884	15,091	15,117	99.83
1854	8,515	8,747	97.35	1885	14,324	14,377	99.63
1855	8,767	8,947	97.99	1886	14,125	14,232	99.25
1856	9,107	9,243	98.53	1887	14,083	14,189	99.25
1857	9,614	9,709	99.02	1888	14,282	14,370	99.39
1858	9,906	9,960	99.46	1889	13,756	13,809	99.62
1859	10,069	10,180	98.91	1890	14,868	14,911	99.71
1860	10,225	10,400	98.32	1891	14,444	14,488	99.70
1861	10,725	10,963	97.83	1892	14,261	14,357	99.33
1862	11,638	11,841	98.29	1893	13,647	13,699	99.62
1863	13,943	14,225	98.02	1894	12,880	12,910	99.77
1864	13,678	13,985	97.80	1895	10,583	10,612	99.73
1865	13,868	14,123	98.19	1896	7,657	7,663	99.92
1866	13,601	13,785	98.67	1897	7,120	7,120	100.00
1867	10,989	11,067	99.30	1898	7,379	7,380	99.99
1868	11,530	11,584	99.53	1899	6,389	6,392	99.95
1869	13,757	13,884	99.09	1900	6,793	6,801	99.88
1870	15,191	15,235	99.71	1901	7,280	7,288	99.89
1871	16,759	16,898	99.18	1902	7,552	7,560	99.89
1872	13,247	13,381	99.00	1903	7,354	7,382	99.62
1873	14,735	14,838	99.31	1904	6,733	6,742	99.87
1874	13,633	13,762	99.06	1905	7,174	7,201	99.63
1875	13,870	14,005	99.04	1906	7,115	7,154	99.45
1876	13,602	13,649	99.66	1907	6,987	7,080	98.69
1877	14,071	14,092	99.85	1908	7,816	8,008	97.60
1878	15,077	15,124	99.69	1909	8,166	8,289	98.52
1879	14,248	14,280	99.78	1910	7,483	7,572	98.82
1880	13,344	13,376	99.76	1911	7,666	7,746	98.97
1881	12,883	12,898	99.88	1912	8,065	8,166	98.76
1882	13,995	14,045	99.64	1913	8,258	8,355	98.84
				Total	696,001	702,205	99.12

Table 6a

Cornwall arsenic ore production and its share of total U K output 1854 to 1913

Year	Corn. Ore(ton)	UK Ore(ton)	%UK Prod	Year	Corn. Ore(ton)	UK Ore(ton)	%UK Prod
1854	477	477	100.00	1884	3,347	7,906	42.33
1855	443	787	56.29	1885	3,889	8,129	47.84
1856	-	-	0.00	1886	1,791	5,027	35.63
1857	477	477	100.00	1887	1,661	4,618	35.97
1858	396	396	100.00	1888	1,584	4,624	34.26
1859	445	465	95.70	1889	1,927	4,758	40.50
1860	515	545	94.50	1890	3,143	7,276	43.20
1861	539	1,211	44.51	1891	3,048	6,049	50.39
1862	901	901	100.00	1892	2,567	5,114	50.20
1863	721	721	100.00	1893	1,751	5,976	29.30
1864	633	633	100.00	1894	1,900	4,801	39.58
1865	827	827	100.00	1895	1,821	4,798	37.95
1866	1,117	1,117	100.00	1896	1,366	3,616	37.78
1867	1,200	1,256	95.54	1897	1,014	4,165	24.35
1868	1,267	1,741	72.77	1898	1,062	4,174	25.44
1869	1,189	2,562	46.41	1899	1,361	3,829	35.54
1870	1,813	4,050	44.77	1900	1,160	4,081	28.42
1871	1,927	4,147	46.47	1901	1,258	3,361	37.43
1872	2,950	5,172	57.04	1902	1,029	2,131	48.29
1873	3,480	5,449	63.86	1903	690	902	76.50
1874	2,286	4,768	47.94	1904	827	976	84.73
1875	2,412	4,569	52.79	1905	1,523	1,528	99.67
1876	2,557	4,228	60.48	1906	1,599	1,599	100.00
1877	1,718	4,810	35.72	1907	1,368	1,497	91.38
1878	1,843	4,991	36.93	1908	1,409	1,919	73.42
1879	1,659	5,492	30.21	1909	1,817	2,865	63.42
1880	2,044	5,739	35.62	1910	1,817	2,153	84.39
1881	2,775	6,156	45.08	1911	1,912	2,121	90.15
1882	3,473	7,469	46.50	1912	2,011	2,266	88.75
1883	3,380	7,622	44.35	1913	1,449	1,652	87.71
				Total	98,565	202,689	48.63

National total for 1871 corrected from Devon volume.

Table 6b

Cornwall arsenical pyrites ore production and its share of total U K output 1854 to 1913

Year	Corn. Ore(ton)	UK Ore(ton)	%UK Prod	Year	Corn. Ore(ton)	UK Ore(ton)	%UK Prod
1854	-	-	0.00	1884	926	1,762	52.55
1855	151	495	30.51	1885	145	1,911	7.59
1856	-	-	0.00	1886	2,415	4,919	49.10
1857	-	-	0.00	1887	101	4,363	2.31
1858	160	320	50.00	1888	1,443	5,325	27.10
1859	-	-	0.00	1889	2,443	7,688	31.78
1860	-	-	0.00	1890	1,536	5,114	30.04
1861	-	-	0.00	1891	1,104	5,095	21.67
1862	-	-	0.00	1892	1,086	4,497	24.15
1863	-	-	0.00	1893	835	3,036	27.50
1864	-	-	0.00	1894	1,519	3,288	46.20
1865	-	-	0.00	1895	1,557	2,931	53.12
1866	-	-	0.00	1896	6,204	8,808	70.44
1867	-	-	0.00	1897	11,514	13,137	87.65
1868	-	-	0.00	1898	9,729	10,823	89.89
1869	-	2,444	0.00	1899	9,679	13,429	72.08
1870	-	1,497	0.00	1900	9,019	9,573	94.21
1871	-	1,731	0.00	1901	2,334	2,578	90.54
1872	-	195	0.00	1902	535	829	64.54
1873	-	-	0.00	1903	-	57	0.00
1874	959	1,936	49.54	1904	43	43	100.00
1875	6,688	7,489	89.30	1905	47	641	7.33
1876	2,693	8,011	33.62	1906	333	640	52.03
1877	13,879	14,929	92.97	1907	207	689	30.04
1878	2,954	3,844	76.85	1908	-	1,925	0.00
1879	895	2,843	31.48	1909	-	93	0.00
1880	4,847	5,920	81.88	1910	623	684	91.08
1881	12,941	14,338	90.26	1911	1,150	1,170	98.29
1882	10,579	12,564	84.20	1912	1,778	1,778	100.00
1883	765	1,300	58.85	1913	35	35	100.00
				Total	125,851	196,717	63.98

National total for 1892 and 1898 corrected from Devon volume.

Table 7

Cornwall lead ore production and its share
of total U K output 1845 to 1913

Year	Corn. Ore(ton)	UK Ore(ton)	%UK Prod	Year	Corn. Ore(ton)	UK Ore(ton)	%UK Prod
1845	10,110	78,267	12.92	1880	754	72,245	1.04
1846	8,228	74,551	11.04	1881	765	64,702	1.18
1847	11,674	83,747	13.94	1882	624	65,002	0.96
1848	10,494	78,944	13.29	1883	830	50,980	1.63
1849	10,326	86,823	11.89	1884	529	54,485	0.97
1850	10,386	92,958	11.17	1885	241	51,302	0.47
1851	9,515	92,312	10.31	1886	227	53,420	0.42
1852	8,999	91,198	9.87	1887	-	51,563	0.00
1853	6,680	85,043	7.85	1888	-	51,259	0.00
1854	7,460	90,554	8.24	1889	-	48,465	0.00
1855	8,963	92,038	9.74	1890	-	45,651	0.00
1856	9,974	101,998	9.78	1891	-	43,859	0.00
1857	9,560	96,820	9.87	1892	15	40,024	0.04
1858	9,710	95,856	10.13	1893	-	40,808	0.00
1859	7,843	91,382	8.58	1894	-	40,600	0.00
1860	6,401	88,791	7.21	1895	-	38,412	0.00
1861	6,691	90,666	7.38	1896	-	41,069	0.00
1862	6,030	95,312	6.33	1897	-	35,338	0.00
1863	6,260	91,283	6.86	1898	-	32,985	0.00
1864	5,302	94,463	5.61	1899	-	30,999	0.00
1865	6,547	90,452	7.24	1900	-	32,010	0.00
1866	6,737	91,048	7.40	1901	-	27,976	0.00
1867	8,645	93,432	9.25	1902	-	24,606	0.00
1868	8,416	95,236	8.84	1903	9	26,567	0.03
1869	9,023	96,866	9.31	1904	-	26,374	0.00
1870	8,481	98,177	8.64	1905	248	27,649	0.90
1871	7,566	93,965	8.05	1906	200	30,795	0.65
1872	5,463	81,619	6.69	1907	190	32,533	0.58
1873	3,909	73,500	5.32	1908	87	29,249	0.30
1874	3,120	76,202	4.09	1909	-	29,744	0.00
1875	2,566	77,746	3.30	1910	-	28,534	0.00
1876	2,727	79,096	3.45	1911	-	23,910	0.00
1877	2,167	80,850	2.68	1912	-	25,409	0.00
1878	1,350	77,351	1.75	1913	-	24,282	0.00
1879	725	66,877	1.08				
				Total	252,767	4,408,229	5.73

Table 8

Cornwall silver production and its share of total U K output 1851 to 1913

Year	Corn. (ozs)	UK (ozs)	%UK Prod	Year	Corn. (ozs)	UK (ozs)	%UK Prod
1851	255,640	674,458	37.90	1883	9,445	344,053	2.75
1852	250,008	818,325	30.55	1884	5,000	325,718	1.54
1853	165,670	496,475	33.37	1885	2,500	320,520	0.78
1854	179,675	562,659	31.93	1886	2,200	325,427	0.68
1855	211,348	561,906	37.61	1887	-	320,345	0.00
1856	248,436	614,188	40.45	1888	-	321,425	0.00
1857	224,277	532,866	42.09	1889	-	306,149	0.00
1858	223,189	569,345	39.20	1890	-	291,724	0.00
1859	215,964	576,027	37.49	1891	-	279,792	0.00
1860	180,757	549,720	32.88	1892	-	271,259	0.00
1861	173,344	569,530	30.44	1893	-	274,100	0.00
1862	205,662	686,123	29.97	1894	-	275,696	0.00
1863	206,312	634,004	32.54	1895	-	280,434	0.00
1864	192,232	641,088	29.99	1896	-	283,826	0.00
1865	214,659	724,856	29.61	1897	-	249,156	0.00
1866	195,218	636,688	30.66	1898	-	211,403	0.00
1867	314,326	805,394	39.03	1899	-	191,927	0.00
1868	303,033	841,328	36.02	1900	-	187,842	0.00
1869	315,714	831,891	37.95	1901	-	178,324	0.00
1870	292,045	784,562	37.22	1902	-	145,873	0.00
1871	267,324	761,490	35.11	1903	234	152,855	0.15
1872	207,710	628,920	33.03	1904	-	141,592	0.00
1873	129,509	524,307	24.70	1905	7,936	163,399	4.86
1874	85,304	509,277	16.75	1906	7,000	147,647	4.74
1875	25,681	487,358	5.27	1907	7,220	150,521	4.80
1876	37,650	483,422	7.79	1908	3,219	135,154	2.38
1877	23,035	497,375	4.63	1909	-	142,006	0.00
1878	16,456	397,471	4.14	1910	-	136,192	0.00
1879	9,435	333,674	2.83	1911	-	118,395	0.00
1880	11,790	295,518	3.99	1912	-	118,540	0.00
1881	14,396	308,398	4.67	1913	-	128,154	0.00
1882	11,460	372,449	3.08				
				Total	5,452,013	25,630,540	21.27

Table 9

Cornwall zinc ore production and its share of total U K output 1856 to 1913

Year	Corn. Ore(ton)	UK Ore(ton)	%UK Prod	Year	Corn. Ore(ton)	UK Ore(ton)	%UK Prod
1856	2,316	9,004	25.72	1885	1,031	24,668	4.18
1857	1,675	9,290	18.03	1886	749	23,156	3.23
1858	2,014	5,659	35.59	1887	649	25,445	2.55
1859	2,423	13,039	18.58	1888	134	26,408	0.51
1860	4,773	15,553	30.69	1889	129	23,202	0.56
1861	5,695	15,770	36.11	1890	-	22,041	0.00
1862	1,232	7,498	16.43	1891	-	22,216	0.00
1863	1,700	13,699	12.41	1892	12	23,880	0.05
1864	890	15,232	5.84	1893	21	23,754	0.09
1865	1,085	17,843	6.08	1894	57	21,821	0.26
1866	1,165	12,770	9.12	1895	-	17,478	0.00
1867	1,731	13,489	12.83	1896	55	19,319	0.28
1868	2,061	12,782	16.12	1897	57	19,278	0.30
1869	848	15,533	5.46	1898	-	23,553	0.00
1870	927	13,587	6.82	1899	42	23,135	0.18
1871	952	17,737	5.37	1900	20	24,675	0.08
1872	1,086	18,543	5.86	1901	237	23,752	1.00
1873	1,428	15,969	8.94	1902	-	25,060	0.00
1874	764	16,830	4.54	1903	10	24,888	0.04
1875	3,087	23,978	12.87	1904	-	27,655	0.00
1876	4,414	23,613	18.69	1905	4	23,909	0.02
1877	4,991	24,406	20.45	1906	38	22,824	0.17
1878	4,483	25,438	17.62	1907	450	20,082	2.24
1879	3,202	22,200	14.42	1908	50	15,225	0.33
1880	4,440	27,548	16.12	1909	824	9,902	8.32
1881	7,793	35,527	21.94	1910	-	11,238	0.00
1882	4,608	32,539	14.16	1911	300	17,652	1.70
1883	2,807	29,728	9.44	1912	-	17,704	0.00
1884	1,290	25,563	5.05	1913	-	17,294	0.00
				Total	80,749	1,151,581	7.01

Table 10

Cornwall manganese ore production and its share of total U K output 1867 to 1890

Year	Corn. Ore(ton)	UK Ore(ton)	%UK Prod	Year	Corn. Ore(ton)	UK Ore(ton)	%UK Prod
1867	-	808	0.00	1879	110	816	13.48
1868	-	1,750	0.00	1880	833	2,839	29.34
1869	-	1,558	0.00	1881	450	2,887	15.59
1870	-	4,839	0.00	1882	-	1,548	0.00
1871	-	5,548	0.00	1883	-	1,287	0.00
1872	-	7,773	0.00	1884	-	909	0.00
1873	17	8,671	0.20	1885	-	1,688	0.00
1874	95	5,778	1.64	1886	-	12,763	0.00
1875	41	3,206	1.28	1887	387	13,777	2.81
1876	21	2,797	0.75	1888	-	4,342	0.00
1877	37	3,039	1.22	1889	-	8,852	0.00
1878	144	1,586	9.08	1890	83	12,444	0.67
				Total	2,218	111,505	1.99

Table 11

Cornwall iron ore production and its share of total U K output 1855 to 1913

Year	Corn. (tons)	UK (tons)	%UK Prod	Year	Corn. (tons)	UK (tons)	%UK Prod
1855	24,057	9,553,741	0.25	1885	-	15,417,982	0.00
1856	22,650	10,483,309	0.22	1886	75	14,110,013	0.00
1857	19,359	9,372,781	0.21	1887	-	13,098,041	0.00
1858	55,150	8,040,959	0.69	1888	-	14,590,713	0.00
1859	35,214	7,876,582	0.45	1889	-	14,546,105	0.00
1860	23,953	8,035,306	0.30	1890	-	13,780,767	0.00
1861	26,262	7,215,518	0.36	1891	-	12,777,689	0.00
1862	24,627	7,562,240	0.33	1892	691	11,312,675	0.01
1863	18,976	8,613,951	0.22	1893	124	11,203,476	0.00
1864	34,210	10,064,891	0.34	1894	955	12,367,308	0.01
1865	36,112	9,910,046	0.36	1895	800	12,615,414	0.01
1866	18,684	9,665,013	0.19	1896	833	13,700,764	0.01
1867	6,427	10,021,058	0.06	1897	-	13,787,878	0.00
1868	8,310	10,169,231	0.08	1898	-	14,176,936	0.00
1869	4,619	11,508,526	0.04	1899	-	14,461,330	0.00
1870	11,214	14,496,427	0.08	1900	-	14,028,208	0.00
1871	21,948	16,470,010	0.13	1901	-	12,275,198	0.00
1872	49,400	15,755,675	0.31	1902	-	13,426,004	0.00
1873	32,655	15,583,669	0.21	1903	-	13,715,645	0.00
1874	45,056	14,844,936	0.30	1904	-	13,774,282	0.00
1875	11,404	15,821,060	0.07	1905	240	14,590,703	0.00
1876	18,390	16,841,584	0.11	1906	60	15,500,406	0.00
1877	4,963	16,692,802	0.03	1907	50	15,731,604	0.00
1878	1,309	15,726,370	0.01	1908	133	15,031,025	0.00
1879	400	14,379,735	0.00	1909	43	14,979,979	0.00
1880	15,865	18,026,050	0.09	1910	688	15,226,015	0.00
1881	7,460	17,446,065	0.04	1911	312	15,519,424	0.00
1882	5,749	18,031,957	0.03	1912	-	13,790,391	0.00
1883	670	17,383,046	0.00	1913	-	15,997,328	0.00
1884	950	16,137,887	0.01				
				Total	591,047	787,263,728	0.08

Table 12

Cornwall tungsten ore production and its share of total U K output 1870 to 1913

Year	Corn. Ore(ton)	UK Ore(ton)	%UK Prod	Year	Corn. Ore(ton)	UK Ore(ton)	%UK Prod
1870	50	50	100.00	1892	125	125	100.00
1871	20	20	100.00	1893	22	22	100.00
1872	88	88	100.00	1894	-	-	0.00
1873	47	47	100.00	1895	-	-	0.00
1874	32	32	100.00	1896	43	43	100.00
1875	46	46	100.00	1897	125	125	100.00
1876	23	23	100.00	1898	324	324	100.00
1877	15	15	100.00	1899	89	89	100.00
1878	10	10	100.00	1900	7	8	87.50
1879	13	13	100.00	1901	18	18	100.00
1880	1	1	100.00	1902	9	9	100.00
1881	54	54	100.00	1903	269	272	98.90
1882	58	58	100.00	1904	156	156	100.00
1883	111	111	100.00	1905	158	163	96.94
1884	64	64	100.00	1906	232	262	88.71
1885	374	374	100.00	1907	250	314	79.62
1886	151	151	100.00	1908	220	229	96.07
1887	55	55	100.00	1909	323	367	87.98
1888	62	62	100.00	1910	251	271	92.73
1889	-	-	100.00	1911	245	260	94.12
1890	104	104	100.00	1912	189	189	99.79
1891	138	138	100.00	1913	179	182	98.63
				Total	4,756	4,950	96.09

1

APPENDIX A

Cornwall's Principal Mines: Divided into Categories of the Total Value of Output of All Minerals 1845–1913 *

Mines with a Total Value of Output in excess of £2 million

Dolcoath, Carn Brea & Tincroft, East Pool & Agar

Total No. Mines 3
Total Value of Output £14,510,881
Output as % of Cornish Total 23.5%

Mines with a Total Value of Output of between £1 and £2 million

Basset, West Basset, Botallack, South Caradon, Grenville, Levant, Phoenix

Total No. Mines 7
Total Value of Output £10,575,450
Output as % of Cornish Total 17.1%

Mines with a Total value of Output of between £1 million and £500,000

Buller, West Caradon, Clifford Amalgamated, South Condurrow, Cooks Kitchen, South Crofty, Eliza Consols, Fowey Consols, South Frances, Kitty, West Kitty, Owles, Par Consols, North Roskear, Seton, West Seton, United Mines, Great Vor

Total No. Mines 18
Total Value of Output £12,445,118
Output as % of Cornish Total 20.1%

Mines with a Total Value of Output between £500,000 and £250,000

Alfred Consols, Balleswidden, Great Busy, East Caradon, Charlestown United, Clifford, Condurrow, Consolidated Mines, Drakewalls, West Frances, Hingston Down Consols, Holmbush, Jane, North Levant, Mark Valley,

* Value of lead output not included before 1873

Pednandrea United, Polberro, Providence Mines, Sisters, St. Day United, St. Ives Consols, South Tolgus, West Tolgus, Trumpet Consols, Uny, Great Work

Total No. Mines 26
Total Value of Output £8,813,790
Output as % of Cornish Total 14.2%

Mines with a Total Value of Output between £250,000 and £100,000

Great Alfred, Basset & Grylls, East Basset, North Basset, South Basset, Blue Hills, Boscean, Camborne Vean, Glasgow Caradon, East Carn Brea, West Chiverton, Craddock Moor, Creegbrawse, South Crinnis, East Crofty, North Crofty, West Damsel, Ding Dong, Great Fortune, West Fowey Consols, South Frances United, Gunnislake Clitters, Killefreth, Kitty, East Lovell, Margaret, Mary, Mellanear, Peevor, Pendarves United, Penhalls, Perran St. George, Poldice, Great Polgooth, North Pool, Prince of Wales, Prosper, Prosper United, Rosewarne United, Spearne Moor, Great South Tolgus, Tremayne, Tresavean, North Treskerby, Treviskey, Tywarnhaile, Wendron Consols.

Total No. Mines 47
Total Value of Output £7,187,442
Output as % of Cornish Total 11.6%

APPENDIX B

List of Mines arranged by Grid Reference

SS603485 Girt Down

SW322325 Yankee Boy
SW344704 Balmynheer
SW350318 Cape Cornwall
SW352316 St. Just Amalgamated
SW352316 St. Just United
SW353319 Porthledden
SW355322 Boswedden
SW355326 Castle
SW356312 Ballowal
SW357307 Hermon
SW358298 St. Just Consols
SW359308 Bosorne
SW359308 St. Just United, East
SW359311 Bellan
SW360328 Edward
SW362317 Cunning United
SW363309 Bosorne, East
SW364321 Boscean
SW364331 Botallack
SW366289 Trevegean, Great
SW366325 Owles
SW368303 Kelynack
SW368339 Spearne Moor
SW368346 Levant
SW371338 Spearne Consols
SW372334 Carnyorth
SW374528 Blue Hills, North
SW375345 Geevor
SW376341 Levant, North
SW380302 Augusta
SW380350 Boscaswell, North
SW381337 Bal
SW381359 Pendeen Consols
SW383339 Carne
SW383344 Boscaswell

SW387312 Balleswidden
SW389309 Balleswidden, North
SW389313 Bostraze
SW390340 Boscaswell, East
SW390340 Hearle
SW393304 Balleswidden, East
SW393304 Balleswidden, New
SW393304 Leswidden
SW396312 Botallack, East
SW396312 Metal, West
SW397275 Margaret, West
SW397312 Botrea
SW407359 Morvah Consols
SW412290 Boswarthen Penzance
          Consols
SW412290 Bosworthan
SW412290 Penzance Consols
SW415301 Ding Dong, West
SW416356 Garden
SW417296 Argus
SW417357 Morvah Hill
SW420363 Rosemergy
SW420403 Providence, North
SW421364 Carn Galver
SW421365 Bosigran Consolidated
SW436298 Balleswidden United
SW436383 Treen Downs
SW437346 Ding Dong
SW442388 Cornello
SW458379 Rosevale
SW461368 Trewey Downs
SW467352 Carnaquidden
SW469382 Sperries
SW470279 Tolvadden, West
SW482404 Brea Consols
SW482404 St. Ives, West
SW482405 Trevega

liv

SW562311 Hallamanning, East
SW563315 Anna
SW563315 Guskus
SW564312 Hallamanning
SW565293 Great Western
SW565293 Greenberry Moor
SW565293 Grylls, East
SW565329 Gurlyn
SW565338 Tremayne, West
SW566284 Keneggy
SW566363 Alfred Consols, West
SW566363 Anne
SW568290 Grylls, Great
SW568290 St. Aubyn & Grylls
SW568301 Mill Pool
SW569309 Croft Gothal
SW571305 Tindene, South
SW571319 Tregembo
SW572314 Tindene
SW572318 Relubbus
SW573332 Bosworgy
SW573342 Treven
SW573374 Alfred, North
SW573402 Trenwith
SW574306 Leeds
SW574333 Lewis Mines
SW574333 Nut
SW577282 Sydney Cove
SW577312 Work, West Great
SW578371 Alfred, Great
SW578397 Boiling Well
SW579286 Mounts Bay
SW579286 Mounts Bay Consols
SW579286 Sydney Godolphin
SW579299 Lemon
SW579341 Cuskease
SW580341 Gilmar
SW580415 Emily
SW581337 Lewis, West
SW582292 Leeds & St. Aubyn
SW582317 Godolphin, West

SW582317 Work, North Great
SW582325 Gilbert
SW582345 Providence, West
SW583353 Kayle
SW585369 Alfred Consols
SW586327 Osborne
SW586327 Tregembo, East
SW586353 Carpenter
SW587381 Angarrack Consols
SW588302 Lady Gwendolin
SW588302 Work, New Great
SW588341 Carzise, West
SW588362 Alfred, South
SW590307 Reeth
SW590307 Reeth
SW591281 Hendra, New
SW591348 Tremayne
SW591370 Herland
SW592272 Prosper
SW593375 Trungle
SW593375 Trungle
SW594577 Prosper, Great
SW594577 St. Dennis Consols
SW595293 Grey
SW595342 Carzise
SW595348 Providence
SW595575 Mary
SW595575 St. Dennis Crown
SW596307 Work, Great
SW596403 St. Andrew
SW597324 Godolphin Bridge, Great
SW597324 Godolphin, East
SW598345 Treasury, West
SW598372 Rosewarne
SW599324 Godolphin
SW600265 Penrose, New
SW600625 Trewavas
SW601331 Paul Downs
SW603335 Leedstown Consolidated
SW604283 Anne
SW604368 Relistian Consols

SW604381 Alfred Consols, East
SW607358 Unity, North
SW607368 Gwinear Consols
SW607368 Relistian, East
SW607368 Rosewarne United, West
SW607368 Rosewarne, New West
SW608288 Trevena
SW608355 Unity Consols
SW609316 Work, East Great
SW609372 Rosewarne, North
SW611366 Parbola, North
SW612376 Annie
SW613310 Polrose
SW614298 Carleen Vor, New
SW614298 Vor, West
SW614363 Jennings
SW614363 Parbola
SW614363 Parbola, South
SW614368 Rosewarne, New
SW615335 Crowan Consols
SW616334 Curtis
SW616359 Bosparra
SW619367 Rosewarne United
SW620385 Dolcoath, West
SW621292 Emma
SW621302 Vor, Great
SW621361 Rosewarne Consols
SW622307 Breage
SW622307 Gwin & Singer
SW623295 Metal & Flow
SW623295 Metal, West
SW623343 Clowance Wood
SW624309 Penhale Wheal Vor
SW624309 Vor, Penhale Wheal
SW624402 Roskear, West
SW625387 Hartley
SW627276 Hanson Mines
SW627282 Fortune, South
SW627282 Pembroke Copper
SW627290 Fortune, Great
SW627314 Vor, North

SW629324 Millett
SW630338 Abraham Consols
SW630338 Crenver & Abraham
SW630366 Copper Bottom
SW631300 Buller, Sithney
SW631300 Metal, Sithney Wheal
SW631378 Pendarves & St. Aubyn
SW631378 Rosewarne, East
SW632306 Metal, North
SW632306 Wallis
SW632337 Oatfields
SW632413 Seton, Violet
SW633306 Vor, New
SW634252 Penrose
SW634296 Carnmeal
SW634334 Crenver, South
SW634385 Dolcoath, North
SW635293 Sithney
SW635310 Harriet
SW635310 Vor, Great East
SW637406 Seton, South
SW637497 Bejawsa
SW638247 Lomax
SW638247 Rose
SW638365 Gernick
SW638400 Crane
SW640373 Grenville, West
SW640373 Trevoole
SW640382 Nelson
SW641323 Nancygollen
SW641323 Pengelly
SW641323 Worthy, Great
SW641373 Maxwell
SW641373 Tolcarne, West
SW642317 Prospidnick
SW645405 Gustavus Mines
SW646305 Vor, East
SW646410 Seton, New
SW647393 Stray Park, West
SW648331 Carbona
SW648331 Polcrebo

SW648388 Condurrow, West
SW648395 Frances
SW650363 Frances, New
SW650414 Seton, West
SW652398 Camborne Vean
SW652398 Condurrow, Great
SW652398 Dolcoath, New
SW652401 Camborne Consols
SW654389 Tryphena
SW654398 Stray Park
SW654421 Emily Henrietta
SW654421 Henrietta
SW655324 Christopher Consols
SW655381 Tolcarne
SW656318 St. Erth Alluvials
SW656386 Tolcarne, South
SW656393 Carn Camborne
SW656410 Pendarves Consols
SW656410 Roskear
SW656410 Roskear, South
SW656413 Roskear, North
SW656415 Seton
SW657358 Frances, New South
SW659395 Harriet
SW660392 Condurrow
SW660392 Pendarves United
SW660392 Tolcarne, North
SW661296 Trannack
SW661418 Seton, East
SW662405 Dolcoath
SW663386 Grenville
SW663389 Condurrow, South
SW663389 King Edward
SW663411 Crofty
SW664408 Cooks Kitchen, New
SW665407 Cooks Kitchen
SW665414 Crofty, East
SW665414 Crofty, North
SW667399 Carmarthen
SW668382 Bolenowe
SW668382 Grenville, South

SW668388 Grenville, East
SW669317 Fursden
SW669413 Crofty, South   ⅄
SW669421 Tolgus Consols
SW670409 Tincroft
SW671397 Dolcoath, South
SW671417 Union
SW672291 Treworlis
SW673297 Trenithick
SW673392 Frances, West
SW674173 Trenance
SW674302 Trumpet Consols
SW674415 Pool, East
SW674417 Agar
SW675422 Pool, North
SW676410 Carn Brea
SW677414 Illogan Mines
SW678296 Tremenhere
SW678396 Basset, West
SW679304 Trumpet, East
SW679427 Tolgus, West
SW679469 Sally
SW680297 Trevenen
SW680393 Frances United, South
SW680393 Frances, South
SW680437 Robartes
SW681280 Lovell, West
SW681280 Lovell, West Wheal
SW681317 Wendron Consols, New
SW681402 Frances, North
SW682417 Tehidy
SW683473 Towan, West
SW684298 Trevennen, New
SW684437 Maude
SW685420 Tehidy Tin Streams
SW685426 Tolgus, South
SW686431 Tolgus
SW687422 Tolgus, Great South
SW688397 Basset
SW688401 Basset, North
SW688407 Carn Brea, South

SW688440 Tolgus, Great North
SW689320 Wendron Consols
SW689431 Gilbert
SW689435 Tolgus United, Old
SW690305 Lovell, Great
SW691355 Enys
SW692394 Basset, South
SW692473 Basset Consols
SW692473 Clarence
SW693328 Basset & Grylls
SW693328 Porkellis United
SW693328 Porkellis, New
SW693442 Mary, West
SW693443 Tolgus, North
SW693443 Treleigh Wood United
SW693470 Nancekuke
SW694367 Polgear
SW694401 Basset, East
SW694409 Uny
SW694426 Alice
SW694426 Redruth Consols
SW694426 Tolgus, East
SW694432 Mary
SW694475 Lydia
SW695340 Boswin
SW695388 Buller & Basset United
SW695405 Buller, North
SW696439 Treleigh Consols, New
SW696475 Towan, South
SW696481 Towan
SW697342 Wendron United
SW697347 White Alice
SW697414 Carn Brea, East
SW697474 United Hills
SW697492 Charlotte
SW699308 Lovell, New
SW699329 Garlidna
SW699329 Lovell, North
SW699348 Calvadnack
SW699401 Basset & Buller Consols
SW699401 Buller

SW699433 Treleigh Wood
SW699472 Towan, New
SW700501 Coates
SW701322 Ruby
SW701399 Buller, West
SW701435 Harmony
SW701598 Polberro, West
SW702301 Mengern
SW702411 Bucketts
SW702411 Clijah & Wentworth
SW702411 Perseverance
SW702411 Uny, East
SW702467 Basset, Old
SW702472 Tywarnhaile
SW702501 Polberro, East
SW703391 Buller, South
SW703391 Penstruthal, West
SW703420 Pednandrea United
SW703471 Music, East
SW703488 Charlotte United
SW703488 Charlotte, New
SW704304 Combellack
SW704318 Lovell, East
SW704417 Sparnon
SW704469 Ellen
SW705301 Lovell, South
SW705301 Wendron, South
SW705304 Lovell
SW705353 Polhigey Moor
SW705440 Peevor, West
SW705445 North Downs
SW705492 Charlotte, East
SW706304 Tregunstic
SW706335 Medlyn Moor
SW706428 Louisa
SW707398 Buller, Penstruthal
SW707398 Penstruthal Consols
SW707405 Copper Hill
SW707410 Trefusis
SW707447 Prussia
SW707447 Treleigh Wood, East

SW707476 Tywarnhaile, East
SW707483 Tallack
SW708312 Lovell, New East
SW708393 Penstruthal, South
SW708414 Trefusis, East
SW708436 Cardrew
SW708442 Peevor
SW708442 Peevor United
SW708466 Ellen, South
SW709318 Penstruthal, North
SW709458 Stencoose
SW711419 Grambler & St. Aubyn
SW711419 St. Aubyn
SW711508 Trevaunance
SW712391 Carvennal
SW712423 Cupid
SW712438 Boys
SW712453 Hallenbeagle, North
SW713406 Buller, East
SW713406 Pennance
SW713443 North Downs, Great
SW713446 Briggan, Great
SW714386 Trethellan, West
SW714423 Grambler, North
SW714442 St. Aubyn United
SW714468 Ellen, East
SW714501 Polbreen, West
SW715397 Bell Vean
SW715397 Tresavean, West
SW715515 Polberro
SW716329 Retanna Hill
SW716390 Brewer
SW717390 Trethellan
SW717392 Tretharup
SW717437 Treskerby
SW717504 Polbreen, New
SW717513 Turnavore
SW717513 Turnavore Stamps
SW718395 Comford
SW718395 Tresavean, North
SW718416 Cathedral

SW718449 Rose
SW718462 Tregullow Consols
SW718503 Kitty, New
SW718503 Livingstone Consols
SW718503 Polbreen
SW718513 Coit
SW718513 Coit Stamps
SW718513 Coit, West
SW719508 Kitty, West
SW719514 Friendly Stamps
SW720451 Treskerby, North
SW720512 Friendly
SW721394 Tresavean
SW722422 Gorland, West
SW723430 Parkanchy
SW723515 St. Agnes Consols
SW723516 Penhalls
SW724414 Damsel United
SW724414 Damsel, West
SW724427 Pink
SW724456 Scorrier Consols
SW724505 Gooninnis
SW725449 East Downs
SW725810 Rock
SW726422 Damsel, North
SW726422 Jewell, West
SW726509 Kitty
SW727396 Treviskey
SW727428 Clinton
SW727449 Hallenbeagle
SW728417 Damsel
SW728458 Britain
SW728460 Treskerby, East
SW728517 Blue Hills
SW730402 Edgecombe
SW730410 Clifford, West
SW730410 Ting Tang Consols
SW730427 Gorland
SW730427 Gorland
SW731402 Clifford, New
SW731407 Moyle

SW732443 Black Dog
SW732452 Boscawen
SW733419 Damsel, East
SW733433 Unity Wood, West
SW733518 Betsy
SW734294 Vyvyan
SW734458 Busy, North
SW734528 Ocean
SW735421 Jewell
SW735425 St. Day United
SW735435 Poldice, West
SW735435 Tolgullow United
SW735435 Tolgullow United Mines
SW735435 Unity Wood
SW735443 Killefreth
SW736430 Unity No.1
SW736430 Unity No.2
SW736528 Perran Great St. George, West
SW736528 Prudence
SW736528 St. George, West
SW738420 Carharrach
SW738452 Seymour
SW738518 Blue Hills, East
SW738536 Prudence, South
SW739448 Busy, Great
SW739448 Busy, Great
SW739490 Burrow & Butson
SW740427 Poldice
SW740427 Poldice & East Maid
SW740427 Poldice (Behenna)
SW740427 Poldice (H.T.Trythall)
SW740427 Poldice (Letcher)
SW740427 Poldice (R.Martin & Son)
SW740427 Poldice (W.J.Trythall)
SW740427 Poldice (Waters & Teague)
SW740427 Poldice Halvaners
SW740427 Poldice, Part of
SW740427 Poldice, Old (Part of)
SW742423 Maid

SW742423 Maiden
SW742534 Good Fortune
SW743419 Virgin, West
SW745412 Ale & Cakes, South
SW745412 Gwennap United
SW745412 United Mines
SW745435 Creegbrawse
SW745435 Creegbrawse (Behenna)
SW746281 Constantine
SW746535 Perran St. George
SW746535 St. George, Great
SW747420 Virgin
SW747466 Burra Burra
SW747496 Chiverton, Great West
SW748407 Britannia
SW748421 Consolidated Mines
SW748437 Penkevill
SW748522 Prince Royal
SW749449 Daniel
SW749515 St. George
SW750487 Silverwell
SW751495 Perran Wheal Vyvyan
SW751542 Perran No.1
SW753426 Henry
SW754539 Perran United
SW754545 Droskin
SW755402 Clifford United, South
SW756417 Clifford
SW756417 Clifford Amalgamated
SW756417 Clifford United
SW757417 Poldory
SW757526 Penwortha
SW757542 Leisure
SW758282 Anna Maria
SW758519 Shepherds, West
SW758542 Leisure, East
SW761378 Roskrow United
SW761378 Tresavean, South
SW761419 Andrew
SW761511 Lambriggan
SW761529 Leisure, South

SW761529 Truro
SW761545 Leisure, North
SW761580 Perran Silver-lead Consols
SW761590 Golden
SW762548 Ramoth
SW762580 Penhale
SW762580 Penhale & Lomax
SW763534 Alfred
SW763534 Leisure, New
SW764420 Nangiles
SW764574 Penhale
SW765377 Magdalen
SW765575 Gravel Hill
SW765581 Phoenix
SW767558 Budnick, North
SW767558 Perran Consols
SW767558 Vlow
SW767581 Golden, East
SW769430 Hope
SW770433 Falmouth & Sperries
SW772493 Chiverton United
SW772493 Perran Wheal Virgin
SW772494 Callestock, Great
SW773547 Budnick Consols
SW773547 Perran, Wheal
SW775565 Halwyn
SW776413 Bissoe Pool
SW776529 Anna, West
SW776529 Chiverton, New
SW777421 Baddern, Great
SW777427 Jane, West
SW779563 Perran No.2
SW779563 Trebisken
SW781430 Falmouth, East
SW781565 Mount No.1
SW781565 Mount No.2
SW781565 Perran Iron
SW782501 Chiverton Valley
SW783434 Jane
SW783454 Penhaldarva, South
SW785504 Callestock Moors, Great

SW785504 Chiverton Moor
SW785553 Budnick, East
SW785570 Trebisken Green
SW785571 Cubert
SW786434 Kea Tremayne
SW786548 Hope
SW786559 Treamble
SW787514 Wentworth Consols
SW787562 Retallack, North
SW788530 Anna
SW788530 Chiverton, North
SW789439 Basore
SW789440 Jane, North
SW789461 Penhaldarva
SW792501 Chiverton, Great South
SW792558 Retallack, Great
SW793508 Chiverton, West
SW794515 Mineral Bottom
SW795510 Chiverton, East
SW796556 Duchy & Peru
SW797512 Chiverton
SW797512 Cornubian
SW797537 Albert
SW797537 Prince Albert Consols
SW800492 Chiverton, South
SW801312 Swanpool
SW803313 Kimberley
SW803313 Lord Kimberly
SW807555 Deer Park
SW808385 Restrongnet
SW808385 Restrongnet Stream
SW810613 Newquay
SW811547 Shepherds, North
SW814336 Clinton
SW814581 Trerew
SW817541 Shepherds, Old
SW818536 Rose & Chiverton
SW819477 Garras, South
SW819623 Rosecliff
SW820512 Chiverton Consols, Great
SW823611 Trelogan

SW826552 Penhallow Moor
SW833546 Constance
SW834541 Cargol, South
SW835542 Cargol
SW837554 Rose, East
SW838560 Rose, North
SW849616 Trewollack Wheal Rose, Great
SW849616 Treworlack
SW849690 Carnewas
SW870558 Pencorse Consols
SW870561 Trevissa
SW891576 Troon
SW894606 Trugo
SW896773 Padstow Silver Lead
SW902561 Trefulach
SW905564 Burthy Row
SW905564 Chyprase Consols
SW905578 Penhale & Barton
SW905578 Penhale Moor
SW907568 Bennallack
SW911558 Toldish
SW911572 Trewheela
SW911588 Park of Mines
SW911588 Parka Consols
SW912768 Treleather, North
SW913731 Creddis
SW915525 Ladock
SW917538 Alice
SW917557 Burthy
SW917558 Chytane
SW918586 Fatwork & Virtue
SW918586 Indian Queens Consols
SW920571 Edith
SW920571 Retew
SW921606 Treliver
SW921695 Trelow
SW923598 Toldis
SW923601 Indian Queens No.1
SW923601 Indian Queens No.2
SW923601 Ruthers

SW923601 Ruthers (Iron)
SW927766 Porthelly, North
SW929523 Blencowe Consols
SW929523 Lambert
SW932528 Terras
SW934537 Treweetha
SW935524 Terras, South
SW935524 Tolgarrick
SW935524 Uranium Mines
SW935592 Gavrigan
SW937509 Crow Hill, New
SW938791 Polzeath
SW941799 Pentire Glaze
SW942530 Park Wynn
SW946519 Bodinnick
SW946554 Trethosa
SW948592 Alfred James
SW949528 Gwindra
SW950600 Goss Moor
SW950616 Castle An Dinas
SW952505 Down Derry Consols
SW952505 Grenville
SW952627 Tregonetha
SW952701 Pawton
SW955514 Fortescue
SW955717 Carthew Consols
SW955717 St. Issey
SW957543 Tin Hill
SW962504 Ventonwyn
SW963514 Dowgas, Great
SW967512 St. Austell Consols
SW968502 Kingsdown
SW970528 Lanjeth
SW972625 Belovely
SW972625 Belowda
SW972625 Belowda Hill
SW972625 Roche Consols
SW974654 Skews, Little
SW976502 Hewas, Great
SW977617 Tregoss
SW978546 Alviggan

SW978618	Brynn Royalton	SX008799	Bodannon
SW978618	Royalton, Great	SX009582	Goonbarrow
SW978701	Molesworth	SX009587	Beam
SW980509	Ninnis Downs	SX013582	Carnesmerry
SW980600	Dyehouse	SX013795	Poltravorgie
SW980600	Trerank	SX014582	Rock Hill
SW981501	Polgooth, West	SX014582	Rocks
SW981542	Burngullow	SX014605	Cariggan Rock
SW983503	Woodclose	SX016595	Hallew
SW985499	Hewas, New	SX016595	Hallow
SW985584	Coldvreath	SX016659	Mulberry
SW985584	Killivreath	SX017594	Tresibble
SW986653	Lanjew	SX018577	Kerrow Moor
SW987497	Commerce	SX019579	Anne
SW987657	Blackhay	SX019658	Mulberry, East
SW988597	Magnetic	SX020613	Savath
SW988597	Tower Consols	SX020800	Treore
SW989499	Polgooth, South	SX021648	Tremoor
SW989499	Treloweth	SX021670	Nantallon
SW991623	Brynn	SX024591	Mollinis
SW994508	Polyear	SX025542	Carclaze
SW998528	Gover	SX027636	Bigbees
SW998597	Cornubia	SX027636	Lanivet
		SX027636	West Down
SX001504	Polgooth Tin Mine	SX027636	Woodley
SX001504	Polgooth United	SX028819	Sampson, Royal
SX001504	Polgooth, Great	SX029518	Carvath United
SX001504	Polgooth, New	SX029518	Polmear, West
SX001504	Polgooth, North Great	SX030517	Polmear
SX001504	Polgooth, Old	SX030562	Carbis
SX001736	Tregordon	SX030642	Prosper
SX006645	Retire	SX031594	Menadue
SX006645	Withiel	SX033564	Treverbyn
SX006653	James	SX033593	Canna
SX007568	Blue Barrow	SX034560	Knightor
SX007571	Bonney, North	SX035515	Polmear, South
SX007571	Bunnie	SX035554	Pentruff
SX007571	Shelton	SX035554	Trethurgy
SX007621	Burney House	SX035557	Ruby
SX007637	Colbiggon	SX037528	Bucklers
SX007784	Trevinnick	SX037673	Boscarne

SX038546	Garker	SX067641	Tretoil
SX038625	William	SX071554	Prideaux
SX040528	Cuddra	SX072539	Fowey Consols, West
SX040531	Charlestown United	SX072557	Prideaux Wood
SX040531	Charlestown, New	SX073533	Par Consols, West
SX040531	Eliza, West	SX074566	Fowey Consols, New
SX040537	Boscoppa	SX080551	Fowey Consols, South
SX041653	Mary Consols	SX083558	Fowey Consols
SX041839	Tregardock	SX084567	Palharmon
SX044634	Reperry	SX084567	Polharman
SX045522	Apple Tree	SX084624	Maudlin
SX045522	Crinnis, South	SX094843	Trethin
SX045531	Boscundle	SX096777	Onslow Consols, Great
SX045540	Tregrehan Consols	SX096777	St. Breward Consols
SX046523	Crinnis, West	SX097583	Pelyn Wood
SX046532	Eliza Consols	SX098614	Restormel Royal
SX047527	Cuddra, South	SX098634	Respryn
SX047530	Boscundle, New	SX107629	Duke of Cornwall
SX049519	Crinnis Consols	SX111676	Glynn
SX049519	Crinnis, Great	SX122596	Silver Vein
SX050520	Carlyon	SX122606	Fortescue, North
SX050523	Regent	SX136651	Sicily
SX051790	Pengenna	SX136655	Jane, East
SX051890	King Arthur	SX149718	Hardhead
SX054558	Par	SX151695	Cabella
SX055554	Restinnis	SX151695	Treveddoe, Great
SX055633	Trebell Consols	SX151706	Esther United
SX056554	Treffry	SX152696	Whisper
SX057530	Pembroke, New	SX158685	Carn Vivian
SX057797	Treburgett, Old	SX164575	Rashleigh
SX058526	Pembroke	SX170698	Gazeland
SX058530	Par Consols	SX174647	Allen
SX059562	Lady Rashleigh	SX174647	Bodithiel
SX063640	Messer	SX175687	Tregeagle
SX065528	Crinnis, East	SX179599	Botelet
SX065641	Bodmin United	SX179599	Herodsfoot, West
SX066553	St. Blazey	SX179678	Goonzion
SX066553	St. Blazey Consols	SX181685	Trevenna
SX066553	St. Blazey, New	SX184646	St. Neot
SX067641	Lanivet Consols	SX184646	Trevenna, South
SX067641	Lanivet Tin	SX184684	Robins

SX186694	Hobbs Hill	SX247690	Caradon, West South
SX187672	Mary Great Consols	SX248699	Norris
SX187672	Tin Valley	SX249691	Caradon United
SX188693	Hammet Consols	SX253855	West Down End
SX188788	Tresellyn	SX255716	Silver Valley
SX188788	Treselyan & Scaddick Consols	SX255776	Trebartha Lemarne
		SX256697	Caradon, New West
SX194674	Ambrose Lake	SX256721	Phoenix, West
SX196789	Dorothy	SX256721	Phoenix, West
SX198789	Jane	SX256776	Luskey
SX198790	Canaframe	SX257698	Caradon Consols
SX198823	Bray	SX259702	Craddock Moor
SX202699	Northwood	SX260732	Sharp Tor, West
SX204670	Coryton	SX262697	Agar, East
SX204670	Killham	SX262700	Caradon, West
SX204670	Tamworth	SX262706	Gonamena
SX205690	Bowden	SX262716	Phoenix, South
SX205690	Carpuan	SX267724	Phoenix
SX207795	St. Vincent	SX268698	Caradon, South
SX209747	Goodevere	SX268718	Phoenix Dunsley
SX209747	Tin Era	SX271756	Caradon & Phoenix Consols
SX209794	Horse Burrow		
SX209794	Wilhelmina	SX272697	Hooper, Caradon South Wheal
SX209795	Vincent		
SX211787	Halvinna	SX273713	Rose Down, West
SX212594	Herodsfoot, South	SX273734	Phoenix, North
SX212600	Herodsfoot	SX274722	Phoenix, East
SX212604	Herodsfoot, North	SX275636	Mary Ann, West
SX212800	Trewint Consols	SX277704	Caradon, East
SX219603	Herodscoombe	SX280718	Marke Valley
SX224797	Tregune Consols, Great	SX282702	Caradon Consols, Glasgow
SX226688	Dreynes & Netherton, East		
		SX286751	Rodd, Great
SX229717	Trewitten	SX287635	Trelawney
SX230692	Caradon, New South	SX287638	Trehane
SX235794	Phoenix, New	SX287666	Butterdon
SX237794	Annie	SX288634	Mary Anne
SX237794	Treburland	SX289646	Hony
SX241850	Lidcott	SX291654	Treweatha
SX242689	Penhale & Larkholes	SX291708	Tokenbury
SX245700	Pollard	SX292675	Penhanger

SX292680 Gill
SX292680 Gill, Glasgow
SX292680 Hayford Ltd.
SX293663 Venton
SX293820 Botathan
SX297656 Trelawney, North
SX297665 Wrey Consols
SX297757 Coads Green
SX298661 Ludcott
SX298702 Caradon
SX298702 Slade
SX298707 Caradon, Great
SX302679 Wrey, North
SX302697 Trebeigh Consols
SX303693 Ida
SX303726 Plusheys
SX305723 Beneathwood
SX305857 Truscott
SX305861 Atway
SX308696 Bicton
SX311683 Trelawney, New
SX328788 Trebullett
SX329790 Trenute
SX345695 Haye Valley
SX356677 Callington, South
SX356677 Colquite & Callington
         United
SX356711 Redmoor
SX357710 Callington
SX358721 Holmbush
SX359719 Emmens United
SX360597 Tredinnick
SX361714 Kelly Bray
SX362757 Kingston Consols
SX362794 Greystone Wood
SX363706 Callington Consols
SX363706 Florence
SX364831 Lawhitton Consols
SX365731 Downgate Consols
SX365731 Holmbush, North
SX368702 Lady Ashburton

SX369705 Tonkin
SX372737 Martha, West
SX372737 Sheba Consols, Great
SX372737 Trehill
SX374710 Kit Hill, South
SX375713 Kit Hill United
SX377696 Silver Hill
SX378705 Princess of Wales
SX381723 Kit Hill Tunnel
SX381724 Excelsior
SX383695 Baring
SX383695 Cornwall & St. Vincent
SX383695 Langford
SX385702 Silver Valley
SX388737 Consols, New Great
SX388737 Martha
SX388737 Martha, New
SX389707 Prince of Wales, West
SX389711 Kit Hill, East
SX391701 Brothers
SX391701 Emily
SX391730 Deer Park
SX394701 Fortune
SX394701 Queen
SX395738 Devon Consols, West
SX397731 Benny
SX399699 Barnard
SX399699 Newton
SX399699 Newton Silver Mine
SX400702 Calstock United
SX401705 Prince of Wales
SX402701 George
SX408738 Williams
SX409708 Drakewalls, West
SX409714 Hingston Downs Consols
SX409732 Cornish, New
SX409732 Duchy Great Consols
SX409732 Latchley
SX413734 Maria, South
SX417723 River Tamar
SX417723 Tamar River

SX421720 Clitters United
SX421720 Gunnislake Clitters
SX423692 Cotehele Consols
SX423718 Gunnislake, East
SX424707 Drakewalls
SX426696 Calstock & Danescombe
SX426696 Calstock Consols
SX426696 Calstock, East
SX426696 Danescombe Valley
SX427718 Dimson
SX428699 Edward
SX428724 Hawkmoor, West
SX430719 Gunnislake, Old
SX431696 Zion
SX433629 Tamar, New South
SX433771 Collacombe

SX434700 Arthur
SX434727 Hawkmoor
SX435697 Morshead
SX435718 Bedford, South
SX437698 Slimeford
SX444689 Okel Tor
SX450695 Harewood Consols
SX463701 Devon & Cornwall
SX463820 Bowden Down
SX483598 Whitleigh
SX483826 Langstone
SX512823 Friendship, North
SX630366 Hender
SX773771 Haytor
SX836837 Adams

Tables of Mine

Production

Ownership

Management

and Employment

ABRAHAM CONSOLS   CROWAN   SW 630338 0001

Production: Copper   Ore(tons) Metal(tons)   Value(£)
            1854   324.00   5.50   169.20
            Comment 1854 (C) ABRAM
            Tin No detailed return
Management: Manager 1866—1867 JOS.VIVIAN

ADAMS   CHRISTOW, DEVON   SX 836837 0003

Production: Zinc   Ore(tons) Metal(tons)   Value(£)
            1859   206.00   0.00   721.00
            Comment 1859 ADDAMS

AGAR   ILLOGAN   SW 674417 0004

Production: Copper   Ore(tons) Metal(tons)   Value(£)
            1847   593.00   41.90   2839.50
            1848   997.00   66.20   3749.60
            1849   1493.00   91.50   5894.90
            1850   866.00   57.30   3839.60
            1851   806.00   70.40   4872.90
            1852   652.00   42.40   3404.30
            1854   69.00   5.70   583.80
            1855   88.00   6.00   617.70
            1856   158.00   9.00   828.40
            1857   161.00   10.50   1069.20
            1858   280.00   22.80   2130.80
            1859   195.00   14.50   1400.70
            1860   330.00   22.80   2078.30
            1861   364.00   26.90   2476.70
            1862   667.00   57.30   5062.70
            1863   145.00   9.90   787.50
            1864   82.00   5.50   454.90
            1865   42.00   2.80   262.10
            1866   28.00   1.80   122.50
            1867   19.00   1.10   72.70
            1868   10.00   0.70   42.20
            1870   12.00   1.00   65.40
            1871   6.00   0.50   29.10
            1876   65.00   3.80   237.10
            1877   23.20   2.00   106.50
            1879   16.00   1.40   74.50
            1880   13.00   1.00   52.50
            1884   16.00   0.00   71.00
            1885   25.00   0.00   89.00
            1886   21.00   0.00   41.00
            1887   87.00   0.00   95.00
            1888   276.00   0.00   414.00
            1889   9.00   0.00   33.00
            1890   93.00   0.00   261.00
            1891   43.00   0.00   175.00
            1892   259.00   0.00   1020.00
            1893   43.00   0.00   155.00
            1894   76.00   0.00   139.00

                                    1

Copper   Ore(tons) Metal(tons)   Value(£)
1897–1899 No detailed return
1901–1913 No detailed return
Comment 1847–1852 (C); 1854–1868 (C); 1870–1871 (C); 1876
(C); 1887 TWO RETURNS AGGREGATED; 1888 ORE EST; 1891–1892
UNDRESSED ORE; 1897–1899 SEE EAST POOL; 1901–1913 SEE EAST
POOL

Tin	Black(tons)	Stuff(tons)	Tin(tons)	Value(£)
1855	0.30	0.00	0.00	11.50
1856	0.20	0.00	0.00	22.40
1857	0.00	0.00	0.00	2.80
1858	0.00	0.00	0.00	42.20
1859	0.00	0.00	0.00	224.40
1860	2.20	0.00	0.00	141.50
1861	6.80	0.00	0.00	477.20
1862	8.00	0.00	0.00	481.20
1863	3.30	0.00	0.00	310.60
1864	4.30	0.00	0.00	365.30
1865	3.00	0.00	0.00	420.80
1866	5.70	0.00	0.00	961.20
1867	2.30	0.00	0.00	579.30
1868	1.80	0.00	0.00	241.60
1869	0.00	0.00	0.00	1676.20
1870	3.00	0.00	0.00	2123.00
1871	2.20	0.00	0.00	645.20
1872	0.50	0.00	0.00	38.10
1876	0.00	175.50	0.00	939.00
1877	0.00	2784.80	0.00	4268.30
1878	0.00	2791.80	0.00	4687.00
1879	5.60	4205.80	0.00	6560.00
1880	108.90	0.00	0.00	5435.20
1881	93.10	0.00	65.20	5242.70
1882	34.70	0.00	24.30	2080.00
1883	222.60	0.00	0.00	11704.00
1884	582.90	0.00	0.00	26287.00
1885	637.40	0.00	0.00	32301.00
1886	641.70	0.00	0.00	35750.00
1887	675.30	0.00	0.00	42251.00
1888	514.50	0.00	0.00	33914.00
1889	160.70	56.00	0.00	8106.00
1890	261.20	0.00	0.00	13287.00
1891	339.80	0.00	0.00	17149.00
1892	413.00	0.00	0.00	20278.00
1893	389.40	0.00	0.00	17128.00
1894	252.60	73.80	0.00	9721.00
1895	81.40	6.10	0.00	2795.00
1896	0.00	45.00	0.00	70.00

1897–1913 No detailed return
Comment 1856 VALUE INC.TIN STUFF; 1857–1859 TIN STUFF;
1863–1868 VALUE INC.TIN STUFF; 1869 TIN STUFF; 1870–1871
VALUE INC.TIN STUFF; 1879 VALUE INC.TIN STUFF; 1889 TWO
RETURNS AGGREGATED; 1897–1913 SEE EAST POOL

Arsenic	Ore(tons)	Metal(tons)	Value(£)
1880	42.60	0.00	294.80
1881	54.80	0.00	264.90

2

Arsenic	Ore(tons)	Metal(tons)	Value(£)
1882	12.00	0.00	60.00
1883	116.10	0.00	728.00
1884	190.10	0.00	1177.00
1885	297.70	0.00	1885.00
1886	275.00	0.00	1778.00
1887	263.00	0.00	1608.00
1888	311.00	0.00	1897.00
1889	303.00	0.00	1676.00
1890	254.00	0.00	1693.00
1891	214.00	0.00	1615.00
1892	331.00	0.00	2167.00
1893	230.00	0.00	1680.00
1894	174.00	0.00	1506.00
1895	82.00	0.00	818.00
1896	15.00	0.00	76.00

1897-1913 No detailed return
Comment 1884-1885 SOOT; 1886 CRUDE; 1897-1913 SEE EAST POOL

Tungsten	Ore(tons)	Metal(tons)	Value(£)
1884	1.10	0.00	15.00
1885	No detailed return		
1886	2.00	0.00	30.00

1898-1913 No detailed return
Comment 1898-1913 SEE EAST POOL

Ownership: 1877-1886 WHEAL AGAR MINING CO.; 1887-1890 WHEAL AGAR
ADVENTURERS; 1891-1893 WHEAL AGAR ADVENTURERS C.B.; 1894-1896
WHEAL AGAR ADVENTURERS
Comment 1897-1913 SEE EAST POOL

Management: Manager 1859-1865 WM.ROBERTS; 1866-1874 EDM.ROGERS; 1881
W.C.TREVENA
Chief Agent 1859-1869 JAS.CHAMPION; 1875 WM.TEAGUE JNR.
ED.MOYLE; 1876 ED.MOYLE; 1877-1879 WM.HAMBLY ED.MOYLE; 1880
WM.HAMBLY; 1881 RALPH DANIELL; 1882-1884 W.C.TREVENA;
1885-1888 WM.T.WHITE; 1889 HY.CARTER; 1892-1896 RALPH
DANIELL
Secretary 1859-1861 RICH.LYLE (P); 1862-1869 GEO.LIGHTLEY;
1870-1876 CORN.BAWDEN; 1877-1881 CORN.BAWDEN (S); 1887-1891
CORN.BAWDEN; 1892-1896 CORN.BAWDEN (P)

Employment:	Underground	Surface	Total
1878	59	34	93
1879	51	60	111
1880	66	81	147
1881	58	86	144
1882	42	85	127
1883	136	118	254
1884	198	134	332
1885	222	129	361
1886	146	176	322
1887	189	211	400
1888	153	173	326
1889	129	135	264
1890	139	138	277
1891	154	142	296
1892	177	152	329
1893	142	140	282

AGAR   ILLOGAN   Continued

                    Underground   Surface   Total
          1894   123   139   262
          1895   5   5   10
          1896   10   13   23
          1897-1913
          Comment 1897-1913 SEE EAST POOL

AGAR,EAST   ILLOGAN   0005

Production: Copper No detailed return
Management: Manager 1859-1861 WM.ROBERTS
            Chief Agent 1859-1861 JAS.CHAMPION
            Secretary 1860-1861 RICH.LYLE (P)

AGAR,EAST   ST.CLEER   SX 262697 0006

Production: Copper   Ore(tons) Metal(tons)   Value(£)
            1865   24.00   1.60   124.20
            Comment 1865 (C)
Management: Manager 1861-1866 FRAN.PRYOR; 1867 WM.JOHNS
            Chief Agent 1860-1864 WM.JOHNS; 1865-1867 JAS.WILLIAMS
            Secretary 1860-1864 FRAN.PRYOR (P); 1865-1867 G.B.COLLINS

ALBERT   PERRANZABULOE   SW 797537 0007

Production: Lead & Silver   Ore(tons) Lead(tons) Silver(ozs)   Value(£)
            1852   14.90   11.20   0.00   0.00
            1853   No detailed return
            1854   22.50   17.00   68.00   0.00
Ownership: Comment 1880 SEE SHEPHERDS WHEAL ROSE UNITED

ALE AND CAKES,SOUTH   GWENNAP   SW 745412 0008

Production: Tin   Black(tons) Stuff(tons)   Tin(tons)   Value(£)
            1905   0.00   15.00   0.00   54.00
            1906   4.60   0.00   0.00   396.00
Ownership: 1905 R.H.MOORE
Employment: Underground   Surface   Total
            1905   4   5   9

ALFRED   PERRANPORTH   SW 763534 0009

Production: Tin No detailed return
Ownership: 1905 ALFRED MINES LTD.
            Comment 1905 PREVIOUSLY NEW LEISURE
Management: Chief Agent 1905 F.MARSHALL

                                4

ALFRED CONSOLS   PHILLACK   SW 585369 0011

Production: Lead   Ore(tons) Metal(tons)   Value(£)
            1857   0.70   0.50   0.00
            1858   2.40   1.80   0.00
            1859   No detailed return
            Comment 1858 FOR AG.SEE GREAT ALFRED
            Zinc   Ore(tons) Metal(tons)   Value(£)
            1857   20.00   0.00   35.00
            1858   54.40   0.00   126.00
            1859   52.00   0.00   120.90
            1860   115.30   0.00   245.10
            Copper   Ore(tons) Metal(tons)   Value(£)
            1846   147.00   10.50   744.30
            1847   1660.00   95.40   6477.80
            1848   1148.00   57.20   3125.30
            1849   256.00   12.10   687.50
            1850   1595.00   154.50   10916.90
            1851   2819.00   271.00   19210.70
            1852   3892.00   347.20   29561.20
            1853   3977.00   398.10   40386.20
            1854   3712.00   337.40   35767.80
            1855   3467.00   276.70   29036.30
            1856   3568.00   295.40   28657.40
            1857   4505.00   322.80   32885.60
            1858   4214.00   282.00   25556.20
            1859   3538.00   211.00   19633.60
            1860   1764.00   100.20   9260.90
            1861   1538.00   95.00   8500.30
            1862   1752.00   89.20   7249.90
            1863   1104.00   50.60   3734.10
            1864   61.00   4.20   362.00
            Comment 1846-1864 (C)
            Tin   Black(tons) Stuff(tons) Tin(tons)   Value(£)
            1860   0.00   0.00   0.00   21.40
            1863   0.00   0.00   0.00   50.90
            Comment 1860 TIN STUFF; 1863 TIN STUFF
Ownership: Comment 1863-1865 SUSPENDED
Management: Manager 1859 THOS.TRELEASE; 1861-1862 JOHN THOMAS
            Chief Agent 1859 THOS.HOSKING; 1860 THOS.TRELEASE
            THOS.HOSKING; 1861 THOS.HOSKING THOS.BAWDEN; 1862
            THOS.BAWDEN
            Secretary 1859-1861 HY.NOELL (P); 1862 T.ROBINSON

ALFRED CONSOLS,EAST   PHILLACK   SW 604381 0012

Production: Zinc   Ore(tons) Metal(tons)   Value(£)
            1857   4.00   0.00   7.90
            1858   14.00   0.00   38.50
            1859   18.00   0.00   50.00
            Copper   Ore(tons) Metal(tons)   Value(£)
            1857   288.00   25.20   2620.30
            1858   291.00   21.00   2023.20
            1859   152.00   10.80   1041.40
            1860   184.00   11.70   1131.50
            1861   440.00   26.90   2519.90

                                5

ALFRED CONSOLS,EAST   PHILLACK   Continued

Copper	Ore(tons)	Metal(tons)	Value(£)
1862	545.00	27.20	2198.70
1863	124.00	6.40	488.20
1864	5.00	0.20	14.60

Comment 1857-1864 (C)
Ownership: Comment 1863-1865 SUSPENDED
Management: Manager 1859-1862 JOS.VIVIAN
Chief Agent 1859-1862 HY.SKEWIS
Secretary 1859-1862 WM.PAINTER (P)

ALFRED CONSOLS,WEST   PHILLACK   SW 566363 0013

Production:
Lead & Silver	Ore(tons)	Lead(tons)	Silver(ozs)	Value(£)
1856	1.60	1.10	0.00	0.00
1861	3.00	1.80	28.00	0.00
1863	0.30	0.20	0.00	0.00
1865	0.10	0.05	0.00	0.00

Zinc	Ore(tons)	Metal(tons)	Value(£)
1862	22.50	0.00	23.50
1863	5.00	0.00	3.80
1864	2.00	0.00	2.00

Comment 1863 WEST ALFRED

Copper	Ore(tons)	Metal(tons)	Value(£)
1851	130.00	7.80	502.00
1852	623.00	35.90	2952.60
1853	471.00	21.40	2017.50
1854	692.00	29.70	2863.70
1855	840.00	32.20	2995.50
1856	639.00	26.80	2266.40
1857	1118.00	48.20	4438.90
1858	810.00	36.70	3080.40
1859	833.00	33.80	2993.80
1860	769.00	33.40	2925.50
1861	807.00	30.50	2517.10
1862	434.00	11.20	732.00
1863	102.00	2.50	135.00
1864	111.00	2.20	123.50
1865	51.00	0.90	35.50

Comment 1851-1865 (C)

Tin	Black(tons)	Stuff(tons)	Tin(tons)	Value(£)
1862	0.00	0.00	0.00	5.10

Comment 1862 TIN STUFF - PART YEAR ONLY
Ownership: Comment 1864-1865 SUSPENDED
Management: Manager 1859-1861 STEP.LEAN
Chief Agent 1859-1861 RICH.STEPHENS; 1862 W.PAULL
RICH.STEPHENS
Secretary 1859 RICH.NICHOLLS (P); 1860-1861 JOHN TREBILCOCK
(P); 1862 H.NISBETT (P)

ALFRED JAMES   ST.DENNIS   SW 948592 0014

Production: Tin No detailed return
Management: Chief Agent 1861-1862 MICHELL; 1863-1865 MICHELL H.GROSE

ALFRED JAMES   ST.DENNIS   Continued

Secretary 1861—1865 JAS.HAWKER(P)

ALFRED,GREAT   PHILLACK   SW 578371 0015

Production: Lead & Silver   Ore(tons) Lead(tons) Silver(ozs)   Value(£)
            1855   2.70   1.70   0.00   0.00
            1858   1.20   0.70   320.00   0.00
            1859   5.20   3.70   0.00   0.00
            1860   1.90   1.30   1200.00   0.00
            1861   1.20   0.80   12.00   0.00
            1863   0.20   0.10   0.00   0.00
            Comment 1858 AG.INC.4 OTHER MINES; 1860 AG.INC.5 OTHER MINES
            Copper   Ore(tons) Metal(tons)   Value(£)
            1852   273.00   16.20   1444.50
            1853   727.00   51.20   4920.70
            1854   1011.00   52.40   5095.90
            1855   1965.00   98.70   9728.80
            1856   2264.00   130.10   11953.40
            1857   2311.00   127.70   12578.40
            1858   2782.00   158.30   14038.20
            1859   2734.00   147.10   13617.00
            1860   2570.00   139.20   12703.30
            1861   3800.00   194.70   16784.50
            1862   1015.00   41.40   3097.40
            1863   119.00   4.10   272.70
            1864   55.00   2.60   203.70
            1865   30.00   1.60   137.30
            Comment 1852—1865 (C)
            Tin   Black(tons) Stuff(tons) Tin(tons)   Value(£)
            1855   0.80   0.00   0.00   41.30
            1856   0.00   0.00   0.00   92.80
            1857   0.00   0.00   0.00   191.70
            1858   0.00   0.00   0.00   447.20
            1859   0.00   0.00   0.00   738.30
            1860   0.00   0.00   0.00   1412.70
            1861   0.00   0.00   0.00   668.30
            1862   0.00   0.00   0.00   744.60
            1863   0.00   0.00   0.00   42.50
            1864   0.00   0.00   0.00   12.30
            Comment 1856—1864 TIN STUFF
Ownership: Comment 1863—1865 SUSPENDED
Management: Manager 1859—1860 MICH.MICHELL; 1862 WM.BUGLEHOLE
            Chief Agent 1859 WM.BUGLEHOLE; 1860—1861 WM.BUGLEHOLE
            WM.ARTHUR; 1862 WM.ARTHUR
            Secretary 1859—1862 WM.PAINTER (P)

ALFRED,NORTH   PHILLACK   SW 573374 0016

Production: Copper No detailed return
Management: Manager 1859—1865 JOHN VIVIAN
            Secretary 1859 R.DALTON (P)

7

ALFRED,SOUTH   PHILLACK   SW 588362 0017

Production: Copper   Ore(tons) Metal(tons)   Value(£)
            1866   3.00   0.10   7.40
            Comment 1866 (C)

ALFRED,WEST   ST.ERTH   SW 552358 0018

Production: Copper   Ore(tons) Metal(tons)   Value(£)
            1852   154.00   9.30   847.40
            1853   303.00   14.90   1396.20
            1854   403.00   14.70   1383.20
            Comment 1852-1854 (C)

ALICE   REDRUTH   SW 694426 0019

Production: Copper   Ore(tons) Metal(tons)   Value(£)
            1845   154.00   13.80   1033.90
            Comment 1845 (C)

ALICE   GRAMPOUND ROAD   SW 917538 0020

Production: Tin   Black(tons) Stuff(tons)   Tin(tons)   Value(£)
            1888   7.40   0.00   0.00   483.00
Ownership: 1886 WHEAL ALICE ADVENTURES; 1887 J.P.STEPHENS
Management: Chief Agent 1886-1887 ED.PASCOE
Employment: Underground   Surface   Total
            1886   6   5   11
            1887   8   6   14

ALLEN   ST.PINNOCK   SX 174647 0021

Production: Lead & Silver No detailed return
Ownership: Comment 1873-1874 FORMERLY BODITHIEL; 1875 SUSPENDED
Management: Manager 1873 JOHN TREDINNICK
            Secretary 1874-1875 JOHN TREDINNICK (S)

ALMA   LANIVET   0022

Production: Iron   Ore(tons)   Iron(%)   Value(£)
            1872   50.00   0.00   30.00
            Comment 1872 BH.
Ownership: 1870-1871 FAITHFUL,COOKSON & CO.
Management: Chief Agent 1870-1871 W.H.HOSKING

ALVIGGAN   ST.STEPHEN-IN-BRANNEL   SW 978546 0023

Production: Tin No detailed return
Ownership: Comment 1877 ALVOGGAN & BURNGULLOW SUSP.MAY 1877; 1880 SEE
            PAR GREAT CONSOLS; 1881 SEE POLGOOTH UNITED
Management: Secretary 1877 G.S.KIRKMAN

8

AMBROSE LAKE   ST.NEOT   SX 194674 0024

Production: Lead & Silver   Ore(tons)   Lead(tons) Silver(ozs)   Value(£)
            1876   80.10   64.00   320.00   280.70
            Copper   Ore(tons)   Metal(tons)   Value(£)
            1875   50.00   2.70   209.40
            1876   100.00   5.20   357.90
            Comment 1875-1876 (C)
            Tin   Black(tons) Stuff(tons) Tin(tons)   Value(£)
            1875   0.50   0.00   0.00   24.00
            Arsenic   Ore(tons)   Metal(tons)   Value(£)
            1875   18.50   0.00   81.00
            1876   17.60   0.00   95.70
Ownership: 1876 JOHN TAYLOR & CO.; 1877 JOS.TAYLOR & CO.
Management: Manager 1872-1877 PETER TEMBY
            Chief Agent 1869-1870 PETER TEMBY; 1873-1874 JOS.TAYLOR
            Secretary 1872 JOS.TAYLOR; 1873-1875 JOS.TAYLOR WM.EATON

ANDREW   ST.DAY   SW 761419 0025

Production: Copper   Ore(tons) Metal(tons)   Value(£)
            1845   166.00   12.90   993.60
            1846   382.00   29.30   2087.90
            1847   442.00   33.50   2389.60
            1848   459.00   34.60   2006.70
            Comment 1845-1848 (C)INC.NANGILES

ANGARRACK CONSOLS   PHILLACK   SW 587381 0026

Production: Copper No detailed return
Ownership: Comment 1862-1865 SUSPENDED
Management: Manager 1859-1860 W.BARRATT; 1861 S.M.ROGERS
            Chief Agent 1860-1861 THOS.NINNIS
            Secretary 1859-1860 W.BARRATT (P); 1861-1865 S.M.ROGERS

ANNA   PERRANZABULOE   SW 788530 0027

Production: Lead   Ore(tons) Metal(tons)   Value(£)
            1855-1856 No detailed return
            1858-1859 No detailed return
            1860   2.50   1.90   0.00
            1861-1862 No detailed return
            1863   1.00   0.70   0.00
            1864-1865 No detailed return
            Comment 1855-1856 SEE ANNA (ST.HILARY); 1858-1859 SEE ANNA
            (ST.HILARY); 1860 FOR SILVER SEE GREAT ALFRED; 1861-1862 SEE
            ANNA (ST.HILARY); 1864-1865 SEE ANNA (ST.HILARY)
            Zinc   Ore(tons) Metal(tons)   Value(£)
            1855   No detailed return
            1856   106.70   0.00   312.30
            1857   155.10   0.00   438.10
            1858   102.70   0.00   283.20
            1860   42.70   0.00   85.40
            1862   44.30   0.00   122.00

Zinc	Ore(tons)	Metal(tons)	Value(£)
1863	12.80	0.00	24.00

Comment 1855 ANNA CONSOLS. SEE GREAT BADDERN. COULD BE ANNA
(ST.HILARY)
Copper No detailed return

Ownership: Comment 1863-1864 SEE NORTH CHIVERTON; 1865 SUSPENDED. SEE
ALSO NORTH CHIVERTON; 1866-1868 SEE NORTH CHIVERTON
Management: Manager 1864 JOHN TAYLOR & SONS
Chief Agent 1859-1860 CHAS.OATES; 1861 CHAS.OATES
J.S.MICHELL
Secretary 1859 CHAS.OATES (P); 1860 THOS.TREGASKIS (P); 1861
RICH.TREGASKIS (P); 1862-1865 JAS.HAMPTON

ANNA   ST.HILARY   SW 563315 0028

Production: Lead & Silver

	Ore(tons)	Lead(tons)	Silver(ozs)	Value(£)
1855	19.20	11.00	0.00	0.00
1856	0.80	0.50	0.00	0.00
1858	1.40	1.00	0.00	0.00
1859	1.30	1.00	0.00	0.00
1860	3.20	2.50	0.00	0.00
1861	No detailed return			
1862	4.10	2.60	53.00	0.00
1863	8.40	5.80	77.00	0.00
1864	3.50	2.30	60.00	0.00
1865	1.80	1.30	28.00	0.00

Comment 1855-1856 COULD BE ANNA (PERRANZABULOE); 1858 FOR
SILVER SEE GREAT ALFRED.COULD BE ANNA (PERRANZABULOE); 1859
COULD BE ANNA (PERRANZABULOE); 1860 INC.GUSKUS.FOR SILVER SEE
GREAT ALFRED; 1861-1862 COULD BE ANNA (PERRANZABULOE);
1864-1865 COULD BE ANNA (PERRANZABULOE)

Zinc	Ore(tons)	Metal(tons)	Value(£)
1855	No detailed return		
1860	145.40	0.00	368.40
1863	24.20	0.00	13.60

Comment 1855 SEE ANNA (PERRANZABULOE)

Copper	Ore(tons)	Metal(tons)	Value(£)
1845	215.00	11.80	822.80
1856	458.00	28.20	2554.70
1857	1115.00	64.20	5982.30
1858	893.00	51.30	4411.80
1859	955.00	59.10	5420.50
1860	868.00	59.50	5412.60
1861	672.00	32.60	2708.50
1862	607.00	35.10	2768.00
1863	708.00	49.30	3802.40
1864	223.00	13.40	1132.30
1865	37.00	1.90	122.90

Comment 1845 (C); 1856-1865 (C)

Tin	Black(tons)	Stuff(tons)	Tin(tons)	Value(£)
1852	2.00	0.00	0.00	0.00
1856	8.40	0.00	0.00	601.90
1857	119.20	0.00	0.00	9386.10
1858	99.90	0.00	0.00	6496.20

ANNA   ST.HILARY   Continued

Tin	Black(tons)	Stuff(tons)	Tin(tons)	Value(£)
1859	81.80	0.00	0.00	6035.90
1860	67.30	0.00	0.00	5407.50
1861	68.90	0.00	0.00	5042.50
1862	51.20	0.00	0.00	3426.20
1863	15.10	0.00	0.00	1002.80
1864	15.50	0.00	0.00	950.70
1865	2.90	0.00	0.00	225.60

Comment 1852 ANNA (WATER STAMP); 1856 ANNA (WATER STAMP);
1860 INC.GUSKUS; 1861 ANNA (WATER STAMP); 1862 ANNE TWO
RETURNS AGGREGATED; 1863-1864 ANNE; 1865 VALUE INC.TIN STUFF

Arsenic	Ore(tons)	Metal(tons)	Value(£)
1859	11.50	0.00	17.30
1860	11.40	0.00	15.70
1861	8.40	0.00	10.60
1862	14.00	0.00	12.30

Comment 1859-1862 CRUDE
Ownership: Comment 1860 INC.GUSCUS; 1864-1865 SUSPENDED
Management: Manager 1859-1861 JOHN NINNIS; 1862-1863 J.ROGERS
Chief Agent 1859-1860 T.FORD; 1861-1863 R.FORD; 1867-1868
H.FORD
Secretary 1859-1865 JOHN TAYLOR & SON (P); 1867-1868 JOHN
BENNETT

ANNA MARIA   CONSTANTINE   SW 758282 0030

Production: Copper No detailed return
Ownership: 1907-1908 CONSTANTINE COPPER SYNDICATE
Management: Chief Agent 1907-1908 WM.TEAGUE
Secretary 1907-1908 W.M.MARTIN (S)

Employment:	Underground	Surface	Total
1907	10	9	19
1908	15	5	20

ANNA,EAST   PERRANZABULOE   0031

Ownership: Comment 1880 SEE SHEPHERDS WHEAL ROSE UNITED

ANNA,WEST   PERRANZABULOE   SW 776529 0033

Production: Lead & Silver No detailed return
Ownership: Comment 1863-1879 SEE NEW CHIVERTON

ANNE   BREAGE   SW 604283 0034

Production: Tin No detailed return
Ownership: 1907 WHEAL ANNE CONSOLS MINING CO.; 1908 WM.TRELOAR
Comment 1908 ABANDONED
Management: Chief Agent 1907-1908 JOS.HARVEY
Secretary 1907 P.R.PERRING (S)

11

ANNE   BREAGE   Continued

Employment: Underground   Surface   Total
                 1907   14   11   25
                 1908   12   12   24

ANNE   PHILLACK   SW 566363 0035

Production: Copper No detailed return
Management: Manager 1859—1861 JAS.PAULL
            Secretary 1859—1861 T.V.GURNEY (P)

ANNE   ST.AUSTELL   SX 019579 0036

Production: Tin No detailed return
Ownership: Comment 1862—1865 SUSPENDED
Management: Manager 1861 H.B.GROSE
            Chief Agent 1861—1865 T.HOOPER
            Secretary 1861 H.B.GROSE (P); 1863—1865 WM.WEST

ANNIE   ALTARNUN   SX 237794 0037

Production: Tin   Black(tons) Stuff(tons)   Tin(tons)   Value(£)
                 1902   2.30   0.00   0.00   168.00
                 1903   3.80   0.00   0.00   295.00
            Comment 1903 OPEN WORK
Ownership: 1880—1881 J.M.B.DAINTY & OTHERS; 1901—1902 WHEAL ANNIE TIN
            MINE LTD.
            Comment 1901 OR TREBURLAND; 1902 OR TREBURLAND OPEN WORKING
            JAN.1903
Management: Manager 1880—1881 J.M.B.DAINTY
            Chief Agent 1901—1902 WM.WADGE
Employment: Underground   Surface   Total
                 1880   4   4
                 1901   6   3   9
                 1902   12   12

ANNIE   GWINEAR   SW 612376 0038

Production: Copper No detailed return
Ownership: Comment 1865 SUSPENDED
Management: Manager 1862—1864 JAS.PAULL
            Chief Agent 1861 WM.C.VIVIAN JAS.PAULL
            Secretary 1861—1865 WM.C.VIVIAN (P)

APPLE TREE   CHARLESTOWN   SX 045522 0039

Production: Copper No detailed return
            Tin No detailed return
Ownership: 1901 E.A.TREGILGAS & JAS.TUCKER; 1902—1910 JAS.TUCKER
            Comment 1901—1903 INC.SOUTH CRINNIS; 1904—1909 INC.SOUTH
            CRINNIS SUSPENDED; 1910 INC.SOUTH CRINNIS NOT WORKED

APPLE TREE   CHARLESTOWN   Continued

Employment: Underground   Surface   Total
             1901   2   2
             1902   2   2
             1903   1   1   2
             1905   1   1   2
             1906   2   2
             1907   2   2
             Comment 1901-1903 INC.SOUTH CRINNIS; 1905-1907 INC.SOUTH
             CRINNIS

ARCHIE   ADVENT   0040

Production: Lead No detailed return
            Copper No detailed return
Ownership: Comment 1872 SEE GREAT NORTH CARADON
Management: Manager 1869-1871 JOHN SPARGO
            Secretary 1869-1871 JOHN PETERS

ARGUS   SANCREED   SW 417296 0041

Production: Tin   Black(tons) Stuff(tons)   Tin(tons)   Value(£)
            1873   9.20   0.00   0.00   690.50
            1874   6.50   0.00   0.00   363.20
            1875   3.40   0.00   0.00   132.40
            Comment 1874 VALUE EST
Ownership: 1875 WHEAL ARGUS MINING CO.
Management: Manager 1875 THOS.TREHAIR
            Chief Agent 1872-1874 THOS.TREHAIR
            Secretary 1872-1874 G.EUSTICE JNR.

ARTHUR   CALSTOCK   SX 434700 0042

Production: Copper   Ore(tons) Metal(tons)   Value(£)
            1852   483.00   39.90   3559.50
            1853   1270.00   95.50   9481.00
            1854   2417.00   127.50   12528.10
            1855   2674.00   119.60   11228.60
            1856   1657.00   64.30   5211.70
            1857-1858 No detailed return
            1860   136.00   5.20   409.10
            1861   194.00   8.40   633.30
            1862   642.00   25.60   1814.20
            1863   394.00   19.40   1398.00
            1864   303.00   11.80   850.50
            1865   214.00   7.70   533.20
            1884   27.00   0.00   59.00
            1885   51.00   0.00   156.00
            Comment 1852-1856 (C); 1857-1858 SEE ARTHUR, DEVON; 1860-1865
            (C); 1884-1885 ARTHUR MINE STAMPS
            Tin   Black(tons) Stuff(tons)   Tin(tons)   Value(£)
            1855   No detailed return
            1856   5.20   0.00   0.00   367.40

                                      13

Tin	Black(tons)	Stuff(tons)	Tin(tons)	Value(£)
1857	15.30	0.00	0.00	1186.80
1858	1.90	0.00	0.00	124.20
1861	2.10	0.00	0.00	145.80
1871	29.70	0.00	0.00	2212.60
1872	48.60	0.00	0.00	4132.40
1873	22.50	0.00	0.00	1755.00
1874	5.40	0.00	0.00	269.40
1875	3.10	0.00	0.00	119.40
1877	4.00	0.00	0.00	114.70
1878	5.40	0.00	0.00	133.50
1880	2.60	0.00	0.00	119.50
1881	5.80	0.00	3.30	270.60
1882	1.70	0.00	0.80	72.00
1884	2.20	0.00	0.00	116.00
1885	1.90	0.00	0.00	94.00

Comment 1855 SEE ARTHUR, DEVON; 1873 VALUE EST; 1884–1885
ARTHUR MINE STAMPS

Arsenic	Ore(tons)	Metal(tons)	Value(£)
1872	10.50	0.00	20.00
1873	18.40	0.00	36.80
1880	3.40	0.00	21.30
1881	36.60	0.00	182.70
1884	18.00	0.00	113.00
1885	33.00	0.00	212.00

Comment 1880 SOOT; 1884–1885 SOOT.ARTHUR MINE STAMPS

Ownership: 1878 WHEAL ARTHUR CO.; 1880–1888 WHEAL ARTHUR MINING CO.;
1895–1897 MOSES BAWDEN; 1898 MOSES BAWDEN & CO.
Comment 1874 STOPPED; 1875 SUSPENDED; 1880 SEE ARTHUR, DEVON;
1894 SEE ARTHUR, DEVON

Management: Manager 1859 C.CARPENTER; 1869 MOSES BAWDEN; 1870–1875
WM.SKEWIS; 1880–1881 MOSES BAWDEN
Chief Agent 1860–1863 F.C.HARPER THOS.CARPENTER; 1864–1866
THOS.CARPENTER; 1870–1871 JAS.MAYNE; 1872–1875 R.TREVARTHEN;
1878 MOSES BAWDEN
Secretary 1859–1866 W.WATSON (P); 1869 WM.SKEWIS; 1870–1875
MOSES BAWDEN; 1882–1888 MOSES BAWDEN

Employment:

	Underground	Surface	Total
1880	3	7	10
1881	4	13	17
1882		3	3
1883		5	5
1884		2	2
1885		3	3
1886		5	5
1887		3	3
1888		2	2
1894			
1896–1897		2	2
1898		1	1

Comment 1894 SEE ARTHUR, DEVON

14

ARTHUR   CAMBORNE   0043

Production: Zinc   Ore(tons) Metal(tons)   Value(£)
            1893   21.00   10.00   59.00
            1894   55.00   0.00   172.00
            1895   No detailed return
            1896   50.00   0.00   150.00
            1897   30.00   0.00   13.00
            Comment 1894 VALUE INC.WEST JANE
Ownership: 1892-1898 DANIELL,THOMAS & TEMBY; 1913 C.V.THOMAS
            Comment 1894-1895 NOT WORKED; 1913 NOT WORKED
Management: Chief Agent 1892-1898 PETER TEMBY
            Secretary 1897-1898 JOHN R.DANIELL (P)
Employment: Underground   Surface   Total
            1892   1   1   2
            1893   5   3   8
            1894   1   1
            1896   3   3   6
            1897   1   1

ATWAY   LAUNCESTON   SX 305861 0002

Ownership: Comment 1880-1881 SEE ATWAY, DEVON

AUGUSTA   ST.JUST   SW 380302 0044

Production: Tin   Black(tons) Stuff(tons)   Tin(tons)   Value(£)
            1852   1.30   0.00   0.00   0.00
            1853   9.40   0.00   0.00   592.50
            1854   11.50   0.00   0.00   727.60
            Comment 1852 TIN ORE

BADDERN,GREAT   KEA   SW 777421 0045

Production: Lead & Silver   Ore(tons) Lead(tons) Silver(ozs)   Value(£)
            1850   309.00   216.00   0.00   0.00
            1851   346.00   244.50   0.00   0.00
            1852   628.00   447.00   9400.00   0.00
            1853   227.00   136.00   3128.00   0.00
            1854   437.20   335.50   5795.00   0.00
            1855   397.00   280.00   6160.00   0.00
            1856   392.00   274.00   5918.00   0.00
            1857   312.00   214.00   5350.00   0.00
            1858   217.00   152.50   3496.00   0.00
            1859   80.00   62.00   1860.00   0.00
            1860-1861 No detailed return
            1867   8.20   6.00   0.00   0.00
            1868   16.00   12.00   0.00   0.00
            1869   7.70   5.40   0.00   0.00
            1870   2.10   1.50   0.00   0.00
            Comment 1850 BADDERN
            Zinc   Ore(tons) Metal(tons)   Value(£)
            1854   57.90   0.00   0.00
            1855   935.00   0.00   0.00

                                15

BADDERN,GREAT   KEA   Continued

Zinc	Ore(tons)	Metal(tons)	Value(£)	
1856	110.30	0.00	88.70	
1857	33.70	0.00	26.50	

Comment 1855 INC.4 OTHER MINES

Tin	Black(tons)	Stuff(tons)	Tin(tons)	Value(£)
1852	4.00	0.00	0.00	0.00
1853	13.50	0.00	0.00	820.40
1854	3.70	0.00	0.00	271.10
1855	3.20	0.00	0.00	192.10
1856	1.30	0.00	0.00	70.50
1857	0.00	0.00	0.00	33.60
1858	0.00	0.00	0.00	53.80
1859	0.00	0.00	0.00	12.00
1867	0.60	0.00	0.00	167.00
1868	0.00	0.00	0.00	166.60
1869	0.00	0.00	0.00	148.00
1870	0.00	0.00	0.00	65.60
1871	0.00	0.00	0.00	42.00
1872	0.00	0.00	0.00	52.30

Comment 1852 TIN ORE; 1855 VALUE EST; 1857-1859 TIN STUFF;
1867 VALUE INC.TIN STUFF; 1868 TIN STUFF; 1870-1872 TIN
STUFF

Arsenic	Ore(tons)	Metal(tons)	Value(£)
1854	2.60	0.00	0.00
1869	11.50	0.00	5.80

Management: Manager 1859-1863 JAS.HAMPTON; 1865-1868 RICH.PRYOR
Chief Agent 1859-1863 JOHN JENKIN; 1864 RICH.PRYOR JOHN
JENKIN; 1865-1868 H.TREGONNING L.N.VISICK
Secretary 1859-1868 T.B.LAWS (P)

BAL   ST.JUST   SW 381337 0046

Production: Tin	Black(tons)	Stuff(tons)	Tin(tons)	Value(£)
1852	10.70	0.00	0.00	0.00
1855	88.10	0.00	0.00	5759.00
1856	71.20	0.00	0.00	5287.60
1857	52.70	0.00	0.00	4200.70
1858	73.00	0.00	0.00	4638.70
1859	67.80	0.00	0.00	4940.60
1860	48.80	0.00	0.00	3856.90
1861	24.50	0.00	0.00	1740.10
1862	2.60	0.00	0.00	414.70

Comment 1852 TIN ORE
Ownership: Comment 1859 BALL
Management: Manager 1859 HY.TREZISE; 1860 JOHN BENNETT
Chief Agent 1859-1860 JOHN BREWETH
Secretary 1859-1860 JOHN T.WHITE (P)

BALDHU   KEA   0047

Production: Lead & Silver No detailed return

Tin	Black(tons)	Stuff(tons)	Tin(tons)	Value(£)
1880	0.00	280.80	0.00	88.80

BALDHU   KEA   Continued

Ownership: 1880-1881 BALDHU MINING CO.
             Comment 1869-1873 FORMERLY BALDHU, NOW GREAT JANE
Management: Chief Agent 1880-1881 JOHN SMITH

BALDHU   LUDGVAN   0048

Ownership: 1907 BALDHU SYNDICATE
             Comment 1907 BALDHU & TINCROFT
Management: Chief Agent 1907 A.M.FOWLRAKER
Employment: Underground   Surface   Total
             1907   2   3   5
             Comment 1907 INC.TINCROFT (TOWEDNACK)

BALLESWIDDEN   ST.JUST   SW 387312 0049

Production: Tin   Black(tons) Stuff(tons)   Tin(tons)   Value(£)
            1852   405.10   0.00   0.00   0.00
            1853   16.00   0.00   0.00   1181.80
            1854   73.60   0.00   0.00   5748.90
            1855   219.20   0.00   0.00   15292.30
            1856   267.60   0.00   0.00   21264.70
            1857   175.40   0.00   0.00   14805.40
            1858   238.90   0.00   0.00   16105.60
            1859   212.50   0.00   0.00   16849.10
            1860   202.10   0.00   0.00   16141.70
            1861   277.20   0.00   0.00   20017.60
            1862   320.10   0.00   0.00   20479.70
            1863   313.40   0.00   0.00   21286.50
            1864   309.80   0.00   0.00   19701.50
            1865   268.60   0.00   0.00   15074.40
            1866   268.50   0.00   0.00   13253.10
            1867   284.70   0.00   0.00   15242.40
            1868   315.50   0.00   0.00   17837.50
            1869   262.40   0.00   0.00   19005.20
            1870   270.60   0.00   0.00   20755.20
            1871   215.00   0.00   0.00   17507.70
            1872   182.20   0.00   0.00   16051.90
            1873   102.80   0.00   0.00   7629.30
            1874   3.00   0.00   0.00   126.00
            1875   15.00   0.00   0.00   623.70
            Comment 1852 TIN ORE; 1862 TWO RETURNS AGGREGATED; 1874 EST.
            SURFACE WORKING ONLY
Ownership: Comment 1872-1873 CEASED WORKING JAN.1873; 1874 BURROWS ONLY
            WORKED IN 1874
Management: Manager 1859-1863 THOS.TREHAIR; 1865-1873 JOHN TAYLOR; 1874
            G.EUSTICE JNR.
            Chief Agent 1859-1861 WM.CLEMENS; 1862-1867 WM.CLEMENS JOHN
            BENNETTS THOS.VEAL; 1868-1872 WM.CLEMENS JOHN BENNETTS
            T.P.ROWE; 1875 THOS.TREHAIR
            Secretary 1859-1867 R.V.DAVEY (P); 1868-1872 EDM.DAVEY; 1873
            EDM.DAVEY & OTHERS; 1875 G.EUSTICE JNR.

BALLESWIDDEN UNITED   ST.JUST   SW 436298 0050

Production: Tin   Black(tons) Stuff(tons)   Tin(tons)   Value(£)
            1854   16.50   0.00   0.00   1039.00

BALLESWIDDEN,EAST   ST.JUST   SW 393304 0051

Production: Tin   Black(tons) Stuff(tons)   Tin(tons)   Value(£)
            1853   3.30   0.00   0.00   0.00
            1873   7.90   0.00   0.00   624.00
            1874   3.70   0.00   0.00   157.50
            Comment 1853 EAST BALSWIDDEN; 1873 VALUE EST
Management: Chief Agent 1872-1874 THOS.TREHAIR
            Secretary 1872-1873 G.V.MORGAN (S)

BALLESWIDDEN,NEW   ST.JUST   SW 393304 0052

Production: Tin   Black(tons) Stuff(tons)   Tin(tons)   Value(£)
            1889   14.00   0.00   0.00   740.00
            1892   4.30   0.00   0.00   240.00
            1893   27.20   0.00   0.00   1304.00
            1894   3.20   0.00   0.00   128.00
            1895   No detailed return
Ownership: 1888-1890 NEW BALLESWIDDEN TIN MINE CO.; 1891-1893 NEW
            BALLESWIDDEN TIN MINE CO.LTD.
            Comment 1889-1890 SUSPENDED; 1894 BALLESWIDDEN ABANDONED
Management: Chief Agent 1888-1890 JOHN HOLLOW; 1891 RICHARDS; 1892-1893
            JOHN HOLLOW
            Secretary 1891-1893 G.J.GRAY (S)
Employment: Underground   Surface   Total
            1888   14   10   24
            1892   36   26   62
            1893   8   37   45
            1894   9   26   35

BALLESWIDDEN,NORTH   ST.JUST   SW 389309 0053

Production: Tin No detailed return
Ownership: 1905-1906 NORTH BALLESWIDDEN MINING SYNDICATE
Management: Chief Agent 1905 L.R.THOMAS; 1906 WM.THOMAS
Employment: Underground   Surface   Total
            1905   8   2   10

BALLOWAL   ST.JUST   SW 356312 0054

Production: Tin No detailed return
Ownership: Comment 1859-1865 SEE BOSORNE

BALMYNHEER   WENDRON   SW 344704 0055

Production: Tin   Black(tons) Stuff(tons)   Tin(tons)   Value(£)
            1870   23.80   0.00   0.00   1437.90

18

BALMYNHEER   WENDRON   Continued

Tin	Black(tons)	Stuff(tons)	Tin(tons)	Value(£)
1871	28.40	0.00	0.00	1878.90
1872	27.20	0.00	0.00	2120.40
1873	84.60	0.00	0.00	5690.00
1874	10.90	0.00	0.00	570.40
1875	47.80	0.00	0.00	2483.50
1876	24.30	0.00	0.00	963.30
1877	2.00	0.00	0.00	76.00
1878	10.70	0.00	0.00	420.50
1879	2.10	0.00	0.00	82.00

Comment 1870 BALMYNHEAR
Ownership: 1877–1878 BALMYNHEER MINING CO.LTD.; 1879–1882 BALMYNHEER TIN
MINING CO.LTD.
Comment 1872 NEW BALMYNHEER; 1880–1881 NOT WORKED
Management: Manager 1871–1873 JOHN TONKIN; 1874 JOHN BURGAN
Chief Agent 1869–1874 J.S.HARRIS; 1875 R.S.COATES; 1876
F.MICHELL; 1877–1882 FRAN.W.MICHELL
Secretary 1869–1873 J.TAYLOR; 1874–1878 WM.BATTYE

Employment:	Underground	Surface	Total
1878	4		4
1879	3	2	5

BALNOON CONSOLS   LELANT   SW 508381 0056

Production: Tin	Black(tons)	Stuff(tons)	Tin(tons)	Value(£)
1852	8.40	0.00	0.00	0.00
1853	18.00	0.00	0.00	1071.00
1855	12.70	0.00	0.00	808.70
1856	1.20	0.00	0.00	79.70

Comment 1852 TIN ORE; 1853 BALNOON; 1856 BALNOON
Ownership: Comment 1905–1909 SEE WORVAS DOWNS

Employment:	Underground	Surface	Total
1905–1909			

Comment 1905–1909 SEE WORVAS DOWNS

BARING   CALLINGTON   SX 383695 0057

Ownership: Comment 1878–1879 SEE LANGFORD

BARNARD   CALSTOCK   SX 399699 0058

Production: Lead & Silver No detailed return

Arsenic	Ore(tons)	Metal(tons)	Value(£)
1873	2.00	0.00	3.30

Ownership: 1873 WHEAL BARNARD SILVER LEAD CO.
Comment 1874–1875 SEE NEWTON
Management: Chief Agent 1873 THOS.J.BARNARD

19

BASORE   CHACEWATER   SW 789439 0059

Production: Lead   Ore(tons) Metal(tons)   Value(£)
           1856   1.10   0.70   0.00
           1857–1858 No detailed return

BASSET   ILLOGAN   SW 688397 0060

Production: Copper   Ore(tons) Metal(tons)   Value(£)
           1851   5442.00   487.90   33324.40
           1852   5966.00   503.70   42933.60
           1853   7207.00   608.10   60638.60
           1854   8378.00   646.50   66024.90
           1855   7713.00   680.60   71095.70
           1856   7600.00   594.90   56324.20
           1857   No detailed return
           1858   4754.00   414.20   38507.20
           1859   4258.00   365.70   34850.30
           1860   3894.00   300.70   27922.80
           1861   2931.00   231.90   21079.10
           1862   2633.00   213.40   17728.90
           1863   1998.00   154.00   12042.00
           1864   1726.00   138.30   11721.80
           1865   2036.00   183.60   14697.50
           1866   2108.00   174.50   12202.10
           1867   1700.00   132.40   9183.90
           1868   1539.00   115.60   7570.80
           1869   1572.00   130.40   8104.10
           1870   1341.00   106.60   6239.60
           1871   442.00   35.10   2186.00
           1872   528.00   38.40   3010.30
           1873   488.00   67.10   4548.50
           1874   396.00   41.40   2847.40
           1875   714.00   62.90   4789.70
           1876   759.00   67.70   4607.70
           1877   448.40   40.50   2437.90
           1878   227.90   23.40   1237.30
           1879   99.10   8.00   387.30
           1880   29.60   2.00   107.70
           1890   3.00   0.00   15.00
           1903   11.00   0.00   107.00
           1912   13.00   0.00   115.00
           Comment 1851–1856 (C); 1857 SEE WEST BASSET; 1858–1861 (C);
           1862 (C)TWO RETURNS AGGREGATED; 1863–1876 (C); 1912 BASSET
           MINES LTD.
           Tin   Black(tons) Stuff(tons) Tin(tons)   Value(£)
           1852   21.00   0.00   0.00   0.00
           1853   31.00   0.00   0.00   2011.20
           1854   46.00   0.00   0.00   3220.00
           1855   25.60   0.00   0.00   1610.40
           1856   20.10   0.00   0.00   1414.40
           1857   32.20   0.00   0.00   2396.60
           1858   45.20   0.00   0.00   2869.40
           1859   42.50   0.00   0.00   3116.10
           1860   60.10   0.00   0.00   4863.30
           1861   51.80   0.00   0.00   3734.70

                                    20

ILLOGAN

Tin	Black(tons)	Stuff(tons)	Tin(tons)	Value(£)
1862	64.20	0.00	0.00	4300.70
1863	77.80	0.00	0.00	5254.10
1864	117.40	0.00	0.00	7333.20
1865	158.50	0.00	0.00	9052.40
1866	167.20	0.00	0.00	8233.00
1867	292.70	0.00	0.00	15917.10
1868	264.90	0.00	0.00	15030.20
1869	313.20	0.00	0.00	21928.80
1870	353.50	0.00	0.00	26439.10
1871	335.10	0.00	0.00	27232.10
1872	255.40	0.00	0.00	22598.00
1873	286.30	0.00	0.00	22827.80
1874	310.90	0.00	0.00	17199.50
1875	248.00	0.00	0.00	12618.20
1876	147.30	0.00	0.00	6514.10
1877	164.40	0.00	0.00	6741.40
1878	208.30	0.00	0.00	7459.30
1879	23.00	0.00	0.00	854.90
1880	7.80	0.00	0.00	334.10
1881	8.20	0.00	5.90	345.80
1882	60.50	0.00	40.80	3632.00
1883	173.50	0.00	0.00	9550.00
1884	267.10	0.00	0.00	12672.00
1885	395.00	0.00	0.00	19535.00
1886	379.20	0.00	0.00	22303.00
1887	370.00	0.00	0.00	24230.00
1888	416.40	0.00	0.00	28420.00
1889	386.50	0.00	0.00	21320.00
1890	396.00	0.00	0.00	22203.00
1891	402.00	138.00	0.00	22448.00
1892	458.00	16.00	0.00	24949.00
1893	432.50	150.00	0.00	21439.00
1894	470.20	0.00	0.00	19060.00
1895	536.20	105.00	0.00	21115.00
1896	554.00	0.00	0.00	20399.00
1897	539.00	132.00	0.00	18863.00
1898	356.00	46.00	0.00	14001.00
1899	268.00	72.00	0.00	19265.00
1900	29.00	0.00	0.00	1769.00
1901	575.00	0.00	0.00	38062.00
1902	794.00	0.00	0.00	51565.00
1903	728.00	0.00	0.00	49384.00
1904	613.00	0.00	0.00	41161.00
1905	768.00	0.00	0.00	59369.00
1906	696.00	0.00	0.00	66949.00
1907	734.00	0.00	0.00	72551.00
1908	760.00	0.00	0.00	56850.00
1909	844.00	0.00	0.00	63923.00
1910	700.00	0.00	0.00	65549.00
1911	586.00	0.00	0.00	71358.00
1912	502.00	0.00	0.00	66311.00
1913	575.00	0.00	0.00	67763.00

Comment 1852 TIN ORE; 1854—1856 BASSETT; 1862 TWO RETURNS AGGREGATED; 1896—1912 BASSET MINES LTD.

Ownership: 1876 WHEAL BASSET CO.; 1877–1890 WHEAL BASSET MINING CO.;
              1891 WHEAL BASSET MINING CO. C.B.; 1892–1893 WHEAL BASSET
              TIN MINING CO. C.B.; 1894–1895 WHEAL BASSET TIN MINING CO.;
              1896–1913 BASSET MINES LTD.
              Comment 1862–1864 GREAT BASSET; 1895 INC.WEST BASSET & SOUTH
              FRANCES NOW BASSET LTD; 1896–1913 BASSET MINES LTD.
Management: Manager 1859–1864 JAS.POPE; 1868–1869 W.W.MARTIN; 1870–1871
              AB.T.JAMES; 1872 W.TREVENA; 1873–1881 W.C.TREVENA
              Chief Agent 1859–1860 WM.JULIFF; 1861–1864 JAS.JULIFF
              S.A.POPE; 1865–1866 W.MARTIN W.H.PASCOE W.FREEMAN; 1867
              W.W.MARTIN W.H.PASCOE E.RICHARDS; 1868–1871 W.H.PASCOE
              W.TREVENA E.RICHARDS; 1872 W.H.PASCOE JOHN OPIE W.REYNOLDS;
              1873–1880 W.J.REYNOLDS JOHN OPIE; 1881 JOHN OPIE; 1882–1887
              W.C.TREVENA; 1888–1889 JOHN H.JAMES; 1890–1891 JAS.THOMAS;
              1892–1913 WM.JAMES
              Secretary 1859–1862 WM.RICHARDS (P); 1863–1868 F.W.DABB;
              1869–1872 G.A.MICHELL; 1873–1881 RICH.MARTIN; 1890–1891
              RICH.RENDLE; 1892–1895 RICH.RENDLE (P); 1896–1902 RICH.RENDLE
              (S)

Employment:

	Underground	Surface	Total
1878	90	67	157
1879	24	37	61
1880	49	37	86
1881	62	37	99
1882	118	137	255
1883	144	81	225
1884	171	86	257
1885	205	133	338
1886	222	132	354
1887	201	156	357
1888	186	190	376
1889	156	201	357
1890	150	192	342
1891	142	182	324
1892	183	168	351
1893	163	152	315
1894	151	144	295
1895	147	138	285
1896	210	180	390
1897	195	199	394
1898	157	151	308
1899	252	185	437
1900	253	186	439
1901	231	188	419
1902	245	244	489
1903	295	236	531
1904	270	229	499
1905	321	239	560
1906	287	252	539
1907	282	267	549
1908	321	241	562
1909	344	194	538
1910	284	184	468
1911	306	191	497
1912	263	192	455

BASSET   ILLOGAN   Continued

                    Underground   Surface   Total
          1913   253   191   444

BASSET AND BULLER CONSOLS   REDRUTH   SW 699401 0061

Production: Tin No detailed return
Ownership: 1880 BULLER AND BASSET CONSOLS MINING CO.; 1881–1887 BASSET
            AND BULLER CONSOLS MINING CO.; 1888–1890 BASSET & BULLER TIN
            & COPPER CO.LTD.
            Comment 1880 BULLER AND BASSET CONSOLS; 1883 NOT WORKED IN
            1883; 1884–1885 NOT WORKED IN 1884; 1887 NOT WORKING; 1888
            NOT WORKING INC.COPPER HILL; 1889 INC.COPPER HILL; 1890
            INC.COPPER HILL NOW ABANDONED
Management: Manager 1881 RICH.PRYOR
            Chief Agent 1880 RICH.PRYOR; 1882–1887 RICH.PRYOR; 1888–1890
            JOHN H.JAMES
Employment: Underground   Surface   Total
          1881   6   6
          1882   10   18   28
          1886   5   2   7
          1888   16   4   20
          1889   4   6   10
          1890   2   2
          Comment 1888–1890 INC.COPPER HILL

BASSET AND GRYLLS   WENDRON   SW 693328 0062

Production: Tin   Black(tons) Stuff(tons)   Tin(tons)   Value(£)
          1861   153.70   0.00   0.00   10729.70
          1862   336.10   0.00   0.00   21227.90
          1863   299.60   0.00   0.00   18603.70
          1864   281.10   0.00   0.00   16645.10
          1865   314.70   0.00   0.00   16659.30
          1866   234.00   0.00   0.00   10636.50
          1867   135.10   0.00   0.00   6873.70
          1868   167.90   0.00   0.00   8834.80
          1869   195.10   0.00   0.00   13129.00
          1870   139.80   0.00   0.00   9620.20
          1871   157.80   0.00   0.00   11865.60
          1872   154.70   0.00   0.00   12761.90
          1873   109.20   0.00   0.00   8064.40
          1874   135.60   0.00   0.00   6775.70
          1875   56.80   0.00   0.00   2405.30
          1876   35.00   0.00   0.00   1341.50
          1877   34.00   0.00   0.0C   1029.60
          1878   11.90   0.00   0.00   305.50
          1879   4.40   0.00   0.00   138.60
          1880   2.90   0.00   0.00   145.00
          1881   2.80   0.00   1.40   110.00
          1882   3.10   0.00   2.00   180.00
          1907   3.40   0.00   0.00   243.00
          1908   15.60   0.00   0.00   1087.00
          1909   5.70   0.00   0.00   408.00

                                        23

BASSET AND GRYLLS   WENDRON   Continued

Tin	Black(tons)	Stuff(tons)	Tin(tons)	Value(£)
1910	17.30	1833.00	0.00	1598.00
1911	24.00	0.00	0.00	2670.00
1912	1.80	0.00	0.00	191.00
1913	29.00	0.00	0.00	2925.00

Comment 1862 TWO RETURNS AGGREGATED

Ownership: 1877–1878 BASSET & GRYLLS MINING CO.; 1879–1882 GRAN.SHARPE; 1907–1911 BASSET & GRYLLS LTD.; 1912–1913 RAYFIELD TIN SYNDICATE LTD.
Comment 1909–1910 IN LIQUIDATION 1ST.JAN.1910; 1911 SUSPENDED MAY 1911; 1912–1913 RESTARTED AUG.1912

Management: Manager 1859–1864 J.B.WILKIN; 1865–1866 J.B.WILKIN WM.TEAGUE; 1867 CHAS.FREEMAN; 1868–1873 W.OATES JNR.; 1874–1876 PAUL PRISK; 1877–1879 JOS.PRISK
Chief Agent 1859 WALT.HARRIS; 1860–1861 JOHN WILLIAMS BENT.JOHNS; 1862 WALT.HARRIS BENT.JOHNS; 1863–1866 WALT.HARRIS W.OATES JNR.; 1867 PAUL PRISK W.OATES JNR.; 1868–1873 PAUL PRISK; 1877–1882 PAUL PRISK
Secretary 1859–1871 T.P.TYACKE (P); 1872–1877 J.WALKER TYACKE; 1878 GRAN.SHARPE; 1907–1911 R.ART.THOMAS

Employment:	Underground	Surface	Total
1878		10	10
1879		6	6
1880		7	7
1881–1882		6	6
1907	23	16	39
1908–1909	18	13	31
1910	35	39	74
1911	32	29	61
1912	32	56	88
1913	49	53	102

BASSET CONSOLS   PORTH TOWAN   SW 692473 0063

Production:	Lead & Silver	Ore(tons)	Lead(tons)	Silver(ozs)	Value(£)
	1861	14.20	9.20	144.00	0.00
	Copper	Ore(tons)	Metal(tons)	Value(£)	
	1858	195.00	10.30	923.50	
	1859	139.00	7.90	756.10	

Comment 1858–1859 (C)

BASSET VALLEY   PORTH TOWAN   0064

Production:	Zinc	Ore(tons)	Metal(tons)	Value(£)
	1879	39.00	0.00	136.50
	1880	70.00	0.00	245.00

Ownership: 1878 RICHARDS POWER & CO.; 1879–1882 R.W.RICKARD & CO.
Comment 1881 NOT WORKED
Management: Chief Agent 1878 HY.DALE; 1879–1882 NICH.VIVIAN

Employment:	Underground	Surface	Total
. 1878	2	3	5
1879	10	4	14
1880	4	3	7

BASSET VALLEY   PORTH TOWAN   Continued

                        Underground   Surface   Total
          1881-1882   2   2

BASSET,EAST   REDRUTH   SW 694401 0065

Production: Barytes   Ore(tons)   Value(£)
            1876   1.00   0.00
            Copper   Ore(tons) Metal(tons)   Value(£)
            1858   591.00   54.30   5081.80
            1859   2039.00   308.40   30290.60
            1860   2023.00   232.70   22106.50
            1861   1799.00   184.10   16736.30
            1862   1535.00   114.40   9214.60
            1863   1249.00   125.80   9992.70
            1864   1111.00   98.50   8442.50
            1865   632.00   48.50   3830.50
            1866   631.00   41.10   2632.40
            1867   190.00   16.40   1153.80
            1868   176.00   15.10   990.10
            1869   176.00   15.60   998.90
            1870   218.00   14.90   836.20
            1871   24.00   1.60   94.90
            1872   27.00   1.10   55.00
            1874   270.00   18.40   1180.50
            1875   218.00   15.20   1089.30
            1876   209.00   15.00   1000.80
            1877   14.00   1.20   62.50
            Comment 1858-1872 (C); 1874-1876 (C)
            Tin   Black(tons) Stuff(tons)   Tin(tons)   Value(£)
            1857   1.50   0.00   0.00   109.40
            1858   0.00   0.00   0.00   995.40
            1859   0.00   0.00   0.00   1537.00
            1860   0.00   0.00   0.00   1310.90
            1861   0.00   0.00   0.00   647.40
            1862   0.00   0.00   0.00   1261.30
            1863   0.00   0.00   0.00   2008.40
            1864   0.80   0.00   0.00   2308.10
            1865   5.00   0.00   0.00   1339.60
            1866   0.00   0.00   0.00   570.40
            1867   0.00   0.00   0.00   1066.20
            1868   0.00   0.00   0.00   1135.70
            1869   0.00   0.00   0.00   1145.20
            1870   0.00   0.00   0.00   442.00
            1871   0.00   0.00   0.00   493.10
            1872   0.00   0.00   0.00   1887.50
            1873   33.20   200.00   0.00   2971.40
            1874   12.30   0.00   0.00   744.10
            1875   0.00   11.80   0.00   36.10
            1879   0.00   50.00   0.00   8.00
            Comment 1858-1863 TIN STUFF; 1864-1865 INC.GRYLLS (ILLOGAN)
            TIN STUFF; 1866-1872 TIN STUFF; 1873 TIN STUFF EST
Ownership: 1875-1876 EAST BASSET MINING CO.; 1879-1881 EAST BASSET
            SYNDICATE
            Comment 1864 INC.GRYLLS (ILLOGAN); 1865 INC.GRYLLS (ILLOGAN),

                                    25

BASSET,EAST   REDRUTH   Continued

SUSPENDED
Management: Manager 1859–1864 JAS.POPE; 1870–1872 J.LEAN; 1873–1874
            J.JENNINGS; 1875–1876 RICH.PRYOR; 1880–1881 RICH.PRYOR
            Chief Agent 1859–1864 WM.NANCARROW; 1869 WM.NANCARROW R.LEAN;
            1873–1874 N.BOTTLE; 1875 ED.ADAMS; 1876 HY.L.PHILLIPS; 1879
            RICH.PRYOR
            Secretary 1859–1862 WM.RICHARDS (P); 1863–1864 F.W.DABB; 1869
            F.W.DABB; 1870–1872 SML.ABBOT; 1873 RICH.PRYOR; 1874
            RICH.PRYOR & SON
Employment: Underground   Surface   Total
            1879   8   8

BASSET,NORTH   ILLOGAN   SW 688401 0066

Production: Copper   Ore(tons) Metal(tons)   Value(£)
            1846   226.00   16.60   1131.10
            1847   569.00   44.90   3164.10
            1848   344.00   28.80   1800.00
            1852   2638.00   181.30   15275.50
            1853   2528.00   174.50   17622.70
            1854   2752.00   222.10   22868.00
            1855   4574.00   433.50   45628.10
            1856   5104.00   552.40   53081.00
            1857   3996.00   328.40   33195.40
            1858   3293.00   201.40   17629.20
            1859   1498.00   83.70   7602.70
            1860   526.00   26.90   2336.60
            1861   566.00   31.70   2721.80
            1862   483.00   26.60   2076.20
            1863   358.00   18.40   1336.80
            1864   180.00   10.00   819.50
            1865   287.00   14.90   1132.20
            1866   220.00   11.30   748.40
            1869   15.00   0.90   50.20
            Comment 1846–1848 (C); 1852–1859 (C); 1860 (C)NORTH BASSETT;
            1861 (C); 1866 NORTH BASSETT
            Tin   Black(tons) Stuff(tons)   Tin(tons)   Value(£)
            1855   5.80   0.00   0.00   309.00
            1856   18.20   0.00   0.00   1129.80
            1857   38.60   0.00   0.00   2831.90
            1858   0.00   0.00   0.00   2325.20
            1859   0.00   0.00   0.00   1914.90
            1860   0.00   0.00   0.00   2258.40
            1861   0.00   0.00   0.00   1084.80
            1862   9.40   0.00   0.00   1015.10
            1863   21.70   0.00   0.00   1428.80
            1864   25.70   0.00   0.00   1590.00
            1865   26.80   0.00   0.00   1512.50
            1866   21.80   0.00   0.00   1108.00
            Comment 1855 VALUE EST; 1858–1861 TIN STUFF; 1862
            VAL.INCS.TIN STUFF. 3 RETURNS.
Management: Manager 1859–1862 THOS.GLANVILLE; 1864 THOS.GLANVILLE
            Chief Agent 1859–1864 GEO.DAVEY; 1865–1866 WM.ROBERTS
            Secretary 1859–1864 RICH.LYLE (P); 1865–1867 GEO.LIGHTLEY

BASSET,OLD   PORTH TOWAN   SW 702467 0067

Production: Copper   Ore(tons) Metal(tons)   Value(£)
            1854   160.00   10.00   1018.30
            1855   114.00   8.00   827.60
            1856   93.00   6.90   653.80
            1857   126.00   8.60   879.10
            1858   67.00   5.10   487.40
            1865   9.00   0.30   15.50
            1867   38.00   0.90   39.00
            Comment 1854-1855 (C); 1856 (C)OLD BASSETT; 1857-1858 (C);
            1865 (C); 1867 (C)

BASSET,SOUTH   ILLOGAN   SW 692394 0068

Production: Copper   Ore(tons) Metal(tons)   Value(£)
            1845   3390.00   267.30   19961.80
            1846   2914.00   211.10   14920.90
            1847   1648.00   133.40   9469.10
            1848   2111.00   201.10   11999.20
            1849   3188.00   322.30   22117.40
            1850   4409.00   427.70   30826.40
            1860   138.00   5.50   444.70
            1861   201.00   8.20   653.60
            1862   238.00   7.70   504.20
            1863   96.00   3.80   254.00
            1864   54.00   2.80   228.90
            1865   36.00   1.50   100.30
            1866   31.00   1.90   117.00
            1867   90.00   5.10   336.40
            1868   27.00   2.60   170.90
            Comment 1845-1850 (C); 1860-1868 (C)
Management: Manager 1859-1864 JAS.POPE
            Chief Agent 1859 JOHN REED; 1860-1864 WM.MORCOM; 1865-1867
            S.POPE F.FAY
            Secretary 1859-1862 WM.RICHARDS; 1863-1867 J.W.DABBS

BASSET,W & DOLCOATH CONSOLS   WENDRON   0069

Production: Copper No detailed return
            Tin No detailed return
Ownership: Comment 1865 SUSPENDED
Management: Manager 1861-1864 W.H.BISHOP
            Secretary 1861-1865 BENT.JOHNS

BASSET,WEST   ILLOGAN   SW 678396 0070

Production: Copper   Ore(tons) Metal(tons)   Value(£)
            1852   574.00   35.50   3059.60
            1853   1438.00   106.80   10709.90
            1854   5231.00   376.70   38277.50
            1855   7219.00   490.40   50211.10
            1856   7673.00   532.40   50102.90
            1857   12839.00   892.00   89945.90

CM-D   27

Copper	Ore(tons)	Metal(tons)	Value(£)
1858	7275.00	476.40	42070.50
1859	6766.00	443.50	40865.70
1860	6641.00	502.20	46166.90
1861	5998.00	406.80	35766.00
1862	5479.00	384.20	31301.80
1863	5401.00	402.40	31269.30
1864	4219.00	332.50	28127.40
1865	3637.00	263.40	20721.80
1866	1956.00	147.20	10035.70
1867	1143.00	77.00	5279.20
1868	1357.00	95.10	6227.80
1869	1113.00	70.00	4151.80
1870	1413.00	110.50	6623.00
1871	927.00	76.70	4616.60
1872	601.00	47.40	3583.40
1873	594.00	70.20	4871.30
1874	370.00	34.30	2287.60
1875	640.00	40.20	2972.20
1876	777.00	42.10	2635.80
1877	207.40	15.00	808.40
1878	121.10	7.90	377.00
1879	62.20	3.70	161.20
1880	56.50	3.40	177.50
1882	9.60	0.70	43.00
1885	4.00	0.00	17.00

Comment 1852-1856 (C); 1857 (C)INC.BASSET; 1858-1876 (C)

Tin	Black(tons)	Stuff(tons)	Tin(tons)	Value(£)
1855	1.30	0.00	0.00	82.50
1856	1.40	0.00	0.00	101.40
1857	16.10	0.00	0.00	1206.40
1858	7.80	0.00	0.00	495.10
1859	12.20	0.00	0.00	844.80
1860	12.50	0.00	0.00	958.40
1861	13.40	0.00	0.00	968.50
1862	14.60	0.00	0.00	924.40
1863	4.80	0.00	0.00	306.00
1864	8.00	0.00	0.00	1308.90
1865	9.10	0.00	0.00	517.60
1866	9.70	0.00	0.00	2862.20
1867	82.40	0.00	0.00	4526.30
1868	32.80	0.00	0.00	1799.20
1869	124.30	0.00	0.00	11525.60
1870	790.90	0.00	0.00	8037.60
1871	138.60	0.00	0.00	11277.90
1872	140.30	0.00	0.00	20501.30
1873	360.20	207.00	0.00	32603.60
1874	300.60	12272.00	0.00	59127.00
1875	206.90	33376.00	0.00	38169.50
1876	521.00	413.00	0.00	28495.90
1877	679.70	0.00	0.00	26854.20
1878	690.90	0.00	0.00	22812.60
1879	1000.30	0.00	0.00	38169.80
1880	867.40	0.00	0.00	43440.10
1881	862.00	0.00	560.30	44626.00

BASSET,WEST   ILLOGAN   Continued

Tin	Black(tons)	Stuff(tons)	Tin(tons)	Value(£)
1882	729.40	0.00	459.50	42303.00
1883	526.60	0.00	0.00	25865.00
1884	479.70	0.00	0.00	20320.00
1885	450.00	0.00	0.00	19779.00
1886	336.40	0.00	0.00	17008.00
1887	297.60	0.00	0.00	17510.00
1888	326.10	0.00	0.00	19564.00
1889	309.00	0.00	0.00	16147.00
1890	310.10	0.00	0.00	16471.00
1891	283.60	0.00	0.00	14421.00

Comment 1862 TWO RETURNS AGGREGATED; 1864 VALUE INC.TIN
STUFF; 1866 VALUE INC.TIN STUFF; 1869 VALUE INC.TIN STUFF;
1870 TONNAGE FIGURE CLEARLY WRONG; 1872 VALUE INC.TIN STUFF
Ownership: 1877-1878 WEST WHEAL BASSET MINING CO.; 1879-1890 WEST BASSET
MINE ADVENTURERS
Comment 1891-1894 SEE SOUTH FRANCES UNITED; 1895 SEE SOUTH
FRANCES UNITED & BASSET
Management: Manager 1859-1869 WM.ROBERTS; 1870-1871 JOHN GILBERT;
1872-1873 JAS.NICHOLAS; 1875 JAS.NICHOLAS; 1878-1881
JAS.NICHOLAS
Chief Agent 1859 GEO.LIGHTLEY; 1860-1869 W.PRYOR N.CUNDY;
1870-1873 N.CUNDY FRAN.EVANS; 1874 JAS.NICHOLAS FRAN.EVANS;
1875 JAS.REED FRAN.EVANS; 1876-1877 JAS.NICHOLAS JAS.REED;
1878 FRAN.HODGE J.HARRIS; 1879-1881 FRAN.HODGE; 1882-1888
JAS.NICHOLAS; 1889-1890 JOHN JENNINGS
Secretary 1859 RICH.LYLE; 1860-1861 RICH.LYLE (P); 1862-1869
GEO.LIGHTLEY; 1870-1871 W.A.BUCKLEY; 1872-1873 GEO.LIGHTLEY;
1875-1876 JAS.EVANS; 1878-1881 C.BAWDEN (S); 1887-1890
CORN.BAWDEN
Employment:

	Underground	Surface	Total
1878	288	238	526
1879	304	261	565
1880	332	264	596
1881	300	258	558
1882	263	258	521
1883	287	195	482
1884	197	137	334
1885	229	123	352
1886	206	114	320
1887	181	103	284
1888	223	116	339
1889	172	114	286
1890	192	245	437
1892-1895			

Comment 1891-1895 SEE SOUTH FRANCES UNITED

BEAM   ROCHE   SX 009587 0071

Production: Tin

	Black(tons)	Stuff(tons)	Tin(tons)	Value(£)
1852	11.20	0.00	0.00	0.00
1853	29.70	0.00	0.00	2408.20
1854	64.00	0.00	0.00	4609.20
1855	54.70	0.00	0.00	3689.80

BEAM   ROCHE   Continued

            Tin   Black(tons) Stuff(tons)   Tin(tons)   Value(£)
            1856   115.60   0.00   0.00   8902.60
            1868   38.90   0.00   0.00   2280.20
            1869   22.80   0.00   0.00   1693.30
            1870   18.50   0.00   0.00   1382.70
            1871   19.20   0.00   0.00   1539.10
            Comment 1852 GREAT BEAM TIN ORE; 1853-1856 GREAT BEAM;
            1870-1871 OLD BEAM
Ownership: Comment 1872-1873 GREAT BEAM; 1874 GREAT BEAM SUSPENDED
Management: Manager 1869-1871 JOS.KNIGHT; 1872 JAS.KNIGHT; 1873-1874
            DAVID COCK
            Secretary 1872-1873 J.K.PEER

BEAM,EAST   ROCHE   0072

Production: Tin   Black(tons) Stuff(tons)   Tin(tons)   Value(£)
            1863   1.20   0.00   0.00   80.30
            1864   0.90   0.00   0.00   53.40
            1865   1.40   0.00   0.00   77.80
            1866   0.30   0.00   0.00   13.50
Ownership: Comment 1863-1865 SUSPENDED
Management: Chief Agent 1861-1862 JOHN WEBB JOHN WEBB JNR.; 1863-1865
            JOHN WEBB JNR.
            Secretary 1861-1865 JOHN WEBB

BEAM,NORTH   ROCHE   0073

Production: Tin   Black(tons) Stuff(tons)   Tin(tons)   Value(£)
            1896   1.00   0.00   0.00   22.00
            1897   1.10   0.00   0.00   42.00
Management: Manager 1873 DAVID COCK
            Chief Agent 1872 DAVID COCK

BEAM,WEST   ROCHE   0074

Production: Copper   Ore(tons) Metal(tons)   Value(£)
            1866   24.00   3.30   250.20
            Comment 1866 (C)
            Tin No detailed return

BEDFORD,SOUTH   TAVISTOCK, DEVON   SX 435718 0075

Production: Copper No detailed return
Ownership: Comment 1863-1866 SEE EAST GUNNISLAKE

BEJAWSA   CAMBORNE   SW 637497 0076

Production: Copper   Ore(tons) Metal(tons)   Value(£)
            1851-1853 No detailed return
            Comment 1851-1853 SEE CRANE

                                    30

BELL   ST.ERTH   0077

Production: Copper No detailed return
            Tin No detailed return
Management: Chief Agent 1862–1866 R.WILLIAMS
            Secretary 1862–1866 THOS.W.FIELD

BELL VEAN   GWENNAP   SW 715397 0079

Production: Tin	Black(tons)	Stuff(tons)	Tin(tons)	Value(£)
1873	10.60	0.00	0.00	790.60
1879	0.00	7.00	0.00	14.00
1882	3.50	0.00	2.30	210.00

Comment 1873 BELL TIN MINE
Ownership: 1878 BELL TIN MINING CO.; 1879–1881 BELL TIN & COPPER MINING
            CO.LTD.; 1882 DAVID BURNS
            Comment 1877 BELL TIN MINE; 1882 IN LIQUIDATION
Management: Manager 1873 JOHN HARRIS; 1880–1881 JAS.BRAY
            Chief Agent 1872 JOHN HARRIS; 1878 JOHN BROKENSHIRE; 1880
            DAVID BURNS
            Secretary 1872–1873 R.J.CUNNACK; 1879 DAVID BURNS (S);
            1880–1881 T.S.SMITH (S)

Employment:	Underground	Surface	Total
1878	1	7	8
1879	4		4
1880	7		7
1881	27	54	81
1882	14	32	46

BELLAN   ST.JUST   SW 359311 0080

Production: Tin	Black(tons)	Stuff(tons)	Tin(tons)	Value(£)
1895	12.00	0.00	0.00	480.00
1896	12.00	0.00	0.00	360.00
1910	7.90	0.00	0.00	740.00
1911	6.70	0.00	0.00	768.00
1912	6.90	0.00	0.00	880.00
1913	7.20	0.00	0.00	921.00

Ownership: 1910 JAS.CHENHALLS; 1911–1913 WHEAL BELLAN TIN SYNDICATE
            LTD.
            Comment 1894–1902 SEE ST.JUST UNITED
Management: Chief Agent 1913 R.BLACKWELL

Employment:	Underground	Surface	Total
1911	10	15	25
1912	12	26	38
1913	24	25	49

BELOVELY   ROCHE   SW 972625 0081

Production: Copper	Ore(tons)	Metal(tons)	Value(£)	
1899	1.00	0.00	42.00	

Tin	Black(tons)	Stuff(tons)	Tin(tons)	Value(£)
1894	3.50	0.00	0.00	124.00
1902	0.00	75.00	0.00	27.00

BELOVELY   ROCHE   Continued

Ownership: 1894-1895 BEALE,BEALE & COCK; 1899-1900 ROYSE,COCK,SHANN;
            1901-1904 ROYSE & SHANN
            Comment 1903-1904 SUSPENDED
Management: Chief Agent 1899-1900 DAVID COCK
Employment: Underground   Surface   Total
            1894   5   5   10
            1899   2   2   4
            1900   9   10   19
            1901-1902   2   2   4

BELOWDA   ROCHE   SW 972625 0082

Production: Tin   Black(tons) Stuff(tons) Tin(tons)   Value(£)
            1873   0.20   0.00   0.00   12.10
            1880   7.00   0.00   0.00   315.00
            Comment 1873 BELLOWDA BEACON
Ownership: 1880-1881 BELOWDA TIN MINING CO.LTD.; 1909-1912 BELOWDA TIN &
            CHINA CLAY MINES; 1913 THOS.JEFFERY
            Comment 1873 BELOWDA BEACON. STOPPED IN 1873 & EARLY IN 1874;
            1911-1913 SUSPENDED AUG.1911
Management: Manager 1909-1913 TAMBLYN
            Chief Agent 1872-1873 GEO.STEPHENS; 1880 J.H.COLLINS;
            1909-1912 KINGSFORD,DORMAN & CO.
            Secretary 1909-1912 JAS.C.INGLIS
Employment: Underground   Surface   Total
            1880   4   24   28
            1909   11   11
            1910   6   4   10

BELOWDA HILL   ROCHE   SW 972625 0084

Production: Tin   Black(tons) Stuff(tons) Tin(tons)   Value(£)
            1873   8.80   0.00   0.00   644.50
            1874   4.60   0.00   0.00   283.00
            1875   2.70   0.00   0.00   162.10
            Comment 1873-1875 BELLOWDA HILL
Management: Chief Agent 1874 H.TREGANOWAN
            Secretary 1872 J.DUNSTAN; 1873 J.DUNSTAN (S); 1874 J.DUNSTAN;
            1875 J.P.DUNSTAN (S)

BELOWDO   PAR   0085

Production: Tin No detailed return
Ownership: Comment 1880 SEE PAR GREAT CONSOLS; 1881 SEE POLGOOTH UNITED

BENEATHWOOD   LINKINHORNE   SX 305723 0086

Production: Lead   Ore(tons) Metal(tons)   Value(£)
            1859   5.50   4.00   0.00
Ownership: Comment 1860-1861 BENCATHWOOD; 1862 BENCATHWOOD SUSPENDED;
            1863-1865 SUSPENDED

                                  32

BENEATHWOOD   LINKINHORNE   Continued

Management: Chief Agent 1860–1861 J.LEAN
            Secretary 1860–1865 J.D.DYMOND (P)

BENNALLACK   ST.ENODER   SW 907568 0087

Production: Iron   Ore(tons)   Iron(%)   Value(£)
            1860–1861 No detailed return
            Comment 1860–1861 BH.SEE TREWHEELA

BENNY   CALSTOCK   SX 397731 0088

Production: Tin   Black(tons) Stuff(tons)   Tin(tons)   Value(£)
            1884   6.20   0.00   0.00   252.00
            1885   4.10   0.00   0.00   192.00
            Arsenic   Ore(tons) Metal(tons)   Value(£)
            1885   12.50   0.00   62.00
            Comment 1885 SOOT
Ownership: 1883–1886 WHEAL BENNY MINING CO.; 1901–1904 KIT TIN MINES
            LTD.
            Comment 1886 NOT WORKED; 1903–1904 SUSPENDED
Management: Chief Agent 1883–1886 THOS.COCKING; 1901–1902 ALF.THOMAS
Employment: Underground   Surface   Total
            1883   8   4   12
            1884   16   9   25
            1885   15   10   25
            1901   10   15   25
            1902   10   14   24
            1904   1   1

BETSY   ST.AGNES   SW 733518 0089

Production: Tin   Black(tons) Stuff(tons)   Tin(tons)   Value(£)
            1855   39.20   0.00   0.00   2548.10
            1856   29.60   0.00   0.00   2185.60
            1857   23.00   0.00   0.00   1767.90
            1858   5.70   0.00   0.00   353.00
            1859   13.10   0.00   0.00   868.80
            1860   5.90   0.00   0.00   453.90
Ownership: Comment 1863–1865 SEE EAST POLBERRO
Management: Manager 1860–1862 ART.GRIPE
            Secretary 1859–1862 JAS.GRIPE (P)

BICTON   ST.IVE   SX 308696 0090

Production: Lead & Silver   Ore(tons) Lead(tons) Silver(ozs)   Value(£)
            1881   4.00   1.50   0.00   14.00
            1882   4.00   3.00   15.00   14.00
            1883   1.40   1.00   5.00   10.00
            Manganese No detailed return
Ownership: 1879–1881 BICTON SILVER LEAD & MANGANESE CO.LTD.; 1882–1886
            BICKTON SILVER LEAD & MANGANESE CO.LTD.; 1887 NEW BICKTON

                                    33

BICTON   ST.IVE   Continued

SILVER, LEAD & MANGANESE CO.LTD.; 1896 PERCY R.DENNY; 1897
BRITISH TRUST AND AGENCY LTD.
Comment 1880 BICTON SILVER LEAD; 1881–1886 BICKTON; 1887 NEW
BICKTON; 1896–1897 NEW BICKTON
Management: Manager 1880–1881 WM.HARRIS
Chief Agent 1879 WM.HARRIS; 1882–1887 WM.HARRIS; 1897
WM.SIMPSON
Secretary 1881 B.GREGORY (S)

Employment:	Underground	Surface	Total
1879		9	9
1880	15	9	24
1881	8	7	15
1882	11	5	6
1883	8	3	11
1884	6	2	8
1885	6	4	10
1886		3	3
1887		2	2
1896–1897	2	1	3

BIGBEES   LANIVET   SX 027636 0091

Production: Iron
	Ore(tons)	Iron(%)	Value(£)
1906	60.00	44.00	10.00
Ownership: 1906 BALDWINS LTD.
Comment 1906 ABANDONED SEPT.1906
Employment:	Underground	Surface	Total
1906	8	3	11

BILLIA CONSOLS   TOWEDNACK   0092

Production: Tin
	Black(tons)	Stuff(tons)	Tin(tons)	Value(£)
1865	74.90	0.00	0.00	4020.00
1866	51.90	0.00	0.00	2435.70
1867	17.80	0.00	0.00	812.30

BISSOE CONSOLS   DEVORAN   0094

Production: Lead No detailed return
Tin No detailed return
Ownership: 1882 PAINTER & TRYTHALL
Management: Secretary 1882 PAINTER & TRYTHALL
Employment:	Underground	Surface	Total
1882	7	5	12

BISSOE POOL   ST.DAY   SW 776413 0095

Production: Tin
	Black(tons)	Stuff(tons)	Tin(tons)	Value(£)
1896	0.00	21.00	0.00	7.00
Arsenic	Ore(tons)	Metal(tons)	Value(£)	
1894	47.00	0.00	236.00	

BISSOE POOL   ST.DAY   Continued

        Arsenic   Ore(tons) Metal(tons)   Value(£)
        1896   4.00   0.00   38.00
        Comment 1894 INC.PORTHTOWAN & ROSEWARNE DOWNS
        Ochre No detailed return
Ownership: 1895-1896 W.J.TRYTHALL
        Comment 1895 NOT WORKED
Employment: Underground   Surface   Total
        1896   1   1   2

BLACK DOG   SCORRIER   SW 732443 0096

Production: Tin No detailed return
Ownership: 1912-1913 PEEVOR UNITED MINES LTD.
Management: Chief Agent 1912-1913 J.A.TEMBY
Employment: Underground   Surface   Total
        1912-1913
        Comment 1912-1913 SEE PEEVOR UNITED

BLACKHAY   WITHIEL   SW 987657 0097

Production: Iron   Ore(tons)   Iron(%)   Value(£)
        1872   1200.00   0.00   720.00
        1873   780.00   0.00   585.00
        1874   1000.00   0.00   800.00
        1879   400.00   0.00   240.00
        1880   950.00   0.00   570.00
        Comment 1872-1874 BH.; 1879-1880 BH.
Ownership: 1863-1865 EDW.CARTER; 1873-1874 AGRA BANKING CO.; 1875-1879
        AGRA BANK AND CO.; 1880 AGRA BANKING CO.; 1881-1882 AGRA BANK
        CO.LTD.
        Comment 1882 NOT WORKED
Management: Manager 1873-1878 A.J.WHALLEY; 1879-1880 PHIL.MICHELL JNR.
        Chief Agent 1863-1865 THOS.HAMBLY; 1875-1878 PHIL.MICHELL
        JNR.; 1881-1882 PHIL.MICHELL JNR.
Employment: Underground   Surface   Total
        1878   4   4
        1879   3   1   4

BLAMEY CONSOLS   ST.BLAZEY   0098

Production: Tin No detailed return
Ownership: 1898-1901 BLAMEY CONSOLS TIN MINE LTD.; 1904-1905 TASMAN
        SYNDICATE LTD.
        Comment 1900-1901 SUSPENDED; 1905 IN LIQUIDATION
Management: Chief Agent 1898-1901 S.BENNETT; 1904-1905 W.WILLIAMS
Employment: Underground   Surface   Total
        1898   6   6   12
        1899   4   5   9
        1904   6   3   9
        1905   1   1

CM-D*   35

BLENCOWE CONSOLS   GRAMPOUND ROAD   SW 929523 0099

Production: Tin

	Black(tons)	Stuff(tons)	Tin(tons)	Value(£)
1873	29.00	0.00	0.00	2023.00
1874	5.00	0.00	0.00	314.60

         Comment 1873 BLENCOWE; 1874 BLENCOWE—VALUE EST

Ownership: Comment 1873 STOPPED IN 1873 & EARLY IN 1874; 1875 BLENCOWE SEE LAMBERT

Management: Manager 1872—1873 JOHN EDWARDS
         Chief Agent 1872—1873 JOHN TREDINNICK; 1874 IS.THOMAS DAVID COCK
         Secretary 1872—1873 JOBLING & BATTYE; 1874 M.E.JOBLING

BLUE BARROW   ST.AUSTELL   SX 007568 0100

Production: Tin

	Black(tons)	Stuff(tons)	Tin(tons)	Value(£)
1901	1.20	0.00	0.00	75.00
1903	2.30	0.00	0.00	170.00
1905	1.80	0.00	0.00	167.00
1906	1.20	0.00	0.00	137.00

         Comment 1901 OPEN WORK; 1903 OPEN WORK; 1905—1906 OPEN WORK

BLUE HILLS   ST.AGNES   SW 728517 0101

Production: Copper No detailed return

Tin

	Black(tons)	Stuff(tons)	Tin(tons)	Value(£)
1869	2.70	0.00	0.00	115.30
1870	32.40	0.00	0.00	2358.80
1871	83.90	0.00	0.00	6447.70
1872	76.00	0.00	0.00	6322.50
1873	75.90	0.00	0.00	5626.00
1874	66.20	0.00	0.00	3671.10
1875	52.30	0.00	0.00	2602.50
1876	40.70	0.00	0.00	1709.80
1877	36.70	0.00	0.00	1440.40
1878	45.20	0.00	0.00	1581.50
1879	64.50	0.00	0.00	2855.00
1880	103.50	0.00	0.00	5413.20
1881	80.90	0.00	55.70	4567.20
1882	76.20	0.00	50.20	4542.00
1883	64.00	0.00	0.00	3347.00
1884	53.00	0.00	0.00	2425.00
1885	98.00	0.00	0.00	4900.00
1886	118.10	0.00	0.00	6904.00
1887	94.50	0.00	0.00	5942.00
1888	105.60	0.00	0.00	6837.00
1889	114.20	0.00	0.00	6283.00
1890	142.00	0.00	0.00	7975.00
1891	133.50	0.00	0.00	7218.00
1892	86.60	0.00	0.00	4724.00
1893	50.00	0.00	0.00	2523.00
1894	74.50	0.00	0.00	2993.00
1895	103.00	0.00	0.00	3895.00
1896	29.00	0.00	0.00	1038.00
1897	13.00	0.00	0.00	425.00

BLUE HILLS   ST.AGNES   Continued

Tin	Black(tons)	Stuff(tons)	Tin(tons)	Value(£)
1898	No detailed return			

Arsenic	Ore(tons)	Metal(tons)	Value(£)
1894	3.00	0.00	21.00
1895	12.00	0.00	76.00
1897	2.00	0.00	24.00

Ownership: 1876-1890 BLUE HILLS MINING CO.; 1891-1893 BLUE HILLS MINING
CO. C.B.; 1894-1895 BLUE HILLS MINE ADVENTURERS; 1896-1897
BLUE HILLS MINING CO.
Comment 1863-1865 SEE EAST POLBERRO; 1897 ABANDONED 1897
Management: Manager 1869-1881 SML.BENNETTS
Chief Agent 1868 JAS.GRIPE; 1869-1871 ART.GRIPE; 1872-1875
ART.GRIPE PHIL.BENNETTS; 1876-1879 ART.GRIPE; 1880-1881
RICH.HARRIS; 1882-1888 SML.BENNETTS; 1889-1891 WM.RICH;
1892-1897 JOS.RICHARDS
Secretary 1868 ED.KING R.H.PIKE & SON; 1869-1870 ROBT.H.PIKE
& SON; 1871 JAS.HICKEY; 1872-1875 J.HICKEY R.PIKE & SON;
1876-1877 ROBT.H.PIKE & SON; 1878-1880 WALT.PIKE; 1891
WALT.PIKE; 1892-1897 WALT.PIKE (P)

Employment:	Underground	Surface	Total
1878	38	27	65
1879	37	16	53
1880	54	32	86
1881	50	36	86
1882	71	45	116
1883	72	43	115
1884	47	49	96
1885	57	53	110
1886	61	49	110
1887	64	43	107
1888	72	50	122
1889	75	47	122
1890	88	53	141
1891	78	52	130
1892	56	40	96
1893	76	28	104
1894	77	32	109
1895	71	37	108
1896	53	36	89
1897		1	1

BLUE HILLS,EAST   ST.AGNES   SW 738518 0102

Production: Tin	Black(tons)	Stuff(tons)	Tin(tons)	Value(£)
1881	3.00	0.00	2.10	193.80
1882	17.30	0.00	11.20	990.00
1883	28.00	0.00	0.00	1468.00
1884	19.80	0.00	0.00	885.00
1885	57.00	0.00	0.00	2702.00
1886	62.00	0.00	0.00	3410.00
1887	30.60	0.00	0.00	1891.00
1888	29.70	0.00	0.00	1974.00
1889	27.70	0.00	0.00	1526.00
1890	20.50	0.00	0.00	1062.00

BLUE HILLS,EAST   ST.AGNES   Continued

            Tin   Black(tons) Stuff(tons)   Tin(tons)   Value(£)
            1891   17.10   0.00   0.00   844.00
            1893   26.10   0.00   0.00   1276.00
            1912   57.00   0.00   0.00   7727.00
            1913   21.90   0.00   0.00   2853.00
Ownership: 1881-1890 EAST BLUE HILLS MINING CO.; 1891-1893 EAST BLUE
            HILLS MINING CO.C.B.; 1911-1913 ST.AGNES CONSOLIDATED MINES
            LTD.
            Comment 1893 ABANDONED IN LIQUIDATION
Management: Manager 1881 SML.BENNETTS
            Chief Agent 1881 ART.GRIPE; 1882-1887 SML.BENNETTS; 1888
            ART.GRIPE; 1889-1890 J.THOMAS; 1891-1893 JAS.WHITE
            Secretary 1881 WALT.PIKE (P); 1889-1890 ART.GRIPE; 1891-1893
            C.B.PARRY (S&P)
Employment: Underground   Surface   Total
            1881   15   5   20
            1882   28   24   52
            1883   20   14   34
            1884   22   26   48
            1885   32   34   66
            1886   37   24   61
            1887   28   23   51
            1888   24   18   42
            1889   23   15   38
            1891   38   28   66
            1892   27   23   50
            1893   18   16   34
            1911   31   10   41
            1912   36   6   42
            1913   15   3   18

BLUE HILLS,NORTH   ST.AGNES   SW 374528 0103

Production: Tin No detailed return
Ownership: 1882-1886 NORTH BLUE HILLS MINING CO.; 1907 J.DEFRIES & SONS
            LTD.
            Comment 1885-1886 NOT WORKED
Management: Chief Agent 1882-1886 SML.BENNETTS
Employment: Underground   Surface   Total
            1882   4   4
            1883   5   1   6

BODANNON   ENDELLION   SX 008799 0104

Production: Antimony   Ore(tons) Metal(tons)   Value(£)
            1884   4.00   0.00   40.00
Ownership: 1884-1885 W.G.FORSTER
Management: Chief Agent 1884-1885 E.L.OWEN
Employment: Underground   Surface   Total
            1884   4   4   8

                                     38

BODINNICK   ST.STEPHEN   SW 946519 0105

Production: Iron   Ore(tons)   Iron(%)   Value(£)
            1858   2150.70   0.00   919.20
            1859   550.60   0.00   183.30
            1860   207.40   0.00   51.80
            1861   205.00   0.00   51.20
            1863   250.00   0.00   135.00
            Comment 1858-1861 BH.; 1863 BH.
Ownership: 1863-1865 EDW.CARTER & CO.
Management: Chief Agent 1863-1865 WM.BUCKTHOUGHT

BODITHIEL   ST.PINNOCK   SX 174647 0106

Production: Lead & Silver No detailed return
Ownership: Comment 1873-1874 SEE ALLEN

BODMIN UNITED   LANIVET   SX 065641 0108

Production: Lead & Silver   Ore(tons)   Lead(tons) Silver(ozs)   Value(£)
            1853   10.00   7.00   78.00   0.00
            Comment 1853 BODMIN CONSOLS
            Copper   Ore(tons) Metal(tons)   Value(£)
            1853   266.00   23.70   2363.20
            1854   505.00   32.60   3319.30
            1855   172.00   10.70   1093.50
            Comment 1853-1855 (C)

BOILING WELL   PHILLACK   SW 578397 0109

Production: Lead & Silver   Ore(tons)   Lead(tons) Silver(ozs)   Value(£)
            1855   46.50   30.70   0.00   0.00
            1856   49.00   37.50   321.00   0.00
            1857   248.00   186.00   2560.00   0.00
            1858   281.00   211.00   10870.00   0.00
            1859   64.80   48.50   1760.00   0.00
            Zinc   Ore(tons) Metal(tons)   Value(£)
            1856   54.10   0.00   81.10
            1858   179.30   0.00   651.10
            1859   281.30   0.00   697.20
            Copper   Ore(tons) Metal(tons)   Value(£)
            1854   304.00   19.40   1974.80
            1855   877.00   51.40   5296.60
            1856   931.00   65.80   6231.60
            1857   848.00   55.40   5617.70
            1858   168.00   9.10   802.80
            1862   25.00   0.70   45.60
            1863   15.00   0.40   24.00
            1864   28.00   0.70   56.60
            Comment 1854-1858 (C); 1862-1864 (C)
            Tin   Black(tons) Stuff(tons)   Tin(tons)   Value(£)
            1857   0.00   0.00   0.00   20.50
            Comment 1857 TIN STUFF
Ownership: Comment 1863-1865 SUSPENDED

                                     39

BOLENOWE   CAMBORNE   SW 668382 0110

```
Production: Tin No detailed return
Ownership:  1899-1900 BOLENOWE MINING CO.
Management: Secretary 1899-1900 WM.ROWE
Employment:           Underground    Surface      Total
            1899          4                        4
            1900          2                        2
```

BONNEY,NORTH   BUGLE   SX 007571 0112

```
Production: Tin         Black(tons) Stuff(tons)  Tin(tons)   Value(£)
            1873           41.00       0.00         0.00      3336.50
            Comment 1873 NORTH BONNY
Ownership:  Comment 1874 SUSPENDED
Management: Chief Agent 1872-1874 DAVID COCK
```

BOSCARNE   BODMIN   SX 037673 0113

```
Production: Iron           Ore(tons)    Iron(%)    Value(£)
            1861             65.00        0.00       16.20
            Comment 1861 BH.
Ownership:  1863-1865 EDW.CARTER & CO.
Management: Chief Agent 1863-1865 THOS.HAMBLY
```

BOSCASTLE   BOSCASTLE   0114

```
Production: Iron           Ore(tons)    Iron(%)    Value(£)
            1858             51.30        0.00       25.50
            1859     No detailed return
            Comment 1858-1859 MO.
```

BOSCASWELL   ST.JUST   SW 383344 0115

```
Production: Copper         Ore(tons) Metal(tons)   Value(£)
            1852            123.00      10.50        877.20
            1853            109.00       6.90        593.80
            1862             42.00       4.50        376.20
            1863              1.00       0.60         41.40
            1864            101.00       7.30        566.20
            1865            143.00      16.20       1313.60
            1866             76.00       6.60        459.20
            1867             90.00       9.30        661.80
            1869              6.00       0.40         18.90
            1871              3.00       1.60         99.10
            Comment 1852-1853 (C)BOSCASWELL DOWNS; 1862-1867 (C); 1869
            (C); 1871 (C)BOSCASWELL DOWNS
            Tin         Black(tons) Stuff(tons)  Tin(tons)   Value(£)
            1855           19.20       0.00         0.00      1129.80
            1856            2.30       0.00         0.00       142.60
            1857           11.40       0.00         0.00       839.30
            1858           81.90       0.00         0.00      5310.70
            1859           20.50       0.00         0.00      1265.10
```

BOSCASWELL   ST.JUST   Continued

Tin	Black(tons)	Stuff(tons)	Tin(tons)	Value(£)
1861	0.00	0.00	0.00	5245.60
1862	39.10	0.00	0.00	2483.30
1863	168.50	0.00	0.00	10701.30
1864	102.40	0.00	0.00	6001.60
1865	91.40	0.00	0.00	4744.70
1866	71.50	0.00	0.00	3368.30
1867	24.40	0.00	0.00	1103.70
1868	65.80	0.00	0.00	3455.10
1869	95.20	0.00	0.00	6475.30
1870	100.90	0.00	0.00	7120.10
1873	152.70	0.00	0.00	12303.30
1874	65.00	0.00	0.00	3404.00
1877	2.70	0.00	0.00	100.00

Comment 1855–1857 BOSCASWELL DOWNS; 1858 TWO RETURNS
AGGREGATED; 1861 TIN STUFF; 1862 PART YEAR ONLY; 1873–1874
BOSCASWELL DOWNS; 1877 BOSCASWELL DOWNS
Ownership: 1877 WM.EDDY
             Comment 1860–1872 OR BOSCASWELL DOWNS; 1873–1874 BOSCASWELL
             DOWNS; 1877–1879 ABOVE ADIT & HEAPS ONLY
Management: Manager 1859–1867 WM.NOY; 1868 WM.EDDY WM.T.WHITE; 1869–1870
             WM.EDDY; 1871–1873 R.WILLIAMS; 1876–1879 WM.EDDY
             Chief Agent 1862–1866 JOHN NOY; 1867 WM.EDDY; 1869 WM.NOY
             WM.EDDY JNR.; 1870 WM.NOY; 1871 WM.EDDY; 1872–1873 WM.EDDY
             ED.RODDA JOHN EDDY; 1874 R.WILLIAMS
             Secretary 1859–1866 SML.YORK (P); 1867 E.J.BOYNS; 1868–1873
             EDW.S.BOYNS; 1876–1879 WM.EDDY

BOSCASWELL,EAST   ST.JUST   SW 390340 0117

Production:	Tin	Black(tons)	Stuff(tons)	Tin(tons)	Value(£)
	1873	36.00	0.00	0.00	2774.40
	1874	10.90	0.00	0.00	532.70
	1877	3.00	0.00	0.00	122.50

Ownership: 1876–1877 THOS.RICHARDS & CO.
             Comment 1874–1875 SUSPENDED; 1877 ABOVE ADIT & HEAPS ONLY
Management: Manager 1872–1873 WM.EDDY JNR.; 1876 WM.EDDY
             Chief Agent 1872–1873 JAS.ELLIS; 1874–1875 WM.EDDY; 1877
             THOS.RICHARDS
             Secretary 1872–1873 F.WARWICK; 1875 WM.WARWICK

BOSCASWELL,NORTH   ST.JUST   SW 380350 0118

Production:	Tin	Black(tons)	Stuff(tons)	Tin(tons)	Value(£)
	1911	19.60	0.00	0.00	1818.00
	1912	4.10	0.00	0.00	351.00

Comment 1911–1912 BOSCASWELL UNITED
Ownership: 1906–1911 NORTH BOSCASWELL MINING CO.LTD.; 1912–1913
             BOSCASWELL UNITED TIN & COPPER MINESLTD.
             Comment 1911 SUSPENDED OCT.1911; 1913 SUSPENDED JULY 1913
Management: Chief Agent 1913 S.JAMES

Employment:		Underground	Surface	Total
	1906	6	11	17

41

BOSCASWELL,NORTH   ST.JUST   Continued

	Underground	Surface	Total
1907	10	12	22
1908	10	9	19
1909	16	25	41
1910	10	20	30
1911	2	25	27
1912	15	14	29
1913	20	10	30

BOSCAWEN   KENWYN   SW 732452 0119

Production: Zinc   Ore(tons) Metal(tons)   Value(£)
            1864   6.60   0.00   7.00
            1870   12.00   0.00   41.40
            1880   15.00   0.00   49.50
            Copper   Ore(tons) Metal(tons)   Value(£)
            1863   319.00   22.80   1871.60
            1864   1235.00   71.40   5882.20
            1865   555.00   36.70   3862.10
            1866   13.00   0.50   30.60
            Comment 1863-1866 (C)
            Tin   Black(tons) Stuff(tons) Tin(tons)   Value(£)
            1893   0.00   25.00   0.00   19.00
            1894   0.00   20.00   0.00   30.00
            1895   No detailed return
            1898   0.00   8.00   0.00   4.00
            Arsenic No detailed return
            Arsenic Pyrite Ore(tons)   Value(£)
            1893   20.00   3.00
            1894   30.00   4.00
            1898   78.00   11.00
            1899   11.00   2.00
Ownership: 1893-1895 JOHN BRENTON; 1898-1899 W.J.TRYTHALL THOS.BEHENNA;
            1900 W.J.TRYTHALL; 1907 RENNY-TAILYOUR
            Comment 1865 SUSPENDED; 1893-1895 TRIBUTER UNDER LORD
            FALMOUTH; 1898-1899 TRYTHALL ARSENIC BEHENNA TIN
Management: Manager 1863-1864 JOHN EDWARDS; 1869-1871 THOS.WESLEY
            Chief Agent 1863-1864 R.GILES; 1898-1900 THOS.TONKIN
            Secretary 1863-1865 ED.KING
Employment: Underground   Surface   Total
            1893-1894   2   2
            1898   2   2   4
            1899   2   2
            Comment 1898 TRYTHALL ABOVE BEHENNA BELOW; 1899 TRYTHALL

BOSCEAN   ST.JUST   SW 364321 0120

Production: Tin   Black(tons) Stuff(tons) Tin(tons)   Value(£)
            1852   23.10   0.00   0.00   0.00
            1853   91.20   0.00   0.00   6439.40
            1854   147.50   0.00   0.00   10295.00
            1855   97.20   0.00   0.00   6452.60
            1856   284.40   0.00   0.00   21565.10

42

BOSCEAN   ST.JUST   Continued

Tin	Black(tons)	Stuff(tons)	Tin(tons)	Value(£)
1857	217.10	0.00	0.00	18070.40
1858	226.10	0.00	0.00	15214.10
1859	187.00	0.00	0.00	16519.50
1860	163.70	0.00	0.00	13635.00
1861	0.00	0.00	0.00	9265.70
1862	34.30	0.00	0.00	2427.80
1863	121.20	0.00	0.00	8352.20
1864	158.00	0.00	0.00	10317.70
1865	124.30	0.00	0.00	7112.40
1866	92.50	0.00	0.00	4649.80
1867	33.60	0.00	0.00	1727.60
1870	99.60	0.00	0.00	7400.30
1871	77.50	0.00	0.00	5990.80

Comment 1852 TIN ORE; 1861 TIN STUFF; 1862 PART YEAR ONLY
Ownership: Comment 1872–1873 SEE CUNNING UNITED
Management: Manager 1859 JOHN CARTHEW; 1860–1864 RICH.BERRYMAN; 1869–1871
             JAS.TREZISE
             Chief Agent 1859 RICH.BERRYMAN; 1860–1864 JAS.TREZISE JOHN
             ROWE; 1865–1868 JAS.TREZISE; 1869–1871 WM.T.WHITE
             Secretary 1859–1866 SML.YORK (P); 1867 E.J.BOYNS; 1868–1871
             SML.YORK

BOSCENCE   SITHNEY   0121

Production: Copper   Ore(tons) Metal(tons)   Value(£)
             1851–1852 No detailed return
             Comment 1851–1852 SEE TRANNACK
             Tin   Black(tons) Stuff(tons)   Tin(tons)   Value(£)
             1855   No detailed return
             Comment 1855 BOSENCE SEE TRANNACK

BOSCOPPA   ST.AUSTELL   SX 040537 0122

Production: Tin	Black(tons)	Stuff(tons)	Tin(tons)	Value(£)
1872	8.50	0.00	0.00	731.30
1873	17.10	0.00	0.00	1345.50
1874	1.90	0.00	0.00	101.30
1876	0.10	0.00	0.00	3.20

Management: Manager 1873–1877 FRAN.BARRATT
             Chief Agent 1872 FRAN.BARRATT; 1874–1877 WALT.J.NICHOLLS
             Secretary 1872 FRAN.BARRATT JNR.; 1873–1877 FRAN.BARRATT

BOSCUNDLE   PAR   SX 045531 0123

Production: Copper	Ore(tons)	Metal(tons)	Value(£)
1851	57.00	9.70	686.90
1852	30.00	5.50	430.60
1854	118.00	10.80	1075.50

Comment 1851–1852 (C); 1854 (C)

Tin	Black(tons)	Stuff(tons)	Tin(tons)	Value(£)
1852	67.40	0.00	0.00	0.00

43

BOSCUNDLE   PAR   Continued

Tin   Black(tons) Stuff(tons)   Tin(tons)   Value(£)
1853   217.00   0.00   0.00   14507.70
1854   148.90   0.00   0.00   9400.00
1855   55.80   0.00   0.00   3458.50
1856   139.90   0.00   0.00   9120.40
1857   119.90   0.00   0.00   9023.60
1858   88.20   0.00   0.00   5894.60
1859   125.80   0.00   0.00   8728.20
1860   135.40   0.00   0.00   10405.70
1861   11.00   0.00   0.00   816.70
1862   2.10   0.00   0.00   132.20
1863   3.90   0.00   0.00   240.40
             Comment 1852 TIN ORE; 1862 PART YEAR ONLY
Ownership: Comment 1863 SUSPENDED; 1865 SUSPENDED
Management: Manager 1859-1862 WM.VIVIAN; 1864 RICH.WILLIAMS
             Chief Agent 1859-1862 JOHN MORCOM WM.ALLEN; 1863 WM.ALLEN
             Secretary 1859-1865 JOS.MORCOM (P)

BOSCUNDLE,NEW   PAR   SX 047530 0124

Production: Copper   Ore(tons) Metal(tons)   Value(£)
            1878   158.10   17.50   632.00
            Tin   Black(tons) Stuff(tons)   Tin(tons)   Value(£)
            1880   1.50   0.00   0.00   82.00
            1881   1.00   0.00   0.60   59.40
Ownership: 1877-1883 RICH.H.WILLIAMS & CO.
Management: Manager 1877 RICH.H.WILLIAMS; 1879-1881 RICH.H.WILLIAMS
            Chief Agent 1877-1879 CHAS.FAULE; 1880-1881 CHAS.C.FAULE;
            1882-1883 RICH.H.WILLIAMS
            Secretary 1877 RICH.H.WILLIAMS; 1878-1879 RICH.H.WILLIAMS &
            CO; 1880 C.W.BROADHURST
Employment: Underground   Surface   Total
            1878   16   9   25
            1879   14   7   21
            1880   11   5   16
            1881   13   5   18
            1882   9   4   13
            1883   6   5   11

BOSIGRAN CONSOLIDATED   MORVAH   SW 421365 0125

Production: Tin No detailed return
Ownership: 1907 BOSIGRAN CONSOLS MINES
            Comment 1907 MORVAH HILL, CARN GALVER, ROSEMERGY.
Management: Chief Agent 1907 G.D.MCGREGOR

BOSKILLAN   ST.AUSTELL   0126

Production: Tin   Black(tons) Stuff(tons)   Tin(tons)   Value(£)
            1903   2.00   0.00   0.00   150.00
            1904   2.50   0.00   0.00   157.00
            1905   1.50   0.00   0.00   130.00

                                    44

Tin	Black(tons)	Stuff(tons)	Tin(tons)	Value(£)
1906	1.00	0.00	0.00	105.00
1908	2.00	0.00	0.00	150.00
1909	2.50	0.00	0.00	205.00
1910	1.40	0.00	0.00	132.00

Ownership: 1903–1913 PETER KENDALL
Comment 1911–1913 WORKING DUMPS
Employment:

	Underground	Surface	Total
1903–1911		2	2
1912–1913		1	1

BOSORNE ST.JUST SW 359308 0128

Production:
Tin	Black(tons)	Stuff(tons)	Tin(tons)	Value(£)
1852	2.70	0.00	0.00	0.00
1853	10.60	0.00	0.00	768.10
1854	10.90	0.00	0.00	714.00

Comment 1852 BOSOM TIN ORE
Ownership: Comment 1859–1864 BOSORNE AND BALLOWAL UNITED.; 1865 NOW EAST
ST. JUST UNITED
Management: Manager 1859 BRY.LAWRY; 1860–1864 JOHN BOYNS; 1865 JOHN
CARTHEW
Chief Agent 1859 JOHN BOYNS; 1862–1864 JOHN ROWE; 1865 PETER
CASLEY
Secretary 1859 SML.YORK (P); 1860–1862 BRY.LAWRY (P);
1863–1864 J.CARTHEW; 1865 W.AUGWIN

BOSORNE,EAST ST.JUST SW 363309 0129

Production:
Tin	Black(tons)	Stuff(tons)	Tin(tons)	Value(£)
1854	1.00	0.00	0.00	74.00

BOSPARRA GWINEAR SW 616359 0130

Ownership: 1911–1912 W.MIDDLIA; 1913 GWINEAR TIN MINES LTD.
Comment 1911 PROSPECTING STOPPED SEPT 1911; 1912 RESUMED
28TH.MAY 1912; 1913 NOT WORKED
Employment:

	Underground	Surface	Total
1911	2	1	3
1912	6	6	12

BOSSINEY TINTAGEL 0131

Production: Lead & Silver No detailed return

Iron	Ore(tons)	Iron(%)	Value(£)
1872	40.00	0.00	24.00

Comment 1872 SP.
Ownership: 1871–1872 J.OLI.MASON; 1874–1875 J.OLI.MASON
Management: Chief Agent 1871–1872 SML.TUCKER; 1874–1875 SML.TUCKER

BOSSOW   TOWEDNACK   SW 494385 0132

Production: Tin No detailed return
Ownership: 1880–1881 M.G.BROWNE WM.BROWNE
            Comment 1880 FORMERLY TYRINGHAM CONSOLS; 1881 FORMERLY
            TYRINGHAM CONSOLS SUSPENDED
Management: Chief Agent 1880–1881 WM.BUGLEHOLE

BOSTRAZE   ST.JUST   SW 389313 0133

Production: Tin   Black(tons) Stuff(tons)   Tin(tons)   Value(£)
            1888   3.60   0.00   0.00   220.00
            1889   15.00   786.00   0.00   743.00
            1890   2.50   0.00   0.00   144.00
Ownership: 1888–1890 RICH.ROBERTS & OTHERS; 1891 WHEAL BOSTRAZE MINING
            CO.
            Comment 1890–1891 SUSPENDED
Management: Chief Agent 1888–1891 RICH.ROBERTS
Employment: Underground   Surface   Total
            1888   17   9   26
            1889   12   8   20
            1890   16   6   22

BOSWARTHEN PENZANCE CONSOLS   SANCREED   SW 412290 0134

Production: Tin No detailed return
Ownership: Comment 1866 SEE ALSO BOSWORTHAN
Management: Chief Agent 1866 FRAN.HOSKING
            Secretary 1866 RICH.PRYOR

BOSWEDDEN   ST.JUST   SW 355322 0135

Production: Copper   Ore(tons) Metal(tons)   Value(£)
            1864   9.00   1.50   115.20
            1865   22.00   2.60   196.90
            1866   10.00   1.10   77.00
            1872   147.00   11.30   858.20
            Comment 1864–1866 (C); 1872 (C)
            Tin   Black(tons) Stuff(tons)   Tin(tons)   Value(£)
            1853   22.00   0.00   0.00   1135.10
            1855   60.10   0.00   0.00   3894.50
            1856   31.90   0.00   0.00   2397.60
            1857   52.40   0.00   0.00   4182.10
            1858   87.70   0.00   0.00   5429.30
            1859   109.70   0.00   0.00   7304.00
            1860   105.50   0.00   0.00   7935.30
            1861   117.60   0.00   0.00   8667.50
            1862   109.90   0.00   0.00   7072.20
            1863   70.20   0.00   0.00   4500.20
            1864   64.80   0.00   0.00   3995.30
            1865   68.10   0.00   0.00   4169.50
            1866   89.70   0.00   0.00   4408.70
            1867   68.80   0.00   0.00   3585.20
            1868   76.00   0.00   0.00   4184.40

BOSWEDDEN   ST.JUST   Continued

Tin   Black(tons) Stuff(tons)   Tin(tons)   Value(£)
1869   27.00   0.00   0.00   1862.60
1870   54.80   0.00   0.00   4174.90
1871   57.20   0.00   0.00   4396.40
1872   45.80   0.00   0.00   3906.30
1874   No detailed return
Comment 1853 INC.CASTLE; 1856–1857 INC.CASTLE; 1861
INC.CASTLE; 1862 INC.CASTLE 2 RETURNS AGGREGATED; 1863–1872
INC.CASTLE; 1874 SEE CUNNING UNITED
Ownership: Comment 1861–1871 INC.CASTLE; 1872–1876 SEE CUNNING UNITED
Management: Manager 1859–1861 R.GOLDSWORTHY
             Chief Agent 1859–1861 R.GOLDSWORTHY JNR.; 1862–1871 JAS.ROACH
             WM.HATTAM
             Secretary 1859–1861 RICH.PEARCE (P); 1862–1871 RICH.BOYNS

BOSWIN   WENDRON   SW 695340 0136

Production: Tin   Black(tons) Stuff(tons)   Tin(tons)   Value(£)
1908   0.00   40.00   0.00   83.00
1911   61.00   0.00   0.00   6700.00
1912   90.10   0.00   0.00   10558.00
1913   108.30   0.00   0.00   12392.00
Ownership: 1907–1908 HUGH S.GORDON; 1909–1913 BOSWIN MINES LTD.
Management: Secretary 1913 HUGH S.GORDON (S)
Employment: Underground   Surface   Total
1907   2   2   4
1908   22   8   30
1911   50   30   80
1912   60   47   107
1913   72   33   105

BOSWORGY   ST.ERTH   SW 573332 0137

Production: Tin   Black(tons) Stuff(tons)   Tin(tons)   Value(£)
1872   0.00   0.00   0.00   432.20
1873   0.00   586.50   0.00   1037.30
1874   10.20   75.50   0.00   734.20
1875   14.10   0.00   0.00   740.90
1876   0.30   0.00   0.00   28.50
Comment 1872–1873 TIN STUFF; 1874 TIN STUFF EST; 1876 VALUE
INC.TIN STUFF
Ownership: 1876 BOSWORGY MINING CO.
             Comment 1871 SEE CROWAN MINES
Management: Chief Agent 1872–1876 WM.ROSEWARNE
             Secretary 1872 WALT.PIKE; 1873–1875 COMMITTEE

BOSWORTHAN   SANCREED   SW 412290 0138

Production: Tin No detailed return
Ownership: Comment 1866 SEE ALSO BOSWORTHEN PENZANCE CONSOLS
Management: Manager 1860–1862 JOHN CARTHEW; 1863–1866 J.JONES
             Secretary 1860–1862 JOHN CARTHEW (P); 1863–1866 J.JONES (P)

Production: Copper

	Ore(tons)	Metal(tons)	Value(£)
1845	1274.00	140.70	9656.50
1846	513.00	58.80	3338.10
1847	251.00	23.90	1551.10
1848	75.00	7.00	370.80
1850	147.00	16.40	1099.40
1851	212.00	21.90	1457.70
1852	629.00	60.90	5036.50
1853	1001.00	93.40	9248.40
1854	1453.00	203.80	21635.50
1855	973.00	133.80	14311.20
1856	651.00	83.30	8201.60
1857	985.00	110.10	11427.00
1858	576.00	70.20	6680.80
1859	330.00	42.00	4038.50
1860	331.00	46.70	4423.90
1861	509.00	51.80	4642.90
1862	539.00	48.00	3871.70
1863	541.00	48.90	3783.00
1864	217.00	16.00	1307.10
1865	369.00	40.30	3244.70
1866	467.00	61.70	4296.30
1867	434.00	50.80	3586.30
1868	434.00	53.90	3596.30
1869	83.00	9.10	584.80
1870	32.00	3.60	213.60
1872	56.00	10.20	740.30
1873	55.00	12.70	844.00
1874	161.00	26.10	1942.50
1875	428.00	62.90	4597.30
1876	430.00	53.20	3686.10
1877	136.70	18.60	1101.60
1878	158.10	19.40	963.10
1879	131.20	16.10	897.20
1880	129.00	14.30	847.00
1881	187.60	21.50	1178.10
1882	32.00	3.10	238.00
1883	64.00	6.30	575.00
1884	18.00	0.00	175.00
1885	7.00	0.00	72.00
1886	9.00	0.00	76.00
1887	10.00	0.00	107.00
1888	15.00	0.00	188.00
1889	43.00	0.00	312.00
1891	9.00	0.00	136.00
1894	12.00	0.00	136.00
1895	18.00	0.00	90.00

Comment 1845–1848 (C); 1850–1870 (C); 1872–1876 (C)

Tin

	Black(tons)	Stuff(tons)	Tin(tons)	Value(£)
1853	147.40	0.00	0.00	8656.80
1854	147.10	0.00	0.00	10842.10
1855	142.70	0.00	0.00	9590.20
1856	149.00	0.00	0.00	13285.50
1857	172.30	0.00	0.00	14254.60
1858	202.20	0.00	0.00	12861.60

Tin	Black(tons)	Stuff(tons)	Tin(tons)	Value(£)
1859	153.10	0.00	0.00	12308.30
1860	183.10	0.00	0.00	15302.90
1861	201.60	0.00	0.00	15233.60
1862	440.10	0.00	0.00	30645.20
1863	413.50	0.00	0.00	28967.60
1864	425.70	0.00	0.00	28857.20
1865	390.10	0.00	0.00	22107.40
1866	342.80	0.00	0.00	17620.90
1867	202.50	0.00	0.00	10632.50
1868	375.30	0.00	0.00	21381.70
1869	525.20	0.00	0.00	38268.60
1870	446.90	0.00	0.00	33760.50
1871	497.70	0.00	0.00	39552.00
1872	391.70	0.00	0.00	34224.10
1873	352.50	0.00	0.00	27392.20
1874	401.50	0.00	0.00	22717.40
1875	358.30	0.00	0.00	18556.20
1876	323.50	0.00	0.00	14318.00
1877	433.70	0.00	0.00	17577.50
1878	477.80	0.00	0.00	16882.00
1879	501.10	0.00	0.00	20355.50
1880	389.00	0.00	0.00	20557.00
1881	245.40	0.00	162.60	13572.50
1882	210.00	0.00	147.00	12659.00
1883	192.00	0.00	0.00	10226.00
1884	231.10	0.00	0.00	10389.00
1885	271.20	0.00	0.00	12886.00
1886	318.80	0.00	0.00	17960.00
1887	338.50	0.00	0.00	21586.00
1888	284.20	0.00	0.00	18259.00
1889	334.20	0.00	0.00	17757.00
1890	415.50	0.00	0.00	22476.00
1891	361.50	0.00	0.00	18882.00
1892	350.00	0.00	0.00	18468.00
1893	402.00	0.00	0.00	19164.00
1894	390.00	0.00	0.00	15424.00
1895	65.50	0.00	0.00	2070.00
1896	No detailed return			
1899	36.00	0.00	0.00	2208.00
1900	29.00	0.00	0.00	1769.00
1901	62.00	0.00	0.00	3410.00
1902	53.00	0.00	0.00	3445.00
1903	48.40	0.00	0.00	3144.00
1904	54.00	0.00	0.00	2921.00
1905	50.60	0.00	0.00	3872.00
1906	49.50	0.00	0.00	4784.00
1908	44.70	0.00	0.00	3346.00
1909	135.40	0.00	0.00	10398.00
1910	86.90	0.00	0.00	7506.00
1911	95.50	0.00	0.00	10453.00
1912	121.50	0.00	0.00	15669.00
1913	156.40	0.00	0.00	18498.00

Comment 1862 TWO RETURNS AGGREGATED; 1872-1875 INC.CARNYORTH

BOTALLACK   ST.JUST   Continued

Arsenic	Ore(tons)	Metal(tons)	Value(£)
1875	16.20	0.00	71.60
1876	97.00	0.00	353.20
1877	135.40	0.00	472.80
1878	72.60	0.00	97.90
1879	114.90	0.00	378.00
1880	60.00	0.00	306.00
1881	69.30	0.00	264.00
1882	72.00	0.00	266.00
1883	69.00	0.00	254.00
1884	78.60	0.00	312.00
1885	41.80	0.00	167.00
1886	35.00	0.00	140.00
1887	35.00	0.00	155.00
1888	40.00	0.00	178.00
1889	95.00	0.00	382.00
1890	108.00	0.00	512.00
1891	79.00	0.00	390.00
1892	94.00	0.00	462.00
1893	95.00	0.00	486.00
1894	98.00	0.00	635.00
1895	20.00	0.00	100.00

Comment 1875 INC. CARNYORTH; 1884-1885 SOOT; 1886 CRUDE
Ownership: 1876-1886 BOTALLACK MINING CO.; 1887-1890 BOTALLACK MINE
ADVENTURERS; 1891-1893 BOTALLACK MINE ADVENTURERS C.B.;
1894-1895 BOTALLACK MINE ADVENTURERS; 1898-1902 CHELLEW &
TROON; 1903-1905 CHELLEW,TROON & CO.; 1906-1911 BOTALLACK
MINES LTD.; 1912-1913 BOTALLACK LTD.
Comment 1868-1878 INC.CARNYORTH; 1895 ABANDONED
Management: Manager 1859-1868 S.H.JAMES; 1869-1870 HY.BOYNS; 1871-1880
FRAN.BENNETTS; 1881 JAS.ROACH
Chief Agent 1859 NICH.HOCKING; 1860-1861 NICH.HOCKING JOHN
ROWE HY.BOYNS; 1862-1868 N.HOCKING J.ROWE H.BOYNS J.BOYNS;
1869 JOHN ROWE JOHN BOYNS; 1870 JOHN ROWE F.OATS JOHN BOYNS;
1871 JOHN ROWE F.OATS HY.HOCKING; 1872-1873 F.OATS
HY.HOCKING; 1874-1875 J.HOLLOW HY.HOCKING; 1876 F.OATES
HY.HOCKING; 1877 F.P.ROWE HY.HOCKING; 1878-1879 F.P.ROWE
RALPH WILLIAMS; 1880-1881 F.P.ROWE RALPH WILLIAMS HY.HOCKING;
1888 ART.H.JAMES; 1891-1895 ART.H.JAMES; 1906-1908 WM.THOMAS;
1909-1912 W.R.THOMAS; 1913 A.B.CLIMAS
Secretary 1859-1887 S.H.JAMES (P); 1889-1895 ART.H.JAMES

Employment:	Underground	Surface	Total
1878	251	158	409
1879	254	153	407
1880	202	110	312
1881	184	105	289
1882	200	97	297
1883	135	80	215
1884	130	66	196
1885	166	90	256
1886	182	97	279
1887	188	104	292
1888	189	126	315
1889	237	172	409
1890	237	168	405

BOTALLACK   ST.JUST   Continued

	Underground	Surface	Total
1891	196	153	349
1892	213	156	369
1893	206	161	367
1894	133	116	249
1895		6	6
1899	7	33	40
1900	8	38	46
1901	5	38	43
1902	5	35	40
1904		42	42
1905		37	37
1906	8	65	73
1907	75	125	200
1908	108	132	240
1909	109	133	242
1910	122	111	233
1911	175	101	276
1912	158	110	268
1913	181	94	275

Comment 1906 INC.44 EMPL.AT SURFACE BY CHELLEW

BOTALLACK,EAST   ST.JUST   SW 396312 0140

Production: Tin
	Black(tons)	Stuff(tons)	Tin(tons)	Value(£)
1881	0.50	0.00	0.20	25.80
1883	5.20	0.00	0.00	270.00
1885	1.00	0.00	0.00	40.00

Ownership: 1880—1885 EAST BOTALLACK MINING CO.
Management: Manager 1880 GEO.H.EUSTACE
             Chief Agent 1882 GEO.H.EUSTICE; 1885 JOHN HOLLOW
             Secretary 1881 GEO.H.EUSTICE; 1883—1884 GEO.H.EUSTICE
Employment:
	Underground	Surface	Total
1880	11	8	19
1881	10	6	16
1882—1883	8	4	12
1885	3	2	5

BOTATHAN   SOUTH PETHERWIN   SX 293820 0141

Production: Manganese No detailed return
Ownership: 1884—1885 C.T.H.BENNETT
Management: Chief Agent 1884—1885 C.T.H.BENNETT
Employment:
	Underground	Surface	Total
1884		2	2
1885		6	6

BOTELET   LANREATH   SX 179599 0142

Production: Lead No detailed return
Management: Chief Agent 1862—1865 THOS.TREVILLION JAS.TREVILLION
            Secretary 1862—1865 RICH.CLOGG

51

BOTREA   ST.JUST   SW 397312 0143

Production: Tin No detailed return
Ownership: 1907-1908 BOTREA MINING CO.
Management: Chief Agent 1907-1908 WM.THOMAS
Employment: Underground   Surface   Total
                1908   6   6

BOWDEN   ST.NEOT   SX 205690 0144

Production: Tin No detailed return
Ownership: Comment 1913 SEE CARPUAN
Employment: Underground   Surface   Total
                1913
            Comment 1913 SEE CARPUAN

BOWDEN DOWN   BRENTOR, DEVON   SX 463820 0145

Production: Iron   Ore(tons)   Iron(%)   Value(£)
                1877   26.00   0.00   15.60
            Comment 1877 BH.

BOYS   REDRUTH   SW 712438 0146

Production: Copper   Ore(tons) Metal(tons)   Value(£)
                1873   10.00   0.40   15.00
                1882   12.20   1.30   74.00
            Comment 1873 (C); 1882 (C)
            Tin   Black(tons) Stuff(tons)   Tin(tons)   Value(£)
                1877   0.00   8.60   0.00   6.50
                1878   0.00   10.00   0.00   120.00
                1881   0.00   119.80   0.00   43.40
Ownership: 1877-1883 FRAN.W.MICHELL THOS.PRYOR
            Comment 1880 NOT WORKED
Management: Chief Agent 1878 THOS.PRYOR; 1879-1880 JOS.PRYOR; 1881
            W.J.WHITE
            Secretary 1877-1883 THOS.PRYOR
Employment: Underground   Surface   Total
                1878   5   5
                1879   2   2
                1881   7   7
                1882   5   5

BRAY   ALTARNUN   SX 198823 0147

Production: Copper   Ore(tons) Metal(tons)   Value(£)
                1857   318.00   9.20   714.30
                1858   190.00   5.20   443.20
            Comment 1857-1858 (C)
Ownership: Comment 1862-1865 SUSPENDED
Management: Manager 1859-1861 RICH.KITTO
            Chief Agent 1859-1864 SML.BENNETTS
            Secretary 1859-1865 RICH.KITTO (P)

52

BREA CONSOLS   TOWEDNACK   SW 482404 0149

Production: Tin   Black(tons) Stuff(tons)   Tin(tons)   Value(£)
            1862   11.80   0.00   0.00   825.70
            1863   47.90   0.00   0.00   3369.40
            Comment 1862 PART YEAR ONLY
Ownership: 1906 KINGSWAY MINING SYNDICATE LTD.
            Comment 1863 SUSPENDED; 1867 SUSPENDED; 1906 BREA
Management: Manager 1859-1862 CHAS.CRAZE
            Chief Agent 1859-1860 CHAS.CRAZE; 1864-1866 C.T.CRAZE
            Secretary 1859 C.THOMAS (P); 1860-1862 E.STRICKLAND (P);
            1864-1867 J.T.MESSER; 1906 M.MONROE (S)
Employment: Underground   Surface   Total
            1906   7   7

BREAGE   BREAGE   SW 622307 0150

Production: Tin   Black(tons) Stuff(tons)   Tin(tons)   Value(£)
            1892   0.00   29.00   0.00   54.00
            1893   0.00   50.00   0.00   70.00
            1894   0.00   44.00   0.00   31.00
            1895   No detailed return
            1896   15.00   0.00   0.00   300.00
Ownership: 1892-1898 BREAGE TIN MINE LTD.
            Comment 1892-1896 LATE GWIN AND SINGER; 1897-1898 LATE GWIN &
            SINGER IN LIQUIDATION
Management: Manager 1892-1898 THOS.B.PROVIS
            Chief Agent 1892-1893 CHAS.ROACH; 1894-1898 STEP.HARRIS
            Secretary 1892-1898 THOS.B.PROVIS (S)
Employment: Underground   Surface   Total
            1892   18   6   24
            1893   16   6   22
            1894   8   1   9

BREWER   GWENNAP   SW 716390 0151

Production: Copper   Ore(tons) Metal(tons)   Value(£)
            1845   1142.00   78.60   5765.80
            1846   432.00   21.40   1348.10
            1847   465.00   17.40   1016.30
            1848   124.00   5.20   240.80
            1852   116.00   6.40   573.40
            1853   133.00   5.60   468.20
            Comment 1845-1848 (C); 1852-1853 (C)
            Tin No detailed return
Ownership: Comment 1900-1901 SEE TRESAVEAN UNITED

BRIGGAN,GREAT   REDRUTH   SW 713446 0152

Production: Copper   Ore(tons) Metal(tons)   Value(£)
            1861   140.00   9.10   802.20
            1862   235.00   17.20   1483.70
            1863   506.00   31.50   2433.80
            1864   137.00   9.70   825.90

53

BRIGGAN,GREAT   REDRUTH   Continued

            Copper   Ore(tons) Metal(tons)   Value(£)
            1865   282.00   21.30   1729.10
            1866   89.00   7.00   488.90
            Comment 1861-1866 (C)
            Tin   Black(tons) Stuff(tons)   Tin(tons)   Value(£)
            1863   0.00   0.00   0.00   2.90
            Comment 1863 TIN STUFF
Management: Chief Agent 1860 THOS.TRELEASE; 1861-1862 THOS.TRELEASE
            GEO.OATES; 1863 JOHN TREDENNICK GEO.OATES; 1864-1865 JOHN
            TREDINNICK
            Secretary 1860 J.G.GOATLEY (P); 1861-1865 ED.KING (P)

BRITAIN   ST.AGNES   SW 728458 0153

Production: Tin   Black(tons) Stuff(tons)   Tin(tons)   Value(£)
            1873   0.00   430.00   0.00   634.00
            1874   0.00   274.60   0.00   404.80
            1875   0.00   39.50   0.00   36.50
            Comment 1874 VALUE EST.
Ownership: 1875 WHEAL BRITAIN CO.
            Comment 1875 SUSPENDED MAY 1875
Management: Manager 1872-1875 JOEL PHILLIPS
            Secretary 1874 J.D.JENKIN

BRITAIN,EAST   ST.AGNES   0154

Production: Tin No detailed return
Management: Secretary 1872 JOEL PHILLIPS & CO.

BRITANNIA   GWENNAP   SW 748407 0155

Production: Copper No detailed return
Ownership: 1882 WHEAL BRITANNIA MINING CO.
Management: Chief Agent 1882 JOS.MICHELL
Employment: Underground   Surface   Total
            1882   6   6

BRITON,WEST   CROWAN   0156

Production: Copper   Ore(tons) Metal(tons)   Value(£)
            1867   20.00   1.20   89.60
            1868   80.00   4.40   310.20
            Comment 1867-1868 (C)
Management: Chief Agent 1867 WM.ROSEWARNE

BROTHERS   CALSTOCK   SX 391701 0157

Production: Lead & Silver   Ore(tons) Lead(tons) Silver(ozs)   Value(£)
            1852-1854 No detailed return
            1874   0.00   0.00   1570.00   461.20

                                        54

BROTHERS   CALSTOCK   Continued

Lead & Silver   Ore(tons) Lead(tons) Silver(ozs)   Value(£)
1878   0.90   12.40   153.00   9.00
Comment 1852–1854 SEE SILVER VALLEY
Copper   Ore(tons) Metal(tons)   Value(£)
1875   1.50   0.00   90.00
1876   1.00   0.00   0.00
Comment 1875 CU & AG PRECIPITATE.VALUE EST.; 1876 CU & AG
PRECIPITATE
Tin   Black(tons) Stuff(tons) Tin(tons)   Value(£)
1878   0.60   0.00   0.00   22.50
Arsenic   Ore(tons) Metal(tons)   Value(£)
1875   13.80   0.00   41.30
Silver   Ore(tons) Metal(tons)   Value(£)
1878   0.40   0.00   2.80
Ownership: 1878–1880 ROBT.HENDERSON; 1881–1882 G.BECKINSALE
             Comment 1879–1882 NOT WORKED
Management: Manager 1873–1876 J.W.DOBLE
             Chief Agent 1875–1876 ROBT.HAYE; 1878 J.W.DOBLE; 1881–1882
             J.W.DOBLE
             Secretary 1873–1874 J.W.DOBLE; 1875 ROBT.HENDERSON; 1876
             ROBT.HENDERSON & CO.; 1878 ROBT.HENDERSON & CO.
Employment: Underground   Surface   Total
             1878   4   10   14

BRYNN   ROCHE   SW 991623 0158

Production: Tin   Black(tons) Stuff(tons) Tin(tons)   Value(£)
            1870   6.30   0.00   0.00   478.90
            1871   9.70   0.00   0.00   748.60
            1872   2.20   0.00   0.00   161.00
Management: Manager 1869–1871 THOS.PARKYN; 1872–1873 MART.RICKARD
            Secretary 1872–1873 S.TRUDGEON

BRYNN ROYALTON   ROCHE   SW 978618 0159

Production: Tin No detailed return
Ownership: Comment 1872–1879 BRYNN ROYALTON CONSOLS; 1880 BRYNN ROYALTON
            CONSOLS.SEE PAR GREAT CONSOLS; 1881 BRYNN ROYALTON
            CONSOLS.SEE POLGOOTH UNITED.
Management: Manager 1869–1881 THOS.PARKYN
            Secretary 1869–1881 H.T.WHITEFIELD

BUCKETTS   REDRUTH   SW 702411 0160

Production: Copper   Ore(tons) Metal(tons)   Value(£)
            1846   119.00   8.10   560.00
            1847   1281.00   85.00   5935.20
            1848   1069.00   68.40   4004.20
            1849   189.00   10.40   630.70
            Comment 1846–1849 (C)

55

BUCKLERS   ST.AUSTELL   SX 037528 0161

Production: Tin	Black(tons)	Stuff(tons)	Tin(tons)	Value(£)
1880	1.40	0.00	0.00	75.60

BUDNICK CONSOLS   PERRANZABULOE   SW 773547 0163

Production: Lead & Silver	Ore(tons)	Lead(tons)	Silver(ozs)	Value(£)
1855	15.30	10.00	444.00	0.00
1856	43.10	28.10	1273.00	0.00
1857	112.60	84.00	3315.00	0.00
1858	57.00	41.50	1266.00	0.00
1859	38.00	29.00	435.00	0.00
1860	16.80	12.60	190.00	0.00
1861	No detailed return			
1862	0.90	0.60	13.00	0.00
1873	0.70	0.50	0.00	6.60
1903	9.00	6.80	234.00	80.00

Comment 1903 BUDNIC CONSOLS

Zinc	Ore(tons)	Metal(tons)	Value(£)
1856	812.80	0.00	1983.30
1857	520.40	0.00	1522.20
1858	410.00	0.00	1080.70
1859	361.00	0.00	1080.10
1860	618.40	0.00	1454.20
1896	5.00	0.00	14.00
1903	9.00	4.00	29.00

Comment 1903 ZINC INC.PERRAN NO.1

Tin	Black(tons)	Stuff(tons)	Tin(tons)	Value(£)
1854	17.50	0.00	0.00	1282.20
1855	43.40	0.00	0.00	2837.70
1856	9.40	0.00	0.00	671.40
1857	8.20	0.00	0.00	647.50
1858	11.70	0.00	0.00	706.00
1859	7.70	0.00	0.00	524.50
1860	16.80	0.00	0.00	1258.10
1861	66.40	0.00	0.00	4615.60
1862	47.80	0.00	0.00	3089.10
1866	4.10	0.00	0.00	189.60
1867	4.60	0.00	0.00	233.80
1868	2.40	0.00	0.00	120.80
1869	3.40	0.00	0.00	232.00
1870	12.30	0.00	0.00	876.20
1871	12.20	0.00	0.00	912.50
1872	10.30	0.00	0.00	900.00
1873	7.70	0.00	0.00	225.10
1874	2.30	0.00	0.00	430.00
1875	2.50	4.60	0.00	129.50
1876	15.60	0.00	0.00	652.50
1877	17.50	0.00	0.00	700.70
1878	5.00	0.00	0.00	162.70
1879	2.80	0.00	0.00	110.00
1880	3.80	0.00	0.00	182.20
1881	3.50	0.00	2.60	174.30
1883	2.30	0.00	0.00	116.00
1887	4.00	0.00	0.00	171.00

Tin	Black(tons)	Stuff(tons)	Tin(tons)	Value(£)
1888	31.20	0.00	0.00	1875.00
1889	30.70	0.00	0.00	1704.00
1890	23.90	0.00	0.00	1293.00
1891	32.20	0.20	0.00	1734.00
1892	22.00	0.00	0.00	1204.00
1893	12.70	0.00	0.00	630.00
1894	8.30	0.00	0.00	392.00
1895	5.20	0.00	0.00	187.00
1896	0.60	0.00	0.00	30.00
1897	5.50	0.00	0.00	205.00
1898	1.00	0.00	0.00	42.00
1899	1.80	0.00	0.00	104.00
1900	2.10	0.00	0.00	114.00
1901	0.30	0.00	0.00	22.00
1902	0.10	0.00	0.00	2.00
1903	1.20	0.00	0.00	74.00
1904	2.00	0.00	0.00	136.00
1905	2.90	0.00	0.00	213.00
1906	2.60	0.00	0.00	240.00
1907	2.40	0.00	0.00	219.00
1908	1.20	0.00	0.00	93.00
1909	1.10	0.00	0.00	82.00
1910	0.30	0.00	0.00	25.00
1911	0.30	0.00	0.00	35.00
1912	0.20	0.00	0.00	20.00

Comment 1862 SEE ALSO PERRAN; 1872 BUDNICH CONSOLS; 1874
VALUE EST; 1877 BUDNICK; 1887 HALVANS; 1888 BUDNICK;
1889-1895 BUDNICK & BUDNICK CONSOLS AGGR.; 1897-1898 BUDNICK
& BUDNICK CONSOLS AGGR.; 1899-1900 BUDNICK AGGREGATED

Ownership: 1877-1900 F.GILB.ENNYS; 1901 JAS.MICHELL; 1902 JAS.MICHELL &
DAVID ZIMAN; 1903-1904 JAS.MICHELL; 1905-1911 DAVIS &
MENADUE
Comment 1861 SUSPENDED; 1867-1872 SEE NEW CHIVERTON CONSOLS;
1873 STOPPED IN 1873; 1875 WINDING UP; 1880-1881 ABOVE ADIT &
HEAPS ONLY; 1882 NOT WORKED; 1890 BUDNICK; 1891-1900 BUDNICK
(PART); 1902 SUSPENDED; 1903-1905 BUDNIC CONSOLS ETC.;
1906-1912 BUDNIC CONSOLS; 1913 BUDNIC CONSOLS NOT WORKED

Management: Manager 1859-1860 WM.BURROWS; 1861 JAS.MICHELL; 1862-1865
JAS.EVANS; 1873 JOHN RAWLINGS; 1880-1881 TRIBUTERS
Chief Agent 1859 JAS.EVANS JNR.JOHN EVANS; 1860 JAS.EVANS
JNR.; 1866 WM.BRYANT; 1874-1875 JOHN RAWLINGS; 1876
R.GLANVILLE; 1879 JOHN RAWLINGS; 1903-1904 JAS.T.ELLERY
Secretary 1859-1860 JOSH.DUNSTAN (P); 1861 J.POLKINGHORNE
(P); 1862-1866 THOS.WOODWARD; 1873 J.PEARCE; 1874-1875
WM.J.COCK; 1876 F.S.ENNYS; 1877-1880 R.GLANVILLE; 1881-1900
REG.C.GLANVILLE; 1901-1902 JAS.T.ELLERY

Employment:	Underground	Surface	Total
1878-1879	5	2	7
1880-1881	3		3
1883	4		4
1884	4	6	10
1888	5	2	7
1889	5	3	8
1890-1892	4	3	7

BUDNICK CONSOLS   PERRANZABULOE   Continued

	Underground	Surface	Total
1893	4	2	6
1894	3	1	4
1895	4		4
1896		3	3
1897	3	1	4
1898	3		3
1899	3	1	4
1900	2		2
1901	4	1	5
1902	2		2
1903	6	1	7
1904—1905	4		4
1906—1912	3		3

Comment 1879 BUDNICK; 1888—1890 BUDNICK; 1891—1900 BUDNICK PART

BUDNICK CONSOLS (CO)   PERRANZABULOE   SW 773547 0164

Production: Lead & Silver No detailed return
            Zinc No detailed return
            Tin   Black(tons) Stuff(tons)   Tin(tons)   Value(£)
            1889—1898 No detailed return
            Comment 1889—1898 FOR PRODUCTION SEE BUDNICK CONSOLS
Ownership: 1889—1890 W.MICHELL & CO.; 1891—1898 BUDNICK CONSOLS MINING
            CO.
Management: Chief Agent 1889—1891 W.MICHELL
            Secretary 1891—1898 RICH.TRENNERY
Employment: Underground   Surface   Total

	Underground	Surface	Total
1889	5	3	8
1890	4	2	6
1891—1892	6	2	8
1893	5	3	8
1894	5	1	6
1895	4	1	5
1896	3	1	4
1897—1898	2		2

BUDNICK CONSOLS (S.M.)   PERRANZABULOE   SW 773547 0165

Production: Lead & Silver No detailed return
            Zinc No detailed return
            Tin   Black(tons) Stuff(tons)   Tin(tons)   Value(£)
            1889—1900 No detailed return
            Comment 1889—1900 FOR PRODUCTION SEE BUDNICK CONSOLS
Ownership: 1889—1890 JAS.MICHELL & OTHERS; 1891—1900 SAMP.MICHELL & CO.
Management: Chief Agent 1889—1891 JAS.MICHELL
            Secretary 1891—1900 JOS.MORRISH
Employment: Underground   Surface   Total

	Underground	Surface	Total
1889	2		2
1890	5	1	6
1891	4	3	7
1892	3	4	7

BUDNICK CONSOLS (S.M.)   PERRANZABULOE   Continued

	Underground	Surface	Total
1893	4	2	6
1894	3	1	4
1895	4	4	8
1896	2	1	3
1897—1899	3	1	4
1900	2		2

BUDNICK,EAST   PERRANZABULOE   SW 785553 0166

Production: Lead No detailed return

Tin	Black(tons)	Stuff(tons)	Tin(tons)	Value(£)
1889	1.00	0.00	0.00	53.00
1890	1.00	0.00	0.00	79.00
1891	4.60	0.00	0.00	236.00
1892	2.00	0.00	0.00	106.00
1893	0.30	0.00	0.00	14.00

Ownership: 1889—1890 T.J.DAVIS & CO.; 1891—1893 T.J.DAVIS & CO. P.;
1894—1900 T.J.DAVIS & CO.
Comment 1860—1868 INC.MOUNT NO.2; 1897—1900 NOT WORKING
Management: Manager 1860 JOHN TONKIN; 1861—1865 W.H.REYNOLDS
Chief Agent 1860 WM.JOHNS; 1866—1868 JAS.EVANS; 1889—1900
T.J.DAVIS
Secretary 1860 JOHN COWLING (P); 1861—1868 H.E.CROKER (P);
1891—1900 T.J.DAVIS

Employment:	Underground	Surface	Total
1889	2	2	4
1890	3	2	5
1891	3	3	6
1892—1893	2	1	3
1894—1895	2		2
1898		1	1
1900	2		2

BUDNICK,NORTH   PERRANZABULOE   SW 767558 0167

Production: Copper No detailed return

Tin	Black(tons)	Stuff(tons)	Tin(tons)	Value(£)
1864	No detailed return			

Comment 1864 SEE VLOW
Management: Manager 1859—1865 JOHN TONKIN
Chief Agent 1859—1865 WM.JOHNS
Secretary 1859—1865 RICH.COWLING (P)

BUDNICK,WEST   PERRANZABULOE   0168

Production: Tin	Black(tons)	Stuff(tons)	Tin(tons)	Value(£)
1878	0.40	0.00	0.00	12.30
1879	0.90	0.00	0.00	31.30
1880	0.20	0.00	0.00	13.40
1883	0.80	0.00	0.00	42.00
1884	1.90	0.00	0.00	82.00

BUDNICK,WEST   PERRANZABULOE   Continued

             Tin   Black(tons) Stuff(tons)   Tin(tons)   Value(£)
             1904   0.90   0.00   0.00   46.00
             1905   1.10   0.00   0.00   79.00
             1906   0.50   0.00   0.00   41.00
             1907   0.70   0.00   0.00   89.00
             1908   0.60   0.00   0.00   42.00
             1909   0.30   0.00   0.00   25.00
Ownership: 1904 JAS.MICHELL; 1905 JOHN MICHELL; 1906-1910 MICHELL,PENNA
             & MICHELL
             Comment 1904-1909 WEST BUDNIC; 1910 WEST BUDNIC NOT WORKED
Management: Chief Agent 1904-1905 JAS.T.ELLERY
Employment: Underground   Surface   Total
             1904   2   1   3
             1905   2   3   5
             1906   1   1   2
             1907-1908   2   1   3
             1909   2   2   4

BULLEN   ST.DAY   0169

Production: Tin   Black(tons) Stuff(tons)   Tin(tons)   Value(£)
            1904   No detailed return
            Comment 1904 SEE WOOD

BULLER   REDRUTH   SW 699401 0170

Production: Copper   Ore(tons) Metal(tons)   Value(£)
            1845   1313.00   76.20   5211.60
            1846   343.00   17.20   1091.90
            1847   No detailed return
            1850   3345.00   328.20   23409.50
            1851   5747.00   489.80   33913.90
            1852   9310.00   632.80   53257.80
            1853   13562.00   850.80   83307.80
            1854   12208.00   742.10   74734.60
            1855   9693.00   575.60   57834.90
            1856   7482.00   430.90   39761.60
            1857   6590.00   361.40   34592.50
            1858   5362.00   299.10   26052.90
            1859   3073.00   158.40   14381.50
            1860   1725.00   117.00   10965.10
            1861   1246.00   84.50   7605.70
            1862   982.00   70.60   5944.70
            1863   710.00   44.40   3386.90
            1864   499.00   21.60   1721.90
            1865   395.00   18.70   1418.50
            1866   228.00   13.40   900.30
            1867   204.00   12.10   807.60
            1868   156.00   11.40   739.00
            1869   102.00   7.90   491.80
            1870   51.00   3.10   117.80
            1871   58.00   7.30   465.60
            1872   122.00   16.30   1301.20

Copper	Ore(tons)	Metal(tons)	Value(£)
1873	117.00	12.50	871.90
1874	38.00	3.70	226.30
1875	22.00	1.00	60.30

Comment 1845–1847 (C); 1850–1875 (C)

Tin	Black(tons)	Stuff(tons)	Tin(tons)	Value(£)
1856	0.00	0.00	0.00	63.70
1857	0.00	0.00	0.00	460.40
1858	0.00	0.00	0.00	1183.00
1859	64.60	0.00	0.00	4780.60
1860	102.00	0.00	0.00	8102.40
1861	137.20	0.00	0.00	9945.10
1862	145.40	0.00	0.00	9524.20
1863	102.30	0.00	0.00	6593.70
1864	73.20	0.00	0.00	4395.20
1865	130.90	0.00	0.00	7137.70
1866	157.00	0.00	0.00	7524.90
1867	167.90	0.00	0.00	8699.10
1868	105.20	0.00	0.00	5777.10
1869	91.70	0.00	0.00	6352.50
1870	61.40	0.00	0.00	4350.40
1871	40.30	0.00	0.00	3253.50
1872	29.10	0.00	0.00	2463.50
1873	20.20	69.80	0.00	1706.10
1874	3.00	71.30	0.00	323.30
1875	0.00	7.40	0.00	25.00

1905–1906 No detailed return

1908	0.00	4.00	0.00	6.00
1909	0.00	87.00	0.00	198.00
1910	0.00	75.00	0.00	150.00

Comment 1856–1858 TIN STUFF; 1862 TWO RETURNS AGGREGATED;
1865 VALUE INC.TIN STUFF; 1905–1906 SEE CUSGARNE

Ownership: 1875–1876 WHEAL BULLER MINING CO.; 1905–1907 T.H.LETCHER;
1908–1909 THOS.PRYOR; 1910–1912 WHITFORD & SONS SOLICITORS
Comment 1905–1907 T.H.LETCHER = LORDS AGENT; 1911–1912 NOT
WORKED

Management: Manager 1864 JOHN DAVEY; 1866 JAS.INCH; 1867 FRAN.PRYOR; 1872
CHAS.THOMAS; 1874 JAS.BROWNE
Chief Agent 1859–1860 JOHN DAVEY JOHN DAVEY JNR.; 1861–1863
JOHN DAVEY JAS.JOHNS; 1864 JAS.JOHNS; 1865 JAS.INCH
JAS.JOHNS; 1866 JAS.JOHNS; 1867 JAS.BROWNE; 1868–1869
JAS.INCH JAS.BROWNE; 1870–1871 JAS.INCH J.R.W.JANE; 1872
CLARKE J.R.W.JANE; 1873 J.A.PRYOR CLARKE; 1874–1876
JAS.PRYOR
Secretary 1859–1864 S.DAVEY R.DAVEY (P); 1865–1866 JOHN
MICHELL; 1867–1874 THOS.PRYOR; 1875 THOS.PRYOR (S)

Employment:	Underground	Surface	Total
1905		1	1
1906		2	2
1908		2	2
1909	4		4
1910	3		3

BULLER & BASSET UNITED   WENDRON   SW 695388 0172

Production: Copper No detailed return
Ownership: Comment 1864-1865 SUSPENDED
Management: Manager 1859 GEO.REYNOLDS; 1860 WM.BOSKEAN; 1861 S.BICE
            Chief Agent 1859 WM.BOSKEAN; 1860 GEO.REYNOLDS; 1862-1863
            WM.PASCOE S.BICE
            Secretary 1859-1865 WM.CHARLES (P)

BULLER,EAST   REDRUTH   SW 713406 0173

Production: Copper No detailed return
            Tin No detailed return
Ownership: 1880-1881 EAST WHEAL BULLER CO.
            Comment 1880-1881 LATE PENNANCE CONSOLS
Management: Chief Agent 1863 J.NANCARROW; 1865 W.HOSKING; 1880-1881
            WM.TREGAY
            Secretary 1864 J.NANCARROW; 1865 THOS.RICHARDS & SON;
            1880-1881 WM.TREGAY
Employment: Underground   Surface   Total
            1880   6   6

BULLER,NORTH   REDRUTH   SW 695405 0174

Production: Copper   Ore(tons) Metal(tons)   Value(£)
            1852   91.00   5.50   463.50
            Comment 1852 (C)
Management: Manager 1859-1863 JOHN DELBRIDGE
            Chief Agent 1859-1863 JOHN DELBRIDGE JNR.; 1864-1865
            RICH.PRYOR WM.HARVEY
            Secretary 1859 ROBT.H.PIKE (P); 1860 W.J.DUNSFORD (P);
            1861-1863 ROBT.H.PIKE; 1864-1865 ROBT.H.PIKE & SON

BULLER,PENSTRUTHAL   GWENNAP   SW 707398 0175

Production: Copper No detailed return
            Tin No detailed return
Management: Manager 1870-1871 JOS.MICHELL
            Chief Agent 1870-1871 RICH.MORCOM
            Secretary 1869 JOS.MICHELL; 1870-1871 MAT.GREEN

BULLER,SITHNEY   SITHNEY   SW 631300 0176

Production: Tin   Black(tons) Stuff(tons)   Tin(tons)   Value(£)
            1853   7.20   0.00   0.00   488.40
            1858   18.40   0.00   0.00   1343.80
Management: Manager 1859-1862 THOS.GILL
            Chief Agent 1859-1862 JOHN TRELOAR
            Secretary 1859-1862 WM.TRURAN (P)

BULLER,SOUTH   REDRUTH   SW 703391 0177

Production: Copper   Ore(tons) Metal(tons)   Value(£)
            1863   21.00   1.20   89.10
            1864   15.00   0.90   78.00
            1865   4.00   0.20   11.10
            Comment 1863—1865 (C)
Ownership: Comment 1861—1864 INC.WEST PENSTRUTHAL; 1865 INC.WEST
            PENSTRUTHAL SUSPENDED
Management: Chief Agent 1861—1863 S.BICE; 1864 JOS.RULE
            Secretary 1861—1865 WM.PASCOE (P)

BULLER,WEST   REDRUTH   SW 701399 0178

Production: Copper   Ore(tons) Metal(tons)   Value(£)
            1849   1165.00   125.40   8968.10
            Comment 1849 (C)
            Tin   Black(tons) Stuff(tons)   Tin(tons)   Value(£)
            1853   0.20   0.00   0.00   9.90

BUNNIE   BUGLE   SX 007571 0179

Production: Tin   Black(tons) Stuff(tons)   Tin(tons)   Value(£)
            1873   36.00   0.00   0.00   2808.00
            1874   4.00   0.00   0.00   180.00
            1902   0.00   21.00   0.00   21.00
            1903   2.60   0.00   0.00   136.00
            1904   61.80   0.00   0.00   4253.00
            1905   83.00   0.00   0.00   6144.00
            1906   52.60   0.00   0.00   4310.00
            1907   20.10   0.00   0.00   2016.00
            Comment 1873—1874 VALUE EST; 1902—1907 BUNNEY
            Tungsten   Ore(tons) Metal(tons)   Value(£)
            1903   8.00   0.00   197.00
            1904   31.50   0.00   1573.00
            1905   30.30   0.00   2176.00
            1906   7.10   0.00   560.00
            1907   3.60   0.00   391.00
            Comment 1903—1907 BUNNY
Ownership: 1899—1907 BUNNEY WOLFRAM & TIN MINE SYNDICATE LTD.
            Comment 1869—1872 BONNEY.NOW SHELTON; 1874 SUSPENDED; 1875
            SUSPENDED SEPT.1874; 1899—1906 BUNNEY; 1907 BUNNEY MINE
            CLOSED OCT.1907
Management: Manager 1872—1875 RICH.HANCOCK
            Chief Agent 1869—1871 RICH.HANCOCK; 1899—1907 GEO.J.BOWKERS
            Secretary 1872 JAS.KNIGHT; 1873—1875 RICH.HANCOCK
Employment: Underground   Surface   Total
            1899   7   11   18
            1900   11   6   17
            1901   19   7   26
            1902   21   7   28
            1903   55   50   105
            1904   76   49   125
            1905   59   50   109
            1906   26   44   70

63

BUNNIE   BUGLE   Continued

                       Underground   Surface   Total
            1907   15   24   39

BURNEY HOUSE   ROCHE   SX 007621 0180

Production: Tin   Black(tons) Stuff(tons)   Tin(tons)   Value(£)
            1874   8.00   0.00   0.00   362.60
            Iron No detailed return
Ownership: 1873 MID CORNWALL MINING CO.LTD.
Management: Chief Agent 1873 DAVID COCK

BURNGULLOW   ST.MEWAN   SW 981542 0181

Production: Tin   Black(tons) Stuff(tons)   Tin(tons)   Value(£)
            1897   1.50   0.00   0.00   45.00
            1898   2.00   0.00   0.00   60.00
            1899   2.80   0.00   0.00   207.00
            1900   1.40   0.00   0.00   103.00
            1901   No detailed return
            Comment 1897-1898 OPEN WORK GARKER & PENTRUFF
Ownership: Comment 1877 SEE ALVIGGAN; 1880 SEE PAR GREAT CONSOLS; 1881
            SEE POLGOOTH UNITED

BURRA BURRA   KENWYN   SW 747466 0182

Production: Lead & Silver   Ore(tons) Lead(tons) Silver(ozs) Value(£)
            1862   1.70   1.20   23.00   0.00
            1863   1.00   0.70   0.00   0.00
            1876   0.50   8.20   32.00   0.00
            Comment 1876 LEAD FIGURE WRONG
            Zinc   Ore(tons) Metal(tons) Value(£)
            1862   24.00   0.00   23.80
            1863   56.00   0.00   56.80
            1873   103.20   0.00   1119.70
            1874   26.90   0.00   93.50
            1875   72.00   0.00   250.00
            1876   35.00   0.00   122.00
            Comment 1862 BURRA; 1863 ORE EST
            Copper   Ore(tons) Metal(tons) Value(£)
            1862   224.00   10.10   792.80
            1877   3.00   0.20   12.70
            Comment 1862 (C)
            Tin No detailed return
Ownership: 1877 G.F.BEVILLE JOHN FAWCETT
            Comment 1863-1865 SUSPENDED; 1873 STOPPED IN 1873
Management: Manager 1860-1862 JOHN DAVEY JNR.; 1872-1877 JAS.BROWNE
            Chief Agent 1861-1862 SIMON TOY; 1872-1873 THOS.JENKIN;
            1875-1877 THOS.JENKIN
            Secretary 1860 BEN.MATTHEWS (P); 1861-1865 W.G.SPARKE (P);
            1872-1873 JAS.JOHNS, 1875-1876 G.BEVILLE T.FAWCETT; 1877
            G.F.BEVILLE

BURROW AND BUTSON   ST.AGNES   SW 739490 0183

Production: Lead & Silver   Ore(tons) Lead(tons) Silver(ozs)   Value(£)
            1873   7.00   5.20   27.00   84.00
            1875   41.80   32.30   160.00   594.00
            Zinc   Ore(tons) Metal(tons)   Value(£)
            1873   150.00   0.00   450.00
            1874   101.10   0.00   176.70
            1875   446.60   0.00   717.60
            1876   45.00   0.00   135.00
            Copper   Ore(tons) Metal(tons)   Value(£)
            1875   21.00   0.70   33.40
            Comment 1875 (C)
            Tin No detailed return
Ownership: 1875—1876 BURROW AND BUTSON MINING CO.LTD.
            Comment 1875 IN LIQUIDATION
Management: Manager 1872 D.R.STRICKLAND; 1873—1876 JOHN CHRISTOPHER
            Chief Agent 1872 JOHN CHRISTOPHER
            Secretary 1872—1874 HY.VON AUSTOR

BURTHY   ST.ENODER   SW 917557 0184

Production: Tin   Black(tons) Stuff(tons)   Tin(tons)   Value(£)
            1900   1.30   0.00   0.00   100.00
            Comment 1900 OPEN WORK

BURTHY ROW   ST.ENODER   SW 905564 0185

Production: Tin No detailed return
Ownership: Comment 1880 SEE PAR GREAT CONSOLS; 1881 SEE POLGOOTH UNITED

BUSY,EAST   CHACEWATER   0186

Production: Lead No detailed return
            Copper No detailed return
            Tin   Black(tons) Stuff(tons)   Tin(tons)   Value(£)
            1898   0.00   26.00   0.00   16.00
Ownership: 1898 JAS.MOYLE & JAS.ROBERTS
            Comment 1865 SUSPENDED
Management: Manager 1860—1864 JOHN TONKIN
            Secretary 1860—1864 JOHN MOYLE (P)

BUSY,GREAT   KENWYN   SW 739448 0187

Production: Lead & Silver   Ore(tons) Lead(tons) Silver(ozs)   Value(£)
            1862   3.50   2.20   44.00   0.00
            Copper   Ore(tons) Metal(tons)   Value(£)
            1845   420.00   16.40   988.60
            1846   231.00   10.10   543.20
            1847   184.00   8.70   537.60
            1848   193.00   9.10   466.50
            1849   208.00   9.80   579.60
            1850   118.00   6.40   395.00

65

Copper	Ore(tons)	Metal(tons)	Value(£)
1851	173.00	8.90	538.20
1852	211.00	10.20	791.70
1853	230.00	11.50	994.60
1854	205.00	8.80	826.90
1855	236.00	12.50	1200.60
1856	209.00	9.30	764.70
1857	1236.00	44.20	3582.30
1858	5172.00	188.20	14856.70
1859	5006.00	188.90	16185.60
1860	4722.00	185.20	15376.30
1861	6279.00	237.70	18963.30
1862	5882.00	234.70	17288.20
1863	5139.00	212.30	14773.10
1864	4230.00	169.70	12720.50
1865	2361.00	91.40	6064.90
1866	1630.00	63.60	3559.80
1867	99.00	3.10	157.90
1868	55.00	2.20	120.20
1869	32.00	1.50	70.30
1909	7.00	0.00	8.00

Comment 1845–1855 (C)BUSY; 1856–1867 (C); 1868–1869 (C)BUSY

Tin	Black(tons)	Stuff(tons)	Tin(tons)	Value(£)
1855	36.20	0.00	0.00	112.60
1856	1.30	0.00	0.00	91.70
1857	2.60	0.00	0.00	204.70
1858	102.90	0.00	0.00	6009.00
1859	163.30	0.00	0.00	10741.30
1860	182.20	0.00	0.00	13170.50
1861	186.40	0.00	0.00	12284.20
1862	193.40	0.00	0.00	11375.90
1863	240.50	0.00	0.00	14460.90
1864	197.20	0.00	0.00	11218.30
1865	284.80	0.00	0.00	14347.80
1866	215.80	0.00	0.00	9277.50
1867	28.70	0.00	0.00	1241.30
1870	1.40	413.50	0.00	696.20
1871	2.50	388.20	0.00	1234.60
1872	3.80	66.60	0.00	508.20
1873	35.80	299.00	0.00	3134.40
1874	1.60	135.00	0.00	159.90
1894	0.00	916.30	0.00	250.00
1895	0.00	2995.00	0.00	748.00
1896	No detailed return			
1900	0.00	8.00	0.00	8.00
1901	0.00	5118.00	0.00	1919.00
1902	0.00	1884.00	0.00	1441.00
1903	10.00	0.00	0.00	700.00
1904	7.70	0.00	0.00	500.00
1905	0.00	143.00	0.00	275.00
1906	1.70	0.00	0.00	120.00
1907	0.00	68.00	0.00	92.00
1908	0.00	2.00	0.00	13.00
1909	15.60	0.00	0.00	1114.00
1911	0.00	72.10	0.00	77.00

Comment 1855–1856 BUSY; 1862 GT.BUSY UNITED.TWO RETURNS AGG.;
1867 VALUE INC.TIN STUFF; 1894–1896 (T.TONKIN)

Iron	Ore(tons)	Iron(%)	Value(£)
1860	0.00	0.00	3.70
1861	0.00	0.00	7.40

Comment 1860–1861 OXIDE OF IRON

Arsenic	Ore(tons)	Metal(tons)	Value(£)
1858	43.10	0.00	77.50
1859	5.00	0.00	9.90
1861	8.50	0.00	12.00
1862	17.50	0.00	24.50
1863	50.70	0.00	73.40
1864	29.00	0.00	22.00
1873	1.70	0.00	0.80
1903	43.00	0.00	260.00
1909	66.50	0.00	584.00
1910	23.00	0.00	243.00
1911	4.50	0.00	21.00

Comment 1859–1864 CRUDE

Arsenic Pyrite	Ore(tons)	Value(£)
1893	171.00	56.00
1896	2986.00	746.00
1897	5778.00	1444.00
1898	5949.00	1487.00
1899	5494.00	1961.00
1900	6115.00	2293.00
1904	43.00	150.00
1905	47.00	21.00

Comment 1893 BUSY

Ownership: 1892 JOHN RUSSELL; 1893–1895 THOS.TONKIN; 1896–1897
W.J.TRYTHALL; 1898–1900 W.J.TRYTHALL & JOHN TONKIN; 1901–1905
THOS.TONKIN; 1906–1907 CONSOLIDATED NI,SN,& CU MINES LTD.;
1908–1913 COMPAGNIE ANGLO–BELGE
Comment 1873 SUSPENDED; 1892 BUSY RHODES SHAFT ONLY; 1893
BUSY TRIBUTER ABANDONED; 1894–1895 HODGERS SHAFT TRIBUTER;
1896–1902 HODGERS SHAFT; 1911 SUSPENDED; 1912 SUSPENDED
RESUMED PUMPING DEC.1912; 1913 UNWATERING

Management: Manager 1859–1861 JOHN DELBRIDGE; 1862–1863 THOS.TRELEASE
Chief Agent 1859–1860 WM.EDWARDS; 1861–1863 JOHN PETTERICK;
1864 JOHN EDWARDS JOHN PETTERICK; 1865 C.RICH JOHN
TREDINNICK; 1872–1873 CHAS.BISHOP; 1874–1875 J.C.LANYON;
1906–1913 THOS.TONKIN
Secretary 1859–1860 ED.KING (P); 1861 MOSES BAWDEN; 1862–1865
ED.KING; 1872 J.LEAN & SON; 1873 J.LEAN J.JOSE & CO.;
1874–1875 J.C.LANYON & SON

Employment:

	Underground	Surface	Total
1892–1893	3		3
1894	26	5	31
1895	36	3	39
1896	63	7	70
1897	82	10	92
1898	86	13	99
1899	75	9	84
1900	60	5	65
1901	43	4	47

CM-E*   67

BUSY,GREAT   KENWYN   Continued

	Underground	Surface	Total
1902	18	2	20
1903	11	2	13
1904	11		11
1905	4	2	6
1906	8	4	12
1907	25	11	36
1908	45	85	130
1909	25	55	80
1910	13	31	44
1911	2	4	6
1912		5	5
1913	4	4	8

BUSY,GREAT   KENWYN   SW 739448 0188

Production: Tin No detailed return
            Arsenic No detailed return
            Arsenic Pyrite   Ore(tons)   Value(£)
            1900   126.00   126.00
Ownership: 1899-1900 ANGLO PENINSULA MINING CO.LTD.
Management: Chief Agent 1899 CHAS.PENGELLY
            Secretary 1899 JOHN SANDOW
Employment: Underground   Surface   Total
            1899   3   3
            1900   4   4

BUSY,NORTH   CHACEWATER   SW 734458 0189

Production: Lead   Ore(tons) Metal(tons)   Value(£)
            1885   0.40   0.00   0.00
            Comment 1885 NORTH BUSY UNITED
            Zinc   Ore(tons) Metal(tons)   Value(£)
            1858   35.00   0.00   56.50
            1859   382.00   0.00   897.00
            1860   26.90   0.00   48.00
            1863   6.30   0.00   3.50
            1876   10.00   0.00   30.00
            1877   37.60   0.00   178.50
            1878   310.90   0.00   478.50
            1879   120.70   0.00   362.10
            1880   83.50   0.00   129.70
            1881   13.60   7.00   44.00
            1882   24.60   11.10   69.00
            1885   6.00   0.00   12.00
            Comment 1859 ORE & VALUE EST; 1860 TWO RETURNS AGGREGATED;
            1863 ORE ESTIMATED; 1876-1880 NORTH BUSY UNITED
            Copper   Ore(tons) Metal(tons)   Value(£)
            1854   139.00   11.40   1155.10
            1855   111.00   7.40   767.90
            1856   264.00   17.80   1672.30
            1857   192.00   11.80   1168.50
            1858   129.00   8.00   789.20

68

Copper	Ore(tons)	Metal(tons)	Value(£)
1859	122.00	5.50	500.30
1862	40.00	3.20	259.30
1875	253.00	13.80	975.60
1876	113.00	5.90	375.90
1877	44.60	2.20	188.50
1878	141.60	11.90	597.10
1879	82.90	5.20	286.40
1880	18.00	1.20	57.60
1881	3.90	0.30	17.50
1883	2.30	0.10	8.00
1884	52.00	0.00	204.00
1885	43.00	0.00	150.00

Comment 1854-1859 (C); 1862 (C); 1875-1876 (C); 1877 NORTH
BUSY UNITED; 1879 NORTH BUSY UNITED; 1881 NORTH BUSY UNITED

Tin	Black(tons)	Stuff(tons)	Tin(tons)	Value(£)
1859	0.00	0.00	0.00	2345.20
1860	0.00	0.00	0.00	3161.80
1861	0.00	0.00	0.00	3273.10
1862	0.00	0.00	0.00	361.20
1863	0.00	0.00	0.00	264.70
1871	1.50	0.00	0.00	90.00
1872	3.40	0.00	0.00	214.90
1873	0.60	21.80	0.00	86.60
1874	0.00	50.10	0.00	22.60
1875	0.00	394.70	0.00	570.00
1876	0.00	592.30	0.00	708.70
1877	33.00	833.80	0.00	2300.50
1878	32.30	0.00	0.00	1128.50
1879	29.00	0.00	0.00	1158.00
1880	26.50	1577.40	0.00	3410.50
1881	28.50	0.00	19.80	1230.50
1882	6.90	0.00	4.50	356.00
1883	3.20	0.00	0.00	148.00
1884	3.70	0.00	0.00	131.00
1885	2.20	0.00	0.00	100.00

Comment 1859-1861 TIN STUFF; 1862 TIN STUFF PART YEAR ONLY;
1863 TIN STUFF; 1872 VALUE INC.TIN STUFF; 1873 NORTH BUSSY;
1877-1885 NORTH BUSY UNITED

Arsenic	Ore(tons)	Metal(tons)	Value(£)
1883	5.00	0.00	15.00

Arsenic Pyrite	Ore(tons)	Value(£)
1878	21.70	11.60
1879	76.20	45.00
1880	73.60	44.50
1885	19.10	18.00

Comment 1878 VALUE INC. 2.6 TONS IRON PYRITE; 1879-1880 NORTH
BUSY UNITED; 1885 NORTH BUSY UNITED
Ownership: 1877-1886 NORTH BUSY UNITED MINING CO.
Comment 1863-1865 SUSPENDED; 1877-1881 NORTH BUSY UNITED;
1886 NOT WORKED
Management: Manager 1859-1861 JAS.CRAZE JNR.; 1874-1876 NICH J.O.ROGERS;
1880-1881 JOHN JAMES
Chief Agent 1859-1861 JAS.CRAZE JNR.; 1862 JAS.W.CRAZE;
1872-1873 JOHN PETERS HY.TREVETHAN; 1877 HY.TREVETHAN;

BUSY,NORTH   CHACEWATER   Continued

          1878-1879 HY.TREVETHAN ED.KNIGHT; 1880-1881 J.PRISK;
          1882-1886 JOHN JAMES
          Secretary 1859-1865 WM.PAINTER (P); 1872-1875 THOS.WOODWARD;
          1876-1880 T.NORM.WOODWARD; 1881 T.NORM.WOODWARD (P)
Employment: Underground   Surface   Total
          1878   17   21   38
          1879   22   19   41
          1880   50   22   72
          1881   35   12   47
          1882   34   12   46
          1883   13   5   18
          1884   20   6   26
          1885   10   3   13

BUTTERDON   MENHENIOT   SX 287666 0190

Production: Lead   Ore(tons) Metal(tons)   Value(£)
          1857   2.10   1.50   0.00

BUTTERS TOR   CAMELFORD   0191

Production: Tin   Black(tons) Stuff(tons)   Tin(tons)   Value(£)
          1906   0.30   0.00   0.00   40.00
Ownership: 1905-1906 BUTTERS TOR & GENERAL SYNDICATE LTD.
Employment: Underground   Surface   Total
          1906   10   10

BUZZA   TOWEDNACK   SW 494385 0192

Production: Tin No detailed return
Ownership: 1880 M.G.BROWN & OTHERS
Management: Chief Agent 1880 WM.BUGLEHOLE
Employment: Underground   Surface   Total
          1880   18   35   53

CABELLA   WARLEGGAN   SX 151695 0193

Production: Tin   Black(tons) Stuff(tons)   Tin(tons)   Value(£)
          1862   No detailed return
          1864   No detailed return
          1868-1871 No detailed return
          Comment 1862 SEE GREAT TREVEDDOE; 1864 SEE GREAT TREVEDDOE;
          1868-1871 CABILLA SEE GREAT TREVEDDOE
Ownership: Comment 1864-1871 SEE GREAT TREVEDDOE; 1872 SEE WHISPER

CALLESTOCK MOORS,GREAT   PERRANZABULOE   SW 785504 0194

Production: Lead   Ore(tons) Metal(tons)   Value(£)
          1847   109.00   64.00   0.00
          1849-1854 No detailed return

CALLESTOCK MOORS,GREAT   PERRANZABULOE   Continued

       Comment 1854 ABANDONED

CALLESTOCK,GREAT   PERRANZABULOE   SW 772494 0195

Production: Lead   Ore(tons) Metal(tons)   Value(£)

	Ore(tons)	Metal(tons)	Value(£)
1847	116.00	72.00	0.00
1848	179.00	110.00	0.00
1849	No detailed return		

       Comment 1847 CALLESTOCK
       Tin No detailed return
Management: Manager 1872 J.NANCARROW; 1873-1874 RICH.NANCARROW

CALLINGTON   CALLINGTON   SX 357710 0196

Production: Lead & Silver

	Ore(tons)	Lead(tons)	Silver(ozs)	Value(£)
1845	950.00	570.00	0.00	0.00
1846	1138.00	683.00	0.00	0.00
1847	1249.00	824.00	0.00	0.00
1848	957.00	632.00	0.00	0.00
1849	625.00	422.00	0.00	0.00
1850	516.00	348.90	0.00	0.00
1851	597.50	417.90	0.00	0.00
1852	243.00	154.50	7700.00	0.00
1853	118.00	71.00	4118.00	0.00
1854	71.00	43.50	2595.00	0.00

CALLINGTON CONSOLS   CALLINGTON   SX 363706 0197

Production: Copper

	Ore(tons)	Metal(tons)	Value(£)
1881	10.00	0.70	35.00

       Tin No detailed return
Ownership: 1881 J.W.SHARP & CO.; 1882 CALLINGTON CONSOLS MINING CO.
Management: Chief Agent 1881 HY.JAMES; 1882 THOS.GREGORY
       Secretary 1881 THOS.GREGORY
Employment:

	Underground	Surface	Total
1881		12	12

CALLINGTON UNITED   CALLINGTON   0198

Production: Copper

	Ore(tons)	Metal(tons)	Value(£)
1889	221.00	0.00	418.00
1890	1417.00	0.00	4055.00
1891	712.00	0.00	900.00
1892	198.00	0.00	304.00

Tin	Black(tons)	Stuff(tons)	Tin(tons)	Value(£)
1889	58.80	0.00	0.00	3270.00
1890	114.20	10.00	0.00	6256.00
1891	115.10	0.00	0.00	6044.00
1892	69.40	0.00	0.00	3721.00
1893	No detailed return			

CALLINGTON UNITED   CALLINGTON   Continued

Arsenic	Ore(tons)	Metal(tons)	Value(£)
1889	153.00	0.00	1224.00
1890	1117.00	0.00	12283.00
1891	960.00	0.00	10080.00
1892	410.00	0.00	4100.00

Arsenic Pyrite	Ore(tons)	Value(£)
1889	551.00	800.00
1893	200.00	250.00

Ownership: 1888–1892 CALLINGTON UNITED MINES LTD.; 1893 CORNISH MUNDIC
SYNDICATE LTD.
Comment 1888–1890 INC.HOLMBUSH,KELLY BRAY & REDMOOR; 1891
INC.HOLMBUSH,KELLY BRAY & REDMOOR & SOUTH KELLY BRAY; 1892
INC.HOLMBUSH,KELLY BRAY & REDMOOR & SOUTH KELLY
BRAY.SUSPENDED PART YEAR; 1893 INC.HOLMBUSH,KELLY BRAY &
REDMOOR ABANDONED
Management: Chief Agent 1888–1892 H.BENNETT; 1893 W.F.WILKINSON
Secretary 1888–1890 G.EMMENS; 1891–1893 P.A.LATHAM

Employment:	Underground	Surface	Total
1888	42	23	65
1889	169	113	282
1890	243	107	350
1891	272	141	413
1892	19	28	47
1893	20	28	48

Comment 1888–1890 INC.HOLMBUSH,KELLY BRAY & REDMOOR;
1891–1893 INC.HOLMBUSH,KELLY BRAY & REDMOOR & SOUTH KELLY
BRAY

CALLINGTON,SOUTH   CALLINGTON   SX 356677 0199

Management: Chief Agent 1866 W.SPARGO

CALSTOCK AND DANESCOMBE   CALSTOCK   SX 426696 0200

Production: Copper	Ore(tons)	Metal(tons)	Value(£)
1875	No detailed return		
1882	36.30	2.20	127.00
1883	134.90	4.90	350.00
1884	138.00	0.00	151.00
1887	80.00	0.00	102.00

Comment 1875 SEE CALSTOCK CONSOLS; 1882–1884 INC.DANESCOMBE
VALLEY; 1887 INC.DANESCOMBE VALLEY
Tin No detailed return
Arsenic No detailed return

Arsenic Pyrite	Ore(tons)	Value(£)
1881	1003.50	759.80
1882	388.60	300.00
1883	760.00	570.00
1884	903.20	151.00
1886	83.10	73.00
1887	102.00	97.00

Ownership: 1880–1887 CALSTOCK & DANESCOMBE CONSOLIDATED CO.
Management: Manager 1880–1881 W.B.COLLOM

CALSTOCK AND DANESCOMBE   CALSTOCK   Continued

          Chief Agent 1882-1887 WM.SOWDEN
Employment: Underground   Surface   Total
          1880   6   1   7
          1881   30   11   41
          1882   24   13   37
          1883   19   10   29
          1884   12   13   25
          1886   6   4   10
          1887   6   6   12

CALSTOCK CONSOLS   CALSTOCK   SX 426696 0201

Production: Lead & Silver   Ore(tons) Lead(tons) Silver(ozs)   Value(£)
          1860   18.00   13.50   400.00   0.00
          1861   11.10   7.00   140.00   0.00
          Comment 1860 AG.INC.GREAT CRINNIS & TREVISSA
          Copper   Ore(tons) Metal(tons)   Value(£)
          1856-1858 No detailed return
          1859   714.00   40.50   3798.30
          1860   868.00   48.60   4277.30
          1861   561.00   23.10   1869.80
          1862   437.00   20.20   1515.00
          1863   144.00   5.40   396.40
          1864   105.00   4.80   379.90
          1875   12.00   0.50   38.40
          1882   No detailed return
          Comment 1856-1858 (C)SEE CALSTOCK CONSOLS, DEVON; 1859-1864
          (C); 1875 (C); 1882 SEE CALSTOCK AND DANESCOMBE
          Arsenic   Ore(tons) Metal(tons)   Value(£)
          1862   25.40   0.00   15.20
          Comment 1862 CRUDE
Management: Manager 1859-1868 WM.B.COLLOM; 1869-1874 WM.SKEWIS; 1875
          WM.B.COLLOM
          Chief Agent 1859-1863 WM.JENNINGS; 1869 MOSES BAWDEN;
          1870-1874 MOSES BAWDEN WM.B.COLLOM
          Secretary 1859-1875 JOHN BAILEY (P)

CALSTOCK,EAST   CALSTOCK   SX 426696 9999

Production: Arsenic   Ore(tons) Metal(tons)   Value(£)
          1878   22.80   0.00   77.40
          Comment 1878 REFINED
          Arsenic Pyrite   Ore(tons)   Value(£)
          1879   160.20   93.90

CALSTOCK EASTERN SETT   CALSTOCK   0202

Production: Copper   Ore(tons) Metal(tons)   Value(£)
          1878   10.90   0.70   35.10
          1879   9.70   0.60   26.00
          Arsenic Pyrite   Ore(tons)   Value(£)
          1878   91.40   65.30

CALSTOCK EASTERN SETT   CALSTOCK   Continued

          Arsenic Pyrite   Ore(tons)   Value(£)
          1879   174.50   130.00
Ownership: 1876-1881 WHEAL CALSTOCK EASTERN SETT CO.LTD.
Management: Manager 1876-1881 WM.B.COLLOM
            Secretary 1876-1878 C.GRANT SMITH
Employment: Underground   Surface   Total
          1878   15   5   20
          1879   4   2   6

CALSTOCK UNITED   CALSTOCK   SX 400702 0203

Production: Copper No detailed return
          Tin   Black(tons) Stuff(tons)   Tin(tons)   Value(£)
          1853   5.30   0.00   0.00   275.20
          1854   3.40   0.00   0.00   150.00
          1856   5.90   0.00   0.00   418.70
          1857   13.00   0.00   0.00   997.10
          1858   16.10   0.00   0.00   903.30
Management: Chief Agent 1860-1863 WM.COOK
            Secretary 1860-1863 P.FISHER (P)

CALSTOCK,WEST   CALSTOCK   0204

Production: Tin No detailed return
          Arsenic Pyrite   Ore(tons)   Value(£)
          1897   220.00   110.00
Ownership: 1896-1901 T.WESTLAKE & CO.
Employment: Underground   Surface   Total
          1897   10   1   11
          1898   4   1   5
          1899   9   1   10
          1900   6   2   8

CALVADNACK   WENDRON   SW 699348 0205

Production: Tin   Black(tons) Stuff(tons)   Tin(tons)   Value(£)
          1855   12.40   0.00   0.00   791.40
          1856   38.10   0.00   0.00   2812.10
          1857   43.80   0.00   0.00   3282.70
          1858   65.70   0.00   0.00   3814.10
          1859   84.50   0.00   0.00   5777.50
          1860   90.90   0.00   0.00   6596.00
          1861   121.60   0.00   0.00   8057.20
          1862   172.80   0.00   0.00   10866.80
          1863   134.50   0.00   0.00   8692.20
          1864   108.80   0.00   0.00   6218.50
          1865   20.90   0.00   0.00   1104.90
          1873   29.20   0.00   0.00   1806.90
          1874   17.30   0.00   0.00   855.10
          1875   16.30   9.00   0.00   1314.50
          Comment 1862 TWO RETURNS AGGREGATED; 1874 VALUE INC.TIN
          STUFF

74

CALVADNACK   WENDRON   Continued

Ownership: 1875 CALVADNACK MINE ADVENTURERS
            Comment 1864–1865 SUSPENDED
Management: Manager 1859 CHAS.THOMAS & SON; 1860–1861 CHAS.THOMAS;
            1862–1863 CHAS.THOMAS & SON; 1873–1874 J.COCK
            Chief Agent 1859 WM.ROWE; 1860–1861 JOH.THOMAS JOHN COCK;
            1862–1863 HY.ARTHUR JOHN COCK; 1872 JOHN TONKIN J.COCK; 1875
            J.TREGONNING
            Secretary 1859–1862 ROBT.H.PIKE (P); 1863–1865 ROBT.H.PIKE &
            SON

CAMBORNE CONSOLS   CAMBORNE   SW 652401 0206

Production: Copper   Ore(tons) Metal(tons)   Value(£)
            1853   62.00   6.70   702.40
            1854   134.00   9.80   1027.60
            1856   164.00   10.20   913.30
            1857   189.00   15.10   1534.80
            1858   132.00   8.20   760.60
            1859   193.00   12.00   1094.40
            1860   100.00   7.40   697.70
            1862   68.00   6.80   575.60
            1863   66.00   6.20   499.00
            1864   23.00   1.30   111.80
            1886   41.00   0.00   160.00
            1887   34.00   0.00   88.00
            1888   23.00   0.00   69.00
            1889   31.00   0.00   60.00
            Comment 1853–1854 (C); 1856–1860 (C); 1862–1864 (C)
            Tin   Black(tons) Stuff(tons) Tin(tons)   Value(£)
            1859   0.00   0.00   0.00   0.80
            1862   0.00   0.00   0.00   3.20
            1865   4.20   0.00   0.00   37.10
            1886   7.90   0.00   0.00   481.00
            1887   28.00   0.00   0.00   1678.00
            1888   15.40   0.00   0.00   938.00
            1889   43.50   0.00   0.00   2207.00
            1890   13.00   0.00   0.00   643.00
            1891   No detailed return
            Comment 1859 TIN STUFF; 1862 TIN STUFF
            Arsenic   Ore(tons) Metal(tons)   Value(£)
            1886   9.00   0.00   48.00
            1887   119.00   0.00   770.00
            1888   64.00   0.00   416.00
            1889   8.00   0.00   35.00
Ownership: 1884–1886 WHEAL CAMBORNE CO.; 1887–1891 CAMBORNE CONSOLS
            LTD.
            Comment 1865 SUSPENDED; 1884–1885 CAMBORNE; 1891 SUSPENDED
Management: Manager 1859–1864 WM.ROBERTS
            Chief Agent 1859–1864 THOS.GUNDRY; 1884–1891 PETER TEMBY
            Secretary 1859–1861 RICH.LYLE (P); 1862–1865 GEO.LIGHTLEY
Employment: Underground   Surface   Total
            1884   10   16   26
            1885   35   17   52
            1886   44   55   99

                             75

CAMBORNE CONSOLS   CAMBORNE   Continued

	Underground	Surface	Total
1887	37	40	77
1888	30	38	68
1889	46	32	78

CAMBORNE VEAN   CAMBORNE   SW 652398 0207

Production: Copper

	Ore(tons)	Metal(tons)	Value(£)
1845	2751.00	185.50	13873.10
1846	3161.00	218.70	15608.40
1847	3297.00	229.50	16289.90
1848	3075.00	211.70	12414.50
1849	2775.00	180.30	11495.80
1850	3046.00	200.30	13613.80
1851	2067.00	120.20	7618.90
1852	1397.00	74.10	5852.20
1853	1590.00	81.30	7718.40
1854	1002.00	48.90	4686.60
1855	841.00	37.20	3507.20
1856	568.00	27.40	2377.10
1857	441.00	24.60	2390.60
1858	584.00	35.60	3193.40
1859	712.00	32.90	2943.90
1860	233.00	14.70	1302.00
1861	451.00	24.00	2068.30
1862	572.00	31.90	2540.00
1863	185.00	11.40	853.00
1864	87.00	3.10	234.20
1865	59.00	2.40	175.10
1866	249.00	13.60	810.30
1867	436.00	22.60	1511.40
1868	110.00	5.20	319.10
1869	9.00	0.30	15.50
1870	5.00	0.10	7.90
1871	17.00	1.50	93.90
1872	12.00	1.00	80.40
1882	13.70	0.80	48.00
1883	74.00	4.80	244.00
1884	82.00	0.00	205.00
1885	46.00	0.00	115.00

Comment 1845-1858 (C); 1859 (C)CAMBORNE VEIN; 1860-1872 (C)

Tin

	Black(tons)	Stuff(tons)	Tin(tons)	Value(£)
1857	No detailed return			
1858	17.20	0.00	0.00	1159.30
1860	9.20	0.00	0.00	705.10
1861	8.50	0.00	0.00	638.10
1862	17.10	0.00	0.00	1119.70
1863	19.70	0.00	0.00	1295.80
1864	22.20	0.00	0.00	1445.40
1865	77.30	0.00	0.00	4374.80
1866	36.40	0.00	0.00	1853.40
1867	40.30	0.00	0.00	2211.80
1868	49.00	0.00	0.00	2700.50
1869	52.70	0.00	0.00	3711.80

Tin	Black(tons)	Stuff(tons)	Tin(tons)	Value(£)
1870	17.80	0.00	0.00	1269.60
1882	1.50	113.00	1.00	75.00
1883	0.30	55.00	0.00	15.00
1884	0.10	14.00	0.00	3.00

Comment 1857 SEE STRAY PARK; 1860—1861 INC.FRANCES; 1862
INC.FRANCES.TWO RETURNS AGG.; 1863—1864 INC.FRANCES; 1870
VALUE INC.TIN STUFF

Arsenic	Ore(tons)	Metal(tons)	Value(£)
1862	10.00	0.00	14.90
1867	3.40	0.00	6.00
1868	15.60	0.00	12.40

Comment 1862 CRUDE; 1867—1868 CRUDE
Ownership: 1881—1886 CAMBORNE VEAN MINING CO.; 1896 PETER TEMBY
Comment 1859—1865 INC.FRANCES
Management: Manager 1859 CHAS.THOMAS; 1860—1864 WM.BAWDEN; 1865—1868
NICH.CLYMO; 1869 JOH.THOMAS; 1870—1871 JOS.VIVIAN & SON; 1881
R.S.TEAGUE
Chief Agent 1860—1864 NICH.CLYMO; 1869 M.T.MINERS; 1870—1871
NICH.CLYMO; 1881 NICH.CLYMO
Secretary 1859 W.H.M.BLUES (P); 1860—1862 ROBT.H.PIKE (P);
1863—1871 ROBT.H.PIKE & SON; 1881—1883 EDW.BOND; 1884—1886
R.S.TEAGUE

Employment:	Underground	Surface	Total
1881	15	4	19
1882	17	4	21
1883	12	5	17
1884	10	4	14

CAMBORNE,WEST   CAMBORNE   9998

Production: Arsenic Pyrite	Ore(tons)	Value(£)
1898	240.00	110.00
1899	350.00	240.00
1900	225.00	112.00

CAMELFORD   CAMELFORD   0208

Production: Lead	Ore(tons)	Metal(tons)	Value(£)
1845	180.00	108.00	0.00

CANAFRAME   ALTARNUN   SX 198790 0209

Ownership: 1912—1913 H.H.KING
Comment 1912 PROSPECTING FOR SN & WO JUNE 1912; 1913 NOT
WORKED

Employment:	Underground	Surface	Total
1912		5	5

CANNA   LUXULYAN   SX 033593 0210

Production: Iron No detailed return
Ownership: 1872—1874 EAST ROCKS CO.
           Comment 1872—1874 CANNA & EAST ROCKS
Management: Chief Agent 1872—1874 WOOD.HORE
            Secretary 1872—1874 J.W.COOM (P)

CAPE CORNWALL   ST.JUST   SW 350318 0211

Production: Tin   Black(tons) Stuff(tons)   Tin(tons)   Value(£)
            1866   14.60   0.00   0.00   702.30
            1867   2.40   0.00   0.00   108.00
            1868   0.20   0.00   0.00   10.20
            Comment 1867 INC.ST JUST CONSOLS
Ownership: Comment 1875 SUSPENDED
Management: Manager 1864 F.GOLDSWORTHY; 1865—1866 RALPH GOLDSWORTHY;
            1867—1873 RICH.PRYOR; 1874—1875 RICH.PRYOR & SON
            Chief Agent 1866 RICH.PRYOR; 1867 FRAN.HOSKING W.WHITE; 1868
            W.TREGLOWN W.WHITE; 1869—1875 JOHN DAVEY
            Secretary 1864—1867 FRED.HOLMAN; 1868 RICH.PRYOR; 1869—1875
            HY.L.PHILLIPS

CARADON   ST.IVE   SX 298702 0212

Production: Lead No detailed return
            Copper No detailed return
            Tin   Black(tons) Stuff(tons)   Tin(tons)   Value(£)
            1883   No detailed return
            Comment 1883 SEE SOUTH PHOENIX
Management: Manager 1861—1866 FRAN.PRYOR
            Chief Agent 1861—1866 JAS.BROWNE
            Secretary 1861—1866 E.LILLY (P)

CARADON & PHOENIX CONSOLS   NORTH HILL   SX 271756 0213

Production: Zinc   Ore(tons) Metal(tons)   Value(£)
            1870   7.00   0.00   24.20
            1871   67.00   0.00   251.00
            1873   18.00   0.00   45.00
            1874   No detailed return
            1877   11.60   0.00   51.70
            Copper No detailed return
Ownership: 1877 CARADON & PHOENIX CONSOLS MINING CO.
Management: Manager 1868—1869 WM.JOHNS & SON
            Chief Agent 1868—1877 JAS.KELLY
            Secretary 1869—1871 CHAS.PEARSON; 1872—1873 HY.SHORT;
            1874—1877 CHAS.PEARSON

CARADON CONSOLS   ST.CLEER   SX 257698 0214

Production: Copper   Ore(tons) Metal(tons)   Value(£)
            1866   34.00   2.00   122.40

CARADON CONSOLS   ST.CLEER   Continued

            Copper   Ore(tons) Metal(tons)   Value(£)
            1867   185.00   12.70   871.70
            1868   89.00   4.10   242.20
            1870   40.00   2.10   115.80
            Comment 1866-1868 (C); 1870 (C)
Management: Manager 1859-1866 WM.RICH; 1867-1870 SML.BENNETTS
            Chief Agent 1866-1868 SML.BENNETTS
            Secretary 1859 E.A.CROUCH (P); 1860 H.M.RICE (P); 1861-1870
            ED.KING (P)

CARADON CONSOLS,GLASGOW   ST.IVE   SX 282702 0215

Production: Copper   Ore(tons) Metal(tons)   Value(£)
            1864   862.00   38.30   3130.70
            1865   888.00   37.80   2909.60
            1866   757.00   38.50   2470.30
            1867   515.00   28.30   1991.00
            1868   1575.00   93.30   6047.70
            1869   1703.00   100.60   6180.20
            1870   1338.00   92.70   5591.00
            1871   1531.00   106.60   6852.90
            1872   2195.00   182.80   14267.20
            1873   2966.00   214.00   13897.30
            1874   2950.00   211.50   14388.20
            1875   2962.00   205.30   15876.80
            1876   2980.00   212.10   14586.60
            1877   2978.20   196.60   11427.30
          . 1878   2562.00   191.60   9848.90
            1879   2299.60   162.10   8322.50
            1880   1926.10   126.20   7138.80
            1881   835.00   55.50   3049.20
            1882   979.30   69.80   4711.00
            1883   932.80   67.60   3294.00
            1884   481.00   0.00   1223.00
            1885   1802.00   0.00   5816.00
            Comment 1864-1876 (C)GLASGOW CARADON; 1877-1878 GLASGOW
            CARADEN; 1882 GLASGOW CARADON
Ownership: 1876 GLASGOW CARADON CONSOLS CORP.CO.LTD.; 1877-1887 GLASGOW
            CARADON CONSOLIDATED CO.LTD.; 1888 GLASGOW CARADON CO.LTD.
            Comment 1861 GLASGOW CONSOLS; 1886-1887 NOT WORKING; 1888 INC
            SOUTH CARADON & EAST CARADON.
Management: Manager 1871-1881 WM.TAYLOR
            Chief Agent 1861-1870 WM.TAYLOR; 1872-1881 W.J.TAYLOR;
            1882-1888 WM.TAYLOR
            Secretary 1861-1868 WM.TAYLOR (P); 1869-1870 DAVID DUNLOP;
            1871 T.E.WATSON; 1872 DAVID DUNLOP; 1873-1881 T.E.WATSON
Employment: Underground   Surface   Total
            1878   93   50   143
            1879   92   48   140
            1880   74   46   120
            1881   45   27   72
            1882   64   32   96
            1883   51   33   84
            1884   35   25   60

                               79

CARADON CONSOLS,GLASGOW   ST.IVE   Continued

                  Underground   Surface   Total
          1885   60   44   104
          1888   10   33   43
          Comment 1888 INC SOUTH CARADON & EAST CARADON.

CARADON UNITED   ST.CLEER   SX 249691 0216

Production: Copper   Ore(tons) Metal(tons)   Value(£)
            1864   63.00   1.70   130.70
            Comment 1864 (C)
            Tin   Black(tons) Stuff(tons)   Tin(tons)   Value(£)
            1864   6.20   0.00   0.00   331.10
Management: Secretary 1859—1865 RICH.CLOGG (P)

CARADON,EAST   LINKINHORNE   SX 277704 0218

Production: Copper   Ore(tons) Metal(tons)   Value(£)
            1860   836.00   65.00   6010.60
            1861   2773.00   228.50   20419.20
            1862   5265.00   432.30   35600.20
            1863   5911.00   426.20   33215.20
            1864   5933.00   406.80   34791.60
            1865   5098.00   285.60   22417.10
            1866   3761.00   213.80   14302.70
            1867   3279.00   182.40   12958.90
            1868   2202.00   133.80   8854.50
            1869   2527.00   162.60   10268.20
            1870   2766.00   197.70   11919.90
            1871   2349.00   168.50   10607.40
            1872   2133.00   146.20   11316.90
            1873   2096.00   164.80   10747.60
            1874   1755.00   135.20   9242.10
            1875   1515.00   114.30   8947.00
            1876   1390.00   99.70   6896.00
            1877   1126.00   78.30   4560.50
            1878   249.20   18.50   1000.40
            1879   83.60   8.70   467.80
            1880   64.10   6.70   381.40
            1881   118.90   9.20   533.80
            1882   280.10   19.60   1260.00
            1883   303.30   19.00   1065.00
            1884   183.00   0.00   641.00
            1885   52.00   0.00   205.00
            Comment 1860—1876 (C)
            Iron No detailed return
Ownership: 1876—1886 EAST CARADON MINING CO.; 1907—1908 BUDD & CO.
            Comment 1888 SEE GLASGOW CARADON CONSOLS; 1907—1908 EAST
            CARADON & MARKE VALLEY
Management: Manager 1859—1864 JAS.SECCOMBE; 1867—1868 JOHN TRUSCOTT;
            1869—1871 WM.THORNE (P); 1872—1880 JAS.KELLOW; 1881 JOHN
            KELLOW
            Chief Agent 1859—1860 JOHN TRUSCOTT; 1861—1866 JAS.KELLOW;
            1869—1871 JAS.KELLOW NOAH COWARD; 1872 WM.THORPE (P);

1873–1878 THOS.TRELEASE; 1879–1880 WM.THORNE; 1881–1886
GEO.SECCOMBE WM.GEORGE; 1907–1908 ALT.E.FLEWIN
Secretary 1859–1868 C.R.NORTON (P); 1869–1881 C.R.NORTON (S)

Employment:	Underground	Surface	Total
1878	28	11	39
1879	24	10	34
1880	34	11	45
1881	50	11	61
1882	40	23	63
1883	30	8	38
1884	17	5	22
1885	9	4	13
1886	9	3	12
1888			
1907–1908		2	2

Comment 1888 SEE GLASGOW CARADON CONSOLS; 1907–1908 INC.MARKE
VALLEY

CARADON,GREAT   ST.IVE   SX 298707 0219

Production: Copper No detailed return
Ownership: Comment 1859–1860 INC.SLADE; 1876 SUSPENDED DURING 1876
Management: Manager 1867–1868 F.C.HARPER; 1869–1876 WM.TAYLOR
Chief Agent 1860–1864 F.C.HARPER; 1865–1866 F.C.HARPER
SML.BENNETTS; 1867–1868 SML.BENNETTS; 1871 THOS.KITTO;
1872–1875 THOS.TRELEASE
Secretary 1860–1868 ED.KING (P); 1869–1876 GREN.SHARP

CARADON,GREAT NORTH   ADVENT   0220

Production: Lead & Silver No detailed return
Copper No detailed return
Ownership: Comment 1872 FORMERLY ARCHIE
Management: Manager 1872–1873 GEO.RICKARD

CARADON,NEW   ST.CLEER   0221

Production: Copper No detailed return
Ownership: 1882–1886 NEW CARADON MINING CO.; 1887 NEW CARADON MINE
ADVENTURERS
Management: Chief Agent 1885 NICH.RICHARDS; 1886–1887 ALF.STEVENS
Secretary 1882–1884 W.H.WATSON

Employment:	Underground	Surface	Total
1882	5		5
1883	5	4	9
1884	9	3	12
1885	10	1	11
1886	7	1	8
1887	3		3

CARADON,NEW SOUTH   ST.CLEER   SX 230692 0222

Production: Copper No detailed return
            Tin No detailed return
Ownership: 1883-1887 NEW SOUTH CARADON MINING CO.
            Comment 1861 SUSPENDED; 1864-1865 SUSPENDED
Management: Manager 1860 ROBT.KNAPP
            Chief Agent 1862-1863 ROBT.KNAPP; 1883-1886 JOHN HOLMAN
            Secretary 1860 ROBT.KNAPP (P); 1861-1865 JOHN O.HARRIS (P);
            1887 WM.H.RULE
Employment: Underground   Surface   Total
            1883   15   15
            1884   9   10   19
            1885   16   8   24
            1886   13   8   21
            1887   10   8   18

CARADON,NEW WEST   ST.CLEER   SX 256697 0223

Production: Copper   Ore(tons) Metal(tons)   Value(£)
            1881   59.20   3.60   189.50
            1882   52.80   3.30   187.00
            1883   46.00   3.10   154.00
            1884   46.00   0.00   156.00
            1885   63.00   0.00   188.00
            1886   26.00   0.00   65.00
Ownership: 1880-1885 JOHN WATSON & OTHERS; 1886-1889 NEW WEST CARADON
            MINES ADVENTURERS
Management: Chief Agent 1880-1881 JOHN WATSON NICH.RICHARDS; 1885
            NICH.RICHARDS; 1886-1889 JAS.KELLY
            Secretary 1880-1884 W.H.WATSON
Employment: Underground   Surface   Total
            1880   13   13
            1881   20   8   28
            1882   12   6   18
            1883   16   5   21
            1884   12   6   18
            1885   10   3   13
            1886-1887   6   1   7
            1888   5   5   10
            1889   2   2   4

CARADON,SOUTH   ST.CLEER   SX 268698 0224

Production: Copper   Ore(tons) Metal(tons)   Value(£)
            1845   4631.00   394.80   27266.50
            1846   4159.00   382.90   24739.80
            1847   4570.00   440.20   30679.00
            1848   3473.00   373.40   21836.70
            1849   2965.00   304.80   20508.40
            1850   2999.00   318.90   22648.10
            1851   2818.00   296.50   20208.00
            1852   2834.00   276.80   23965.70
            1853   2871.00   340.40   35040.30
            1854   3006.00   377.90   39986.80

Copper	Ore(tons)	Metal(tons)	Value(£)
1855	3679.00	383.80	40857.40
1856	4694.00	417.80	46488.40
1857	4538.00	452.80	46893.70
1858	4995.00	528.90	50093.20
1859	5164.00	523.10	50587.80
1860	5232.00	543.50	50571.10
1861	4777.00	502.20	45071.10
1862	5460.00	597.90	50031.00
1863	5846.00	649.10	52069.50
1864	5744.00	658.10	57070.40
1865	6306.00	668.70	54033.80
1866	5785.00	671.30	46760.90
1867	5993.00	705.20	50293.40
1868	6416.00	724.90	48826.20
1869	6433.00	763.80	48767.00
1870	6706.00	717.90	42699.10
1871	6125.00	647.90	41234.40
1872	5195.00	551.70	44285.80
1873	5293.00	569.90	38676.70
1874	5467.00	542.10	36471.80
1875	6006.00	597.80	44966.00
1876	6157.00	607.20	40875.30
1877	6467.80	665.60	39188.80
1878	6102.50	616.50	33130.40
1879	6048.80	552.00	29647.10
1880	5871.60	518.10	30249.00
1881	5185.40	475.50	27609.20
1882	5484.90	356.50	25984.00
1883	3016.30	19.00	13725.00
1884	4096.00	0.00	14278.00
1885	3436.00	0.00	11174.00
1886	162.00	0.00	83.00

Comment 1845-1876 (C)
Ownership: 1876-1887 SOUTH CARADON MINING CO.
Comment 1887 NOT WORKING; 1888 SEE GLASGOW CARADON CONSOLS
Management: Manager 1859-1868 PETER CLYMO; 1869-1870 WM.RULE JOHN HOLMAN;
1871-1881 JOHN HOLMAN
Chief Agent 1859 WM.RULE; 1860 WM.RULE J.PEARCE JOHN HOLMAN
W.WILLS; 1861-1867 WM.RULE J.PEARCE JOHN HOLMAN; 1868 WM.RULE
JOHN HOLMAN; 1869-1874 WM.TYACKE; 1875 WM.TYACKE JAS.DYMOND;
1876-1878 WM.TYACKE; 1879-1881 WM.TYACKE WM.CLOGG
FRED.WADDLETON; 1882 JOHN HOLMAN; 1883-1886 WM.GEORGE &
GEO.SECCOMBE
Secretary 1859-1868 THOS.KITTO (P); 1872-1875 THOS.KITTO;
1876-1878 JAS.DYMOND; 1879 JAS.G.DYMOND; 1880-1881 WM.H.RULE

Employment:	Underground	Surface	Total
1878	275	124	399
1879	242	156	398
1880	245	177	422
1881	213	148	361
1882	178	125	303
1883	263	132	395
1884	176	122	298
1885	162	127	289

CARADON,SOUTH   ST.CLEER   Continued

                 Underground   Surface   Total
          1886   2   2
          1888
          Comment 1888 SEE GLASGOW CARADON CONSOLS

CARADON,WEST   ST.CLEER   SX 262700 0225

Production: Copper   Ore(tons) Metal(tons)   Value(£)
          1845   4457.00   475.00   33272.10
          1846   4736.00   503.90   32801.10
          1847   4135.00   413.90   29056.10
          1848   3668.00   424.30   24726.70
          1849   3966.00   439.70   29873.00
          1850   4049.00   458.20   32646.20
          1851   4128.00   443.00   30329.90
          1852   4048.00   407.30   34813.00
          1853   4355.00   417.00   42252.60
          1854   4018.00   351.00   36563.80
          1855   3941.00   323.50   33596.00
          1856   4313.00   348.10   33371.70
          1857   4217.00   313.10   31482.60
          1858   3702.00   290.40   26508.60
          1859   3778.00   377.80   36205.90
          1860   3936.00   421.00   39711.80
          1861   3166.00   312.80   28185.60
          1862   3090.00   265.10   21862.80
          1863   2520.00   204.20   15835.30
          1864   1834.00   141.90   12199.30
          1865   1270.00   88.10   6794.20
          1866   1002.00   76.00   5048.10
          1867   728.00   53.20   3667.80
          1868   600.00   43.40   2704.60
          1869   794.00   55.30   3277.70
          1870   720.00   51.20   2912.40
          1871   782.00   51.40   2990.70
          1872   853.00   47.40   3408.80
          1873   638.00   36.50   1969.70
          1874   172.00   9.00   558.80
          1881   227.10   15.80   804.70
          1882   331.60   27.30   1721.00
          1883   417.60   43.80   2254.00
          1884   277.00   0.00   855.00
          1885   122.00   0.00   294.00
          1886   20.00   0.00   70.00
          Comment 1845-1874 (C)
Ownership: 1880-1885 JOHN WATSON & OTHERS; 1886-1891 WEST CARADON MINES
          ADVENTURERS
          Comment 1874 STOPPED; 1875 SUSPENDED; 1876 ABANDONED
          OCT.1876; 1891 SUSPENDED
Management: Manager 1859 FRAN.PRYOR; 1861-1863 FRAN.PRYOR; 1864-1870
          WM.JOHNS; 1871-1877 NICH.RICHARDS; 1880-1881 JOHN WATSON
          Chief Agent 1859 WM.JOHNS; 1860 WM.JOHNS W.TRATHAN; 1861-1863
          WM.JOHNS ROBT.TRATHAN J.WILLIAMS; 1864-1867 ROBT.TRATHAN
          J.WILLIAMS; 1868-1870 NICH.RICHARDS; 1871 THOS.PRYOR;

                                    84

1872—1877 J.PRYOR; 1880—1881 NICH.RICHARDS; 1885
NICH.RICHARDS; 1886—1891 JAS.KELLY
Secretary 1859 E.A.CROUCH (P); 1860—1863 FRAN.PRYOR (P); 1864
H.LAVINGTON; 1865—1877 W.J.LAVINGTON; 1880—1884 W.H.WATSON

Employment:	Underground	Surface	Total
1880	16	8	24
1881	28	26	54
1882	32	22	54
1883	42	23	65
1884	24	20	44
1885	10	10	20
1886	7	3	10
1887	9	3	12
1888	6	5	11
1889	2	2	4
1890	4	1	5

CARADON,WEST SOUTH   ST.CLEER   SX 247690 0226

Production: Copper No detailed return
Management: Chief Agent 1860 JAS.WILLIAMS; 1861—1865 JAS.WILLIAMS
WM.JOHNS
Secretary 1860—1865 FRAN.PRYOR

CARBIS   ST.AUSTELL   SX 030562 0227

Production:	Tin	Black(tons)	Stuff(tons)	Tin(tons)	Value(£)
	1891	No detailed return			

Comment 1891 OPEN WORK AT WORK
Ownership: 1891 WM.NICHOLLS

CARBOLA   0228

Production:	Tin	Black(tons)	Stuff(tons)	Tin(tons)	Value(£)
	1873	0.00	304.00	0.00	616.20

CARBONA   CROWAN   SW 648331 0229

Production:	Tin	Black(tons)	Stuff(tons)	Tin(tons)	Value(£)
	1852	6.90	0.00	0.00	0.00

Comment 1852 TIN ORE

CARCLAZE   ST.AUSTELL   SX 025542 0230

Production:	Lead	Ore(tons)	Metal(tons)	Value(£)
	1874	0.50	0.30	26.80
	Zinc	Ore(tons)	Metal(tons)	Value(£)
	1874	0.90	0.00	58.80

Comment 1874 MISTAKE?

CARCLAZE   ST.AUSTELL   Continued

Tin	Black(tons)	Stuff(tons)	Tin(tons)	Value(£)
1865	0.00	0.00	0.00	595.90
1869	8.60	0.00	0.00	616.70
1870	7.30	0.00	0.00	543.00
1871	5.90	0.00	0.00	482.30
1872	7.60	0.00	0.00	696.50
1873	7.60	0.00	0.00	613.60
1874	4.90	0.00	0.00	281.80
1875	6.20	0.00	0.00	329.40
1876	2.00	0.00	0.00	90.00
1880	1.20	0.00	0.00	58.00
1906	0.50	0.00	0.00	52.00
1910	0.10	0.00	0.00	8.00

Comment 1865 TIN STUFF; 1906 OPEN WORK; 1910 OPEN WORK
Ownership: 1876–1881 J.LOVERING & CO.
Management: Manager 1873–1881 ROBT.MARTIN
Chief Agent 1872 ROBT.MARTIN
Secretary 1860–1875 J.LOVERING

CARDELE   LEEDSTOWN   0231

Ownership: Comment 1871 SEE CROWAN MINES

CARDREW   REDRUTH   SW 708436 0232

Production: Copper   Ore(tons) Metal(tons)   Value(£)
            1883   No detailed return
Comment 1883 SEE PRUSSIA
Tin   Black(tons) Stuff(tons)   Tin(tons)   Value(£)
1879–1883 No detailed return
Comment 1879–1883 SEE PRUSSIA
Ownership: Comment 1879–1883 SEE PRUSSIA
Employment: Underground   Surface   Total
1880–1883
Comment 1880–1883 SEE PRUSSIA

CARGOL   NEWLYN EAST   SW 835542 0233

Production: Lead & Silver   Ore(tons)   Lead(tons) Silver(ozs)   Value(£)

1845	55.00	33.00	0.00	0.00
1846	306.00	183.00	0.00	0.00
1847	951.00	570.00	0.00	0.00
1848	964.00	577.00	0.00	0.00
1849	505.10	303.00	0.00	0.00
1850	200.00	118.80	0.00	0.00
1851	309.90	184.90	0.00	0.00
1852	70.60	40.00	1800.00	0.00
1853	23.90	13.50	600.00	0.00
1854	133.00	89.50	1556.00	0.00
1855	378.00	260.00	2600.00	0.00
1856	363.00	232.50	6277.00	0.00
1857	458.00	300.00	8700.00	0.00

Lead & Silver	Ore(tons)	Lead(tons)	Silver(ozs)	Value(£)
1858	497.00	303.00	8910.00	0.00
1859	455.50	334.00	9088.00	0.00
1860	489.90	360.00	7165.00	0.00
1861	1026.00	667.00	12670.00	0.00
1862	1086.00	814.00	11400.00	0.00
1863	895.50	671.00	9394.00	0.00
1864	730.20	540.00	11880.00	0.00
1865	890.90	675.60	14860.00	0.00
1866	579.50	434.80	12152.00	0.00
1867	574.50	430.60	12056.00	0.00
1868	633.50	475.00	13300.00	0.00
1869	600.70	450.60	11200.00	0.00
1870	474.80	356.50	9982.00	0.00
1871	389.70	292.20	8176.00	0.00
1872	172.10	129.00	4820.00	0.00
1873	13.60	10.20	285.00	154.20
1875	7.50	5.70	0.00	172.40
1876	47.70	36.50	180.00	881.50
1877	78.00	58.50	298.00	1053.00
1878	20.10	15.00	76.00	240.00
1879	6.50	4.80	25.00	68.20
1880	41.00	30.70	145.00	389.00
1881	14.50	10.50	49.00	145.00
1882	9.20	7.00	20.00	65.00
1883	3.00	2.20	7.00	30.00
1884	6.00	0.00	0.00	31.00
1885	3.00	0.00	0.00	19.00

Comment 1869-1873 CARGOLL; 1879-1880 CARGOLL

Zinc	Ore(tons)	Metal(tons)	Value(£)
1864	103.40	0.00	431.70
1865	304.90	0.00	1258.90
1866	447.60	0.00	1809.80
1867	537.40	0.00	1866.30
1868	955.00	0.00	3301.80
1869	584.70	0.00	1724.40
1870	517.60	0.00	1426.60
1871	271.70	0.00	972.80
1872	47.10	0.00	117.50
1873	25.60	0.00	61.40
1875	25.00	0.00	65.00
1876	28.00	0.00	127.50
1880	30.00	0.00	121.50
1885	10.00	0.00	39.00
1892	12.00	6.00	50.00
1893	No detailed return		

Comment 1864-1867 CARGOLL; 1868-1869 CARGOLL.TWO RETURNS
AGGREGATED; 1870 CARGOLL; 1871 CARGOLL.TWO RETURNS
AGGREGATED; 1885 CARGOLL; 1892-1893 NEW CARGOLL

Copper	Ore(tons)	Metal(tons)	Value(£)
1864	120.00	18.20	1690.40
1865	40.00	6.60	619.00
1867	55.00	8.90	641.90
1868	21.00	3.70	257.20
1869	55.00	10.20	646.20

Copper	Ore(tons)	Metal(tons)	Value(£)
1870	12.00	2.40	147.90
1871	8.00	1.30	81.40
1885	1.00	0.00	1.00

Comment 1864 (C); 1865 (C)CARGOLL; 1867-1871 (C); 1885
CARGOLL

Ownership: 1877 CARGOLL MINING CO.; 1882-1885 CHAS.TAMBLYN JOHN GROSE;
1891-1893 NEW CARGOLL MINING CO. C.B.; 1894-1895 NEW CARGOLL
MINING CO.
Comment 1891-1892 NEW CARGOL; 1893-1895 NEW CARGOL.
SUSPENDED

Management: Manager 1868-1873 JOHN GROSE; 1874-1881 JOHN JENNINGS
Chief Agent 1860-1861 JOHN GROSE; 1862-1865 JOHN GROSE
H.TYZZER; 1866-1867 JOHN GROSE R.TYZZER; 1869-1873 R.TYZZER;
1879-1885 CHAS.TAMBLYN JOHN GROSE; 1891-1895 JOHN GROSE
Secretary 1860-1867 ED.MICHELL (P); 1868 ED.MICHELL R.TYZZER;
1869-1878 ED.MICHELL; 1879-1881 CHAS.TAMBLYN J.GROSE;
1891-1895 F.W.THOMAS (P)

Employment:	Underground	Surface	Total
1878	30	20	50
1879-1880	2	15	17
1881-1882	2	6	8
1883	2	3	5
1884	2	2	4
1885	7	8	15
1891	9	6	15
1892	10	8	18

CARGOL,SOUTH NEWLYN EAST SW 834541 0235

Production: Lead & Silver	Ore(tons)	Lead(tons)	Silver(ozs)	Value(£)
1851	42.00	30.00	0.00	0.00
1852	70.60	40.00	1800.00	0.00
1853	44.00	30.00	345.00	0.00
1854	212.00	105.00	978.00	0.00
1855	57.00	28.20	210.00	0.00
1856	363.00	232.00	6728.00	0.00

1857-1861 No detailed return

Management: Manager 1859 JOHN CHAMPION
Chief Agent 1859-1861 JOHN GROSE
Secretary 1859-1861 ED.MICHELL (P)

CARHARRACH ST.DAY SW 738420 0236

Production: Tin	Black(tons)	Stuff(tons)	Tin(tons)	Value(£)
1855	1.30	0.00	0.00	80.80

CARIGGAN ROCK BUGLE SX 014605 0237

Production: Tin	Black(tons)	Stuff(tons)	Tin(tons)	Value(£)
1882	27.00	0.00	17.80	1620.00
1883	5.00	0.00	0.00	263.00

CARIGGAN ROCK   BUGLE   Continued

        Tin   Black(tons) Stuff(tons)   Tin(tons)   Value(£)
        1884   9.00   0.00   0.00   435.00
        1885   8.00   0.00   0.00   348.00
        Comment 1882 OPEN WORK
Ownership: 1873-1874 JOHN FLETCHER PAGEN; 1882-1885 JOHN MIDDLETON &
            CO.
            Comment 1873-1874 CARRIGAN ROCK; 1882 OPEN WORKS
Management: Chief Agent 1883-1885 W.WILLIAMS
Employment: Underground   Surface   Total
        1883   22   29   51
        1884   22   27   49
        1885   18   23   41

CARLEEN VOR,NEW   BREAGE   SW 614298 0238

Production: Tin   Black(tons) Stuff(tons)   Tin(tons)   Value(£)
        1876   1.80   75.00   0.00   159.50
        Comment 1876 INC.WEST METAL (BREAGE)
Ownership: 1876 JOHN HALL & OTHERS
            Comment 1876 INC.WEST METAL (BREAGE)
Management: Chief Agent 1876 JOHN HALL

CARLYON   ST.AUSTELL   SX 050520 0239

Production: Copper   Ore(tons) Metal(tons)   Value(£)
        1877-1881 No detailed return
        Comment 1877-1881 SEE GREAT CRINNIS
        Silver   Ore(tons) Metal(tons)   Value(£)
        1877   No detailed return
        1879-1881 No detailed return
        Comment 1877 SEE CRINNIS; 1879-1881 SEE GREAT CRINNIS
Ownership: Comment 1872-1881 SEE GREAT CRINNIS
Employment: Underground   Surface   Total
        1878-1881
        Comment 1878-1881 SEE GREAT CRINNIS

CARMARTHEN   ILLOGAN   SW 667399 0240

Production: Copper No detailed return
Ownership: Comment 1859-1865 SEE SOUTH DOLCOATH; 1866-1875 CARNARTHEN
            SEE SOUTH DOLCOATH

CARN BREA   ILLOGAN   SW 676410 0241

Production: Copper   Ore(tons) Metal(tons)   Value(£)
        1845   6674.00   532.50   39462.70
        1846   7288.00   600.10   41477.30
        1847   10372.00   1028.80   73445.40
        1848   10432.00   1051.20   62641.40
        1849   10086.00   986.10   66632.30
        1850   9318.00   822.80   57366.80

                                89

Copper	Ore(tons)	Metal(tons)	Value(£)
1851	7663.00	632.40	42181.30
1852	6616.00	472.30	39049.30
1853	7543.00	453.00	43598.10
1854	6231.00	358.80	35455.30
1855	5938.00	329.60	32553.90
1856	6632.00	318.90	27286.70
1857	4116.00	178.30	16088.70
1858	3345.00	1682.50	14237.80
1859	3421.00	192.40	16871.40
1860	3475.00	180.60	15382.20
1861	3014.00	133.30	10685.90
1862	2291.00	101.70	7397.90
1863	1563.00	94.00	7151.40
1864	1956.00	124.40	10327.40
1865	1590.00	89.60	6877.00
1866	1717.00	104.20	6646.30
1867	1581.00	106.70	7361.70
1868	2349.00	175.40	11465.30
1869	1855.00	122.10	7382.00
1870	2010.00	151.40	8723.30
1871	1114.00	78.60	4703.10
1872	1039.00	77.80	6036.80
1873	1137.00	67.40	4048.20
1874	1080.00	84.60	5458.10
1875	469.00	30.80	2213.80
1876	1174.00	65.90	4329.20
1877	714.70	85.40	1541.70
1878	428.80	20.00	1020.30
1879	152.10	8.80	436.00
1880	110.00	6.00	314.50
1882	90.00	27.00	300.00
1883	42.00	2.50	132.00
1884	57.00	0.00	161.00
1885	96.00	0.00	204.00
1889	14.00	0.00	18.00
1892-1893 No detailed return			
1896	618.00	0.00	1688.00
1897	621.00	0.00	1436.00
1898	1533.00	0.00	1792.00
1899	1294.00	0.00	2549.00
1900	836.00	0.00	1947.00
1901	474.00	0.00	1120.00
1902	786.00	0.00	936.00
1903	724.00	0.00	1475.00
1904	863.00	0.00	1454.00
1905	814.00	0.00	1881.00
1906	672.00	0.00	1744.00
1907	384.00	0.00	611.00
1908	21.00	0.00	104.00
1909	61.00	0.00	221.00
1910	77.00	0.00	327.00
1911	38.00	0.00	149.00
1913	17.00	0.00	31.00

Comment 1845-1876 (C); 1892-1893 SEE TINCROFT; 1896-1911

INC.TINCROFT; 1913 INC.TINCROFT

Tin	Black(tons)	Stuff(tons)	Tin(tons)	Value(£)
1853	80.00	0.00	0.00	4560.00
1855	333.00	0.00	0.00	20768.20
1856	294.40	0.00	0.00	20508.20
1857	433.00	0.00	0.00	32459.80
1858	509.40	0.00	0.00	30796.60
1859	515.30	0.00	0.00	35412.00
1860	653.50	0.00	0.00	47802.40
1861	535.90	0.00	0.00	38595.10
1862	694.50	0.00	0.00	44953.50
1863	569.00	0.00	0.00	37651.30
1864	511.50	0.00	0.00	31897.00
1865	437.90	0.00	0.00	25767.80
1866	424.70	0.00	0.00	20709.80
1867	317.00	0.00	0.00	16729.10
1868	420.90	0.00	0.00	22884.00
1869	448.20	0.00	0.00	30850.20
1870	476.80	0.00	0.00	34626.80
1871	402.00	0.00	0.00	32410.50
1872	559.30	0.00	0.00	50374.60
1873	545.00	0.00	0.00	40559.50
1874	652.50	0.00	0.00	37780.10
1875	668.10	0.00	0.00	34432.60
1876	852.60	0.00	0.00	37243.60
1877	1038.10	0.00	0.00	42540.60
1878	1195.00	0.00	0.00	42258.60
1879	1047.80	0.00	0.00	41910.00
1880	1063.00	0.00	0.00	54233.50
1881	974.30	0.00	649.10	58561.90
1882	707.00	0.00	459.00	41678.00
1883	719.00	0.00	0.00	37650.00
1884	463.90	0.00	0.00	20248.00
1885	392.60	0.00	0.00	17609.00
1886	535.20	0.00	0.00	29635.00
1887	578.10	0.00	0.00	37130.00
1888	760.20	0.00	0.00	50694.00
1889	1130.00	0.00	0.00	61950.00
1890	1670.80	0.00	0.00	90218.00
1891	1819.80	0.00	0.00	90943.00
1892	1614.00	0.00	0.00	80543.00
1893	1397.30	0.00	0.00	64455.00
1894	1142.50	0.00	0.00	40118.00
1895	780.40	0.00	0.00	24687.00
1896	846.50	0.00	0.00	26946.00
1897	696.20	0.00	0.00	23846.00
1898	618.50	0.00	0.00	24364.00
1899	492.40	0.00	0.00	33284.00
1900	662.90	0.00	0.00	51745.00
1901	652.00	0.00	0.00	42132.00
1902	584.80	0.00	0.00	39000.00
1903	549.70	0.00	0.00	38900.00
1904	657.10	0.00	0.00	45346.00
1905	827.30	0.00	0.00	63774.00
1906	873.50	0.00	0.00	83834.00

Tin	Black(tons)	Stuff(tons)	Tin(tons)	Value(£)
1907	813.40	0.00	0.00	78523.00
1908	1027.10	0.00	0.00	72003.00
1909	1106.90	0.00	0.00	77439.00
1910	963.90	0.00	0.00	76076.00
1911	820.60	0.00	0.00	87522.00
1912	919.80	0.00	0.00	113197.00
1913	733.90	0.00	0.00	82532.00

Comment 1862 TWO RETURNS AGGREGATED; 1896-1913 INC.TINCROFT

Arsenic	Ore(tons)	Metal(tons)	Value(£)
1854	145.40	0.00	254.30
1855	140.00	0.00	0.00
1857	116.50	0.00	204.80
1858	122.60	0.00	214.60
1859	219.10	0.00	389.00
1860	302.30	0.00	556.50
1861	218.30	0.00	351.40
1862	171.70	0.00	206.80
1863	166.00	0.00	165.90
1864	229.40	0.00	160.00
1865	174.70	0.00	132.40
1866	146.60	0.00	153.50
1867	77.30	0.00	130.90
1868	94.80	0.00	231.40
1869	173.90	0.00	424.30
1870	198.80	0.00	614.00
1871	101.10	0.00	167.70
1872	237.20	0.00	477.30
1873	185.40	0.00	444.30
1874	87.00	0.00	324.50
1875	83.30	0.00	566.90
1876	188.70	0.00	933.00
1877	93.90	0.00	298.10
1878	78.70	0.00	203.50
1879	75.90	0.00	197.50
1880	27.40	0.00	123.20
1881	20.90	0.00	95.00
1882	76.00	0.00	283.00
1883	16.00	0.00	79.00
1884	27.00	0.00	81.00
1885	20.00	0.00	60.00
1886	48.00	0.00	147.00
1887	35.00	0.00	51.00
1888	25.00	0.00	59.00
1889	19.00	0.00	68.00
1890	7.00	0.00	32.00
1896	370.00	0.00	4472.00
1897	298.00	0.00	5307.00
1898	421.00	0.00	4116.00
1899	278.00	0.00	3001.00
1900	312.00	0.00	4457.00
1901	339.00	0.00	3113.00
1902	313.00	0.00	2007.00
1903	255.00	0.00	1529.00
1904	323.00	0.00	1757.00

Arsenic	Ore(tons)	Metal(tons)	Value(£)
1905	233.00	0.00	1352.00
1906	262.00	0.00	3982.00
1907	174.00	0.00	4421.00
1908	222.00	0.00	1805.00
1909	362.30	0.00	2969.00
1910	452.00	0.00	3041.00
1911	372.70	0.00	1851.00
1912	337.80	0.00	2456.00
1913	287.50	0.00	1648.00

Comment 1855 WHITE ARSENIC; 1859-1871 CRUDE; 1884-1885 SOOT; 1886 CRUDE; 1896-1913 INC.TINCROFT

Tungsten	Ore(tons)	Metal(tons)	Value(£)
1896	22.00	0.00	371.00
1897	24.00	0.00	392.00
1898	239.00	0.00	11614.00
1899	44.00	0.00	2333.00
1901	15.00	0.00	262.00
1903	24.00	0.00	858.00
1904	20.00	0.00	2125.00
1905	1.50	0.00	95.00
1906	12.10	0.00	1089.00
1907	23.70	0.00	2890.00
1908	4.70	0.00	201.00
1909	4.40	0.00	258.00
1910	10.60	0.00	898.00
1911	21.70	0.00	2047.00
1912	13.60	0.00	947.00
1913	18.40	0.00	1612.00

Comment 1896-1899 INC.TINCROFT; 1901 INC.TINCROFT; 1903-1913 INC.TINCROFT

Arsenic Pyrite	Ore(tons)	Value(£)
1894	32.00	78.00

Ownership: 1876-1890 CARN BREA MINING CO.; 1891 CARN BREA MINE ADVENTURERS C.B.; 1892-1893 CARN BREA MINING CO. C.B.; 1894-1895 CARN BREA MINING CO.; 1896-1913 CARN BREA & TINCROFT MINING CO.LTD.
Comment 1896-1913 INC.TINCROFT

Management: Manager 1859-1869 JOHN DAWE; 1870-1879 WM.TEAGUE JNR.; 1880-1881 WM.TEAGUE
Chief Agent 1859-1866 EDW.HOSKING; 1867-1869 EDW.HOSKING JOHN JAMES THOS.MINERS; 1870-1871 J.THOMAS EDW.HOSKIN WM.POLKINGHORNE; 1872 J.QUENTRALL J.THOMAS WM.POLKINGHORNE; 1873 J.QUENTRALL WM.POLKINGHORNE J.DADDOW; 1874 J.QUENTRALL F.TRYTHALL WM.TEAGUE JNR; 1875-1879 J.QUENTRALL F.TRYTHALL JOHN DADDOW; 1880-1881 J.QUENTRALL J.DADDOW WM.TEAGUE JNR.; 1882-1888 WM.TEAGUE; 1889-1904 WM.T.WHITE; 1905-1910 JOHN PENHALL; 1911-1913 E.S.KING
Secretary 1859-1862 ROBT.H.PIKE (P); 1863-1864 ROBT.H.PIKE & SON; 1865-1871 ROBT.H.PIKE; 1872-1881 WM.TEAGUE; 1891 JOHN TREVETHAN; 1892-1902 JOHN TREVETHAN (P)

Employment:	Underground	Surface	Total
1878	337	382	719
1879	441	348	789
1880	389	371	760

CARN BREA   ILLOGAN   Continued

	Underground	Surface	Total
1881	356	321	677
1882	366	359	725
1883	338	312	650
1884	226	247	473
1885	230	231	461
1886	335	292	627
1887	394	336	730
1888	388	347	735
1889	437	348	785
1890	431	399	830
1891	475	421	896
1892	451	417	868
1893	432	398	830
1894	407	398	805
1895	333	361	694
1896	466	607	1073
1897	379	463	842
1898	388	482	870
1899	451	411	862
1900	491	433	924
1901	491	439	930
1902	438	427	865
1903	391	381	772
1904	410	401	811
1905	455	427	832
1906	502	428	930
1907	516	464	980
1908	474	460	934
1909	500	478	978
1910	496	488	984
1911	513	542	1055
1912	526	544	1070
1913	495	530	1025

Comment 1896-1913 INC.TINCROFT

CARN BREA,EAST   REDRUTH   SW 697414 0242

Production: Copper	Ore(tons)	Metal(tons)	Value(£)
1859	124.00	7.80	750.50
1860	205.00	16.80	1575.60
1861	802.00	61.20	5691.70
1862	2470.00	181.40	15239.50
1863	2824.00	192.70	15267.20
1864	2953.00	181.40	15103.40
1865	4021.00	230.80	17961.50
1866	2996.00	172.90	11478.20
1867	2509.00	151.20	10363.50
1868	1646.00	80.50	4955.80
1869	890.00	50.90	2949.20
1870	569.00	33.20	1826.30
1871	67.00	4.20	250.90

Comment 1859-1871 (C)

94

CARN BREA,EAST   REDRUTH   Continued

Tin	Black(tons)	Stuff(tons)	Tin(tons)	Value(£)
1857	0.00	0.00	0.00	21.40
1859	0.00	0.00	0.00	4.60
1860	1.40	0.00	0.00	103.90
1864	0.00	0.00	0.00	12.30
1865	2.50	0.00	0.00	123.20
1866	2.10	0.00	0.00	114.60
1867	0.00	0.00	0.00	132.30
1868	0.00	0.00	0.00	57.70
1869	0.00	0.00	0.00	398.70
1870	0.00	0.00	0.00	127.10

Comment 1857 TIN STUFF; 1859 TIN STUFF; 1864 TIN STUFF; 1866
VALUE INC.TIN STUFF; 1867–1870 TIN STUFF
Management: Manager 1859–1865 THOS.GLANVILLE; 1866 J.SCOLLER; 1869 JOHN
RODDA; 1870–1871 THOS.PRYOR
Chief Agent 1859–1865 J.SCOLLER; 1866 THOS.JAMES THOS.MINERS;
1867–1868 IS.RICHARDS JOHN PRYOR; 1869 THOS.PRYOR
Secretary 1859–1861 RICH.LYLE (P); 1862–1863 GEO.LIGHTLEY;
1864–1869 WM.A.BUCKLEY; 1870–1871 THOS.PRYOR

CARN BREA,SOUTH   ILLOGAN   SW 688407 0243

Production: Copper	Ore(tons)	Metal(tons)	Value(£)
1858	232.00	13.00	1140.50
1859	214.00	13.00	1230.60
1862	110.00	6.30	501.70
1863	117.00	6.50	468.30
1864	118.00	6.80	552.30
1865	191.00	10.10	726.50
1866	117.00	6.40	401.50
1870	5.00	0.20	13.60
1871	112.00	13.90	877.10
1872	628.00	87.70	7489.60
1873	492.00	56.50	3789.30
1874	642.00	51.90	3199.60
1875	165.00	12.20	892.80
1876	294.00	29.30	2074.20
1877	188.00	12.50	759.80

Comment 1858–1859 (C); 1862–1866 (C); 1870–1876 (C)

Tin	Black(tons)	Stuff(tons)	Tin(tons)	Value(£)
1856	2.30	0.00	0.00	144.00
1857	1.60	0.00	0.00	108.20
1858	0.00	0.00	0.00	77.30
1859	0.00	0.00	0.00	377.10
1860	15.40	0.00	0.00	1159.70
1861	62.00	0.00	0.00	4118.10
1862	153.20	0.00	0.00	9724.70
1863	106.60	0.00	0.00	6711.50
1864	58.00	0.00	0.00	3431.90
1865	66.10	0.00	0.00	2548.80
1866	33.50	0.00	0.00	1686.40
1867	8.40	0.00	0.00	291.50
1868	8.00	0.00	0.00	439.00
1869	7.70	0.00	0.00	515.20

CARN BREA,SOUTH   ILLOGAN   Continued

Tin	Black(tons)	Stuff(tons)	Tin(tons)	Value(£)
1870	26.20	0.00	0.00	1843.30
1871	16.50	0.00	0.00	1283.60
1872	20.70	0.00	0.00	1794.60
1873	36.50	0.00	0.00	2706.80
1874	119.40	0.00	0.00	6486.50
1875	49.50	0.00	0.00	2583.00
1876	21.10	0.00	0.00	1259.80
1877	15.30	0.00	0.00	626.50

Comment 1858–1859 TIN STUFF; 1862 TWO RETURNS AGGREGATED;
1868 VALUE INC.TIN STUFF

Arsenic	Ore(tons)	Metal(tons)	Value(£)
1877	4.00	0.00	2.90

Ownership: 1876–1877 SOUTH CARN BREA MINING CO.
             Comment 1865 SUSPENDED
Management: Manager 1859–1863 THOS.GLANVILLE; 1864 IS.RICHARDS; 1869 JOHN
             DAWE; 1871–1874 WM.RICH; 1876–1877 WM.RICH
             Chief Agent 1859–1864 GEO.BROWNE; 1868 JOHN NICHOLL; 1869
             JOHN MICHELL JAS.CUSTWELL; 1870 WM.RICH JAS.KNOTWELL;
             1871–1877 JAS.KNOTWELL
             Secretary 1859–1862 RICH.LYLE (P); 1863–1865 GEO.LIGHTLY;
             1868–1877 ROBT.H.PIKE & SON

CARN CAMBORNE   CAMBORNE   SW 656393 0244

Production: Zinc	Ore(tons)	Metal(tons)	Value(£)
1880	2.50	0.00	3.70

Copper	Ore(tons)	Metal(tons)	Value(£)
1862	193.00	11.40	889.10
1863	103.00	4.80	330.00
1864	124.00	7.00	569.80
1865	522.00	24.90	1796.80
1866	1247.00	62.70	3938.40
1867	1076.00	47.00	2980.40
1868	538.00	20.50	1178.40
1869	292.00	11.40	578.60
1870	296.00	12.80	652.20
1871	362.00	19.20	1056.80
1872	283.00	18.20	1284.60
1873	396.00	19.50	1109.10
1874	210.00	9.60	461.40
1875	269.00	13.50	959.60
1876	323.00	16.20	1041.70
1877	33.00	1.40	82.30
1881	37.00	3.20	162.20
1883	17.00	1.30	63.00

Comment 1862–1876 (C)

Tin	Black(tons)	Stuff(tons)	Tin(tons)	Value(£)
1870	38.80	0.00	0.00	726.10
1871	42.40	0.00	0.00	3382.30
1872	25.20	0.00	0.00	2201.50
1875	2.20	0.00	0.00	124.90
1876	0.00	81.70	0.00	99.60
1879	0.00	4.00	0.00	6.50

Tin	Black(tons)	Stuff(tons)	Tin(tons)	Value(£)
1880	0.00	10.50	0.00	9.30
1883	1.00	80.00	0.00	25.00
1886	0.20	30.00	0.00	8.00

Comment 1870 VALUE INC.TINSTUFF, WRONG

Arsenic	Ore(tons)	Metal(tons)	Value(£)
1872	15.00	0.00	12.00
1873	24.00	0.00	19.20
1875	33.30	0.00	99.80

Arsenic Pyrite	Ore(tons)	Value(£)
1879	28.90	14.60
1881	10.00	3.50

Ownership: 1879-1882 CARN CAMBORNE MINING CO.; 1883-1886 CARNE CAMBORNE
             MINING CO.
             Comment 1883-1885 CARNE CAMBORNE; 1886 CARNE CAMBORNE NOT
             WORKED
Management: Manager 1869-1871 JOHN TRUSCOTT; 1877 WM.MARTIN
             Chief Agent 1861 CAPT.BISHOP; 1862-1863 W.H.BISHOP; 1864
             RICH.BENNETTS; 1865-1866 H.T.BENNETTS; 1867-1868 JOHN
             TRUSCOTT H.T.BENNETTS; 1869-1871 H.T.BENNETTS; 1872
             WM.PENBERTHY; 1873-1874 GEO.ROWE; 1875-1876 WM.MARTIN; 1879
             WM.C.VIVIAN; 1880-1881 WM.C.VIVIAN RICH.SOUTHEY; 1882-1886
             WM.C.VIVIAN
             Secretary 1861-1863 BENT.JOHNS (P); 1864 JAS.SECCOMBE (P);
             1865-1868 WM.THORNE; 1869 WM.THOMAS; 1870-1871 JOHN HARDING;
             1872 GEO.ROWE; 1873-1874 R.WATSON; 1875-1877 THOS.PRYOR; 1880
             W.H.PRYOR; 1881 W.H.PRYOR (S)

Employment:	Underground	Surface	Total
1880	12	7	19
1881	14	7	21
1882	18	12	30
1883	15	8	23
1884	13	7	20
1885	7	5	12

CARN GALVER   MORVAH & ZENNOR   SW 421364 0245

Production: Tin	Black(tons)	Stuff(tons)	Tin(tons)	Value(£)
1871	1.60	0.00	0.00	130.10
1872	11.80	0.00	0.00	962.10
1873	30.70	0.00	0.00	2252.70
1874	38.40	0.00	0.00	1795.40
1875	31.20	0.00	0.00	1511.50
1876	8.70	0.00	0.00	398.30
1877	7.70	0.00	0.00	290.00

Ownership: 1875-1877 BOLITHO & SONS AND FIELD
             Comment 1877 SUSPENDED; 1907 SEE BOSIGRAN CONSOLIDATED
Management: Manager 1859 JOHN CARTHEW; 1860 HY.WILLIAMS; 1861-1867
             THOS.WILLIAMS; 1873-1874 JOHN ROACH; 1876-1877 JOHN ROACH
             Chief Agent 1859 HY.WILLIAMS; 1872 JOHN ROACH; 1875
             WM.WILLIAMS JOHN ROACH
             Secretary 1859-1867 JOHN COULSON (P); 1871-1873 THOS.W.FIELD;
             1874 BOLITHO & FIELD; 1875-1877 R.WELLINGTON (S)

CARN PERRAN   PERRANUTHNOE   SW 534296 0246

Production: Copper   Ore(tons) Metal(tons)   Value(£)
            1845   158.00   8.00   518.10
            1846   197.00   10.30   645.20
            1847   106.00   7.80   566.10
            Comment 1845-1847 (C)

CARN VIVIAN   WARLEGGAN   SX 158685 0247

Production: Lead & Silver   Ore(tons) Lead(tons) Silver(ozs)   Value(£)
            1856   15.00   7.00   210.00   0.00
            1857-1860 No detailed return
            1861   1052.00   758.00   7580.00   0.00
            1862   19.00   11.00   138.00   0.00
            Copper No detailed return
            Tin No detailed return
Ownership: Comment 1863-1865 SUSPENDED
Management: Manager 1859-1862 WM.EUSTICE
            Secretary 1859-1860 R.J.MORESHEAD (P); 1861-1865 JAS.POLGLAZE
            (P)

CARNAQUIDDEN   GULVAL   SW 467352 0248

Production: Tin No detailed return
Ownership: 1909-1912 CARNAQUIDDEN MINING CO.; 1913 CARNAQUIDDEN MINING
            CO.LTD.
            Comment 1913 NOT WORKED IN LIQUIDATION
Employment: Underground   Surface   Total
            1909   14   14
            1910-1911   6   6
            1912   100   99   199

CARNE   ST.JUST   SW 383339 0249

Production: Tin   Black(tons) Stuff(tons)   Tin(tons)   Value(£)
            1852   19.00   0.00   0.00   0.00
            1853   9.20   0.00   0.00   562.60
            1855   28.50   0.00   0.00   1820.60
            1856   20.60   0.00   0.00   1418.70
            1857   4.20   0.00   0.00   335.90
            Comment 1852 TIN ORE

CARNESMERRY   ST.AUSTELL   SX 013582 0250

Production: Tin   Black(tons) Stuff(tons)   Tin(tons)   Value(£)
            1883   No detailed return
            Comment 1883 SEE ROCKS
Ownership: 1881 ROCK MINING CO.
Management: Chief Agent 1881 DAVID COCK

CARNEWAS   ST.EVAL   SW 849690 0251

Production: Lead & Silver No detailed return
Copper No detailed return

Iron	Ore(tons)	Iron(%)	Value(£)
1868–1870 No detailed return			
1871	350.00	0.00	262.50
1873	100.00	0.00	75.00
1874	486.00	0.00	388.80

Comment 1868–1870 BH.SEE LANIVET; 1871 BH.; 1873–1874
BH.CARNEWIS
Ownership: Comment 1864–1865 SUSPENDED
Management: Manager 1859–1860 GEO.REYNOLDS; 1863 WM.RUTTER JNR.
Chief Agent 1861–1862 WM.RUTTER JNR.T.DREW; 1863 T.DREW
Secretary 1859–1865 RICH.TREDINNICK (P)

CARNMEAL   SITHNEY   SW 634296 0253

Production: Tin No detailed return
Ownership: Comment 1859–1865 SEE SITHNEY

CARNWINNICK   PROBUS   0254

Ownership: 1912–1913 PITCHBLENDE,URANIUM & RADIUM SYNDICATE
Comment 1912–1913 DRIVING ADIT

Employment:	Underground	Surface	Total
1913	2	2	4

CARNYORTH   ST.JUST   SW 372334 0255

Production: Tin	Black(tons)	Stuff(tons)	Tin(tons)	Value(£)
1853	9.00	0.00	0.00	163.30
1855	80.60	0.00	0.00	5101.90
1856	101.30	0.00	0.00	7483.90
1857	115.20	0.00	0.00	8927.20
1858	137.50	0.00	0.00	8589.10
1859	94.40	0.00	0.00	6398.60
1860	120.00	0.00	0.00	9149.20
1861	99.30	0.00	0.00	6906.50
1862	33.70	0.00	0.00	2174.60
1863	103.60	0.00	0.00	6572.10
1864	70.30	0.00	0.00	4714.80
1865	76.00	0.00	0.00	4905.10
1866	40.80	0.00	0.00	1952.50
1872–1875 No detailed return				

Comment 1862 PART YEAR ONLY; 1872–1875 SEE BOTALLACK

Arsenic	Ore(tons)	Metal(tons)	Value(£)
1875	No detailed return		

Comment 1875 SEE BOTALLACK
Ownership: Comment 1865 SUSPENDED; 1866–1867 WORKED BY THE BOTALLACK
ADVENTURERS; 1868–1878 SEE BOTALLACK
Management: Manager 1859–1861 JOHN CARTHEW
Chief Agent 1859 JOHN WALLIS; 1860–1861 JOHN WALLIS
WM.TREMBATH; 1862–1864 JOHN WALLIS

CM-F*   99

Secretary 1859–1861 RICH.PEARCE (P); 1862–1865 SML.YORK

CAROLINE   PERRANUTHNOE   SW 543305 0256

Production: Tin No detailed return
Ownership: 1881 CAROLINE MINING CO.
            Comment 1860–1865 SUSPENDED; 1874–1876 NEW CAROLINE; 1881 OLD
            CAROLINE; 1907–1908 SEE VERRANT; 1909–1913 SEE GOLDSITHNEY
Management: Manager 1859 AB.BENNETT; 1874–1876 ABS.BENNETT; 1881
            R.POWNING
            Chief Agent 1859–1861 FRAN.GUNDRY
            Secretary 1859–1865 AB.BENNETT; 1874–1876 ABS.BENNETT
Employment: Underground   Surface   Total
            1907
            Comment 1907 SEE VERRANT

CARPENTER   0258

Production: Copper   Ore(tons) Metal(tons)   Value(£)
            1853   72.00   7.80   781.50
            1854   205.00   19.40   2047.30
            1855   251.00   23.90   2548.40
            Comment 1853–1855 (C). MAY BE CARPENTER, DEVON

CARPENTER   GWINEAR   SW 586353 0259

Production: Copper   Ore(tons) Metal(tons)   Value(£)
            1854   488.00   16.60   1444.40
            1855   212.00   8.30   745.00
            Comment 1854–1855 (C)

CARPUAN   ST.NEOT   SX 205690 0260

Production: Tin No detailed return
Ownership: 1889–1891 R.CUMMING; 1913 SOUTH MOUNT BOPPY GOLD CO.LTD.
            Comment 1890–1891 SUSPENDED; 1913 PROSPECTING INC.BOWDEN
Management: Manager 1913 W.A.PASCOE
            Chief Agent 1889–1891 R.KNOTT
Employment: Underground   Surface   Total
            1889   3   3
            1890   2   2
            1913   18   4   22
            Comment 1913 INC.BOWDEN

CARRACK DEWS   ST.IVES   SW 507413 0261

Production: Copper   Ore(tons) Metal(tons)   Value(£)
            1856   99.00   7.00   659.30
            1857   398.00   26.30   2660.80
            1858   348.00   22.50   2089.00

Copper	Ore(tons)	Metal(tons)	Value(£)
1859	253.00	15.40	1501.10

Comment 1856-1859 (C)

Tin	Black(tons)	Stuff(tons)	Tin(tons)	Value(£)
1859	0.00	0.00	0.00	33.20
1860	4.70	0.00	0.00	377.40

Comment 1859 TIN STUFF
Ownership: Comment 1861-1865 ABANDONED
Management: Manager 1859-1860 WM.HOLLOW
             Chief Agent 1859 MART.DUNN; 1860 JAS.BERRYMAN
             Secretary 1859 EALEY (P); 1860-1861 JAS.HOLLOW (P)

CARROCK,NORTH   0263

Production: Lead	Ore(tons)	Metal(tons)	Value(£)
1862	45.50	27.20	0.00

CARTHEW CONSOLS   ST.ISSEY   SW 955717 0264

Production: Lead & Silver	Ore(tons)	Lead(tons)	Silver(ozs)	Value(£)
1849	45.00	31.00	0.00	0.00
1850	44.00	30.50	0.00	0.00
1851	33.00	23.20	0.00	0.00
1852	16.50	11.50	120.00	0.00

1853-1854 No detailed return

Copper	Ore(tons)	Metal(tons)	Value(£)
1850	92.00	6.50	379.50
1851	113.00	7.10	443.60

Comment 1850-1851 (C)

CARVATH UNITED   ST.AUSTELL   SX 029518 0265

Production: Copper No detailed return

Tin	Black(tons)	Stuff(tons)	Tin(tons)	Value(£)
1858	71.30	0.00	0.00	4256.70
1859	15.40	0.00	0.00	953.70
1860	24.50	0.00	0.00	1758.70

Ownership: Comment 1859-1863 OR WEST POLMEAR; 1864-1865 OR WEST POLMEAR.
             SUSPENDED
Management: Manager 1859 RICH.HANCOCK; 1860-1863 WM.BODY
             Chief Agent 1863 WM.POLKINGHORNE
             Secretary 1859 WM.CHARLES (P); 1860-1862 W.HIGMAN (P);
             1863-1865 C.E.TREFFRY

CARVENNAL   GWENNAP   SW 712391 0266

Production: Copper	Ore(tons)	Metal(tons)	Value(£)
1851	102.00	6.50	416.90
1852	214.00	18.60	1512.80
1853	189.00	14.70	1478.20
1854	410.00	34.60	3587.20

CARVENNAL   GWENNAP   Continued

          Copper   Ore(tons) Metal(tons)   Value(£)
          1855   474.00   37.80   3867.90
          1856   393.00   27.00   2493.00
          1857   346.00   23.20   2193.80
          1858   343.00   19.20   1690.20
          1859   247.00   16.10   1451.10
          Comment 1851 (C)CARVANNAL; 1852-1853 (C)CARVANNALL; 1854-1859
          (C)CARVANNAL
          Tin   Black(tons) Stuff(tons)   Tin(tons)   Value(£)
          1860   0.00   0.00   0.00   131.00
          1861   0.00   0.00   0.00   91.10
          Comment 1860-1861 CARVENNEL TIN STUFF
Ownership: Comment 1860-1865 SUSPENDED
Management: Manager 1859 WM.ROBERTS
          Secretary 1859 RICH.LYLE (P)

CARZISE   CROWAN   SW 595342 0267

Production: Copper No detailed return
          Tin   Black(tons) Stuff(tons)   Tin(tons)   Value(£)
          1872   0.00   0.00   0.00   70.80
          1873   0.00   44.30   0.00   162.20
          Comment 1872 TIN STUFF
Ownership: Comment 1873 SUSPENDED
Management: Manager 1872-1873 CHAS.CARKEEK
          Chief Agent 1872-1873 JOHN TREDINNICK
          Secretary 1871-1873 W.PAGE CARDOZA
Employment: Underground   Surface   Total
          1888   20   14   34
          1889   8   9   17
          1890   6   3   9

CARZISE,WEST   CROWAN   SW 588341 0268

Production: Tin   Black(tons) Stuff(tons)   Tin(tons)   Value(£)
          1889   2.00   29.00   0.00   173.00
          1890   0.00   16.70   0.00   43.00
Ownership: 1887-1890 WEST CARZISE MINE CO.
          Comment 1890 NOW ABANDONED
Management: Chief Agent 1887-1890 JOHN CURTIS

CASTLE   ST.JUST   SW 355326 0269

Production: Tin   Black(tons) Stuff(tons)   Tin(tons)   Value(£)
          1853   No detailed return
          1856-1857 No detailed return
          1861-1872 No detailed return
          1874   No detailed return
          1885   0.70   0.00   0.00   32.00
          Comment 1853 SEE BOSWEDDEN; 1856-1857 SEE BOSWEDDEN;
          1861-1872 SEE BOSWEDDEN; 1874 SEE CUNNING UNITED

```
        Iron            Ore(tons)      Iron(%)    Value(£)
        1886              75.00         43.00       25.00
        Comment 1886 BH.
Ownership:  1883 WHEAL CASTLE MINING CO.; 1884-1886 WHEAL CASTLE CO.
        Comment 1861-1871 SEE BOSWEDDEN; 1872-1876 SEE CUNNING
        UNITED
Management: Chief Agent 1883-1886 JOHN BOYNS
Employment:              Underground     Surface      Total
        1883                15            21           36
        1884                22             5           27
        1885                18             9           27
        1886                15             9           24
```

CASTLE AN DINAS   ST.COLUMB MAJOR   SW 950616 0270

```
Production: Tin           Black(tons) Stuff(tons)  Tin(tons)   Value(£)
        1852              0.40         0.00         0.00         0.00
        1873              6.00         0.00         0.00       495.40
        Comment 1852 CASTLE DINAS
Management: Manager 1870-1871 H.REYNOLDS; 1873 CHAS.G.KESSELL
        Chief Agent 1870-1871 D.TONKIN; 1872 D.TONKIN JOHN PARKYN;
        1874 CHAS.G.KESSELL
        Secretary 1870-1871 WM.WARD; 1872-1873 S.SYMONS
```

CATHEDRAL   GWENNAP   SW 718416 0272

```
Production: Copper          Ore(tons) Metal(tons)    Value(£)
        1874               89.00       8.20         606.00
        1875              312.00      21.60        1618.00
        1876              122.00       5.80         379.10
        1877               52.00       2.90         171.60
        1880                7.00       0.90          28.70
        1883                7.30       0.20          14.00
        Comment 1874-1876 (C); 1880 CATHEDRAL CONSOLS; 1883 CATHEDRAL
        CONSOLS
        Tin           Black(tons) Stuff(tons)  Tin(tons)   Value(£)
        1873              0.00         2.20         0.00        12.40
        1874              0.00         2.20         0.00         9.40
        1883              0.60         0.00         0.00        31.00
        1884              1.10         0.00         0.00        49.00
        Comment 1883-1884 CATHEDRAL CONSOLS
Ownership:  1878 NEW CATHEDRAL MINING CO.LTD.; 1879-1880 NEW CATHEDRAL
        COPPER & TIN CO.LTD.; 1881-1885 CATHEDRAL CONSOLS MINING CO.
        Comment 1878-1880 NEW CATHEDRAL; 1881-1885 CATHEDRAL CONSOLS
Management: Manager 1870-1878 JOS.MICHELL
        Chief Agent 1870-1871 THOS.WILLIAMS; 1872-1873 STEP.DAVEY;
        1874-1875 MAT.GREEN STEP.DAVEY; 1876-1877 MAT.GREEN;
        1878-1880 STEP.DAVEY; 1881-1882 STEP.DAVEY JNR.; 1883-1885
        STEP.DAVEY
        Secretary 1870-1873 MAT.GREEN; 1874-1877 GEO.CORFIELD; 1881
        ED.ASHMEAD (S)
Employment:              Underground     Surface      Total
        1878                 2             3            5
```

	Underground	Surface	Total
1879	8	7	15
1880	24	8	32
1881	19	9	28
1882	18	9	27
1883	23	13	36
1884	21	7	28
1885	16	9	25

CHARITY LANDS   0273

Production: Lead & Silver

	Ore(tons)	Lead(tons)	Silver(ozs)	Value(£)
1851	38.70	25.00	0.00	0.00
1852	No detailed return			
1853	28.40	19.00	380.00	0.00
1854	No detailed return			

Comment 1853 TRURO CHARITY LANDS

CHARLESTOWN UNITED   ST.AUSTELL   SX 040531 0274

Production: Copper

	Ore(tons)	Metal(tons)	Value(£)
1847	198.00	47.20	3615.50
1848	208.00	26.30	1528.40
1849	255.00	27.00	1840.10
1850	292.00	23.20	1537.20

Comment 1847-1848 (C)CHARLES TOWN UNITED; 1849-1850 (C)

Tin

	Black(tons)	Stuff(tons)	Tin(tons)	Value(£)
1852	80.80	0.00	0.00	0.00
1853	55.00	0.00	0.00	3423.80
1854	2.00	0.00	0.00	149.60
1855	4.60	0.00	0.00	272.40
1856	26.40	0.00	0.00	1573.80
1857	15.30	0.00	0.00	934.00
1858	1.30	0.00	0.00	64.10
1860	69.50	0.00	0.00	5352.70
1861	296.40	0.00	0.00	20970.80
1862	517.50	0.00	0.00	33121.90
1863	480.10	0.00	0.00	31392.50
1864	382.10	0.00	0.00	24113.70
1865	429.30	0.00	0.00	23169.40
1866	509.10	0.00	0.00	24450.10
1867	247.20	0.00	0.00	12625.80
1868	308.40	0.00	0.00	16857.00
1869	262.50	0.00	0.00	18224.40
1870	204.40	0.00	0.00	14752.80
1871	213.70	0.00	0.00	17847.30
1872	158.70	0.00	0.00	13719.70
1873	202.50	0.00	0.00	15714.40
1874	120.70	0.00	0.00	6887.20
1875	16.80	0.00	0.00	940.40

Comment 1852 TIN ORE; 1854 CHARLESTON; 1855-1856 CHARLESTOWN
MINES; 1858 CHARLESTOWN MINES; 1862 TWO RETURNS AGGREGATED
Ownership: Comment 1872-1874 INC.CUDDRA; 1875 SUSPENDED SEPT.1874

CHARLESTOWN UNITED   ST.AUSTELL   Continued

Management: Manager 1859-1874 RICH.H.WILLIAMS
             Chief Agent 1859-1864 SML.NORTHEY CHAS.SIMMONS; 1865-1874
             SML.NORTHEY CHAS.FAULE
             Secretary 1859-1862 JOHN WILLIAMS; 1863 JOHN H.WILLIAMS;
             1864-1868 RICH.H.WILLIAMS; 1869-1874 C.W.BROADHURST

CHARLESTOWN,NEW   CHARLESTOWN   SX 040531 0275

Ownership: Comment 1865 SUSPENDED
Management: Manager 1859 KEMP
            Secretary 1859 J.B.STEPHENS (P); 1865 J.B.STEPHENS (P)

CHARLOTTE   ST.AGNES   SW 697492 0276

Production: Copper   Ore(tons) Metal(tons)   Value(£)
            1854   570.00   46.30   4817.70
            1855   894.00   68.20   7085.50
            1856   658.00   44.80   4274.40
            1857   436.00   33.90   3376.90
            1858   855.00   67.20   6267.60
            1859   1323.00   94.00   8790.20
            1860   1189.00   84.80   7857.20
            1861   741.00   61.40   5599.10
            1862   644.00   48.00   3946.40
            1863   268.00   19.90   1590.40
            Comment 1854-1863 (C)
            Tin   Black(tons) Stuff(tons)   Tin(tons)   Value(£)
            1860   0.00   0.00   0.00   214.60
            1861   0.00   0.00   0.00   159.70
            Comment 1860-1861 TIN STUFF
Management: Manager 1859-1862 RICH.KENDALL; 1869-1872 JOHN TONKIN
            Secretary 1859 JAS.HOLLOW (P); 1860-1862 ED.KING (P); 1869
            JOHN HITCHINGS; 1870-1872 JOHN.HITCHINGS

CHARLOTTE UNITED   ST.AGNES   SW 703488 0277

Production: Copper   Ore(tons) Metal(tons)   Value(£)
            1864   708.00   48.00   4023.20
            1865   63.00   3.10   227.20
            Comment 1864-1865 (C)
            Tin   Black(tons) Stuff(tons)   Tin(tons)   Value(£)
            1861   0.00   0.00   0.00   125.90
            1862   0.00   0.00   0.00   38.50
            1863   0.00   0.00   0.00   17.90
            1864   0.00   0.00   0.00   130.60
            1865   0.00   0.00   0.00   9.30
            Comment 1861-1865 TIN STUFF
Ownership: 1876-1877 CHARLOTTE UNITED MINING CO.
            Comment 1865 SUSPENDED
Management: Manager 1860-1862 RICH.KENDALL; 1863-1864 PETER FLOYD
            Chief Agent 1860-1861 JAS.PENBERTHY; 1862-1864 RICH.JOHNS;
            1876-1877 SML.BENNETTS

                                       105

CHARLOTTE UNITED   ST.AGNES   Continued

        Secretary 1860–1861 ED.KING (P); 1862 JOHN HOSKING; 1863–1865
        F.P.TYACKE

CHARLOTTE,EAST   ST.AGNES   SW 705492 0278

Production: Copper   Ore(tons) Metal(tons)   Value(£)
            1862   14.00   0.40   29.00
            Comment 1862 (C)
            Tin   Black(tons) Stuff(tons)   Tin(tons)   Value(£)
            1908   8.80   0.00   0.00   780.00
            1909   5.20   0.00   0.00   433.00
Ownership: 1907–1913 EAST WHEAL CHARLOTTE SYNDICATE
            Comment 1911–1912 SUSPENDED; 1913 NOT WORKED
Management: Chief Agent 1908–1910 WM.WEARNE; 1911–1913 J.W.KEAST
            Secretary 1908–1913 W.MIDDLIN
Employment: Underground   Surface   Total
            1907   11   12   23
            1908   23   7   30
            1909   15   6   21
            1910   10   1   11

CHARLOTTE,NEW   ST.AGNES   SW 703488 0279

Production: Tin   Black(tons) Stuff(tons)   Tin(tons)   Value(£)
            1870   2.30   0.00   0.00   176.00
            1871   2.50   0.00   0.00   187.20
            1872   0.30   0.00   0.00   25.10
            1873   0.50   0.00   0.00   39.00
            Comment 1873 VALUE EST
Ownership: Comment 1873 STOPPED IN 1873 & 1874
Management: Manager 1872–1873 JOHN TONKIN
            Chief Agent 1872–1873 JOHN WATERS

CHARLOTTE,SOUTH   ST.AGNES   0280

Production: Tin   Black(tons) Stuff(tons)   Tin(tons)   Value(£)
            1852   11.00   0.00   0.00   0.00
            1855   6.60   0.00   0.00   443.20
            Comment 1852 TIN ORE

CHIVERTON   PERRANZABULOE   SW 797512 0281

Production: Lead & Silver   Ore(tons) Lead(tons) Silver(ozs)   Value(£)
            1866   55.40   42.00   1533.00   0.00
            1867   294.80   220.80   8030.00   0.00
            1868   118.50   88.80   3212.00   0.00
            1869   327.00   245.20   8942.00   0.00
            1870   28.00   21.00   0.00   0.00
            1871   7.20   5.40   0.00   0.00
            1872   15.00   11.20   401.00   0.00

CHIVERTON   PERRANZABULOE   Continued

           Tin   Black(tons) Stuff(tons)   Tin(tons)   Value(£)
           1853   7.00   0.00   0.00   443.50
Ownership: Comment 1863-1871 THE OLD CORNUBIAN
Management: Manager 1863 J.JULIFF; 1864 J.JULIFF JNR.; 1865-1868
            J.JULIFF; 1869-1871 GEO.TREMAYNE
            Chief Agent 1864-1871 JOHN BORLASE
            Secretary 1863 ED.BURGESS J.JULIFF; 1864-1868 RICH.CLOGG;
            1869-1871 R.MICHELL

CHIVERTON CENTRAL   PERRANZABULOE   0282

Production: Lead & Silver No detailed return
Management: Chief Agent 1864-1865 H.GREY(SURVEYOR)

CHIVERTON CONSOLS   PERRANZABULOE   0283

Production: Lead & Silver No detailed return
Management: Manager 1864-1865 J.OATES

CHIVERTON CONSOLS,GREAT   ST.ALLEN   SW 820512 0284

Production: Lead No detailed return
Management: Manager 1869-1871 CYRUS SYMONS
            Secretary 1869-1871 JAS.EVANS

CHIVERTON CONSOLS,NEW   PERRANZABULOE   0285

Production: Lead No detailed return
            Zinc   Ore(tons) Metal(tons)   Value(£)
            1870   8.00   0.00   27.60
            Tin No detailed return
Ownership: Comment 1867-1872 FORMERLY BUDNICK CONSOLS
Management: Manager 1867-1871 JAS.EVANS; 1872 JOHN RAWLINGS
            Secretary 1867-1868 THOS.WOODWARD; 1869-1872 J.PEARCE

CHIVERTON MOOR   PERRANZABULOE   SW 785504 0286

Production: Lead & Silver   Ore(tons) Lead(tons) Silver(ozs)   Value(£)
            1866   45.10   33.80   507.00   0.00
            1867   262.80   197.10   2955.00   0.00
            1868   285.50   214.20   3210.00   0.00
            1869   360.20   270.00   4050.00   0.00
            1870   508.90   381.00   5715.00   0.00
            1871   330.50   247.70   3706.00   0.00
            1872   263.00   195.50   2925.00   0.00
            1873   71.10   53.30   798.00   1344.40
Management: Manager 1865-1867 J.JULIFF; 1868-1871 GEO.E.TREMAYNE
            Chief Agent 1864 J.JULIFF; 1865-1871 WM.BENNETTS
            Secretary 1864-1867 RICH.CLOGG; 1868-1871 PRYOR (P) WARD (S)

                                  107

CHIVERTON POND MOOR   PERRANZABULOE   0287

Production: Lead & Silver No detailed return
Ownership: Comment 1867 SUSPENDED
Management: Chief Agent 1864-1866 J.NANKIVELL
            Secretary 1864-1867 WM.HOSKING

CHIVERTON ST.GEORGE   PERRANZABULOE   0288

Production: Lead & Silver No detailed return
Management: Chief Agent 1881 R.NANCARROW
            Secretary 1881 J.H.DINGLE

CHIVERTON UNITED   PERRANZABULOE   SW 772493 0289

Production: Lead No detailed return
Management: Manager 1869-1871 GEO.E.TREMAYNE
            Secretary 1869-1871 P.PRYN

CHIVERTON VALLEY   PERRANZABULOE   SW 782501 0290

Production: Lead & Silver

	Ore(tons)	Lead(tons)	Silver(ozs)	Value(£)
1870	38.60	29.90	0.00	0.00
1871	46.70	34.90	170.00	0.00

Management: Manager 1865-1871 J.JULIFF
            Chief Agent 1864 J.JULIFF JNR.; 1867-1871 JAS.TREVELYAN
            Secretary 1865-1871 RICH.CLOGG

CHIVERTON,EAST   PERRANZABULOE   SW 795510 0291

Production: Lead & Silver

	Ore(tons)	Lead(tons)	Silver(ozs)	Value(£)
1870	1.80	1.30	0.00	0.00
1871	8.50	6.30	31.00	0.00
1872	7.00	5.50	28.00	0.00
1873	13.30	10.00	50.00	271.90
1874	6.10	4.10	36.00	87.80
1875	5.80	4.00	20.00	0.00
1876	1.80	1.20	5.00	0.00
1877	10.00	7.00	28.00	160.70
1878	8.80	6.50	24.00	90.00
1879	8.60	6.50	24.00	89.00
1880	10.00	15.00	60.00	135.50
1881	105.00	77.00	420.00	1000.00
1882	100.00	75.00	2250.00	1260.00
1883	50.00	37.50	1100.00	528.00

Zinc	Ore(tons)	Metal(tons)	Value(£)
1881	105.00	47.30	341.30

Ownership: 1877-1883 EAST CHIVERTON MINING CO.
Management: Manager 1862-1863 J.JULIFF; 1864 JAS.JULIFF; 1866
            JAS.NANCARROW; 1869-1871 JOHN GROSE; 1872-1881 RICH.SOUTHEY
            Chief Agent 1859-1861 J.ROGERS; 1862-1863 JOHN NANCARROW;
            1864-1865 JAS.NANCARROW; 1866 JOHN GROSE; 1867-1868
            RICH.SOUTHEY JOHN GROSE; 1869-1871 RICH.SOUTHEY; 1882-1883

RICH.SOUTHEY
Secretary 1862-1864 J.JULIFF JNR.; 1866 J.WOODWARD; 1867-1872
WM.BARTLETT; 1873-1881 THOS.WOODWARD

Employment:	Underground	Surface	Total
1878	9	7	16
1879	12	5	17
1880	23	5	28
1881	18	16	34
1882	18	14	32
1883	18	12	30

CHIVERTON,GREAT SOUTH   PERRANZABULOE   SW 792501 0292

Production: Lead & Silver	Ore(tons)	Lead(tons)	Silver(ozs)	Value(£)
1870	23.10	17.20	0.00	0.00
1871	No detailed return			
1872	23.60	16.50	0.00	0.00
1873	22.10	16.50	99.00	375.10

Zinc	Ore(tons)	Metal(tons)	Value(£)
1870	4.50	0.00	18.00
1872	2.60	0.00	5.00

Arsenic	Ore(tons)	Metal(tons)	Value(£)
1873	22.20	0.00	44.00

Ownership: Comment 1873 STOPPED IN 1873
Management: Manager 1868-1873 JOHN NANCARROW
            Chief Agent 1864-1867 JOHN NANCARROW JOHN GEORGE
            Secretary 1864-1873 H.CHAPMAN

CHIVERTON,GREAT WEST   PERRANZABULOE   SW 747496 0293

Production: Lead No detailed return
Ownership: 1882-1884 GREAT WEST CHIVERTON MINING CO.
            Comment 1869 GREAT WHEAL CHIVERTON; 1884 NOT WORKED
Management: Manager 1869-1871 S.NANCARROW
            Chief Agent 1882-1884 JOHN CURTIS
            Secretary 1869-1871 DAVID STRICKLAND

Employment:	Underground	Surface	Total
1881	4		4
1882-1883	6		6

CHIVERTON,NEW   PERRANZABULOE   SW 776529 0294

Production: Lead & Silver	Ore(tons)	Lead(tons)	Silver(ozs)	Value(£)
1864	6.00	3.90	70.00	0.00
1865	10.20	7.50	130.00	0.00
1866	2.90	2.20	0.00	0.00
1867	13.10	9.20	0.00	0.00
1868	5.60	4.20	80.00	0.00
1869	5.00	3.70	0.00	0.00
1870	6.10	4.50	22.00	0.00
1872	30.40	22.80	112.00	0.00
1873	24.10	18.00	90.00	321.90

CHIVERTON,NEW   PERRANZABULOE   Continued

Lead & Silver   Ore(tons) Lead(tons) Silver(ozs)   Value(£)
1874   81.00   60.70   302.00   1012.80
1875   115.90   86.90   516.00   1251.70
1876   92.10   70.00   420.00   920.00
1877   17.00   12.70   65.00   125.00
1878   1.10   0.80   4.00   12.00
Comment 1869 FOR AG.SEE GREAT RETALLACK
Zinc   Ore(tons) Metal(tons)   Value(£)
1864   18.90   0.00   41.20
1865   61.70   0.00   72.20
1866   34.10   0.00   103.20
1867   264.80   0.00   764.80
1868   234.20   0.00   691.10
1872   26.40   0.00   19.90
Arsenic   Ore(tons) Metal(tons)   Value(£)
1873   24.20   0.00   316.80

Ownership: 1877-1878 NEW CHIVERTON MINING CO.LTD.
             Comment 1863-1877 THE OLD WEST ANNA; 1878-1879 THE OLD WEST
             ANNA IN LIQUIDATION
Management: Manager 1866 J.JULIFF; 1867-1876 JOHN TREWARTHA; 1877-1879
             JAS.TREWARTHA
             Chief Agent 1863 J.JULIFF; 1864-1865 J.JULIFF JOHN TREWARTHA;
             1866 JOHN TREWARTHA; 1869-1871 WM.TREGAY
             Secretary 1863-1864 W.PAGE CARDOZA; 1865-1873 G.CARDOZA;
             1874-1877 W.PAGE CARDOZA
Employment: Underground   Surface   Total
             1878   3   3

CHIVERTON,NORTH   PERRANZABULOE   SW 788530 0295

Production: Lead & Silver   Ore(tons) Lead(tons) Silver(ozs)   Value(£)
            1864   5.50   3.60   65.00   0.00
            1865   6.00   3.80   75.00   0.00
            1866   2.40   1.80   0.00   0.00
            1867   10.70   7.70   250.00   0.00
            1869   0.70   0.40   0.00   0.00
            1870   No detailed return
            Comment 1869 FOR AG.SEE GREAT RETALLACK
            Zinc   Ore(tons) Metal(tons)   Value(£)
            1864   97.30   0.00   441.20
            1865   332.90   0.00   1009.30
            1866   133.30   0.00   440.60
            1867   56.70   0.00   120.70
            1868   37.40   0.00   75.90
            1869   28.50   0.00   27.00
            Copper   Ore(tons) Metal(tons)   Value(£)
            1867   3.60   0.30   23.70
            Comment 1867 (P)
Ownership: 1875 NORTH CHIVERTON CO.
            Comment 1863-1868 THE OLD ANNA; 1880 SEE SHEPHERDS WHEAL ROSE
            UNITED
Management: Manager 1874-1875 JOHN COOK
            Chief Agent 1863 CAPT.HAMPTON; 1864 WM.HAMPTON; 1865-1866
            WM.HANCOCK; 1867-1868 WM.HANCOCK W.T.BRYANT

CHIVERTON,NORTH   PERRANZABULOE   Continued

        Secretary 1864—1866 WM.WATSON; 1867—1868 G.T.NOAKES

CHIVERTON,SOUTH   PERRANZABULOE   SW 800492 0296

Production: Lead & Silver   Ore(tons) Lead(tons) Silver(ozs)   Value(£)
            1866   77.00   55.40   3245.00   0.00
            1867   52.20   39.10   2301.00   0.00
            1868   27.00   20.20   1180.00   0.00
            1869   7.00   5.20   309.00   0.00
Management: Manager 1865—1867 ROBT.NANCARROW
            Chief Agent 1864 JAS.NANCARROW ROBT.NANCARROW; 1865—1867
            ROBT.R.NANCARROW
            Secretary 1864—1867 RICH.CLOGG

CHIVERTON,WEST   PERRANZABULOE   SW 793508 0297

Production: Lead & Silver   Ore(tons) Lead(tons) Silver(ozs)   Value(£)
            1859   162.00   143.00   3729.00   0.00
            1860   217.20   162.00   4230.00   0.00
            1861   706.00   450.00   13700.00   0.00
            1862   914.50   585.00   21035.00   0.00
            1863   1181.60   779.50   27269.00   0.00
            1864   1533.30   937.00   39900.00   0.00
            1865   2809.50   1797.70   73670.00   0.00
            1866   3166.00   1970.00   66950.00   0.00
            1867   4082.20   3061.60 130624.00   0.00
            1868   4516.10   3387.00 144512.00   0.00
            1869   4707.80   3529.00 150504.00   0.00
            1870   4777.50   3582.70 161190.00   0.00
            1871   3983.70   2987.70 127444.00   0.00
            1872   2874.00   2155.70   90510.00   0.00
            1873   2224.00   1668.30   70056.00   29929.80
            1874   1557.90   1167.50   24574.00   19296.40
            1875   1613.70   1209.70   18135.00   0.00
            1876   1594.80   1197.00   29925.00   20683.80
            1877   1296.60   1007.50   15700.00   17504.10
            1878   895.00   671.00   10842.00   10740.00
            1879   425.80   319.50   5170.00   3834.00
            1880   238.50   179.00   2685.00   2261.00
            1881   177.20   87.70   1747.00   2120.00
            1882   20.00   15.00   300.00   240.00
            1883   36.00   14.40   720.00   213.00
            1884   54.00   27.00   0.00   249.00
            1885   31.00   0.00   0.00   155.00
            1886   3.00   0.00   0.00   0.00
            Comment 1860 AG.INC.TREGARDOCK & WEST CRINNIS
            Zinc   Ore(tons) Metal(tons)   Value(£)
            1860   187.70   0.00   253.30
            1863   206.50   0.00   242.10
            1864   232.30   0.00   598.50
            1865   117.60   0.00   142.40
            1866   224.40   0.00   324.40
            1867   550.60   0.00   827.90

                                     111

Zinc	Ore(tons)	Metal(tons)	Value(£)
1868	465.60	0.00	617.40
1869	181.30	0.00	320.10
1870	318.80	0.00	782.20
1871	529.60	0.00	858.60
1872	807.80	0.00	2161.00
1873	1002.50	0.00	2966.80
1874	509.60	0.00	1244.70
1875	2078.70	0.00	7271.50
1876	3004.10	0.00	10514.00
1877	3660.40	0.00	10980.00
1878	3284.00	0.00	9852.00
1879	2781.00	0.00	9733.50
1880	1568.10	0.00	5488.00
1881	523.30	235.50	1726.90
1882	88.00	44.00	264.00
1883	189.00	66.10	274.00
1884	164.00	0.00	204.00
1885	68.00	0.00	59.00
1886	39.00	0.00	39.00

Copper	Ore(tons)	Metal(tons)	Value(£)
1876	15.00	0.70	40.50

Comment 1876 (C)

Ownership: 1877–1883 WEST CHIVERTON MINING CO.; 1884–1885 RICH.SOUTHEY
Management: Manager 1859–1873 JAS.JULIFF; 1874–1881 RICH.SOUTHEY
Chief Agent 1859–1864 RICH.NANCARROW; 1865–1879
RICH.NANCARROW WM.NICHOLLS; 1880 WM.NICHOLLS; 1881 JAS.MOYCE;
1882–1885 RICH.SOUTHEY
Secretary 1859–1861 JAS.JULIFF (P); 1862–1873 RICH.CLOGG (P);
1874–1878 T.R.WOODWARD; 1879–1881 GREN.SHARP (S)

Employment:	Underground	Surface	Total
1878	152	166	318
1879	121	137	258
1880	24	47	71
1881	19	60	79
1882		6	6
1883		17	17
1884		12	12
1885		7	7

CHRISTOPHER CONSOLS   SITHNEY   SW 655324 0298

Production: Tin	Black(tons)	Stuff(tons)	Tin(tons)	Value(£)
1857	58.90	0.00	0.00	3791.00
1858	2.20	0.00	0.00	165.90

Comment 1857 CHRISTOPHER
Ownership: Comment 1863–1865 SUSPENDED
Management: Manager 1861–1862 J.WEBB
Chief Agent 1860–1862 WM.TEAGUE
Secretary 1859–1860 R.DALTON (P); 1861–1865 W.H.MORTIMER (P)

CHYPRASE CONSOLS   ST.ENODER   SW 905564 0300

Production: Tin   Black(tons) Stuff(tons)   Tin(tons)   Value(£)
            1852   21.50   0.00   0.00   0.00
            1853   3.30   0.00   0.00   205.70
            Comment 1852 TIN ORE

CHYTANE   ST.ENODER   SW 917558 0301

Production: Tin   Black(tons) Stuff(tons)   Tin(tons)   Value(£)
            1873   4.40   0.00   0.00   328.30
Ownership: Comment 1873 STOPPED IN 1873 & EARLY IN 1874
Management: Manager 1872-1873 JOHN EDWARDS
            Chief Agent 1872-1873 JOHN TONKIN
            Secretary 1872-1873 M.E.JOBLING

CLARENCE   PORTH TOWAN   SW 692473 0303

Production: Copper   Ore(tons) Metal(tons)   Value(£)
            1865   14.00   0.60   38.10
            1866   18.00   0.80   48.00
            Comment 1865-1866 (C)
Ownership: 1898 CAPT.HAMPTON

CLIFFORD   GWENNAP   SW 756417 0304

Production: Copper   Ore(tons) Metal(tons)   Value(£)
            1845   348.00   29.50   2254.00
            1846   345.00   24.40   1790.80
            1847   409.00   27.10   1922.20
            1848   305.00   17.60   949.90
            1851   174.00   14.10   964.20
            1852   618.00   48.00   4294.30
            1853   650.00   59.10   6237.50
            1854   566.00   46.10   4844.30
            1855   1651.00   124.50   12938.10
            1856   3583.00   258.40   24810.50
            1857   3234.00   193.80   19660.80
            1858   4836.00   315.20   28254.30
            1859   7450.00   529.70   49869.80
            1860   8466.00   540.40   49118.60
            1861   6301.00   442.50   39227.90
            Comment 1845-1848 (C); 1851-1861 (C)
            Tin   Black(tons) Stuff(tons)   Tin(tons)   Value(£)
            1856   0.40   0.00   0.00   28.90
            1858   3.20   0.00   0.00   159.70
            1859   15.10   0.00   0.00   923.30
            1860   15.70   0.00   0.00   1055.20
            1861   15.50   0.00   0.00   987.60
Ownership: Comment 1862-1868 SEE CLIFFORD AMALGAMATED
Management: Manager 1859-1860 JOHN RICHARDS
            Chief Agent 1859-1860 LEAN
            Secretary 1859-1860 WM.WILLIAMS & SON (P)

113

CLIFFORD AMALGAMATED GWENNAP SW 756417 0305

Production:

Zinc	Ore(tons)	Metal(tons)	Value(£)	
1862	12.60	0.00	12.60	
1865	12.00	0.00	24.10	
1870	12.00	0.00	41.40	
Copper	Ore(tons)	Metal(tons)	Value(£)	
1861	3315.00	205.80	18930.60	
1862	14322.00	855.80	68475.60	
1863	14273.00	896.80	68667.20	
1864	15180.00	965.40	79702.50	
1865	15111.00	884.00	66999.80	
1866	13961.00	784.70	50320.00	
1867	12460.00	820.80	55825.10	
1868	10732.00	639.20	40878.20	
1869	8309.00	587.90	35168.30	
1870	2080.00	120.60	6743.60	
1871	184.00	9.10	497.00	
1872	40.00	1.60	110.00	

Comment 1861-1870 (C); 1871 (C)CLIFFORD; 1872 (C)

Tin	Black(tons)	Stuff(tons)	Tin(tons)	Value(£)
1862	11.90	0.00	0.00	706.80
1863	32.40	0.00	0.00	1811.50
1864	15.80	0.00	0.00	827.10
1865	7.70	0.00	0.00	346.90
1866	1.80	0.00	0.00	440.40
1867	26.50	0.00	0.00	1592.00
1868	62.80	0.00	0.00	3420.20
1869	58.40	0.00	0.00	3933.10
1870	48.80	0.00	0.00	3261.80
1872	13.30	0.00	0.00	1576.10
1873	22.80	87.80	0.00	1577.80
1874	19.00	0.00	0.00	908.00
1875	1.10	0.00	0.00	64.40
1879	0.00	21.60	0.00	20.00

Comment 1862 PART YEAR ONLY; 1866-1867 VALUE INC.TIN STUFF; 1872 VALUE INC.TIN STUFF

Arsenic	Ore(tons)	Metal(tons)	Value(£)
1870	4.00	0.00	7.90
1874	3.00	0.00	3.80

Comment 1870 CRUDE

Ownership: Comment 1862-1868 CLIFFORD CONSOLIDATED & UNITED; 1869-1871 ABANDONED JUNE 29TH.1870; 1880 ABOVE ADIT & HEAPS ONLY

Management: Manager 1862-1867 JOHN RICHARDS; 1868-1869 JOHN GILBERT Chief Agent 1861 JOHN RICHARDS; 1868-1869 JOHN MAYNE JOS.HIGGINS JOS.JEWELL; 1879 JOHN MICHELL Secretary 1861-1867 WM.WILLIAMS & SON (P); 1868-1871 WM.WILLIAMS & SON; 1879-1880 JAS.BOSKEAN

Employment:

	Underground	Surface	Total
1879	5		5

CLIFFORD UNITED GWENNAP SW 756417 0306

Production:

Tin	Black(tons)	Stuff(tons)	Tin(tons)	Value(£)
1892	0.00	50.00	0.00	25.00
1893	0.00	120.00	0.00	100.00

114

CLIFFORD UNITED   GWENNAP   Continued

Tin	Black(tons)	Stuff(tons)	Tin(tons)	Value(£)
1894	0.00	42.40	0.00	51.00
1895	0.00	171.80	0.00	144.00
1896	No detailed return			

Comment 1895 AGGREGATED
Ownership: 1892–1893 ART.STRAUSS P; 1894–1895 W.J.TRYTHALL
Management: Chief Agent 1892–1895 JOHN ODGERS
             Secretary 1892–1893 JAS.HICKS

Employment:	Underground	Surface	Total
1892	7		7
1893	4	1	5
1894	3		3
1895	2		2

CLIFFORD UNITED(PT. OF)   GWENNAP   SW 756417 0307

Ownership: 1895–1896 JOHN ODGERS

Employment:	Underground	Surface	Total
1895	6		6

CLIFFORD UNITED,SOUTH   GWENNAP   SW 755402 0308

Production: Copper No detailed return
Ownership: Comment 1860–1865 SUSPENDED
Management: Manager 1859 WM.BRENTON JNR.
             Chief Agent 1859 WM.BRENTON JNR.
             Secretary 1859–1860 JOHN PASCOE (P)

CLIFFORD,EAST   GWENNAP   0309

Production: Copper No detailed return
             Tin No detailed return
Management: Manager 1885 W.LANLEN

CLIFFORD,NEW   GWENNAP   SW 731402 0310

Production: Copper No detailed return

Tin	Black(tons)	Stuff(tons)	Tin(tons)	Value(£)
1870	0.00	5.10	0.00	10.50
1871	0.00	4.30	0.00	4.70

             Comment 1870–1871 TIN STUFF
Ownership: Comment 1863–1864 SUSPENDED
Management: Manager 1860–1862 WM.KITTO; 1870–1873 JOS.MICHELL
             Chief Agent 1860–1862 HY.TREVETHAN; 1865–1866 J.MICHELL; 1867
             W.MICHELL; 1868–1869 JOS.MICHELL
             Secretary 1860–1864 WM.BRAY (P); 1865–1866 JOHN GREEN;
             1867–1873 MAT.GREEN

CLIFFORD,WEST   GWENNAP   SW 730410 0311

Production: Copper   Ore(tons) Metal(tons)   Value(£)
            1865   73.00   2.90   206.00
            Comment 1865 (C)

CLIJAH AND WENTWORTH   REDRUTH   SW 702411 0312

Production: Copper   Ore(tons) Metal(tons)   Value(£)
            1854   428.00   26.20   2644.00
            1855   874.00   50.20   5074.30
            1856   562.00   32.40   2975.30
            1857   432.00   29.20   2925.50
            1858   230.00   14.70   1327.00
            1860   122.00   7.10   637.50
            1861   171.00   8.50   712.30
            1862   32.00   1.00   73.30
            1866   12.00   0.90   54.60
            Comment 1854—1858 (C); 1860—1862 (C); 1866 (C)
            Tin   Black(tons) Stuff(tons)   Tin(tons)   Value(£)
            1856   3.70   0.00   0.00   268.60
            1857   0.00   0.00   0.00   961.60
            1858   0.00   0.00   0.00   805.50
            1859   0.00   0.00   0.00   1488.20
            1860   0.00   0.00   0.00   1487.30
            1861   0.00   0.00   0.00   2206.80
            1862   0.00   0.00   0.00   2618.20
            1863   0.00   0.00   0.00   4079.50
            1864   0.00   0.00   0.00   3537.60
            1865   0.00   0.00   0.00   2267.70
            1866   0.00   0.00   0.00   667.50
            Comment 1857—1861 TIN STUFF; 1862 TIN STUFF.TWO RETURNS
            AGGREGATED; 1863—1866 TIN STUFF
Management: Manager 1859—1865 FRAN.PRYOR
            Chief Agent 1859—1865 CHAS.GLASSON
            Secretary 1859—1862 ROBT.H.PIKE (P); 1863—1865 ROBT.H.PIKE &
            SONS

CLINTON   MYLOR   SW 814336 0313

Production: Lead   Ore(tons) Metal(tons)   Value(£)
            1858   17.50   10.50   0.00
            1859   No detailed return
            Comment 1858 FOR AG.SEE GREAT ALFRED

CLINTON   ST.DAY   SW 727428 0314

Production: Tin   Black(tons) Stuff(tons)   Tin(tons)   Value(£)
            1856   0.40   0.00   0.00   24.60
            Comment 1856 INC.EDGECOMBE
Ownership: Comment 1864—1868 CLINTON AND PINK UNITED
Management: Chief Agent 1864—1868 WM.WILLIAMS JOEL HIGGINS JAS.HIGGINS
            Secretary 1864—1868 C.REDRUTH

116

CLITTERS UNITED   GUNNISLAKE   SX 421720 0315

Production: Copper   Ore(tons) Metal(tons)   Value(£)
            1905   80.00   0.00   104.00
            Tin   Black(tons) Stuff(tons)   Tin(tons)   Value(£)
            1902   108.30   0.00   0.00   6004.00
            1903   111.60   0.00   0.00   8817.00
            1904   104.30   0.00   0.00   8121.00
            1905   107.00   0.00   0.00   9359.00
            1906   62.00   0.00   0.00   6881.00
            1907   35.00   0.00   0.00   4023.00
            1908   18.70   0.00   0.00   1589.00
            1909   27.90   0.00   0.00   2127.00
            Arsenic   Ore(tons) Metal(tons)   Value(£)
            1905   123.00   0.00   1368.00
            1906   60.00   0.00   1343.00
            Tungsten   Ore(tons) Metal(tons)   Value(£)
            1903   230.00   0.00   11548.00
            1904   91.00   0.00   9120.00
            1905   62.50   0.00   4868.00
            1906   26.70   0.00   2365.00
            1907   18.50   0.00   2565.00
            1908   4.90   0.00   426.00
            1909   28.30   0.00   2212.00
            Wolfram No detailed return
Ownership: 1900—1909 CLITTERS UNITED MINES CO.LTD.; 1910—1912 HINGSTON &
            CLITTERS MINES LTD.
            Comment 1900—1902 INC.GUNNISLAKE CLITTERS, OLD GUNNISLAKE,
            HAWKMOOR & HINGSTON DOWN CONSOLS; 1910 NO MINERAL WORKED
            1910; 1911 NO MINERAL WORKED 1911; 1912 NO MINERAL WORKED
            1912
Management: Chief Agent 1900—1901 W.H.WILLIAMS; 1902—1908 J.PAULL
Employment: Underground   Surface   Total
            1900   45   9   54
            1901   68   68   136
            1902   42   47   89
            1903   40   118   158
            1904   77   121   198
            1905   58   90   148
            1906   56   87   143
            1907   57   89   146
            1908   60   82   142
            1909   14   14
            1911   3   3

CLOWANCE WOOD   CROWAN   SW 623343 0316

Production: Copper No detailed return
            Tin No detailed return
Management: Chief Agent 1863—1865 E.CHEGWIN
            Secretary 1863—1865 W.PAGE CARDOZA

                                    117

COADS GREEN   NORTH HILL   SX 297757 0317

Production: Manganese   Ore(tons) Metal(tons)   Value(£)
            1876   21.00   0.00   94.50
Management: Manager 1875-1876 JOHN GOLDSWORTHY
            Chief Agent 1873-1874 JOHN GOLDSWORTHY
            Secretary 1875-1876 JOHN GOLDSWORTHY

COATES   ST.AGNES   SW 700501 0318

Production: Copper   Ore(tons) Metal(tons)   Value(£)
            1883   69.00   6.90   371.00
            1884   265.00   0.00   1060.00
            Tin   Black(tons) Stuff(tons)   Tin(tons)   Value(£)
            1855   1.30   0.00   0.00   80.10
            1856   1.40   0.00   0.00   88.20
            1863   1.40   0.00   0.00   94.10
            1864   7.60   0.00   0.00   497.90
            1865   11.50   0.00   0.00   643.00
            1866   4.00   0.00   0.00   204.50
            1869   0.90   0.00   0.00   61.90
            1870   9.00   0.00   0.00   633.70
            1871   12.90   0.00   0.00   985.90
            1872   8.10   0.00   0.00   689.40
            1873   19.80   0.00   0.00   1542.50
            1874   20.30   0.00   0.00   1126.90
            1875   40.60   0.00   0.00   1944.40
            1876   95.80   0.00   0.00   4264.10
            1877   24.90   0.00   0.00   1070.40
            1878   65.50   0.00   0.00   2312.20
            1879   16.40   0.00   0.00   573.20
            1880   35.50   0.00   0.00   1887.30
            1881   101.10   0.00   70.00   5896.30
            1882   80.00   0.00   56.00   4800.00
            1883   63.00   0.00   0.00   3614.00
            1884   13.00   0.00   0.00   588.00
            1888   4.20   0.00   0.00   209.00
            1889   3.90   0.00   0.00   184.00
            1901   0.00   13.00   0.00   15.00
            1902   0.00   2.00   0.00   11.00
            1913   16.00   0.00   0.00   1884.00
            Comment 1881 COATES UNITED
Ownership: 1876 WHEAL COATES TIN MINING CO.; 1877-1882 WHEAL COATES TIN
            MINING CO.LTD.; 1883-1886 WHEAL COATES TIN MINING CO.;
            1901-1902 DOIDGE,TEAGUE & TEAGUE; 1911-1913 CORNISH MINES
            LTD.
            Comment 1864-1865 STAMPING BURROWS ONLY; 1880-1881 COATES
            UNITED; 1886 NOT WORKED; 1911-1913 UNWATERING
Management: Manager 1859-1861 JOHN TAYLOR & SONS; 1873-1881 W.H.MARTIN;
            1913 WM.THOMAS
            Chief Agent 1859-1861 JOHN HANCOCK; 1862-1865 WM.WATERS; 1871
            T.KEMPTHORNE; 1872 W.H.MARTIN; 1873-1875 T.BROWN; 1876
            J.S.HARRIS T.BROWN; 1877-1881 J.S.HARRIS; 1882-1886
            WM.VIVIAN
            Secretary 1859-1861 MICH.MORCOM (P); 1862-1865 EDW.CARTER;
            1871-1875 M.BRYANT; 1876-1881 WM.BATTYE (S)

                                      118

Employment:

	Underground	Surface	Total
1878	38	18	56
1879	47	27	74
1880	64	42	106
1881	89	49	138
1882	82	46	128
1883	46	16	62
1884	30	15	45
1901	2	2	4
1902	3	4	7
1911	6	6	12
1912	10	20	30
1913	39	21	60

COIT   ST.AGNES   SW 718513 0319

Production: Tin

	Black(tons)	Stuff(tons)	Tin(tons)	Value(£)
1855	2.60	0.00	0.00	156.70
1856	2.50	0.00	0.00	183.50
1857	5.50	0.00	0.00	402.80
1858	3.80	0.00	0.00	239.20
1859	2.70	0.00	0.00	193.40
1860	5.90	0.00	0.00	456.00
1861	4.50	0.00	0.00	330.90
1862	12.10	0.00	0.00	747.00
1863	7.90	0.00	0.00	506.70
1864	6.50	0.00	0.00	412.00
1865	6.40	0.00	0.00	330.50
1866	11.40	0.00	0.00	474.10
1867	6.20	0.00	0.00	301.50
1868	21.60	0.00	0.00	1156.10
1869	27.50	0.00	0.00	1790.20
1870	19.20	0.00	0.00	1230.70
1871	26.40	0.00	0.00	1891.40
1872	16.80	0.00	0.00	1405.20
1873	10.50	0.00	0.00	699.90
1874	13.10	0.00	0.00	585.50
1875	16.00	0.00	0.00	825.50
1884	1.40	0.00	0.00	63.00

Comment 1884 INC.FRIENDLY
Management: Manager 1861-1875 JAS.EVANS
Chief Agent 1861-1868 JOHN EVANS; 1869-1871 FRAN.EVANS
Secretary 1861-1872 HY.BORROW (P); 1875 JAS.EVANS & OTHERS

COIT STAMPS   ST.AGNES   SW 718513 0320

Production: Tin

	Black(tons)	Stuff(tons)	Tin(tons)	Value(£)
1868	6.20	0.00	0.00	311.30
1869	24.00	0.00	0.00	1519.20
1870	13.10	0.00	0.00	835.10
1871	20.10	0.00	0.00	1364.30
1872	24.30	0.00	0.00	1703.90
1873	14.10	0.00	0.00	993.40

COIT STAMPS   ST.AGNES   Continued

Tin	Black(tons)	Stuff(tons)	Tin(tons)	Value(£)
1874	26.80	0.00	0.00	1320.70

COIT,WEST   ST.AGNES   SW 718513 0321

Production: Tin No detailed return
Management: Manager 1870–1872 NICH.BRYANT

COLBIGGON   ROCHE   SX 007637 0322

Production:	Iron	Ore(tons)	Iron(%)	Value(£)
	1874	4931.40	0.00	3081.70

Comment 1874 BH.
Ownership: 1874–1877 YNISCEDWYN IRON CO.
Management: Chief Agent 1874–1876 RICH.THOMAS; 1877 THOS.LOBB

COLDVREATH   ROCHE   SW 985584 0323

Production:	Iron	Ore(tons)	Iron(%)	Value(£)
	1858	890.50	0.00	289.40
	1859	2294.70	0.00	745.00
	1860	546.60	0.00	144.30
	1861	561.60	0.00	140.40
	1862	140.00	0.00	98.00
	1863	No detailed return		
	1864	289.00	0.00	72.20
	1865	305.00	0.00	75.20
	1870	1116.30	0.00	371.80
	1871	194.20	0.00	48.50
	1872	4740.60	0.00	2527.10
	1873	2039.00	0.00	1325.20
	1874	440.00	0.00	385.00
	1875	328.00	0.00	215.00
	1881	No detailed return		
	1882	550.00	49.00	275.00

Comment 1858–1865 BH.; 1870 BH.; 1871 BH.HIGHER & LOWER
COLDVREATH; 1872 BH.TWO RETURNS AGG.; 1873–1875 BH.HIGHER &
LOWER COLDVREATH; 1881 BH.SEE RETEW INC.SOME RH.; 1882 BH.
Ownership: 1863–1865 EDW.CARTER & CO.; 1870–1874 THE IRONMASTERS CO.;
1875 THE IRONMASTERS CO.LTD.; 1880–1881 WEST OF ENGLAND IRON
ORE CO.; 1882 WEST OF ENGLAND IRON ORE CO.LTD.
Comment 1870–1875 HIGHER & LOWER COLDVREATH
Management: Manager 1872–1875 W.H.HOSKING; 1880–1881 DAVID COCK
Chief Agent 1863–1865 WM.BUCKTHOUGHT; 1870–1871 W.H.HOSKING;
1882 DAVID COCK

Employment:	Underground	Surface	Total
1880	6	6	12
1881	5		5
1882	4	1	5

COLLACOMBE   SYDENHAM DAMEREL, DEVON   SX 433771 0324

Production: Zinc   Ore(tons) Metal(tons)   Value(£)
         1868   100.00   0.00   250.00

COLLINS   TRURO   0325

Ownership: Comment 1860–1865 SUSPENDED
Management: Secretary 1859 GEO.R.ODGERS (P)

COLQUITE & CALLINGTON UNITED   CALLINGTON   SX 356677 0326

Management: Manager 1867–1869 JAS.EVANS; 1870–1871 THOS.DOIDGE
            Chief Agent 1867–1868 I.DOIDGE; 1869 THOS.DOIDGE
            Secretary 1869–1871 G.HILL

COMBE AND DANESCOMBE   CALSTOCK   0327

Production: Copper   Ore(tons) Metal(tons)   Value(£)
         1898   24.00   0.00   168.00
         1900   20.00   0.00   63.00
         Arsenic No detailed return
         Arsenic Pyrite   Ore(tons)   Value(£)
         1896   628.00   1366.00
         1897   2060.00   4879.00
         1898   1443.00   2164.00
         1899   797.00   1192.00
         1900   671.00   1927.00
         1901   686.00   515.00
Ownership: 1896–1903 COMBE & DANESCOMBE ARSENIC CO.LTD.
            Comment 1902 IN LIQUIDATION
Management: Manager 1901–1902 JOHN H.JAMES
            Chief Agent 1903 J.CROKER
            Secretary 1896–1902 J.CROKER (S)
Employment: Underground   Surface   Total
         1896   40   30   70
         1897   54   53   107
         1898   25   19   44
         1899   27   13   40
         1900   23   17   40
         1901   6   6
         1902   7   7
         1903   1   1

COMBELLACK   WENDRON   SW 704304 0328

Production: Tin   Black(tons) Stuff(tons)   Tin(tons)   Value(£)
         1878   18.40   0.00   0.00   648.00
         1880   6.00   0.00   0.00   288.30
         1881   13.60   0.00   11.90   778.40
         1882   39.00   0.00   26.50   2342.00
         1883   35.90   0.00   0.00   1964.00
         1884   23.20   0.00   0.00   1158.00

COMBELLACK   WENDRON   Continued

            Comment 1878 SUSPENDED TEMPORARILY; 1880-1884 INC.MENGERN
Ownership: 1875-1876 COMBELLACK CO.; 1877-1879 COMBELLACK MINING
            CO.LTD.; 1880-1884 COMBELLACK & MINGERN MINING CO.
            Comment 1879 NOT WORKED; 1880-1884 INC.MENGERN
Management: Manager 1878-1879 STEP.ROBERTS; 1881 C.M.BLAINBERG
            Chief Agent 1875-1877 JOHN NANCARROW; 1880-1884 JOHN CURTIS
Employment: Underground   Surface   Total
            1878   14   15   29
            1879   1   1
            1880-1884
            Comment 1880-1884 SEE MENGERN

COMFORD   GWENNAP   SW 718395 0329

Production: Copper   Ore(tons) Metal(tons)   Value(£)
            1845   122.00   5.20   313.50
            1846   267.00   16.20   1058.00
            1847   1436.00   69.20   4507.40
            1848   2264.00   104.30   5393.20
            1849   3084.00   128.50   7405.00
            1850   2263.00   82.50   4649.60
            1851   1573.00   52.80   2698.90
            1852   1158.00   39.50   2900.20
            1853   1286.00   37.70   2919.00
            1854   1353.00   38.80   3170.80
            1855   1294.00   32.70   2534.60
            1856   586.00   16.40   1060.90
            1857   316.00   8.20   580.10
            1871   2.00   0.10   5.80
            1872   3.00   0.20   12.40
            1875   6.00   0.40   33.90
            1876   52.00   5.20   338.60
            1877   51.40   4.10   331.10
            1878   90.00   6.80   349.20
            1879   66.60   4.10   240.00
            1880   26.00   1.90   106.60
            1881   113.00   7.60   396.20
            1882   43.00   2.90   172.00
            1883   58.40   3.50   198.00
            1884   121.00   0.00   372.00
            1885   242.00   0.00   354.00
            Comment 1845 (C)COMFORT; 1846-1857 (C); 1871-1872 (C);
            1875-1876 (C); 1879-1885 INC.NORTH TRESAVEAN
            Tin   Black(tons) Stuff(tons)   Tin(tons)   Value(£)
            1873   0.00   345.80   0.00   420.70
            1874   19.60   94.80   0.00   1030.50
            1875   6.20   320.90   0.00   619.00
            1876   0.00   309.00   0.00   298.80
            1877   1.30   159.70   0.00   190.50
            1880   0.00   114.00   0.00   120.00
            1881   18.00   0.00   0.00   820.00
            1882   26.00   0.00   0.00   1560.00
            1883   13.40   0.00   0.00   742.00
            1884   14.40   0.00   0.00   640.00

                                     122

```
Tin          Black(tons) Stuff(tons)   Tin(tons)    Value(£)
1893              1.00        0.00        0.00       41.00
```
Comment 1880–1884 INC.NORTH TRESAVEAN; 1893 INC.NORTH
TRESAVEAN
```
Arsenic Pyrite  Ore(tons)    Value(£)
1884            13.00        22.00
```
Ownership: 1877 WHEAL COMFORD MINING CO.; 1879–1886 COMFORD & NORTH
TRESAVEAN ADVENTURERS; 1893–1894 WM.CORNISH & CO.
Comment 1879–1886 INC.NORTH TRESAVEAN; 1893–1894 TRIBUTERS;
1900–1901 SEE TRESAVEAN UNITED
Management: Manager 1873 PETER PHILLIPS
Chief Agent 1872 PETER PHILLIPS; 1874 GEO.LIGHTLEY; 1875–1881
JOH.JAMES; 1882–1886 H.TREGANOWAN
Secretary 1872–1873 GEO.LIGHTLEY; 1874–1879 JOHN L.PETERS;
1880–1881 JOHN L.PETERS (S); 1893–1894 WM.ROSEWARNE
Employment: Underground   Surface   Total
```
1878              21            1          22
1879              25            4          29
1880              30           15          45
1881              26           23          49
1882              19           37          56
1883              16           22          38
1884              21           20          41
1885              10            5          15
1893               4                        4
```
Comment 1880–1885 INC.NORTH TRESAVEAN

COMMERCE   ST.EWE   SW 987497 0330

```
Production: Tin          Black(tons) Stuff(tons)   Tin(tons)    Value(£)
1901              0.60        0.00        0.00       43.00
1902             19.10        0.00        0.00     1536.00
1903             10.70        0.00        0.00      835.00
1904             16.10        0.00        0.00     1341.00
1905              7.80        0.00        0.00      680.00
1906             21.50        0.00        0.00     2446.00
1907              1.20        0.00        0.00      244.00
1908             30.80        0.00        0.00     2526.00
1909             47.40        0.00        0.00     3976.00
```
Ownership: 1899 E.A.TREGILGAS; 1900–1913 COMMERCE TIN MINE LTD.
Comment 1899 INC.PENTRASSOE; 1910–1912 CLOSED; 1913 ABANDONED
JUNE 1910
Management: Manager 1901–1902 P.J.PARRY
Chief Agent 1903–1905 H.BLAKE; 1906–1913 RICH.H.WILLIAMS
Secretary 1900–1902 WM.COOPER (S)
```
Employment:              Underground    Surface      Total
1899               2            2           4
1900              23            9          32
1901              26            4          30
1902              16            7          23
1903              16           12          28
1904              12           11          23
1905              20           18          38
1906              24            7          31
```

COMMERCE   ST.EWE   Continued

	Underground	Surface	Total
1907	40	30	70
1908	36	34	70
1909	36	22	58

CONCORD   ST.AGNES   0331

Production: Lead   Ore(tons) Metal(tons)   Value(£)
            1846   30.00   18.00   0.00
            1847   30.00   18.00   0.00
            1849—1852 No detailed return
            1854   No detailed return

CONDURROW   CAMBORNE   SW 660392 0332

Production: Copper   Ore(tons) Metal(tons)   Value(£)
            1845   166.00   10.70   785.70
            1846   609.00   36.00   2395.50
            1847   1235.00   72.80   4905.50
            1848   1550.00   102.30   5736.90
            1849   1493.00   115.00   7647.80
            1850   1664.00   113.20   7614.00
            1851   1707.00   121.50   8057.20
            1852   1633.00   126.90   10939.80
            1853   1844.00   125.90   12533.50
            1854   1490.00   110.10   10003.40
            1855   1696.00   116.80   12022.90
            1856   1941.00   128.50   12012.50
            1857   1589.00   97.60   9831.40
            1858   1465.00   84.60   7483.00
            1859   1752.00   89.80   7962.90
            1860   1926.00   80.30   6612.30
            1861   1705.00   71.90   5897.30
            1862   1800.00   78.70   5931.70
            1863   377.00   19.20   1406.10
            1864   238.00   13.90   1153.20
            1865   213.00   11.80   892.70
            1876   199.00   6.20   326.00
            Comment 1845—1865 (C); 1876 (C)
            Tin   Black(tons) Stuff(tons)   Tin(tons)   Value(£)
            1852   44.50   0.00   0.00   0.00
            1853   17.70   0.00   0.00   1057.10
            1854   122.40   0.00   0.00   8718.00
            1855   108.00   0.00   0.00   7097.00
            1856   77.70   0.00   0.00   5759.70
            1857   121.80   0.00   0.00   9772.90
            1858   102.80   0.00   0.00   6737.60
            1859   73.00   0.00   0.00   5290.70
            1860   49.00   0.00   0.00   3900.60
            1861   45.90   0.00   0.00   3430.10
            1862   158.20   0.00   0.00   10824.00
            1863   193.90   0.00   0.00   13032.00
            1864   244.10   0.00   0.00   15924.00

124

CONDURROW   CAMBORNE   Continued

Tin	Black(tons)	Stuff(tons)	Tin(tons)	Value(£)
1865	367.50	0.00	0.00	21107.70
1866	404.60	0.00	0.00	20110.60
1907	2.60	0.00	0.00	244.00
1908	2.00	0.00	0.00	138.00
1910	1.60	0.00	0.00	103.00
1911	4.30	0.00	0.00	445.00
1912	0.70	0.00	0.00	62.00

Comment 1852 TIN ORE; 1862 TWO RETURNS AGGREGATED; 1865 9
MONTHS ONLY; 1907-1908 CONDURROW UNITED

Arsenic	Ore(tons)	Metal(tons)	Value(£)
1857	6.00	0.00	12.00
1858	8.00	0.00	16.00
1860	10.00	0.00	16.30
1861	6.10	0.00	8.00
1862	6.00	0.00	3.00

Comment 1860-1862 CRUDE

Ownership: 1905-1910 CONDURROW UNITED MINES LTD.; 1911-1913 CONDURROW
            MINES LTD.
            Comment 1865-1872 SEE PENDARVES UNITED; 1905-1910 CONDURROW
            UNITED
Management: Manager 1859-1861 NICH.VIVIAN; 1862-1864 J.MOYLE
            REG.T.GRYLLS
            Chief Agent 1859-1860 JOS.VIVIAN JNR.; 1861 CHAS.DAVEY; 1905
            JAS.NEGUS; 1906-1909 S.BENNETT; 1910-1913 J.TAMBLYN
            Secretary 1859-1860 NICH.VIVIAN (P); 1861-1864 REG.T.GRYLLS
            (P)

Employment:	Underground	Surface	Total
1906	9	43	52
1907	30	34	64
1908	29	27	56
1909	25	23	48
1910	13	27	40
1911	17	20	37
1912	11	19	30
1913	12	18	30

CONDURROW,GREAT   CAMBORNE   SW 652398 0333

Production: Tin No detailed return
Ownership: 1894 E.C.MURDOCK
            Comment 1894 OR NEW DOLCOATH

Employment:	Underground	Surface	Total
1894	3	2	5

CONDURROW,SOUTH   CAMBORNE   SW 663389 0334

Production: Copper	Ore(tons)	Metal(tons)	Value(£)
1864	20.00	2.80	237.50
1865	65.00	7.30	631.20
1866	340.00	25.70	1457.50
1867	247.00	30.60	2187.90
1868	106.00	11.10	734.50

CONDURROW,SOUTH   CAMBORNE   Continued

Copper	Ore(tons)	Metal(tons)	Value(£)
1869	43.00	4.70	325.90
1871	10.00	1.40	86.00
1873	7.00	0.70	48.10
1877	51.00	5.60	296.90
1878	66.00	8.30	448.60
1879	7.60	0.70	30.50
1880	28.80	2.50	138.40
1881	50.90	3.90	213.70
1882	20.00	1.20	50.00

Comment 1864-1869 (C); 1871 (C); 1873 (C)

Tin	Black(tons)	Stuff(tons)	Tin(tons)	Value(£)
1865	13.50	0.00	0.00	751.40
1866	6.90	0.00	0.00	336.00
1867	7.30	0.00	0.00	386.90
1868	14.70	0.00	0.00	867.50
1869	50.90	0.00	0.00	3523.20
1870	219.30	0.00	0.00	16109.00
1871	297.00	0.00	0.00	21510.50
1872	158.40	0.00	0.00	13393.40
1873	300.70	0.00	0.00	22572.40
1874	475.30	0.00	0.00	26896.30
1875	439.50	0.00	0.00	22981.20
1876	588.00	0.00	0.00	26536.50
1877	621.00	0.00	0.00	26563.60
1878	861.00	0.00	0.00	30371.80
1879	762.00	0.00	0.00	31372.30
1880	551.20	0.00	0.00	29391.40
1881	457.60	0.00	319.20	26392.70
1882	444.00	0.00	290.40	27414.00
1883	520.00	0.00	0.00	28833.00
1884	546.00	0.00	0.00	26032.00
1885	440.00	0.00	0.00	21786.00
1886	439.00	0.00	0.00	25350.00
1887	458.00	0.00	0.00	30037.00
1888	445.00	0.00	0.00	30383.00
1889	407.20	0.00	0.00	23307.00
1890	347.00	0.00	0.00	20066.00
1891	329.50	0.00	0.00	18463.00
1892	345.50	0.00	0.00	19609.00
1893	268.00	0.00	0.00	14165.00
1894	208.50	0.00	0.00	9047.00
1895	181.80	0.00	0.00	7294.00
1896	131.70	0.00	0.00	5080.00
1899	1.30	0.00	0.00	75.00
1900	0.00	741.00	0.00	756.00
1901	14.00	175.00	0.00	1110.00
1902	74.90	0.00	0.00	5854.00

Arsenic	Ore(tons)	Metal(tons)	Value(£)
1877	4.00	0.00	2.00

Ownership: 1876-1890 SOUTH CONDURROW MINING CO.; 1891-1893 SOUTH
CONDURROW MINING CO. C.B.; 1894-1896 SOUTH CONDURROW MINE
ADVENTURERS; 1899-1902 SOUTH CONDURROW MINING CO.; 1903-1910
WHEAL GRENVILLE MINING CO.
Comment 1903-1910 WORKED WITH GRENVILLE,CAMBORNE

CONDURROW,SOUTH   CAMBORNE   Continued

Management: Manager 1859 JOS.VIVIAN; 1860—1861 JOS.VIVIAN JNR.; 1862—1863
            JOS.VIVIAN WM.THOMAS; 1864—1873 JOS.VIVIAN & SON; 1874—1881
            WM.RICH
            Chief Agent 1859—1861 WM.RICHARDS; 1862—1863 WM.RICHARDS
            WM.DUNSFORD; 1864—1868 WM.RICHARDS W.WILLIAMS; 1869
            W.WILLIAMS; 1870—1871 H.ABRAHAMS; 1872—1879 H.ABRAHAMS
            C.W.WILLIAMS; 1880—1881 H.KING W.WILLIAMS; 1882—1896 WM.RICH;
            1899—1902 WM.THOMAS
            Secretary 1859 WM.DARKE (P); 1860—1861 WM.DUNSFORD (P);
            1862—1873 JOS.VIVIAN JNR.; 1874—1881 ROBT.H.PIKE & SON; 1891
            CHAS.CLARKE; 1892—1896 CHAS.CLARKE (S); 1899—1902 CHAS.CLARKE
            (S)
Employment: Underground   Surface   Total
            1878   182   205   387
            1879   174   204   378
            1880   180   225   405
            1881   139   165   304
            1882   150   175   325
            1883   166   178   344
            1884   172   183   355
            1885   141   161   302
            1886   141   158   299
            1887   140   151   291
            1888   140   167   307
            1889   116   175   291
            1890   128   157   285
            1891   159   161   320
            1892   168   154   322
            1893   147   149   296
            1894   110   120   230
            1895   81   83   164
            1896   53   75   128
            1897   62   18   80
            1898   80   5   85
            1899   16   11   27
            1900   46   37   83
            1901   62   58   120
            1902   50   43   93

CONDURROW,SOUTH (SCHOOL)   CAMBORNE   SW 663389 0335

Production: Tin No detailed return
Ownership: 1897—1900 CAMBORNE MINING SCHOOL
Management: Chief Agent 1897—1900 WM.THOMAS; 1901—1902 SEE KING EDWARD
Employment: Underground   Surface   Total
            1899   30   6   36
            1900   35   7   42

CONDURROW,WEST   CAMBORNE   SW 648388 0336

Production: Copper   Ore(tons) Metal(tons)   Value(£)
            1862   4.00   0.30   23.00
            1865   3.00   0.50   39.90

                                127

Copper	Ore(tons)	Metal(tons)	Value(£)
1866	61.00	4.50	286.70
1886	14.00	0.00	36.00

Comment 1862 (C); 1865-1866 (C); 1886 LATE SOUTH TOLCARNE

Tin	Black(tons)	Stuff(tons)	Tin(tons)	Value(£)
1859	0.00	0.00	0.00	126.30
1861	0.00	0.00	0.00	52.50
1863	0.00	0.00	0.00	560.20
1864	0.00	0.00	0.00	605.20
1865	0.00	0.00	0.00	171.70
1866	0.00	0.00	0.00	186.10
1886	4.50	137.90	0.00	227.00
1887	No detailed return			
1888	0.50	12.00	0.00	24.00

Comment 1859 TIN STUFF; 1861 TIN STUFF; 1863-1866 TIN STUFF
Ownership: 1886-1888 WEST CONDURROW MINING CO.
Management: Manager 1859-1864 GEO.BENNETTS; 1865 J.HOSKING
             Chief Agent 1859-1861 GEO.JEWELL; 1862-1865 F.GILBERT JNR.;
             1872-1874 STEP.TERRILL; 1886-1888 JOHN JENNINGS
             Secretary 1859-1865 ALM.E.PAULL (P)

Employment:	Underground	Surface	Total
1886	20	8	28
1887	18	10	28
1888	23	9	32

CONSOLIDATED MINES   GWENNAP   SW 748421 0337

Production: Zinc	Ore(tons)	Metal(tons)	Value(£)
1857	59.50	0.00	200.00

Comment 1857 ORE ESTIMATED

Copper	Ore(tons)	Metal(tons)	Value(£)
1845	8798.00	685.70	51146.70
1846	9057.00	746.10	54386.10
1847	9849.00	812.10	59122.00
1848	9140.00	754.70	45898.10
1849	7906.00	626.30	41864.00
1850	7592.00	576.20	40159.80
1851	5661.00	407.80	27285.30
1852	5147.00	345.00	28526.70
1853	4724.00	291.80	28637.60
1854	3225.00	171.40	16965.90
1855	2655.00	135.30	13296.10
1856	2276.00	127.70	11568.70
1857	1105.00	59.20	6038.00

Comment 1845-1857 (C)

Tin	Black(tons)	Stuff(tons)	Tin(tons)	Value(£)
1855	7.30	0.00	0.00	461.40
1856	0.00	0.00	0.00	1439.20
1857	0.00	0.00	0.00	1391.80
1858	0.00	0.00	0.00	154.60

Comment 1856-1858 TIN STUFF

Arsenic	Ore(tons)	Metal(tons)	Value(£)
1858	4.00	0.00	6.80

Ownership: Comment 1862-1868 SEE CLIFFORD AMALGAMATED

Management: Chief Agent 1861 JOHN RICHARDS
            Secretary 1861 WM.WILLIAMS & SONS (P)

CONSOLS,NEW GREAT   STOKE CLIMSLAND   SX 388737 0339

Production: Copper   Ore(tons) Metal(tons)   Value(£)
            1874   No detailed return
            1875   52.90   0.00   3117.80
            1877   9.40   0.40   39.90
            1878   53.20   1.30   53.20
            1879   2.20   0.10   7.50
            Comment 1874 SEE NEW CONSOLS, DEVON; 1875 NEW CONSOLS.COPPER
            & SILVER PRECIPITATE; 1878 NEW CONSOLS
            Tin   Black(tons) Stuff(tons)   Tin(tons)   Value(£)
            1870   49.40   0.00   0.00   3322.30
            1871   76.30   0.00   0.00   5159.10
            1872   191.20   0.00   0.00   15326.10
            1873   147.50   0.00   0.00   10434.10
            1874   50.70   0.00   0.00   2632.40
            1875   17.50   0.00   0.00   750.90
            1876   27.50   0.00   0.00   983.30
            1877   5.40   0.00   0.00   227.90
            1878   5.20   0.00   0.00   162.60
            1879   23.20   0.00   0.00   734.80
            Comment 1873 TWO RETURNS FOR 1873. OTHER RETURN = 195 ORE,
            £13295
            Arsenic   Ore(tons) Metal(tons)   Value(£)
            1870   180.10   0.00   450.10
            1871   610.50   0.00   1387.60
            1872   1300.90   0.00   3614.60
            1873   1117.80   0.00   4261.60
            1874   558.60   0.00   2361.80
            1875   492.70   0.00   3152.20
            1876   330.50   0.00   1585.00
            1878   95.30   0.00   330.30
            Comment 1870-1871 CRUDE; 1875 NEW CONSOLS TIN & ARSENIC
            WORKS; 1876 NEW CONSOLS.SOOT; 1878 NEW CONSOLS SILVER &
            ARSENIC WORKS
            Arsenic Pyrite   Ore(tons)   Value(£)
            1877   2273.00   1015.00
            1878   2005.80   1000.00
            1879   206.30   99.70
            Comment 1877 ARSENICAL MUNDIC; 1878 NEW CONSOLS SILVER &
            ARSENIC WORKS
Ownership: 1877-1879 NEW CONSOLS SILVER & ARSENIC WORKS LTD.; 1880-1882
            HY.L.PHILLIPS
            Comment 1869-1870 SEE NEW GREAT CONSOLS, DEVON; 1877 NEW
            CONSOLS; 1878-1879 NEW CONSOLS IN LIQUIDATION; 1880-1881 NOT
            WORKED; 1882 NEW CONSOLS NOT WORKED
Management: Manager 1869-1873 RICH.PRYOR; 1875-1877 RICH.PRYOR & SON,
            Chief Agent 1869 RICH.FRATHER; 1870-1871 J.THOMAS HY.VIAL;
            1872 J.PRYOR JOHN GREEN; 1873 J.PRYOR; 1874 RICH.PRYOR & SON;
            1875-1877 THOS.JENKIN HY.VIAL; 1878 WM.J.COCK THOS.NEAL;
            1879-1881 THOS.NEAL

CONSOLS,NEW GREAT   STOKE CLIMSLAND   Continued

                Secretary 1869-1877 HY.L.PHILLIPS
Employment: Underground   Surface   Total
                1878   12   9   21
                1879   5   5

CONSOLS,WEST GREAT   CALSTOCK   0340

Production: Copper No detailed return
Management: Manager 1872-1873 W.VIAL
            Secretary 1872-1873 H.PRYOR

CONSTANCE   NEWLYN EAST   SW 833546 0341

Production: Lead & Silver   Ore(tons)   Lead(tons) Silver(ozs) Value(£)
            1852   20.00   12.70   400.00   0.00
            1853   21.00   16.00   510.00   0.00
            1854   42.00   28.00   1148.00   0.00
            1855   24.00   14.00   574.00   0.00
            1856-1857 No detailed return
            1859   9.00   6.20   0.00   0.00
            Tin No detailed return
Ownership: Comment 1860-1865 SUSPENDED
Management: Chief Agent 1859 A.CUNDY
            Secretary 1859 WM.WEST (P)

CONSTANTINE   CONSTANTINE   SW 746281 0342

Production: Iron   Ore(tons)   Iron(%)   Value(£)
            1866   150.00   0.00   45.00
            1869   1000.00   0.00   300.00
            1870   1000.00   0.00   300.00
            1871   881.40   0.00   661.00
            1872   2241.00   0.00   636.50
            1873   2000.00   0.00   1500.40
            1874   1060.00   0.00   856.00
            1875   1276.00   0.00   638.00
            Comment 1866 BH.; 1869-1875 BH.
Ownership: 1865-1867 WM.BROWNE; 1869-1875 J.BROGDEN & SON
Management: Manager 1874-1875 GEO.NOBLE
            Chief Agent 1869-1873 GEO.NOBLE

COOKS KITCHEN   ILLOGAN   SW 665407 0343

Production: Copper   Ore(tons) Metal(tons)   Value(£)
            1845   502.00   24.20   1522.40
            1846   134.00   7.20   468.60
            1851   226.00   16.30   1068.80
            1853   240.00   9.10   753.50
            1855   414.00   10.40   789.20
            1856   219.00   7.70   562.40
            1862   53.00   1.30   80.30

Copper	Ore(tons)	Metal(tons)	Value(£)
1863	10.00	0.70	50.70
1864	4.00	0.20	19.70
1866	11.00	0.40	23.90
1867	11.00	0.40	18.10
1877	5.00	0.40	17.60
1878	6.10	0.10	7.40
1879	9.50	0.60	30.00
1880	10.00	0.40	22.70
1881	102.10	7.10	357.50
1882	2.60	0.10	4.00
1884	2.00	0.00	2.00

Comment 1845–1846 (C); 1851 (C); 1853 (C); 1855–1856 (C); 1862–1864 (C); 1866–1867 (C)

Tin	Black(tons)	Stuff(tons)	Tin(tons)	Value(£)
1854	50.40	0.00	0.00	3544.60
1855	79.90	0.00	0.00	5253.50
1856	97.50	0.00	0.00	7312.60
1857	102.10	0.00	0.00	8433.60
1858	124.50	0.00	0.00	8302.30
1859	162.30	0.00	0.00	12444.40
1860	186.40	0.00	0.00	15357.00
1861	231.00	0.00	0.00	17154.50
1862	331.50	0.00	0.00	22663.30
1863	271.50	0.00	0.00	18758.10
1864	222.00	0.00	0.00	14511.10
1865	259.50	0.00	0.00	14724.00
1866	264.00	0.00	0.00	13159.00
1867	235.60	0.00	0.00	12798.10
1868	302.10	0.00	0.00	16910.30
1869	297.30	0.00	0.00	21604.20
1870	347.20	0.00	0.00	25244.60
1871	349.50	0.00	0.00	27863.60
1872	319.70	0.00	0.00	27951.10
1873	259.50	0.00	0.00	20302.50
1874	239.90	0.00	0.00	14069.70
1875	229.70	0.00	0.00	11922.00
1876	210.70	0.00	0.00	9431.10
1877	213.20	0.00	0.00	8892.50
1878	231.30	0.00	0.00	8274.00
1879	212.80	0.00	0.00	8472.00
1880	179.70	0.00	0.00	9245.60
1881	188.60	0.00	126.70	11192.10
1882	230.40	0.00	156.70	13822.00
1883	241.00	0.00	0.00	12842.00
1884	175.30	0.00	0.00	8035.00
1885	86.90	0.00	0.00	3989.00
1886	157.40	0.00	0.00	9104.00
1887	179.30	0.00	0.00	11772.00
1888	155.70	0.00	0.00	10612.00
1889	216.00	0.00	0.00	11976.00
1890	205.90	0.00	0.00	11638.00
1891	239.60	0.00	0.00	13165.00
1892	152.00	0.00	0.00	8344.00
1893	155.10	0.00	0.00	7996.00

COOKS KITCHEN   ILLOGAN   Continued

Tin	Black(tons)	Stuff(tóns)	Tin(tons)	Value(£)
1894	183.00	0.00	0.00	7671.00
1895	161.60	0.00	0.00	6260.00
1896	No detailed return			

Comment 1862 TWO RETURNS AGGREGATED

Arsenic	Ore(tons)	Metal(tons)	Value(£)
1868	7.50	0.00	6.30
1875	1.00	0.00	5.00
1878	13.10	0.00	35.80
1879	34.60	0.00	94.00
1880	11.80	0.00	40.30
1881	9.00	0.00	37.30
1882	6.30	0.00	25.00
1883	12.60	0.00	36.00
1885	6.00	0.00	15.00
1886	9.60	0.00	19.00
1888	13.00	0.00	26.00
1890	11.00	0.00	39.00
1894	9.00	0.00	13.00

Comment 1868 CRUDE; 1881 CRUDE; 1885-1886 CRUDE

Arsenic Pyrite	Ore(tons)	Value(£)
1892	10.00	15.00

Ownership: 1877-1890 COOKS KITCHEN MINING CO.; 1891-1893 COOKS KITCHEN
              MINING CO.C.B.; 1894-1895 COOKS KITCHEN MINE ADVENTURERS
              Comment 1895 NOW INCORP.WITH TINCROFT
Management: Manager 1859-1866 CHAS.THOMAS; 1867-1874 JOH.THOMAS;
              1876-1881 JOH.THOMAS
              Chief Agent 1859-1861 SML.DAVEY; 1865-1866 C.T.CRAZE
              CHAS.THOMAS JNR.; 1867-1873 CHAS.THOMAS JNR.F.GILBERT JNR.;
              1874 CHAS.THOMAS FRAN.GILBERT; 1875 JOH.THOMAS FRAN.GILBERT;
              1876 FRAN.GILBERT; 1877-1879 CHAS.THOMAS FRAN.GILBERT; 1880
              WM.THOMAS FRAN.GILBERT; 1881-1895 CHAS.THOMAS
              Secretary 1859-1862 ROBT.H.PIKE (P); 1863-1881 ROBT.H.PIKE &
              SON; 1891-1895 ROBT.H.PIKE & SONS (P)

Employment:	Underground	Surface	Total
1879	113	111	224
1880	100	121	221
1881	105	122	227
1882	104	122	226
1883	113	115	228
1884	91	105	196
1885	83	73	156
1886	114	92	206
1887	109	89	198
1888	125	97	222
1889	116	103	219
1890	121	102	223
1891	115	98	213
1892	92	100	192
1893	112	94	206
1894	94	86	180
1895	66	67	133

Production: Copper

	Ore(tons)	Metal(tons)	Value(£)
1877	116.40	8.10	441.50
1878	95.00	7.80	395.20
1879	158.70	12.80	646.00
1880	154.00	15.50	976.40
1881	336.30	29.40	1705.20
1882	522.00	40.40	2349.00
1883	452.00	30.50	1633.00
1884	367.00	0.00	1284.00
1885	364.00	0.00	1154.00
1886	214.00	0.00	622.00
1887	65.00	0.00	111.00
1888	103.00	0.00	411.00
1889	8.00	0.00	27.00
1890	8.00	0.00	34.00
1893	8.00	0.00	19.00

Tin	Black(tons)	Stuff(tons)	Tin(tons)	Value(£)
1877	0.00	48.20	0.00	45.00
1878	0.00	172.00	0.00	105.00
1879	0.00	273.00	0.00	264.00
1880	2.00	468.00	0.00	258.70
1881	0.00	698.00	0.00	397.70
1882	18.00	911.00	11.90	900.00
1883	22.00	0.00	0.00	759.00
1884	16.60	866.00	0.00	453.00
1885	16.90	678.00	0.00	530.00
1886	20.70	1035.50	0.00	757.00
1887	15.80	783.00	0.00	598.00
1888	21.00	1052.50	0.00	1059.00
1889	34.00	1905.00	0.00	1691.00
1890	46.80	0.00	0.00	1620.00
1891	41.20	0.00	0.00	1336.00
1892	38.30	0.00	0.00	1217.00
1893	0.00	1377.00	0.00	932.00

Ownership: 1877-1890 NEW COOKS KITCHEN MINING CO.; 1891-1893 NEW COOKS
KITCHEN MINING CO.C.B.; 1894 NEW COOKS KITCHEN MINING CO.
Comment 1894 NOT WORKED
Management: Manager 1872-1873 JOH.THOMAS; 1877-1878 JOH.THOMAS; 1892-1894
P)
Chief Agent 1872-1873 CHAS.THOMAS; 1877-1879 WM.THOMAS;
1880-1881 WM.THOMAS CHAS.THOMAS; 1882-1894 WM.THOMAS
Secretary 1872-1873 ROBT.H.PIKE & SON; 1878 WALT.PIKE;
1879-1881 ROBT.H.PIKE & SON; 1891 ROBT.H.PIKE & SONS;
1892-1894 ROBT.H.PIKE & SONS (P)

Employment:

	Underground	Surface	Total
1878	15	16	31
1879	28	17	45
1880	25	15	40
1881	26	21	47
1882	22	29	51
1883	22	27	49
1884	20	23	43
1885-1886	15	20	35
1887	15	17	32
1888	20	16	36

COOKS KITCHEN,NEW   ILLOGAN   Continued

	Underground	Surface	Total
1889	22	16	38
1890	22	14	36
1891—1892	21	14	35
1893	20	13	33

COPPER BOTTOM   GWINEAR   SW 630366 0345

Production: Copper   Ore(tons) Metal(tons)   Value(£)
            1851   104.00   8.80   610.30
            Comment 1851 (C)

COPPER HILL   REDRUTH   SW 707405 0346

Production: Copper   Ore(tons) Metal(tons)   Value(£)
            1855   246.00   15.80   1619.60
            1856   514.00   38.90   3713.30
            1857   841.00   59.40   6074.50
            1858   690.00   44.40   3942.50
            1859   460.00   49.80   4784.20
            1860   586.00   42.90   4008.20
            1861   1285.00   86.20   7665.90
            1862   1447.00   82.10   6549.30
            1863   1258.00   69.10   5067.00
            1864   1103.00   59.90   4816.80
            1865   839.00   49.60   3788.70
            1866   606.00   30.20   1912.40
            1867   624.00   35.50   2332.00
            1868   531.00   29.80   1818.40
            1869   401.00   23.10   1342.70
            1870   332.00   24.10   1423.70
            1871   21.00   1.90   119.70
            Comment 1855 (C)COPPER HALL; 1856—1871 (C)
            Tin   Black(tons) Stuff(tons)   Tin(tons)   Value(£)
            1856   0.00   0.00   0.00   172.00
            1857   0.00   0.00   0.00   267.60
            1858   0.00   0.00   0.00   48.00
            1859   0.00   0.00   0.00   18.50
            1864   1.20   0.00   0.00   72.50
            Comment 1856—1859 TIN STUFF
Ownership: Comment 1888—1890 SEE BASSET & BULLER CONSOLS
Management: Manager 1862—1868 JOHN DAVEY & SON; 1869 JOHN DAVEY JNR.
            Chief Agent 1859—1860 JOHN DAVEY; 1861 JOHN DAVEY AND.JOHNS;
            1862—1868 AND.JOHNS JAS.INCH; 1869 AND.JOHNS
            Secretary 1859—1865 R.DAVEY S.DAVEY (P); 1866—1868 J.MICHELL;
            1869 THOS.DAVEY
Employment: Underground   Surface   Total
            1888—1890
            Comment 1888—1890 SEE BASSET & BULLER CONSOLS

COPPER HOUSE SLAG   0347

Production: Copper   Ore(tons) Metal(tons)   Value(£)
            1845   218.00   9.10   332.10
            1846   158.00   7.60   306.90
            Comment 1845-1846 (C)

COPPLEY   ST.ERNE   0348

Production: Antimony No detailed return
Ownership: 1893-1895 THOS.H.GEAKE
Management: Chief Agent 1893-1895 WM.GEORGE
Employment: Underground   Surface   Total
            1893   6   4   10
            1894   11   3   14
            1895   10   5   15

CORNELLO   ZENNOR   SW 442388 0349

Production: Copper No detailed return
            Tin   Black(tons) Stuff(tons) Tin(tons)   Value(£)
            1872   4.30   0.00   0.00   350.70
            1873   1.30   0.00   0.00   103.30
            Comment 1872-1873 CARNELLOE
Ownership: Comment 1865 CERNELLOA; 1866 CERNELLOW; 1867-1869 CORNELLOW;
            1871-1873 CARNELLOE; 1874 CARNELLOE SUSPENDED
Management: Manager 1862-1863 E.BENNETTS; 1864 F.BENNETTS; 1871 WM.RICH;
            1873-1874 WM.RICH
            Chief Agent 1862-1869 JOHN ROACH; 1871 ENNOR
            Secretary 1862-1868 THOS.W.FIELD; 1869 R.R.MICHELL; 1873-1874
            W.GEO.NETTLE

CORNISH CONSOLIDATED   0350

Production: Lead & Silver   Ore(tons) Lead(tons) Silver(ozs)   Value(£)
            1872   23.00   16.80   83.00   0.00
            1873   20.00   15.00   75.00   174.20
            1874   13.00   9.70   50.00   0.00
            Comment 1873 CORNISH CONSOLS
            Iron   Ore(tons)   Iron(%)   Value(£)
            1872   2261.50   0.00   1846.40
            1873   904.00   0.00   744.60
            Comment 1872-1873 BH.

CORNISH TIN STREAMS   9997

Production: Arsenic   Ore(tons) Metal(tons)   Value(£)
            1891   4.00   0.00   15.00

CORNISH,NEW   CALSTOCK   SX 409732 0351

Production: Copper   Ore(tons) Metal(tons)   Value(£)
            1863   100.00   4.60   334.70
            1864   87.00   3.40   237.30
            1865   386.00   16.20   1189.70
            1866   274.00   16.50   1043.40
            1867   57.00   2.70   181.50
            Comment 1863-1867 (C)

CORNUBIA   ROCHE   SW 998597 0352

Production: Tin   Black(tons) Stuff(tons)   Tin(tons)   Value(£)
            1863   15.10   0.00   0.00   1062.80
            1873   2.20   0.00   0.00   161.80
            1874   6.20   0.00   0.00   273.30
            Comment 1874 VALUE EST
Ownership: 1912 CORNISH EXPLORATION CO.LTD.; 1913 CORNUBIA TIN CO.LTD.
            Comment 1863-1865 SUSPENDED; 1912 PROSPECTING OCT.1912; 1913
            PROSPECTING
Management: Chief Agent 1859 H.B.GROSE; 1860-1862 W.H.GRAY; 1872-1873
            DAVID COCK; 1874 H.LANCASTER
            Secretary 1859-1860 P.WATSON (P); 1861-1865 W.H.GRAY (P)
Employment: Underground   Surface   Total
            1912   6   6
            1913   28   22   50

CORNUBIAN   PERRANZABULOE   SW 797512 0353

Production: Lead & Silver   Ore(tons) Lead(tons) Silver(ozs)   Value(£)
            1845   420.00   252.00   0.00   0.00
Ownership: Comment 1863-1871 SEE CHIVERTON

CORNWALL AND ST.VINCENT   CALLINGTON   SX 383695 0354

Production: Tin No detailed return
Management: Manager 1871 J.PENGELLY

CORNWALL GREAT CONSOLS   CALLINGTON   0355

Production: Tin   Black(tons) Stuff(tons)   Tin(tons)   Value(£)
            1880   19.80   0.00   0.00   964.40
            1881   25.80   0.00   16.80   1392.00
            1882   3.70   0.00   2.40   221.00
Ownership: 1879-1882 CORNWALL GREAT CONSOLS MINING CO.
Management: Manager 1879 H.MINERS
            Chief Agent 1879 THOS.DOIDGE; 1880-1882 WM.DOIDGE
            Secretary 1879-1881 THOS.HORSWILL (P)
Employment: Underground   Surface   Total
            1879   6   7   13
            1880   22   15   37
            1881   28   19   47
            1882   8   6   14

136

Production: Lead & Silver No detailed return
Ownership: 1876 NORTH CORNWALL MINING CO.; 1877—1880 NORTH CORNWALL
            SILVER & LEAD CO.LTD.
            Comment 1879—1880 NOT WORKED
Management: Chief Agent 1876—1880 THOS.DOIDGE
Employment: Underground   Surface   Total
            1878   2   2

CORYTON   ST.NEOT   SX 204670 0357

Ownership: 1899 P.B.HENWOOD; 1900 P.B.HENWOOD & JOHNS

COTEHELE CONSOLS   CALSTOCK   SX 423692 0358

Production: Copper   Ore(tons) Metal(tons)   Value(£)
            1880   6.00   0.40   21.90
            1883   No detailed return
            Comment 1880 COTHELE CONSOLS INC.OKEL TOR; 1883 SEE OKEL TOR
            Tin   Black(tons) Stuff(tons)   Tin(tons)   Value(£)
            1883—1885 No detailed return
            Comment 1883—1885 SEE OKEL TOR
            Arsenic   Ore(tons) Metal(tons)   Value(£)
            1883—1887 No detailed return
            Comment 1883—1887 SEE OKEL TOR
            Arsenic Pyrite   Ore(tons)   Value(£)
            1880   60.00   45.00
            1881   150.00   110.30
            1882   112.00   84.00
Ownership: 1880 THOS.S.TREBY JAS.START; 1881—1882 COTEHELE MINING
            CO.LTD.
            Comment 1883—1886 SEE OKEL TOR
Management: Manager 1881 HY.BULFORD
            Chief Agent 1880—1881 THOS.S.TREBY; 1882 HY.BULFORD
Employment: Underground   Surface   Total
            1880   6   6
            1881   17   5   22
            1882   9   9
            1883—1886
            Comment 1883—1886 SEE OKEL TOR

CRADDOCK MOOR   ST.CLEER   SX 259702 0360

Production: Copper   Ore(tons) Metal(tons)   Value(£)
            1856   623.00   57.20   5627.60
            1857   989.00   93.10   9676.30
            1858   1252.00   109.80   10410.60
            1859   1433.00   125.90   12200.70
            1860   1434.00   125.50   11573.20
            1861   1755.00   149.30   13479.30
            1862   1879.00   145.10   11951.90
            1863   2083.00   154.00   11828.80
            1864   1663.00   124.10   10317.40

Copper	Ore(tons)	Metal(tons)	Value(£)
1865	1247.00	86.80	6835.80
1866	933.00	66.70	4268.40
1867	854.00	63.10	4333.90
1868	930.00	76.30	4950.90
1869	990.00	69.80	4214.60
1870	708.00	52.30	2976.20
1871	516.00	46.00	2846.30
1872	479.00	35.40	2648.30
1873	342.00	25.00	1548.40
1874	31.00	2.90	211.00

Comment 1856-1874 (C)
Ownership: 1907-1913 HOLMES & GUTTRIDGE
Comment 1873 STOPPED IN 1873; 1911-1913 PROSPECTING
Management: Manager 1859-1871 HY.TAYLOR; 1872-1873 HY.PHILLIPS
Chief Agent 1859-1860 JOHN TAYLOR; 1861-1866 HY.PHILLIPS JOHN
TAYLOR; 1867-1871 HY.PHILLIPS
Secretary 1859 E.A.CROUCH (P); 1860 RICH.HINGSTON (P);
1861-1871 JOHN TAYLOR (P); 1872-1873 JOHN TAYLOR; 1907-1913
R.KNOTT (S)

Employment:	Underground	Surface	Total
1907		5	5
1908	3	1	4
1909	1	3	4
1910-1911		3	3
1912		2	2
1913		3	3

CRANE   CAMBORNE   SW 638400 0361

Production: Lead & Silver	Ore(tons)	Lead(tons)	Silver(ozs)	Value(£)
1862	1.10	0.70	13.00	0.00
1866	0.80	0.60	0.00	0.00

Copper	Ore(tons)	Metal(tons)	Value(£)
1851	103.00	10.40	739.70
1852	326.00	28.90	2373.40
1853	242.00	21.40	2105.40
1862	77.00	6.80	578.90
1863	36.00	3.70	296.60
1864	108.00	11.70	1011.20
1865	131.00	12.40	1012.20
1866	25.00	1.50	107.60
1868	8.00	0.70	63.40
1870	8.00	0.60	40.60

Comment 1851-1853 (C)INC.BEJAWASA; 1862-1866 (C); 1868 (C);
1870 (C)
Management: Manager 1860 J.SQUIRE (P); 1862-1864 HY.SKEWIS
Chief Agent 1860-1861 HY.SKEWIS JOS.RULE; 1862-1864 JOS.RULE;
1865-1866 WM.ROWE C.DUNN; 1868 RICH.ROWE
Secretary 1860 BEN.MATTHEWS; 1861-1866 BEN.MATTHEWS (P); 1868
BEN.MATTHEWS (P)

CREDDIS   LITTLE PETHERICK   SW 913731 0366

Production: Copper   Ore(tons) Metal(tons)   Value(£)
            1867   54.00   4.30   0.00
            Comment 1867 (P)CREDIS

CREEGBRAWSE   KENWYN   SW 745435 0362

Production: Zinc   Ore(tons) Metal(tons)   Value(£)
            1875   3.60   0.00   4.50
            Copper   Ore(tons) Metal(tons)   Value(£)
            1845   546.00   39.10   2902.40
            1846   817.00   59.50   4297.80
            1847   1314.00   86.50   5984.10
            1848   1299.00   86.70   5002.90
            1849   747.00   53.30   3371.40
            1852   1220.00   66.70   5665.90
            1853   2062.00   119.40   11473.20
            1854   1339.00   73.80   7243.40
            1855   815.00   40.20   3903.40
            1856   350.00   16.90   1516.50
            1857   266.00   13.70   1299.00
            1858   234.00   11.80   1047.40
            1859   152.00   7.50   680.90
            1862   57.00   3.40   268.00
            1863   9.00   0.50   39.10
            1864   14.00   0.60   51.60
            1865   18.00   0.80   55.00
            1873   9.00   0.40   25.90
            1898   731.00   0.00   1608.00
            Comment 1845–1849 (C); 1852–1854 (C); 1855 (C)INC.PENKEVILL;
            1856 (C)CREEGBRAWSE & PENKEVILL UTD.; 1857–1859 (C);
            1862–1865 (C); 1873 (C)
            Tin   Black(tons) Stuff(tons)   Tin(tons)   Value(£)
            1852   4.00   0.00   0.00   0.00
            1853   3.10   0.00   0.00   268.70
            1855   5.30   0.00   0.00   299.70
            1856   63.00   0.00   0.00   3764.50
            1857   54.70   0.00   0.00   3568.00
            1858   42.00   0.00   0.00   2370.00
            1859   44.00   0.00   0.00   2752.00
            1860   47.00   0.00   0.00   3323.20
            1861   61.90   0.00   0.00   3534.50
            1862   74.60   0.00   0.00   4344.20
            1863   47.90   0.00   0.00   3108.50
            1864   93.00   0.00   0.00   5139.90
            1865   66.00   0.00   0.00   3127.50
            1866   75.50   0.00   0.00   3252.60
            1867   79.10   0.00   0.00   3761.00
            1868   134.20   0.00   0.00   6795.10
            1869   41.00   0.00   0.00   2331.30
            1870   78.30   0.00   0.00   5370.30
            1871   39.00   0.00   0.00   2773.70
            1872   63.00   0.00   0.00   5219.10
            1873   34.10   457.00   0.00   4074.50
            1874   31.80   92.00   0.00   1777.20

                              139

Tin	Black(tons)	Stuff(tons)	Tin(tons)	Value(£)
1875	22.50	0.00	0.00	1055.70
1876	21.30	0.00	0.00	775.50
1877	16.00	0.00	0.00	672.20
1878	0.00	201.00	0.00	230.70
1879	8.00	215.00	0.00	490.50
1880	275.30	0.00	0.00	426.30
1881	7.20	0.00	4.80	360.00
1882	3.60	71.00	2.40	175.00
1883	8.30	346.00	0.00	414.00
1884	5.20	134.30	0.00	207.00
1885	1.00	19.90	0.00	45.00
1893	0.00	56.00	0.00	42.00
1894	0.00	27.50	0.00	14.00
1895	No detailed return			
1896	20.00	0.00	0.00	600.00
1899	0.00	22.00	0.00	42.00
1900	0.00	41.00	0.00	84.00
1901	0.60	0.00	0.00	43.00
1902	0.00	11.00	0.00	23.00
1903	0.00	41.00	0.00	46.00
1904	0.00	6.00	0.00	10.00
1905	0.00	4.80	0.00	7.00
1912	0.00	50.00	0.00	55.00
1913	0.00	20.00	0.00	25.00

Comment 1852 TIN ORE; 1856 INC.PENKEVILL; 1857 CREEGBRAWSE &
PENKEVILL UNITED; 1858-1861 INC.PENKEVILL; 1862
INC.PENKEVILL.TWO RETURNS AGG.; 1863 VALUE.INC.TIN STUFF;
1864-1866 INC.PENKEVILL; 1867 CREEGBRAWSE & PENKEVILL UNITED;
1868-1873 INC.PENKEVILL; 1874 CREEGBRAWSE UNITED; 1875-1876
INC.PENKEVILL; 1904-1905 OPEN WORK; 1912 VALUE EST

Arsenic	Ore(tons)	Metal(tons)	Value(£)
1873	0.20	0.00	0.40
1874	9.50	0.00	20.50
1875	11.60	0.00	56.00
1876	15.80	0.00	44.00
1877	6.00	0.00	17.50

Comment 1873-1874 INC.PENKEVILL

Arsenic Pyrite	Ore(tons)	Value(£)
1875	48.80	21.50
1900	40.00	15.00

Comment 1900 INC.UNITY WOOD

Ownership: 1877-1886 CREEGBRAWSE TIN MINING CO.; 1893-1894 THOS.BEHENNA;
1895-1896 F.GRIBBLE; 1898 OLI.NORTHEY W.J.TRYTHALL; 1899-1901
OLI.NORTHEY; 1902-1903 F.DEITZSCH; 1904 WM.BURROWS; 1911-1913
WELLINGTON BROS.
Comment 1859-1877 INC.PENKEVILL; 1886 NOT WORKED; 1893-1894
TADPOLE PART TRIBUTER; 1895 NOT WORKED; 1904 HELD ON TRIBUTE
Management: Manager 1859-1869 FRAN.PRYOR; 1872 THOS.FIELD; 1873
WM.BUCKINGHAM; 1874-1877 SML.BAILEY; 1878-1881 WM.GILES
Chief Agent 1859-1869 J.BLIGHT; 1870 J.BLIGHT JOHN JAMES;
1871 JOHN JAMES J.BLIGHT; 1872 JOHN JAMES J.BLIGHT THOS.GOUGH
JOHN JOHNS; 1873 JOHN NINNIS JNR.JOHN JAMES J.BLIGHT;
1874-1875 JOHN JAMES J.BLIGHT; 1903-1904 WM.BURROWS
Secretary 1859-1870 W.H.TREGONNING (P); 1871 THOS.W.FIELD;

1872 WM.BUCKINGHAM; 1873 T.W.FIELD T.GOOGH; 1874
THOS.W.FIELD; 1875–1878 T.C.GREGORY; 1880–1881 JOHN L.SMITH
(P); 1882–1886 JOHN L.SMITH

Employment:

	Underground	Surface	Total
1878	11	2	13
1879	10	14	14
1880	10	3	13
1881	3	7	10
1882	3	4	7
1883	5	6	11
1884	6	6	12
1885	2	4	6
1893	3		3
1894	2		2
1896	9	3	12
1898–1900	2		2
1901	14		14
1902	11		11
1903	2		2
1911–1913	3		3

CREEGBRAWSE (BEHENNA)   KENWYN   SW 745435 0363

Production: Tin No detailed return
Ownership: 1898–1900 THOS.BEHENNA
          Comment 1898–1900 INC. UNITY WOOD
Employment:

	Underground	Surface	Total
1898	9	4	13
1899	7	1	8
1900	2	1	3

Comment 1898–1900 INC. UNITY WOOD

CRENVER AND ABRAHAM   CROWAN   SW 630338 0364

Production:

Lead & Silver	Ore(tons)	Lead(tons)	Silver(ozs)	Value(£)
1867	0.50	0.30	0.00	0.00
1870	0.20	0.10	0.00	0.00
1871	No detailed return			
1872	4.00	3.00	15.00	0.00
1873	0.20	0.10	0.00	2.50

Comment 1867 FOR AG.SEE GREAT RETALLACK; 1873 CRENVER

Copper	Ore(tons)	Metal(tons)	Value(£)
1866	396.00	22.40	1391.40
1867	668.00	32.10	2127.10
1868	2120.00	97.80	5709.90
1869	2321.00	109.90	6025.60
1870	1895.00	95.20	5065.70
1871	1568.00	105.40	6392.00
1872	1363.00	76.60	5835.30
1873	1441.00	137.30	9033.70
1874	2725.00	205.50	13293.40
1875	3644.00	202.40	14003.50
1876	3543.00	190.90	12024.40

CRENVER AND ABRAHAM   CROWAN   Continued

Comment 1866-1876 (C)
Tin   Black(tons) Stuff(tons)   Tin(tons)   Value(£)
1868   0.00   0.00   0.00   18.80
1869   0.00   121.60   0.00   165.00
1870   0.00   45.30   0.00   21.10
1871   0.00   905.60   0.00   2127.60
1872   0.00   804.60   0.00   1969.00
1873   0.00   751.30   0.00   1752.10
1874   38.70   0.00   0.00   1637.00
1875   85.20   0.00   0.00   4320.70
1876   27.20   0.00   0.00   915.50
Comment 1868 TIN STUFF
Arsenic   Ore(tons) Metal(tons)   Value(£)
1873   0.30   0.00   2.60
Ownership: Comment 1864 INC.OATFIELDS; 1865 INC.OATFIELDS SUSPENDED;
              1868-1871 CRENVER WHEAL ABRAHAM; 1872-1874 CRENVER WHEAL
              ABRAHAM UNITED; 1875-1876 CRENVER AND ABRAHAM UNITED
Management: Manager 1864 JAS.VIVIAN; 1868 JAS.VIVIAN; 1869-1872 WM.KITTO;
              1876 WM.THOMAS
              Chief Agent 1869 JAS.GARLAND; 1870-1871 WM.THOMAS W.J.PAULL;
              1872 JOHN VIVIAN W.J.PAULL; 1873 WM.THOMAS W.J.PAULL JOHN
              VIVIAN; 1874-1875 WM.THOMAS JOHN VIVIAN JOHN EDWARDS; 1876
              JOHN EDWARDS JOHN VIVIAN
              Secretary 1864 W.PAGE CARDOZA; 1868-1876 W.PAGE CARDOZA

CRENVER,SOUTH   CROWAN   SW 634334 0365

Production: Copper   Ore(tons) Metal(tons)   Value(£)
1853   834.00   35.80   3124.60
1854   1629.00   58.40   5258.00
1855   1809.00   61.20   5384.50
1856   935.00   35.80   2894.00
1857   1115.00   41.80   3618.40
1858   735.00   30.20   2504.80
1859   559.00   24.10   2125.90
1860   415.00   18.50   1568.20
1861   287.00   14.20   1209.70
1862   462.00   17.70   1329.90
1863   291.00   14.00   1045.20
1864   49.00   2.10   169.20
Comment 1853-1864 (C)
Tin   Black(tons) Stuff(tons)   Tin(tons)   Value(£)
1862   0.00   0.00   0.00   67.80
Comment 1862 TIN STUFF
Ownership: Comment 1865 SUSPENDED
Management: Manager 1859 JOHN DELBRIDGE
              Chief Agent 1859-1864 ED.CHEGWIN
              Secretary 1859-1864 W.PAGE CARDOZA (P)

CRINNIS CONSOLS   ST.AUSTELL   SX 049519 0367

Production: Copper   Ore(tons) Metal(tons)   Value(£)
1869   80.00   6.90   433.80

CRINNIS CONSOLS   ST.AUSTELL   Continued

        Copper   Ore(tons) Metal(tons)   Value(£)
        1873   15.00   0.70   30.40
        Comment 1869 (C).SEE ALSO GREAT CRINNIS; 1873 (C).SEE ALSO
        GREAT CRINNIS

CRINNIS,EAST   ST.BLAZEY   SX 065528 0368

Production: Lead   Ore(tons) Metal(tons)   Value(£)
            1859   0.05   0.04   0.00
            Zinc   Ore(tons) Metal(tons)   Value(£)
            1859   No detailed return
            Comment 1859 SEE PEMBROKE
            Copper   Ore(tons) Metal(tons)   Value(£)
            1852-1859 No detailed return
            1860   890.00   54.60   4678.60
            1861   909.00   57.60   4988.90
            1862   1014.00   63.10   4888.00
            Comment 1852-1859 SEE PEMBROKE; 1860-1861 (C)INC.SOUTH PAR;
            1862 (C)
            Tin   Black(tons) Stuff(tons)   Tin(tons)   Value(£)
            1856-1857 No detailed return
            1882   3.60   0.00   2.30   122.00
            1884   2.40   0.00   0.00   114.00
            1886   3.40   0.00   0.00   172.00
            1887   1.80   0.00   0.00   116.00
            1888   1.00   0.00   0.00   67.00
            Comment 1856-1857 SEE PEMBROKE
Ownership: Comment 1861 INC.SOUTH PAR; 1862-1865 INC.SOUTH PAR
            SUSPENDED
Management: Manager 1859-1860 REV.J.J.JEFFREY
            Chief Agent 1859 FRAN.GILL; 1861 CHAS.MARRATT
            Secretary 1859-1860 MAJOR DAVIS (P); 1861-1862 JOHN
            POLKINGHORNE; 1863-1864 WM.POLKINGHORNE; 1865 JOHN
            POLKINGHORNE

CRINNIS,GREAT   CHARLESTOWN   SX 049519 0369

Production: Lead & Silver   Ore(tons) Lead(tons) Silver(ozs)   Value(£)
            1854   10.80   4.00   950.00   0.00
            1856   148.10   80.80   0.00   0.00
            1857   6.30   12.00   0.00   0.00
            1858   0.40   0.30   0.00   0.00
            1859   No detailed return
            1860   6.60   4.90   0.00   0.00
            1861   No detailed return
            Comment 1858 FOR AG.SEE GREAT ALFRED; 1860 FOR AG.SEE
            CALSTOCK CONSOLS
            Copper   Ore(tons) Metal(tons)   Value(£)
            1854   157.00   9.20   879.40
            1855   629.00   47.00   4806.00
            1856   928.00   58.40   5237.40
            1857   1037.00   52.60   4919.20
            1858   207.00   11.30   1080.10

                                    143

Copper	Ore(tons)	Metal(tons)	Value(£)
1859	230.00	13.00	1158.50
1860	260.00	13.80	1121.80
1861	183.00	8.80	745.40
1862	56.00	2.40	183.30
1863	6.00	1.00	79.60
1864	3.00	0.80	55.20
1866	1.00	0.20	7.60
1868	7.00	0.10	2.60
1869	2.00	0.10	6.50
1877	53.30	5.50	295.10
1878	176.90	16.20	818.70
1879	33.40	3.60	167.10
1880	136.90	11.40	608.90
1881	58.10	5.50	263.80

Comment 1854–1864 (C); 1866 (C); 1868 (C); 1869 (C).SEE ALSO
CRINNIS CONSOLS; 1873 SEE ALSO CRINNIS CONSOLS; 1877
INC.CARLYON; 1878–1881 CRINNIS & CARLYON CONSOLS

Tin	Black(tons)	Stuff(tons)	Tin(tons)	Value(£)
1864	0.00	0.00	0.00	21.30

Comment 1864 TIN STUFF

Silver	Ore(tons)	Metal(tons)	Value(£)
1877	2.10	0.00	0.00
1879	10.70	0.00	440.00
1880	12.60	0.00	273.00
1881	4.30	0.00	183.40

Comment 1877 CRINNIS & CARLYON CONSOLS; 1879–1881
INC.CARLYON

Ownership: 1875 GREAT CRINNIS CO.; 1876–1878 RICH.H.WILLIAMS; 1879 GREAT
CRINNIS & CARLYON CONSOLS CO.; 1880–1881 RICH.H.WILLIAMS
Comment 1863–1865 SUSPENDED; 1872–1874 INC.CARLYON; 1875–1881
GREAT CRINNIS AND CARLYON CONSOLS
Management: Manager 1859–1860 JOHN WEBB; 1872–1881 RICH.H.WILLIAMS
Chief Agent 1859–1862 WM.WILCOCK; 1877–1881 WM.TURNER
Secretary 1859–1861 WM.CHARLES (P); 1862–1865
WM.POLKINGHORNE; 1873–1874 COMMITTEE; 1876–1881
C.W.BROADHURST

Employment:	Underground	Surface	Total
1878	18	14	32
1879	19	11	30
1880	22	9	31
1881	17	9	26

Comment 1878–1881 INC.CARLYON

CRINNIS,SOUTH   CHARLESTOWN   SX 045522 0370

Production: Lead	Ore(tons)	Metal(tons)	Value(£)
1859	0.05	0.04	0.00

1860–1861 No detailed return

Zinc	Ore(tons)	Metal(tons)	Value(£)
1859	103.90	0.00	222.70
1860	129.80	0.00	324.50
1864	36.80	0.00	38.10

CRINNIS,SOUTH   CHARLESTOWN   Continued

	Copper Ore(tons)	Metal(tons)	Value(£)
1849	156.00	12.70	817.90
1850	75.00	6.20	419.00
1851	95.00	5.30	302.20
1853	550.00	50.00	4873.30
1854	1925.00	144.30	14532.80
1855	1772.00	138.10	13943.50
1856	1524.00	108.50	10061.40
1857	2369.00	154.70	15201.70
1858	1622.00	118.10	10990.90
1859	1405.00	85.50	7834.70
1860	1250.00	78.10	6920.00
1861	1481.00	110.00	9742.50
1862	1020.00	56.00	4329.80
1863	214.00	10.40	727.30
1864	204.00	13.20	1106.30
1865	108.00	7.90	605.20
1866	34.00	1.20	77.50
1868	60.00	2.30	136.00

Comment 1849-1851 (C); 1853-1866 (C); 1868 (C)
Ownership: Comment 1901-1910 SEE APPLE TREE
Management: Manager 1859-1865 FRAN.BARRETT
             Chief Agent 1859-1860 FRAN.BARRETT JNR.; 1861-1865
             CHAS.ROWETT
             Secretary 1859-1865 WM.PETHERICK (P)
Employment: Underground   Surface   Total
             1901-1907
             Comment 1901-1907 SEE APPLE TREE

CRINNIS,WEST   CHARLESTOWN   SX 046523 0371

Production: Lead & Silver   Ore(tons) Lead(tons) Silver(ozs) Value(£)
            1860   1.60   1.10   0.00   0.00
            1861   1.60   1.00   16.00   0.00
            Comment 1860 FOR AG.SEE WEST CHIVERTON

Copper	Ore(tons)	Metal(tons)	Value(£)
1855	311.00	21.70	2180.80
1856	188.00	14.90	1414.20
1857	134.00	9.60	914.60
1858	568.00	37.60	3418.00
1859	467.00	251.00	2300.10

            Comment 1855-1859 (C)
Ownership: Comment 1859 WEST CRINNIS & REGENT UNITED; 1860-1865 WEST
            CRINNIS & REGENT UNITED SUSPENDED
Management: Manager 1859 JOHN WEBB
            Secretary 1859-1861 WM.CHARLES (P); 1862-1865 G.J.SOPER

CROFT GOTHAL   ST.HILARY   SW 569309 0373

Production: Copper   Ore(tons) Metal(tons)   Value(£)
            1853   No detailed return
            1855-1858 No detailed return
            Comment 1853 SEE HALLAMANNING; 1855-1858 SEE HALLAMANNING

CROFTY   ILLOGAN   SW 663411 0374

Production: Copper   Ore(tons) Metal(tons)   Value(£)
            1863   34.00   1.40   99.30
            1864   42.00   1.80   145.70
            1865   13.00   0.70   58.80
            Comment 1863-1865 (C)

CROFTY,EAST   ILLOGAN   SW 665414 0375

Production: Copper   Ore(tons) Metal(tons)   Value(£)
            1845   6173.00   483.20   36302.70
            1846   4127.00   312.40   22698.70
            1847   3580.00   273.00   19540.80
            1848   3100.00   228.00   13453.30
            1849   3217.00   230.70   14541.70
            1850   3261.00   221.20   14863.60
            1851   2267.00   152.80   10033.60
            1852   1771.00   119.30   10177.90
            1853   1655.00   92.50   9288.90
            1854   620.00   33.60   3359.90
            Comment 1845-1854 (C)
            Tin   Black(tons) Stuff(tons)   Tin(tons)   Value(£)
            1855   2.70   0.00   0.00   163.50
            1861   38.00   0.00   0.00   2936.40

CROFTY,NORTH   ILLOGAN   SW 665414 0376

Production: Copper   Ore(tons) Metal(tons)   Value(£)
            1854   653.00   40.50   4089.30
            1855   1582.00   100.70   10385.50
            1856   1478.00   91.90   8552.40
            1857   918.00   51.40   5156.70
            1858   575.00   30.00   2635.40
            1859   468.00   27.20   2523.80
            1860   544.00   28.20   2546.30
            1861   497.00   20.70   1772.40
            1862   525.00   25.40   1998.00
            1863   389.00   18.60   1348.70
            1864   316.00   16.70   1345.10
            1865   237.00   12.00   917.20
            1866   69.00   3.00   219.10
            1867   81.00   5.80   398.00
            1868   49.00   3.40   227.60
            1869   136.00   17.10   1102.10
            1870   165.00   21.40   1353.10
            1871   353.00   30.10   1861.90
            1872   199.00   16.00   1330.80
            1873   60.00   4.80   327.90
            1874   25.00   1.80   108.90
            1910   1.00   0.00   6.00
            Comment 1854-1874 (C); 1910 VALUE EST
            Tin   Black(tons) Stuff(tons)   Tin(tons)   Value(£)
            1855   5.90   0.00   0.00   369.10
            1857   18.00   0.00   0.00   1345.80

                                  146

Tin	Black(tons)	Stuff(tons)	Tin(tons)	Value(£)
1858	56.10	0.00	0.00	3646.60
1859	67.30	0.00	0.00	4911.70
1860	70.80	0.00	0.00	5724.80
1862	42.20	0.00	0.00	2961.00
1863	84.10	0.00	0.00	5772.60
1864	129.50	0.00	0.00	8654.90
1865	92.00	0.00	0.00	5219.40
1866	106.60	0.00	0.00	5385.70
1867	94.90	0.00	0.00	5315.90
1868	171.70	0.00	0.00	9388.30
1869	122.10	0.00	0.00	8943.80
1870	145.40	0.00	0.00	10750.30
1871	123.20	0.00	0.00	9732.70
1872	136.50	0.00	0.00	10003.50
1873	101.80	0.00	0.00	7937.60
1874	51.00	0.00	0.00	2864.00
1899	4.40	0.00	0.00	245.00
1910	12.00	0.00	0.00	888.00

Comment 1862 TWO RETURNS AGGREGATED; 1863 9 MONTHS ONLY; 1872
VALUE INC.TIN STUFF; 1874 VALUE EST

Arsenic	Ore(tons)	Metal(tons)	Value(£)
1857	5.50	0.00	12.40
1866	109.10	0.00	153.90
1867	55.30	0.00	82.90
1868	131.90	0.00	366.50
1869	87.20	0.00	259.80
1870	222.80	0.00	783.60
1871	309.70	0.00	613.10
1872	147.60	0.00	283.30
1873	380.00	0.00	840.00
1874	17.20	0.00	18.50
1899	15.00	0.00	174.00
1909	4.20	0.00	26.00
1910	81.00	0.00	486.00
1911	1.50	0.00	12.00

Comment 1866-1868 CRUDE; 1870 CRUDE; 1871 CRUDE & SOOT

Arsenic Pyrite	Ore(tons)	Value(£)
1897	545.00	259.00
1898	410.00	193.00
1900	37.00	37.00

Ownership: 1897-1900 ROSKEAR CALCINING CO.; 1907-1913 SCOTT-THOMPSON
SYNDICATE LTD.
Comment 1911-1912 SUSPENDED; 1913 ABANDONED
Management: Manager 1859 JOS.PURAN; 1860-1865 JOS.VIVIAN; 1866-1868
JOS.VIVIAN & SON; 1869-1872 JOS.VIVIAN; 1873 JOS.VIVIAN &
SON
Chief Agent 1859 JOHN IVEY; 1860-1864 WM.THOMAS JNR.;
1865-1873 WM.THOMAS JNR.GEO.BENNETTS; 1874 JOHN VIVIAN & SON;
1897-1900 J.A.TEMBY; 1907-1913 PETER TEMBY
Secretary 1859-1865 ALM.E.PAULL (P); 1866-1873 WM.WATSON

Employment:	Underground	Surface	Total
1897	8	12	20
1898	10	11	21
1899	5	36	41

	Underground	Surface	Total
1907	8	3	11
1908	16	28	44
1909	22	52	74
1910	39	48	87

CROFTY,SOUTH   ILLOGAN   SW 669413 0377

Production: Copper	Ore(tons)	Metal(tons)	Value(£)
1854	492.00	23.30	2213.70
1855	851.00	40.10	3881.50
1856	866.00	40.40	3587.30
1857	980.00	56.20	5563.40
1858	785.00	39.80	3513.40
1859	624.00	33.50	3047.80
1860	744.00	39.40	3488.00
1861	401.00	18.80	1551.70
1862	213.00	10.10	776.50
1863	325.00	14.70	1057.10
1864	642.00	28.40	2257.80
1865	1459.00	96.20	7430.10
1866	1599.00	99.70	6457.30
1867	2136.00	122.50	8227.90
1868	3131.00	155.00	9537.40
1869	2564.00	146.80	8599.50
1870	1509.00	69.60	3645.60
1871	1782.00	85.20	4769.40
1872	2333.00	119.40	8283.20
1873	3208.00	164.50	9507.10
1874	1590.00	81.40	4688.30
1875	1371.00	72.10	5092.80
1876	1942.00	106.10	6807.40
1877	1880.30	91.40	4650.50
1878	1810.60	93.50	4219.30
1879	112.60	6.20	265.50
1880	102.30	7.50	306.00
1881	97.30	6.80	331.40
1882	13.40	0.90	54.00
1884	4.00	0.00	9.00
1885	7.00	0.00	23.00
1886	2.50	0.00	10.00
1887	11.00	0.00	26.00
1889	12.00	0.00	31.00
1891	9.00	0.00	32.00
1892	4.00	0.00	13.00
1901	12.00	0.00	36.00
1903	6.00	0.00	20.00
1904	8.00	0.00	28.00
1906	7.00	0.00	39.00
1910	12.00	0.00	71.00
1911	27.00	0.00	134.00

Comment 1854-1876 (C)

Tin	Black(tons)	Stuff(tons)	Tin(tons)	Value(£)
1855	1.90	0.00	0.00	87.60

148

Tin	Black(tons)	Stuff(tons)	Tin(tons)	Value(£)
1856	0.30	0.00	0.00	16.00
1857	0.80	0.00	0.00	56.60
1858	4.10	0.00	0.00	226.20
1859	0.00	0.00	0.00	204.80
1860	1.70	0.00	0.00	107.30
1861	0.80	0.00	0.00	47.60
1862	1.90	0.00	0.00	93.80
1863	1.70	0.00	0.00	282.50
1864	1.30	0.00	0.00	904.30
1865	2.30	0.00	0.00	1008.40
1866	1.00	0.00	0.00	227.70
1867	5.60	0.00	0.00	755.60
1868	4.80	0.00	0.00	1179.00
1869	7.40	0.00	0.00	2056.20
1870	10.70	0.00	0.00	3144.40
1871	18.00	0.00	0.00	6471.60
1872	63.60	0.00	0.00	7030.50
1873	207.00	490.00	0.00	14188.80
1874	180.30	0.00	0.00	8624.80
1875	163.40	0.00	0.00	7182.60
1876	144.20	0.00	0.00	5397.30
1877	136.60	0.00	0.00	1070.40
1878	173.50	0.00	0.00	6120.00
1879	1.20	1627.60	0.00	1346.00
1880	0.00	1228.30	0.00	1138.00
1881	117.70	43.90	0.00	6076.00
1882	120.30	0.00	0.00	6977.00
1883	154.40	0.00	0.00	7505.00
1884	169.60	0.00	0.00	7143.00
1885	91.80	0.00	0.00	4073.00
1886	99.00	0.00	0.00	5402.00
1887	125.30	0.00	0.00	7967.00
1888	125.50	0.00	0.00	8248.00
1889	122.10	0.00	0.00	6673.00
1890	147.40	0.00	0.00	8152.00
1891	184.80	0.00	0.00	9900.00
1892	203.10	0.00	0.00	11107.00
1893	258.80	0.00	0.00	13313.00
1894	307.00	0.00	0.00	13148.00
1895	251.00	0.00	0.00	10001.00
1896	34.00	0.00	0.00	1083.00
1898	0.00	150.00	0.00	72.00
1899	0.00	260.00	0.00	142.00
1900	4.20	0.00	0.00	306.00
1901	158.40	0.00	0.00	10758.00
1902	296.00	0.00	0.00	21831.00
1903	348.60	0.00	0.00	27341.00
1904	275.30	0.00	0.00	20272.00
1905	291.40	0.00	0.00	22397.00
1906	153.80	0.00	0.00	14606.00
1907	228.50	0.00	0.00	22530.00
1908	434.10	0.00	0.00	33554.00
1909	682.80	0.00	0.00	52421.00
1910	631.50	0.00	0.00	56619.00

Tin	Black(tons)	Stuff(tons)	Tin(tons)	Value(£)
1911	677.60	0.00	0.00	76234.00
1912	624.80	0.00	0.00	81562.00
1913	629.40	0.00	0.00	78612.00

Comment 1859 TIN STUFF; 1862 TWO RETURNS AGGREGATED;
1863–1872 VALUE INC.TIN STUFF

Iron	Ore(tons)	Iron(%)	Value(£)
1908	133.00	45.00	555.00
1909	43.00	42.40	248.00

Arsenic	Ore(tons)	Metal(tons)	Value(£)
1858	2.00	0.00	3.50
1872	165.20	0.00	276.00
1873	351.10	0.00	1003.00
1874	235.00	0.00	801.70
1875	393.20	0.00	2822.60
1876	252.60	0.00	1136.00
1877	299.60	0.00	898.60
1878	304.60	0.00	913.60
1879	46.10	0.00	138.20
1880	20.00	0.00	60.00
1881	101.40	0.00	532.20
1882	209.70	0.00	1258.00
1883	137.80	0.00	853.00
1884	156.00	0.00	1071.00
1885	148.60	0.00	928.00
1886	42.00	0.00	235.00
1887	34.00	0.00	202.00
1888	17.00	0.00	109.00
1889	14.00	0.00	84.00
1890	19.00	0.00	116.00
1891	8.00	0.00	61.00
1892	60.00	0.00	339.00
1893	20.00	0.00	104.00
1894	6.00	0.00	40.00
1895	16.00	0.00	96.00
1896	26.00	0.00	176.00
1901	36.00	0.00	295.00
1902	37.00	0.00	167.00
1903	66.00	0.00	387.00
1904	74.00	0.00	360.00
1905	87.00	0.00	400.00
1906	92.00	0.00	1397.00
1907	294.00	0.00	7305.00
1908	321.00	0.00	2477.00
1909	707.80	0.00	7247.00
1910	635.00	0.00	6619.00
1911	952.80	0.00	8787.00
1912	1000.30	0.00	8970.00
1913	773.10	0.00	10036.00

Comment 1884–1885 REFINED; 1886 CRUDE

Tungsten	Ore(tons)	Metal(tons)	Value(£)
1873	14.00	0.00	98.00
1887	3.00	0.00	5.00
1891	1.10	0.00	4.00

1892–1895 No detailed return

Tungsten	Ore(tons)	Metal(tons)	Value(£)
1899	1.00	0.00	6.00
1905	0.40	0.00	20.00
1907	51.00	0.00	7597.00
1908	101.70	0.00	9116.00
1909	160.40	0.00	13902.00
1910	131.20	0.00	13997.00
1911	147.30	0.00	14732.00
1912	132.00	0.00	11316.00
1913	115.40	0.00	10805.00

Arsenic Pyrite	Ore(tons)	Value(£)
1879	54.10	32.00
1899	23.00	14.00

Ownership: 1877–1890 SOUTH WHEAL CROFTY MINING CO.; 1891–1893 SOUTH WHEAL CROFTY MINING CO. C.B.; 1894–1905 SOUTH WHEAL CROFTY MINING CO.; 1906–1913 SOUTH CROFTY LTD.
Comment 1896 SUSPENDED FEB.1896; 1897–1898 SUSPENDED
Management: Manager 1859–1864 WM.RUTTER; 1869 FRAN.GILBERT; 1872–1873 JOH.THOMAS; 1874–1879 N.JAMES JOH.THOMAS; 1880 JOH.THOMAS; 1881 JOH.THOMAS N.JAMES
Chief Agent 1859–1865 FRAN.GILBERT; 1866–1868 FRAN.GILBERT SIMON TOY; 1869 SIMON TOY; 1870–1871 JOH.THOMAS SIMON TOY JOHN JORY; 1872–1876 JOHN JORY JAS.JOHNS; 1877–1879 JOHN JORY WM.PASCOE; 1880 JOHN JORY; 1881 JOHN JORY WM.PASCOE; 1882–1900 JOH.THOMAS; 1901–1904 R.ART.THOMAS; 1913 J.PAULL
Secretary 1859–1871 E.H.RODD (P); 1872–1874 E.H.RODD; 1875–1879 HY.J.LEAN; 1881 HY.J.LEAN (S); 1887–1891 HY.J.LEAN; 1892–1902 HY.J.LEAN (P)

Employment:	Underground	Surface	Total
1878	89	113	202
1879	42	23	65
1880	52	45	97
1881	67	55	122
1882	80	69	149
1883	92	71	163
1884	84	76	160
1885	52	49	101
1886	56	66	122
1887	74	68	142
1888	67	64	131
1889	65	62	127
1890	73	61	134
1891	81	64	145
1892	96	74	170
1893	142	81	223
1894	125	75	200
1895	88	68	156
1896	7	6	13
1897	7	7	14
1898		3	3
1899	1	4	5
1900	25	48	73
1901	93	90	183
1902	129	97	226
1903	163	126	289

CROFTY,SOUTH   ILLOGAN   Continued

	Underground	Surface	Total
1904	189	123	312
1905	172	118	290
1906	178	144	322
1907	254	222	476
1908	270	240	510
1909	327	240	567
1910	332	221	553
1911	376	207	583
1912	380	242	622
1913	375	245	620

CROW HILL,NEW   ST.STEPHEN   SW 937509 0378

Production: Lead & Silver

	Ore(tons)	Lead(tons)	Silver(ozs)	Value(£)
1853	20.00	8.70	40.00	0.00
1855	12.70	8.00	0.00	0.00
1856-1857 No detailed return				
1859	1.00	0.80	0.00	0.00
1861	20.50	12.00	175.00	0.00
1863	69.50	46.90	0.00	0.00
1864	16.40	10.80	170.00	0.00
1865	49.30	31.30	500.00	0.00
1866	96.00	60.40	968.00	0.00
1867	68.70	51.50	622.00	0.00
1868	98.20	73.60	876.00	0.00
1869	9.80	7.40	90.00	0.00
1870	No detailed return			

Comment 1853 CROW HILL; 1855-1857 CROW HILL
Plumbago No detailed return

	Ore(tons)	Value(£)
Uranium		
1910	63.00	0.00

Ownership: 1891 TREGREHAN CONSOLS MINING CO.LTD; 1909-1910 PITCH URANIUM
& RADIUM SYNDICATE LTD.; 1911-1913 RADIUM ORE MINES LTD.
Comment 1873 STOPPED IN 1873 & EARLY IN 1874; 1891 CROWN
HILL. LETTER RETURNED UNKNOWN; 1909-1910 CROW HILL; 1911-1913
CROW HILL OR TOLGARRICK REOPENING
Management: Manager 1862-1868 WM.BROWNE; 1870-1872 A.KENT; 1873
THOS.TRELEASE
Chief Agent 1859-1869 STEP.COLLINS; 1870-1872 THOS.TRELEASE;
1913 C.GOLDSWORTHY
Secretary 1859-1861 WM.BROWNE (P); 1864-1866 WM.BROWNE;
1867-1868 JOHN HITCHINGS; 1869-1873 WM.BROWNE; 1891
FRAN.W.MICHELL
Employment:

	Underground	Surface	Total
1909	6	1	7
1910	10	1	11
1911	9	7	16
1913	2	3	5

CROWAN CONSOLS   LEEDSTOWN   SW 615335 0379

Production: Zinc   Ore(tons) Metal(tons)   Value(£)
            1864   100.30   0.00   77.60
            1865   102.20   0.00   99.70
            Copper   Ore(tons) Metal(tons)   Value(£)
            1863   13.00   0.30   15.20
            1864   103.00   3.30   230.80
            1865   35.00   2.10   160.50
            1866   66.00   3.40   243.50
            Comment 1863-1866 (C)
            Tin   Black(tons) Stuff(tons) Tin(tons)   Value(£)
            1864   0.00   0.00   0.00   75.20
            1865   0.00   0.00   0.00   20.20
            Comment 1864-1865 TIN STUFF
Ownership: Comment 1864 CROWN CONSOLS

CROWAN MINES   LEEDSTOWN   0380

Production: Tin No detailed return
Ownership: Comment 1871 INC.OSBORNE, BOSWORGY, CARDELE & GODOLPHIN
Management: Chief Agent 1871 CARKEEK
            Secretary 1871 W.PAGE CARDOZA

CUBERT   CUBERT   SW 785571 0382

Production: Lead & Silver Ore(tons) Lead(tons) Silver(ozs)   Value(£)
            1846   136.00   81.00   0.00   0.00
            1847   354.00   212.00   0.00   0.00
            1848   68.00   41.00   0.00   0.00
            1849   No detailed return
            1852   70.00   50.00   750.00   0.00
            1853   70.00   51.50   852.00   0.00
            1854   160.00   120.00   1900.00   0.00
            1855   221.50   165.00   2640.00   0.00
            1856   135.50   103.00   1380.00   0.00
            1857   118.50   25.00   390.00   0.00
            Comment 1852 CUBERT UNITED SILVER LEAD; 1853 CUBERT UNITED;
            1854 CUBERT UNITED SILVER LEAD; 1855-1857 CUBERT UNITED

CUDDRA   ST.AUSTELL   SX 040528 0383

Production: Copper   Ore(tons) Metal(tons)   Value(£)
            1861   306.00   12.20   952.40
            1862   35.00   1.50   115.50
            Comment 1861-1862 (C)
            Tin   Black(tons) Stuff(tons) Tin(tons)   Value(£)
            1862   20.90   0.00   0.00   1329.30
            1863   25.20   0.00   0.00   1661.00
            1864   33.80   0.00   0.00   2139.70
            1865   51.00   0.00   0.00   2900.70
            1866   60.90   0.00   0.00   3026.60
            1867   78.70   0.00   0.00   4406.10
            1868   72.10   0.00   0.00   4176.30

                                    153

CUDDRA   ST.AUSTELL   Continued

Tin	Black(tons)	Stuff(tons)	Tin(tons)	Value(£)
1869	73.10	0.00	0.00	5324.40
1870	46.90	0.00	0.00	3446.10
1871	36.00	0.00	0.00	2860.90
1872	13.90	0.00	0.00	1190.80

Comment 1862 TWO RETURNS AGG.
Ownership: Comment 1872–1874 SEE CHARLESTOWN UNITED
Management: Manager 1863–1871 FRAN.PUCKEY
             Chief Agent 1860–1861 J.WEBB A.CUNDY; 1862 FRAN.PUCKEY
             A.CUNDY; 1863 A.CUNDY; 1864 A.CUNDY JOHN GREEN (S); 1865–1871
             A.CUNDY JOHN HITCHINGS (S)
             Secretary 1860–1871 WM.POLKINGHORNE (P)

CUDDRA,EAST   ST.BLAZEY   0384

Copper	Ore(tons)	Metal(tons)	Value(£)
1864	17.00	1.00	79.90
1865	9.00	0.50	39.10

Production:
Comment 1864–1865 (C)
Management: Chief Agent 1864–1865 WM.WOOLCOCK
             Secretary 1864–1865 H.C.VIVIAN

CUDDRA,SOUTH   CHARLESTOWN   SX 047527 0385

Copper	Ore(tons)	Metal(tons)	Value(£)
1857	57.00	5.80	673.10

Production:
Comment 1857 (C)
Tin No detailed return
Ownership: Comment 1859 ABANDONED; 1860–1865 SUSPENDED

CUNNING UNITED   ST.JUST   SW 362317 0386

Tin	Black(tons)	Stuff(tons)	Tin(tons)	Value(£)
1873	96.20	0.00	0.00	7488.00
1874	77.80	0.00	0.00	4150.70
1875	75.00	0.00	0.00	3600.00
1876	34.20	0.00	0.00	1574.50

Production:
Comment 1873 VALUE EST; 1874 INC.BOSWEDDEN & CASTLE
Ownership: 1875 RICH.BOYNS; 1876 RICH.BOYNS THOS.BOLITHO
             Comment 1872–1873 FORMERLY BOSCEAN,BOSWEDDEN & CASTLE; 1874
             FORMERLY BOSWEDDEN & CASTLE; 1875–1876 FORMERLY BOSWEDDEN &
             CASTLE SUSPENDED
Management: Manager 1872–1873 JAS.ROACH; 1874–1876 RICH.BOYNS
             Chief Agent 1872–1873 WM.HATTAM CHAS.CLEMENS; 1874 JAS.ROACH
             WM.HATTAM CHAS.CLEMENS; 1875 JAS.ROACH WM.HATTAM; 1876
             JAS.ROACH
             Secretary 1872–1874 RICH.BOYNS

CUPID   REDRUTH   SW 712423 0387

Production:

Copper	Ore(tons)	Metal(tons)	Value(£)	
1862	21.00	1.70	133.50	
1865	1.00	0.10	6.00	

Comment 1862 (C); 1865 (C)

Tin	Black(tons)	Stuff(tons)	Tin(tons)	Value(£)
1855	6.10	0.00	0.00	427.20
1856	0.00	0.00	0.00	161.80
1857	0.00	0.00	0.00	24.30

Comment 1856-1857 TIN STUFF

Fluorspar	Ore(tons)	Value(£)
1896	No detailed return	

Comment 1896 SEE DAMSEL,WEST

Ownership: Comment 1860 SUSPENDED; 1863-1865 SUSPENDED

Management: Manager 1859 RICH.PRYOR
Chief Agent 1861-1862 RICH.PRYOR
Secretary 1859 ROBT.H.PIKE (P); 1861-1862 ROBT.H.PIKE (P)

CURTIS   CAMBORNE   SW 616334 0388

Production:

Zinc	Ore(tons)	Metal(tons)	Value(£)
1864	3.10	0.00	5.40

Copper	Ore(tons)	Metal(tons)	Value(£)
1864	171.00	7.90	613.20
1865	354.00	20.20	1561.10
1866	398.00	19.20	1270.60
1867	151.00	5.70	359.20
1868	23.00	0.80	47.00

Comment 1864 (C)TWO RETURNS AGGREGATED; 1865-1868 (C)

CUSGARNE   ST.DAY   0389

Production:

Tin	Black(tons)	Stuff(tons)	Tin(tons)	Value(£)
1905	0.00	272.00	0.00	170.00
1906	0.00	665.00	0.00	241.00

Comment 1905-1906 INC.BULLER & POLDICE T.H.LETCHER

CUSKEASE   ST.ERTH   SW 579341 0390

Ownership: 1913 GERI RIVER TIN MINES LTD.
Comment 1913 PROSPECTING

Employment:

	Underground	Surface	Total
1913	5	5	10

DAMSEL   ST.DAY   SW 728417 0391

Production:

Copper	Ore(tons)	Metal(tons)	Value(£)
1861	1830.00	89.60	7672.30
1862	29.00	1.80	140.00
1863	47.00	2.30	158.20
1864	12.00	0.70	52.20
1873	86.00	4.10	250.90

DAMSEL   ST.DAY   Continued

        Copper   Ore(tons) Metal(tons)   Value(£)
        1876   2.00   0.10   6.50
        Comment 1861-1864 (C); 1873 (C); 1876 (C)NEW DAMSEL
        Tin   Black(tons) Stuff(tons)   Tin(tons)   Value(£)
        1855   17.10   0.00   0.00   1109.90
        1856   18.20   0.00   0.00   1359.30
        1857   20.60   0.00   0.00   1726.90
        1858   29.80   0.00   0.00   2024.90
        1859   5.60   0.00   0.00   439.70
        1860   8.20   0.00   0.00   636.40
        1861   2.00   0.00   0.00   145.30
        1862   0.10   0.00   0.00   6.70
        1872   0.00   0.00   0.00   15.50
        1873   0.00   0.00   2.50   116.10
        1903-1904 No detailed return
        Comment 1856 TWO RETURNS AGGREGATED; 1872 GREAT DAMSEL TIN
        STUFF; 1873 TIN STUFF; 1903-1904 SEE WOOD MINE
        Fluorspar   Ore(tons)   Value(£)
        1860   32.70   36.70
Ownership: Comment 1865 SUSPENDED
Management: Chief Agent 1859-1860 BLANEY; 1861-1862 RICH.PRYOR; 1863
        W.J.DUNSFORD
        Secretary 1859-1860 FRAN.PRYOR (P); 1861 THOS.KING (P); 1862
        W.J.DUNSFORD; 1863-1865 ROBT.H.PIKE & SON

DAMSEL UNITED   ST.DAY   SW 724414 0392

Production: Copper No detailed return
        Tin No detailed return
        Fluorspar   Ore(tons)   Value(£)
        1872   80.60   40.30
Ownership: Comment 1872 LATE WEST DAMSEL
Management: Manager 1872-1873 JOHN WHITBURN
        Secretary 1872-1873 G.A.MICHELL

DAMSEL,EAST   ST.DAY   SW 733419 0393

Production: Copper No detailed return
        Tin   Black(tons) Stuff(tons)   Tin(tons)   Value(£)
        1891   0.00   394.00   0.00   26.00
Ownership: 1903-1904 T.H.LETCHER
        Comment 1864-1865 SUSPENDED; 1903-1904 T.H.LETCHER LORDS
        AGENT
Management: Chief Agent 1860 WM.HUNT; 1861 WM.HUNT JOS.MICHELL; 1862-1863
        JOS.MICHELL J.W.HUNT
        Secretary 1860-1865 JOHN MICHELL (P)
Employment: Underground   Surface   Total
        1903-1904   1   1

156

DAMSEL,NORTH   ST.DAY   SW 726422 0394

Production: Copper   Ore(tons) Metal(tons)   Value(£)
            1852   165.00   12.60   1215.80
            1853   322.00   23.90   2359.40
            1854   91.00   7.90   828.20
            Comment 1852-1854 (C)
            Tin   Black(tons) Stuff(tons)   Tin(tons)   Value(£)
            1855   2.80   0.00   0.00   181.10

DAMSEL,WEST   ST.DAY   SW 724414 0395

Production: Copper   Ore(tons) Metal(tons)   Value(£)
            1852   290.00   15.90   1396.40
            1853   1165.00   66.50   6292.00
            1854   1662.00   96.00   9476.30
            1855   1543.00   92.50   9336.60
            1856   1732.00   100.80   9213.00
            1857   1837.00   108.70   10658.60
            1858   1609.00   87.90   7856.60
            1859   1305.00   65.10   5874.40
            1860   1910.00   98.50   8716.80
            1862   2065.00   101.30   7867.00
            1863   2252.00   111.20   8339.60
            1864   2166.00   117.60   9570.70
            1865   1927.00   101.70   7905.50
            1866   1603.00   86.70   5466.60
            1867   1426.00   83.80   5786.10
            1868   1338.00   78.70   5033.10
            1869   871.00   57.50   3514.90
            1870   613.00   38.50   2250.10
            1871   380.00   21.30   1220.30
            1872   129.00   7.50   532.30
            Comment 1852-1860 (C); 1862-1872 (C)
            Tin   Black(tons) Stuff(tons)   Tin(tons)   Value(£)
            1855   0.90   0.00   0.00   55.10
            1859   0.00   0.00   0.00   17.40
            1860   0.00   0.00   0.00   2.60
            1865   0.10   0.00   0.00   1.00
            1868   0.00   0.00   0.00   139.60
            1869   0.00   537.60   0.00   601.10
            1870   0.00   273.10   0.00   334.60
            1871   0.00   222.40   0.00   116.90
            Comment 1859-1860 TIN STUFF; 1868 TIN STUFF
            Fluorspar   Ore(tons)   Value(£)
            1864   1.00   12.60
            1865   2.80   46.10
            1866   1.60   1.90
            1868   60.00   42.00
            1871   41.00   20.50
            1891   22.00   20.00
            1892   54.00   31.00
            1893   38.00   25.00
            1894   No detailed return
            1896   100.00   37.00
            Comment 1891 VARIOUS WORKERS; 1892-1893 VARIOUS WORKERS,

157

DAMSEL,WEST   ST.DAY   Continued

            OPENWORKS; 1894 SEE ST.AUBYN; 1896 INC.CUPID
Ownership: 1906-1908 T.H.LETCHER; 1909-1910 J.T.LETCHER
            Comment 1872 SEE DAMSEL UNITED; 1906-1908 T.H.LETCHER LORDS
            AGENT; 1909-1910 J.T.LETCHER LORDS AGENT
Management: Manager 1864-1867 WM.THOMAS; 1868-1871 AB.T.JAMES
            Chief Agent 1859 JOHN MOYLE; 1860-1863 WM.THOMAS
            Secretary 1859-1871 G.A.MICHELL (P)
Employment: Underground   Surface   Total
            1906   1   1

DANESCOMBE VALLEY   CALSTOCK   SX 426696 0396

Production: Copper   Ore(tons) Metal(tons)   Value(£)
            1882-1884 No detailed return
            1887   No detailed return
            1888   103.00   0.00   251.00
            1890   200.00   0.00   219.00
            1891   83.00   0.00   135.00
            1892   283.00   0.00   376.00
            1893   158.00   0.00   90.00
            1895   90.00   0.00   37.00
            Comment 1882-1884 SEE CALSTOCK AND DANESCOMBE; 1887 SEE
            CALSTOCK AND DANESCOMBE
            Tin   Black(tons) Stuff(tons)   Tin(tons)   Value(£)
            1892   2.00   0.00   0.00   81.00
            1893   No detailed return
            Arsenic   Ore(tons) Metal(tons)   Value(£)
            1892   287.00   0.00   3160.00
            Arsenic Pyrite Ore(tons)   Value(£)
            1888   182.00   191.00
            1889   510.00   602.00
            1890   964.00   1274.00
            1891   575.00   863.00
            1892   969.00   969.00
            1893   306.00   306.00
            1894   851.00   851.00
            1895   175.00   179.00
Ownership: 1888-1890 DANESCOMBE VALLEY MINING CO.; 1891 DANESCOMBE
            VALLEY MINING CO.LTD.; 1892-1895 HUGHES & WM.SOWDEN
            Comment 1891 DANESCOMBE; 1892-1894 LESSEES DANESCOMBE VALLEY
            MINING CO
Management: Chief Agent 1888-1895 WM.SOWDEN
            Secretary 1891 H.HENDRICKS
Employment: Underground   Surface   Total
            1888   10   12   22
            1889   16   13   29
            1890   29   20   49
            1891-1892   24   17   41
            1893   18   5   23
            1894   25   11   36
            1895   18   10   28

                                   158

DANIEL   KENWYN   SW 749449 0397

Production: Copper No detailed return
            Tin   Black(tons) Stuff(tons)   Tin(tons)   Value(£)
            1871   0.00   37.70   0.00   44.30
            1872   0.00   0.00   0.00   1290.20
            1873   0.00   195.20   0.00   286.40
            Comment 1872 TIN STUFF; 1873 DANIELL
Ownership: Comment 1873 STOPPED IN 1873
Management: Manager 1870 WM.COUCH; 1872-1873 JOHN NINNIS
            Chief Agent 1871 H.GOUGH
            Secretary 1870-1871 JOHN NINNIS; 1872-1873 JOHN JOSE

DARLINGTON   MARAZION   SW 507318 0398

Production: Copper   Ore(tons) Metal(tons)   Value(£)
            1845   649.00   35.30   2397.40
            1846-1847 No detailed return
            Comment 1845-1847 (C)

DAVEY   ST.MEWAN   0399

Production: Tin No detailed return
Ownership: 1902-1904 JOHN J.TONKIN
Employment: Underground   Surface   Total
            1903   2   1   3
            1904   1   1   1

DEAN AND CHAPTER LANDS   0400

Production: Lead   Ore(tons) Metal(tons)   Value(£)
            1851   6.50   5.00   0.00
            1852-1854 No detailed return

DEER PARK   CALLINGTON   SX 391730 0401

Production: Tin   Black(tons) Stuff(tons)   Tin(tons)   Value(£)
            1873   0.10   0.00   0.00   7.50
            Comment 1873 VALUE EST
Management: Manager 1869-1872 JOHN BUCKNELL
            Chief Agent 1867 JOHN TAYLOR; 1868 JOHN BUCKNELL; 1873-1875
            JOHN GOLDSWORTHY
            Secretary 1867-1869 JOHN BUCKNELL; 1870-1871 J.WRIGHT;
            1872-1875 E.T.R.WILD

DEER PARK   NEWLYN EAST   SW 807555 0402

Production: Lead   Ore(tons) Metal(tons)   Value(£)
            1875   2.10   1.50   0.00
            1877   0.50   0.30   0.00
            1878   0.50   0.30   0.00
            1879   0.60   0.50   0.00

                                159

DEER PARK   NEWLYN EAST   Continued

Zinc	Ore(tons)	Metal(tons)	Value(£)
1877	10.00	0.00	40.00
Iron	Ore(tons)	Iron(%)	Value(£)
1875	40.00	0.00	30.00
1876	227.00	0.00	192.90

Comment 1875-1876 BH.
Ownership: 1863-1865 EDW.CARTER & EXECUTORS OF T.WHITFORD; 1875 BARTON &
             CO.; 1876-1879 DEER PARK LEAD & IRON ORE MINING CO.LTD.; 1880
             CORNISH STEEL IRON ORE CO.; 1881-1883 NEWQUAY MINING CO.LTD.
             Comment 1879-1880 NOT WORKED
Management: Manager 1876 JAS.HENDERSON
             Chief Agent 1863-1865 JOHN BALL; 1875 J.G.BARTON; 1876-1879
             CHAS.PARKIN; 1880 JOHN H.JAMES; 1881 R.A.SMITH; 1882-1883
             JOHN H.JAMES

Employment:	Underground	Surface	Total
1881	17	4	21
1882	9	15	24
1883		1	1

DEVON AND CORNWALL   TAVISTOCK, DEVON   SX 463701 0403

Production: Copper	Ore(tons)	Metal(tons)	Value(£)
1856	481.00	24.80	2399.50
1857	1032.00	56.80	5254.00
1858	1877.00	90.10	7785.90
1859	1709.00	72.60	6263.60
1860	1009.00	42.80	3622.80
1861	621.00	23.30	1791.90

Comment 1856-1861 (C)

DEVON CONSOLS,WEST   LAMERTON, DEVON   SX 395738 0404

Production: Copper No detailed return
Ownership: 1881-1884 WEST DEVON GREAT CONSOLS MINING CO.
            Comment 1863-1865 SUSPENDED; 1881-1884 WEST DEVON GREAT
            CONSOLS
Management: Manager 1861-1862 GEO.ROWE
            Chief Agent 1881-1884 GEO.ROWE
            Secretary 1861-1864 W.S.TROTTER (P)

Employment:	Underground	Surface	Total
1881	6	6	12
1882	14	3	17
1883		10	10
1884	3		3

DIMSON   GUNNISLAKE   SX 427718 0405

Production: Tin	Black(tons)	Stuff(tons)	Tin(tons)	Value(£)
1911	0.00	200.00	0.00	300.00
1912	0.00	1053.00	0.00	1224.00
1913	6.40	0.00	0.00	712.00

DIMSON   GUNNISLAKE   Continued

```
        Tungsten        Ore(tons) Metal(tons)    Value(£)
        1913              0.50       0.00          30.00
Ownership:  1911-1912 CONSOLIDATED TAMAR MINES LTD.; 1913 DIMSON MINES
Management: Chief Agent 1911-1913 JOS.CARTER
Employment:             Underground    Surface      Total
        1911              8             7           15
        1912             12             6           18
        1913             22             2           24
```


DING DONG   GULVAL   SW 437346 0406

Production:	Tin	Black(tons)	Stuff(tons)	Tin(tons)	Value(£)
	1855	130.20	0.00	0.00	8849.60
	1856	212.30	0.00	0.00	17214.80
	1857	115.80	0.00	0.00	9420.30
	1858	63.20	0.00	0.00	4312.00
	1859	60.00	0.00	0.00	4902.30
	1860	80.70	0.00	0.00	6722.70
	1861	93.00	0.00	0.00	6553.00
	1862	65.80	0.00	0.00	4357.10
	1863	58.40	0.00	0.00	4042.30
	1864	62.30	0.00	0.00	3923.40
	1865	49.60	0.00	0.00	2691.30
	1866	161.30	0.00	0.00	8219.50
	1867	170.20	0.00	0.00	9371.00
	1868	190.30	0.00	0.00	11636.10
	1869	182.40	0.00	0.00	13413.10
	1870	126.30	0.00	0.00	9818.00
	1871	100.20	0.00	0.00	8236.70
	1872	121.40	0.00	0.00	10500.20
	1873	199.10	0.00	0.00	14469.80
	1874	268.20	0.00	0.00	15206.60
	1875	201.10	0.00	0.00	10276.10
	1876	159.50	0.00	0.00	7002.10
	1877	77.30	0.00	0.00	2781.00
	1878	9.90	0.00	0.00	350.10

```
        Comment 1862 TWO RETURNS AGGREGATED; 1878 MINE CLOSED
        JAN.1878
Ownership:  1875-1876 T.BOLITHE SONS & FIELD; 1877-1878 DING DONG MINING
        CO.
        Comment 1877 CLOSED END OF YEAR
Management: Manager 1859-1864 JOHN TRURAN; 1865-1867 FRAN.BENNETTS;
        1869-1877 WM.WILLIAMS
        Chief Agent 1859 THOS.BLIGHT; 1860-1867 WM.DANIELL; 1868
        WM.WILLIAMS WM.DANIELL; 1869-1871 THOS.DANIELL; 1872-1877
        T.P.ROWE THOS.DANIELL
        Secretary 1859-1874 R.WELLINGTON (P); 1875-1878 R.WELLINGTON
        (S)
```


DING DONG,SOUTH   GULVAL   0407

```
Production: Tin No detailed return
Ownership:  Comment 1863-1865 SUSPENDED
```

DING DONG,SOUTH   GULVAL   Continued

Management: Chief Agent 1860–1862 JOHN PRINCE
            Secretary 1860–1862 JOHN PRINCE (P)

DING DONG,WEST   GULVAL   SW 415301 0408

Production: Tin   Black(tons) Stuff(tons)   Tin(tons)   Value(£)
            1853   12.70   0.00   0.00   821.60
            1855   7.60   0.00   0.00   524.10

DO EM   ST.JUST   0409

Production: Tin   Black(tons) Stuff(tons)   Tin(tons)   Value(£)
            1862   6.90   0.00   0.00   439.50
Management: Chief Agent 1862–1864 JAS.TREZISE
            Secretary 1862–1864 RICH.QUICK

DOLCOATH   CAMBORNE   SW 662405 0410

Production: Zinc   Ore(tons) Metal(tons)   Value(£)
            1897   12.00   0.00   20.00
            Copper   Ore(tons) Metal(tons)   Value(£)
            1845   3504.00   233.70   16996.80
            1846   2156.00   138.60   9546.50
            1847   2057.00   135.40   9414.80
            1848   1254.00   95.10   5587.80
            1849   1028.00   77.50   5197.40
            1850   1115.00   72.60   4909.20
            1851   801.00   55.40   3625.90
            1852   832.00   42.70   3344.60
            1853   1040.00   51.90   4920.00
            1854   992.00   45.50   4313.50
            1855   711.00   28.30   2634.10
            1856   617.00   24.40   1998.40
            1857   566.00   25.90   2429.90
            1858   593.00   34.40   3085.40
            1859   757.00   38.70   3531.70
            1860   712.00   28.70   2426.70
            1861   417.00   18.90   1589.00
            1862   508.00   29.80   2357.70
            1863   636.00   39.80   3029.20
            1864   621.00   39.20   3288.80
            1865   607.00   44.70   3510.40
            1866   688.00   52.10   3512.20
            1867   267.00   15.90   1068.10
            1868   153.00   12.60   863.80
            1869   153.00   10.60   648.60
            1870   57.00   3.80   224.10
            1871   86.00   5.60   326.10
            1872   46.00   3.20   215.50
            1873   16.00   1.10   78.40
            1874   75.00   5.80   420.30
            1876   41.00   2.40   162.60

Copper	Ore(tons)	Metal(tons)	Value(£)
1877	30.30	2.10	112.30
1878	13.60	0.60	27.40
1879	4.20	0.10	12.00
1889	3.00	0.00	10.00
1899	4.00	0.00	23.00
1907	15.00	0.00	96.00
1908	172.00	0.00	806.00
1909	88.00	0.00	519.00
1910	37.00	0.00	133.00

Comment 1845-1874 (C); 1876 (C)

Tin	Black(tons)	Stuff(tons)	Tin(tons)	Value(£)
1853	360.00	0.00	0.00	22680.00
1854	363.50	0.00	0.00	25261.10
1855	352.40	0.00	0.00	23169.80
1856	416.70	0.00	0.00	30727.00
1857	544.00	0.00	0.00	42880.10
1858	635.50	0.00	0.00	41859.20
1859	723.90	0.00	0.00	53506.20
1860	805.10	0.00	0.00	64974.80
1861	864.40	0.00	0.00	63862.50
1862	1246.40	0.00	0.00	83806.00
1863	1026.50	0.00	0.00	69741.70
1864	1029.90	0.00	0.00	66959.20
1865	944.50	0.00	0.00	53238.00
1866	919.40	0.00	0.00	46120.50
1867	847.90	0.00	0.00	46169.90
1868	984.20	0.00	0.00	55847.70
1869	813.40	0.00	0.00	59694.10
1870	1034.80	0.00	0.00	78601.10
1871	1169.90	0.00	0.00	95373.10
1872	1284.80	0.00	0.00	114550.10
1873	1045.30	8.30	0.00	82880.30
1874	1121.00	0.00	0.00	65558.80
1875	1241.50	0.00	0.00	65346.70
1876	1263.30	0.00	0.00	55825.40
1877	1404.70	13.00	0.00	59700.70
1878	1539.10	0.00	0.00	55902.80
1879	1780.40	0.00	0.00	71216.00
1880	1737.30	0.00	0.00	93702.00
1881	1816.30	0.00	1226.00	102039.00
1882	1976.20	0.00	1324.00	120244.00
1883	1875.60	0.00	0.00	101707.00
1884	2423.20	0.00	0.00	113965.00
1885	2555.20	0.00	0.00	124998.00
1886	2383.00	0.00	0.00	134881.00
1887	2366.00	0.00	0.00	152241.00
1888	2239.00	0.00	0.00	148734.00
1889	2125.00	0.00	0.00	114029.00
1890	2023.50	0.00	0.00	110696.00
1891	2131.60	0.00	0.00	114761.00
1892	2535.00	0.00	0.00	139818.00
1893	2421.40	0.00	0.00	124841.00
1894	2126.00	0.00	0.00	89347.00
1895	1766.00	0.00	0.00	68389.00

DOLCOATH   CAMBORNE   Continued

Tin	Black(tons)	Stuff(tons)	Tin(tons)	Value(£)
1896	2039.30	0.00	0.00	75142.00
1897	2095.00	0.00	0.00	79397.00
1898	2302.30	0.00	0.00	99719.00
1899	2078.90	0.00	0.00	151874.00
1900	2004.40	0.00	0.00	164116.00
1901	2035.60	0.00	0.00	143808.00
1902	1828.50	0.00	0.00	131885.00
1903	1739.90	0.00	0.00	133458.00
1904	1705.20	0.00	0.00	129619.00
1905	1696.90	0.00	0.00	146981.00
1906	1813.90	0.00	0.00	198642.00
1907	1708.00	0.00	0.00	184644.00
1908	1782.80	0.00	0.00	142007.00
1909	1904.40	0.00	0.00	154276.00
1910	1730.80	0.00	0.00	161460.00
1911	1705.90	0.00	0.00	198696.00
1912	1665.00	0.00	0.00	217217.00
1913	1525.10	0.00	0.00	185637.00

Comment 1862 TWO RETURNS AGGREGATED; 1880 INC.210 TONS
STOCKED IN 1874

Arsenic	Ore(tons)	Metal(tons)	Value(£)
1854	132.30	0.00	243.90
1855	102.00	0.00	0.00
1867	49.90	0.00	81.40
1868	148.80	0.00	400.10
1869	43.30	0.00	126.00
1870	117.80	0.00	246.30
1871	59.00	0.00	79.80
1872	74.00	0.00	95.80
1873	309.00	0.00	840.00
1874	58.20	0.00	87.30
1875	102.00	0.00	396.20
1876	58.80	0.00	280.00
1877	38.50	0.00	125.10
1878	56.80	0.00	128.90
1879	73.40	0.00	147.00
1881	35.80	0.00	142.40
1882	4.30	0.00	15.00
1886	94.00	0.00	115.00
1907	2.00	0.00	34.00

Comment 1855 WHITE ARSENIC; 1867–1868 CRUDE; 1870–1871 CRUDE;
1886 CRUDE

Arsenic Pyrite	Ore(tons)	Value(£)
1880	178.00	150.50
1894	13.00	6.00
1896	146.00	100.00
1897	276.00	188.00
1898	5.00	3.00
1899	6.00	2.00
1900	10.00	3.00

Bismuth	Ore(tons)	Value(£)
1871	0.10	13.80

Comment 1871 ORE EST.
Ownership: 1875–1889 DOLCOATH MINING CO.; 1890–1893 DOLCOATH MINING

164.

CO.C.B.; 1894 DOLCOATH MINING CO.; 1895–1913 DOLCOATH MINE
LTD.
Comment 1870–1872 INC.STRAY PARK; 1875 INC.STRAY PARK
Management: Manager 1859 CHAS.THOMAS; 1860–1866 CHAS.THOMAS & SON;
1867–1881 JOH.THOMAS
Chief Agent 1859 WM.PROVIS; 1860–1861 WM.PROVIS JOHN TONKIN;
1862–1869 WM.PROVIS JOHN TONKIN JOHN BAWDEN; 1870–1876 JOHN
TONKIN WM.PROVIS JOHN BAWDEN; 1877 JAS.JOHNS WM.PROVIS JOHN
BAWDEN; 1878–1880 JAS.JOHNS RICH.PEARCE JOHN CHENOWETH; 1881
JAS.JOHNS RICH.PEARCE JOS.CHENOWETH; 1882–1900 JOH.THOMAS;
1901–1903 ART.THOMAS; 1904–1913 R.ART.THOMAS
Secretary 1859–1873 COMMITTEE (P); 1874 COMMITTEE JOH.THOMAS;
1891–1897 JOH.THOMAS (P)

Employment:

	Underground	Surface	Total
1878	405	636	1041
1879	432	653	1085
1880	478	458	936
1881	560	637	1197
1882	618	640	1258
1883	613	703	1316
1884	626	685	1311
1885	598	700	1298
1886	496	754	1250
1887	586	695	1281
1888	610	687	1297
1889	605	712	1317
1890	604	698	1302
1891	624	659	1283
1892	591	745	1336
1893	660	696	1356
1894	599	710	1309
1895	566	602	1168
1896	704	603	1307
1897	735	631	1366
1898	650	554	1204
1899	743	595	1338
1900	773	606	1379
1901	700	609	1309
1902	717	550	1267
1903	665	513	1178
1904	635	484	1119
1905	637	464	1101
1906	673	469	1142
1907	718	507	1225
1908	647	481	1128
1909	672	470	1142
1910	593	471	1064
1911	548	475	1023
1912	569	481	1050
1913	618	473	1091

DOLCOATH,NEW   CAMBORNE   SW 652398 0411

Production: Copper   Ore(tons) Metal(tons)   Value(£)
            1872   44.00   3.40   203.50
            1873   112.00   10.20   698.40
            1874   33.00   2.40   139.90
            1875   28.00   2.10   151.10
            Comment 1872—1875 (C)
            Tin   Black(tons) Stuff(tons) Tin(tons) Value(£)
            1872   0.00   245.20   0.00   329.90
            1873   0.80   120.70   0.00   176.30
            1874   0.00   164.70   0.00   67.90
            1875   0.00   32.70   0.00   262.00
Ownership: Comment 1875 IN LIQUIDATION; 1894 SEE GREAT CONDURROW
Management: Manager 1872 NICH.CLYMO; 1873 RICH.PRYOR; 1874 RICH.PRYOR &
            SON
            Chief Agent 1875—1876 RICH.PRYOR
            Secretary 1872 J.VIVIAN & SON; 1874—1876 CHAS.INNES
Employment: Underground   Surface   Total
            1894
            Comment 1894 SEE GREAT CONDURROW

DOLCOATH,NORTH   CAMBORNE   SW 634385 0412

Production: Lead & Silver No detailed return
            Copper   Ore(tons) Metal(tons)   Value(£)
            1862   12.00   1.00   77.10
            1863   16.00   1.40   104.80
            1864   10.00   0.60   43.50
            1867   5.00   0.40   27.70
            Comment 1862—1864 (C); 1867 (C)
Ownership: 1907—1910 WHEAL GERSLER SYNDICATE; 1913 RAYFIELD TIN
            SYND.LTD.
            Comment 1873 SEE STRAY PARk; 1910 NOT WORKED
Management: Manager 1859—1866 JOS.VIVIAN
            Chief Agent 1859—1866 JAS.PAULL
            Secretary 1859 T.V.GURNEY (P); 1860—1866 JOS.VIVIAN JNR (P);
            1907—1910 W.MIDDLIN (S)
Employment: Underground   Surface   Total
            1909   2   2
            1913   14   6   20

DOLCOATH,SOUTH   CAMBORNE   SW 671397 0413

Production: Copper   Ore(tons) Metal(tons)   Value(£)
            1862   61.00   8.00   690.40
            1863   79.00   10.70   890.50
            1864   118.00   15.40   1334.80
            1865   175.00   18.00   1473.90
            1866   225.00   18.60   1213.70
            1867   236.00   16.90   1180.50
            1868   66.00   3.90   254.20
            1869   36.00   1.90   107.70
            1870   22.00   1.50   85.80
            1871   31.00   1.50   87.70

                              166

DOLCOATH,SOUTH   CAMBORNE   Continued

          Copper   Ore(tons) Metal(tons)   Value(£)
          1872   82.00   5.70   411.80
          1873   181.00   14.40   932.10
          1874   273.00   19.40   1254.30
          1875   100.00   5.80   405.50
          1882   21.30   1.30   85.00
          Comment 1862-1875 (C)
Ownership: 1881-1882 SOUTH DOLCOATH MINING CO.
          Comment 1859-1871 SOUTH DOLCOATH & CARMARTHEN CONSOLS;
          1872-1875 INC.CARMARTHEN CONSOLS
Management: Manager 1859-1866 WM.ROBERTS; 1869 GEO.LIGHTLEY; 1870-1873
          JOS.PRISK; 1874-1875 JOHN WILLIAMS; 1881 J.NICHOLLS
          Chief Agent 1859-1869 RICH.BENNETTS; 1870-1873 JOHN WILLIAMS;
          1875 CORN.BAWDEN; 1882 J.NICHOLLS
          Secretary 1859-1861 RICH.LYLE (P); 1862-1868 GEO.LIGHTLEY;
          1869 WM.A.BUCKLEY; 1870-1875 GEO.LIGHTLEY
Employment: Underground   Surface   Total.
          1881   16   15   31
          1882   16   10   26

DOLCOATH,WEST   CAMBORNE   SW 620385 0414

Production: Copper No detailed return
Ownership: Comment 1865 SUSPENDED
Management: Manager 1859 GEO.R.ODGERS
          Chief Agent 1859 JOS.ANGOVE; 1860-1864 STEP.LEAN JOS.ANGOVE
          Secretary 1859-1865 C.WESCOMBE (P)

DOROTHY   ALTARNUN   SX 196789 0415

Production: Tin No detailed return
Ownership: 1892-1893 WHEAL DOROTHY TIN MINE LTD.
          Comment 1893 SUSPENDED
Management: Chief Agent 1892-1893 R.B.WILTON
          Secretary 1892-1893 H.CLENCH
Employment: Underground   Surface   Total
          1892   3   3

DOWGAS,GREAT   ST.STEPHEN-IN-BRANNEL   SW 963514 0416

Production: Tin   Black(tons) Stuff(tons)   Tin(tons)   Value(£)
          1856   70.10   0.00   0.00   4744.00
          1857   34.10   0.00   0.00   2249.70
          1858   4.20   0.00   0.00   297.10
          1860   11.80   0.00   0.00   840.50
          1907   3.90   0.00   0.00   343.00
          1908   10.80   0.00   0.00   634.00
          1912   5.90   0.00   0.00   760.00
          1913   26.30   0.00   0.00   3668.00
          Comment 1912 INC.VENTONWYN VALUE EST; 1913 INC.VENTONWYN
          Cobalt   Ore(tons) Metal(tons)   Value(£)
          1856   No detailed return

                               167

DOWGAS,GREAT   ST.STEPHEN—IN—BRANNEL   Continued

        Cobalt   Ore(tons) Metal(tons)   Value(£)
        1857   4.00   0.00   168.80
Ownership: 1903-1904 W.T.ENGLEDUE; 1905-1910 GREAT DOWGAS TIN MINES
        LTD.; 1911-1913 GREAT DOWGAS(1912)TIN MINING CO.LTD.
        Comment 1863-1865 SUSPENDED; 1905-1908 INC.HORSE BURROW
        (ST.AUSTELL); 1909-1910 INC.HORSE BURROW (ST.AUSTELL),
        VENTONWYN ETC.; 1911-1913 INC.HORSE BURROW (ST.AUSTELL),
        VENTONWYN, ETC.SUSPENDED
Management: Manager 1859 J.ROGERS; 1861 J.ROGERS; 1913 G.BARGATE
        Chief Agent 1862 J.ROGERS; 1904 HY.GRIPE
Employment: Underground   Surface   Total
        1903   6   6
        1904   6   3   9
        1905   7   7
        1906   24   63   87
        1907   63   82   145
        1908   21   17   38
        1909   18   15   33
        1910   3   3
        1912   61   58   119
        1913   55   50   105
        Comment 1909-1910 INC.HORSE BURROW (ST.AUSTELL) & VENTONWYN;
        1912-1913 INC.HORSE BURROW (ST.AUSTELL) & VENTONWYN

DOWN DERRY CONSOLS   ST.STEPHEN—IN—BRANNEL   SW 952505 0417

Production: Copper No detailed return
        Tin No detailed return
Ownership: Comment 1862-1865 SUSPENDED
Management: Manager 1860-1861 J.P.MUFFORD
        Secretary 1860-1865 ROBT.SARGEANT (P)

DOWNGATE CONSOLS   STOKE CLIMSLAND   SX 365731 0418

Production: Copper No detailed return
        Tin No detailed return
Ownership: Comment 1874-1875 SEE NORTH HOLMBUSH

DRAKEWALLS   CALSTOCK   SX 424707 0419

Production: Lead & Silver   Ore(tons) Lead(tons) Silver(ozs)   Value(£)
        1862   0.20   0.10   3.00   0.00
        Copper   Ore(tons) Metal(tons)   Value(£)
        1888   117.00   0.00   275.00
        1889   805.00   0.00   1356.00
        1890   697.00   0.00   1066.00
        1891   323.00   0.00   439.00
        1892   54.00   0.00   97.00
        1894   19.00   0.00   30.00
        Tin   Black(tons) Stuff(tons) Tin(tons)   Value(£)
        1852   218.80   0.00   0.00   0.00
        1853   203.20   0.00   0.00   15397.60

                                    168

Tin	Black(tons)	Stuff(tons)	Tin(tons)	Value(£)
1854	236.00	0.00	0.00	18470.00
1855	301.20	0.00	0.00	20224.80
1856	232.00	0.00	0.00	17870.70
1857	230.50	0.00	0.00	18675.80
1858	225.30	0.00	0.00	14762.10
1859	240.90	0.00	0.00	17618.00
1860	176.30	0.00	0.00	14348.50
1861	214.40	0.00	0.00	15916.80
1862	296.10	0.00	0.00	20197.40
1863	247.60	0.00	0.00	17171.50
1864	203.10	0.00	0.00	13626.10
1865	192.10	0.00	0.00	11286.80
1866	133.20	0.00	0.00	6719.30
1867	23.60	0.00	0.00	1251.50
1868	100.50	0.00	0.00	5646.30
1869	161.80	0.00	0.00	11517.80
1870	166.60	0.00	0.00	12646.80
1871	158.40	0.00	0.00	12938.40
1872	82.20	0.00	0.00	7280.20
1873	40.40	0.00	0.00	3146.00
1874	131.10	0.00	0.00	6575.00
1875	223.80	0.00	0.00	12311.00
1876	179.00	0.00	0.00	7655.50
1877	51.90	0.00	0.00	2000.40
1878	29.90	0.00	0.00	1054.70
1879	14.40	0.00	0.00	439.10
1880	1.20	0.00	0.00	51.60
1882	45.20	0.00	29.90	2709.00
1883	125.50	0.00	0.00	6759.00
1884	79.10	0.00	0.00	3585.00
1885	75.60	0.00	0.00	3522.00
1886	25.40	0.00	0.00	1359.00
1887	24.80	0.00	0.00	1478.00
1888	28.40	27.40	0.00	1715.00
1889	14.00	0.00	0.00	756.00
1890	23.00	0.00	0.00	1216.00
1891	24.50	0.00	0.00	1204.00
1892	43.30	0.00	0.00	2099.00
1893	34.60	0.00	0.00	1649.00
1894	47.60	0.00	0.00	1810.00
1895	36.80	0.00	0.00	1317.00
1896	No detailed return			
1897	1.00	0.00	0.00	30.00
1898	No detailed return			
1909	0.00	155.00	0.00	155.00
1910	1.70	49.00	0.00	273.00

Comment 1852 TIN ORE; 1862 TWO RETURNS AGGREGATED; 1873-1874
VALUE EST; 1888 TWO RETURNS AGGREGATED; 1910 VALUE EST

Arsenic	Ore(tons)	Metal(tons)	Value(£)
1865	17.00	0.00	12.80
1873	4.60	0.00	9.00
1874	20.90	0.00	20.00
1875	60.30	0.00	60.00
1876	10.70	0.00	40.30

Arsenic	Ore(tons)	Metal(tons)	Value(£)
1879	4.00	0.00	20.00
1882	11.00	0.00	38.00
1883	11.20	0.00	47.00
1884	25.30	0.00	143.00
1885	41.60	0.00	244.00
1888	64.00	0.00	429.00
1889	316.00	0.00	2477.00
1890	502.00	0.00	4256.00
1891	608.00	0.00	5242.00
1892	417.00	0.00	3241.00
1893	283.00	0.00	2632.00
1894	284.00	0.00	2818.00
1895	333.00	0.00	3081.00

Comment 1865 CRUDE; 1875 SOOT; 1884-1885 SOOT

Tungsten	Ore(tons)	Metal(tons)	Value(£)
1858	No detailed return		
1860	No detailed return		
1910	0.60	0.00	59.00

Comment 1858 NO RETURN AVAILABLE; 1860 4LBS ONLY PRODUCED;
1910 VALUE EST.

Arsenic Pyrite	Ore(tons)	Value(£)
1890	100.00	53.00

Ownership: 1877 DRAKEWALLS MINING CO.; 1878-1882 DRAKEWALLS TIN & COPPER
MINING CO.; 1883 DRAKEWALLS UNITED TIN & COPPER MINING CO;
1884-1890 DRAKEWALLS TIN & COPPER MINING CO.; 1891-1898
DRAKEWALLS MINING CO.LTD.; 1901 JOHN TAYLOR & SONS; 1902-1904
DRAKEWALLS LTD.; 1906-1913 BRITISH MINING & METAL CO.LTD.
Comment 1880-1881 DRAKEWALLS UNITED; 1883-1885 DRAKEWALLS;
1897-1898 IN LIQUIDATION; 1900 SEE DRAKEWALLS,DEVON; 1904
SUSPENDED; 1911-1913 KEEPING LEVELS OPEN

Management: Manager 1859-1871 THOS.GREGORY; 1872-1877 WM.SKEWIS;
1878-1881 MOSES BAWDEN; 1892-1898 MOSES BAWDEN (P)
Chief Agent 1859 J.ANDREWS; 1860-1865 JAS.HOSKING; 1869-1871
JAS.HOSKING; 1872 WM.DUNSTAN; 1873-1876 ED.DUNSTAN; 1879-1881
ED.DUNSTAN; 1889-1892 H.RODDA; 1893 JAS.HOSKING; 1894-1898
MOSES BAWDEN; 1904 J.HOOPER; 1906-1913 J.H.HEAP
Secretary 1859-1862 E.BETTELEY (P); 1863-1869 RICH.CLOGG;
1870-1871 RICH CLOGG & SON; 1872-1890 MOSES BAWDEN; 1891-1898
W.J.LAVINGTON (S)

Employment:	Underground	Surface	Total
1878	7	30	37
1879	6	27	33
1880	10	20	30
1881	44	38	82
1882	51	32	83
1883	68	48	116
1884	70	60	130
1885	25	34	59
1886	20	24	44
1887	54	39	93
1888	50	44	94
1889	46	37	83
1890	44	48	92
1891	48	50	98

DRAKEWALLS   CALSTOCK   Continued

	Underground	Surface	Total
1892	39	52	91
1893	47	63	110
1894	58	51	109
1895	53	59	112
1896–1897		3	3
1898		2	2
1901	20	20	40
1902	26	20	46
1903	10	8	18
1904		2	2
1906	12	16	28
1907	2		2
1908–1909	2	4	6
1910	2	2	4
1911	2	3	5
1913	2	2	4

DRAKEWALLS,WEST   CALSTOCK   SX 409708 0420

Production: Tin No detailed return
Ownership: Comment 1873 STOPPED IN 1873
Management: Manager 1870–1873 THOS.GREGORY
            Chief Agent 1867–1868 THOS.GREGORY; 1869 THOS.GREGORY
            R.P.COATH; 1870–1873 R.P.COATH
            Secretary 1867–1869 RICH.CLOGG; 1870–1873 RICH.CLOGG & SON

DREYNES & NETHERTON,EAST   ST.NEOT   SX 226688 0422

Production: Tin No detailed return
Ownership: 1908 S.S.WHITBURN; 1909–1911 S.S.WHITBURN & CO.
            Comment 1911 ABANDONED
Employment: Underground   Surface   Total
            1909   6   6
            1910   6   2   8
            1911   6   4   10

DROSKIN   PERRANZABULOE   SW 754545 0423

Production: Copper No detailed return
            Tin   Black(tons) Stuff(tons)   Tin(tons)   Value(£)
            1875   No detailed return
            1887–1894 No detailed return
            1899   No detailed return
            Comment 1875 SEE GREAT ST GEORGE; 1887–1894 SEE GREAT ST
            GEORGE; 1899 SEE GREAT ST GEORGE
            Tungsten   Ore(tons) Metal(tons)   Value(£)
            1899   No detailed return
            Comment 1899 SEE GREAT ST.GEORGE
Ownership: Comment 1887 SEE GREAT ST.GEORGE; 1891–1893 SEE GREAT ST
            GEORGE; 1899 SEE GREAT ST GEORGE

DROSKIN   PERRANZABULOE   Continued

Employment: Underground   Surface   Total
            1891-1893
            1899
            Comment 1891-1893 SEE GREAT ST.GEORGE; 1899 SEE GREAT
            ST.GEORGE

DUCHY AND PERU   PERRANZABULOE   SW 796556 0424

Production: Lead & Silver   Ore(tons)   Lead(tons) Silver(ozs)   Value(£)
            1865   0.20   0.10   0.00   0.00
            1884   2.00   0.00   13.00   0.00
            Comment 1884 DUCHY PERU
            Zinc   Ore(tons) Metal(tons)   Value(£)
            1862   482.10   0.00   810.70
            1863   324.10   0.00   753.50
            1864   135.30   0.00   351.10
            1875   6.00   0.00   18.00
            1877   12.40   0.00   54.70
            1878   140.90   0.00   420.00
            1880   2256.00   0.00   5640.00
            1881   7151.00   3218.00   21453.00
            1882   4059.00   1948.30   12177.00
            1883   2369.70   474.00   1991.00
            1884   1060.00   0.00   1530.00
            1885   840.00   0.00   1398.00
            1886   411.00   0.00   776.00
            Comment 1878 DUCHY PERU; 1886 DUCHY PERU
            Copper   Ore(tons) Metal(tons)   Value(£)
            1862   185.00   5.60   348.40
            Comment 1862 (C)
            Iron   Ore(tons)   Iron(%)   Value(£)
            1858   2762.00   0.00   1242.40
            1859   1995.00   0.00   878.00
            1860   4895.50   0.00   1223.90
            1861   No detailed return
            1862   169.00   0.00   42.20
            1863   3950.90   0.00   2568.10
            1864   2366.90   0.00   591.80
            1865   3539.90   0.00   884.90
            1866   150.00   0.00   37.50
            1872   50.00   0.00   30.00
            1873   528.00   0.00   409.60
            1874   2566.00   0.00   2489.40
            1875   412.10   0.00   370.00
            1876   1121.60   0.00   673.00
            1878   17.00   0.00   10.20
            1879   No detailed return
            1880   3238.00   0.00   1942.80
            1881   1616.00   0.00   808.00
            1882   1862.00   0.00   931.00
            1883   300.00   0.00   75.00
            1884   950.00   42.00   237.00
            1885   No detailed return
            Comment 1858-1861 BH.; 1862-1864 SP.; 1865-1866 BH.;

172

1872-1876 BH.& SOME SPATHOSE; 1878 BH.& SOME SPATHOSE;
1879-1885 BH.

Silver	Ore(tons)	Metal(tons)	Value(£)
1860	184.70	0.00	362.30

Comment 1860 SILVER AND COPPER

Ownership: 1872-1877 CORNISH CONSOLIDATED IRON MINES CORP.LTD; 1878-1879
NEW PERRAN MINERALS CO.LTD.; 1880-1882 DUCHY MINING CO.LTD.;
1883-1886 DUCHY PERU MINING CO.
Comment 1871 SEE TREBISKEN; 1875-1876 INC.GRAVEL HILL; 1881
DUCHY MINE; 1882 DUCHY; 1883-1886 DUCHY PERU; 1892-1893 DUCHY
OF PERU SEE LAMBRIGGAN

Management: Manager 1859-1860 W.BALL; 1861-1865 JOHN BALL; 1873 JOHN
BORLASE; 1874 WM.PARKYN; 1875-1877 JOHN PARKYN; 1878-1879
JAS.HENDERSON; 1880 S.H.HARWOOD
Chief Agent 1861-1865 G.PHILLIPS; 1872 R.PALMER; 1877
JAS.HENDERSON; 1879-1880 JOHN H.JAMES; 1881-1883 PHIL.ARGALL;
1884-1886 R.NANCARROW J.NANCARROW
Secretary 1859-1861 THOS.BLENKINSON (P); 1862-1865
EDM.CARTER; 1873-1875 J.S.COLLARD (S); 1876-1878 LEWIS BIDDER
& NOBLE; 1879-1880 W.R.ROEBUCK & OTHERS

Employment:	Underground	Surface	Total
1878	21	30	51
1879	6	6	12
1880	70	68	138
1881	111	71	182
1882	80	56	136
1883	45	35	80
1884	38	28	66
1885	32	24	56
1886	38	29	67

DUCHY GREAT CONSOLS   CALSTOCK   SX 409732 0425

Production: Copper	Ore(tons)	Metal(tons)	Value(£)
1873	54.00	3.00	168.70
1874	301.00	14.40	819.30
1875	111.00	4.10	275.10
1876	20.00	0.70	41.80
1877	22.00	1.20	93.20

Comment 1873-1875 (C); 1876 (C)DUCHY CONSOLS

Arsenic	Ore(tons)	Metal(tons)	Value(£)
1874	253.00	0.00	293.00

Arsenic Pyrite	Ore(tons)	Value(£)
1875	223.20	303.90

Ownership: 1877 DUCHY GREAT CONSOLS CO.LTD.
Comment 1877 STOPPED
Management: Chief Agent 1873-1877 JAS.RICHARDS
Secretary 1873-1877 FRED.EVERY

DUCHY TIN WORKS   ALTARNUN   0426

Production: Tin No detailed return
Ownership: 1885 DUCHY TIN WORKS CO.

DUCHY TIN WORKS   ALTARNUN   Continued

Management: Chief Agent 1885 SML.MAYNE
Employment: Underground   Surface   Total
            1885   10   4   14

DUKE OF CORNWALL   BODMIN   SX 107629 0427

Production: Zinc   Ore(tons) Metal(tons)   Value(£)
            1857   45.00   0.00   101.30
            1858   214.80   0.00   498.00
            Copper   Ore(tons) Metal(tons)   Value(£)
            1855   279.00   11.80   1014.20
            1856   1190.00   56.20   4683.20
            1857   1691.00   66.00   6130.10
            1858   762.00   29.80   2518.60
            1859   207.00   7.90   693.70
            Comment 1855-1859 (C)
Ownership: Comment 1861-1865 SUSPENDED
Management: Manager 1859 JOHN VERCOE
            Chief Agent 1859 JAS.EVANS; 1860 THOS.TREVILLION W.HOSKING
            Secretary 1859 JAS.HENDERSON (P); 1860 THOS.TREVILLIAN (P)

DULTA   ST.STEPHEN-IN-BRANNEL   0428

Production: Tin No detailed return
Ownership: Comment 1865 SUSPENDED
Management: Manager 1863-1864 J.MARTIN
            Chief Agent 1860-1862 J.MARTIN
            Secretary 1860-1865 W.BUTT (P)

DURLO   TOWEDNACK   SW 498370 0429

Production: Tin   Black(tons) Stuff(tons)   Tin(tons)   Value(£)
            1859   55.90   0.00   0.00   4419.60
            1860   99.30   0.00   0.00   7982.50
            1861   95.20   0.00   0.00   6823.10
            1862   82.40   0.00   0.00   5441.00
            1863   83.60   0.00   0.00   5585.40
            1864   66.60   0.00   0.00   4132.80
            Comment 1862 DURLOE.TWO RETURNS AGGREGATED; 1863-1864 DURLOE
Ownership: Comment 1864-1865 SUSPENDED
Management: Manager 1859 BEN.CHAMPION; 1860-1861 RICH.JAMES
            Chief Agent 1859 BEN.MARTIN; 1860 JAS.PENBERTHY; 1861
            BEN.MARTIN; 1862-1863 BEN.MARTIN BLIGHT; 1864-1865 BLIGHT
            Secretary 1859-1860 J.WILLIAMSON (P); 1861 H.WILLIAMSON;
            1862-1863 COMMITTEE

DYEHOUSE   ROCHE   SW 980600 0430

Production: Iron   Ore(tons)   Iron(%)   Value(£)
            1858   344.00   0.00   97.50
            1859-1861 No detailed return

174

ROCHE

Comment 1858 BH.; 1859 SEE GWINDRA

EAST DOWNS   SCORRIER   SW 725449 0431

Production: Copper   Ore(tons) Metal(tons)   Value(£)
            1864   24.00   1.50   125.40
            1865   51.00   2.50   189.60
            Comment 1864-1865 (C)
            Tin   Black(tons) Stuff(tons)   Tin(tons)   Value(£)
            1899   0.10   0.00   0.00   7.00
            Comment 1899 INC.SILVERWELL
            Arsenic No detailed return
            Arsenic Pyrite   Ore(tons)   Value(£)
            1897   42.00   42.00
            1898   65.00   21.00
Ownership: 1897-1900 C.D.TEAGUE
            Comment 1866-1868 SEE ROSE (SCORRIER); 1869-1871 SEE
            HALLENBEAGLE; 1897-1899 MICHAELS SHAFT; 1900 MICHAELS SHAFT
            SUSPENDED
Management: Chief Agent 1865 JOHN WATERS
            Secretary 1865 HY.MICHELL
Employment: Underground   Surface   Total
            1897   11   11
            1898   2   3   5
            1899   2   2

EAST DOWNS,WILLIAMS   SCORRIER   0432

Production: Copper   Ore(tons) Metal(tons)   Value(£)
            1845   242.00   16.20   1160.90
            1846   139.00   10.20   722.00
            1847   No detailed return
            Comment 1845-1846 (C)

EDGECOMBE   ST.DAY   SW 730402 0433

Production: Tin   Black(tons) Stuff(tons)   Tin(tons)   Value(£)
            1856   No detailed return
            Comment 1856 SEE CLINTON(ST DAY)

EDITH   ST.ENODER   SW 920571 0434

Production: Iron No detailed return
Ownership: 1872-1873 CORNISH CONSOLIDATED FE MINES CORP.LTD
Management: Chief Agent 1872 S.J.PITTAR; 1873 ART.PETO

EDWARD   CALSTOCK   SX 428699 0435

Production: Copper   Ore(tons) Metal(tons)   Value(£)
            1855   172.00   14.30   1467.70

175

EDWARD   CALSTOCK   Continued

Copper	Ore(tons)	Metal(tons)	Value(£)
1856	357.00	24.40	2271.60
1857	1600.00	93.10	8722.80
1858	1542.00	82.00	7184.50
1859	784.00	31.50	2682.70
1860	547.00	26.40	2227.90
1861	1363.00	65.30	5314.20
1862	1324.00	62.70	4599.40
1863	1239.00	62.30	4452.20
1864	668.00	27.00	2082.00
1865	34.00	1.10	68.00
1866	102.00	4.30	236.90
1867	38.00	1.20	70.00
1877	9.00	0.40	18.20

Comment 1855–1867 (C)
Ownership: 1877 EMMENS & CO.LTD.
Management: Manager 1859–1862 M.H.EAST; 1863–1865 GEO.ROWE; 1874–1877
            H.BENNETT
            Chief Agent 1877 G.EMMENS
            Secretary 1859 ED.KING (P); 1860–1865 W.E.COMMINS (P);
            1874–1877 R.F.HAWKE

EDWARD   ST.JUST   SW 360328 0436

Production: Tin No detailed return
Ownership: Comment 1878–1881 SEE OWLES

ELIZA CONSOLS   ST.AUSTELL   SX 046532 0437

Production: Copper   Ore(tons) Metal(tons)   Value(£)
1864	28.00	2.80	234.50
1869	9.00	0.60	36.00
1875	11.00	1..50	106.40
1877	48.20	4.40	227.80
1880	54.50	5.00	261.60
1881	33.30	3.20	172.80
1888	28.00	0.00	129.00

Comment 1864 (C)ELIZA; 1869 (C)ELIZA; 1875 (C)ELIZA; 1880
ELIZA
Tin   Black(tons) Stuff(tons)   Tin(tons)   Value(£)
1864	147.60	0.00	0.00	9187.00
1865	232.00	0.00	0.00	12814.60
1866	244.10	0.00	0.00	12180.10
1867	147.50	0.00	0.00	7707.60
1868	175.40	0.00	0.00	10030.10
1869	154.40	0.00	0.00	10550.00
1870	175.00	0.00	0.00	12804.00
1871	142.50	0.00	0.00	11398.10
1872	93.40	0.00	0.00	8341.30
1873	113.60	0.00	0.00	8740.90
1874	221.40	0.00	0.00	12629.00
1875	358.90	0.00	0.00	18924.20
1876	445.40	0.00	0.00	20392.30

Tin	Black(tons)	Stuff(tons)	Tin(tons)	Value(£)
1877	457.80	0.00	0.00	19630.50
1878	626.40	0.00	0.00	23417.40
1879	614.90	0.00	0.00	25922.80
1880	763.30	0.00	0.00	41419.50
1881	524.00	0.00	357.00	31019.40
1882	470.30	0.00	310.50	29940.00
1883	506.80	0.00	0.00	26860.00
1884	483.70	0.00	0.00	23781.00
1885	338.60	0.00	0.00	17081.00
1886	323.00	0.00	0.00	19244.00
1887	287.20	0.00	0.00	19394.00
1888	317.30	0.00	0.00	22137.00
1889	307.00	0.00	0.00	17962.00
1890	226.50	0.00	0.00	13489.00
1891	231.10	0.00	0.00	13226.00
1892	130.20	0.00	0.00	6937.00
1908	0.30	0.00	0.00	15.00

Comment 1908 NEW ELIZA
Ownership: 1876 STEP.BARKER & OTHERS; 1877-1878 WHEAL ELIZA CONSOLS
MINING CO.; 1879-1890 STEP.BARKER & OTHERS; 1891-1892 ELIZA
CONSOLS MINING CO. C.B.; 1908-1912 NEW WHEAL ELIZA CONSOLS
LTD.
Comment 1862-1864 ELIZA; 1869-1871 RETURNS FOR ELIZA & ELIZA
CONSOLS; 1892 ABANDONED JULY 1892; 1908-1910 NEW ELIZA;
1911-1912 NEW ELIZA NOT WORKED
Management: Manager 1869-1871 JOHN TRUSCOTT; 1873-1882 RICH.H.WILLIAMS;
1891 JOHN LEWIS (S)
Chief Agent 1862-1872 RICH.H.WILLIAMS; 1875 JOHN TRUSCOTT;
1877-1878 JOHN TRUSCOTT; 1880-1881 JOHN TRUSCOTT; 1882-1891
RICH.H.WILLIAMS; 1908-1912 WM.WEDLAKE
Secretary 1862-1863 JOHN H.WILLIAMS; 1864-1873
RICH.H.WILLIAMS; 1874-1882 C.W.BROADHURST; 1891
C.W.BROADHURST (P)

Employment:	Underground	Surface	Total
1878	149	85	234
1879	174	86	260
1880	195	112	307
1881	197	133	330
1882	214	105	319
1883	239	110	349
1884	190	111	301
1885	180	89	269
1886	181	96	277
1887	166	93	259
1888	181	107	288
1889	166	102	268
1890	165	86	251
1891	162	83	245
1892	104	71	175
1908	8		8
1909	14	4	18
1910	8		8

ELIZA,WEST   ST.AUSTELL   SX 040531 0438

Production: Copper   Ore(tons) Metal(tons)   Value(£)
            1877   7.00   0.90   50.70
            Tin   Black(tons) Stuff(tons)   Tin(tons)   Value(£)
            1874   8.40   0.00   0.00   477.30
            1875   23.30   0.00   0.00   1211.10
            1876   11.10   0.00   0.00   476.30
            1877   14.80   0.00   0.00   594.90
            1878   83.90   0.00   0.00   2960.90
            1879   59.30   0.00   0.00   2281.00
            1880   1.30   0.00   0.00   75.50
            1881   4.70   0.00   3.10   265.60
            1882   16.80   0.00   11.10   1059.00
            1883   9.50   0.00   0.00   503.00
            1884   3.10   0.00   0.00   133.00
            1885   12.40   0.00   0.00   587.00
            1886   6.60   0.00   0.00   137.00
            1887   4.50   0.00   0.00   224.00
            1888   3.90   0.00   0.00   196.00
            Comment 1880 WEST ELIZA CONSOLS
Ownership: 1876 STEP.BARKER & OTHERS; 1877-1878 WEST WHEAL ELIZA MINING
            CO.; 1879-1888 STEP.BARKER & OTHERS
Management: Manager 1874-1881 RICH.H.WILLIAMS
            Chief Agent 1877-1878 SML.NORTHEY; 1882-1888 RICH.H.WILLIAMS
            Secretary 1874-1881 C.W.BROADHURST
Employment: Underground   Surface   Total
            1878   47   37   84
            1879   19   15   34
            1880   16   9   25
            1881   10   5   15
            1882   25   17   42
            1883   12   9   21
            1884   12   8   20
            1885   9   5   14
            1886-1888   4   4   4

ELIZABETH   ST.MEWAN   0439

Production: Tin No detailed return
Ownership: Comment 1880 SEE NORTH GREAT POLGOOTH; 1881 SEE PAR GREAT
            CONSOLS

ELLEN   PORTH TOWAN   SW 704469 0440

Production: Zinc   Ore(tons) Metal(tons)   Value(£)
            1858   30.00   0.00   75.00
            1860   40.00   0.00   81.70
            Copper   Ore(tons) Metal(tons)   Value(£)
            1845   714.00   57.90   4373.00
            1846   743.00   62.00   4466.60
            1847   614.00   57.00   4129.40
            1848   641.00   60.90   3648.70
            1849   403.00   35.20   2284.70
            1850   405.00   38.00   2690.90

                                      178

ELLEN PORTH TOWAN Continued

Copper	Ore(tons)	Metal(tons)	Value(£)
1851	392.00	30.40	2049.30
1852	366.00	20.90	1690.50
1853	451.00	21.70	2036.90
1854	152.00	6.20	557.10
1857	433.00	21.30	2004.60
1858	944.00	50.80	4553.10
1859	1484.00	77.40	7080.90
1860	1055.00	49.60	4189.90
1861	233.00	10.50	865.40
1863	6.00	0.20	14.70
1865	20.00	0.70	51.80

Comment 1845-1854 (C); 1857-1861 (C); 1863 (C); 1865 (C)ELLEN UNITED

Tin	Black(tons)	Stuff(tons)	Tin(tons)	Value(£)
1861	0.00	0.00	0.00	60.00

Comment 1861 TIN STUFF
Ownership: 1884 J.DOBSON GOOD & CO.
Comment 1861-1864 SUSPENDED; 1884 ELLEN UNITED
Management: Manager 1859-1860 JOHN HOSKING
Chief Agent 1859-1860 NICH.MINERS; 1884 JOS.MICHELL
Secretary 1859-1860 ROBT.H.PIKE (P); 1861-1865 RICH.PRYOR (P)

Employment:	Underground	Surface	Total
1884	6		6

ELLEN,EAST PORTH TOWAN SW 714468 0441

Production: Zinc	Ore(tons)	Metal(tons)	Value(£)
1864	25.10	0.00	25.10
1905	4.00	1.00	18.00
1906	38.00	8.00	114.00
Copper	Ore(tons)	Metal(tons)	Value(£)
1863	10.00	0.50	35.00
1864	198.00	10.40	834.70

Comment 1863-1864 (C)
Ownership: 1903-1905 W.T.ENGLEDUE; 1906-1907 MINES LTD.
Comment 1862-1863 SUSPENDED; 1865 SUSPENDED
Management: Manager 1864 WM.C.VIVIAN
Chief Agent 1859 STEP.THOMAS; 1860-1861 JOS.MICHELL; 1864 THOS.CORFIELD
Secretary 1859-1863 RICH.GREENWOOD (P); 1864 WALT.THOMPSON

Employment:	Underground	Surface	Total
1904	25	17	42
1905	20	16	36
1906	2	5	7
1907	8	6	14

ELLEN,SOUTH PORTH TOWAN SW 708466 0442

Production: Lead	Ore(tons)	Metal(tons)	Value(£)
1857	0.60	0.50	0.00
1858	No detailed return		

179

ELLEN,SOUTH   PORTH TOWAN   Continued

Lead	Ore(tons)	Metal(tons)	Value(£)
1859	1.50	1.00	0.00
1860	2.00	1.40	0.00
1861	No detailed return		

Zinc	Ore(tons)	Metal(tons)	Value(£)
1859	63.30	0.00	88.60
1860	73.70	0.00	95.20
1861	1.00	0.00	1.50

Copper	Ore(tons)	Metal(tons)	Value(£)
1856	537.00	32.20	2965.80
1857	747.00	38.80	3811.90
1858	593.00	33.00	2921.70
1859	347.00	16.10	1437.70
1860	178.00	6.90	553.80
1862	25.00	2.10	171.70

Comment 1856-1860 (C); 1862 (C)NEW SOUTH ELLEN
Ownership: Comment 1860-1861 NEW SOUTH ELLEN; 1862-1865 NEW SOUTH ELLEN
            SUSPENDED
Management: Manager 1859 J.THOMAS; 1860-1861 JAS.GARLAND
            Chief Agent 1859 J.THOMAS
            Secretary 1859 HY.LOWRY (P); 1860-1863 REG.GRYLLS (P)

EMILY   CALSTOCK   SX 391701 0443

Production: Lead & Silver No detailed return
Management: Chief Agent 1867 JOHN ROBY
            Secretary 1867 H.W.WATSON

EMILY   GWITHIAN   SW 580415 0444

Production: Lead	Ore(tons)	Metal(tons)	Value(£)
1860	73.00	54.70	0.00
1861	No detailed return		

Comment 1860 FOR AG.SEE GREAT ALFRED

Zinc	Ore(tons)	Metal(tons)	Value(£)
1860	23.50	0.00	25.10

Copper No detailed return
Ownership: Comment 1860-1865 SUSPENDED
Management: Manager 1859 HUGH STEVENS
            Chief Agent 1859 DAVID HICKLAND
            Secretary 1859 HUGH STEVENS (P)

EMILY HENRIETTA   ILLOGAN   SW 654421 0445

Production: Zinc	Ore(tons)	Metal(tons)	Value(£)
1869	11.70	0.00	14.50

Copper	Ore(tons)	Metal(tons)	Value(£)
1862	51.00	4.00	325.40
1863	54.00	4.40	347.90
1864	109.00	7.60	660.40
1865	8.00	0.30	21.20
1867	188.00	12.20	850.80

EMILY HENRIETTA   ILLOGAN   Continued

        Copper   Ore(tons) Metal(tons)   Value(£)
        1868   627.00   45.70   3033.40
        1869   488.00   37.10   2314.10
        1870   250.00   21.20   1244.80
        1871   23.00   1.80   109.40
        1872   23.00   1.30   114.70
        Comment 1862–1865 (C); 1867–1872 (C)
        Tin   Black(tons) Stuff(tons)   Tin(tons)   Value(£)
        1868   0.00   0.00   0.00   0.50
        1869   0.00   0.00   0.00   15.30
        1871   0.00   0.00   0.00   52.00
        1872   0.00   0.00   0.00   10.70
        Comment 1868–1869 TIN STUFF; 1871–1872 TIN STUFF
Ownership: Comment 1860–1861 SEE HENRIETTA; 1872–1873 INCORPORATED WITH
            EAST SETON
Management: Manager 1861–1869 JOHN DAWE; 1871–1872 SML.G.TRURAN; 1873
            THOS.PRYOR
            Chief Agent 1861–1869 M.HARRIS; 1870 M.HARRIS SML.G.TRURAN;
            1871–1872 M.HARRIS; 1873 WM.ARTHUR
            Secretary 1861–1872 JOHN F.PENROSE (P)

EMMA   SITHNEY   SW 621292 0446

Management: Manager 1860–1861 JOHN DAVEY; 1862–1864 JOHN DAVEY JNR
            Chief Agent 1865 J.SOUTHEY
            Secretary 1860–1865 T.MILLS (P)

EMMENS UNITED   CALLINGTON   SX 359719 0447

Production: Copper   Ore(tons) Metal(tons)   Value(£)
        1874   41.00   2.30   139.40
        1876   673.00   18.60   933.50
        Comment 1874 (C) SEE ALSO EMMENS UNITED, DEVON; 1876 (C)
        Tin   Black(tons) Stuff(tons)   Tin(tons)   Value(£)
        1874   1.90   0.00   0.00   98.30
        1875   0.70   0.00   0.00   27.30
        Comment 1874–1875 FORMERLY REDMOOR
        Arsenic Pyrite   Ore(tons)   Value(£)
        1875   3161.80   1580.00
Ownership: 1876 CORNWALL CHEMICAL CO.LTD.
            Comment 1874 INC.HOLMBUSH,REDMOOR,KELLY BRAY,& FLORENCE AND
            TONKYN.
Management: Manager 1874 H.BENNETT W.VERRAN; 1875–1876 H.BENNETT
            Chief Agent 1875 WM.VERRAN; 1876 WM.VERRAN G.EMMENS
            Secretary 1874–1875 RICH.HAWKE

ENYS   WENDRON   SW 691355 0448

Production: Tin   Black(tons) Stuff(tons)   Tin(tons)   Value(£)
        1853   18.20   0.00   0.00   1245.90
        1854   38.70   0.00   0.00   2503.00
        1855   78.20   0.00   0.00   4661.70

                                     181

ENYS   WENDRON   Continued

Tin	Black(tons)	Stuff(tons)	Tin(tons)	Value(£)
1856	53.40	0.00	0.00	3315.70
1857	28.20	0.00	0.00	1742.00
1859	26.40	0.00	0.00	1746.50
1860	15.80	0.00	0.00	1179.40

Comment 1859—1860 ENYS MINES
Ownership: Comment 1860—1865 SUSPENDED
Management: Manager 1859 CHAS.PARRY
Chief Agent 1859 THOS.WATERS
Secretary 1859 CHAS.PARRY (P)

ESTHER UNITED   BODMIN   SX 151706 0449

Production: Tin No detailed return
Management: Secretary 1864—1869 J.H.DREW

EXCELSIOR   CALLINGTON   SX 381724 0450

Production: Copper No detailed return

Tin	Black(tons)	Stuff(tons)	Tin(tons)	Value(£)
1873	1.50	0.00	0.00	117.00

1884—1885 No detailed return
Comment 1873 VALUE EST; 1884—1885 SEE KIT HILL UNITED
Ownership: Comment 1874 STOPPED
Management: Manager 1869 JOHN BUCKNELL; 1873—1874 SML.G.TRURAN
Chief Agent 1870—1872 GEO.RICKARD
Secretary 1869 W.WARD; 1870—1874 MAT.GREEN

FALMOUTH AND SPERRIES   KENWYN   SW 770433 0451

Production:

Lead & Silver	Ore(tons)	Lead(tons)	Silver(ozs)	Value(£)
1860	10.90	8.00	0.00	0.00
1861	8.10	5.20	84.00	0.00
1862	6.80	4.40	69.00	0.00
1863	9.50	6.60	82.00	0.00
1864	7.10	5.30	68.00	0.00
1865	3.00	1.90	20.00	0.00
1866	8.50	5.30	0.00	0.00
1867	1.80	1.30	0.00	0.00
1868	3.60	2.70	0.00	0.00
1869	2.40	1.80	0.00	0.00

1870—1872 No detailed return
Comment 1860 FOR AG.SEE GREAT ALFRED; 1867—1869 FOR AG.SEE
GREAT RETALLACK

Zinc	Ore(tons)	Metal(tons)	Value(£)
1861	8.00	0.00	16.00
1863	37.40	0.00	37.40
1864	60.40	0.00	62.30
1865	18.30	0.00	19.20
1866	38.90	0.00	31.10

Comment 1861 ORE EST

FALMOUTH AND SPERRIES   KENWYN   Continued

```
Copper        Ore(tons) Metal(tons)    Value(£)
1862           117.00      4.70         369.40
1863           207.00     10.00         779.10
1864           169.00      5.50         410.10
1865           128.00      3.90         278.60
1868           130.00      5.10         728.20
1869            70.00      2.10         134.70
Comment 1862-1865 (C); 1868-1869 (C)
Manganese No detailed return
Tin           Black(tons) Stuff(tons)  Tin(tons)     Value(£)
1856           0.00        0.00         0.00          5.00
1857           0.00        0.00         0.00          3.50
1860           0.00        0.00         0.00         29.70
1861           0.00        0.00         0.00         45.60
1862           0.00        0.00         0.00         66.40
1863           0.00        0.00         0.00        192.50
1869           0.00        7.70         0.00         22.30
Comment 1856-1857 TIN STUFF; 1860-1861 TIN STUFF; 1862 TIN
STUFF.TWO RETURNS AGGREGATED; 1863 TIN STUFF
```
Management: Manager 1859-1864 WM.KITTO; 1865-1868 WM.KITTO JNR
            Chief Agent 1859-1864 WM.KITTO JNR
            Secretary 1859-1863 JOHN TIPPET (P); 1864-1868 COMMITTEE

FALMOUTH,EAST   KENWYN   SW 781430 0452

```
Production: Lead & Silver    Ore(tons) Lead(tons) Silver(ozs)    Value(£)
            1856              51.50      36.00     2016.00         0.00
            1857             133.00     112.00     7280.00         0.00
            1858              69.00      51.00     3087.00         0.00
            1859             156.50     117.30     5372.00         0.00
            1860              72.90      54.70     2430.00         0.00
            1861         No detailed return
            1862               0.15       0.10        0.00         0.00
            Zinc             Ore(tons) Metal(tons)   Value(£)
            1856              28.90       0.00        75.10
            1857             118.30       0.00       418.30
            1858              74.30       0.00       194.60
            1859              68.40       0.00       131.60
            1860              97.40       0.00       208.40
            1863             174.30       0.00       161.20
            Tin            Black(tons) Stuff(tons) Tin(tons)    Value(£)
            1857              0.00        0.00        0.00         7.00
            1863              0.00        0.00        0.00       142.50
            Comment 1857 TIN STUFF; 1863 TIN STUFF
```
Management: Manager 1859-1865 WM.HANCOCK
            Chief Agent 1859-1861 WM.HANCOCK
            Secretary 1859-1865 GEO.DOWN (P)

FANCY   0453

```
Production: Copper         Ore(tons) Metal(tons)    Value(£)
            1877            4.00        0.20         10.70
```

FANNY ADELA   LELANT   SW 535388 0454

Production: Copper   Ore(tons) Metal(tons)   Value(£)
            1868   23.00   1.10   85.70
            1869   13.00   0.40   24.70
            Comment 1868-1869 (C)
            Tin No detailed return
Ownership: Comment 1867-1868 SEE HAWKS POINT; 1869 FORMERLY HAWKS POINT
Management: Manager 1869 JOS.VIVIAN & SON
            Chief Agent 1869 HY.SKEWIS
            Secretary 1869 ALB.VIVIAN

FATWORK AND VIRTUE   INDIAN QUEENS   SW 918586 0455

Production: Tin   Black(tons) Stuff(tons)   Tin(tons)   Value(£)
            1853   9.00   0.00   0.00   594.80

FLORENCE   CALLINGTON   SX 363706 0457

Production: Copper   Ore(tons) Metal(tons)   Value(£)
            1870   31.00   1.60   95.30
            1871   21.00   1.40   84.00
            1872   100.00   4.80   378.60
            1873   103.00   4.20   218.90
            Comment 1870-1873 (C)INC.TONKIN
            Tin No detailed return
            Arsenic No detailed return
            Arsenic Pyrite   Ore(tons)   Value(£)
            1897   76.00   7.00
Ownership: 1875 ADAM MURRAY & OTHERS; 1897-1898 WM.R.FOOTNER; 1901-1905
            FLORENCE SYNDICATE; 1906 LAKE & CURRIE; 1909-1912 FLORENCE
            MINING & EXPLORATION CO.; 1913 WHEAL FLORENCE LTD.
            Comment 1870-1873 FLORENCE & TONKIN UNITED; 1874 SEE EMMENS
            UNITED; 1875 FLORENCE & TONKIN; 1897-1898 SUSPENDED THROUGH
            CHANCERY ACTION; 1905 SUSPENDED; 1910 NOT WORKED; 1912 NOT
            WORKED
Management: Manager 1867 WM.GIFFORD; 1868-1870 WM.JOHNS; 1871-1873
            WM.VERRAN; 1875 WM.VERRAN
            Chief Agent 1867-1870 WM.VERRAN; 1909-1913 JOHN W.HOLT
            Secretary 1868-1869 R.E.KNOWLING; 1870-1872 HY.PEET; 1873
            ADAM MURRAY HY.PEET; 1901-1905 WM.R.FOOTNER
Employment: Underground   Surface   Total
            1897   16   7   23
            1901-1902   4   4
            1903   2   2
            1904   8   1   9
            1905   1   1
            1906   17   12   29
            1909   8   2   10
            1911   2   2   4
            1913   14   2   16

184

Production: Tin No detailed return
Ownership: 1908—1910 J.DEFRIES & SONS; 1911—1912 A.J.ANGOVE PER D.ROGER
JOHNS
Comment 1912 TAKEN OVER BY G.W.M.MARTIN JUNE 1912

Employment:

	Underground	Surface	Total
1908	10	1	11
1911—1912	2		2

FLORENCE   ST.HILARY   SW 556297 0459

Production: Copper

	Ore(tons)	Metal(tons)	Value(£)
1863	42.00	3.90	303.60

Comment 1863 (C)

Tin

	Black(tons)	Stuff(tons)	Tin(tons)	Value(£)
1862	21.20	0.00	0.00	1297.30
1863	4.40	0.00	0.00	284.80
1872	11.90	0.00	0.00	1071.20
1873	41.70	0.00	0.00	3252.00
1874	43.40	0.00	0.00	2541.70
1875	7.60	0.00	0.00	375.00

Comment 1862 TWO RETURNS AGGREGATED; 1873 FLORENCE CONSOLS
VALUE EST; 1874—1875 FLORENCE CONSOLS

Arsenic

	Ore(tons)	Metal(tons)	Value(£)
1875	9.00	0.00	9.00

Comment 1875 FLORENCE CONSOLS
Ownership: Comment 1873—1875 FLORENCE CONSOLS
Management: Manager 1861—1862 JOHN CURTIS; 1872 WM.JOHNS; 1873 PETER
SKEWIS; 1875 PETER SKEWIS
Chief Agent 1872 PETER FLOYD WM.SEARLE; 1873 WM.SEARLE; 1874
PETER SKEWIS
Secretary 1861—1862 WM.VAWDREY; 1872 R.MICHELL IS.THOMAS;
1873 R.MICHELL; 1875 GEO.EVELEIGH

FLORENCE,EAST   ST.HILARY   0460

Production: Copper

	Ore(tons)	Metal(tons)	Value(£)
1864	58.00	2.90	226.00

Comment 1864 (C)

FORTESCUE   ST.STEPHEN   SW 955514 0461

Production: Tin

	Black(tons)	Stuff(tons)	Tin(tons)	Value(£)
1864	0.10	0.00	0.00	0.00
1870	5.30	0.00	0.00	0.00
1871	3.10	0.00	0.00	0.00
1872	0.00	0.00	0.00	21.40
1874	3.90	0.00	0.00	0.00
1880	2.00	2.40	0.00	180.00
1881	0.00	4.40	0.00	0.00

Comment 1870—1871 FORTESCUE CONSOLS; 1872 FORTESCUE CONSOLS
TIN STUFF; 1874 VALUE EST; 1880 FORTESCUE STANNAGWYN
Ownership: 1880—1881 FORTESCUE MINING CO.LTD.

FORTESCUE   ST.STEPHEN   Continued

            Comment 1872–1873 THE FORTESCUE STOPPED IN 1873 & 1874;
            1874–1875 THE FORTESCUE; 1880–1881 FORTESCUE STANNAGWYN
Management: Manager 1869–1872 WM.ARTHUR; 1873–1875 JOHN H.JAMES
            Chief Agent 1869–1875 T.PHILLIPS; 1880–1881 JOHN H.JAMES
            Secretary 1869–1872 R.HOSKING; 1873–1875 H.M.BYERS
Employment: Underground   Surface   Total
            1880   20   27   47

FORTESCUE,NORTH   LOSTWITHIEL   SX 122606 0462

Production: Lead No detailed return
Management: Manager 1859–1872 WM.VERRAN
            Secretary 1859–1872 JAS.HENDERSON (P)

FORTUNE   ST.MELLION   SX 394701 0463

Production: Lead & Silver No detailed return
            Copper No detailed return
            Tin No detailed return
            Arsenic No detailed return
            Silver   Ore(tons) Metal(tons)   Value(£)
            1880   1.80   806.00   2418.00
            1881   1.60   0.00   175.00
            1882   0.40   0.00   17.00
Ownership: 1880–1883 WHEAL FORTUNE MINING CO.LTD.; 1884–1885 BIRMINGHAM
            & HARROWBARROW MINING CO.
            Comment 1880–1881 WHEAL NEWTON WHEAL FORTUNE; 1883–1885 NOT
            WORKED
Management: Manager 1880–1881 CAPT.KNOTT
            Chief Agent 1880–1881 R.W.DOWLING
            Secretary 1880–1881 A.C.COX (S); 1882–1885 R.W.DOWLING
Employment: Underground   Surface   Total
            1882   40   6   46
            1897   7   1   8
            1898   8   2   10

FORTUNE,EAST   PERRANWELL   0464

Production: Tin   Black(tons) Stuff(tons)   Tin(tons)   Value(£)
            1905   0.00   2.00   0.00   3.00
Ownership: 1904–1905 R.H.STAPLE
Employment: Underground   Surface   Total
            1904   3   3
            1905   2   2

FORTUNE,GREAT   BREAGE   SW 627290 0465

Production: Zinc   Ore(tons) Metal(tons)   Value(£)
            1860   13.00   0.00   37.20
            1876   10.00   0.00   45.00
            Comment 1876 FORTUNE

Copper	Ore(tons)	Metal(tons)	Value(£)
1855	47.00	5.20	545.30
1856	46.00	6.70	615.50
1857	126.00	5.70	568.70
1862	10.00	0.80	70.00
1864	19.00	1.80	162.60
1884	3.00	0.00	7.00

Comment 1855-1857 (C); 1862 (C); 1864 (C)

Tin	Black(tons)	Stuff(tons)	Tin(tons)	Value(£)
1855	89.00	0.00	0.00	6108.80
1856	126.80	0.00	0.00	9197.40
1857	122.00	0.00	0.00	9684.40
1858	114.10	0.00	0.00	7312.40
1859	142.30	0.00	0.00	10830.90
1860	181.70	0.00	0.00	15343.50
1861	226.10	0.00	0.00	17572.10
1862	518.50	0.00	0.00	37284.40
1863	447.40	0.00	0.00	32632.40
1864	278.60	0.00	0.00	19284.10
1865	119.20	0.00	0.00	6977.80
1866	103.90	0.00	0.00	5407.00
1867	55.70	0.00	0.00	2688.60
1868	43.00	0.00	0.00	2457.80
1873	43.10	0.00	0.00	3142.20
1874	18.10	0.00	0.00	952.00
1875	14.00	0.00	0.00	634.10
1876	12.40	0.00	0.00	484.10
1877	10.30	0.00	0.00	370.00
1878	7.30	9.00	0.00	347.50
1879	11.90	3.00	0.00	478.30
1880	15.50	0.00	0.00	721.80
1881	16.10	4.70	0.00	760.90
1882	16.50	0.00	11.60	990.00
1883	13.90	0.00	0.00	629.00
1884	15.90	0.00	0.00	578.00
1885	13.30	400.00	0.00	592.00
1886	19.20	0.00	0.00	1111.00
1887	40.30	0.00	0.00	2495.00
1888	46.00	0.00	0.00	2851.00
1889	47.00	0.00	0.00	2482.00
1890	42.10	0.00	0.00	2179.00
1891	13.00	0.00	0.00	589.00
1892	2.10	0.00	0.00	170.00
1895	9.00	250.00	0.00	300.00
1899	4.70	0.00	0.00	321.00
1900	4.90	0.00	0.00	359.00
1901	5.60	0.00	0.00	365.00
1902	4.80	0.00	0.00	285.00
1903	6.20	0.00	0.00	422.00
1904	5.20	2.00	0.00	349.00
1905	2.20	0.00	0.00	188.00
1906	5.00	0.00	0.00	394.00

Comment 1885 TWO RETURNS AGGREGATED; 1886-1887 PART
INC.GRYLLS; 1892 FORTUNE; 1895 FORTUNE OPEN WORK; 1899
FORTUNE OPEN WORK; 1900-1902 NEW FORTUNE OPEN WORK; 1903-1906

FORTUNE,GREAT   BREAGE   Continued

OPEN WORKS
Arsenic   Ore(tons) Metal(tons)   Value(£)
  1866   150.00   0.00   186.70
  1867   138.00   0.00   171.30
  1875   10.40   0.00   45.00
  1891   1.50   0.00   5.00
  1892-1895 No detailed return
  Comment 1866-1867 CRUDE; 1892-1895 FORTUNE
Arsenic Pyrite   Ore(tons)   Value(£)
  1897   42.00   11.00
  1898   90.00   31.00
  Comment 1897-1898 FORTUNE
Ownership: 1877-1890 GREAT WHEAL FORTUNE MINING CO.; 1891 GREAT FORTUNE
             MINING CO. C.B.; 1892-1893 NEW FORTUNE SYNDICATE C.B.;
             1897-1898 W.J.TRYTHALL; 1907-1912 NEW GREAT WHEAL FORTUNE
             MINING CO.
             Comment 1891 GREAT FORTUNE; 1892 FORTUNE FORMERLY GREAT
             FORTUNE; 1893 FORTUNE OPENCASTE; 1897-1898 FORTUNE; 1907-1912
             NEW GREAT FORTUNE
Management: Manager 1859-1860 J.DANIELL J.PRYOR; 1861-1864 JOS.VIVIAN;
             1872-1875 STEP.HARRIS
             Chief Agent 1859-1860 MICHELL; 1861 THOS.GEORGE; 1862-1864
             N.T.MINERS; 1865-1866 WM.BAWDEN THOS.GEORGE N.T.MINERS; 1867
             THOS.GEORGE R.HOSKING; 1872-1875 J.P.BAWDEN; 1877-1878
             STEP.HARRIS; 1879-1880 WM.ARGALL STEP.HARRIS; 1881-1883
             STEP.HARRIS; 1884-1888 WM.ARGALL; 1889-1893 STEP.CURTIS
             Secretary 1859-1867 THOS.W.ROBINSON (P); 1872-1875 WM.ARGALL;
             1881 WM.ARGALL; 1891 ED.ASHMEAD; 1892-1893 GEO.E.COLLINS;
             1907-1912 G.FREEMAN (S)
Employment: Underground   Surface   Total
             1878-1879   4   18   22
             1880   4   15   19
             1881-1882   4   17   21
             1883   25   25
             1884   3   14   17
             1885   10   12   22
             1886   32   41   73
             1887   30   29   59
             1888   40   35   75
             1889   45   32   77
             1890   40   38   78
             1891   8   10   18
             1892   5   8   13
             1897   7   1   8
             1898   8   2   10
             1907   3   3
             1908   10   2   12

FORTUNE,SOUTH   BREAGE   SW 627282 0466

Production: Copper   Ore(tons) Metal(tons)   Value(£)
            1847   82.00   10.30   715.70
            1848   339.00   34.40   2046.60
            1849   238.00   24.10   1610.20

188

FORTUNE,SOUTH   BREAGE   Continued

        Copper   Ore(tons) Metal(tons)   Value(£)
        1850   191.00   15.90   1067.50
        Comment 1847—1850 (C)

FOWEY   ST.BLAZEY   0467

Production: Copper No detailed return
            Tin No detailed return
Ownership: Comment 1859—1863 FOWEY & PAR UNITED
Management: Manager 1859 WM.PASCOE; 1860—1861 J.ROGERS; 1863 JOHN
            TREDINNICK
            Chief Agent 1859 R.T.STEPHENS; 1860—1862 JOHN TREDINNICK
            Secretary 1859 W.S.TROTTER (P); 1860—1861 WM.CHARLES (P);
            1862—1863 E.S.CODD

FOWEY CONSOLS   TYWARDREATH   SX 083558 0468

Production: Zinc   Ore(tons) Metal(tons)   Value(£)
        1858   45.80   0.00   135.70
        Copper   Ore(tons) Metal(tons)   Value(£)
        1845   8976.00   751.80   49233.00
        1846   7189.00   572.80   34760.80
        1847   6726.00   574.90   38820.80
        1848   6015.00   568.10   32146.30
        1849   6302.00   564.00   36449.30
        1850   6169.00   542.00   37335.60
        1851   5203.00   442.20   28967.40
        1852   4622.00   380.60   30853.30
        1853   4173.00   324.00   31040.50
        1854   4440.00   347.30   34316.10
        1855   5189.00   392.70   39385.70
        1856   5899.00   446.80   40687.60
        1857   4981.00   368.60   36112.40
        1858   5365.00   400.20   35044.90
        1859   4890.00   371.90   34047.20
        1860   4822.00   376.90   32742.90
        1861   4919.00   348.10   29552.50
        1862   4061.00   296.70   23098.20
        1863   3499.00   246.60   18259.70
        1864   3654.00   257.60   20744.20
        1865   3078.00   197.20   14677.90
        1866   2269.00   150.30   9649.80
        1867   480.00   28.00   1984.70
        Comment 1845—1867 (C)
        Tin   Black(tons) Stuff(tons)   Tin(tons)   Value(£)
        1861   0.50   0.00   0.00   29.10
        1864   0.50   0.00   0.00   28.90
        1882   11.60   0.00   7.60   706.00
        1883   15.30   0.00   0.00   845.00
        Nickel   Ore(tons) Metal(tons)   Value(£)
        1858   3.00   0.00   110.30
        1860   3.40   0.00   119.90
        1867   1.90   0.00   14.60

                                 189

FOWEY CONSOLS   TYWARDREATH   Continued

Ownership: 1881 STOKES & HALL; 1882-1883 JOHN BAINES
Management: Manager 1860 MAJOR DAVIS (P); 1863-1867 FRAN.PUCKEY
             Chief Agent 1859 PHIL.RICH; 1860 FRAN.PUCKEY; 1861-1862
             FRAN.PUCKEY SML.SAMPSON; 1863-1866 CHAS.MERRETT GEO.JOB; 1867
             CHAS.MERRETT; 1881-1883 W.H.PASCOE
             Secretary 1859-1860 WM.CHARLES (P); 1861-1864 MAJOR DAVIS
             (P); 1865-1867 WM.POLKINGHORNE
Employment: Underground   Surface   Total
             1882   36   20   56
             1883   30   6   36

FOWEY CONSOLS,NEW   TYWARDREATH   SX 074566 0469

Production: Copper   Ore(tons) Metal(tons)   Value(£)
             1880   3.00   0.20   8.20
             Comment 1880 NEW FOWEY
             Tin No detailed return
Ownership: 1880-1881 NEW FOWEY CONSOLS CO.; 1884-1886 JOHN BAINES; 1887
             FRED.R.PIKE
Management: Chief Agent 1879-1881 SML.NORTHEY; 1884-1887 W.H.PASCOE
Employment: Underground   Surface   Total
             1884   7   3   10
             1885   2   2
             1886   3   3
             1887   6   3   9

FOWEY CONSOLS,SOUTH   TYWARDREATH   SX 080551 0470

Production: Copper No detailed return
             Tin   Black(tons) Stuff(tons)   Tin(tons)   Value(£)
             1873   1.40   0.00   0.00   103.10
Ownership: Comment 1873 STOPPED IN 1873 & EARLY IN 1874
Management: Manager 1866-1867 FRAN.PUCKEY; 1868 CHAS.MERRETT; 1869-1873
             FRAN.PUCKEY
             Chief.Agent 1868 FRAN.PUCKEY; 1869-1873 CHAS.MERRETT
             Secretary 1866-1872 WM.POLKINGHORNE; 1873 JOHN POLKINGHORNE

FOWEY CONSOLS,WEST   TYWARDREATH   SX 072539 0471

Production: Copper   Ore(tons) Metal(tons)   Value(£)
             1845   169.00   11.50   739.40
             1846   130.00   10.10   601.40
             1848   109.00   9.75   735.00
             1849   459.00   41.50   2810.50
             1850   315.00   27.30   1860.70
             1851   469.00   39.70   2596.50
             1852   551.00   47.50   3951.40
             1853   256.00   22.20   2181.90
             1854   277.00   23.30   2361.00
             1855   411.00   38.40   4058.40
             1856   853.00   85.10   8374.70
             1857   774.00   70.90   7294.40

FOWEY CONSOLS,WEST   TYWARDREATH   Continued

Copper	Ore(tons)	Metal(tons)	Value(£)
1858	593.00	51.80	4842.20
1859	597.00	54.50	5346.50
1860	544.00	44.90	4383.50
1861	207.00	17.70	1624.40
1862	225.00	20.50	1710.60
1863	269.00	24.20	1866.60
1864	231.00	22.00	1877.10
1865	42.00	4.20	346.50

Comment 1845-1846 (C); 1848-1865 (C)

Tin	Black(tons)	Stuff(tons)	Tin(tons)	Value(£)
1857	2.80	0.00	0.00	223.50
1858	17.10	0.00	0.00	1145.30
1859	56.20	0.00	0.00	4384.80
1860	129.70	0.00	0.00	10313.80
1861	185.00	0.00	0.00	13337.10
1862	232.50	0.00	0.00	15406.90
1863	178.20	0.00	0.00	11912.30
1864	116.20	0.00	0.00	7353.60
1865	45.10	0.00	0.00	2491.10

Comment 1862 TWO RETURNS AGGREGATED
Ownership: Comment 1865 SUSPENDED
Management: Manager 1860-1864 FRAN.PUCKEY
             Chief Agent 1859 WM.STEPHENS; 1861-1863 ED.DUNSTAN
             WM.STEVENS; 1864 WM.STEVENS
             Secretary 1859-1865 MAJOR DAVIS (P)

FOX HOLE   ILLOGAN   0472

Production: Tin	Black(tons)	Stuff(tons)	Tin(tons)	Value(£)
1863	0.20	0.00	0.00	12.30
1864	0.10	0.00	0.00	5.10

FRANCES   CAMBORNE   SW 648395 0473

Production: Copper No detailed return
            Tin   Black(tons) Stuff(tons) Tin(tons)   Value(£)
            1860-1864 No detailed return
            Comment 1860-1864 SEE CAMBORNE VEAN
Ownership: Comment 1859-1865 SEE CAMBORNE VEAN

FRANCES UNITED,SOUTH   ILLOGAN   SW 680393 0474

Production: Copper	Ore(tons)	Metal(tons)	Value(£)
1892	23.00	0.00	26.00
1895	22.00	0.00	74.00

Tin	Black(tons)	Stuff(tons)	Tin(tons)	Value(£)
1892	763.00	0.00	0.00	39559.00
1893	687.10	0.00	0.00	33087.00
1894	727.60	0.00	0.00	29714.00
1895	631.40	0.00	0.00	23842.00
1896	No detailed return			

191

FRANCES UNITED,SOUTH   ILLOGAN   Continued

            Ochre No detailed return
Ownership: 1891–1893 SOUTH FRANCES UNITED ADVENTURERS C.B.; 1894–1895
            SOUTH FRANCES UNITED ADVENTURERS
            Comment 1891–1894 SOUTH FRANCES & WEST BASSET; 1895 SOUTH
            FRANCES & WEST BASSET.NOW BASSET LTD.
Management: Chief Agent 1892–1895 WM.HOOPER
            Secretary 1891 CORN.BAWDEN; 1892–1894 CORN.BAWDEN (P); 1895
            CORN.BAWDEN
Employment: Underground   Surface   Total
            1891   308   416   724
            1892   346   407   753
            1893   331   356   687
            1894   302   360   662
            1895   232   321   553
            Comment 1891–1895 INC.WEST BASSET

FRANCES,NEW   CROWAN   SW 650363 0475

Production: Copper No detailed return
            Tin   Black(tons) Stuff(tons)   Tin(tons)   Value(£)
            1859   0.00   0.00   0.00   13.40
            1860   0.00   0.00   0.00   146.50
            1861   4.70   0.00   0.00   301.80
            1862   7.10   0.00   0.00   485.90
            Comment 1859–1860 TIN STUFF
Ownership: Comment 1862–1865 SUSPENDED
Management: Chief Agent 1859–1861 CHAS.CARKEEK
            Secretary 1859–1862 ROBT.H.PIKE (P)

FRANCES,NEW SOUTH   ILLOGAN   SW 657358 0476

Production: Tin No detailed return
Ownership: 1882 NEW SOUTH WHEAL FRANCES ADVENTURERS
Management: Chief Agent 1882 CHAS.CRAZE
Employment: Underground   Surface   Total
            1882   6   6

FRANCES,NORTH   ILLOGAN   SW 681402 0477

Production: Copper   Ore(tons) Metal(tons)   Value(£)
            1857   167.00   16.00   1639.10
            1858   101.00   7.20   645.10
            1862   40.00   3.40   276.00
            1863   74.00   4.30   314.40
            1864   75.00   3.90   326.30
            Comment 1857–1858 (C); 1862–1864 (C)
            Tin   Black(tons) Stuff(tons)   Tin(tons)   Value(£)
            1865   0.00   0.00   0.00   112.10
            Comment 1865 TIN STUFF
Ownership: Comment 1865 SUSPENDED
Management: Manager 1859–1860 J.MOYLE; 1861–1864 FRAN.PRYOR J.MOYLE
            Secretary 1859–1863 PHILLIPS & DARLINGTON (P); 1864–1865

THOS.PRYOR

FRANCES,SOUTH   ILLOGAN   SW 680393 0478

Production: Copper	Ore(tons)	Metal(tons)	Value(£)
1845	249.00	14.40	979.60
1846	903.00	127.20	9368.20
1847	2294.00	319.80	23879.00
1848	2462.00	305.10	19055.80
1849	1844.00	246.90	17448.80
1850	2678.00	291.20	21388.80
1851	2709.00	292.00	20351.90
1852	2506.00	225.60	19150.70
1853	2652.00	196.10	19517.50
1854	2783.00	221.80	23204.80
1855	4544.00	407.00	42991.40
1856	6463.00	505.50	48263.60
1857	5795.00	414.40	42612.40
1858	6256.00	459.40	41462.20
1859	4654.00	333.20	31059.60
1860	3573.00	258.50	24163.90
1861	2807.00	191.20	17078.20
1862	2445.00	184.50	15216.50
1863	2091.00	163.00	12764.00
1864	1138.00	81.00	6905.50
1865	965.00	51.00	4053.00
1866	914.00	71.60	4919.20
1867	1462.00	135.90	9558.90
1868	1136.00	94.40	6409.20
1869	422.00	38.20	2442.90
1870	697.00	61.80	3824.00
1871	418.00	32.20	1949.20
1872	313.00	31.40	2601.60
1873	57.00	4.00	262.30
1874	27.00	1.90	126.90
1875	53.00	3.00	208.00
1876	6.00	0.30	17.50

Comment 1845-1876 (C)

Tin	Black(tons)	Stuff(tons)	Tin(tons)	Value(£)
1852	5.00	0.00	0.00	0.00
1853	27.20	0.00	0.00	1842.50
1854	22.50	0.00	0.00	1457.50
1855	17.80	0.00	0.00	1089.90
1856	20.80	0.00	0.00	1442.80
1857	17.80	0.00	0.00	1283.90
1858	14.10	0.00	0.00	825.10
1859	14.40	0.00	0.00	966.40
1860	13.90	0.00	0.00	1008.60
1861	10.10	0.00	0.00	696.00
1862	20.90	0.00	0.00	1332.60
1863	33.10	0.00	0.00	2174.30
1864	40.90	0.00	0.00	2860.30
1865	29.20	0.00	0.00	1534.50
1866	20.10	0.00	0.00	997.00

Tin	Black(tons)	Stuff(tons)	Tin(tons)	Value(£)
1867	16.80	0.00	0.00	885.50
1868	26.10	0.00	0.00	1527.60
1869	68.00	0.00	0.00	4924.10
1870	123.20	0.00	0.00	9219.90
1871	137.40	0.00	0.00	10890.60
1872	86.20	0.00	0.00	7509.30
1873	92.60	0.00	0.00	7434.40
1874	97.30	0.00	0.00	5848.40
1875	42.20	0.00	0.00	2127.50
1876	22.90	0.00	0.00	996.80
1877	212.00	0.00	0.00	8320.00
1878	532.10	0.00	0.00	18484.90
1879	728.60	0.00	0.00	28953.30
1880	468.30	0.00	0.00	24509.80
1881	394.20	0.00	261.20	22250.90
1882	290.80	0.00	191.90	17451.00
1883	391.30	0.00	0.00	20155.00
1884	433.40	0.00	0.00	19860.00
1885	399.10	0.00	0.00	19251.00
1886	417.40	0.00	0.00	22811.00
1887	318.10	0.00	0.00	19463.00
1888	343.00	0.00	0.00	21236.00
1889	302.00	0.00	0.00	15677.00
1890	251.20	0.00	0.00	12905.00
1891	409.20	0.00	0.00	20424.00

Comment 1852 TIN ORE; 1862 TWO RETURNS AGGREGATED
Ownership: 1876-1886 SOUTH WHEAL FRANCES MINING CO.; 1887-1890 SOUTH WHEAL FRANCES ADVENTURERS
Comment 1891-1894 SEE SOUTH FRANCES UNITED; 1895 SEE SOUTH FRANCES UNITED & BASSET
Management: Manager 1863-1871 WM.PASCOE; 1872-1879 AB.T.JAMES; 1881 WILLIAMS & BLARNEY
Chief Agent 1859-1860 WM.PASCOE HY.BENNETTS; 1861 WM.PASCOE JOS.PRISK; 1862 WM.PASCOE JOS.PRISK JOHN POPE JNR; 1863-1866 JOHN POPE JNR JOS.PRISK; 1867 JOS.PRISK; 1869-1871 JOS.PRISK; 1872-1875 JOHN JAMES JOHN OPIE; 1876-1877 JOHN OPIE; 1878 JOHN OPIE WM.VINCENT; 1879 JOHN OPIE; 1880 CHAS.CRAZE JOHN OPIE W.JENKIN; 1881-1887 CHAS.CRAZE; 1889-1890 WM.HOOPER
Secretary 1859-1862 JOHN CADY (P); 1863 JAS.LANYON; 1864-1871 COMMITTEE; 1872-1875 JOHN F.PENROSE; 1876-1880 SML.ABBOT; 1881 CORN.BAWDEN (P); 1887-1890 CORN.BAWDEN

Employment:	Underground	Surface	Total
1878	117	150	267
1879	141	165	306
1880	126	154	280
1881	170	220	390
1882	152	230	382
1883	157	196	353
1884	173	209	382
1885	167	187	354
1886	189	188	377
1887	236	210	446
1888	142	186	328
1889	169	189	358

FRANCES, SOUTH   ILLOGAN   Continued

	Underground	Surface	Total
1890	157	190	347

FRANCES, WEST   ILLOGAN   SW 673392 0479

Production: Copper

	Ore(tons)	Metal(tons)	Value(£)
1865	7.00	0.40	24.30
1871	20.00	1.30	76.00

Comment 1865 (C); 1871 (C)

Tin

	Black(tons)	Stuff(tons)	Tin(tons)	Value(£)
1855	3.20	0.00	0.00	210.50
1856	3.00	0.00	0.00	222.00
1858	1.00	0.00	0.00	65.90
1860	2.30	0.00	0.00	192.60
1861	7.60	0.00	0.00	543.10
1862	29.40	0.00	0.00	1980.90
1863	30.30	0.00	0.00	2059.40
1864	55.30	0.00	0.00	3498.60
1865	98.70	0.00	0.00	5521.60
1866	136.50	0.00	0.00	6587.30
1867	125.30	0.00	0.00	6579.00
1868	183.70	0.00	0.00	9829.40
1869	221.50	0.00	0.00	15251.50
1870	208.50	0.00	0.00	15134.30
1871	231.10	0.00	0.00	18220.80
1872	219.00	0.00	0.00	18404.80
1873	206.90	0.00	0.00	16146.00
1874	275.90	0.00	0.00	13676.80
1875	258.70	0.00	0.00	12888.90
1876	299.60	0.00	0.00	13376.20
1877	312.30	0.00	0.00	13202.00
1878	339.10	0.00	0.00	11865.00
1879	317.60	0.00	0.00	12990.90
1880	277.60	0.00	0.00	14356.80
1881	216.90	0.00	162.00	13520.00
1882	139.70	0.00	90.80	8100.00
1883	121.00	0.00	0.00	6276.00
1884	394.50	0.00	0.00	17850.00
1885	343.00	0.00	0.00	16245.00
1886	385.50	0.00	0.00	22032.00
1887	415.60	0.00	0.00	26922.00
1888	422.20	0.00	0.00	28067.00
1889	379.50	0.00	0.00	20575.00
1890	402.10	0.00	0.00	22092.00
1891	364.40	0.00	0.00	19080.00
1892	381.00	0.00	0.00	19967.00
1893	343.70	0.00	0.00	16632.00
1894	386.50	0.00	0.00	15553.00
1895	316.80	0.00	0.00	11563.00
1896	150.40	0.00	0.00	4954.00

Comment 1862 TWO RETURNS AGGREGATED; 1873 VALUE EST
Ownership: 1876—1890 WEST WHEAL FRANCES MINING CO.; 1891—1893 WEST WHEAL
FRANCES MINING CO. C.B.; 1894—1897 WEST WHEAL FRANCES MINING
CO.

CM-I*   195

Comment 1897 NOT WORKING
Management: Manager 1859–1866 CHAS.THOMAS; 1867–1881 JOH.THOMAS
Chief Agent 1859–1863 JAS.MAYNE; 1864–1871 CHAS.CRAZE
H.RABLIN; 1872–1873 WM.ROWE H.RABLIN; 1874–1881 WM.ROWE
WM.THOMAS; 1882–1897 JOH.THOMAS
Secretary 1859–1863 ROBT.H.PIKE (P); 1864–1881 ROBT.H.PIKE &
SON; 1891–1897 ROBT.H.PIKE & SON(P)

Employment:	Underground	Surface	Total
1878	121	154	275
1879	129	141	270
1880	119	115	234
1881	100	110	210
1882	64	94	158
1883	72	95	167
1884	107	109	216
1885	134	120	254
1886	102	135	237
1887	132	158	290
1888	140	165	305
1889	135	156	291
1890	132	165	297
1891	137	161	298
1892	135	168	303
1893	137	171	308
1894	130	149	279
1895	103	129	232
1896	52	74	126
1897		1	1

FRIENDLY   ST.AGNES   SW 720512 0480

Production: Tin	Black(tons)	Stuff(tons)	Tin(tons)	Value(£)
1855	39.80	0.00	0.00	2095.20
1856	24.20	0.00	0.00	1262.50
1857	19.60	0.00	0.00	928.90
1858	26.90	0.00	0.00	1365.80
1859	16.50	0.00	0.00	945.90
1860	19.00	0.00	0.00	1353.20
1861	26.10	0.00	0.00	1690.20
1862	31.60	0.00	0.00	1720.00
1863	28.30	0.00	0.00	1654.60
1864	23.80	0.00	0.00	1366.70
1865	18.30	0.00	0.00	937.70
1866	22.40	0.00	0.00	981.90
1867	19.70	0.00	0.00	946.10
1868	16.40	0.00	0.00	832.50
1869	19.60	0.00	0.00	1263.10
1870	13.40	0.00	0.00	839.30
1871	5.50	0.00	0.00	366.30
1872	1.70	0.00	0.00	110.80
1874	4.40	0.00	0.00	207.50
1875	25.60	0.00	0.00	1125.70
1876	1.50	0.00	0.00	64.50
1884	No detailed return			

FRIENDLY   ST.AGNES   Continued

Tin	Black(tons)	Stuff(tons)	Tin(tons)	Value(£)
1890	1.10	0.00	0.00	62.00
1891	0.50	0.00	0.00	27.00
1892	1.90	0.00	0.00	106.00
1893	6.00	0.00	0.00	303.00
1894	4.40	0.00	0.00	173.00
1895	2.00	0.00	0.00	70.00
1896	No detailed return			

Comment 1857 TWO RETURNS AGGREGATED; 1862 TWO RETURNS
AGGREGATED; 1874 VALUE EST; 1884 SEE COIT
Ownership: 1888-1890 J.P.RICKMAN & OTHERS; 1891 WHEAL FRIENDLY CO. C.B.;
             1892-1895 WHEAL FRIENDLY CO.
Management: Manager 1859-1871 JAS.EVANS; 1872 JAS.BOYNS; 1873-1876
             JAS.EVANS; 1891-1893 F.J.HARVEY (S); 1894-1895 THOS.WILLIAMS
             (S)
             Chief Agent 1859-1865 JOHN EVANS; 1866 JOHN EVANS
             THOS.RICHARDS; 1867-1868 JOHN EVANS; 1869-1871 FRAN.EVANS;
             1888 HENDERSON & SONS; 1892-1895 NICH.VIVIAN
             Secretary 1859-1874 HY.BORROW (P); 1875-1876 HY.BORROW &
             OTHERS; 1888-1890 HENDERSON & SONS; 1891-1895 G.COULTER
             HANCOCK(P)

Employment:	Underground	Surface	Total
1888	9	2	11
1889	6		6
1890	8	2	10
1891	8	1	9
1892	12	2	14
1893	14	3	17
1894	6	2	8
1895	4	2	6

FRIENDLY STAMPS   ST.AGNES   SW 719514 0481

Production: Tin	Black(tons)	Stuff(tons)	Tin(tons)	Value(£)
1866	8.10	0.00	0.00	390.20

FRIENDSHIP   0482

Production: Copper	Ore(tons)	Metal(tons)	Value(£)
1869	434.00	45.80	2914.40

Comment 1869 (C)

FRIENDSHIP   ST.HILARY   SW 553316 0421

Production: Tin	Black(tons)	Stuff(tons)	Tin(tons)	Value(£)
1852	3.00	0.00	0.00	0.00
1853	6.50	0.00	0.00	448.80
1854	2.60	0.00	0.00	127.80
1855	2.10	0.00	0.00	132.80
1856	0.00	0.00	0.00	51.90

Comment 1852 TIN ORE; 1856 TIN STUFF

FRIENDSHIP & PROSPER   ST.HILARY   SW 556314 0456

Production: Copper   Ore(tons) Metal(tons)   Value(£)
         1852   No detailed return
         Comment 1852 SEE FRIENDSHIP & PROSPER, DEVON

FRIENDSHIP,NORTH   MARYTAVY, DEVON   SX 512823 0483

Production: Lead No detailed return
Management: Chief Agent 1860–1865 Z.WILLIAMS F.KENT
         Secretary 1860–1865 JOS.MATTHEWS (P)

FURSDEN   SITHNEY   SW 669317 0484

Production: Tin No detailed return
Ownership: Comment 1863–1865 SUSPENDED
Management: Manager 1859–1862 JOS.RICHARDS
         Chief Agent 1859–1863 THOS.RICHARDS; 1865 THOS.RICHARDS
         Secretary 1859–1865 JOS.RICHARDS (P)

GARDEN   MORVAH   SW 416356 0485

Production: Tin   Black(tons) Stuff(tons)   Tin(tons)   Value(£)
         1864   6.40   0.00   0.00   397.50
         1870   4.90   0.00   0.00   345.60
         1871   1.50   0.00   0.00   104.90
Management: Chief Agent 1860–1861 NICH.WHITE; 1862–1869 JOHN WHITE
         Secretary 1860–1869 RICH.WHITE (P)

GARKER   ST.AUSTELL   SX 038546 0486

Production: Tin   Black(tons) Stuff(tons)   Tin(tons)   Value(£)
         1891   No detailed return
         1897–1898 No detailed return
         Comment 1891 OPEN WORK.AT WORK; ,1897–1898 SEE BURNGULLOW
Ownership: 1891 WM.NICHOLLS
         Comment 1891 OPEN WORKS

GARLIDNA   WENDRON   SW 699329 0488

Production: Copper   Ore(tons) Metal(tons)   Value(£)
         1863   3.00   0.60   48.40
         Comment 1863 (C)
         Tin   Black(tons) Stuff(tons)   Tin(tons)   Value(£)
         1861   19.40   0.00   0.00   1339.60
         1862   79.60   0.00   0.00   5322.00
         1863   42.30   0.00   0.00   2692.00
         1864   5.30   0.00   0.00   241.50
         1865   15.00   0.00   0.00   730.20
         1866   3.90   0.00   0.00   167.50
         1875   2.00   0.00   0.00   107.00
         1880   0.00   20.00   0.00   106.20

GARLIDNA   WENDRON   Continued

Comment 1865—1866 GARLIDNA UNITED; 1880 INC.POLGRAIN
Management: Manager 1861—1865 J.ROWE
Chief Agent 1861—1865 PAUL PRISK
Secretary 1861—1865 J.G.PLOMER (P)

GARRAS,SOUTH   KENWYN   SW 819477 0489

Production:	Lead & Silver	Ore(tons)	Lead(tons)	Silver(ozs)	Value(£)
	1855	319.50	165.20	15440.00	0.00
	1856	506.00	243.00	23455.00	0.00
	1857	429.00	183.00	20100.00	0.00
	1858	560.00	360.00	14789.00	0.00
	1859	373.40	237.00	7117.00	0.00
	1860	250.60	159.00	4770.00	0.00
	1861	54.00	38.00	1047.00	0.00

Ownership: Comment 1860—1865 SUSPENDED
Management: Manager 1859 WM.BURROWS
Chief Agent 1859 ROBT.TYZZER
Secretary 1859 ED.MICHELL (P)

GAVRIGAN   INDIAN QUEENS   SW 935592 0491

Production:	Tin	Black(tons)	Stuff(tons)	Tin(tons)	Value(£)
	1866	0.70	0.00	0.00	40.00

Comment 1866 GAVERIGAN
Iron No detailed return
Ownership: 1872 CORNISH CONSOLIDATED IRON MINES CORP.LTD
Management: Chief Agent 1872 S.J.PITTAR

GAZELAND   ST.NEOT   SX 170698 0492

Production: Tin No detailed return
Ownership: 1913 GAZELAND CHINA CLAY CO.
Comment 1913 WORKING ADITS FOR TIN

Employment:		Underground	Surface	Total
	1913	18	51	69

GEEVOR   ST.JUST   SW 375345 0493

Production:	Copper	Ore(tons)	Metal(tons)	Value(£)
	1895	No detailed return		

Comment 1895 SEE NORTH LEVANT

	Tin	Black(tons)	Stuff(tons)	Tin(tons)	Value(£)
	1892	17.50	0.00	0.00	944.00
	1893	40.30	0.00	0.00	1949.00
	1894—1904	No detailed return			
	1906	14.70	0.00	0.00	1203.00
	1907—1913	No detailed return			

Comment 1892—1893 PART OF NORTH LEVANT OLD MINE; 1894—1904
SEE NORTH LEVANT; 1906 INC.NORTH LEVANT; 1907—1913 SEE NORTH
LEVANT

GEEVOR   ST.JUST   Continued

Ownership: 1905 J.S.RODDA
             Comment 1892-1904 SEE NORTH LEVANT; 1906-1913 SEE NORTH
             LEVANT
Employment: Underground   Surface   Total
             1905   4   8   12
             1906-1913
             Comment 1906-1913 SEE NORTH LEVANT

GEORGE   CALSTOCK   SX 402701 0494

Production: Tin   Black(tons) Stuff(tons)   Tin(tons)   Value(£)
            1854   1.50   0.00   0.00   106.10
            Arsenic No detailed return
            Arsenic Pyrite   Ore(tons)   Value(£)
            1876   897.00   672.50
            1877   420.00   155.80
            1897   72.00   42.00
            Comment 1876-1877 ARSENICAL MUNDIC
Ownership: 1876-1880 T.BULKLEY & CO.; 1881-1883 WHEAL FORTUNE MINING
            CO.LTD.; 1897 THOS.WHITE
            Comment 1879-1883 NOT WORKED
Management: Chief Agent 1876-1881 WM.R.FOOTNER
            Secretary 1881-1883 R.W.DOWLING
Employment: Underground   Surface   Total
            1897   3   3

GEORGIA   ST.HILARY   SW 557293 0495

Production: Copper No detailed return
            Tin   Black(tons) Stuff(tons)   Tin(tons)   Value(£)
            1873   0.00   50.00   0.00   60.00
            Comment 1873 GEORGIA TIN & COPPER
Ownership: Comment 1868-1871 SEE GREAT WESTERN; 1873 STOPPED IN 1873 OR
            1874
Management: Chief Agent 1872-1873 SML.J.REED N.W.JAMES
            Secretary 1872-1873 R.MICHELL J.HOSKING

GEORGIA CONSOLS   TOWEDNACK   SW 487363 0496

Production: Tin   Black(tons) Stuff(tons)   Tin(tons)   Value(£)
            1852   125.30   0.00   0.00   0.00
            1853   7.10   0.00   0.00   426.80
            1854   28.60   0.00   0.00   1996.10
            1855   7.20   0.00   0.00   549.10
            1911-1912 No detailed return
            Comment 1852 TIN ORE; 1855 GEORGIA MINES; 1911-1912
            GEORGIA:SEE GIEW

GERNICK   CROWAN   SW 638365 0497

Production: Copper No detailed return
            Tin   Black(tons) Stuff(tons)   Tin(tons)   Value(£)
            1856   0.00   0.00   0.00   26.10
            Comment 1856 TIN STUFF
Ownership: Comment 1862-1865 SUSPENDED
Management: Chief Agent 1859-1861 CHAS.CARKEEK
            Secretary 1859 ROBT.H.PIKE (P); 1860-1862 WM.CHARLES (P)

GERNICK,SOUTH   CROWAN   0498

Production: Copper No detailed return
            Tin No detailed return
Ownership: Comment 1863-1865 SUSPENDED
Management: Chief Agent 1859-1862 CHAS.CARKEEK
            Secretary 1859 ROBT.H.PIKE (P); 1860-1863 W.J.DUNSFORD (P)

GERSLER   GWINEAR   0499

Production: Tin   Black(tons) Stuff(tons)   Tin(tons)   Value(£)
            1907   0.70   0.00   0.00   56.00
            1908   0.70   0.00   0.00   45.00
Ownership: 1907-1910 WHEAL GERSLER SYNDICATE
            Comment 1910 NOT WORKED
Management: Secretary 1907-1910 WM.MIDDLIN (S)
Employment: Underground   Surface   Total
            1907   10   8   18
            1908   13   5   18
            1909   2   2

GIEW   TOWEDNACK   SW 498370 0500

Production: Copper No detailed return
            Tin   Black(tons) Stuff(tons)   Tin(tons)   Value(£)
            1911   102.00   0.00   0.00   10186.00
            1912   184.00   0.00   0.00   21500.00
            1913   218.50   0.00   0.00   24189.00
            Comment 1911-1912 INC.GEORGIA CONSOLS
Ownership: 1908-1913 ST.IVES CONSOLS MINES LTD.
            Comment 1869 GIEW CONSOLS; 1877 GEW. SEE SOUTH PROVIDENCE,
            LELANT
Management: Manager 1869 THOS.TREWEEKE JNR
            Chief Agent 1910-1912 H.P.ROBERTSON; 1913 F.C.CANN
Employment: Underground   Surface   Total
            1908   40   26   66
            1909   40   27   67
            1910   75   70   145
            1911   118   91   209
            1912   176   73   249
            1913   171   97   268

                                  201

GILBERT   LEEDSTOWN   SW 582325 0501

Production: Tin   Black(tons) Stuff(tons)   Tin(tons)   Value(£)
            1890   7.00   0.00   0.00   385.00
            1891   No detailed return
            Comment 1890 OPEN WORK; 1891 AT WORK
Ownership: 1890—1891 JAS.POPE
            Comment 1890—1891 OPEN WORKS

GILBERT   TOWEDNACK   SW 689431 0502

Production: Tin No detailed return
Management: Manager 1875 JAS.POPE
            Chief Agent 1875 STEP.POPE
            Secretary 1875 JOHN B.REYNOLDS

GILBERT CONSOLS   LUDGVAN   0503

Production: Tin No detailed return
Management: Manager 1875 JAS.POPE
            Chief Agent 1875 STEP.ANDREWS
            Secretary 1875 GEO.STILL

GILL   ST.IVE   SX 292680 0504

Production: Lead No detailed return
            Copper   Ore(tons) Metal(tons)   Value(£)
            1862   3.00   0.10   10.10
            Comment 1862 (C)
Management: Chief Agent 1862—1863 WM.ROWE; 1866—1870 F.KENT
            Secretary 1862—1870 C.ISAACS

GILL,GLASGOW   ST.IVE   SX 292680 0505

Production: Lead No detailed return
Management: Chief Agent 1862—1863 WM.ROWE; 1866—1870 F.KENT
            Secretary 1862—1870 C.ISAACS

GILMAR   ST.ERTH   SW 580341 0506

Production: Tin   Black(tons) Stuff(tons)   Tin(tons)   Value(£)
            1856   7.30   0.00   0.00   606.40
            1857   6.90   0.00   0.00   595.00

GIRT DOWN   COMBE MARTIN, DEVON   SS 603485 0507

Production: Iron   Ore(tons)   Iron(%)   Value(£)
            1874   300.00   0.00   225.00
            Comment 1874 BH.& SOME SPATHOSE.

GLEBELAND   ST.WENN   0508

Production: Iron No detailed return
Ownership: Comment 1876 SEE ROSTIDYON & DEEP LEVEL

GLYNN   LANIVET   SX 111676 0509

Production: Lead   Ore(tons) Metal(tons)   Value(£)
            1859   17.30   12.00   0.00

GODOLPHIN   BREAGE   SW 599324 0510

Production: Copper   Ore(tons) Metal(tons)   Value(£)
            1845   852.00   102.00   7715.30
            1846   635.00   48.70   3330.00
            Comment 1845-1846 (C)
            Tin   Black(tons) Stuff(tons)   Tin(tons)   Value(£)
            1907   8.20   0.00   0.00   809.00
            1908   11.60   10.00   0.00   1036.00
            1909   0.00   500.00   0.00   500.00
            1910   0.00   211.00   0.00   380.00
            Comment 1907-1910 GODOLPHIN ETC.
Ownership: 1876-1877 NEW GODOLPHIN MINING CO.; 1906-1912 SOUTH-WEST
            CORNWALL MINES LTD.
            Comment 1871 SEE CROWAN MINES; 1873 STOPPED IN 1873 OR 1874;
            1876-1877 NEW GODOLPHIN; 1904-1905 SEE GREAT WORK; 1907-1909
            INC.GREAT WORK,LADY GWENDOLIN & REETH (BREAGE); 1910-1912
            INC.GREAT WORK,LADY GWENDOLIN & REETH (BREAGE).SUSPENDED
Management: Manager 1876-1877 JOHN CURTIS
            Chief Agent 1872-1873 FRAN.GILBERT
            Secretary 1872-1873 R.H.CUDE
Employment: Underground   Surface   Total
            1904-1905
            1906   7   7
            1907   12   10   22
            1908   12   11   23
            1909   10   10
            1910   13   5   18
            1911   7   7
            1912   3   2   5
            Comment 1904-1905 SEE GREAT WORK; 1907-1912 INC.GREAT
            WORK,LADY GWENDOLIN & REETH

GODOLPHIN BRIDGE,GREAT   BREAGE   SW 597324 0511

Production: Lead & Silver No detailed return
            Tin   Black(tons) Stuff(tons)   Tin(tons)   Value(£)
            1870   0.20   0.00   0.00   13.00
Ownership: Comment 1868 GODOLPHIN BRIDGE
Management: Manager 1868-1871 T.H.JOHNS
            Secretary 1868-1871 T.H.JOHNS

                                  203

GODOLPHIN,EAST   BREAGE   SW 597324 0512

Production: Copper   Ore(tons) Metal(tons)   Value(£)
           1851   96.00   5.10   311.70
           Comment 1851 (C)

GODOLPHIN,WEST   BREAGE   SW 582317 0514

Production: Copper   Ore(tons) Metal(tons)   Value(£)
           1872   4.00   0.60   43.40
           1875   13.00   2.90   226.20
           1876   8.00   2.60   160.00
           1877   50.00   8.90   565.40
           1878   97.00   15.90   888.50
           1879   29.00   4.70   270.00
           1882   12.40   2.00   141.00
           1883   9.30   1.10   81.00
           1889   4.00   0.00   11.00
           Comment 1872 (C); 1875-1876 (C)
           Tin   Black(tons) Stuff(tons)   Tin(tons)   Value(£)
           1870   26.40   0.00   0.00   1881.20
           1871   36.70   0.00   0.00   2947.60
           1872   70.60   0.00   0.00   6117.60
           1873   95.90   0.00   0.00   7328.00
           1874   110.90   0.00   0.00   6275.60
           1875   135.90   0.00   0.00   6923.40
           1876   177.20   0.00   0.00   7752.80
           1877   195.50   0.00   0.00   7117.80
           1878   182.30   0.00   0.00   6429.80
           1879   36.30   0.00   0.00   1452.00
           1881   14.20   0.00   9.80   810.10
           1882   41.00   0.00   28.90   2530.00
           1883   48.80   0.00   0.00   2750.00
           1884   67.70   0.00   0.00   3357.00
           1885   93.80   0.00   0.00   4874.00
           1886   67.10   0.00   0.00   4046.00
           1887   57.60   0.00   0.00   3871.00
           1888   37.40   0.00   0.00   2618.00
           1889   7.60   0.00   0.00   418.00
Ownership: 1877-1889 WEST GODOLPHIN MINING CO.
            Comment 1889 NOW ABANDONED
Management: Manager 1867-1868 J.VIVIAN & SON; 1871-1875 JOHN POPE;
            1877-1879 JOHN POPE
            Chief Agent 1867-1868 JOHN POPE JNR; 1870 JOHN POPE JNR; 1876
            JOHN POPE WM.T.POPE; 1877-1879 WM.T.POPE; 1880-1888
            THOS.HODGE
            Secretary 1867-1868 CHAS.THOMAS; 1870 JOHN POPE JNR;
            1871-1879 CHAS.THOMAS; 1880-1881 RICH.MICHELL
Employment: Underground   Surface   Total
           1878   96   53   159
           1880   20   20
           1881   28   38   66
           1882   36   39   75
           1883   45   35   80
           1884   50   33   83
           1885   52   37   89

                              204

GODOLPHIN,WEST   BREAGE   Continued

                    Underground   Surface   Total
          1886   53   32   85
          1887   42   26   68
          1888-1889   29   20   49

GOLDEN   PERRANZABULOE   SW 761590 0515

Production: Lead & Silver   Ore(tons) Lead(tons) Silver(ozs)   Value(£)
          1849   80.00   56.00   0.00   0.00
          1850   119.00   72.00   0.00   0.00
          1851   429.00   284.20   0.00   0.00
          1852   586.00   409.20   7700.00   0.00
          1853   457.00   320.00   6720.00   0.00
          1854   494.00   336.00   7728.00   0.00
          1855   185.00   125.00   2750.00   0.00
          1856   15.00   8.20   191.00   0.00
          1857   No detailed return
Ownership: Comment 1873 GOLDEN UNITED MINES. STOPPED IN 1873 OR 1874
Management: Manager 1872-1873 GEO.E.TREMAYNE
            Chief Agent 1872-1873 W.H.BORLASE
            Secretary 1872-1873 R.F.MICHELL

GOLDEN,EAST   PERRANZABULOE   SW 767581 0517

Production: Lead & Silver   Ore(tons) Lead(tons) Silver(ozs)   Value(£)
          1861   15.00   11.50   181.00   0.00
Ownership: Comment 1862-1865 SUSPENDED; 1872 SEE PHOENIX
            (PERRANZABULOE)
Management: Manager 1860-1861 W.BURROWS J.EVANS; 1869-1871 MAT.WORLEY
            Secretary 1860-1865 ED.MICHELL (P); 1869-1871 J.J.LYNCH

GOLDSITHNEY   GOLDSITHNEY   SW 546306 0518

Production: Tin   Black(tons) Stuff(tons) Tin(tons)   Value(£)
          1910   0.00   115.00   0.00   150.00
          1911   0.00   141.70   0.00   54.00
          1912   0.00   30.00   0.00   11.00
          1913   0.00   15.00   0.00   13.00
          Comment 1910 VALUE EST
Ownership: 1909-1911 G.D.MCGRIGOR; 1912-1913 H.SYKES & PARTNERS
            Comment 1909-1913 VERRANT & CAROLINE
Management: Manager 1913 S.C.DICKINSON
Employment: Underground   Surface   Total
          1909   4   4
          1911   3   3
          1912   7   7   14
          1913   3   4   7

                              205

GONAMENA   ST.CLEER   SX 262706 0519

Production: Copper   Ore(tons) Metal(tons)   Value(£)
            1848   75.00   9.40   566.60
            1850   96.00   11.60   853.40
            1851   159.00   16.20   1112.40
            1852   243.00   23.80   2143.50
            1853   391.00   35.90   3609.00
            1854   50.00   5.00   511.30
            1856   780.00   67.80   6692.90
            1857   862.00   67.60   6785.60
            1858   675.00   54.10   5110.20
            1859   784.00   61.40   5819.30
            1860   961.00   58.80   5152.50
            1861   937.00   53.30   4462.70
            1862   31.00   2.10   179.60
            1864   373.00   21.30   1690.80
            1865   678.00   38.60   1963.30
            1866   369.00   15.60   887.00
            1867   50.00   3.50   241.20
            1868   327.00   22.30   1438.20
            1869   701.00   46.80   2774.30
            1870   558.00   34.50   1887.90
            1871   490.00   30.20   1739.60
            1872   95.00   5.60   420.20
            1877   8.00   0.60   29.20
            Comment 1848 (C); 1850—1854 (C); 1856—1862 (C); 1864—1872
            (C)
            Tin   Black(tons) Stuff(tons)   Tin(tons)   Value(£)
            1861   8.30   0.00   0.00   519.30
            1862   1.00   0.00   0.00   56.50
            1866   1.60   0.00   0.00   84.40
Management: Manager 1860 WM.GEORGE JNR; 1869—1870 JOHN TRUSCOTT
            Chief Agent 1860 WM.PASCOE; 1861—1870 R.PASCOE
            Secretary 1859 E.A.CROUCH (P.); 1860—1865 RICH.HINGSTON (P);
            1866—1870 JOHN TAYLOR

GONAMENA,WEST   ST.CLEER   0520

Production: Copper   Ore(tons) Metal(tons)   Value(£)
            1884   20.00   0.00   70.00
Ownership: 1883—1886 WEST GONAMENA MINING CO.
            Comment 1886 NOT WORKED
Management: Chief Agent 1885—1886 NICH.RICHARDS
            Secretary 1883—1884 W.H.WATSON
Employment: Underground   Surface   Total
            1883   6   6
            1884   8   8   16

GOOD FORTUNE   PERRANZABULOE   SW 742534 0521

Production: Copper   Ore(tons) Metal(tons)   Value(£)
            1904   4.00   0.00   38.00
            Tin   Black(tons) Stuff(tons)   Tin(tons)   Value(£)
            1900   6.10   0.00   0.00   464.00

GOOD FORTUNE   PERRANZABULOE   Continued

Tin	Black(tons)	Stuff(tons)	Tin(tons)	Value(£)
1901	9.70	0.00	0.00	684.00
1902	15.40	0.00	0.00	1137.00
1903	7.70	0.00	0.00	631.00
1904	5.60	0.00	0.00	436.00
1905	4.20	0.00	0.00	382.00
1906	3.60	0.00	0.00	417.00
1907	3.90	0.00	0.00	343.00
1908	3.20	0.00	0.00	245.00
1909	5.40	0.00	0.00	437.00
1910	4.80	0.00	0.00	445.00
1911	3.30	0.00	0.00	375.00
1912	1.90	0.00	0.00	236.00
1913	2.10	0.00	0.00	268.00

Tungsten	Ore(tons)	Metal(tons)	Value(£)
1904	0.50	0.00	58.00

Arsenic Pyrite	Ore(tons)	Value(£)
1906	8.00	6.00

Ownership: 1900-1913 NOBELS EXPLOSIVES CO.LTD.
Management: Chief Agent 1900-1913 JOS.TURNER

Employment:	Underground	Surface	Total
1900	27	6	33
1901	16	5	21
1902	19	5	24
1903	9	2	11
1904	6	2	8
1905	5	1	6
1906	17	3	20
1907	20	3	23
1908	8	4	12
1909	14	2	16
1910	9	2	11
1911	6	2	8
1912	4	2	6
1913	4	1	5

GOODEVERE   ST.CLEER   SX 209747 0522

Production: Tin	Black(tons)	Stuff(tons)	Tin(tons)	Value(£)
1882	1.90	0.00	1.30	113.00
1883	2.50	0.00	0.00	111.00
1884	0.80	0.00	0.00	38.00
1885	No detailed return			
1886	1.10	0.00	0.00	63.00
1887	No detailed return			

Comment 1887 NOW KNOWN AS TIN ERA
Ownership: 1876 GOODEVERE CO.; 1877-1887 GOODEVERE MINING CO.LTD.
Comment 1879-1880 NOT WORKED; 1887 NOW TIN ERA
Management: Manager 1881 WM.GEORGE
Chief Agent 1876-1877 CHAS.J.SIMS; 1881 R.KNOTT; 1882
WM.GEORGE; 1883-1887 R.KNOTT
Secretary 1877-1881 CHAS.J.SIMS

Employment:	Underground	Surface	Total
1878	4		4

GOODEVERE   ST.CLEER   Continued

	Underground	Surface	Total
1881	10	7	17
1882	13	8	21
1884	4	2	6
1885	6	3	9
1886	6	2	8
1887	6	5	11

GOOLE PELLAS   TOWEDNACK   SW 498397 0523

Production: Tin   Black(tons) Stuff(tons) Tin(tons) Value(£)
            1876   No detailed return
            1877   139.30   0.00   0.00   5013.50
            1878   148.40   0.00   0.00   5235.30
            1879   101.00   0.00   0.00   4038.50
            1880   120.30   0.00   0.00   5921.20
            1881   18.00   0.00   11.70   875.60
            Comment 1876 SEE ROSEWALL HILL; 1877 FORMERLY ROSEWALL HILL &
            RANSOME
Ownership: 1877—1879 M.G.BROWNE; 1880—1881 M.G.BROWNE WM.BROWNE
            Comment 1876 LATE ROSEWALL HILL & RANSOM; 1880—1881
            SUSPENDED
Management: Chief Agent 1876 THOS.BUGLEHOLE; 1877—1881 WM.BUGLEHOLE
            Secretary 1876—1879 M.G.BROWNE
Employment: Underground Surface   Total
            1878   65   52   117
            1879   94   80   174
            1880   100   70   170
            1881   77   63   140

GOONBARROW   ST.AUSTELL   SX 009582 0524

Production: Tin   Black(tons) Stuff(tons) Tin(tons) Value(£)
            1863   26.10   0.00   0.00   1880.90
            1864   13.40   0.00   0.00   959.50
            1868   23.70   0.00   0.00   1413.00
            1869   6.10   0.00   0.00   447.50
            1873—1874 No detailed return
            1906   2.00   0.00   0.00   227.00
            1907   0.80   0.00   0.00   70.00
            Comment 1863 INC.MOLLINIS; 1864 INC.MOLLINIS ONE QUARTER
            ONLY; 1873—1874 SEE ROCKS; 1906—1907 IMPERIAL GOONBARROW
Ownership: Comment 1859 GUNBARROW; 1863 SUSPENDED; 1864—1865
            INC.MOLLINIS SUSPENDED; 1872—1873 SEE ROCKS
Management: Manager 1862 S.SYMONS
            Chief Agent 1859—1861 T.SIMMONS
            Secretary 1859—1860 W.HODGE (P); 1862—1865 WM.BROWNE

GOONINNIS   ST.AGNES   SW 724505 0526

Production: Copper No detailed return
            Tin No detailed return

GOONINNIS   ST.AGNES   Continued

Ownership: 1898-1906 GOONINNIS MINING CO.
            Comment 1862-1865 SUSPENDED; 1903-1906 SUSPENDED
Management: Manager 1859-1860 RICH.DAVIES; 1861 R.GREGORY
            Chief Agent 1900-1906 JOHN WILLIAMS
            Secretary 1859-1860 N.SETCHER (P); 1861-1863 W.M.HIGGINS (P)
Employment: Underground   Surface   Total
            1899   9   14   23
            1900   12   20   32
            1901   20   15   35
            1902   13   12   25

GOONZION   ST.NEOT   SX 179678 0527

Production: Copper   Ore(tons) Metal(tons)   Value(£)
            1863   8.00   0.60   44.60
            Comment 1863 (C)
            Tin   Black(tons) Stuff(tons) Tin(tons)   Value(£)
            1863   7.00   0.00   0.00   396.30
            1864   10.60   0.00   0.00   528.50
Ownership: Comment 1864-1865 SUSPENDED
Management: Chief Agent 1861-1862 JAS.ROWE; 1863 THOS.TREVILLION
            Secretary 1861-1862 J.DUNSTUN (P); 1864-1865 JAS.ROWE

GOOUGUMPAS   ST.DAY   0528

Production: Tin No detailed return
Ownership: Comment 1891-1896 SEE POLDICE
Employment: Underground   Surface   Total
            1891-1896
            Comment 1891-1896 SEE POLDICE

GORLAND   ST.DAY   SW 730427 0529

Production: Copper   Ore(tons) Metal(tons)   Value(£)
            1845   366.00   23.10   1652.40
            1846   144.00   10.00   677.00
            1847   69.00   5.40   380.80
            1848   105.00   8.20   509.10
            1851   93.00   0.00   571.80
            Comment 1845-1848 (C); 1851 (C)GORLANA
            Tin   Black(tons) Stuff(tons) Tin(tons)   Value(£)
            1888   No detailed return
            1889   13.00   308.00   0.00   660.00
            1890   0.00   439.00   0.00   624.00
            1891   0.00   167.00   0.00   200.00
            1892   0.00   60.00   0.00   52.00
            1893   No detailed return
            1898   0.00   154.00   0.00   66.00
            1899   0.00   25.00   0.00   70.00
            1900   0.00   14.00   0.00   13.00
            1908   5.90   0.00   0.00   406.00
            1909   11.90   0.00   0.00   851.00

                              209

Tin	Black(tons)	Stuff(tons)	Tin(tons)	Value(£)
1910	13.00	0.00	0.00	1430.00
1911	3.00	0.00	0.00	327.00

Comment 1888 SEE ST DAY MANOR; 1889 SEE ALSO ST DAY MANOR;
1890—1893 INC.UNITY; 1910 VALUE EST

Arsenic	Ore(tons)	Metal(tons)	Value(£)
1874	5.30	0.00	5.00
1876	12.30	0.00	52.00
1893	No detailed return		
1906	4.00	0.00	22.00
1907	17.00	0.00	353.00
1908	24.00	0.00	197.00
1909	56.60	0.00	497.00
1910	15.00	0.00	90.00
1911	16.00	0.00	84.00

Comment 1893 SEE POLDICE

Tungsten	Ore(tons)	Metal(tons)	Value(£)
1899	20.50	0.00	6.00
1906	26.70	0.00	2025.00
1907	29.40	0.00	3620.00
1908	36.80	0.00	2334.00
1909	70.50	0.00	6051.00
1910	34.00	0.00	2924.00
1911	11.00	0.00	1158.00

Comment 1910 VALUE EST.

Arsenic Pyrite	Ore(tons)	Value(£)
1893	No detailed return	

Comment 1893 SEE POLDICE

Ownership: 1889—1890 THOS.PRYOR; 1891—1893 THOS.PRYOR P.; 1898—1900
WM.H.JEFFREY; 1905—1912 ALLEN EDGAR & CO.LTD.
Comment 1888—1891 INC.UNITY NO.1 (ST.DAY); 1911—1912
ABANDONED JULY 1911
Management: Chief Agent 1888 THOS.PRYOR; 1909—1912 J.A.TEMBY
Secretary 1889—1890 THOS.PRYOR

Employment:	Underground	Surface	Total
1888	7		7
1889	15	1	16
1890	13	1	14
1891	5	1	6
1892	2	1	3
1898—1899	2		2
1900	1		1
1905	4	20	24
1906	17	37	54
1907	29	51	80
1908	29	59	88
1909	45	54	99
1910	27	39	66
1911—1912	25	36	61

Comment 1888 INC.UNITY NO.1 SEE ALSO POLDICE; 1889—1891
INC.UNITY NO.1

GORLAND   ST.DAY   SW 730427 0487

Production: Tin No detailed return
Ownership: 1905-1906 T.H.LETCHER
            Comment 1905 GARLAND. T.H.LETCHER LORDS AGENT
Employment: Underground   Surface   Total
            1905   2   2
            1906   2   2

GORLAND,WEST   ST.DAY   SW 722422 0530

Production: Copper   Ore(tons) Metal(tons)   Value(£)
            1871   15.00   0.70   38.20
            1872   10.00   1.00   79.20
            1873   25.00   1.70   96.40
            1875   14.00   0.90   64.00
            1876   2.00   0.10   7.00
            Comment 1871-1873 (C); 1875-1876 (C)
            Tin   Black(tons) Stuff(tons) Tin(tons)   Value(£)
            1873   29.70   0.00   0.00   1857.50
            1874   23.00   0.00   0.00   1288.50
            1875   14.70   0.00   0.00   735.60
            1876   1.70   0.00   0.00   72.80
            Comment 1874 VALUE EST
            Arsenic   Ore(tons) Metal(tons)   Value(£)
            1873   2.50   0.00   7.00
Management: Manager 1875-1876 J.MAYNE
            Chief Agent 1870-1872 J.MAYNE; 1873-1876 J.MAYNE JOHN
            HANCOCK
            Secretary 1870-1876 J.H.MAYNE

GOSS MOOR   ST.AUSTELL   SW 950600 0531

Production: Tin   Black(tons) Stuff(tons) Tin(tons)   Value(£)
            1910   10.00   0.00   0.00   1000.00
            1911   22.00   0.00   0.00   2200.00
            1912   55.30   0.00   0.00   6912.00
            1913   28.00   0.00   0.00   3000.00

GOVER   ST.MEWAN   SW 998528 0532

Production: Tin   Black(tons) Stuff(tons) Tin(tons)   Value(£)
            1869   7.10   0.00   0.00   507.10
            1870   9.70   0.00   0.00   704.00
            1871   13.40   0.00   0.00   1080.20
            1872   0.80   0.00   0.00   71.60
            1874   6.40   0.00   0.00   375.30
            1875   9.40   0.00   0.00   482.50
            1876   9.70   0.00   0.00   427.50
            1877   7.30   0.00   0.00   275.20
            1878   4.00   0.00   0.00   140.00
            1881   7.30   0.00   4.80   394.30
            1882   3.20   0.00   2.10   194.00
            1883   6.70   0.00   0.00   351.00

                                    211

Comment 1881 GOVER CONSOLS; 1882–1883 PART OF TREVARRAN
UNITED MINES

Iron	Ore(tons)	Iron(%)	Value(£)
1858	1410.40	0.00	463.50
1859–1861 No detailed return			
1877	314.00	0.00	189.20

Comment 1858–1861 BH.; 1877 BH.

Ownership: 1877–1879 WM.BROWNE & SONS; 1880–1881 GOVER CONSOLS CO.LTD.;
             1882 TREVARREN UNITED MINING CO.LTD.
             Comment 1879 NOT WORKED; 1880–1881 GOVER CONSOLS
Management: Manager 1869–1873 WM.BROWNE; 1880 HY.BROWNE; 1881 WM.HOOPER
             Chief Agent 1875–1879 RICH.HANCOCK; 1882 HY.BROWNE
             Secretary 1874–1878 WM.BROWNE & SONS; 1881 HY.BROWNE

Employment:	Underground	Surface	Total
1878	5	4	9
1880	7	8	15
1881	19	15	34
1882	4	2	6

GRAMBLER AND ST.AUBYN   REDRUTH   SW 711419 0533

Production: Copper	Ore(tons)	Metal(tons)	Value(£)
1845	1494.00	112.10	8201.70
1846	1292.00	94.30	6628.10
1847	1201.00	84.00	5921.10
1848	872.00	59.60	3438.20
1849	398.00	28.40	1827.80
1855	103.00	10.40	1101.50
1856	284.00	49.10	5182.40
1857	309.00	62.70	6740.50
1858	585.00	107.80	10655.70
1859	697.00	99.00	9773.00
1860	693.00	63.40	6000.50
1861	254.00	19.60	1773.90
1862	164.00	11.20	926.60
1863	49.00	3.40	264.60
1864	77.00	6.30	530.00
1865	123.00	8.40	672.00
1866	179.00	15.90	1075.80
1867	94.00	5.70	391.00
1868	17.00	1.30	85.60

Comment 1845–1849 (C); 1855–1868 (C)

Tin	Black(tons)	Stuff(tons)	Tin(tons)	Value(£)
1852	3.00	0.00	0.00	0.00
1857	0.30	0.00	0.00	22.90
1858	0.00	0.00	0.00	363.60
1859	0.00	0.00	0.00	44.50
1860	0.00	0.00	0.00	19.90
1861	0.00	0.00	0.00	8.80
1862	0.00	0.00	0.00	8.00
1863	0.00	0.00	0.00	2.30
1864	0.00	0.00	0.00	149.30
1865	0.00	0.00	0.00	894.40
1866	0.00	0.00	0.00	469.80

GRAMBLER AND ST.AUBYN   REDRUTH   Continued

Tin	Black(tons)	Stuff(tons)	Tin(tons)	Value(£)
1867	0.00	0.00	0.00	498.30
1868	0.40	0.00	0.00	206.80
1871	0.00	532.00	0.00	233.50
1872	0.00	713.20	0.00	513.50
1873	0.00	791.60	0.00	287.50
1874	0.00	910.40	0.00	104.90

Comment 1852 TIN ORE; 1862 TWO RETURNS AGGREGATED; 1868 VALUE INC.TIN STUFF; 1871-1874 GRAMBLER
Ownership: Comment 1865 SUSPENDED; 1870-1875 GRAMBLER
Management: Manager 1859-1864 JOHN DAVEY; 1870-1875 WM.TREGAY
Chief Agent 1859-1864 JOHN MICHELL; 1870-1875 JOHN TREGAY
Secretary 1859-1862 WM.RICHARDS (P); 1863-1864 F.W.DABB;
1870-1875 W.PAGE CARDOZA

GRAMBLER,NORTH   REDRUTH   SW 714423 0534

Production: Copper

	Ore(tons)	Metal(tons)	Value(£)
1859	96.00	6.30	595.50
1860	281.00	30.10	2758.60
1861	511.00	45.00	4125.30
1862	403.00	27.60	2310.30
1863	196.00	13.10	1007.10
1864	151.00	10.40	865.20
1865	120.00	8.10	636.70
1866	258.00	19.10	1639.90
1867	236.00	20.20	1429.30
1868	326.00	25.40	1698.50
1869	70.00	5.90	387.90

Comment 1859-1869 (C)

Tin	Black(tons)	Stuff(tons)	Tin(tons)	Value(£)
1860	0.00	0.00	0.00	38.00
1861	0.00	0.00	0.00	42.20
1862	0.00	0.00	0.00	52.30
1863	0.00	0.00	0.00	121.20
1864	0.00	0.00	0.00	797.30
1865	0.00	0.00	0.00	619.80
1866	0.00	0.00	0.00	461.10
1867	0.00	0.00	0.00	627.10
1868	0.00	0.00	0.00	637.90
1869	0.00	0.00	0.00	116.80

Comment 1860-1869 TIN STUFF
Management: Manager 1859-1861 JOS.VIVIAN; 1863 JOS.VIVIAN
Chief Agent 1859 J.THOMAS; 1860 J.THOMAS RICH.PRYOR;
1861-1863 WM.PASCOE; 1864-1865 WM.PASCOE JOS.VIVIAN;
1866-1867 WM.PASCOE
Secretary 1859 ROBT.H.PIKE (P); 1860 W.J.DUNSFORD (P);
1861-1862 ROBT.H.PIKE (P); 1863-1865 ROBT.H.PIKE; 1866-1867
ROBT.H.PIKE & SON

GRAVEL HILL   PERRANZABULOE   SW 765575 0535

Production: Zinc   Ore(tons) Metal(tons)   Value(£)
            1882   30.00   12.00   45.00
            Iron   Ore(tons)   Iron(%)   Value(£)
            1874   2505.00   0.00   2003.90
            1875   494.40   0.00   395.50
            1876   493.80   0.00   294.40
            1877   806.00   0.00   584.00
            1878   947.00   0.00   662.90
            1879   No detailed return
            1880   2652.00   0.00   1856.40
            1881   581.00   0.00   290.50
            1882   1900.00   47.00   950.00
            Comment 1874-1880 BH.& SOME SPATHOSE; 1881-1882 BH.
Ownership: 1873-1874 CORNISH CONSOLIDATED FE MINES CORP.LTD; 1877 GREAT
            PERRAN MINES ADVENTURERS; 1878 NEW PERRAN MINERAL CO.; 1879
            W.R.ROEBUCK & OTHERS; 1880-1881 CORNISH STEEL IRON ORE CO.;
            1882-1883 NEWQUAY MINING CO.LTD.
            Comment 1875-1876 SEE DUCHY & PERU; 1880 GRAVEL HILLS
Management: Manager 1873 ART.PETO; 1874 WM.PARKYN; 1877-1878
            JAS.HENDERSON
            Chief Agent 1879 JOHN H.JAMES; 1880-1883 J.WHITAKER BUSHE
            Secretary 1874 J.S.COLLARD (S)
Employment: Underground   Surface   Total
            1878   16   18   34
            1879   6   6
            1880   24   21   45
            1881   12   10   22
            1882   18   18   36

GREAT ROCK   ST.AUSTELL   0536

Production: Tin No detailed return
Management: Chief Agent 1870-1871 J.KEMP
            Secretary 1870-1871 J.KEMP

GREAT WESTERN   ST.HILARY   SW 565293 0537

Production: Copper No detailed return
            Tin   Black(tons) Stuff(tons)   Tin(tons)   Value(£)
            1870   143.90   0.00   0.00   10384.30
            1871   134.90   0.00   0.00   10393.80
            1872   97.10   0.00   0.00   8151.40
            1873   130.00   20.00   0.00   10180.30
            1886   3.20   0.00   0.00   151.00
            Comment 1873 VALUE EST
            Arsenic   Ore(tons) Metal(tons)   Value(£)
            1871   20.00   0.00   10.00
            1873   30.00   0.00   75.00
            Comment 1871 CRUDE
Ownership: 1886 GREAT WESTERN ADVENTURERS
            Comment 1868-1871 GRYLLS, GRYLLS WHEAL FLORENCE, E. & W.
            GRYLLS, ST.AUBYN & GRYLLS, GEORGIA, WELLINGTON; 1873 STOPPED
            IN 1873 & EARLY IN 1874

                                      214

GREAT WESTERN   ST.HILARY   Continued

Management: Manager 1868—1871 ED.ROGERS
            Chief Agent 1868—1871 EDM.ROGERS JNR; 1873 ED.ROGERS; 1886
            JAS.THOMAS
            Secretary 1868 W.WATSON; 1869—1871 WATSON J.H.MURCHISON;
            1872—1873 THOS.W.FIELD
Employment: Underground   Surface   Total
            1886   21   16   37

GREENBERRY MOOR   ST.HILARY   SW 565293 0538

Production: Tin   Black(tons) Stuff(tons) Tin(tons)   Value(£)
            1859   0.00   0.00   0.00   2319.30
            1860   0.00   0.00   0.00   727.60
            1861   0.00   0.00   0.00   169.90
            Comment 1859—1860 GREENBERRY TIN STUFF
Ownership: Comment 1868—1871 SEE EAST GRYLLS

GRENVILLE   CAMBORNE   SW 663386 0539

Production: Copper   Ore(tons) Metal(tons)   Value(£)
            1860   247.00   17.00   1540.00
            1861   140.00   8.50   721.50
            1862   522.00   37.90   3053.40
            1863   738.00   67.30   5343.40
            1864   262.00   22.60   2005.10
            1865   89.00   9.20   754.10
            1866   67.00   5.10   330.50
            1867   55.00   3.30   214.80
            1868   35.00   2.20   142.80
            1869   14.00   1.10   66.00
            1870   5.00   6.40   24.70
            1871   14.00   1.20   70.20
            1872   40.00   2.80   214.70
            1873   14.00   1.40   79.10
            1874   7.00   0.40   18.00
            1875   7.00   0.30   16.10
            1876   26.00   2.80   190.00
            1877   45.40   4.60   280.30
            1878   6.70   0.70   37.30
            Comment 1860—1876 (C)
            Tin   Black(tons) Stuff(tons) Tin(tons)   Value(£)
            1860   0.00   0.00   0.00   80.70
            1862   0.00   0.00   0.00   325.70
            1863   9.90   0.00   0.00   1793.70
            1864   111.30   0.00   0.00   7057.90
            1865   168.80   0.00   0.00   9410.60
            1866   150.20   0.00   0.00   7510.00
            1867   107.90   0.00   0.00   5544.70
            1868   132.40   0.00   0.00   7210.30
            1869   135.40   0.00   0.00   9283.20
            1870   133.30   0.00   0.00   9774.20
            1871   187.10   0.00   0.00   12122.40
            1872   124.70   0.00   0.00   10839.30

                               215

Tin	Black(tons)	Stuff(tons)	Tin(tons)	Value(£)
1873	141.50	0.00	0.00	11037.00
1874	123.40	0.00	0.00	6941.90
1875	155.60	0.00	0.00	7784.90
1876	180.90	0.00	0.00	8771.20
1877	155.60	0.00	0.00	6637.70
1878	240.40	0.00	0.00	8811.10
1879	230.50	0.00	0.00	9612.70
1880	333.70	0.00	0.00	17830.70
1881	357.40	0.00	243.50	21083.20
1882	359.90	0.00	247.40	22576.00
1883	327.40	0.00	0.00	17352.00
1884	463.10	0.00	0.00	22174.00
1885	594.80	0.00	0.00	29958.00
1886	589.40	0.00	0.00	34100.00
1887	639.40	0.00	0.00	41882.00
1888	481.00	0.00	0.00	32654.00
1889	506.00	0.00	0.00	27868.00
1890	574.50	0.00	0.00	32332.00
1891	611.90	0.00	0.00	33142.00
1892	772.00	0.00	0.00	42345.00
1893	926.00	0.00	0.00	46570.00
1894	1006.00	0.00	0.00	41979.00
1895	999.00	0.00	0.00	39059.00
1896	889.00	0.00	0.00	33338.00
1897	841.00	0.00	0.00	32561.00
1898	877.00	0.00	0.00	39110.00
1899	677.00	0.00	0.00	51101.00
1900	727.00	0.00	0.00	60965.00
1901	699.00	0.00	0.00	49086.00
1902	723.00	0.00	0.00	55541.00
1903	804.50	0.00	0.00	65590.00
1904	622.90	0.00	0.00	48212.00
1905	793.30	0.00	0.00	68403.00
1906	714.70	0.00	0.00	76319.00
1907	620.90	0.00	0.00	66111.00
1908	637.00	0.00	0.00	49297.00
1909	633.80	0.00	0.00	49549.00
1910	611.70	0.00	0.00	56662.00
1911	596.40	0.00	0.00	70632.00
1912	749.80	0.00	0.00	103365.00
1913	703.00	0.00	0.00	86117.00

Comment 1860 TIN STUFF; 1862 TIN STUFF; 1863 VALUE INC.TIN
STUFF; 1864 TWO RETURNS AGGREGATED; 1873 VALUE EST
Ownership: 1877–1890 WHEAL GRENVILLE MINING CO.; 1891–1893 WHEAL
GRENVILLE MINING CO.C.B.; 1894–1909 WHEAL GRENVILLE MINING
CO.; 1910–1913 GRENVILLE UNITED MINES LTD.
Comment 1903–1910 WORKED WITH SOUTH CONDURROW
Management: Manager 1859–1871 GEO.R.ODGERS; 1872–1875 EDW.HOSKING; 1891
D.JULYAN (S)
Chief Agent 1860–1866 WM.BENNETTS; 1867–1874 WM.BENNETTS JOHN
OSBORNE; 1875 WM.BENNETTS THOS.HODGE JOHN OSBORNE; 1876–1877
THOS.HODGE WM.BENNETTS; 1878–1881 THOS.HODGE JOS.HOCKING;
1882–1888 THOS.HODGE; 1889–1902 CHAS.BISHOP; 1903–1905
JAS.NEGUS; 1906–1909 FRANK WILLIAMS; 1910–1913 H.BATTENS

Secretary 1859 JOHN WATSON (P); 1860–1862 JOHN CADY (P); 1863
H.STEVENS JNR (P); 1864–1870 JOHN WATSON (P); 1871–1874
WATSON SML.STEPHENS; 1875–1877 T.B.LAWS S.STEPHENS; 1878–1879
RICH.MICHELL; 1891 A.J.TANGYE (P); 1892–1902 D.JULYAN (S)

Employment:	Underground	Surface	Total
1878	128	109	237
1879	124	109	233
1880	140	109	249
1881	133	101	234
1882	120	108	228
1883	105	117	222
1884	113	125	238
1885	158	139	297
1886	156	139	295
1887	144	138	282
1888	142	144	286
1889	163	169	332
1890	199	179	378
1891	184	202	386
1892	238	231	469
1893	270	285	555
1894	246	280	526
1895	240	274	514
1896	234	259	493
1897	229	255	484
1898	194	284	478
1899	222	293	515
1900	245	285	530
1901	221	251	472
1902	234	254	488
1903	267	245	512
1904	310	240	550
1905	280	265	545
1906	272	230	502
1907	254	227	481
1908	236	207	443
1909	248	199	447
1910	246	193	439
1911	249	184	433
1912	280	189	469
1913	265	194	459

GRENVILLE   ST.STEPHEN   SW 952505 0540

Production: Tin No detailed return
Management: Manager 1869–1871 JOHN H.JAMES
            Secretary 1869–1871 R.HOSKING

GRENVILLE,EAST   CAMBORNE   SW 668388 0541

Production: Copper	Ore(tons)	Metal(tons)	Value(£)
1862	73.00	3.00	218.50
1863	180.00	8.20	574.20

Copper	Ore(tons)	Metal(tons)	Value(£)
1864	386.00	18.80	1433.70
1865	575.00	27.40	2036.50
1866	546.00	28.00	1638.00
1867	376.00	23.40	1537.60
1868	250.00	17.90	1126.50
1869	850.00	57.30	3450.90
1870	878.00	59.30	3397.70
1871	1184.00	67.20	3826.00
1872	293.00	15.10	1089.80
1873	286.00	18.30	1141.40
1874	225.00	13.70	804.90
1875	261.00	13.10	893.00
1876	60.00	3.20	207.40
1877	6.00	0.20	5.50

Comment 1862-1876 (C)

Tin	Black(tons)	Stuff(tons)	Tin(tons)	Value(£)
1861	0.50	0.00	0.00	33.20
1862	18.10	0.00	0.00	1226.80
1863	33.30	0.00	0.00	2300.40
1864	30.80	0.00	0.00	1972.10
1865	52.60	0.00	0.00	2942.20
1866	37.80	0.00	0.00	1932.20
1867	12.10	0.00	0.00	626.90
1868	10.10	0.00	0.00	559.40
1869	9.00	0.00	0.00	609.50
1870	12.50	0.00	0.00	898.90
1871	22.00	0.00	0.00	1975.30
1872	11.70	0.00	0.00	970.70
1873	16.00	0.00	0.00	1128.90
1874	25.20	25.20	0.00	1412.80
1875	16.70	0.00	0.00	830.30
1876	6.70	22.10	0.00	283.10

Comment 1862 TWO RETURNS AGGREGATED

Iron	Ore(tons)	Iron(%)	Value(£)
1874	1073.00	0.00	536.50

Comment 1874 SP.& SOME BROWN HAEMATITE.
Management: Manager 1859-1871 GEO.R.ODGERS; 1872-1875 EDW.HOSKING; 1877
WM.BENNETTS
Chief Agent 1860-1861 WM.BENNETTS; 1862-1863 WM.BENNETTS JOHN
ODGERS; 1864 WM.BENNETTS JOHN ODGERS W.H.BISHOP; 1865
WM.BENNETTS JOHN ODGERS; 1866-1876 WM.BENNETTS
Secretary 1859-1863 JOHN CADY (P); 1864-1870 JOHN WATSON;
1871-1874 J.WATSON S.STEPHENS; 1875-1877 T.B.LAWS S.STEPHENS

GRENVILLE,SOUTH   CAMBORNE   SW 668382 0542

Production: Copper No detailed return
Management: Chief Agent 1865-1866 GEO.R.ODGERS WM.BENNETTS; 1867-1868
GEO.R.ODGERS WM.BENNETTS J.OSBORNE
Secretary 1865-1868 JOHN WATSON

GRENVILLE,WEST   CROWAN   SW 640373 0543

Production: Tin No detailed return
Ownership: 1886 WEST GRENVILLE ADVENTURERS; 1887-1890 WEST WHEAL
GRENVILLE LTD.; 1891 HARVEY & CO. P.
Management: Chief Agent 1886-1888 THOS.HODGE; 1889-1891 JAS.SKEAT
Secretary 1891 WM.ROSEWARNE

Employment:	Underground	Surface	Total
1886		2	2
1887	3	15	18
1888	13	17	30
1889	28	12	40
1890	26	11	37
1891	16	8	24

GREY   GERMOE   SW 595293 0544

Production: Tin	Black(tons)	Stuff(tons)	Tin(tons)	Value(£)
1876	0.20	0.00	0.00	23.20
1884	0.70	0.00	0.00	22.00
1885	0.30	0.00	0.00	12.00
1888	0.10	0.00	0.10	7.00
1895	2.00	0.00	0.00	55.00
1900	7.20	0.00	0.00	575.00
1901	19.20	0.00	0.00	1325.00
1903	19.10	0.00	0.00	1415.00
1904	1.20	0.00	0.00	81.00
1905	0.40	0.00	0.00	31.00
1906	18.30	0.00	0.00	2054.00
1909	1.10	0.00	0.00	92.00
1910	4.60	0.00	0.00	488.00

Comment 1876 VALUE INC.TIN STUFF; 1884-1885 OPEN WORKS; 1888
OPEN WORKS; 1895 OPEN WORKS; 1900-1901 OPEN WORKS; 1904-1905
OPEN WORK; 1906 TWO RETURNS AGG.; 1909-1910 OPEN WORK
Ownership: 1884-1885 WHEAL GREY CLAY & BRICK CO.; 1887 WHEAL GREY CLAY &
BRICK CO.; 1903-1906 LOVERING & CO.
Comment 1884-1885 OPEN WORK; 1887 OPEN WORK

Employment:	Underground	Surface	Total.
1903	11	8	19
1905	2		2
1906	3	8	11

GREYSTONE WOOD   LEZANT   SX 362794 0545

Production: Lead	Ore(tons)	Metal(tons)	Value(£)
1878	0.60	0.50	136.50

Comment 1878 GREYSTONE

Manganese	Ore(tons)	Metal(tons)	Value(£)
1878	35.00	0.00	70.00
1879	20.00	0.00	20.00
1880	40.00	0.00	60.00

Ownership: 1877-1881 WM.LANGDON; 1907-1908 SID.HAWKYARD
Comment 1877 GREYSTON; 1879 GREYSTONE SILVER LEAD; 1880-1881
GREYSTONE; 1908 CLOSED JULY 1908
Management: Chief Agent 1877 WM.LANGDON; 1878 HY.JAMES; 1879 JOHN H.JAMES

CM-J   219

WM.SKEWIS; 1880–1881 J.WOOD WM.SKEWIS; 1885 FRAN.SPRY;
1907–1908 JOHN H.H.JAMES
Secretary 1878–1881 WM.LANGDON

Employment:	Underground	Surface	Total
1878	2		2
1879	14	10	24
1880		2	2
1881	2		2
1885		6	6
1908	11	4	15

Comment 1879 TOTAL FOR GREYSTONE & GREYSTONE WOOD

GRYLLS   ILLOGAN   0547

Production: Tin   Black(tons) Stuff(tons)   Tin(tons)   Value(£)
1864–1865 No detailed return
Comment 1864–1865 SEE EAST BASSET
Ownership: Comment 1864–1865 SEE EAST BASSET

GRYLLS   ST.HILARY   SW 561295 0548

Production: Tin	Black(tons)	Stuff(tons)	Tin(tons)	Value(£)
1859	0.00	0.00	0.00	5940.30
1860	0.00	0.00	0.00	3975.50
1861	0.00	0.00	0.00	7673.30
1862	92.60	0.00	0.00	14042.10
1863	253.20	0.00	0.00	16750.60
1864	121.80	0.00	0.00	7148.90
1865	131.40	0.00	0.00	6678.60
1866	84.40	0.00	0.00	3810.60
1886–1887 No detailed return				
1910	39.00	0.00	0.00	3401.00
1911	23.00	5584.00	0.00	2672.00
1912	4.30	0.00	0.00	538.00

Comment 1859–1861 TIN STUFF; 1862–1863 VALUE INC.TIN STUFF;
1886 SEE GREAT FORTUNE; 1887 SEE GREAT FORTUNE. SUSPENDED
JUNE 1887
Ownership: 1908–1913 GRYLLS MINES LTD.
Comment 1868–1871 SEE GREAT WESTERN; 1911–1912 TWO LOWEST
LEVELS ABANDONED SEPT.1911; 1913 NOT WORKED
Management: Manager 1859 JOHN MORCOM; 1860–1867 ED.ROGERS
Chief Agent 1859 JOHN CURTIS; 1860–1867 JAS.POPE; 1908–1912
H.S.KEOGH; 1913 J.TREEBY BARRETT
Secretary 1859 R.R.MICHELL (P); 1860–1867 WM.WATSON (P)

Employment:	Underground	Surface	Total
1908	23	22	45
1909	11	20	31
1910	33	31	64
1911	40	28	68
1912	10	8	18

GRYLLS CONSOLS   ST.HILARY   0549

Production: Copper No detailed return
            Tin No detailed return
Management: Manager 1863-1866 W.H.RICHARDS
            Chief Agent 1863-1866 C.GILBERT
            Secretary 1863-1866 W.H.RICHARDS

GRYLLS WHEAL FLORENCE   ST.HILARY   SW 556297 0550

Production: Copper   Ore(tons) Metal(tons)   Value(£)
            1865   6.00   0.30   19.40
            Comment 1865 (C)
            Tin   Black(tons) Stuff(tons)   Tin(tons)   Value(£)
            1865   3.70   0.00   0.00   185.20
            1866   7.10   0.00   0.00   310.40
Ownership: Comment 1868-1871 SEE GREAT WESTERN
Management: Manager 1863-1866 ED.ROGERS
            Chief Agent 1863-1866 JAS.POPE
            Secretary 1863-1866 WM.WATSON

GRYLLS,EAST   ST.HILARY   SW 565293 0552

Production: Tin   Black(tons) Stuff(tons)   Tin(tons)   Value(£)
            1864   42.90   0.00   0.00   2433.30
            1865   75.10   0.00   0.00   3630.60
            1866   53.10   0.00   0.00   2398.60
Ownership: Comment 1868-1871 LATE GREENBERRY MOOR. SEE GREAT WESTERN

GRYLLS,GREAT   ST.HILARY   SW 568290 0553

Production: Copper No detailed return
            Tin   Black(tons) Stuff(tons)   Tin(tons)   Value(£)
            1864   0.00   0.00   0.00 ·   117.50
            1865   0.00   0.00   0.00   1235.00
            1866   0.00   0.00   0.00   215.30
            Comment 1864-1866 TIN STUFF
            Arsenic   Ore(tons) Metal(tons)   Value(£)
            1886   13.90   0.00   64.00
            1887   9.00   0.00   48.00
            Comment 1886 CRUDE
Ownership: Comment 1887 GREAT GRYLLS UNITED SUSPENDED JUNE 1887
Management: Manager 1864-1865 ED.ROGERS
            Chief Agent 1864-1865 JAS.POPE; 1885-1887 J.W.DOBLE
            Secretary 1864-1865 WM.WATSON
Employment: Underground   Surface   Total
            1885   4   16   20
            1886   8   13   21
            1887   6   14   20
            Comment 1887 GREAT GRYLLS UNITED

                                    221

GRYLLS,SOUTH   ST.HILARY   SW 559284 0554

Production: Copper   Ore(tons) Metal(tons)   Value(£)
            1866   17.00   1.40   108.40
            Comment 1866 (C)

GRYLLS,WEST   ST.HILARY   SW 559296 0555

Production: Copper   Ore(tons) Metal(tons)   Value(£)
            1864   5.00   0.40   30.60
            1865   24.00   1.40   104.20
            Comment 1864-1865 (C)
Ownership: Comment 1865 SUSPENDED; 1868-1871 SEE GREAT WESTERN

GUNNISLAKE CLITTERS   GUNNISLAKE   SX 421720 0556

Production: Copper   Ore(tons) Metal(tons)   Value(£)
            1860   462.00   33.00   2952.10
            1862   618.00   39.80   3192.40
            1863   284.00   19.60   1485.20
            1865   225.00   10.20   738.70
            1866   573.00   31.90   2047.50
            1868   579.00   45.80   3017.40
            1869   618.00   59.00   3819.30
            1870   938.00   80.10   4805.50
            1871   1025.00   71.30   4420.70
            1873   939.00   50.90   2737.80
            1874   957.00   86.20   6006.50
            1875   1372.00   116.00   9154.90
            1876   1774.00   147.80   10328.20
            1877   1898.60   148.70   8700.70
            1878   1930.10   154.40   7846.90
            1879   1935.60   134.80   6630.50
            1880   1978.60   183.50   10120.20
            1881   2520.00   274.50   15831.80
            1882   2415.10   205.20   14121.00
            1883   2705.40   236.70   13517.00
            1884   1899.00   0.00   7852.00
            1885   1561.00   0.00   5787.00
            1886   865.00   0.00   4048.00
            1887   137.00   0.00   666.00
            1888   274.00   0.00   2357.00
            1889   163.00   0.00   887.00
            Comment 1859 SEE GUNNISLAKE, DEVON; 1860 (C); 1862-1863 (C);
            1865-1866 (C); 1868-1871 (C); 1873-1876 (C)
            Manganese No detailed return
            Tin   Black(tons) Stuff(tons)   Tin(tons)   Value(£)
            1873   17.40   0.00   0.00   1302.00
Ownership: 1875-1876 T.C.ISSACS & OTHERS; 1877-1889 GUNNISLAKE CLITTERS
            MINING CO.; 1899 JOS.PAULL
            Comment 1859 GUNNISLAKE; 1889 ABANDONED; 1899 CLITTERS;
            1900-1902 SEE CLITTERS UNITED
Management: Manager 1859-1863 N.SECCOMBE; 1869-1872 W.SKEWIS; 1873-1877
            WM.SKEWIS; 1880-1881 JOHN C.SECCOMBE
            Chief Agent 1859-1863 JOHN RODDA; 1864-1865 JOHN RODDA

GUNNISLAKE CLITTERS   GUNNISLAKE   Continued

        WM.SKEWIS; 1866-1868 WM.SKEWIS S.C.SECCOMBE; 1869
        S.C.SECCOMBE; 1870-1871 J.SECCOMBE; 1872-1879 JOHN
        C.SECCOMBE; 1880-1881 WM.SKEWIS T.SKINNER; 1882 WM.SKEWIS;
        1883-1889 JOHN C.SECCOMBE
        Secretary 1859 R.T.SKINNER; 1860-1861 R.T.SKINNER (P);
        1862-1868 R.J.SKINNER; 1869-1874 R.T.SKINNER; 1879
        R.T.SKINNER; 1880-1881 BAW.SKEWIS
Employment: Underground   Surface   Total
        1878-1879   58   63   121
        1880   97   70   167
        1881   98   69   167
        1882   120   87   207
        1883   93   68   161
        1884   86   57   143
        1885   48   41   89
        1886   59   31   90
        1887   36   23   59
        1888   51   30   81
        1889   49   23   72

GUNNISLAKE,EAST   GUNNISLAKE   SX 423718 0557

Production: Copper No detailed return
        Comment 1855-1873 SEE EAST GUNNISLAKE, DEVON
        Tin   Black(tons) Stuff(tons)   Tin(tons)   Value(£)
        1860-1862 No detailed return
        Comment 1860-1862 SEE EAST GUNNISLAKE, DEVON
Ownership: Comment 1859-1862 SEE EAST GUNNISLAKE, DEVON; 1863-1866
        INC.SOUTH BEDFORD.SEE EAST GUNNISLAKE, DEVON; 1867-1875 SEE
        EAST GUNNISLAKE, DEVON
Management: Manager 1863-1866 WM.G.GARD
        Chief Agent 1863-1866 JAS.PHILLIPS
        Secretary 1863-1866 WM.G.GARD

GUNNISLAKE,OLD   GUNNISLAKE   SX 430719 0558

Production: Copper   Ore(tons) Metal(tons)   Value(£)
        1861   548.00   40.40   3448.20
        1864   218.00   15.70   1385.90
        1866   14.00   0.50   22.70
        1867   398.00   32.90   2324.80
        1868   60.00   4.00   250.20
        1872   902.00   46.00   3377.70
        1882   12.00   1.60   27.00
        1885   7.00   0.00   0.00
        Comment 1861 (C)GUNNISLAKE; 1864 (C)GUNNISLAKE; 1866 (C);
        1867 (C)GUNNISLAKE; 1868 (C); 1872 (C)GUNNISLAKE
        Tin   Black(tons) Stuff(tons)   Tin(tons)   Value(£)
        1865   64.60   0.00   0.00   3136.10
        1866   19.00   0.00   0.00   1071.90
        1870   8.50   0.00   0.00   682.70
        1871   22.80   0.00   0.00   1820.10
        1872   16.30   0.00   0.00   2177.60

                                        223

GUNNISLAKE,OLD   GUNNISLAKE   Continued

Tin   Black(tons) Stuff(tons)   Tin(tons)   Value(£)
1885   2.00   11.30   0.00   90.00
1888   19.00   0.00   0.00   83.00
Comment 1865 GUNNISLAKE; 1866 SEE OLD GUNNISLAKE, DEVON;
1870–1872 GUNNISLAKE; 1888 BLACK TIN FIGURE WRONG?

Ownership: 1880–1883 OLD GUNNIS LAKE MINING CO.; 1884–1888 OLD
GUNNISLAKE MINING CO.
Comment 1886–1888 NOT WORKED; 1900–1902 SEE CLITTERS UNITED

Management: Manager 1860–1865 WM.G.GARD; 1881 R.C.SECCOMBE
Chief Agent 1860–1865 JAS.PHILLIPS; 1866–1867 HY.RICKARD;
1868–1870 F.PHILLIPS; 1880–1881 ALF.F.SECCOMBE; 1882–1886
R.C.SECCOMBE; 1887–1888 WM.SKEWIS
Secretary 1860–1865 WM.G.GARD (P); 1866–1868 JOHN HITCHINGS;
1869–1870 JEHU.HITCHINGS

Employment: Underground   Surface   Total
1880   6   5   11
1881   10   5   15
1882   15   4   19
1883   11   4   15
1884   11   1   12
1885   5   5

GURLYN   ST.ERTH   SW 565329 0559

Production: Copper   Ore(tons) Metal(tons)   Value(£)
1848   118.00   9.40   599.00
1862   107.00   9.00   774.70
1863   187.00   12.80   966.60
1864   51.00   3.60   314.90
1865   12.00   0.80   63.00
Comment 1848 (C); 1862–1865 (C)
Tin   Black(tons) Stuff(tons)   Tin(tons)   Value(£)
1861   20.90   0.00   0.00   1426.80
1862   80.50   0.00   0.00   5217.20
1863   61.80   0.00   0.00   4057.30
1864   34.80   0.00   0.00   2160.50
1865   4.50   0.00   0.00   234.80
1904   6.50   0.00   0.00   490.00
1906   2.00   0.00   0.00   170.00
1907   26.00   0.00   0.00   2565.00
1909   0.00   200.00   0.00   200.00
1910   0.00   77.00   0.00   70.00
Comment 1862 TWO RETURNS AGGREGATED

Ownership: 1903–1904 WHEAL GURLYN SYNDICATE; 1906–1913 GURLYN CONSOLS
TIN MINES LTD.
Comment 1864–1865 SUSPENDED; 1909 SHUT & FOR SALE MARCH 1910;
1910 PUMPING; 1911 SUSPENDED MARCH 1911; 1912 SUSPENDED; 1913
NOT WORKED

Management: Chief Agent 1860 RICH.KENDALL; 1862–1863 JOHN CURTIS
J.MARTIN; 1903–1904 T.J.BOLLAND; 1906–1913 ART.BRIGSTOCKE
Secretary 1860 W.MARTIN (P); 1861–1862 WM.VAWDREY; 1863–1865
WM.VAWDREY (P); 1904 JOS.PHILLIPS & CO.

Employment: Underground   Surface   Total
1903   23   13   36

224

GURLYN   ST.ERTH   Continued

	Underground	Surface	Total
1904	30	26	56
1906	31	49	80
1907	25	25	50
1908	4		4
1909	4	2	6
1910	4	4	8
1913		1	1

GUSKUS   ST.HILARY   SW 563315 0560

Production: Lead

Lead	Ore(tons)	Metal(tons)	Value(£)
1859	1.10	0.80	0.00
1860	No detailed return		

Comment 1860 SEE ANNA (ST.HILARY)
Zinc No detailed return

Copper	Ore(tons)	Metal(tons)	Value(£)
1853	236.00	15.60	1545.10
1854	273.00	14.00	1315.60
1855	260.00	11.80	1026.00
1856	178.00	10.00	795.40

Comment 1853-1856 (C)

Tin	Black(tons)	Stuff(tons)	Tin(tons)	Value(£)
1853	22.50	0.00	0.00	1696.50
1854	76.80	0.00	0.00	4950.00
1855	105.60	0.00	0.00	6791.60
1856	58.20	0.00	0.00	4100.20
1857	15.10	0.00	0.00	1079.70
1858	6.60	0.00	0.00	445.80
1859	1.10	0.00	0.00	78.10
1860	No detailed return			
1861	0.90	0.00	0.00	69.70

Comment 1860 SEE ANNA(ST HILARY)
Ownership: Comment 1860 SEE ANNA(ST.HILLARY)

GUSTAVUS MINES   CAMBORNE   SW 645405 0561

Production: Copper

Copper	Ore(tons)	Metal(tons)	Value(£)
1853	89.00	5.80	576.80

Comment 1853 (C)

GWENNAP CONSOLS,WEST   GWENNAP   0563

Production: Copper No detailed return

Tin	Black(tons)	Stuff(tons)	Tin(tons)	Value(£)
1873	0.00	14.20	0.00	3.90

Management: Chief Agent 1873 JOHN NANCARROW
            Secretary 1872-1873 CHAS.PARRY

225

GWENNAP UNITED   GWENNAP   SW 745412 0564

Production: Copper   Ore(tons) Metal(tons)   Value(£)
            1901   6.00   0.00   33.00
            1902   495.00   0.00   211.00
            Tin   Black(tons) Stuff(tons)   Tin(tons)   Value(£)
            1894   0.20   0.00   0.00   7.00
            1901   77.00   0.00   0.00   4607.00
            1902   24.00   26.00   0.00   1375.00
            Arsenic   Ore(tons) Metal(tons)   Value(£)
            1901   117.00   0.00   834.00
            1902   37.00   0.00   226.00
Ownership: 1894-1895 WM.KELLOW & SONS; 1900-1902 COHEN MOSS & CO.;
            1905-1906 WM.KINSMAN & FATHER; 1909-1913 ELIJAH KELLOW
            Comment 1894 TRIBUTERS; 1895 TRIBUTERS NOT WORKED; 1905-1906
            PART OF; 1909-1910 WHITEWORKS; 1911-1913 WHITEWORKS NOT
            WORKED
Employment: Underground   Surface   Total
            1894   2   1   3
            1900   6   25   31
            1901   42   79   121
            1902   9   30   39
            1905-1906   2   2
            Comment 1905-1906 PART OF

GWIN AND SINGER   BREAGE   SW 622307 0565

Production: Tin   Black(tons) Stuff(tons)   Tin(tons)   Value(£)
            1885   0.10   6.00   0.00   6.00
            1886   8.20   137.00   0.00   417.00
            1887   19.00   255.20   0.00   1093.00
            1888   64.10   0.00   0.00   3947.00
            1889   24.50   76.00   0.00   1370.00
            1890   2.60   160.70   0.00   561.00
            1891   2.40   171.00   0.00   543.00
            Comment 1885 GWIN & SINGER UNITED; 1890-1891 GWIN & SINGER
            UNITED
Ownership: 1887 WALT.C.VENNING; 1888-1890 GWIN & SINGER UNITED TIN
            MINING CO.LTD; 1891 WALT.C.VENNING
            Comment 1885 GWINEAR & SINGER; 1892-1898 SEE BREAGE
Management: Chief Agent 1885-1891 CHAS.ROACH
Employment: Underground   Surface   Total
            1885   18   6   24
            1886   15   7   22
            1887   25   51   76
            1888   38   45   83
            1889   37   34   71
            1890   32   14   46
            1891   20   14   34

GWINDRA   ST.STEPHEN   SW 949528 0566

Production: Iron   Ore(tons)   Iron(%)   Value(£)
            1858   276.30   0.00   103.70
            1859   150.00   0.00   95.00

226

GWINDRA   ST.STEPHEN   Continued

        Iron   Ore(tons)   Iron(%)   Value(£)
        1860   No detailed return
        Comment 1858 BH.; 1859 BH.INC DYEHOUSE, LANJEN, PENHALE(SP)
Ownership: 1907 KINGSWAY SYNDICATE LTD.
        Comment 1907 GWENDRA
Management: Secretary 1907 M.MUNROE (S)
Employment: Underground   Surface   Total
        1907   8   8

GWINEAR CONSOLS   GWINEAR   SW 607368 0567

Production: Copper   Ore(tons) Metal(tons)   Value(£)
        1847   249.00   10.40   637.20
        1848   262.00   13.70   709.10

GWINNINGS   COLROGER   0568

Production: Copper No detailed return
Ownership: 1912-1913 GWINNINGS COPPER MINE CO.
        Comment 1912-1913 PROSPECTING DEC.1912
Employment: Underground   Surface   Total
        1912-1913   2   2   4

HADEN   ROCHE   0569

Production: Iron   Ore(tons)   Iron(%)   Value(£)
        1873   400.00   0.00   300.00
        Comment 1873 HE.
Ownership: 1873 MID CORNWALL MINING CO.LTD.
Management: Manager 1873 DAVID COCK

HALL   ST.COLUMB   0570

Production: Tin No detailed return
Ownership: 1876-1877 HALL SYNDICATE
Management: Chief Agent 1876-1877 S.H.F.COX

HALLAMANNING   BREAGE   SW 564312 0571

Production: Copper   Ore(tons) Metal(tons)   Value(£)
        1851   123.00   8.10   552.30
        1852   2947.00   197.20   16952.90
        1853   2541.00   165.50   16356.80
        1854   2508.00   153.70   15481.50
        1855   2543.00   154.90   15626.30
        1856   1754.00   121.30   11375.50
        1857   857.00   73.40   7633.50
        1858   112.00   9.60   903.40
        Comment 1851-1852 (C); 1853 (C)INC.CROFT GOTHAL; 1854 (C);
        1855-1858 (C)INC.CROFT GOTHAL

CM-J*   227

HALLAMANNING,EAST   BREAGE   SW 562311 0572

Production:	Tin	Black(tons)	Stuff(tons)	Tin(tons)	Value(£)
	1852	2.50	0.00	0.00	0.00
	1855	4.50	0.00	0.00	271.60
	1856	1.70	0.00	0.00	122.50
	1857	0.00	0.00	0.00	875.90

Comment 1852 TIN ORE; 1857 TIN STUFF
Management: Manager 1860–1862 S.J.REES
Secretary 1860–1862 WM.PAINTER (P)

HALLENBEAGLE   KENWYN   SW 727449 0573

Production:	Copper	Ore(tons)	Metal(tons)	Value(£)
	1845	2879.00	152.50	10478.90
	1846	1291.00	62.50	4000.50
	1847	No detailed return		
	1864	734.00	51.90	4335.90
	1865	1475.00	88.40	6847.40
	1866	701.00	38.90	2429.30
	1867	332.00	16.20	1089.50

Comment 1845–1847 (C); 1864–1867 (C)

	Tin	Black(tons)	Stuff(tons)	Tin(tons)	Value(£)
	1855	0.30	0.00	0.00	20.60
	1867	0.00	0.00	0.00	16.80

Comment 1867 TIN STUFF
Ownership: 1899–1900 PHIL.MICHELL
Comment 1869–1871 INC.EAST DOWNS
Management: Manager 1864 E.RICHARDS; 1869–1871 WM.RICH
Chief Agent 1864 R.M.KITTO; 1865–1866 WM.BAWDEN
Secretary 1864–1866 ED.KING; 1869–1871 HY.MICHELL

HALLENBEAGLE,NORTH   GWENNAP   SW 712453 0574

Production:	Copper	Ore(tons)	Metal(tons)	Value(£)
	1862	31.00	2.30	201.10
	1863	30.00	2.00	153.70

Comment 1862–1863 (C)
Tin No detailed return
Ownership: Comment 1865 SUSPENDED
Management: Chief Agent 1860–1864 JAS.CRAZE
Secretary 1860–1861 C.THOMAS (P); 1862–1865 FRAN.LISTER

HALLEW   BUGLE   SX 016595 0575

Production: Iron No detailed return
Ownership: 1873 DAVID COCK
Management: Manager 1873 R.DYER
Chief Agent 1872 DAVID COCK

HALLOW   BUGLE   SX 016595 0576

Production: Tin No detailed return
Management: Chief Agent 1872 DAVID COCK; 1873 W.DANDLOCK

HALVINNA   ALTARNUN   SX 211787 0577

Production: Tin No detailed return
Ownership: 1889 C.T.H.BENNETT; 1902-1904 C.H.SWANTON
            Comment 1903-1904 TEMP.ABANDONED 12TH DEC.1903
Management: Chief Agent 1889 J.JOLL; 1902-1904 JOHN JOEL
Employment: Underground   Surface   Total
            1889   20   20
            1902   2   7   9
            1903   2   4   6

HALWYN   PERRANZABULOE   SW 775565 0578

Production: Iron No detailed return
Ownership: 1864-1865 E.CARTER & CO.
Management: Chief Agent 1864-1865 JOHN BALL

HAMMET CONSOLS   ST.NEOT   SX 188693 0579

Production: Copper No detailed return
            Tin   Black(tons) Stuff(tons)   Tin(tons)   Value(£)
            1870   2.10   0.00   0.00   158.60
            Comment 1870 HAMMETT
Ownership: Comment 1862-1865 SUSPENDED; 1870 HAMMETT
Management: Manager 1860-1861 J.ROSKILLEY; 1869-1872 PETER TEMBY
            Secretary 1860-1864 P.ROSKILLEY (P); 1869-1871 C.RULE; 1872
            BEAS.C.RULE & CO.

HAMMIE   GERMOE   0581

Production: Tin   Black(tons) Stuff(tons)   Tin(tons)   Value(£)
            1912   0.00   34.00   0.00   7.00
Ownership: 1912-1913 JOS.WILLIAMS
            Comment 1912-1913 CLEARING FEB.1912 CLOSED JUNE 1912

HAMPTON   ST.HILARY   SW 534313 0582

Production: Tin   Black(tons) Stuff(tons)   Tin(tons)   Value(£)
            1904   16.00   0.00   0.00   1299.00
            1910   70.20   0.00   0.00   5676.00
            1911   104.70   0.00   0.00   10587.00
            1912   95.80   0.00   0.00   11552.00
            1913   66.60   0.00   0.00   7653.00
Ownership: 1902-1903 JOS.PHILLIPS; 1904 WHEAL HAMPTON LTD.; 1908-1913
            ST.STEPHENS SYNDICATE LTD.
            Comment 1909-1911 INC.TREGURTHA DOWNS, OWEN VEAN; 1912-1913
            INC.TREGURTHA DOWNS, OWEN VEAN, RODNEY REOPENING JUNE 1912

                                  229

HAMPTON   ST.HILARY   Continued

Management: Chief Agent 1904 JOS.PHILLIPS & CO. & J.H.GILBERT
            Secretary 1908-1913 JOS.PHILLIPS (S)
Employment: Underground   Surface   Total
            1903   22   14   36
            1904   20   36   56
            1909   32   28   60
            1910   41   41   82
            1911   67   64   131
            1912   60   61   121
            1913   60   47   107
            Comment 1909-1911 INC.OWEN VEAN & TREGURTHA DOWNS; 1912-1913
            INC.OWEN VEAN,TREGURTHA DOWNS & RODNEY

HANSON MINES   BREAGE   SW 627276 0583

Production: Copper   Ore(tons) Metal(tons)   Value(£)
            1846   162.00   10.90   747.70
            1847   101.00   6.30   432.20
            Comment 1846-1847 (C)

HARDHEAD   WARLEGGAN   SX 149718 0584

Production: Tin   Black(tons) Stuff(tons)   Tin(tons)   Value(£)
            1911   0.00   936.00   0.00   291.00
            1912   4.00   0.00   0.00   460.00
            1913   2.60   0.00   0.00   286.00
            Iron No detailed return
Ownership: 1911 E.J.COLWILL; 1912 CAPT.ABBOTT; 1913 ASHFORD & WESTLAKE
            Comment 1874 SUSPENDED
Management: Manager 1873 JOHN KESSELL
            Chief Agent 1872 JOHN KESSELL; 1874 JOHN KESSELL
            Secretary 1873 JOS.KELLOW
Employment: Underground   Surface   Total
            1911   5   4   9
            1912   4   5   9
            1913   4   3   7

HAREWOOD CONSOLS   CALSTOCK   SX 450695 0585

Production: Copper No detailed return
Management: Manager 1869-1870 THOS.NEAL
            Secretary 1869-1870 W.F.PEARCE

HARMONY   REDRUTH   SW 701435 0586

Production: Copper   Ore(tons) Metal(tons)   Value(£)
            1867   10.00   0.20   10.50
            1868   6.00   0.30   18.00
            1880   15.00   0.40   32.60
            Comment 1867-1868 (C)

230

HARRIET   SITHNEY   SW 635310 0587

Production: Tin   Black(tons) Stuff(tons)   Tin(tons)   Value(£)
            1853   0.40   0.00   0.00   19.50
            1856   0.00   0.00   0.00   25.80
            1857   0.00   0.00   0.00   56.30
            1858   0.00   0.00   0.00   26.60
            1859   0.00   0.00   0.00   61.00
            1860   0.00   0.00   0.00   483.60
            1861   0.00   0.00   0.00   2373.70
            1862   0.00   0.00   0.00   1407.80
            1863   0.00   0.00   0.00   6158.20
            1864   0.00   0.00   0.00   1133.70
            1865   0.00   0.00   0.00   128.50
            1866   0.00   0.00   0.00   6.20
            1874   No detailed return
            Comment 1856-1861 TIN STUFF; 1862 TIN STUFF.TWO RETURNS
            AGGREGATED; 1863-1866 TIN STUFF; 1874 SEE NORTH METAL
Ownership: Comment 1872-1874 SEE NORTH METAL

HARRIET   CAMBORNE   SW 659395 0588

Production: Copper   Ore(tons) Metal(tons)   Value(£)
            1845   689.00   41.10   2941.50
            1846   987.00   60.10   4200.30
            1847   546.00   31.00   2144.70
            1848   281.00   13.60   724.70
            1857   161.00   7.30   683.70
            1858   234.00   14.30   1281.50
            1859   221.00   12.40   1116.20
            1860   596.00   35.30   3290.30
            1861   344.00   23.30   2135.90
            1862   191.00   12.00   962.30
            1863   257.00   13.90   1007.20
            1864   142.00   7.20   586.90
            1865   26.00   2.30   182.40
            1866   34.00   1.90   141.30
            Comment 1845-1848 (C); 1857-1866 (C)
Ownership: Comment 1867-1872 SEE PENDARVES UNITED
Management: Manager 1859-1865 STEP.WILLIAMS
            Secretary 1859-1865 ED.KING (P)

HARROWBARROW,SOUTH   ST.MELLION   0589

Production: Copper No detailed return
Ownership: Comment 1862-1865 SUSPENDED
Management: Chief Agent 1861 JOHN GIFFORD
            Secretary 1861-1864 H.E.CROKER (P); 1865 THOS.HORSWILL

HARTLEY   GWINEAR   SW 625387 0590

Production: Copper   Ore(tons) Metal(tons)   Value(£)
            1866   7.00   0.60   50.00
            Comment 1866 (C)

                                  231

HARTLEY   GWINEAR   Continued

        Tin   Black(tons) Stuff(tons)   Tin(tons)   Value(£)
        1863   0.00   0.00   0.00   25.40
        1864   0.00   0.00   0.00   11.50
        1865   0.00   0.00   0.00   27.90
        1866   0.00   0.00   0.00   44.10
        Comment 1863-1865 TIN STUFF
Ownership: 1907-1908 J.A.TEMBY
Management: Manager 1860 JOS.VIVIAN
             Chief Agent 1860 HY.SKEWIS; 1861-1863 HY.SKEWIS GEO.BENNETTS;
             1864-1865 JAS.VIVIAN PHIL.SKEWIS
             Secretary 1860-1865 ALM.E.PAULL (P)
Employment: Underground   Surface   Total
             1907   4   4

HARVEYS DROSS   0592

Production: Copper   Ore(tons) Metal(tons)   Value(£)
            1845   211.00   8.70   314.70
            1846   226.00   12.30   558.80
            Comment 1845-1846 (C)

HARVOSE BLENDE   ST.STEPHEN-IN-BRANNEL   0593

Production: Zinc No detailed return
Ownership: 1899-1904 RICH.H.WILLIAMS
             Comment 1902 WORK IN ABEYANCE; 1903 HARVOSE SUSPENDED
Management: Chief Agent 1904 JOHN HARPER
Employment: Underground   Surface   Total
             1899   6   6   12
             1900   11   11
             1901   8   1   9

HAVEN,EAST AND WEST   0594

Production: Lead   Ore(tons) Metal(tons)   Value(£)
            1845   16.00   9.00   0.00

HAWKER   ST.HILARY   0595

Production: Copper No detailed return
            Tin No detailed return
Management: Chief Agent 1862 JOHN PEARCE
            Secretary 1862 THOS.W.FIELD

HAWKMOOR   CALSTOCK   SX 434727 0596

Production: Copper   Ore(tons) Metal(tons)   Value(£)
            1852   119.00   7.70   637.60
            1853   86.00   5.40   543.20
            1854   243.00   16.80   1698.60

                                  232

Copper	Ore(tons)	Metal(tons)	Value(£)
1855	331.00	20.50	2053.90
1856	238.00	14.40	1307.10
1857	154.00	8.50	806.30
1858	106.00	6.30	550.70
1859	421.00	25.30	2402.10
1860	576.00	33.70	2966.30
1861	462.00	26.20	2227.00
1862	341.00	20.40	1616.60
1863	171.00	10.00	739.50
1864	115.00	6.10	505.90
1865	108.00	5.40	415.10
1866	88.00	4.00	241.40
1912	14.00	0.00	0.00

Comment 1852 (C); 1853 (C)HAWKS MOOR; 1854-1865 (C); 1866 (C)HAWKS MOOR

Tin	Black(tons)	Stuff(tons)	Tin(tons)	Value(£)
1853	1.30	0.00	0.00	44.20
1854	1.00	0.00	0.00	40.10
1862	3.40	0.00	0.00	136.20
1863	7.20	0.00	0.00	346.10
1864	5.60	0.00	0.00	287.10
1865	5.00	0.00	0.00	216.50
1866	5.30	0.00	0.00	128.30
1875	4.10	0.00	0.00	223.30
1912	1.00	0.00	0.00	129.00

Comment 1912 VALUE EST

Arsenic	Ore(tons)	Metal(tons)	Value(£)
1912	19.00	0.00	80.00

Tungsten	Ore(tons)	Metal(tons)	Value(£)
1912	0.50	0.00	0.00

Ownership: 1912-1913 SML.G.KNOTT
Comment 1900-1902 SEE CLITTERS UNITED; 1913 NOT WORKED
Management: Manager 1859 J.T.PHILLIPS; 1860 JOSH.PHILLIPS; 1869 MOSES BAWDEN
Chief Agent 1859 JAS.RICHARDS; 1860 J.T.RICHARDS; 1861-1862 J.T.RICHARDS JOSH.RICHARDS; 1863-1865 JOSH.RICHARDS JAS.KELLY
Secretary 1859-1865 P.FISHER (P); 1869 WM.SKEWIS

Employment:	Underground	Surface	Total
1912	4	1	5

HAWKMOOR,WEST   CALSTOCK   SX 428724 0597

Production: Copper No detailed return

Tin	Black(tons)	Stuff(tons)	Tin(tons)	Value(£)
1872	3.10	0.00	0.00	272.00
1873	62.00	0.00	0.00	4836.00
1874	3.60	0.00	0.00	207.00

Comment 1872-1873 WEST HAWKSMOOR
Ownership: 1880-1888 WEST HAWKMOOR MINING CO.
Comment 1885-1886 NOT WORKED; 1887 NOT WORKING; 1888 SUSPENDED
Management: Manager 1870-1875 WM.SKEWIS; 1880-1881 MOSES BAWDEN

Chief Agent 1870—1871 JAS.MAYNE; 1882 MOSES BAWDEN
Secretary 1870—1875 MOSES BAWDEN; 1880 MOSES BAWDEN;
1883—1888 MOSES BAWDEN

Employment:	Underground	Surface	Total
1880	4		4
1881	2		2

HAWKS POINT   LELANT   SW 535388 0598

Production: Copper

	Ore(tons)	Metal(tons)	Value(£)
1851	136.00	7.20	456.20
1852	349.00	15.40	1203.60
1853	177.00	5.40	489.30
1884	8.00	0.00	16.00

Comment 1851—1853 (C)

Tin

	Black(tons)	Stuff(tons)	Tin(tons)	Value(£)
1883	0.40	0.00	0.00	14.00
1884	0.30	0.00	0.00	6.00

Ownership: 1883—1884 NUGENT & KAY; 1900 N.T.ASHTON; 1901 DANIELL &
THOMAS
Comment 1867—1868 OR FANNY ADELA; 1869 SEE FANNY ADELA; 1901
OR EAST PROVIDENCE
Management: Manager 1867—1868 HY.SKEWIS
Chief Agent 1883 J.C.TREGEAR
Secretary 1867—1868 ALB.VIVIAN; 1884 N.F.ASHTON

Employment:	Underground	Surface	Total
1883	14	9	23
1884	3		3
1901	4		4

HAWKS TOR   0599

Production: Tin

	Black(tons)	Stuff(tons)	Tin(tons)	Value(£)
1864	0.50	0.00	0.00	34.10
1865	0.80	0.00	0.00	50.40
1866	0.50	0.00	0.00	32.10

HAYE VALLEY   CALLINGTON   SX 345695 0600

Production: Tin

	Black(tons)	Stuff(tons)	Tin(tons)	Value(£)
1863	11.90	0.00	0.00	752.50
1864	2.20	0.00	0.00	144.30
1870	2.70	0.00	0.00	163.40
1871	2.30	0.00	0.00	131.80
1872	1.90	0.00	0.00	133.90
1873	3.90	0.00	0.00	235.40

Comment 1873 HAY VALLEY
Management: Chief Agent 1861—1866 THOS.TAYLOR; 1869—1870 WM.TAYLOR
Secretary 1861—1866 THOS.TAYLOR (P); 1869—1870 JOHN TAYLOR

HAYFORD LTD.   ST.IVE   SX 292680 0601

Production: Zinc   Ore(tons) Metal(tons)   Value(£)
            1901   237.00   92.00   488.00
Ownership: 1899-1901 HAYFORD MINES CO.LTD.
Management: Chief Agent 1899-1901 GEO.H.BELLAMY
Employment: Underground   Surface   Total
            1899   10   5   15
            1900   8   14   22
            1901   10   18   28

HAYTOR   ILSINGTON, DEVON   SX 773771 0602

Production: Iron   Ore(tons)   Iron(%)   Value(£)
            1874   1669.00   0.00   1251.50
            Comment 1874 BH.

HEARLE   ST.JUST   SW 390340 0603

Production: Copper   Ore(tons) Metal(tons)   Value(£)
            1862   20.00   2.40   185.00
            Comment 1862 (C)
            Tin   Black(tons) Stuff(tons)   Tin(tons)   Value(£)
            1861   59.70   0.00   0.00   4165.60
            1862   87.30   0.00   0.00   5617.20
            1863   34.90   0.00   0.00   2250.00
            1864   42.70   0.00   0.00   2607.00
            Comment 1862 TWO RETURNS AGGREGATED; 1864 NINE MONTHS ONLY
Ownership: Comment 1865 SUSPENDED
Management: Manager 1859 N.TREDINNICK; 1861 WM.RUTTER
            Chief Agent 1859 S.TREDINNICK; 1860 JOHN WESLEY; 1861
            WM.WESLEY; 1862 THOS.UREN WM.WESLEY; 1863-1864 J.G.PHILLIPS
            WM.WESLEY
            Secretary 1859 WALT.BORLASE (P); 1860-1864 JAS.HOLLOW (P);
            1865 THOS.HOLLOW

HELENA   ST.HILARY   SW 539311 0604

Production: Tin   Black(tons) Stuff(tons)   Tin(tons)   Value(£)
            1891   66.40   0.00   0.00   3788.00
            Comment 1891 FORMERLY TREGURTHA DOWNS
Ownership: 1891 DAVID BISCHOFSWERDER
            Comment 1891 FORMERLY TREGURTHA DOWNS; 1892 SEE TREGURTHA
            DOWNS
Employment: Underground   Surface   Total
            1891   58   55   113

HELLEN   0605

Production: Copper   Ore(tons) Metal(tons)   Value(£)
            1854   90.00   6.20   621.10
            Comment 1854 (C)

                            235

HENDER   GWINEAR   SX 630366 0606

Production: Lead & Silver   Ore(tons) Lead(tons) Silver(ozs)   Value(£)
            1854   100.00   74.00   0.00   0.00
            1855   66.00   47.00   0.00   0.00
            1856   44.20   29.00   0.00   0.00
            1857   15.70   12.00   36.00   0.00
            Comment 1854 HENDRE
            Copper No detailed return
            Tin   Black(tons) Stuff(tons) Tin(tons)   Value(£)
            1855   0.30   0.00   0.00   23.00
            1856   0.00   0.00   0.00   47.20
            Comment 1856 TIN STUFF
Ownership: Comment 1862-1865 ABANDONED
Management: Manager 1860-1861 JOHN WILLIAMS
            Secretary 1860-1861 WM.HUTHNANCE (P)

HENDER,NEW   GWINEAR   0607

Production: Copper   Ore(tons) Metal(tons)   Value(£)
            1861   128.00   10.80   992.50
            1862   24.00   1.40   113.40
            Comment 1861-1862 (C)
            Tin   Black(tons) Stuff(tons) Tin(tons)   Value(£)
            1861   0.00   0.00   0.00   5.30
            Comment 1861 TIN STUFF
Management: Chief Agent 1861 HY.WOOLCOCK JNR.
            Secretary 1861 WM.RICHARDS (P)

HENDRA,NEW   BREAGE   SW 591281 0608

Production: Copper   Ore(tons) Metal(tons)   Value(£)
            1864   3.00   0.20   19.20
            1877   3.20   0.20   11.80
            Comment 1864 (C)
            Tin   Black(tons) Stuff(tons) Tin(tons)   Value(£)
            1873   3.50   0.00   0.00   70.00
Ownership: 1877-1878 NEW HENDRA MINING CO.; 1912-1913 PRAA SYNDICATE
            Comment 1906-1910 SEE HORSE BURROW (FIVE LANES); 1912-1913
            HENDRA
Management: Manager 1865 J.JULIAN; 1872-1875 RICH.KING; 1876-1877
            WM.ROWE
            Chief Agent 1861-1865 RICH.KING; 1872-1875 WM.ROWE; 1876-1878
            RICH.KING
            Secretary 1861-1865 WM.ARGAL; 1872-1876 WM.WARD; 1913
            J.TREEBY BARRATT
Employment: Underground   Surface   Total
            1878   2   2   4
            1907-1908
            Comment 1907-1908 SEE HORSE BURROW (FIVE LANES)

HENRIETTA   ILLOGAN   SW 654421 0609

Production: Copper No detailed return
Ownership: Comment 1861 SEE ALSO EMILY HENRIETTA
Management: Manager 1860—1861 JOHN DAWE
            Secretary 1860—1861 JOHN F.PENROSE (P)

HENRY   KEA   SW 753426 0610

Production: Copper   Ore(tons) Metal(tons)   Value(£)
            1845   148.00   10.30   750.30
            1848   311.00   24.80   1469.80
            1849   338.00   28.50   1855.50
            1850   503.00   39.90   2835.30
            1851   154.00   13.10   899.70
            1854   88.00   7.90   840.00
            1857   67.00   9.80   1045.00
            1858   103.00   10.30   1000.70
            1859   161.00   11.30   1113.10
            1862   13.00   0.70   54.30
            Comment 1845 (C); 1848—1851 (C); 1854 (C); 1857—1859 (C);
            1862 (C)
Ownership: Comment 1865 SUSPENDED
Management: Manager 1859—1864 FRAN.PRYOR
            Chief Agent 1859—1860 WM.KITTO; 1861—1864 EL.RALPH
            Secretary 1859—1865 FRAN.PRYOR (P)

HENRY   ST.HILARY   0611

Production: Copper No detailed return
            Tin No detailed return
Management: Manager 1872 J.RICHARDS
            Secretary 1871—1872 F.C.SASSE

HERLAND   GWINEAR   SW 591370 0612

Production: Copper No detailed return
            Tin   Black(tons) Stuff(tons)   Tin(tons)   Value(£)
            1860   No detailed return
            Comment 1860 SEE ROSEWARNE
Ownership: Comment 1859—1865 SEE ROSEWARNE

HERMON   ST.JUST   SW 357307 0613

Production: Tin   Black(tons) Stuff(tons)   Tin(tons)   Value(£)
            1874   7.00   0.00   0.00   350.00
            1875   2.10   0.00   0.00   107.50
            1877   13.60   0.00   0.00   535.20
            1878   6.80   0.00   0.00   239.00
            1879   5.90   0.00   0.00   236.00
            1880   5.40   0.00   0.00   273.70
Ownership: 1877—1882 WHEAL HERMON MINING CO.; 1913 H.F.LETCHA VEAN OLDS
            Comment 1875 SUSPENDED; 1882 NOT WORKED

237

Management: Manager 1874 JOHN CHENHALLS; 1878-1881 JAS.BENNETTS
            Chief Agent 1875 T.CHENHALLS; 1876 RICH.BOYNS; 1877
            JAS.BENNETTS
            Secretary 1874-1876 JOHN CHENHALLS; 1878-1879 RICH.BOYNS (P);
            1880-1882 RICH.BOYNS (S)

Employment: Underground   Surface   Total
            1878   3   7   10
            1879   6   4   10
            1880   5   6   11
            1881   5   5
            1913   10   10   20

HERODSCOOMBE   DULOE   SX 219603 0614

Production: Lead   Ore(tons) Metal(tons)   Value(£)
            1847   37.00   20.00   0.00
            1848-1854 No detailed return

HERODSFOOT   LANREATH   SX 212600 0615

Production: Lead & Silver	Ore(tons)	Lead(tons)	Silver(ozs)	Value(£)
1847	375.00	300.00	0.00	0.00
1848	721.00	570.00	0.00	0.00
1849	1050.00	830.00	0.00	0.00
1850	933.00	590.00	0.00	0.00
1851	833.00	624.00	0.00	0.00
1852	663.00	505.50	9282.00	0.00
1853	715.00	536.00	9295.00	0.00
1854	572.50	364.50	8607.00	0.00
1855	298.00	212.00	5512.00	0.00
1856	359.00	301.00	4417.00	0.00
1857	513.30	392.50	11516.00	0.00
1858	510.00	405.90	11845.00	0.00
1859	644.50	457.50	22850.00	0.00
1860	536.50	375.50	18750.00	0.00
1861	635.00	331.00	28797.00	0.00
1862	685.00	445.20	38715.00	0.00
1863	606.00	412.00	35844.00	0.00
1864	637.70	433.60	37691.00	0.00
1865	576.60	368.60	32016.00	0.00
1866	547.40	350.20	30450.00	0.00
1867	540.10	405.00	39285.00	0.00
1868	532.40	399.30	38730.00	0.00
1869	537.30	402.90	39090.00	0.00
1870	503.70	377.40	36569.00	0.00
1871	473.20	354.90	34435.00	0.00
1872	460.00	345.00	33465.00	0.00
1873	321.00	240.00	16050.00	7483.10
1874	378.40	283.50	27500.00	6655.60
1875	331.00	248.00	4800.00	5998.50
1876	296.80	222.60	4296.00	5543.60
1877	322.80	257.20	5145.00	4358.10
1878	307.80	231.00	4925.00	3696.00

Lead & Silver	Ore(tons)	Lead(tons)	Silver(ozs)	Value(£)
1879	256.30	192.70	4110.00	3072.00
1880	451.10	338.50	8750.00	4374.70
1881	435.00	217.50	12180.00	4386.40
1882	362.00	271.00	8130.00	4344.00
1883	285.20	199.40	2850.00	2150.00
1884	110.00	55.00	0.00	630.00
Copper	Ore(tons)	Metal(tons)	Value(£)	
1881	8.70	0.60	30.40	
1882	5.80	0.40	24.00	
1884	2.00	0.00	3.00	
Tungsten	Ore(tons)	Metal(tons)	Value(£)	
1881	5.10	0.00	48.70	

Ownership: 1877-1878 HERODSFOOT MINING CO.; 1879-1884 HERODSFOOT MINE
ADVENTURERS
Management: Manager 1859-1860 THOS.TREVILLION; 1861 JOHN H.TREVILLION;
1862-1868 THOS.TREVILLION; 1869-1873 W.L.TREVILLION; 1874
THOS.WILLIAMS; 1875-1877 THOS.TREVILLION; 1878-1879 JOHN
A.TREVILLION; 1880-1881 PETER TEMBY
Chief Agent 1859-1860 THOS.TREVILLION; 1861-1872 JOHN
TREVILLION; 1875 JOHN RUTTER; 1878 PETER TEMBY; 1879 JOHN
TAYLOR & SON; 1881 F.W.DABB; 1882 PETER TEMBY; 1884
THOS.TRELEASE
Secretary 1859-1879 THOS.TREVILLION (P); 1883 FELIX F.WATSON

Employment:	Underground	Surface	Total
1878	30	30	60
1879	44	40	84
1880	55	60	115
1881	67	41	108
1882	56	38	94
1883	62	33	95
1884	25	21	46

HERODSFOOT,EAST   LANREATH   0616

Production: Lead & Silver No detailed return
Ownership: 1880 EAST HERODSFOOT MINING CO.; 1881-1882 EAST HERODSFOOT
SILVER LEAD MINING CO.
Comment 1882 NOT WORKED
Management: Manager 1881 T.H.BENNETT
Chief Agent 1880-1881 J.A.TREVILLION
Secretary 1882 WM.BATTYE

Employment:	Underground	Surface	Total
1881	5		5

HERODSFOOT,NORTH   LANREATH   SX 212604 0617

Production: Lead & Silver	Ore(tons)	Lead(tons)	Silver(ozs)	Value(£)
1881	28.80	14.80	0.00	150.00
1882	63.20	33.50	673.00	458.00

Ownership: 1880-1884 NORTH HERODSFOOT MINING CO.
Comment 1862-1865 SUSPENDED; 1884 NOT WORKED
Management: Manager 1880-1881 THOS.TRELEASE

HERODSFOOT,NORTH   LANREATH   Continued

             Chief Agent 1861 R.PASCOE; 1882-1884 THOS.TRELEASE
             Secretary 1861-1865 R.HINGSTON (P); 1880-1881 F.F.WILSON (S)
Employment: Underground   Surface   Total
             1880   8   7   15
             1881   22   17   39
             1882   17   9   26
             1883   15   10   25

HERODSFOOT,SOUTH   DULOE   SX 212594 0618

Production: Lead No detailed return
Management: Chief Agent 1860-1861 GEO.TREMAYNE; 1862-1865 G.GOLDSWORTHY;
            1866-1870 W.GOLDSWORTHY
            Secretary 1860-1863 JAS.WOLFERSTAN (P); 1867-1870
            THOS.HORSWILL

HERODSFOOT,WEST   LANREATH   SX 179599 0619

Ownership: 1880-1881 WEST HERODSFOOT MINING CO.
Management: Manager 1880-1881 PETER TEMBY
            Secretary 1880 RICH.HAWKE & OTHERS
Employment: Underground   Surface   Total
            1880   3   3

HEWAS,GREAT   ST.MEWAN   SW 976502 0621

Production: Lead & Silver   Ore(tons)   Lead(tons) Silver(ozs)   Value(£)
            1856   19.90   12.00   48.00   0.00
            Tin   Black(tons) Stuff(tons)   Tin(tons)   Value(£)
            1852   No detailed return
            1854   10.50   0.00   0.00   884.00
            1855   135.50   0.00   0.00   8906.50
            1856   88.20   0.00   0.00   6260.70
            1857   166.50   0.00   0.00   13317.40
            1858   257.00   0.00   0.00   17454.90
            1859   98.90   0.00   0.00   6978.90
            1861   110.10   0.00   0.00   9218.50
            Comment 1852 HEWAS.SEE WEST POLGOOTH; 1854 GREAT HEWAS
            UNITED
Ownership: 1880 GREAT HEWAS MINING CO.; 1912-1913 ART.THOMAS
            Comment 1859 GREAT HEWAS UNITED; 1860-1865 GREAT HEWAS UNITED
            SUSPENDED; 1913 NOT WORKED
Management: Manager 1859 RICH.H.WILLIAMS
            Chief Agent 1880 R.GOLDSWORTHY
            Secretary 1859 RICH.H.WILLIAMS (P); 1860-1861 J.WEBB (P);
            1862-1865 RICH.H.WILLIAMS

HEWAS,NEW   ST.MEWAN   SW 985499 0622

Production: Tin No detailed return
Ownership: 1905-1908 JOHN J.TONKIN

Employment:	Underground	Surface	Total
1905–1906	4		4

HINGSTON DOWNS CONSOLS   GUNNISLAKE   SX 409714 0623

Production: Copper	Ore(tons)	Metal(tons)	Value(£)
1850	75.00	11.90	831.30
1851	127.00	14.00	978.00
1852	699.00	66.50	5721.20
1853	1259.00	107.50	11014.80
1854	2659.00	215.30	22335.20
1855	3138.00	204.10	20615.00
1856	2284.00	152.80	14110.80
1857	1492.00	99.10	9923.60
1858	1294.00	74.10	6431.30
1859	1327.00	67.70	5978.30
1860	1763.00	85.10	7137.70
1861	1876.00	96.00	8061.90
1862	2228.00	112.40	8604.40
1863	2440.00	114.90	8109.70
1864	2694.00	142.40	11604.10
1865	2914.00	166.70	12645.60
1866	3473.00	197.20	12867.90
1867	2182.00	116.30	7703.20
1868	1128.00	49.00	3016.70
1869	217.00	11.40	640.70
1870	220.00	9.10	454.30
1871	884.00	77.60	5038.60
1872	2660.00	154.70	11043.80
1873	2611.00	143.80	7970.20
1874	3017.00	143.60	8117.10
1875	3155.00	142.60	9442.40
1876	2503.00	112.10	6531.40
1877	2197.50	104.60	5541.40
1878	1700.70	81.00	3814.60
1879	450.60	24.10	1058.90
1880	27.50	2.10	131.40
1882	35.30	2.40	159.00
1905	267.00	0.00	724.00
1906	177.00	0.00	622.00
1907	122.00	0.00	346.00
1908	106.00	0.00	129.00

Comment 1850–1864 (C)HINGSTON DOWNS; 1865–1866 (C); 1867–1876
(C)HINGSTON DOWNS; 1905–1908 HINGSTON

Tin	Black(tons)	Stuff(tons)	Tin(tons)	Value(£)
1870	0.00	65.20	0.00	176.70
1871	0.00	52.70	0.00	65.80
1881	0.00	21.00	0.00	0.00
1905	20.50	0.00	0.00	1889.00
1906	96.00	0.00	0.00	10936.00
1907	90.20	0.00	0.00	10178.00
1908	42.70	0.00	0.00	3598.00

Comment 1905–1908 HINGSTON

Arsenic	Ore(tons)	Metal(tons)	Value(£)
1905	10.00	0.00	112.00
1906	169.00	0.00	3270.00
1907	15.00	0.00	460.00
1908	5.00	0.00	97.00

Comment 1905-1908 HINGSTON

Tungsten	Ore(tons)	Metal(tons)	Value(£)
1905	14.70	0.00	1090.00
1906	39.90	0.00	3558.00
1907	48.50	0.00	6708.00
1908	14.50	0.00	1261.00

Arsenic Pyrite	Ore(tons)	Value(£)
1878	18.50	9.50
1880	7.50	7.00

Comment 1880 HINGSTON DOWNS

Ownership: 1876-1884 HINGSTON DOWNS CONSOLS MINING CO.LTD.; 1905-1909 CLITTERS UNITED MINES CO.LTD.; 1910-1912 HINGSTON AND CLITTERS MINES LTD.
Comment 1884 HINGSTON DOWNS NOT WORKED; 1900-1902 SEE CLITTERS UNITED; 1905-1910 HINGSTON; 1911-1912 HINGSTON NO MINERAL WORKED

Management: Manager 1859-1860 T.T.RICHARDS; 1861-1871 THOS.RICHARDS; 1872-1877 JAS.RICHARDS; 1878-1881 THOS.RICHARDS
Chief Agent 1859-1865 WM.ROSEWARNE; 1866-1869 JNO.WALTERS; 1870-1871 T.B.LAWS (S); 1872-1877 T.B.LAWS (S) THOS.RICHARDS; 1880-1881 JOHN TAYLOR & SONS; 1882-1884 THOS.RICHARDS
Secretary 1859-1867 W.E.COMMINS (P); 1868-1869 JAS.RICHARDS; 1870-1875 JAS.RICHARDS (P); 1876-1881 T.B.LAWS (S)

Employment:	Underground	Surface	Total
1878	59	34	93
1879	18	7	25
1880	18	4	22
1881	17	6	23
1882	27	7	34
1883	15	7	22
1905	35	24	59
1906	88	34	122
1907	127	53	180
1908	42	34	76
1909	27	25	52
1911		2	2

HOBBS HILL   ST.NEOT   SX 186694 0624

Production: Tin	Black(tons)	Stuff(tons)	Tin(tons)	Value(£)
1872	0.50	817.20	0.00	751.90
1873	2.20	0.00	0.00	127.00
1874	12.30	0.00	0.00	547.80

Comment 1874 VALUE EST
Plumbago No detailed return
Ownership: 1907 KINGSWAY SYNDICATE LTD.
Management: Manager 1872-1873 JAS.PEARCE
Secretary 1872-1873 JAS.COCK; 1874 H.BEAZLEY; 1875 E.BEAZLEY; 1907 M.MONROE (S)

Employment: Underground   Surface   Total
            1907   11   2   13

HOLMBUSH   STOKE CLIMSLAND   SX 358721 0625

Production: Lead & Silver	Ore(tons)	Lead(tons)	Silver(ozs)	Value(£)
1846	12.00	7.00	0.00	0.00
1847	60.00	36.00	0.00	0.00
1848	154.00	90.00	0.00	0.00
1849	102.00	61.00	0.00	0.00
1850	200.00	119.10	0.00	0.00
1851	30.00	17.80	0.00	0.00
1852	25.00	14.00	640.00	0.00
1853	16.50	10.20	368.00	0.00
1854	No detailed return			
1856	18.00	11.00	541.00	0.00
1857	58.00	40.50	1336.00	0.00
1858	116.00	82.00	2187.00	0.00
1859	46.90	30.00	340.00	0.00
1860	20.40	13.00	147.00	0.00
1861	46.30	29.00	1044.00	0.00
1862	47.70	35.50	1020.00	0.00
1863	415.30	282.20	8178.00	0.00
1882	0.05	0.04	72.00	13.00
1883	43.30	33.00	860.00	605.00
1884	62.00	0.00	0.00	542.00
1885	4.00	0.00	0.00	34.00

Copper	Ore(tons)	Metal(tons)	Value(£)
1845	1887.00	208.10	14957.30
1846	1279.00	123.20	8118.00
1847	802.00	70.80	4845.50
1848	398.00	33.60	1926.00
1849	454.00	36.20	2356.20
1850	761.00	62.10	4148.30
1851	1188.00	100.30	6731.60
1852	2122.00	117.40	9389.90
1853	2091.00	134.90	13002.80
1854	1557.00	83.60	8067.30
1855	1601.00	97.40	9790.00
1856	1085.00	87.00	8437.80
1857	1035.00	90.90	9196.60
1858	971.00	94.40	9102.40
1859	812.00	78.20	7471.80
1860	1052.00	85.60	7871.90
1861	1495.00	95.80	8315.70
1862	1507.00	135.90	11356.70
1863	1048.00	98.50	7828.00
1864	167.00	15.10	1420.60
1867	13.00	0.60	33.80
1868	37.00	1.70	104.50
1877	238.70	5.80	171.70
1880	15.00	0.90	47.80
1881	32.30	2.90	159.40
1882	493.80	44.40	1148.00

Copper	Ore(tons)	Metal(tons)	Value(£)
1883	1373.10	35.00	3142.00
1884	1635.00	0.00	1973.00
1885	2756.00	0.00	3308.00
1886	795.00	0.00	1183.00

Comment 1845–1864 (C); 1867–1868 (C); 1880 HOLMBUSH CONSOLS

Tin	Black(tons)	Stuff(tons)	Tin(tons)	Value(£)
1885	1.00	85.00	0.00	39.00
1886	13.60	0.00	0.00	87.00

Arsenic	Ore(tons)	Metal(tons)	Value(£)
1880	909.60	0.00	5273.20
1881	1802.00	0.00	14133.60
1882	1497.00	0.00	11524.00
1883	1204.80	0.00	10149.00
1884	1103.00	0.00	9436.00
1885	1748.00	0.00	14359.00

Comment 1881–1882 GREENHILL WORKS.OBTAINED FROM ARSENICAL
PYRITES GIVEN BELOW; 1884–1885 REFINED ARSENIC.

Fluorspar	Ore(tons)	Value(£)
1884	5.00	6.00
1885	108.00	134.00

Tungsten	Ore(tons)	Metal(tons)	Value(£)
1882	3.10	0.00	30.00

Arsenic Pyrite	Ore(tons)	Value(£)
1877	9649.50	4824.50
1878	692.30	1216.00
1880	4457.70	5786.00
1881	6377.90	7972.00
1882	5491.50	6178.00
1886	2291.40	5800.00

Comment 1877 ARSENICAL MUNDIC; 1880 SOLD AS CRUDE ARSENIC AND
IN STONE. SEE ORIGINAL RETURN.
Ownership: 1877–1879 HOLMBUSH MINING CO.LTD.; 1880–1887 NEW HOLMBUSH
MINING CO.LTD.
Comment 1864–1865 SUSPENDED; 1868–1870 HOLMBUSH & KELLY BRAY
UNITED; 1874 SEE EMMENS UNITED; 1879 INC.KELLY BRAY;
1882–1886 INC.KELLY BRAY; 1887 INC.KELLY BRAY. IN
LIQUIDATION; 1888–1893 SEE CALLINGTON UNITED
Management Manager 1859–1861 FRAN.PRYOR; 1868–1869 J.DAINTY; 1877
S.H.EMMENS; 1878–1881 H.BENNETT
Chief Agent 1859 WM.WOOLCOCK; 1860–1861 THOS.WOOLCOCK;
1862–1863 JOHN BORLASE THOS.WOOLCOCK; 1868–1869 T.ODGERS;
1870 R.CROKER; 1878–1879 J.S.BENNET; 1880–1881 J.S.BENNET
G.EMMENS (S); 1882–1887 H.BENNETT
Secretary 1859–1863 FRAN.PRYOR (P); 1869 THOS.THOMPSON; 1870
WARD & LITTLEWARD; 1878–1879 S.BROOME (S); 1880–1881
GEO.BUTTLER (P)

Employment:	Underground	Surface	Total
1878	32	31	63
1879	7	6	13
1880	126	45	171
1881	150	56	206
1882	97	64	161
1883	119	56	175
1884	165	71	236

HOLMBUSH   STOKE CLIMSLAND   Continued

	Underground	Surface	Total
1885	233	62	295
1886	150	32	182
1888–1893			

Comment 1879–1880 INC.KELLY BRAY; 1883–1886 INC.KELLY BRAY;
1888–1893 SEE CALLINGTON UNITED

HOLMBUSH,NORTH   STOKE CLIMSLAND   SX 365731 0626

Production: Copper No detailed return
            Tin No detailed return
Ownership: Comment 1874–1875 OR DOWNGATE CONSOLS
Management: Manager 1874–1875 H.BENNETT
            Secretary 1874–1875 R.F.HAWKE

HONY   MENHENIOT   SX 289646 0627

Production: Lead   Ore(tons) Metal(tons)   Value(£)
            1883   57.80   42.00   806.00
            1884   30.00   0.00   325.00
            Comment 1883 INC.TRELAWNEY; 1884 " FOR AG SEE ZZ SUNDRIES
Ownership: 1880–1884 WHEAL HONY & TRELAWNY MINING CO.LTD.
            Comment 1880–1883 HONY & TRELAWNY UNITED; 1884 HONY &
            TRELAWNY
Management: Manager 1880–1881 WM.HANCOCK
            Chief Agent 1882 WM.HANCOCK; 1883–1884 WM.DERRY
Employment: Underground   Surface   Total
            1880   5   40   45
            1881   85   85
            1882   52   42   94
            1883   41   46   87
            1884   29   21   50
            Comment 1880–1884 INC.TRELAWNEY

HONY,EAST   MENHENIOT   0628

Production: Lead No detailed return
Ownership: 1882–1884 EAST HONY MINING CO.
            Comment 1884 NOT WORKED
Management: Chief Agent 1882 WM.HANCOCK
            Secretary 1881 WM.HANCOCK
Employment: Underground   Surface   Total
            1882   8   8

HOOPER,CARADON SOUTH WHEAL   ST.CLEER   SX 272697 0217

Production: Copper No detailed return
Ownership: Comment 1865 SUSPENDED
Management: Manager 1859 W.C.COCK; 1860–1863 FRAN.PRYOR
            Chief Agent 1859–1864 W.C.COCK

                                    245

HOPE   PERRANZABULOE   SW 786548 0629

Production: Lead & Silver   Ore(tons)   Lead(tons) Silver(ozs)   Value(£)
            1856   6.10   3.60   0.00   0.00
            1857—1859 No detailed return
            1863   56.50   39.20   0.00   0.00
            1864   58.00   43.50   1420.00   0.00
            1865   20.50   12.80   405.00   0.00
            1866   6.10   4.50   148.00   0.00
            Zinc   Ore(tons) Metal(tons)   Value(£)
            1856   1.20   0.00   1.90
            Tin   Black(tons) Stuff(tons) Tin(tons)   Value(£)
            1865   1.00   0.00   0.00   56.70
Management: Chief Agent 1860—1863 W.H.REYNOLDS; 1864—1865 JOHN NICHOLLS
            Secretary 1860—1863 W.H.REYNOLDS (P); 1864—1865 C.WESCOMBE

HOPE   KEA   SW 769430 0630

Ownership: 1880—1881 WHEAL HOPE ADVENTURERS; 1882—1885 WHEAL HOPE MINING
            CO.
            Comment 1885 NOT WORKED
Management: Manager 1880—1881 WM.GILES
            Chief Agent 1881—1884 HY.MICHELL
            Secretary 1880—1881 HARVEY & CO. & OTHERS
Employment: Underground   Surface   Total
            1881   7   7
            1882   1   11   12
            1883   8   8
            1884   7   1   8

HOPE VALLEY   0631

Production: Lead   Ore(tons) Metal(tons)   Value(£)
            1854   128.00   98.00   0.00

HORSE BURROW   ALTARNUN   SX 209794 0632

Production: Tin No detailed return
Ownership: 1906—1908 HORSEBRO TIN CO.LTD.; 1909—1910 JOHN S.SAWREY
            Comment 1906—1909 INC.TREWINT CONSOLS & HENDRA; 1910
            INC.TREWINT CONSOLS & HENDRA. SHUT
Management: Chief Agent 1906—1910 J.S.SAWREY
Employment: Underground   Surface   Total
            1907   7   7
            1908   6   1   7
            Comment 1907 INC.TREWINT CONSOLS & NEW HENDRA; 1908 INC.NEW
            HENDRA

HORSE BURROW   ST.AUSTELL   0633

Production: Tin No detailed return
Ownership: Comment 1905—1913 SEE GREAT DOWGAS

HORSE BURROW   ST.AUSTELL   Continued

Employment: Underground   Surface   Total
            1909-1910
            1912-1913
            Comment 1909-1910 SEE GREAT DOWGAS; 1912-1913 SEE GREAT
            DOWGAS

IDA   ST.IVE   SX 303693 0634

Production: Lead & Silver No detailed return
Management: Manager 1867-1868 WM.BARTLETT; 1869-1870 WM.TAYLOR
            Secretary 1869-1870 GREN.SHARP

ILLOGAN MINES   ILLOGAN   SW 677414 0635

Production: Copper   Ore(tons) Metal(tons)   Value(£)
            1863   21.00   0.90   64.00
            1864   5.00   0.20   9.50
            Comment 1863-1864 (C)
            Tin No detailed return
Ownership: 1876 ILLOGAN TIN & COPPER CO.LTD.
            Comment 1865 ILLOGAN CONSOLS
Management: Manager 1874 RICH.PRYOR; 1875 RICH.PRYOR & SON; 1876
            RICH.PRYOR
            Chief Agent 1865 WM.TEAGUE
            Secretary 1875-1876 HY.L.PHILLIPS

INDIAN QUEENS CONSOLS   INDIAN QUEENS   SW 918586 0636

Production: Tin   Black(tons) Stuff(tons)   Tin(tons)   Value(£)
            1881   4.00   0.00   2.60   220.00
            1882   31.00   0.00   21.70   1891.00
            Comment 1881 INDIAN QUEENS; 1882 PART OF TREVARREN UNITED
            MINES
            Iron No detailed return
Ownership: 1880-1881 INDIAN QUEENS CONSOLS MINING CO.LTD.; 1882
            TREVARREN UNITED MINING CO.LTD.
Management: Manager 1881 WM.HOOPER
            Secretary 1880-1882 HY.BROWNE
Employment: Underground   Surface   Total
            1880   36   38   74
            1881   18   18   36
            1882   15   7   22

INDIAN QUEENS NO.1   ST.COLUMB MAJOR   SW 923601 0637

Production: Iron   Ore(tons)   Iron(%)   Value(£)
            1858   1739.60   0.00   521.90
            1859   3021.00   0.00   906.30
            1860   4946.40   0.00   1423.90
            1861   234.00   0.00   70.00
            1862   1750.00   0.00   962.50

                                  247

INDIAN QUEENS NO.1   ST.COLUMB MAJOR   Continued

Iron	Ore(tons)	Iron(%)	Value(£)
1863	473.00	0.00	236.00
1864	No detailed return		

Comment 1858-1864 BH.COULD BE INDIAN QUEENS NO.2
Ownership: 1863 THOS.WHITFORD; 1864-1865 THOS.WHITFORD & E.CARTER

INDIAN QUEENS NO.2   ST.COLUMB MAJOR   SW 923601 0638

Production: Iron Ore(tons) Iron(%) Value(£)
1858-1864 No detailed return
Comment 1858-1864 SEE INDIAN QUEENS NO.1
Ownership: 1863-1866 JOS.MORCOM

JAMES   WITHIEL   SX 006653 0640

Production: Iron No detailed return
Ownership: 1864-1865 EDW.CARTER
Management: Chief Agent 1864-1865 THOS.PARKYN

JANE   BROADOAK   SX 198789 0641

Production: Lead & Silver No detailed return
Management: Chief Agent 1861 JOHN VERCOE
Secretary 1861 W.MANSEL (P)

JANE   KEA   SW 783434 0642

Production: Lead & Silver	Ore(tons)	Lead(tons)	Silver(ozs)	Value(£)
1859	35.40	24.70	0.00	0.00
1860	97.50	66.00	3387.00	0.00
1861	34.70	19.00	1102.00	0.00
1862	235.00	168.00	8730.00	0.00
1863	61.80	43.20	2246.00	0.00
1865	17.00	10.80	150.00	0.00
1866	24.80	18.60	279.00	0.00
1867	4.60	3.40	0.00	0.00
1868	19.20	14.40	0.00	0.00
1869	14.30	10.70	0.00	0.00
1870	11.60	8.50	0.00	0.00
1871	3.60	2.70	13.00	0.00
1872	8.00	6.00	30.00	0.00
1873	17.90	6.70	33.00	189.00
1874	5.20	3.90	20.00	112.10
1876	10.60	8.00	0.00	0.00
1877	4.70	3.20	0.00	60.70
1884	5.00	0.00	0.00	61.00

Comment 1867-1868 FOR AG.SEE GREAT RETALLACK

Zinc	Ore(tons)	Metal(tons)	Value(£)
1859	150.70	0.00	327.90
1860	107.50	0.00	110.60
1862	93.40	0.00	71.90

Zinc	Ore(tons)	Metal(tons)	Value(£)
1863	102.00	0.00	103.40
1864	28.50	0.00	31.30
1865	56.10	0.00	111.80
1866	10.90	0.00	10.40
1867	4.90	0.00	5.50
1868	10.10	0.00	10.60
1872	15.00	0.00	30.00

Copper	Ore(tons)	Metal(tons)	Value(£)
1847	219.00	10.60	653.30
1848	286.00	14.50	738.90
1862	80.00	3.90	308.90
1863	9.00	0.40	29.50
1864	5.00	0.20	17.00
1878	5.60	0.50	30.00
1879	8.00	0.60	35.00

Comment 1847–1848 (C); 1862–1864 (C)

Tin	Black(tons)	Stuff(tons)	Tin(tons)	Value(£)
1853	3.40	0.00	0.00	196.40
1855	6.10	0.00	0.00	379.20
1856	14.20	0.00	0.00	1012.00
1857	35.30	0.00	0.00	2528.20
1858	12.70	0.00	0.00	808.60
1859	0.00	0.00	0.00	1232.70
1860	0.00	0.00	0.00	2310.30
1861	0.00	0.00	0.00	8988.30
1862	10.60	0.00	0.00	10916.70
1863	11.60	0.00	0.00	9695.00
1864	3.50	0.00	0.00	7647.40
1865	16.10	0.00	0.00	3586.20
1866	168.70	0.00	0.00	8032.00
1867	203.70	0.00	0.00	10870.60
1868	248.20	0.00	0.00	13144.40
1869	251.90	0.00	0.00	17327.60
1870	260.70	0.00	0.00	18578.60
1871	229.00	0.00	0.00	18365.90
1872	206.00	0.00	0.00	17507.50
1873	243.10	0.00	0.00	18954.00
1874	249.70	0.00	0.00	14255.90
1875	213.60	0.00	0.00	5910.00
1876	70.70	0.00	0.00	3091.70
1877	164.00	0.00	0.00	6888.50
1878	197.00	0.00	0.00	6950.20
1879	344.00	0.00	0.00	13760.00
1880	113.00	0.00	0.00	5714.40
1881	90.80	0.00	60.70	5006.90
1882	117.00	0.00	76.00	7225.00
1883	208.00	1862.00	0.00	9448.00
1884	90.50	692.00	0.00	3927.00
1885	26.30	788.00	0.00	818.00
1893	0.00	135.00	0.00	55.00
1894	0.00	54.00	0.00	41.00
1895	0.00	504.00	0.00	201.00
1896	0.00	143.00	0.00	49.00
1899	0.00	596.00	0.00	261.00

Tin	Black(tons)	Stuff(tons)	Tin(tons)	Value(£)
1900	0.00	619.00	0.00	283.00
1901	0.00	166.00	0.00	66.00
1905	4.50	5.00	0.00	392.00
1906	18.80	0.00	0.00	1988.00
1907	13.90	0.00	0.00	1495.00
1908	25.00	0.00	0.00	1853.00
1909	20.50	0.00	0.00	1402.00
1910	112.00	0.00	0.00	12672.00
1911	122.40	0.00	0.00	12378.00
1912	234.20	0.00	0.00	30096.00
1913	293.20	0.00	0.00	34615.00

Comment 1859-1861 TIN STUFF; 1862 VALUE INC.TIN STUFF.3
RTNS.AG.; 1863-1865 VALUE INC.TIN STUFF; 1873 VALUE EST;
1883-1885 TWO RETURNS AGGREGATED

Iron	Ore(tons)	Iron(%)	Value(£)
1864	1.00	0.00	5.00

Comment 1864 OXIDE OF IRON

Arsenic	Ore(tons)	Metal(tons)	Value(£)
1859	4.00	0.00	6.00
1861	5.00	0.00	7.00
1862	5.50	0.00	2.80
1863	4.00	0.00	2.40
1866	1.00	0.00	0.50
1867	19.00	0.00	9.50
1868	19.00	0.00	19.00
1869	30.50	0.00	24.80
1870	37.50	0.00	25.50
1871	0.80	0.00	0.80
1872	30.00	0.00	15.20
1873	19.00	0.00	9.50
1874	15.70	0.00	14.50
1875	2.70	0.00	16.10
1876	5.50	0.00	21.00
1877	1.50	0.00	2.30
1879	13.00	0.00	26.00
1880	7.00	0.00	21.00
1882	13.00	0.00	14.00
1884	5.00	0.00	5.00
1913	16.30	0.00	38.00

Comment 1859 CRUDE; 1861-1863 CRUDE; 1866-1868 CRUDE;
1870-1871 CRUDE; 1884 SOOT

Arsenic Pyrite	Ore(tons)	Value(£)
1894	105.00	79.00
1895	88.00	22.00

Ownership: 1877-1885 WHEAL JANE MINING CO.; 1893-1896 WM.BURROWS;
1898-1904 WM.BURROWS; 1905 DEITZSCH & SHIFF; 1906
S.G.MINERALS SYNDICATE LTD.; 1907-1913 FALMOUTH CONSOLS MINES
LTD.
Comment 1869-1873 GREAT WHEAL JANE FORMERLY BALDHU; 1893-1896
TIPPETTS SHAFT.TRIBUTER
Management: Manager 1859-1860 JOHN BRAY; 1861 THOS.BRAY; 1862-1867
THOS.BRAY WM.GILES; 1868-1875 WM.GILES; 1876-1880
RICH.SOUTHEY
Chief Agent 1859-1860 JAS.NICHOLAS; 1861-1864 WM.BRAY;

1865-1869 WM.BRAY JOHN HALL; 1870-1871 WM.GILES JNR. JOHN
HALL; 1872-1875 WM.GILES JNR.THOS.WILLIAMS; 1876-1877
WM.GILES THOS.WILLIAMS; 1878-1880 JOHN HOCKING JNR.; 1881
JOHN REED JAS.REED; 1882 JAS.REED; 1883-1885 RICH.SOUTHEY;
1903-1904 WM.BURROWS; 1906 H.C.JENKIN; 1908-1910 JOHN
EDWARDS; 1911 W.J.OATES; 1912-1913 A.R.SHUTES
Secretary 1859-1863 JOHN TIPPETT (P); 1864-1867 NICH.GILBERT;
1868-1876 CHAS.HAWKE

Employment:

	Underground	Surface	Total
1878	68	82	150
1879	65	57	122
1880	68	80	148
1881	119	96	215
1882	95	107	202
1883	92	78	170
1884	3	39	42
1885		13	13
1893-1894	3		3
1895	9	1	10
1896	1		1
1899	12	2	14
1900-1901	7	2	9
1902	3		3
1903	2		2
1905	10	18	28
1906	26	20	46
1907	121	61	182
1908	118	183	301
1909	26	36	62
1910	76	154	230
1911	146	188	334
1912	173	228	401
1913	175	188	363

JANE STAMPS,NORTH   KENWYN   0643

Production: Tin   Black(tons) Stuff(tons)   Tin(tons)   Value(£)
            1869   1.20   0.00   0.00   67.10

JANE,EAST   BROADOAK   SX 136655 0644

Production: Lead & Silver   Ore(tons) Lead(tons) Silver(ozs)   Value(£)
            1862   156.90   119.00   1010.00   0.00
            1863   96.40   71.50   636.00   0.00
            1864   48.30   36.00   324.00   0.00
            1865   23.80   15.30   138.00   0.00
Ownership: Comment 1863 TWO RETURNS AGGREGATED
Management: Manager 1863-1865 JAS.SECCOMBE
            Chief Agent 1862 JOHN VERCOE; 1863 JOHN VERCOE THOS.HODGE;
            1864-1865 J.HODGE
            Secretary 1862 W.W.MANSEL; 1863 MANSEL & SECCOMBE; 1864-1865
            JAS.SECCOMBE

JANE,NORTH   KENWYN   SW 789440 0645

Production: Lead & Silver   Ore(tons)   Lead(tons) Silver(ozs)   Value(£)
            1862   0.50   7.20   144.00   0.00
            Tin   Black(tons)   Stuff(tons)   Tin(tons)   Value(£)
            1862   12.70   0.00   0.00   1005.00
            1863   19.20   0.00   0.00   1221.00
            1864   18.20   0.00   0.00   1279.00
            1865   12.90   0.00   0.00   675.60
            1869   0.00   140.80   0.00   111.50
            1870   0.00   29.80   0.00   35.40
            1871   0.00   17.40   0.00   29.20
            1873   2.10   137.20   0.00   237.50
            1874   0.80   3.10   0.00   47.20
            1875   8.50   8.50   0.00   143.30
            Comment 1862 THREE RETURNS AGGREGATED; 1864 VALUE INC.TIN
            STUFF; 1875 LOW DRESSED ORE
Ownership: 1875 NORTH JANE CO.
            Comment 1874 STOPPED IN 1874
Management: Manager 1860 CHAS.PENGELLY; 1866-1874 JOHN ROWE; 1875
            T.NORM.WOODWARD
            Chief Agent 1860 THOS.WASLEY; 1861-1863 JAS.EVANS
            THOS.WASLEY; 1864-1865 JOHN ROWE THOS.WASLEY; 1866-1874
            THOS.WASLEY
            Secretary 1860 CHAS.PENGELLY (P); 1861-1866 WOODWARD (P);
            1867-1868 C.THOMAS; 1869-1874 THOR.WOODWARD

JANE,SOUTH   BROADOAK   0646

Production: Lead & Silver No detailed return
Management: Manager 1863-1865 JAS.SECCOMBE
            Chief Agent 1863-1865 FRAN.HODGE
            Secretary 1863-1865 JAS.SECCOMBE

JANE,WEST   KEA   SW 777427 0647

Production: Lead & Silver   Ore(tons)   Lead(tons) Silver(ozs)   Value(£)
            1864   24.90   18.70   158.00   0.00
            1868   0.60   0.50   0.00   0.00
            Comment 1868 FOR AG.SEE GREAT RETALLACK
            Zinc   Ore(tons)   Metal(tons)   Value(£)
            1856   23.20   0.00   39.00
            1857   8.20   0.00   14.00
            1858   0.40   0.00   0.70
            1859   27.30   0.00   54.30
            1861   76.50   0.00   76.00
            1862   42.50   0.00   22.30
            1863   43.40   0.00   26.10
            1864   24.40   0.00   24.40
            1865   53.30   0.00   19.20
            1876   63.00   0.00   212.50
            1877   25.70   0.00   105.00
            1888   3.00   0.00   4.00
            1894   2.50   0.00   0.00
            1895-1896 No detailed return

                                 252

Comment 1863 ORE EST.TWO RETURNS AGGREGATED; 1894 FOR VALUE
SEE ARTHUR

Copper	Ore(tons)	Metal(tons)	Value(£)
1856	100.00	7.80	750.50

Comment 1856 (C)

Tin	Black(tons)	Stuff(tons)	Tin(tons)	Value(£)
1854	15.10	0.00	0.00	903.10
1855	25.60	0.00	0.00	1595.70
1856	17.70	0.00	0.00	1209.60
1857	14.10	0.00	0.00	994.50
1858	25.50	0.00	0.00	1445.40
1859	0.00	0.00	0.00	1902.50
1860	0.00	0.00	0.00	2501.50
1861	0.00	0.00	0.00	3456.00
1862	71.30	0.00	0.00	8772.00
1863	54.00	0.00	0.00	5215.70
1864	37.10	0.00	0.00	5219.50
1865	58.30	0.00	0.00	3091.80
1866	0.00	0.00	0.00	1571.60
1867	0.30	0.00	0.00	1420.00
1868	0.00	0.00	0.00	1785.60
1869	0.00	0.00	0.00	1889.40
1870	0.00	0.00	0.00	2101.30
1871	0.00	0.00	0.00	3938.00
1872	0.00	0.00	0.00	5234.20
1873	0.00	4269.00	0.00	2567.30
1874	0.00	3672.80	0.00	1150.00
1875	0.00	1876.00	0.00	666.30
1876	1.00	564.00	0.00	190.00
1877	0.00	457.60	0.00	157.50
1878	0.00	132.00	0.00	49.00
1879	0.00	192.00	0.00	96.00
1880	0.00	403.00	0.00	180.00
1881	2.10	0.00	1.40	113.90
1882	4.60	470.00	3.00	275.00
1883	6.40	552.00	0.00	318.00
1884	3.40	320.00	0.00	136.00
1885	6.00	450.00	0.00	241.00
1886	7.50	541.00	0.00	368.00
1888	0.80	330.00	0.00	52.00
1893	0.00	11.00	0.00	7.00
1900	0.00	10.00	0.00	11.00

Comment 1855 VALUE EST; 1859-1861 TIN STUFF; 1862 THREE
RETURNS AGGREGATED; 1863-1864 VALUE INC.TIN STUFF; 1866 TIN
STUFF; 1867 VALUE INC.TIN STUFF; 1868-1872 TIN STUFF; 1874
VALUE EST

Arsenic	Ore(tons)	Metal(tons)	Value(£)
1857	34.00	0.00	55.70
1858	25.70	0.00	40.30
1862	4.50	0.00	6.00
1864	52.00	0.00	38.80
1865	27.60	0.00	35.00
1868	6.30	0.00	5.00
1869	30.00	0.00	25.60
1875	34.90	0.00	50.50

```
            Arsenic        Ore(tons) Metal(tons)    Value(£)
            1876             7.00        0.00         31.50
            Comment 1862 CRUDE; 1864-1865 CRUDE; 1868 CRUDE
            Wolfram No detailed return
Ownership:  1877-1888 JOHN SMITH; 1893-1894 JAS.COLLINS; 1899-1900
            NORTHEY & FOWLER; 1902-1905 JAS.COLLINS
            Comment 1865 SUSPENDED; 1893-1894 SMITHS SHAFT.TRIBUTER
Management: Manager 1859 JOHN TREGONNING; 1860-1863 JOHN TONKIN;
            1869-1881 JOHN SMITH
            Chief Agent 1859-1863 JOHN SMITH; 1864 JOHN SMITH WM.TEAGUE;
            1878-1881 WM.GILES; 1882-1888 JOHN SMITH
            Secretary 1859-1863 WM.PAINTER (P); 1864-1865 ED.KING;
            1869-1881 JOHN SMITH
Employment:              Underground      Surface        Total
            1878              4             2              6
            1879              9                            9
            1880              6             2              8
            1881              7             3             10
            1882              9             1             10
            1883              6             1              7
            1884             10             4             14
            1885              6             4             10
            1886              7             2              9
            1887                            2              2
            1888                            3              3
            1893              2                            2
            1894              1                            1
```

JANEY   PERRANUTHNOE   SW 540298 0648

```
Production: Copper         Ore(tons) Metal(tons)    Value(£)
            1862            24.00       0.70         39.00
            Comment 1862 (C)
```

JENNINGS   GWINEAR   SW 614363 0649

```
Production: Tin            Black(tons) Stuff(tons)    Tin(tons)    Value(£)
            1875             16.80       0.00           0.00        870.00
            1876             64.30       0.00           0.00       2318.20
            1877             48.50       0.00           0.00       1854.50
            1882             26.50       0.00          17.20       1590.00
            1883             52.60       0.00           0.00       2787.00
            1884             10.20       0.00           0.00        443.00
            Comment 1875 WAS PARBOLA.RETURN FROM 31 AUG
Ownership:  1875 WHEAL JENNINGS ADVENTURERS; 1876-1877 WHEAL JENNINGS
            MINING CO.; 1882-1884 WILLIAMS,HARVEY & CO.
            Comment 1875 SEE PARBOLA
Management: Chief Agent 1875-1877 JOHN TREGONNING
            Secretary 1877 JOHN TREGONNING; 1882-1884 JOHN TREGONNING
Employment:              Underground      Surface        Total
            1882             50             36             86
            1883             56             30             86
            1884             38                            38
```

JEWELL   0650

Production: Copper   Ore(tons) Metal(tons)   Value(£)
            1871   169.00   10.20   573.60
            1872   290.00   14.40   1083.40
            Comment 1871 (C)JEWELL; 1872 (C)
            Tin   Black(tons) Stuff(tons)   Tin(tons)   Value(£)
            1871   0.00   0.00   0.00   558.10
            1872   0.00   0.00   0.00   137.00
            Comment 1871-1872 TIN STUFF

JEWELL   PERRANZABULOE   0651

Production: Lead & Silver No detailed return
Ownership: Comment 1862-1865 SUSPENDED
Management: Chief Agent 1859-1861 G.SIMMONS
            Secretary 1859-1863 WALT.TRESIDER (P)

JEWELL   ST.DAY   SW 735421 0652

Production: Copper   Ore(tons) Metal(tons)   Value(£)
            1845   1476.00   109.30   7892.90
            1846   844.00   57.80   4035.30
            1847   806.00   57.60   4065.40
            1848   495.00   37.10   2278.80
            1849   100.00   6.80   436.70
            1850   132.00   7.30   467.00
            1851   114.00   7.40   481.80
            1852   142.00   8.20   668.60
            1853   136.00   7.50   695.40
            Comment 1845-1853 (C)
            Tin   Black(tons) Stuff(tons)   Tin(tons)   Value(£)
            1855   5.10   0.00   0.00   318.00
            1856   3.10   0.00   0.00   202.60
            1857   1.10   0.00   0.00   73.10
            1858   0.90   0.00   0.00   50.20
            1859   1.00   0.00   0.00   66.50
            Comment 1855-1856 JEWEL

JEWELL   MARAZION   SW 534306 0653

Production: Lead   Ore(tons) Metal(tons)   Value(£)
            1883   20.00   14.00   174.00
            Copper   Ore(tons) Metal(tons)   Value(£)
            1880   35.00   1.50   63.00
            1881   385.60   22.60   1028.20
            1882   636.80   34.20   229.00
            1883   368.50   18.40   1022.00
            Tin   Black(tons) Stuff(tons)   Tin(tons)   Value(£)
            1880   0.00   4.20   0.00   2.50
            1882   0.10   5.00   0.00   2.00
Ownership: 1880-1881 WHEAL JEWELL MINE ADVENTURERS; 1882-1884 WHEAL
            JEWEL MINE ADVENTURERS
            Comment 1882 JEWEL; 1884 NOT WORKED

255

JEWELL   MARAZION   Continued

Management: Manager 1859 AB.BENNETTS; 1880 FRAN.GUNDRY
             Chief Agent 1859 FRAN.GUNDRY; 1880 J.W.DABB; 1881
             JOS.TREGONNING
             Secretary 1859 AB.BENNETTS (P); 1881-1884 F.W.DABB
Employment: Underground   Surface   Total
             1880   41   14   15
             1881   51   26   77
             1882   60   22   82
             1883   21   20   41

JEWELL,EAST   PERRANZABULOE   0654

Production: Copper No detailed return
             Tin   Black(tons) Stuff(tons)   Tin(tons)   Value(£)
             1858   12.10   0.00   0.00   780.40
Ownership: Comment 1873 STOPPED IN 1873 OR 1874
Management: Manager 1869-1875 FRAN.GUNDRY
             Secretary 1869-1875 A.BENNETT

JEWELL,NEW   PERRANZABULOE   0655

Production: Lead No detailed return
             Tin   Black(tons) Stuff(tons)   Tin(tons)   Value(£)
             1878   3.00   2.00   0.00   30.00
Ownership: 1878-1879 WM.JEWELL
Management: Chief Agent 1878-1879 WM.JEWELL
Employment: Underground   Surface   Total
             1878   4   4
             1879   2   2

JEWELL,WEST   ST.DAY   SW 726422 0656

Production: Copper   Ore(tons) Metal(tons)   Value(£)
             1845   1883.00   113.70   8045.10
             1846   1520.00   88.40   5908.70
             1847   975.00   57.20   3875.50
             1848   689.00   38.80   2128.00
             1849   400.00   25.00   1617.20
             1850   379.00   23.70   1609.20
             1851   327.00   26.40   1814.60
             1852   162.00   15.00   1207.70
             Comment 1845-1847 (C)WEST JEWEL; 1848-1852 (C)
             Tin   Black(tons) Stuff(tons)   Tin(tons)   Value(£)
             1873   4.60   440.90   0.00   1713.60
             1874   0.00   39.70   0.00   156.30
             Comment 1873 VALUE INC.TIN STUFF
Management: Manager 1870-1874 JOS.MICHELL
             Chief Agent 1870-1871 GEO.JOHNS JOHN MAYNE; 1872-1874
             GEO.JOHNS; 1875 GEO.T.CORFIELD
             Secretary 1870-1873 MAT.GREEN; 1874 GEO.T.CORFIELD

                                256

KAYLE   ST.ERTH   SW 583353 0657

Production: Copper   Ore(tons) Metal(tons)   Value(£)
            1846   237.00   20.30   1383.30
            1847   No detailed return
            Comment 1846-1847 (C)

KEA TREMAYNE   KEA   SW 786434 0658

Production: Lead No detailed return
            Tin No detailed return
Management: Manager 1861-1864 WM.KITTO; 1865 WM.KIBLE
            Secretary 1861-1865 JOHN TIPPETT (P)

KEHELLAND CONSOLS   CAMBORNE   0659

Production: Copper No detailed return
Management: Chief Agent 1873 WM.PASCOE
            Secretary 1873 J.DANIELL

KELLY BRAY   KELLY BRAY   SX 361714 0660

Production: Copper   Ore(tons) Metal(tons)   Value(£)
            1851   147.00   11.50   754.60
            1852   289.00   14.90   1175.40
            1853   376.00   19.90   1876.20
            1854   430.00   28.80   2850.10
            1855   306.00   15.40   1500.30
            1856   1356.00   75.70   6822.10
            1857   1188.00   59.20   5638.80
            1858   1489.00   69.50   5920.30
            1859   1908.00   94.80   8473.10
            1860   1747.00   81.50   6980.40
            1861   988.00   37.60   2912.70
            1862   843.00   37.80   2811.10
            1863   937.00   53.70   3921.80
            1864   484.00   21.90   1706.20
            1865   386.00   19.40   1480.00
            1866   97.00   4.60   312.10
            1867   20.00   0.90   56.50
            1868   116.00   7.60   493.00
            1869   1400.00   90.70   5759.20
            1870   1628.00   86.40   4758.50
            1871   260.00   10.50   564.00
            1872   44.00   1.40   97.00
            1877   5.70   0.40   23.20
            Comment 1851 (C)CALLINGTON KELLEBRAY; 1852-1853 (C)KILLY
            BRAY; 1854 (C)KILLEBRAY; 1855 (C)CALLINGTON KILLYBRAY;
            1856-1872 (C)
            Arsenic No detailed return
            Arsenic Pyrite   Ore(tons)   Value(£)
            1877   283.00   211.80
            1878   17.10   12.80
            Comment 1877 ARSENICAL MUNDIC
                               257

KELLY BRAY   KELLY BRAY   Continued

Ownership: 1877-1878 EMMENS AND CO.LTD.
            Comment 1868-1870 SEE HOLMBUSH; 1874 SEE EMMENS UNITED; 1879
            SEE HOLMBUSH; 1882-1887 SEE HOLMBUSH; 1888-1893 SEE
            CALLINGTON UNITED
Management: Manager 1859-1862 SILAS JAMES; 1863-1865 GEO.ROWE; 1877
            S.H.EMMENS; 1878 H.BENNETT
            Chief Agent 1859-1860 SILAS JAMES; 1863-1865 SILAS JAMES;
            1878 J.S.BENNETT
            Secretary 1859-1865 WM.WATSON (P); 1878 S.BROOME (S)
Employment: Underground   Surface   Total
            1878   10   2   12
            1879-1880
            1883-1886
            1888-1893
            Comment 1879-1880 SEE HOLMBUSH; 1883-1887 SEE HOLMBUSH;
            1888-1893 SEE CALLINGTON UNITED

KELLY BRAY,SOUTH   KELLY BRAY   0661

Ownership: Comment 1891-1893 SEE CALLINGTON UNITED
Employment: Underground   Surface   Total
            1891-1893
            Comment 1891-1893 SEE CALLINGTON UNITED

KELYNACK   ST.JUST   SW 368303 0662

Production: Tin   Black(tons) Stuff(tons)   Tin(tons)   Value(£)
            1901   0.30   0.00   0.00   18.00
            1910   23.70   0.00   0.00   2101.00
            1911   1.10   0.00   0.00   97.00
            Comment 1911 VALUE EST
Ownership: 1900-1901 N.T.WILLIAMS; 1906-1910 STANNARIES SYNDICATE LTD.;
            1911-1913 KELYNACK TIN MINES LTD.
            Comment 1901 ABANDONED MARCH 1901; 1911-1912 SUSPENDED IN
            LIQUIDATION; 1913 SUSPENDED IN LIQUIDATION CLOSED DOWN
Management: Chief Agent 1909-1913 N.T.WILLIAMS
Employment: Underground   Surface   Total
            1900   8   8
            1901   6   6
            1906   6   6
            1907   14   6   20
            1908   17   3   20
            1909   27   27   54
            1910   40   35   75

KENEGGY   BREAGE   SW 566284 0663

Production: Copper   Ore(tons) Metal(tons)   Value(£)
            1854   144.00   10.60   1087.90
            1858   156.00   7.30   618.50
            Comment 1854 (C); 1858 (C)

                                     258

KERROW MOOR   BUGLE   SX 018577 0664

Production: Tin   Black(tons) Stuff(tons)   Tin(tons)   Value(£)
            1874   3.50   0.00   0.00   199.80
            1903   0.50   0.00   0.00   37.00
            1905   0.40   0.00   0.00   38.00
            1906   0.40   0.00   0.00   39.00
            1907   0.80   0.00   0.00   97.00
            1908   0.70   0.00   0.00   54.00
            1909   0.80   0.00   0.00   70.00
            1910   0.60   0.00   0.00   62.00
            1911   0.60   0.00   0.00   73.00
            1912   0.60   0.00   0.00   80.00
            1913   0.70   0.00   0.00   90.00
            Comment 1903 KERROW OPEN WORKS; 1905-1913 KERROW OPEN WORKS

KIDNEY,EAST   GOLDSITHNEY   0665

Production: Copper No detailed return
            Tin   Black(tons) Stuff(tons)   Tin(tons)   Value(£)
            1901   0.00   3.00   0.00   6.00
Ownership: 1901 EAST WHEAL KIDNEY MINING CO.
Management: Secretary 1901 ROBT.BURN
Employment: Underground   Surface   Total
            1901   6   6

KILLEFRETH   KENWYN   SW 735443 0666

Production: Copper   Ore(tons) Metal(tons)   Value(£)
            1858   118.00   12.70   1236.30
            1859   128.00   11.40   1130.70
            1866   8.00   0.70   51.60
            1876   22.00   1.50   83.60
            1877   306.60   34.50   2187.50
            1878   110.00   6.50   305.00
            1880   5.00   0.30   16.40
            1882   8.20   0.50   29.00
            1883   3.60   0.30   25.00
            1884   5.00   0.00   34.00
            Comment 1858-1859 (C); 1866 (C)KILLIFREETH; 1876 (C)
            Tin   Black(tons) Stuff(tons)   Tin(tons)   Value(£)
            1858   0.00   0.00   0.00   18.20
            1860   0.00   0.00   0.00   139.60
            1873   54.60   1.00   0.00   4260.00
            1874   20.20   394.50   0.00   2085.30
            1875   18.50   170.00   0.00   974.60
            1876   87.70   0.00   0.00   3744.50
            1877   97.10   0.00   0.00   3880.70
            1878   0.00   156.00   0.00   950.00
            1879   37.90   0.00   0.00   1516.00
            1880   9.40   0.00   0.00   480.00
            1881   15.00   5.00   0.00   815.00
            1882   245.80   528.00   27.20   14528.00
            1883   259.00   0.00   0.00   13260.00
            1884   153.20   0.00   0.00   7059.00

CM-K*   259

Tin	Block(tons)	Stuff(tons)	Tin(tons)	Value(£)
1885	181.50	0.00	0.00	8495.00
1886	154.00	0.00	0.00	8985.00
1887	115.00	0.00	0.00	7131.00
1888	154.70	0.00	0.00	8030.00
1889	151.00	0.00	0.00	7545.00
1890	136.00	0.00	0.00	8212.00
1891	258.00	0.00	0.00	14596.00
1892	309.70	0.00	0.00	16846.00
1893	351.60	0.00	0.00	17617.00
1894	436.10	0.00	0.00	15938.00
1895	321.50	86.00	0.00	11873.00
1896	273.60	0.00	0.00	9847.00
1897	135.60	44.20	0.00	5089.00
1898	0.00	876.00	0.00	808.00
1899	0.00	687.00	0.00	726.00
1900	0.00	633.00	0.00	853.00
1901	0.00	942.00	0.00	797.00
1902	0.00	579.00	0.00	559.00
1903	0.00	445.00	0.00	299.00
1904	0.00	383.00	0.00	217.00
1905	0.00	104.00	0.00	212.00
1906	0.00	97.00	0.00	78.00
1907	0.00	4.00	0.00	6.00
1911	0.00	98.00	0.00	160.00
1912	0.00	20.00	0.00	50.00

Comment 1858 TIN STUFF; 1860 TIN STUFF; 1873–1874 VALUE INC.TIN STUFF; 1875 VALUE NOT INC.TIN STUFF; 1882 TWO RETURNS AGGREGATED

Arsenic	Ore(tons)	Metal(tons)	Value(£)
1873	6.70	0.00	13.00
1874	4.50	0.00	7.40
1876	21.10	0.00	109.90
1877	34.00	0.00	80.00
1880	10.00	0.00	30.00
1882	12.00	0.00	36.00
1883	17.60	0.00	40.00
1884	14.00	0.00	39.00
1886	4.30	0.00	9.00
1893	31.00	0.00	239.00
1894	55.00	0.00	471.00
1895	61.00	0.00	551.00
1896	45.00	0.00	447.00
1897	43.00	0.00	543.00

Comment 1884 SOOT; 1886 CRUDE

Arsenic Pyrite	Ore(tons)	Value(£)
1876	16.90	13.90
1892	12.00	18.00

Comment 1876 ARSENICAL MUNDIC

Ownership: 1877–1881 KILLIFRETH MINING CO.; 1882–1883 KILLEFRETH MINING
CO.; 1884–1890 KILLIFRETH MINING CO.; 1891–1893 KILLIFRETH
MINING CO. C.B.; 1894–1895 KILLIFRETH MINING CO.; 1896–1897
KILLIFRETH MINING CO.LTD.; 1898–1901 W.J.TRYTHALL; 1902–1907
WM.BURROWS; 1911–1912 WM.ELFORD & PAUL FRANCIS; 1913
KILLIFRETH MINES LTD.

Comment 1884–1896 KILLIFRETH; 1897 KILLIFRETH ABOVE ADIT
ONLY; 1898–1900 ABOVE ADIT ONLY; 1901 ABOVE ADIT ONLY IN
LIQUIDATION; 1902 ABOVE ADIT ONLY; 1907 GAVE UP THE MINE
MARCH 1907

Management: Manager 1859–1862 JOEL HIGGINS; 1863–1868 GEO.E.TREMAYNE;
1880–1881 JOHN MICHELL
Chief Agent 1859–1862 JOHN WHITBURN; 1863–1864 ALEX.NANCARROW
JOHN WHITBURN; 1865–1868 ALEX.NANCARROW; 1872–1874 THOS.GOOGH
WM.BUCKINGHAM JOH.PAUL; 1875–1876 THOS.GOOGH JOH.PAUL;
1877–1878 GEO.TREMAYNE JOH.PAUL; 1879 GEO.TREMAYNE JOHN
MICHELL; 1881 JOHN TREVETHAN; 1882–1893 JOHN MICHELL;
1894–1897 R.A.JAMES; 1898–1907 WM.BURROWS
Secretary 1859–1862 ED.H.HAWKE (P); 1863–1868 ED.H.HAWKE
JNR.; 1871 THOS.GOUGH; 1872–1877 JOHN TREGONNING; 1878–1881
JOHN TREGONNING (P); 1891 JOHN TREGONNING; 1892–1897
THOS.F.TROWNSON (P)

Employment:	Underground	Surface	Total
1878	61	29	90
1879	52	25	77
1880	42	15	57
1881	51	26	77
1882	90	68	158
1883	119	64	183
1884	106	51	157
1885	105	49	154
1886	99	44	143
1887	102	36	138
1888	111	47	158
1889	108	45	153
1890	103	41	144
1891	122	55	177
1892	120	83	203
1893	164	60	224
1894	183	98	281
1895	185	82	267
1896	138	75	213
1897	94	59	153
1898	33	8	41
1899	28	5	33
1900	19	1	20
1901	23	5	28
1902	8	1	9
1903	4		4
1904	2	2	4
1905	5	1	6
1906	2	2	4
1907		1	1
1911	4		4
1912	2		2
1913	18	43	61

KILLHAM   ST.NEOT   SX 204670 0667

Production: Tin	Black(tons)	Stuff(tons)	Tin(tons)	Value(£)
1908	0.20	0.00	0.00	18.00

Ownership: 1908 ASSOCIATED TAMWORTH MINES LTD.
         Comment 1908 IN LIQUIDATION 1ST JAN 1909
Management: Secretary 1908 DAVID BLOTTON (S)

Employment:	Underground	Surface	Total
1908	4	4	8

KILLICOR   CHACEWATER   9999

Production: Arsenic Pyrite	Ore(tons)	Value(£)
1894	3.00	3.00

Comment 1894 OPENWORKS.INC.TWELVE HEADS

KILLIVREATH   ROCHE   SW 985584 0668

Production: Iron	Ore(tons)	Iron(%)	Value(£)
1872	1967.00	0.00	786.80

         Comment 1872 BH.
Ownership: 1870-1872 THE IRONMASTERS CO.
         Comment 1872-1874 KILLIFRETH
Management: Chief Agent 1870-1871 W.H.HOSKING; 1872-1874 H.KESSELL

KIMBERLEY   FALMOUTH   SW 803313 0669

Production: Lead No detailed return
Ownership: 1880-1881 LORD KIMBERLEY MINING CO.LTD.
         Comment 1881 NOT WORKED
Management: Chief Agent 1880-1881 JOS.MICHELL

Employment:	Underground	Surface	Total
1880	9	1	10

KING ARTHUR   TINTAGEL   SX 051890 0670

Production: Lead	Ore(tons)	Metal(tons)	Value(£)
1871	2.40	1.80	0.00

Tin No detailed return
Management: Manager 1870-1873 SML.TUCKER
         Secretary 1870 J.H.TILLY; 1871-1872 KENDRICK & MASSEY; 1873
         MASON,FULLER & MASSEY

KING EDWARD   CAMBORNE   SW 663389 0671

Production: Copper	Ore(tons)	Metal(tons)	Value(£)	
1912	3.00	0.00	113.00	
Tin	Black(tons)	Stuff(tons)	Tin(tons)	Value(£)
1904	3.00	0.00	0.00	216.00
1905	4.00	0.00	0.00	324.00
1906	10.40	0.00	0.00	915.00
1907	8.00	0.00	0.00	708.00

KING EDWARD   CAMBORNE   Continued

Tin	Black(tons)	Stuff(tons)	Tin(tons)	Value(£)
1908	10.60	0.00	0.00	794.00
1909	15.20	0.00	0.00	857.00
1910	13.50	0.00	0.00	1156.00
1911	11.40	0.00	0.00	1288.00
1912	14.80	0.00	0.00	1842.00
1913	7.80	0.00	0.00	855.00

Ownership: 1901-1913 CAMBORNE MINING SCHOOL
           Comment 1901-1902 SOUTH CONDURROW(SCHOOL); 1903-1913
           INSTRUCTION ONLY
Management: Manager 1912-1913 JOHN C.SHEPHERD
            Chief Agent 1901-1905 WM.THOMAS; 1906-1912 B.ANGWIN
Employment: Underground   Surface   Total
	Underground	Surface	Total
1901	48	11	59
1902	52	14	66
1903	14	16	30
1904	60	21	81
1905	53	17	70
1906	6	16	22
1907-1908	6	20	26
1909	9	18	27
1910	8	18	26
1911-1913	9	17	26

KING,THE   CALSTOCK   0672

Production: Copper No detailed return
            Tin No detailed return
            Silver No detailed return
Management: Manager 1870-1871 WM.KNOTT
            Secretary 1870-1871 J.T.BARNARD

KINGSDOWN   ST.EWE   SW 968502 0673

Ownership: 1913 KINGSDOWN TIN MINE
           Comment 1913 SINKING SHAFTS
Employment: Underground   Surface   Total
	Underground	Surface	Total
1913	8	3	11

KINGSTON CONSOLS   STOKE CLIMSLAND   SX 362757 0674

Production: Lead & Silver	Ore(tons)	Lead(tons)	Silver(ozs)	Value(£)
1876	83.50	62.50	200.00	0.00
1877	157.90	118.50	590.00	1895.90
1878	63.00	47.20	235.00	693.00
1879	7.00	5.20	25.00	78.10

Zinc	Ore(tons)	Metal(tons)	Value(£)
1872	23.00	0.00	46.00
1876	222.80	0.00	666.00
1877	709.30	0.00	2100.00
1878	130.00	0.00	390.00
1879	40.00	0.00	120.60

KINGSTON CONSOLS   STOKE CLIMSLAND   Continued

          Comment 1872 NORTH KINGSTON; 1879 TWO RETURNS AGGREGATED
          Copper   Ore(tons) Metal(tons)   Value(£)
          1877   2.00   0.20   8.50
          1878   1.00   0.10   4.00
Ownership: 1875-1876 KINGSTON CONSOLS SILVER LEAD CO.; 1877-1878
          KINGSTON CONSOLS CO.LTD.
          Comment 1870-1871 KINGSTON; 1872-1874 KINGSTON UNITED;
          1880-1881 NOT WORKED
Management: Manager 1870-1872 W.WETHERELL; 1878-1881 JAS.CHENOWETH
          Chief Agent 1873-1875 G.F.RICHARDS; 1876 D.FORREST; 1877-1881
          D.FORREST WM.HANCOCK
          Secretary 1870-1871 G.PENNINGTON; 1873-1874 PORTER &
          INGLEBACH; 1878-1881 RODEN
Employment: Underground   Surface   Total
          1878   33   17   50
          1879   7   16   23

KINGSTON VALLEY   STOKE CLIMSLAND   0675

Production: Lead No detailed return
Management: Chief Agent 1874 JAS.CHENOWETH

KIT HILL TUNNEL   CALSTOCK   SX 381723 0677

Ownership: 1877-1878 KIT HILL TUNNEL CO.LTD.
Management: Manager 1877-1878 S.H.EMMENS
Employment: Underground   Surface   Total
          1878   14   14

KIT HILL UNITED   STOKE CLIMSLAND   SX 375713 0678

Production: Copper   Ore(tons) Metal(tons)   Value(£)
          1862   15.00   0.10   5.60
          Comment 1862 (C)
          Tin   Black(tons) Stuff(tons)   Tin(tons)   Value(£)
          1860   19.30   0.00   0.00   1658.10
          1861   44.40   0.00   0.00   3128.00
          1862   13.30   0.00   0.00   867.90
          1864   6.10   0.00   0.00   382.50
          1870   6.30   0.00   0.00   462.20
          1871   4.60   0.00   0.00   345.70
          1874   4.70   0.00   0.00   209.60
          1879   0.90   0.00   0.00   11.40
          1883   3.80   300.00   0.00   150.00
          1884   3.50   300.00   0.00   150.00
          1885   No detailed return
          Comment 1864 3 MONTHS ONLY; 1870-1871 KIT HILL; 1883 KIT HILL
          GREAT CONSOLS; 1884-1885 INC.EXCELSIOR
          Tungsten   Ore(tons) Metal(tons)   Value(£)
          1858   No detailed return
          1859   26.60   0.00   292.70
          1860   19.00   0.00   0.00

                                    264

KIT HILL UNITED   STOKE CLIMSLAND   Continued

              Tungsten   Ore(tons) Metal(tons)   Value(£)
              1870   10.00   0.00   118.80
              1872   12.90   0.00   146.60
              1879   2.10   0.00   16.90
              Comment 1858 KIT HILL.NO RETURNS AVAILABLE; 1870 KIT HILL
Ownership: 1879-1880 KIT HILL UNITED MINING CO.; 1881-1885 KIT HILL
              GREAT CONSOLS MINING CO.
              Comment 1860 KIT HILL; 1869-1871 KIT HILL; 1874 STOPPED;
              1881-1885 KIT HILL GREAT CONSOLS
Management: Manager 1869-1877 WM.SKEWIS; 1878-1881 MOSES BAWDEN
              Chief Agent 1860 WM.B.COLLOM THOS.ODGERS; 1861 WM.B.COLLOM
              THOS.ODGERS WM.SKEWIS; 1862-1865 WM.B.COLLOM WM.SKEWIS
              THOS.JURY; 1880-1885 IS.RICHARDS
              Secretary 1860-1865 JOHN BAYLY (P); 1869-1871 JOHN BAYLY (P);
              1872-1877 MOSES BAWDEN; 1878 JOHN BAYLY; 1881 MOSES BAWDEN
Employment: Underground   Surface   Total
              1879   2   2   4
              1880   2   2
              1881   7   18   25
              1882   38   19   57
              1883   49   16   65
              1884   37   14   51
              1885   21   7   28

KIT HILL,EAST   STOKE CLIMSLAND   SX 389711 0679

Production: Copper No detailed return
            Tin   Black(tons) Stuff(tons)   Tin(tons)   Value(£)
            1853   8.70   0.00   0.00   473.50
            1898   9.00   0.00   0.00   400.00
            1899   No detailed return
            1902   14.30   0.00   0.00   1097.00
            1903   117.00   0.00   0.00   9601.00
            1904   58.00   0.00   0.00   4819.00
            1905   8.00   0.00   0.00   553.00
            1907   0.70   0.00   0.00   69.00
            1908   0.80   0.00   0.00   48.00
            1909   2.40   0.00   0.00   150.00
            Arsenic   Ore(tons) Metal(tons)   Value(£)
            1905   2.00   0.00   13.00
            Tungsten   Ore(tons) Metal(tons)   Value(£)
            1898   11.00   0.00   528.00
            1899   No detailed return
            Wolfram No detailed return
Ownership: 1897-1908 KIT HILL WOLFRAM MINING CO.LTD.; 1909-1910 KIT HILL
            MINING CO.LTD.
            Comment 1910 ABANDONED JAN.1910
Management: Manager 1869 WM.SKEWIS
            Chief Agent 1897-1910 FRANK WRIGHT
            Secretary 1869 JOHN BAYLY
Employment: Underground   Surface   Total
            1897   16   13   29
            1898   15   10   25
            1899   2   1   3

KIT HILL,EAST   STOKE CLIMSLAND   Continued

	Underground	Surface	Total
1901		1	1
1902	17	12	29
1903	49	38	87
1904	51	33	84
1905	71	25	96
1906	2	1	3
1907	2	2	4
1908		2	2
1909	2	3	5

KIT HILL,SOUTH   STOKE CLIMSLAND   SX 374710 0680

Production: Copper No detailed return

Tin	Black(tons)	Stuff(tons)	Tin(tons)	Value(£)
1870	12.10	0.00	0.00	754.30
1871	8.50	0.00	0.00	463.60
1873	17.90	0.00	0.00	1384.00
1874	4.70	0.00	0.00	209.60
1880	15.60	0.00	0.00	752.90
1882	3.70	0.00	2.40	200.00
1883	0.40	0.00	0.00	12.00

Comment 1873 VALUE EST
Ownership: Comment 1874 STOPPED
Management: Manager 1869 MOSES BAWDEN; 1870 MOSES BAWDEN (P); 1873—1875
WM.SKEWIS
Chief Agent 1870—1871 WM.JORY; 1872 WM.SKEWIS T.PROWSE;
1873—1875 T.PROWSE
Secretary 1869 JOHN BAYLY; 1870 JOHN BAYLY (S); 1872—1875
E.NICHOLLS

KITTY   LELANT   SW 510362 0681

Production: Tin	Black(tons)	Stuff(tons)	Tin(tons)	Value(£)
1852	3.20	0.00	0.00	0.00
1854	18.10	0.00	0.00	1187.10
1855	18.60	0.00	0.00	1163.80
1856	148.60	0.00	0.00	11416.80
1857	146.50	0.00	0.00	11540.60
1858	71.20	0.00	0.00	4488.70
1859	163.70	0.00	0.00	11653.50
1860	97.70	0.00	0.00	8013.10
1861	115.20	0.00	0.00	8585.60
1862	132.10	0.00	0.00	8810.70
1863	134.50	0.00	0.00	9047.70
1864	145.70	0.00	0.00	9220.50
1865	120.50	0.00	0.00	6726.90
1866	134.50	0.00	0.00	6486.60
1867	136.50	0.00	0.00	7063.10
1869	175.40	0.00	0.00	12216.70
1870	175.80	0.00	0.00	12194.40
1871	139.30	0.00	0.00	11005.30
1872	46.40	0.00	0.00	3897.90

KITTY   LELANT   Continued

Tin   Black(tons) Stuff(tons)   Tin(tons)   Value(£)
1873   68.20   0.00   0.00   4926.60
1874   No detailed return
Comment 1852 TIN ORE; 1862 TWO RETURNS AGGREGATED; 1873
CHANGED TO POLPEOR 17.4.73; 1874 SEE POLPEOR
Ownership: Comment 1873-1874 NOW POLPEOR; 1875-1881 SEE SISTERS
Management: Manager 1859-1866 THOS.RICHARDS; 1867-1871 THOS.RICHARDS &
SON; 1873-1874 WM.ROSEWARNE
Chief Agent 1859 HY.PEARCE; 1860-1861 WM.WILLIAMS; 1862-1866
WM.WILLIAMS A.ANTHONY; 1867-1872 WM.ROSEWARNE; 1873-1874
SIMON THOMAS
Secretary 1859-1871 THOS.RICHARDS (P); 1872-1874
JAS.B.COULSON

KITTY   REDRUTH   0682

Production: Tin   Black(tons) Stuff(tons)   Tin(tons)   Value(£)
1860-1862 No detailed return
Comment 1860-1862 SEE MARY (REDRUTH)
Ownership: Comment 1860 SEE MARY (REDRUTH)

KITTY   ST.AGNES   SW 726509 0683

Production: Copper   Ore(tons) Metal(tons)   Value(£)
1856   63.00   6.30   616.20
1857   96.00   9.90   1050.70
1858   65.00   6.80   644.10
1862   13.00   1.20   107.80
1863   27.00   2.40   201.70
1864   18.00   1.70   144.00
1865   40.00   3.50   281.50
1866   35.00   2.90   217.00
1867   14.00   1.20   86.80
1869   15.00   1.20   86.60
1870   115.00   9.40   568.50
1872   19.00   1.70   135.40
1874   45.00   4.00   238.50
1876   14.00   1.20   89.20
1878   13.90   1.30   67.60
1879   367.20   36.70   1927.00
1881   15.00   1.30   70.10
1885   12.00   0.00   50.00
1888   15.00   0.00   115.00
1891   26.00   0.00   177.00
1896   5.00   0.00   40.00
Comment 1856-1858 (C); 1862-1867 (C); 1869-1870 (C); 1872
(C); 1874 (C); 1876 (C)
Tin   Black(tons) Stuff(tons)   Tin(tons)   Value(£)
1853   27.10   0.00   0.00   1895.00
1854   83.40   0.00   0.00   2826.60
1855   133.10   0.00   0.00   8322.60
1856   173.90   0.00   0.00   12321.30
1857   221.50   0.00   0.00   15971.30

                            267

Tin	Black(tons)	Stuff(tons)	Tin(tons)	Value(£)
1858	174.60	0.00	0.00	10417.90
1859	182.70	0.00	0.00	13408.80
1860	166.50	0.00	0.00	12906.60
1861	71.50	0.00	0.00	4983.10
1862	151.70	0.00	0.00	9496.40
1863	213.90	0.00	0.00	14150.50
1864	239.40	0.00	0.00	14982.60
1865	236.30	0.00	0.00	11985.90
1866	218.20	0.00	0.00	10529.90
1867	168.30	0.00	0.00	8839.40
1868	192.10	0.00	0.00	10439.70
1869	189.80	0.00	0.00	13821.50
1870	200.70	0.00	0.00	15431.60
1871	181.40	0.00	0.00	15128.40
1872	224.40	0.00	0.00	20525.80
1873	235.20	0.00	0.00	18752.00
1874	198.00	0.00	0.00	11515.30
1875	183.70	0.00	0.00	9567.50
1876	188.40	0.00	0.00	8637.80
1877	195.30	0.00	0.00	8122.50
1878	219.80	0.00	0.00	8211.60
1879	270.20	0.00	0.00	12157.00
1880	213.00	0.00	0.00	11724.80
1881	122.80	0.00	81.80	6968.10
1882	112.40	0.00	73.00	6770.00
1883	134.80	0.00	0.00	7458.00
1884	175.30	0.00	0.00	8340.00
1885	181.20	0.00	0.00	9152.00
1886	159.30	0.00	0.00	9457.00
1887	195.70	0.00	0.00	13487.00
1888	167.80	0.00	0.00	11806.00
1889	147.20	0.00	0.00	8496.00
1890	172.50	0.00	0.00	9903.00
1891	157.60	0.00	0.00	8744.00
1892	121.00	0.00	0.00	6727.00
1893	112.50	0.00	0.00	5901.00
1894	97.10	0.00	0.00	4129.00
1895	63.80	0.00	0.00	2530.00
1896	172.90	0.00	0.00	6684.00
1897	213.20	0.00	0.00	8569.00
1898	132.00	0.00	0.00	5685.00
1899	138.00	0.00	0.00	10423.00
1900	143.40	0.00	0.00	12007.00
1901	183.10	0.00	0.00	13103.00
1902	187.90	0.00	0.00	14122.00
1903	112.00	0.00	0.00	8972.00
1904	31.30	0.00	0.00	2341.00
1905	32.00	0.00	0.00	2809.00
1906	45.00	0.00	0.00	4789.00
1907	73.00	0.00	0.00	7457.00
1908	221.10	0.00	0.00	19101.00
1909	241.00	0.00	0.00	20479.00
1910	231.30	0.00	0.00	22442.00
1911	172.70	0.00	0.00	20076.00

Tin	Black(tons)	Stuff(tons)	Tin(tons)	Value(£)
1912	155.00	0.00	0.00	19701.00
1913	175.90	0.00	0.00	20780.00

Comment 1862 TWO RETURNS AGGREGATED

Ownership: 1877–1890 WHEAL KITTY MINING CO.; 1891–1893 WHEAL KITTY
MINING CO. C.B.; 1894–1895 WHEAL KITTY MINES ADVENTURERS;
1896–1905 WHEAL KITTY MINING CO.; 1906–1913 WHEAL KITTY &
PENHALLS UNITED LTD.

Management: Manager 1859 MART.EDWARDS; 1860–1862 RICH.PRYOR; 1863–1881
WM.TEAGUE
Chief Agent 1859–1861 JAMES NICHOLAS T.M.THOMAS; 1862
STEP.DAVEY WM.HIGGINS; 1863–1869 STEP.DAVEY WM.POLKINGHORNE;
1870–1871 STEP.DAVEY JAS.WILLIAMS; 1872–1874 STEP.DAVEY JOHN
WILLIAMS; 1875–1876 STEP.DAVEY; 1877–1879 STEP.DAVEY
RICH.HARRIS; 1880–1881 RICH.HARRIS; 1882–1887 WM.TEAGUE;
1888–1889 HY.GRIPE; 1890–1897 WM.TEAGUE; 1898–1903 JOHN
HARPER; 1905–1910 J.H.COLLINS; 1911–1913 J.H.PRISK
Secretary 1859–1863 ROBT.H.PIKE (P); 1864 D.STEPHENS;
1865–1873 ROBT.H.PIKE & SON; 1874–1879 PIKE & SON J.HICKEY;
1880–1881 W.PIKE(P)J.HICKEY(S); 1891–1897 WM.TEAGUE;
1898–1902 GEO.A.HANCOCK (P)

Employment:

	Underground	Surface	Total
1878	102	81	183
1879	114	84	198
1880	108	64	172
1881	87	69	156
1882	95	65	160
1883	107	66	173
1884	94	68	162
1885	90	61	151
1886	124	76	200
1887	117	76	193
1888	111	77	188
1889	98	91	189
1890	103	82	185
1891	74	71	145
1892	86	75	161
1893	79	73	152
1894	50	40	90
1895	68	38	106
1896	93	60	153
1897	85	50	130
1898	84	55	139
1899	130	58	188
1900	115	56	171
1901	100	58	158
1902	98	57	155
1903	90	55	145
1904	25	19	44
1905	55	50	105
1906	80	50	130
1907	87	60	147
1908	109	71	180
1909	101	93	194
1910	109	98	207

KITTY   ST.AGNES   Continued

```
            Underground    Surface      Total
    1911        120          112         232
    1912        122           97         219
    1913        114           89         203
```

KITTY,NEW   ST.AGNES   SW 718503 0684

Production: Copper   Ore(tons) Metal(tons)   Value(£)
    1886   36.00   0.00   247.00
    1888   25.00   0.00   233.00
    1889   6.00   0.00   35.00
    Tin   Black(tons) Stuff(tons)   Tin(tons)   Value(£)
    1886   1.50   0.00   0.00   73.00
    1887   0.70   0.00   0.00   48.00
    1888   4.00   0.00   0.00   203.00
    1889   2.70   0.00   0.00   150.00
Ownership: 1880-1881 JOHN B.REYNOLDS & CO.; 1882-1886 JOHN B.REYNOLDS &
            OTHERS; 1887-1889 NEW KITTY MINE CO.
            Comment 1879 SEE LIVINGSTONE CONSOLS; 1889 SUSPENDED
Management: Chief Agent 1880-1889 WM.VIVIAN
            Secretary 1881 J.COULTER HANCOCK(P)
Employment: Underground   Surface   Total
    1880   3   11   14
    1881   19   7   26
    1882   14   7   21
    1883   15   6   21
    1884   14   8   22
    1885   12   8   20
    1886-1887   14   7   21
    1888   18   6   24

KITTY,SOUTH   LELANT   0685

Production: Tin No detailed return
Ownership: Comment 1863-1865 SUSPENDED
Management: Chief Agent 1860 SML.MICHELL; 1861-1862 SML.MICHELL JNR.
            Secretary 1860-1865 J.WEBB (P)

KITTY,SOUTH WHEAL   LELANT   0686

Production: Tin No detailed return
Ownership: Comment 1863 FORMERLY ROCK
Management: Chief Agent 1861-1862 S.MICHELL
            Secretary 1861-1862 C.R.WEBB

KITTY,WEST   ST.AGNES   SW 719508 0687

Production: Copper   Ore(tons) Metal(tons)   Value(£)
    1884   66.00   0.00   396.00
    1886   21.50   0.00   145.00
    1888   11.00   0.00   100.00

270

Copper	Ore(tons)	Metal(tons)	Value(£)	
1892	12.00	0.00	61.00	
1903	6.00	0.00	38.00	
1904	1.00	0.00	9.00	
1905	9.00	0.00	75.00	
1906	9.00	0.00	97.00	
1907	24.00	0.00	291.00	
1908	2.00	0.00	17.00	
1909	22.00	0.00	159.00	
1910	2.00	0.00	17.00	

Tin	Black(tons)	Stuff(tons)	Tin(tons)	Value(£)
1863	2.20	0.00	0.00	133.40
1864	2.70	0.00	0.00	165.40
1865	2.10	0.00	0.00	186.00
1866	0.00	0.00	0.00	119.70
1867	2.50	267.30	0.00	145.30
1868	8.80	0.00	0.00	421.00
1869	4.80	0.00	0.00	295.40
1870	4.10	0.00	0.00	286.50
1871	2.70	0.00	0.00	182.30
1880	10.20	0.00	0.00	539.50
1881	84.50	0.00	59.10	4912.10
1882	206.70	0.00	144.70	12613.00
1883	327.00	0.00	0.00	18439.00
1884	372.00	0.00	0.00	16833.00
1885	349.00	0.00	0.00	17737.00
1886	318.00	0.00	0.00	18734.00
1887	312.90	0.00	0.00	21400.00
1888	318.00	0.00	0.00	22693.00
1889	333.50	0.00	0.00	19711.00
1890	384.20	0.00	0.00	23127.00
1891	356.90	0.00	0.00	20664.00
1892	320.80	0.00	0.00	18339.00
1893	294.30	0.00	0.00	3252.00
1894	336.20	0.00	0.00	14555.00
1895	371.00	0.00	0.00	14827.00
1896	389.40	0.00	0.00	15104.00
1897	415.00	3.70	0.00	16606.00
1898	420.90	0.00	0.00	19277.00
1899	444.00	0.00	0.00	34634.00
1900	467.00	0.00	0.00	40102.00
1901	432.50	0.00	0.00	31915.00
1902	450.90	0.00	0.00	34502.00
1903	413.00	0.00	0.00	33611.00
1904	363.00	0.00	0.00	28981.00
1905	300.00	0.00	0.00	26358.00
1906	337.00	0.00	0.00	36476.00
1907	362.30	0.00	0.00	38198.00
1908	325.00	0.00	0.00	25524.00
1909	230.00	0.00	0.00	17213.00
1910	82.00	0.00	0.00	7229.00
1911	138.00	0.00	0.00	16243.00
1912	132.00	0.00	0.00	18037.00
1913	160.00	0.00	0.00	19809.00

Comment 1865 VALUE INC.TIN STUFF; 1866 TIN STUFF; 1867

FIGURES FROM 1868 FOOTNOTE; 1893 VALUE DOUBTFUL
Ownership: 1879-1884 JOHN B.REYNOLDS & CO.; 1885-1886 JOHN B.REYNOLDS &
            OTHERS; 1887-1890 WEST KITTY MINE CO.LTD.; 1891-1893 WEST
            KITTY MINING CO. C.B.; 1894-1912 WEST KITTY MINING CO.; 1913
            WEST KITTY MINING CO.LTD.
            Comment 1864-1871 FORMERLY ROCK
Management: Manager 1864-1871 JOS.VIVIAN; 1891-1902 FRAN.W.MICHELL (P)
            Chief Agent 1864 HY.SKEWIS; 1865-1871 WM.VIVIAN; 1879-1890
            WM.VIVIAN; 1891-1904 JOEL HOOPER JOHN WILLIAMS; 1905-1908
            JOHN WILLIAMS
            Secretary 1864-1865 ALM.E.PAULL; 1866-1871 W.WATSON; 1881
            H.W.MICHELL (P); 1891-1902 F.J.HARVEY (S)

Employment:	Underground	Surface	Total
1879	22	8	30
1880	28	22	50
1881	42	45	87
1882	45	54	99
1883	52	73	125
1884	54	73	127
1885	60	67	127
1886	60	70	130
1887	62	74	136
1888-1889	62	73	135
1890	62	78	140
1891	62	75	137
1892	73	84	157
1893	99	92	191
1894	115	96	211
1895	149	99	248
1896	118	103	221
1897	138	100	238
1898	161	121	282
1899	211	133	344
1900	181	124	305
1901	202	130	332
1902	200	141	341
1903	213	131	344
1904	233	130	363
1905	238	140	378
1906	236	133	369
1907	237	136	373
1908	172	117	289
1909	118	82	200
1910	82	62	144
1911	178	156	334
1912	116	122	238
1913	100	135	235

KITTY,WEST(SETT ADJ.)   ST.AGNES   0688

Production: Tin	Black(tons)	Stuff(tons)	Tin(tons)	Value(£)
1887	23.70	0.00	0.00	1507.00
1888	64.90	0.00	0.00	4639.00
1889	49.30	0.00	0.00	2951.00

KITTY,WEST(SETT ADJ.)   ST.AGNES   Continued

         Tin   Black(tons) Stuff(tons)   Tin(tons)   Value(£)
        1890   43.00   0.00   0.00   2602.00

KNIGHTOR   ST.AUSTELL   SX 034560 0689

Production: Iron   Ore(tons)   Iron(%)   Value(£)
           1864-1871 No detailed return
           1872   3901.70   0.00   1170.50
           1873   329.40   0.00   246.00
           1874   350.00   0.00   320.00
           1875   No detailed return
           Comment 1864-1871 SEE RUBY; 1872 BH.INC.TREVERBYN & RESURGY;
           1873-1875 BH.INC.TREVERBYN
Ownership: 1872 FRAN.BARRATT & CO.; 1873-1874 G.B.SANDEMAN
           Comment 1864-1871 SEE RUBY (ST.AUSTELL)
Management: Chief Agent 1872 FRAN.BARRATT JNR.

LADDER WAY   PERRANUTHNOE   SW 540298 0690

Production: Tin No detailed return
Ownership: 1902 G.D.MCGREGOR
           Comment 1903-1905 SEE PERRAN

LADOCK   LADOCK   SW 915525 0691

Production: Iron   Ore(tons)   Iron(%)   Value(£)
           1860   1254.50   0.00   326.00
           1861   59.30   0.00   15.00
           1862   762.00   0.00   420.00
           1863   2413.00   0.00   844.50
           1864   1177.60   0.00   356.30
           1865   121.00   0.00   30.20
           1872   844.00   0.00   506.00
           1873   512.00   0.00   384.00
           1874   250.00   0.00   212.50
           1875   200.00   0.00   160.00
           Comment 1860-1865 BH.; 1872-1875 BH.
Ownership: 1863-1866 JOS.MORCOM; 1870-1876 FAITHFUL,COOKSON & CO.
           Comment 1871 SEE ALSO LADOCK,DEVON; 1872 LADOCK & PAWTON;
           1873-1876 LADOCK & TREVERBYN
Management: Manager 1872-1876 W.H.HOSKING
           Chief Agent 1863-1866 THOS.DANIELL; 1870-1871 W.H.HOSKING

LADY ASHBURTON   CALLINGTON   SX 368702 0692

Production: Lead & Silver   Ore(tons)   Lead(tons) Silver(ozs)   Value(£)
           1873   2.00   1.50   7.00   40.00
           1874   0.80   0.60   0.00   0.00
           Copper No detailed return
           Tin No detailed return
Ownership: 1880 LADY ASHBURTON SILVER LEAD MINING CO.; 1881-1882 LADY

                                     273

LADY ASHBURTON   CALLINGTON   Continued

ASHBURTON MINING CO.LTD.
Comment 1875 SUSPENDED SEPT.1874; 1882 NOT WORKED
Management: Manager 1873 ROBT.SARGEANT; 1874-1875 H.BENNETT; 1880-1882
JAS.WILLCOCK
Chief Agent 1874-1875 ROBT.SARGEANT
Secretary 1873-1875 F.NEVILL; 1880-1882 J.J.STANSFIELD
Employment: Underground   Surface   Total
1880   8   8
1881   8   2   10

LADY GWENDOLIN   GERMOE   SW 588302 0693

Production: Tin No detailed return
Ownership: 1905 ROSEWARNE & CO.; 1906 SOUTH-WEST CORNWALL MINES LTD.;
1913 LADY GWEN TIN MINE LTD.
Comment 1905 LADY GWENDOLEN; 1906 FORMERLY NEW GREAT WORK;
1907-1910 FORMERLY NEW GRT.WORK SEE GODOLPHIN; 1911-1912 SEE
GODOLPHIN; 1913 COMMENCED DEC.1913
Employment: Underground   Surface   Total
1905   2   2
1907-1912
1913   13   13
Comment 1907-1912 SEE GODOLPHIN

LADY RASHLEIGH   LUXULYAN   SX 059562 0694

Production: Tin No detailed return
Ownership: 1881 LADY RASHLEIGH CONSOLS TIN MINING CO.LTD.; 1907-1910
MINES AND COMMERCE LTD.
Comment 1881 RASHLEIGH LADY; 1909 IN LIQUIDATION; 1910
SUSPENDED IN LIQUIDATION
Management: Manager 1881 PHIL.RICH
Chief Agent 1907-1908 A.A.HUMPHRIES; 1909-1910 JOHN PENHALL
Secretary 1881 J.J.ALLBROOK; 1908 C.V.THOMAS
Employment: Underground   Surface   Total
1881   6   4   10
1909   9   2   11

LAMBERT   GRAMPOUND ROAD   SW 929523 0695

Production: Tin   Black(tons) Stuff(tons)   Tin(tons)   Value(£)
1875   0.00   50.00   0.00   6.00
Ownership: 1875 THE LAMBERT CO.
Comment 1875 FORMERLY BLENCOWE CONSOLS
Management: Chief Agent 1875 WM.WEST

LAMBRIGGAN   PERRANZABULOE   SW 761511 0696

Production: Zinc No detailed return
Ownership: 1892-1893 EDW.CARTER (P)
Comment 1892 FORMERLY DUCHY OF PERU; 1893 FORMERLY DUCHY OF

LAMBRIGGAN   PERRANZABULOE   Continued

                 PERU SUSPENDED
Employment: Underground   Surface   Total
                 1892   2   2

LAMTREROVE   0697

Production: Tin   Black(tons) Stuff(tons)   Tin(tons)   Value(£)
            1852   0.80   0.00   0.00   0.00
            Comment 1852 TIN ORE

LANGEATH   ST.STEPHEN-IN-BRANNEL   0698

Production: Iron   Ore(tons)   Iron(%)   Value(£)
            1862   1000.00   0.00   620.00
            Comment 1862 HE.LANGETH
Ownership: 1863-1865 T.CLARE
Management: Chief Agent 1863-1865 W.BECK

LANGFORD   CALLINGTON   SX 383695 0699

Production: Lead & Silver   Ore(tons) Lead(tons) Silver(ozs)   Value(£)
            1852   7.50   5.60   0.00   0.00
            1853-1854 No detailed return
            1855   7.20   4.00   0.00   0.00
            1856   33.00   10.00   1660.00   0.00
            1857   10.00   7.50   500.00   0.00
            1858-1859 No detailed return
            1884   4.00   0.00   0.00   32.00
            1886   23.00   0.00   0.00   0.00
            Comment 1884 NEW LANGFORD; 1886 NEW LANGFORD
            Zinc   Ore(tons) Metal(tons)   Value(£)
            1884   25.00   0.00   45.00
            1885   No detailed return
            Comment 1884-1885 NEW LANGFORD
            Copper No detailed return
Ownership: 1877 EMMENS AND CO.LTD.; 1878 BARING & LANGFORD LTD.; 1881
            LANGFORD MINING CO.; 1882-1888 WHEAL LANGFORD MINING CO.
            Comment 1878-1879 BARING AND LANGFORD; 1881 FORMERLY QUEEN;
            1884-1887 NEW LANGFORD; 1888 NEW LANGFORD SUSPENDED
Management: Manager 1877-1879 S.H.EMMENS
            Chief Agent 1882 R.GOLDSWORTHY; 1883-1884 THOS.GREGORY
            Secretary 1878-1879 S.BROOME (S); 1885-1888 MOSES BAWDEN
Employment: Underground   Surface   Total
            1878   4   4   8
            1882   9   20   29
            1883   16   3   19
            1884   8   4   12
            1885   6   6   12
            1886   7   4   11
            1887   7   2   9

                                 275

LANGSTONE   LAMERTON, DEVON   SX 483826 0700

Production: Manganese No detailed return
Management: Chief Agent 1874 WM.HOOPER

LANIVET   LANIVET   SX 027636 0701

Production: Iron   Ore(tons)   Iron(%)   Value(£)
            1867   468.00   0.00   152.50
            1868   1256.00   0.00   314.00
            1869   1942.00   0.00   582.60
            1870   2341.00   0.00   585.20
            1871   1827.00   0.00   1370.20
            1872   6132.00   0.00   3679.20
            1873   2200.00   0.00   1650.00
            1874   103.50   0.00   77.70
            1875   2089.00   0.00   1566.70
            1876   4625.00   0.00   3468.00
            1880   810.00   0.00   486.00
            Comment 1867 BH.; 1868-1870 BH.INC.CARNEWAS; 1871-1872 BH.;
            1873 BH.OR WEST DOWN; 1874-1875 BH.; 1876 BH.& SOME SPATHOSE;
            1880 BH.
Ownership: 1873-1874 BEGBIE & CO.
            Comment 1873-1874 OR WEST DOWN
Management: Manager 1873 ART.J.BEGBIE; 1874 ART.L.BEGBIE

LANIVET CONSOLS   LANIVET   SX 067641 0702

Production: Copper   Ore(tons) Metal(tons)   Value(£)
            1845   1125.00   91.60   6081.70
            1846   1159.00   100.90   6235.10
            1847   1126.00   82.50   5312.60
            1848   138.00   8.10   409.80
            Comment 1845-1848 (C)

LANIVET TIN   LANIVET   SX 067641 0703

Production: Tin No detailed return
Management: Secretary 1863-1866 E.CARTER

LANJETH   ST.STEPHEN-IN-BRANNEL   SW 970528 0704

Production: Tin No detailed return
Ownership: 1913 UNITED NIGERIAN TIN LTD.
            Comment 1913 IN LIQUIDATION
Employment: Underground   Surface   Total
            1913   6   4   10

LANJEW   WITHIEL   SW 986653 0705

Production: Iron   Ore(tons)   Iron(%)   Value(£)
            1858   196.90   0.00   91.60

                              276

Iron	Ore(tons)	Iron(%)	Value(£)
1859—1861	No detailed return		
1874	No detailed return		

Comment 1858 BH.; 1859 BH.SEE GWINDRA; 1860—1861 BH.; 1874
SEE MID—CORNWALL MINES
Ownership: 1873 MID CORNWALL MINING CO.LTD.
Management: Chief Agent 1872—1873 DAVID COCK

LANLIVERY   LANLIVERY   0706

Production: Tin	Black(tons)	Stuff(tons)	Tin(tons)	Value(£)
1913	3.20	0.00	0.00	359.00

Comment 1913 OPENWORK

LATCHLEY   CALSTOCK   SX 409732 0707

Production: Copper No detailed return
Ownership: Comment 1865—1866 SEE SOUTH MARIA

LAWHITTON CONSOLS   LAWHITTON   SX 364831 0708

Production: Lead No detailed return
             Copper No detailed return
Ownership: Comment 1862—1865 SUSPENDED
Management: Chief Agent 1861 JAS.NICHOLLS
             Secretary 1860—1863 ED.BATTELEY (P)

LEEDS   BREAGE   SW 574306 0709

Production: Tin	Black(tons)	Stuff(tons)	Tin(tons)	Value(£)
1870	0.00	0.00	0.00	69.40
1873	0.00	0.00	0.00	40.00

Comment 1870 TIN STUFF; 1873 LEEDS TIN MINE TIN STUFF
Ownership: Comment 1873 LEEDS TIN STOPPED IN 1873 OR 1874
Management: Manager 1869—1871 THOS.M.EUSTICE; 1873 WM.PASCOE
             Secretary 1869—1871 JOHN R.DANIELL

LEEDS AND ST.AUBYN   GERMOE   SW 582292 0710

Production: Tin	Black(tons)	Stuff(tons)	Tin(tons)	Value(£)
1853	0.00	0.00	0.00	19.30
1854	63.10	0.00	0.00	3780.00
1856	64.50	0.00	0.00	4757.60
1857	31.10	0.00	0.00	2404.00
1858	45.70	0.00	0.00	2889.70
1859	31.80	0.00	0.00	2231.00
1860	23.00	0.00	0.00	1809.30
1861	36.00	0.00	0.00	2571.80
1862	49.80	0.00	0.00	3139.10
1863	44.50	0.00	0.00	3030.50

Tin	Black(tons)	Stuff(tons)	Tin(tons)	Value(£)
1864	51.40	0.00	0.00	3278.70
1865	64.00	0.00	0.00	3561.50
1866	102.00	0.00	0.00	5150.20
1867	106.20	0.00	0.00	5634.70
1868	86.50	0.00	0.00	4701.50
1869	12.20	0.00	0.00	783.30
1870	10.30	0.00	0.00	722.50
1871	39.30	0.00	0.00	3009.20
1872	43.90	0.00	0.00	3599.10
1873	44.30	0.00	0.00	3199.90
1874	6.10	0.00	0.00	416.00

Comment 1862 TWO RETURNS AGGREGATED; 1874 VALUE EST.
Ownership: Comment 1873 STOPPED IN 1873 OR 1874
Management: Manager 1871-1873 JOHN CURTIS
Chief Agent 1859 JOHN CURTIS; 1860-1862 JOHN CURTIS
WM.VAWDREY JNR.; 1863-1870 JOHN CURTIS JOHN BEARE; 1871 JOHN
BEARE; 1874-1875 JOHN CURTIS
Secretary 1859-1870 WM.VAWDREY (P); 1871-1873 F.R.WILSON

LEEDSTOWN CONSOLIDATED   LEEDSTOWN   SW 603335 0711

Production: Tin

	Black(tons)	Stuff(tons)	Tin(tons)	Value(£)
1901	0.40	0.00	0.00	24.00
1902	1.30	0.00	0.00	82.00
1903	0.00	21.00	0.00	48.00
1904	No detailed return			

Arsenic No detailed return

Arsenic Pyrite

	Ore(tons)	Value(£)
1900	37.00	74.00
1901	16.00	33.00
1902	12.00	12.00

Ownership: 1899-1903 ANGLO-PENINSULA MINING & CHEM.CO.LTD.
Management: Chief Agent 1899-1903 WM.ROSEWARNE
Employment:

	Underground	Surface	Total
1899-1900	2	4	6
1901	2	2	4
1902	1	3	4
1903		3	3

LEISURE   PERRANZABULOE   SW 757542 0712

Production: Copper

	Ore(tons)	Metal(tons)	Value(£)
1852	111.00	5.60	415.20
1853	159.00	7.10	653.20
1854-1855	No detailed return		
1862	17.00	0.60	44.30
1863	20.00	0.70	46.50
1865	10.00	0.40	24.00
1867	10.00	0.40	24.50

Comment 1852-1853 (C)GREAT LEISURE; 1854-1855 (C)SEE PERRAN
UNITED; 1862-1863 (C); 1865 (C); 1867 (C)
Ownership: Comment 1865 SUSPENDED

LEISURE   PERRANZABULOE   Continued

Management: Chief Agent 1862–1864 THOS.PILL
            Secretary 1862–1865 M.T.HITCHINGS

LEISURE,EAST   PERRANZABULOE   SW 758542 0713

Production: Copper   Ore(tons) Metal(tons)   Value(£)
            1851   355.00   17.10   993.90
            1852   315.00   13.20   907.80
            Comment 1851–1852 (C)
Ownership: Comment 1864–1865 SUSPENDED

LEISURE,NEW   PERRANZABULOE   SW 763534 0714

Production: Zinc   Ore(tons) Metal(tons)   Value(£)
            1907   450.00   158.00   1700.00
            1908   50.00   17.00   250.00
            1909   824.00   321.00   4944.00
            1911   300.00   70.00   450.00
            Copper No detailed return
            Tin No detailed return
Ownership: 1877–1879 FRAN.BLIGHT; 1904 NEW WHEAL LEISURE MINING CO.;
            1905–1913 ALFRED MINES LTD.
            Comment 1905 SEE ALFRED; 1909 CLOSED AT END OF 1909;
            1910–1913 CLOSED JAN.1910
Management: Manager 1905–1913 J.H.HARRIS JAMES
            Chief Agent 1877–1879 FRAN.BLIGHT; 1904–1905 J.H.HARRIS
            JAMES
            Secretary 1905–1913 F.MARSHALL (S)
Employment: Underground   Surface   Total
            1878–1879   6   6
            1905   20   10   30
            1906   35   45   80
            1907   43   52   95
            1908   22   26   48
            1909   49   56   105
            1910   4   4
            1911   2   2
            1912   3   3
            1913   2   2

LEISURE,NORTH   PERRANZABULOE   SW 761545 0715

Production: Tin No detailed return
Ownership: Comment 1901–1902 SEE RAMOTH
Employment: Underground   Surface   Total
            1901–1902
            Comment 1901–1902 SEE RAMOTH

LEISURE,SOUTH   PERRANZABULOE   SW 761529 0716

Production: Zinc No detailed return
           Tin   Black(tons) Stuff(tons)   Tin(tons)   Value(£)
           1910   3.40   0.00   0.00   361.00
           1911   0.40   0.00   0.00   45.00
           1913   0.10   0.00   0.00   5.00
Ownership: 1909—1913 WM.RICHARDS
            Comment 1910 NEW SINKING MAY 1910; 1912 NOT WORKED
Employment: Underground   Surface   Total
           1910   4   2   6
           1911   3   1   4
           1913   5   3   8

LELANT CONSOLS   LELANT   SW 504359 0717

Production: Tin   Black(tons) Stuff(tons)   Tin(tons)   Value(£)
           1853   14.00   0.00   0.00   1021.60
           1857   85.80   0.00   0.00   6355.60
           1858   17.80   0.00   0.00   1198.10
           1859 .   11.00   0.00   0.00   726.70
           1860   15.50   0.00   0.00   1208.70
Ownership: Comment 1865 SUSPENDED
Management: Manager 1859—1864 JAS.WILLIAMS
            Secretary 1859—1865 WM.RICHARDS (P)

LEMAN   0718

Production: Lead   Ore(tons) Metal(tons)   Value(£)
           1846   30.00   18.00   0.00
           1847   73.00   44.00   0.00
           1849   No detailed return

LEMON   GERMOE   SW 579299 0719

Production: Tin   Black(tons) Stuff(tons)   Tin(tons)   Value(£)
           1855   29.70   0.00   0.00   2034.00
           1856   1.30   0.00   0.00   97.70

LESWIDDEN   ST.JUST   SW 393304 0720

Production: Tin No detailed return
Ownership: 1913 SCRIP LTD.
Employment: Underground   Surface   Total
           1913   4   4   8

LEVANT   ST.JUST   SW 368346 0721

Production: Copper   Ore(tons) Metal(tons)   Value(£)
           1845   1088.00   106.60   7154.90
           1846   1293.00   130.30   8199.90
           1847   851.00   98.80   6849.10

                                          280

Copper	Ore(tons)	Metal(tons)	Value(£)
1848	1776.00	177.60	9857.20
1849	1904.00	170.90	10972.70
1850	2668.00	217.10	14156.80
1851	1539.00	118.10	7350.30
1852	1333.00	101.60	7993.20
1853	1627.00	91.30	8141.30
1854	1589.00	87.70	8201.20
1855	1578.00	94.90	9060.80
1856	1641.00	94.00	7859.40
1857	1587.00	81.60	7446.10
1858	1473.00	96.80	8267.30
1859	1567.00	90.30	7940.10
1860	957.00	57.20	5126.00
1861	967.00	61.80	5260.70
1862	1069.00	71.30	5508.10
1863	1203.00	85.20	6252.50
1864	839.00	53.60	4235.70
1865	609.00	42.20	3112.40
1866	349.00	25.60	1543.10
1867	376.00	37.20	2493.00
1868	380.00	34.70	2278.20
1869	142.00	18.60	1185.80
1870	100.00	16.20	1029.90
1871	72.00	9.00	572.50
1873	74.00	8.00	518.90
1874	61.00	12.50	912.60
1875	123.00	18.40	1411.80
1876	338.00	41.60	2767.30
1877	419.80	52.40	2897.10
1878	741.10	100.70	5386.40
1879	385.60	45.20	2312.10
1880	519.00	70.30	3990.70
1881	1002.40	123.90	6949.60
1882	1173.00	117.30	7800.00
1883	1477.70	155.00	7357.00
1884	1637.00	0.00	8014.00
1885	1364.00	0.00	6073.00
1886	634.00	0.00	3419.00
1887	727.00	0.00	3698.00
1888	1209.00	0.00	14187.00
1889	1505.00	0.00	7293.00
1890	1979.00	0.00	6798.00
1891	2189.00	0.00	5994.00
1892	1719.00	0.00	3788.00
1893	1230.00	0.00	3441.00
1894	1882.00	0.00	7577.00
1895	4333.00	0.00	15730.00
1896	4897.00	0.00	16299.00
1897	3455.00	0.00	11009.00
1898	2981.00	0.00	12057.00
1899	3627.00	0.00	17827.00
1900	5064.00	0.00	22097.00
1901	3550.00	0.00	17430.00
1902	3056.00	0.00	10540.00

Copper	Ore(tons)	Metal(tons)	Value(£)
1903	3381.00	0.00	10961.00
1904	2884.00	0.00	8440.00
1905	3387.00	0.00	11788.00
1906	2140.00	0.00	8819.00
1907	1962.00	0.00	9007.00
1908	807.00	0.00	4215.00
1909	1202.00	0.00	3737.00
1910	683.00	0.00	2049.00
1911	900.00	0.00	3062.00
1912	510.00	0.00	5004.00
1913	383.00	0.00	3303.00

Comment 1845-1871 (C); 1873-1876 (C); 1892-1895 INC.SOME PRECIPITATE

Tin	Black(tons)	Stuff(tons)	Tin(tons)	Value(£)
1852	57.50	0.00	0.00	0.00
1853	41.00	0.00	0.00	2861.50
1854	287.40	0.00	0.00	19540.00
1855	310.00	0.00	0.00	20099.70
1856	218.40	0.00	0.00	15659.80
1857	192.20	0.00	0.00	15161.90
1858	151.30	0.00	0.00	9587.20
1859	210.50	0.00	0.00	15479.50
1860	227.30	0.00	0.00	18138.60
1861	178.50	0.00	0.00	13404.10
1862	241.30	0.00	0.00	15906.60
1863	170.30	0.00	0.00	11178.30
1864	184.00	0.00	0.00	11920.80
1865	183.30	0.00	0.00	9913.90
1866	94.90	0.00	0.00	9657.70
1867	187.10	0.00	0.00	9916.90
1868	134.90	0.00	0.00	7434.00
1869	174.10	0.00	0.00	12073.30
1870	138.80	0.00	0.00	9914.10
1871	86.30	0.00	0.00	6368.80
1872	6.80	0.00	0.00	456.60
1873	83.90	0.00	0.00	6154.80
1874	150.10	0.00	0.00	8024.20
1875	195.60	0.00	0.00	9592.90
1876	247.80	0.00	0.00	10518.00
1877	260.10	0.00	0.00	10569.50
1878	233.20	0.00	0.00	8323.60
1879	273.60	0.00	0.00	10944.00
1880	227.40	0.00	0.00	11452.50
1881	232.60	0.00	158.50	12603.60
1882	230.20	0.00	196.70	13475.00
1883	283.10	0.00	0.00	14984.00
1884	316.70	0.00	0.00	14498.00
1885	452.20	0.00	0.00	21664.00
1886	456.40	0.00	0.00	25244.00
1887	492.40	0.00	0.00	30598.00
1888	436.90	0.00	0.00	28182.00
1889	451.20	0.00	0.00	24769.00
1890	475.00	0.00	0.00	26730.00
1891	578.40	0.00	0.00	31436.00

Tin	Black(tons)	Stuff(tons)	Tin(tons)	Value(£)
1892	498.80	0.00	0.00	27265.00
1893	665.00	0.00	0.00	33775.00
1894	628.90	0.00	0.00	26109.00
1895	517.30	0.00	0.00	19630.00
1896	421.50	0.00	0.00	16201.00
1897	571.80	0.00	0.00	23162.00
1898	556.00	0.00	0.00	24428.00
1899	551.40	0.00	0.00	38940.00
1900	464.00	0.00	0.00	37852.00
1901	514.20	0.00	0.00	37075.00
1902	573.20	0.00	0.00	43726.00
1903	561.80	0.00	0.00	43925.00
1904	540.40	0.00	0.00	43005.00
1905	528.50	0.00	0.00	46450.00
1906	456.90	0.00	0.00	49031.00
1907	338.80	0.00	0.00	37044.00
1908	361.60	0.00	0.00	28670.00
1909	422.00	0.00	0.00	33857.00
1910	370.00	0.00	0.00	33490.00
1911	431.00	0.00	0.00	49035.00
1912	363.30	0.00	0.00	46593.00
1913	381.00	0.00	0.00	45583.00

Comment 1852 TIN ORE; 1862 TWO RETURNS AGGREGATED; 1866 SHOULD PRODUCTION BE 194 TONS?

Arsenic	Ore(tons)	Metal(tons)	Value(£)
1881	20.00	0.00	50.00
1882	20.00	0.00	80.00
1883	11.30	0.00	43.00
1884	110.00	0.00	467.00
1885	207.80	0.00	861.00
1886	142.80	0.00	625.00
1887	144.00	0.00	773.00
1888	146.00	0.00	762.00
1889	173.00	0.00	935.00
1890	75.00	0.00	553.00
1891	186.00	0.00	1267.00
1892	246.00	0.00	1143.00
1893	189.00	0.00	1083.00
1894	192.00	0.00	1243.00
1895	216.00	0.00	1299.00
1896	250.00	0.00	1609.00
1897	231.00	0.00	2019.00
1898	132.00	0.00	434.00
1899	200.00	0.00	912.00
1900	216.00	0.00	1579.00
1904	105.00	0.00	368.00
1905	650.00	0.00	1787.00
1906	234.00	0.00	2267.00
1907	196.00	0.00	4222.00
1908	106.00	0.00	512.00
1909	100.00	0.00	569.00
1910	99.00	0.00	586.00
1911	163.00	0.00	522.00
1912	168.80	0.00	563.00

Arsenic   Ore(tons) Metal(tons)   Value(£)
1913   122.00   0.00   555.00
Comment 1884–1885 SOOT; 1886 CRUDE
Ownership: 1875–1890 LEVANT MINING CO.; 1891–1893 LEVANT MINING CO.
C.B.; 1894–1913 LEVANT MINING CO.
Management: Manager 1859 JAS.EVANS; 1860–1868 JOHN NANKERRIS; 1869–1870
JAS.NICHOLLS; 1871–1874 HY.BOYNS; 1879–1880 HY.TREZISE; 1881
JAS.NEWTON
Chief Agent 1859 RICH.WHITE; 1860 JOHN WHITE JOH.MARTIN
JAS.EVANS; 1861 JOH.MARTIN JAS.EVANS; 1862–1864 JOH.MARTIN
HY.TREZISE JAS.EVANS; 1865–1868 JAS.NICHOLLS HY.TREZISE
JAS.EVANS; 1869 JOHN NANKERRIS HY.TREZISE JAS.EVANS; 1870
JOHN NANKERRIS JAS.EVANS; 1871–1873 JAS.THOMAS; 1874
JAS.THOMAS NAT.WHITE; 1875–1878 HENRY TREZISE JAS.THOMAS
NAT.WHITE; 1879–1880 JAS.THOMAS NAT.WHITE; 1881 HY.NANKERRIS
NAT.WHITE; 1891–1908 RICH.WHITE; 1909–1913 BEN.NICHOLAS
Secretary 1860–1868 HY.BORROW (P); 1869–1870 L.DAUBUZ;
1871–1874 RICH.WHITE; 1875–1881 RICH.WHITE (S); 1882–1890
RICH.WHITE; 1891–1896 RICH.WHITE (P); 1897–1902 RICH.WHITE

Employment:	Underground	Surface	Total
1878	192	102	294
1879	192	87	279
1880	205	94	299
1881	212	96	308
1882	221	105	326
1883	246	116	362
1884	244	114	358
1885	279	127	406
1886	296	134	430
1887	315	133	448
1888	355	150	505
1889	350	159	509
1890	354	159	513
1891	365	154	519
1892	332	157	489
1893	339	162	501
1894	383	179	562
1895	412	192	604
1896	401	201	602
1897	407	202	609
1898	409	198	607
1899	427	203	630
1900	479	213	692
1901	498	226	724
1902	509	201	710
1903	484	201	685
1904	479	200	679
1905	435	199	634
1906	359	188	547
1907	320	181	501
1908	352	171	523
1909	322	172	494
1910	292	165	457
1911	305	171	476
1912	302	172	474

	Underground	Surface	Total
1913	272	167	439

LEVANT,NORTH   ST.JUST   SW 376341 0722

Production: Copper

	Ore(tons)	Metal(tons)	Value(£)
1875	12.00	1.80	141.60
1876	22.00	2.40	156.70
1877	10.70	1.20	73.90
1879	15.00	1.90	93.40
1886	1.50	0.00	5.00
1895	10.00	0.00	17.00

Comment 1875–1876 (C); 1895 NORTH LEVANT WHEAL GEEVOR

Tin

	Black(tons)	Stuff(tons)	Tin(tons)	Value(£)
1854	33.50	0.00	0.00	2318.80
1857	19.70	0.00	0.00	1471.70
1858	57.70	0.00	0.00	3737.80
1859	50.10	0.00	0.00	3614.00
1860	44.40	0.00	0.00	3319.60
1861	43.10	0.00	0.00	3125.90
1862	53.80	0.00	0.00	3541.10
1863	27.50	0.00	0.00	1791.80
1864	35.00	0.00	0.00	2208.00
1865	68.50	0.00	0.00	3904.20
1866	97.20	0.00	0.00	4954.30
1867	69.80	0.00	0.00	3682.60
1868	94.90	0.00	0.00	5324.70
1869	134.20	0.00	0.00	9615.40
1870	176.90	0.00	0.00	13268.30
1871	182.50	0.00	0.00	13585.30
1872	172.10	0.00	0.00	14942.10
1873	133.30	0.00	0.00	10245.80
1874	126.80	0.00	0.00	703.30
1875	168.70	0.00	0.00	8566.50
1876	199.10	0.00	0.00	8670.60
1877	153.30	0.00	0.00	6582.00
1878	82.00	0.00	0.00	2842.90
1879	57.00	0.00	0.00	2278.00
1880	88.60	0.00	0.00	4483.20
1881	63.90	0.00	43.10	3377.40
1882	98.10	0.00	64.70	5880.00
1883	73.00	0.00	0.00	3744.00
1884	72.20	0.00	0.00	3204.00
1885	83.80	0.00	0.00	3996.00
1886	100.50	0.00	0.00	5715.00
1887	136.00	0.00	0.00	8899.00
1888	130.00	0.00	0.00	8557.00
1889	93.00	0.00	0.00	5112.00
1890	52.10	0.00	0.00	2892.00
1891	21.50	0.00	0.00	1098.00
1892–1893 No detailed return				
1894	32.90	0.00	0.00	1317.00
1895	8.00	0.00	0.00	300.00
1896	15.50	0.00	0.00	540.00

Tin	Black(tons)	Stuff(tons)	Tin(tons)	Value(£)
1897	20.90	0.00	0.00	770.00
1898	19.00	0.00	0.00	779.00
1899	13.00	0.00	0.00	841.00
1900	11.30	0.00	0.00	805.00
1901	13.80	0.00	0.00	896.00
1902	8.40	0.00	0.00	550.00
1903	2.70	0.00	0.00	175.00
1904	4.30	0.00	0.00	273.00
1906	No detailed return			
1907	52.20	0.00	0.00	5129.00
1908	39.00	0.00	0.00	2757.00
1909	100.70	0.00	0.00	7468.00
1910	98.80	0.00	0.00	8510.00
1911	116.10	0.00	0.00	12631.00
1912	95.90	0.00	0.00	11795.00
1913	291.40	0.00	0.00	34795.00

Comment 1862 TWO RETURNS AGGREGATED; 1892–1893 SEE GEEVOR; 1894–1898 OTHERWISE KNOWN AS GEEVOR; 1899–1904 NORTH LEVANT (GEEVOR); 1906 SEE GEEVOR; 1907–1913 NORTH LEVANT AND GEEVOR

Arsenic	Ore(tons)	Metal(tons)	Value(£)
1885	2.00	0.00	2.00

Comment 1885 CRUDE

Ownership: 1876–1891 NORTH LEVANT MINING CO.; 1892–1904 WHEAL GEEVOR CO.; 1906–1908 WEST AUSTRALIAN GOLD FIELDS CO.LTD.; 1909–1910 NORTH LEVANT & GEEVOR LTD.; 1911–1913 GEEVOR TIN MINES LTD. Comment 1891 ABANDONED; 1892–1904 GEEVOR PART; 1906–1913 INC.GEEVOR

Management: Manager 1859 RICH.JAMES; 1869–1881 JAS.BENNETTS; 1894–1902 J.DOIDGE
Chief Agent 1859 JAS.BENNETTS; 1860–1864 JAS.BENNETTS JAS.THOMAS; 1865–1866 JAS.BENNETTS; 1867–1868 JAS.BENNETTS JAS.THOMAS; 1869 JAS.THOMAS; 1870 JAS.THOMAS HY.EDDY; 1871 HY.EDDY JAS.THOMAS; 1872–1873 HY.EDDY; 1874–1879 HY.EDDY JAS.M.ARTHUR; 1880–1881 W.H.HUMPHRIES JAS.M.ARTHUR; 1883–1884 R.E.BOYNS; 1891–1904 RICH.WHITE; 1906–1909 W.T.MAY; 1910–1913 W.C.WILLIAMS
Secretary 1859–1861 GEO.HIGGS (P); 1862–1872 SML.HIGGS & SON; 1873–1882 RICH.BOYNS; 1885–1890 RICH.WHITE; 1894–1902 RICH.WHITE

Employment:
	Underground	Surface	Total
1878	69	49	118
1879	46	35	81
1880	55	41	96
1881	51	43	94
1882	54	42	96
1883	48	44	92
1884	58	37	95
1885	71	51	122
1886	81	60	141
1887	90	64	154
1888	106	70	176
1889	56	43	99
1890	39	34	73
1891		12	12

	Underground	Surface	Total
1892	11	5	16
1893	21	11	32
1894	12	5	17
1895	9	2	11
1896–1897	10	4	14
1898	10	5	15
1899	10	2	12
1900	8	2	10
1901–1902	9	2	11
1903	2	2	4
1904		2	2
1906	109	47	156
1907	123	59	182
1908	28	49	77
1909	61	45	106
1910	113	48	161
1911	127	63	190
1912	139	107	246
1913	147	102	249

Comment 1906–1913 INC.GEEVOR

LEWIS MINES   ST.ERTH   SW 574333 0723

Production: Copper No detailed return

Tin	Black(tons)	Stuff(tons)	Tin(tons)	Value(£)
1852	204.50	0.00	0.00	0.00
1853	262.40	0.00	0.00	17816.80
1854	199.40	0.00	0.00	13754.00
1855	65.00	0.00	0.00	4146.30
1856	76.50	0.00	0.00	5483.20
1857	88.60	0.00	0.00	6933.60
1858	133.50	0.00	0.00	8555.90
1859	92.10	0.00	0.00	6992.40
1860	136.60	0.00	0.00	11298.80
1861	16.40	0.00	0.00	1368.20

Comment 1852 LEWIS TIN ORE; 1853 LEWIS
Ownership: Comment 1861–1865 SUSPENDED
Management: Manager 1859–1860 WM.BISHOP
             Chief Agent 1859 W.H.MARTIN; 1860 W.TREDINNICK
             Secretary 1859–1861 JOHN LITTLE (P)

LEWIS,WEST   ST.ERTH   SW 581337 0724

Production: Copper No detailed return
            Tin No detailed return
Ownership: Comment 1861–1865 SUSPENDED
Management: Manager 1860–1861 WM.BISHOP
             Secretary 1860–1865 WM.VAWDREY (P)

LIDCOTT   EGLOSKERRY   SX 241850 0744

Production: Manganese   Ore(tons) Metal(tons)   Value(£)
            1875   16.00   0.00   80.00
            1877   37.00   0.00   74.50
            1878   60.00   0.00   120.00
            1879   45.00   0.00   92.50
            1881   150.00   0.00   187.00
            Comment 1875 LUDCOTT. INC.WEST DOWN END; 1877-1879 LIDCOLT
            INC.WEST DOWN END; 1881 LYDCOCK INC.WEST DOWN END
Ownership: 1877-1881 J.C.PRAWLE; 1882 J.C.PRAWLE JOS.ROBINSON
            Comment 1875-1877 LUDCOTT AND WEST DOWN END; 1878 INC.WEST
            DOWN END; 1879-1881 LYDCOTT AND WEST DOWN END; 1882 LYDCOTT
            AND WEST DOWN END NOT WORKED
Management: Manager 1875-1878 J.C.PRAWLE
            Chief Agent 1874-1878 H.J.NICHOLLS; 1879-1881 WM.LITTLEJOHN;
            1882 JOS.ROBINSON
            Secretary 1875-1881 JOS.ROBINSON
Employment: Underground   Surface   Total
            1878   8   3   11
            1879   2   4   6
            1880   3   4   7
            1881   8   7   15
            Comment 1878-1881 INC.WEST DOWN

LILLY OF THE VALLEY   0725

Production: Tin   Black(tons) Stuff(tons)   Tin(tons)   Value(£)
            1868   25.30   0.00   0.00   1469.90
            1869   4.60   0.00   0.00   327.20
            1870   1.20   0.00   0.00   89.70
            1871   7.10   0.00   0.00   569.40
            1872   0.10   0.00   0.00   0.60
            Comment 1869-1872 LILY OF THE VALLEY

LIVINGSTONE CONSOLS   ST.AGNES   SW 718503 0726

Production: Tin No detailed return
Ownership: 1878-1879 REYNOLDS AND CO.
            Comment 1879 (NEW KITTY) NOT WORKED
Management: Manager 1878-1879 WM.VIVIAN
Employment: Underground   Surface   Total
            1878   2   2

LOCKE   LUDGVAN   0727

Production: Tin   Black(tons) Stuff(tons)   Tin(tons)   Value(£)
            1894   2.10   0.00   0.00   83.00
            1895-1896 No detailed return
Ownership: 1893-1894 WHEAL LOCKE MINING CO.
            Comment 1894 SUSPENDED
Management: Chief Agent 1893-1894 CHRIS.EDWARDS
Employment: Underground   Surface   Total
            1893   6   9   15

                                288

LOCKE   LUDGVAN   Continued

                  Underground   Surface   Total
          1894   6   5   11

LOMAX   SITHNEY   SW 638247 0728

Production: Lead & Silver No detailed return
Ownership: 1880 WALT.THOMPSON & SON; 1881 LOMAX SILVER LEAD MINING CO.
            Comment 1880-1881 FORMERLY ROSE(SITHNEY)
Management: Manager 1881 WM.ARGALL
            Chief Agent 1880 WM.ARGALL
Employment: Underground   Surface   Total
          1881   3   3

LOO POOL STREAM   0729

Production: Tin   Black(tons) Stuff(tons) Tin(tons)   Value(£)
          1871   4.60   0.00   0.00   292.90
          1872   1.50   0.00   0.00   103.50
Management: Chief Agent 1872 JOHN OLD

LORD KIMBERLY   FALMOUTH   SW 803313 0730

Production: Copper No detailed return
Ownership: 1880-1881 LORD KIMBERLY EXPLORATION MINING CO.
Management: Manager 1880-1881 JOS.MICHELL

LOUISA   REDRUTH   SW 706428 0731

Production: Copper No detailed return
            Tin No detailed return
Ownership: Comment 1865 SUSPENDED
Management: Chief Agent 1860-1864 JOHN DELBRIDGE
            Secretary 1860 ED.KING (P); 1861-1865 W.H.PAULL (P)

LOVELL   WENDRON   SW 705304 0732

Production: Copper   Ore(tons) Metal(tons)   Value(£)
          1865   3.00   0.10   5.40
          Comment 1865 (C)
          Tin   Black(tons) Stuff(tons) Tin(tons)   Value(£)
          1852   130.20   0.00   0.00   0.00
          1854   132.20   0.00   0.00   9601.90
          1855   110.80   0.00   0.00   7616.70
          1856   85.40   0.00   0.00   6677.50
          1857   74.30   0.00   0.00   6264.60
          1858   64.40   0.00   0.00   4293.20
          1860   76.70   0.00   0.00   6217.50
          1861   37.60   0.00   0.00   2824.10
          1862   44.90   0.00   0.00   2945.70
          1863   5.00   0.00   0.00   335.20

                              289

Tin	Black(tons)	Stuff(tons)	Tin(tons)	Value(£)
1864	5.60	0.00	0.00	354.10
1865	15.50	0.00	0.00	799.40
1870	0.00	0.00	0.00	32.30
1871	8.10	0.00	0.00	656.50
1873	49.50	0.00	0.00	3766.60
1874	57.10	0.00	0.00	3232.70
1875	44.90	0.00	0.00	2349.60
1876	60.00	0.00	0.00	2665.90
1877	64.90	0.00	0.00	2629.50
1878	20.20	0.00	0.00	739.70
1879	20.20	0.00	0.00	773.00
1880	19.60	0.00	0.00	1104.20
1881	19.30	0.00	12.80	1094.60
1882	8.90	0.00	6.00	476.00
1883	5.50	5.00	0.00	302.00
1884	11.60	0.00	0.00	531.00
1885	7.90	0.00	0.00	386.00
1886	12.40	0.00	0.00	696.00
1887	4.30	0.00	0.00	252.00
1888	2.00	0.00	0.00	113.00
1889	1.20	0.00	0.00	57.00
1890	66.80	0.00	0.00	4024.00
1891	73.80	0.00	0.00	4320.00
1892	8.30	0.00	0.00	470.00
1893	11.90	0.00	0.00	619.00
1894	5.40	0.00	0.00	223.00
1895	1.80	0.00	0.00	66.00
1896	1.80	2.00	0.00	80.00

Comment 1852 TIN ORE; 1862 TWO RETURNS AGGREGATED; 1863-1864
VALUE INC.TIN STUFF; 1870 TIN STUFF; 1873-1881 THE LOVELL;
1883 TWO RETURNS AGGREGATED
Ownership: 1876-1890 THE LOVELL MINING CO.; 1891-1893 THE LOVELL MINING
             CO. C.B.; 1894-1896 THE LOVELL MINING CO.; 1912 WENDRON
             CORNWALL TIN MINES LTD.; 1913 NIGERIAN FINANCE SYNDICATE
             LTD.
             Comment 1865 SUSPENDED; 1872-1896 THE LOVELL; 1912-1913
             REOPENING ADIT
Management: Manager 1859-1862 HY.JAMES; 1863-1864 JOS.PHILLIPS; 1868-1873
             J.NANCARROW; 1874-1881 JOS.PRISK
             Chief Agent 1861-1862 JOS.PHILLIPS; 1869 M.KEMPTHORNE;
             1870-1873 ED.KEMPTHORNE; 1874-1875 ED.KEMPTHORNE J.NANCARROW;
             1876-1877 ED.KEMPTHORNE; 1882-1896 JOS.PRISK
             Secretary 1859-1862 WM.CARNE (P); 1863-1865 T.P.TYACKE;
             1869-1875 BARTLETT & CHAPMAN; 1876-1881 GREN.SHARP; 1891-1896
             JOHN BIDDER (S); 1913 J.S.ALLEN (S)
Employment: Underground   Surface   Total
             1878   30   17   47
             1879   25   14   39
             1880   31   11   42
             1881   33   13   46
             1882   26   15   41
             1883   16   10   26
             1884   24   8   32
             1885   24   7   31

LOVELL   WENDRON   Continued

	Underground	Surface	Total
1886	30	7	37
1887	20	4	24
1888	14	3	17
1889	10	3	13
1890	28	18	46
1891	35	16	51
1892	22	9	31
1893	18	7	25
1894	12	5	17
1895	9	4	13
1896	6	2	8
1913	5	5	10

Comment 1912 SEE WEST VOR

LOVELL CONSOLS   WENDRON   0733

Production: Tin   Black(tons) Stuff(tons)   Tin(tons)   Value(£)
            1870   No detailed return
            1871   0.50   0.00   0.00   38.40
            1872   68.80   0.00   0.00   5861.90
Management: Manager 1868-1871 J.NANCARROW
            Chief Agent 1870-1871 ED.KEMPTHORNE
            Secretary 1870-1871 BARTLETT & CHAPMAN

LOVELL,EAST   WENDRON   SW 704318 0734

Production: Tin	Black(tons)	Stuff(tons)	Tin(tons)	Value(£)
1859	154.80	0.00	0.00	11853.20
1860	55.80	0.00	0.00	4668.10
1861	39.70	0.00	0.00	3033.40
1862	4.10	0.00	0.00	242.30
1863	45.20	0.00	0.00	2873.00
1864	70.10	0.00	0.00	4269.40
1865	72.80	0.00	0.00	4063.80
1866	97.60	0.00	0.00	4790.80
1867	139.70	0.00	0.00	7450.00
1868	93.80	0.00	0.00	5328.40
1869	189.90	0.00	0.00	14230.00
1870	215.10	0.00	0.00	16662.70
1871	172.00	0.00	0.00	13897.00
1872	32.60	0.00	0.00	2766.10
1873	78.80	0.00	0.00	5896.50
1874	140.00	0.00	0.00	8193.60
1875	71.70	0.00	0.00	3616.60
1876	30.40	0.00	0.00	1284.40
1877	23.50	0.00	0.00	893.80
1878	28.50	0.00	0.00	941.90
1879	4.00	0.00	0.00	160.00
1880	5.80	0.00	0.00	276.60
1881	14.10	0.00	9.90	750.50
1882	21.30	0.00	13.90	1111.00
1883	13.60	0.00	0.00	745.00

CM-L*   291

LOVELL,EAST   WENDRON   Continued

Tin	Black(tons)	Stuff(tons)	Tin(tons)	Value(£)
1884	7.20	0.00	0.00	306.00
1885	5.00	0.00	0.00	222.00
1886	4.50	0.00	0.00	224.00
1887	4.60	0.00	0.00	276.00
1888	4.60	0.00	0.00	284.00
1889	2.00	0.00	0.00	93.00
1890	1.90	0.00	0.00	103.00
1891	2.00	0.00	0.00	73.00

Ownership: 1876 EAST WHEAL LOVELL ADVENTURERS; 1877-1886 EAST WHEAL
LOVELL MINING CO.; 1887-1891 EAST WHEAL LOVELL MINE
ADVENTURERS
Comment 1891 STOPPED
Management: Manager 1859-1865 JOHN BURGAN; 1871-1879 RICH.QUENTRALL;
1880-1881 RICH.QUENTRALL & SON
Chief Agent 1859-1862 J.BRYANT; 1865-1868 RICH.QUENTRALL;
1869-1870 RICH.QUENTRALL NICH.PETERS; 1871-1881 NICH.PETERS;
1882-1884 RICH.QUENTRALL & SONS; 1885-1888 THOS.QUENTRALL
Secretary 1859-1870 HY.ROGERS (P); 1871-1876 HY.ROGERS;
1877-1881 HY.ROGERS (S); 1889-1890 HY.ROGERS; 1891 HY.ROGERS
(P)

Employment:		Underground	Surface	Total
	1878	31	16	47
	1879	14	11	25
	1880	34	9	43
	1881	45	10	55
	1882	45	13	58
	1883	30	7	37
	1884	18	10	28
	1885-1886	11	3	14
	1887	10	2	12
	1888	9	1	10
	1889	7	3	10
	1891	4		4

LOVELL,GREAT   WENDRON   SW 690305 0735

Production: Tin	Black(tons)	Stuff(tons)	Tin(tons)	Value(£)
1871	4.70	0.00	0.00	343.30
1872	3.10	0.00	0.00	239.30
1873	3.60	0.00	0.00	256.30
1874	9.20	0.00	0.00	522.60
1875	15.50	0.00	0.00	815.50
1876	0.90	0.00	0.00	35.00

Management: Manager 1873-1877 JOS.PRISK
Chief Agent 1870-1871 C.BAWDEN; 1872 THOS.BRAY
Secretary 1870-1871 C.BAWDEN; 1872-1877 GREN.SHARP

LOVELL,NEW   WENDRON   SW 699308 0736

Production: Copper	Ore(tons)	Metal(tons)	Value(£)
1872	51.00	2.30	149.30
1873	8.00	0.30	18.60

LOVELL,NEW   WENDRON   Continued

        Comment 1872-1873 (C)
        Tin   Black(tons) Stuff(tons)   Tin(tons)   Value(£)
        1867   1.60   0.00   0.00   87.00
        1868   22.10   0.00   0.00   1515.10
        1870   49.10   0.00   0.00   3556.70
        1871   79.90   0.00   0.00   6376.00
        1872   38.90   0.00   0.00   3886.00
        1873   23.30   0.00   0.00   1723.30
        1874   10.80   0.00   0.00   624.70
Ownership: Comment 1873 STOPPED IN 1873 OR 1874; 1874 STOPPED IN 1874;
            1875 IN LIQUIDATION
Management: Manager 1868-1869 C.BAWDEN; 1870-1873 JOS.PRISK
            Chief Agent 1863-1866 J.PRISK; 1867 J.PRISK C.BAWDEN; 1869
            JOS.PRISK; 1874-1875 JOS.PRISK
            Secretary 1863-1866 G.HILL; 1867-1869 COMMITTEE; 1870-1871
            T.P.TYACKE; 1872-1873 THOS.W.FIELD

LOVELL,NEW EAST   CONSTANTINE   SW 708312 0737

Production: Tin No detailed return
Ownership: Comment 1873 STOPPED IN 1873 OR 1874
Management: Manager 1869 CHAS.BAWDEN
            Chief Agent 1872-1873 CHAS.BAWDEN
            Secretary 1869 ED.COOK JNR.

LOVELL,NORTH   WENDRON   SW 699329 0738

Production: Tin   Black(tons) Stuff(tons)   Tin(tons)   Value(£)
        1869   0.00   2.80   0.00   6.40
        1873   3.20   0.00   0.00   275.40
        1874   2.00   0.00   0.00   113.80
Ownership: 1880-1881 NORTH LOVELL MINING CO.
Management: Manager 1869-1871 JAS.ROWE; 1880-1881 JOS.PRISK
            Chief Agent 1869-1871 JOS.PRISK; 1873 JAS.ROWE
            Secretary 1869-1871 THOR.WOODWARD; 1873 GREN.SHARP

LOVELL,SOUTH   WENDRON   SW 705301 0739

Production: Tin No detailed return
Management: Chief Agent 1862 J.PRISK
            Secretary 1862 W.SLEEMAN

LOVELL,WEST   WENDRON   SW 681280 0740

Production: Tin No detailed return
Ownership: Comment 1859-1865 SEE EAST TREVENNEN

LOVELL,WEST WHEAL   WENDRON   SW 681280 0741

Production: Tin No detailed return
Ownership: Comment 1863-1865 SUSPENDED
Management: Chief Agent 1860-1862 H.HARRIS
            Secretary 1860-1861 T.CLARKE (P); 1862-1865 W.PAGE CARDOZA
            (P)

LUCY   PHILLACK   SW 558389 0742

Production: Tin   Black(tons) Stuff(tons)   Tin(tons)   Value(£)
            1872   10.00   0.00   0.00   889.50
            1873   3.20   6.50   0.00   275.70
            1874   0.00   10.00   0.00   37.50
            1893   0.00   15.00   0.00   45.00
            1895   11.60   0.00   0.00   459.00
            1896   No detailed return
Ownership: 1893-1896 WM.C.VIVIAN (P)
            Comment 1873 STOPPED IN 1873 OR 1874; 1893-1895 BATTS ADIT;
            1896 BATTS ADIT SUSPENDED
Management: Manager 1872-1873 WM.HARRIS
            Chief Agent 1874 WM.HARRIS; 1893-1896 HOLMAN
            Secretary 1872 G.EUSTICE JNR.; 1873 JEHU.HITCHINGS
Employment: Underground   Surface   Total
            1893   3   1   4
            1894   9   5   14
            1895   16   20   36
            1896   6   6

LUCY,WEST   PHILLACK   SW 546380 0743

Production: Tin   Black(tons) Stuff(tons)   Tin(tons)   Value(£)
            1873   0.00   10.50   0.00   24.80
Ownership: Comment 1873 STOPPED IN 1873 OR 1874
Management: Manager 1872-1873 WM.HARRIS
            Secretary 1872 G.EUSTICE JNR.; 1873 JEHU.HITCHINGS; 1874
            GEO.EUSTICE

LUDCOTT   ST.IVE   SX 298661 0745

Production: Lead & Silver   Ore(tons) Lead(tons) Silver(ozs)   Value(£)
            1856   194.00   135.00   4995.00   0.00
            1857   211.00   142.00   5396.00   0.00
            1858   357.00   241.00   8767.00   0.00
            1859   528.10   353.70   12708.00   0.00
            1860   589.70   390.00   14040.00   0.00
            1861   493.00   368.00   24068.00   0.00
            1862   735.30   551.20   35830.00   0.00
            1863   749.60   510.00   33150.00   0.00
            1864   578.60   399.50   25970.00   0.00
            1865   472.10   302.00   19630.00   0.00
            1866   68.70   43.50   2827.00   0.00
            Comment 1863-1866 INC.WREY

                              294

LUDCOTT   ST.IVE   Continued

          Silver   Ore(tons) Metal(tons)   Value(£)
          1861   12.10   0.00   1300.40
Ownership: Comment 1862 LUDCOTT & WREY UNITED; 1863-1864 LUDCOTT & WREY
          CONSOLS; 1865 LUDCOTT & WREY CONSOLS.SUSPENDED
Management: Manager 1859-1864 ROBT.KNAPP
          Chief Agent 1859 ROBT.KNAPP; 1860 JOSH HUSBAND; 1861-1864
          JOSH.HUSBAND WM.KNAPP
          Secretary 1859 E.A.CROUCH (P); 1860-1865 JOHN TAYLOR (P)

LUSKEY   NORTH HILL   SX 256776 0746

Production: Lead No detailed return
Ownership: 1881-1884 WHEAL LUSKEY MINING CO.
Management: Manager 1880-1881 E.SKEWIS
          Chief Agent 1880-1881 JOHN G.MAY; 1883-1884 WM.SKEWIS
          Secretary 1880-1881 E.NICHOLLS (P); 1882 E.NICHOLLS
Employment: Underground   Surface   Total
          1881   6   6
          1883   6   6
          1884   2   2

LYDIA   PORTH TOWAN   SW 694475 0747

Production: Copper   Ore(tons) Metal(tons)   Value(£)
          1845   2267.00   136.70   9652.70
          1846   1460.00   87.00   5839.30
          1847   161.00   9.90   645.80
          Comment 1845-1847 (C)

MAGDALEN   POLSANOOTH   SW 765377 0748

Production: Tin   Black(tons) Stuff(tons)   Tin(tons)   Value(£)
          1913   2.50   0.00   0.00   193.00
Ownership: 1913 F.W.MARKHAM A.C.CLARKE
          Comment 1913 DRIVING ADIT
Employment: Underground   Surface   Total
          1913   6   9   15

MAGNETIC   ROCHE   SW 988597 0749

Production: Iron   Ore(tons)   Iron(%)   Value(£)
          1873   200.00   0.00   150.00
          1874   No detailed return
          Comment 1873 MO.; 1874 SEE MID-CORNWALL MINES
Ownership: 1873 MID-CORNWALL MINING CO.LTD.
          Comment 1873 FORMERLY TOWER CONSOLS
Management: Manager 1873 DAVID COCK

                                   295

MAID   ST.DAY   SW 742423 0750

Production: Tin   Black(tons) Stuff(tons)   Tin(tons)   Value(£)
           1888–1889 No detailed return
           Comment 1888–1889 SEE ST. DAY MANOR
           Arsenic   Ore(tons) Metal(tons)   Value(£)
           1893   No detailed return
           Comment 1893 SEE POLDICE
           Arsenic Pyrite   Ore(tons)   Value(£)
           1893   No detailed return
           Comment 1893 SEE POLDICE
Ownership: Comment 1888–1896 SEE POLDICE
Employment: Underground   Surface   Total
           1888–1896
           Comment 1888–1896 SEE POLDICE

MAIDEN   ST.DAY   SW 742423 0751

Production: Copper   Ore(tons) Metal(tons)   Value(£)

Year	Ore(tons)	Metal(tons)	Value(£)
1845	484.00	32.40	2333.20
1846	361.00	21.60	1456.50
1847	326.00	20.10	1371.00
1848	194.00	12.20	702.60
1849	171.00	11.20	742.20
1850	168.00	11.30	763.20
1851	134.00	10.70	728.70
1852	112.00	8.80	748.20

           Comment 1845–1852 (C)

MARGARET   LELANT   SW 505362 0752

Production: Tin   Black(tons) Stuff(tons)   Tin(tons)   Value(£)

Year	Black(tons)	Stuff(tons)	Tin(tons)	Value(£)
1855	207.80	0.00	0.00	11986.30
1856	245.20	0.00	0.00	15854.90
1857	253.10	0.00	0.00	18485.60
1858	286.90	0.00	0.00	17242.70
1859	362.90	0.00	0.00	25182.10
1860	219.90	0.00	0.00	16882.40
1861	103.70	0.00	0.00	7519.00
1862	302.70	0.00	0.00	18547.80
1863	178.70	0.00	0.00	11345.70
1864	87.40	0.00	0.00	5672.10
1865	113.20	0.00	0.00	6102.10
1866	148.20	0.00	0.00	6916.00
1867	74.70	0.00	0.00	3814.60
1868	99.30	0.00	0.00	5243.00
1869	105.50	0.00	0.00	7107.80
1870	118.90	0.00	0.00	8239.90
1871	130.60	0.00	0.00	10135.00
1872	104.20	0.00	0.00	8545.10
1873	112.30	0.00	0.00	8174.00
1874	106.50	0.00	0.00	5409.30
1875	16.00	0.00	0.00	840.00

           1877   No detailed return
           Comment 1862 TWO RETURNS AGGREGATED; 1877 SEE SISTERS

MARGARET   LELANT   Continued

Ownership: Comment 1875–1881 SEE SISTERS
Management: Manager 1859–1871 THOS.TREWEEKE; 1872 THOS.TREWEEKE JNR.;
            1873–1874 THOS.MICHELL JNR.
            Chief Agent 1859 JOHN UREN; 1860–1866 JOHN UREN JOHN
            WILLIAMS; 1867–1872 THOS.MICHELL JNR.
            Secretary 1859–1871 THOS.TREWEEKE (P); 1872–1874
            THOS.TREWEEKE & SON

MARGARET ANN   ST.JUST   0753

Production: Tin   Black(tons) Stuff(tons)   Tin(tons)   Value(£)
            1874   1.10   0.00   0.00   48.00
Ownership: Comment 1874 SUSPENDED
Management: Chief Agent 1873 JOHN HOLLOW
            Secretary 1873–1874 EDW.S.BOYNS

MARGARET,EAST   LELANT   SW 525388 0754

Production: Copper No detailed return
            Tin   Black(tons) Stuff(tons)   Tin(tons)   Value(£)
            1852   7.30   0.00   0.00   0.00
            1853   63.00   0.00   0.00   4076.90
            1854   23.00   0.00   0.00   1676.90
            1855   92.20   0.00   0.00   5450.10
            1856   84.90   0.00   0.00   5534.90
            1857   80.00   0.00   0.00   5797.30
            1858   67.60   0.00   0.00   4054.00
            1859   60.00   0.00   0.00   4072.00
            1860   21.50   0.00   0.00   1665.90
            1861   58.40   0.00   0.00   4454.70
            1862   60.30   0.00   0.00   3637.60
            1863   63.40   0.00   0.00   3925.20
            1864   65.50   0.00   0.00   3726.60
            1865   41.50   0.00   0.00   2033.30
            1902   1.80   0.00   0.00   110.00
            Comment 1852 TIN ORE; 1862 TWO RETURNS AGGREGATED; 1902 EAST
            MARGARET PRAED ESTATE
Ownership: 1901–1902 EAST WHEAL MARGARET CO.
Management: Manager 1859–1861 THOS.TREWEEKE; 1864–1865 BEN.MARTIN
            Chief Agent 1859–1861 BEN.MICHELL; 1862 RICH.JAMES
            BEN.MICHELL; 1863 RICH.JAMES WM.WILLIAMS F.H.BIRCH; 1864–1865
            F.H.BIRCH
            Secretary 1859–1861 THOS.TREWEEKE JNR(P); 1862 SML.HIGGS;
            1863–1865 SML.HIGGS & SON; 1901–1902 WM.THOMAS
Employment: Underground   Surface   Total
            1902   8   2   10

MARGARET,WEST   LELANT   SW 397275 0755

Production: Tin   Black(tons) Stuff(tons)   Tin(tons)   Value(£)
            1869   0.00   0.00   0.00   18.30
            1870   0.00   0.00   0.00   165.60

                                297

MARGARET,WEST   LELANT   Continued

      Tin   Black(tons) Stuff(tons)   Tin(tons)   Value(£)
      1871   0.00   0.00   0.00   69.10
      Comment 1869-1871 TIN STUFF
Ownership: Comment 1862-1865 SUSPENDED
Management: Manager 1859 WM.HOLLOW
             Chief Agent 1859 THOS.UREN; 1860-1861 THOS.UREN WM.WHITE;
             1869-1871 JAS.EVANS RICH.STEPHENS
             Secretary 1859-1861 JAS.HOLLOW (P); 1869-1871 CHRIS.STEPHENS

MARGERY   ST.IVES   SW 525394 0756

Production: Lead & Silver   Ore(tons) Lead(tons) Silver(ozs)   Value(£)
            1861   2.00   1.20   20.00   0.00
            1866   2.40   1.80   0.00
            Copper   Ore(tons) Metal(tons)   Value(£)
            1855   141.00   9.30   960.20
            1856   905.00   61.50   5793.50
            1857   848.00   44.20   4475.40
            1858   1385.00   79.20   7011.30
            1859   1658.00   93.50   8732.40
            1860   1300.00   75.40   6931.80
            1861   1094.00   68.10   6155.50
            1862   1179.00   66.10   5497.70
            1863   998.00   50.50   3831.00
            1864   1932.00   96.90   7921.00
            1865   1907.00   86.50   6611.30
            1866   1242.00   55.50   3713.90
            1867   776.00   40.00   2727.20
            1868   988.00   43.30   2707.40
            1870   12.00   0.60   33.60
            Comment 1855-1868 (C); 1870 (C)
            Tin   Black(tons) Stuff(tons)   Tin(tons)   Value(£)
            1854   2.30   0.00   0.00   135.90
            1856   7.40   0.00   0.00   447.90
            1858   29.60   0.00   0.00   1643.50
            1859   14.60   0.00   0.00   882.50
            1860   18.20   0.00   0.00   1313.60
            1861   8.90   0.00   0.00   497.00
            1862   8.10   0.00   0.00   417.00
            1863   5.70   0.00   0.00   310.30
            1864   3.70   0.00   0.00   213.70
            1865   2.40   0.00   0.00   112.30
            1866   8.70   0.00   0.00   347.70
            1868   5.70   0.00   0.00   226.80
Ownership: Comment 1863-1865 WHEAL MARGERY (TRELYON)
Management: Manager 1859-1868 RICH.JAMES
             Chief Agent 1859 JOHN BENBOW; 1860 JOHN BENBOW WM.ROGERS;
             1861-1866 WM.ROGERS
             Secretary 1859-1862 SML.HIGGS (P); 1863-1865 SML.HIGGS & SON
             (P); 1866-1868 SML.HIGGS

MARIA   CROWAN   0757

Production: Tin   Black(tons) Stuff(tons)   Tin(tons)   Value(£)
            1874   0.00   128.40   0.00   91.70
            1875   0.00   24.50   0.00   0.00
            Arsenic Pyrite   Ore(tons)   Value(£)
            1876   6.40   5.00
            Comment 1876 ARSENICAL MUNDIC
Ownership: Comment 1873 STOPPED IN 1873 OR 1874
Management: Chief Agent 1874—1875 S.S.NOELL

MARIA,SOUTH   CALSTOCK   SX 413734 0758

Production: Copper No detailed return
Ownership: Comment 1865—1866 INC.LATCHLEY
Management: Manager 1865—1866 JAS.RICHARDS
            Secretary 1866 BARLOW & SMITH

MARKE VALLEY   LINKINHORNE   SX 280718 0759

Production: Copper   Ore(tons) Metal(tons)   Value(£)
            1845   165.00   7.80   521.90
            1846   853.00   41.40   2583.50
            1847   1016.00   49.90   3251.90
            1848   1209.00   61.00   3295.90
            1849   1359.00   66.40   4182.80
            1850   1612.00   75.80   4873.60
            1851   1701.00   91.00   5923.50
            1852   2021.00   93.30   7630.90
            1853   2466.00   92.10   8664.90
            1854   2330.00   80.50   7676.30
            1855   2318.00   101.60   10135.30
            1856   2316.00   89.30   7826.50
            1857   1884.00   73.20   6767.80
            1858   1377.00   66.40   5923.80
            1859   1533.00   81.10   7591.00
            1860   3114.00   173.20   15754.20
            1861   3983.00   228.50   20342.10
            1862   4821.00   243.70   19407.00
            1863   4912.00   217.60   15979.10
            1864   5020.00   207.70   16726.10
            1865   5174.00   228.80   17590.40
            1866   5041.00   256.20   16713.20
            1867   5240.00   283.70   20038.80
            1868   5430.00   343.00   22729.10
            1869   5884.00   358.00   22259.70
            1870   5896.00   369.10   21847.40
            1871   5926.00   317.10   19362.80
            1872   4427.00   221.70   16586.60
            1873   3945.00   222.80   13408.10
            1874   4283.00   239.50   14818.70
            1875   4243.00   233.70   16963.80
            1876   4357.00   245.90   16070.90
            1877   4822.20   276.40   15488.00
            1878   4379.90   270.00   12632.70

                               299

Copper	Ore(tons)	Metal(tons)	Value(£)
1879	3399.80	216.70	10532.00
1880	2693.90	135.10	7366.10
1881	2693.90	124.80	6900.30
1882	2379.20	133.40	8327.00
1883	1776.00	85.00	4683.00
1884	685.00	0.00	1495.00
1885	50.00	0.00	114.00
1886	4.50	0.00	62.00

Comment 1845—1876 (C)

Tin	Black(tons)	Stuff(tons)	Tin(tons)	Value(£)
1872	0.00	6.10	0.00	5.90
1873	2.10	0.00	0.00	147.90
1874	1.50	0.00	0.00	74.00
1881	0.40	0.00	0.30	22.00
1882	1.30	0.00	0.80	476.00
1883	5.70	0.00	0.00	297.00
1884	41.60	0.00	0.00	1831.00
1885	52.40	0.00	0.00	2432.00
1886	12.10	0.00	0.00	635.00
1887	58.80	0.00	0.00	3529.00
1888	83.00	0.00	0.00	4885.00
1889	102.30	0.00	0.00	4826.00
1890	27.20	0.00	0.00	1221.00

Comment 1882 VALUE FIGURE WRONG?

Ownership: 1876—1890 MARKE VALLEY MINING CO.
              Comment 1890 ABANDONED; 1907—1908 SEE EAST CARADON
Management: Manager 1859—1864 JAS.SECCOMBE; 1869—1871 JOHN TRUSCOTT;
              1872—1875 JAS.SECCOMBE; 1876—1881 WM.GEORGE
              Chief Agent 1859—1868 JOHN TRUSCOTT; 1869—1871 JAS.STANLAKE
              WM.THORNE (P); 1872 JAS.STANLAKE J.HARDING (S); 1873
              JAS.STANLAKE F.RENALS J.HARDING (P); 1874—1875 JAS.STANLAKE
              F.RENALS J.HARDING (S); 1876 JAS.STANLAKE; 1877—1881
              JAS.STANLAKE JAS.SECCOMBE; 1882—1890 GEORGE AND SECCOMBE
              Secretary 1859 JAS.SECCOMBE (P); 1860 JOHN HARDING (P);
              1861—1862 JAS.SECCOMBE (P); 1863—1868 JOHN HARDING; 1869—1871
              JOHN HARDING (S); 1872—1881 SML.SECCOMBE (P)

Employment:	Underground	Surface	Total
1878	124	106	230
1879	82	62	144
1880	91	69	160
1881	82	68	150
1882	94	64	158
1883	56	54	110
1884	31	34	65
1885	44	41	85
1886	19	46	65
1887	74	58	132
1888	71	62	133
1889	72	84	156
1890	33	37	70
1907—1908			

Comment 1907—1908 SEE EAST CARADON

MARSHALL   ST.STEPHEN—IN—BRANNEL   0760

Production: Tin   Black(tons) Stuff(tons)   Tin(tons)   Value(£)
            1855   4.10   0.00   0.00   243.40
            1856   1.30   0.00   0.00   77.60

MARTHA   STOKE CLIMSLAND   SX 388737 0761

Production: Copper   Ore(tons) Metal(tons)   Value(£)
            1861   No detailed return
            1862   2965.00   88.90   5448.30
            Comment 1861 SEE GREAT MARTHA, DEVON; 1862 (C)GREAT MARTHA
Ownership: Comment 1860—1863 SEE GREAT MARTHA, DEVON; 1864—1865 GREAT
            MARTHA
Management: Chief Agent 1860—1865 HY.RICKARD
            Secretary 1861—1863 ROBT.SARGEANT; 1864—1865 HY.RICKARD

MARTHA,EAST   CALLINGTON   0762

Ownership: Comment 1864—1865 SUSPENDED
Management: Chief Agent 1861—1863 JOS.RICHARDS
            Secretary 1861—1863 JOS.RICHARDS (P)

MARTHA,NEW   STOKE CLIMSLAND   SX 388737 0763

Production: Copper   Ore(tons) Metal(tons)   Value(£)
            1863   2255.00   78.20   5346.50
            1864   3384.00   93.00   7193.40
            1865   3005.00   81.80   6007.50
            1866   1686.00   36.10   2856.40
            1867   116.00   2.30   158.20
            Comment 1863—1867 (C)
Management: Chief Agent 1864—1865 HY.RICKARD
            Secretary 1864—1865 J.WRIGHT

MARTHA,WEST   STOKE CLIMSLAND   SX 372737 0764

Production: Copper   Ore(tons) Metal(tons)   Value(£)
            1863—1864 No detailed return
            Comment 1863—1864 SEE WEST MARTHA, DEVON
Ownership: Comment 1864—1865 SEE GREAT SHEBA CONSOLS

MARY   LELANT   SW 506366 0765

Production: Tin   Black(tons) Stuff(tons)   Tin(tons)   Value(£)
            1855   111.40   0.00   0.00   7237.10
            1856   121.50   0.00   0.00   9043.30
            1857   151.20   0.00   0.00   12039.70
            1858   148.30   0.00   0.00   9960.40
            1859   184.70   0.00   0.00   14147.40
            1860   131.70   0.00   0.00   10553.20
            1861   95.50   0.00   0.00   6884.20

                                     301

MARY   LELANT   Continued

Tin	Black(tons)	Stuff(tons)	Tin(tons)	Value(£)
1862	142.60	0.00	0.00	9511.10
1863	144.80	0.00	0.00	9612.70
1864	116.00	0.00	0.00	7236.70
1865	132.00	0.00	0.00	7090.10
1866	111.90	0.00	0.00	5416.60
1867	134.80	0.00	0.00	6941.30
1868	157.70	0.00	0.00	8547.00
1869	154.70	0.00	0.00	10737.90
1870	154.50	0.00	0.00	10983.40
1871	194.50	0.00	0.00	14979.80
1872	184.40	0.00	0.00	15642.10
1873	204.10	0.00	0.00	15344.80
1874	176.80	0.00	0.00	9940.80
1875	120.20	0.00	0.00	5863.60
1876	230.20	0.00	0.00	8771.00
1877	227.90	0.00	0.00	8253.20

              Comment 1877 INC.TRENCROM SEE SISTERS
Ownership: 1875-1876 WHEAL MARY ADVENTURERS
              Comment 1875-1876 SEE SISTERS; 1877-1881 INC.TRENCROM. SEE
              SISTERS
Management: Manager 1859-1866 THOS.ROBERTS; 1869-1874 MAT.CURNOW; 1876
              MAT.CURNOW
              Chief Agent 1859 MAT.CURNOW; 1860-1868 MAT.CURNOW PAUL ROACH;
              1869-1871 PAUL ROACH; 1872 MAT.CURNOW A.O.MICHELL; 1873
              MAT.CURNOW JNR. A.O.MICHELL; 1874 MAT.CURNOW JNR.; 1875
              THOS.D.FIELD; 1876 N.RICHARDS THOS.F.FIELD
              Secretary 1859-1870 R.R.MICHELL (P); 1871-1874 THOS.W.FIELD;
              1876 A.O.MICHELL

MARY   REDRUTH   SW 694432 0766

Production: Copper   Ore(tons) Metal(tons) Value(£)
            1848   270.00   21.40   1225.60
            1849   1197.00   86.40   5670.80
            1850   1270.00   78.80   5282.20
            1851   513.00   31.00   1969.30
            Comment 1848-1851 (C)

Tin	Black(tons)	Stuff(tons)	Tin(tons)	Value(£)
1860	11.30	0.00	0.00	958.50
1861	20.00	0.00	0.00	1406.90
1862	24.60	0.00	0.00	1689.80

            Comment 1860 MARY & KITTY UNITED; 1861-1862
            INC.KITTY(REDRUTH)
Ownership: Comment 1860 WHEAL MARY & KITTY UNITED

MARY   ST.DENNIS   SW 595575 0767

Production: Tin   Black(tons) Stuff(tons) Tin(tons) Value(£)
            1874   4.00   0.00   0.00   161.80
Management: Chief Agent 1872-1874 THOS.PARKYN

                                  302

MARY ANN,WEST   MENHENIOT   SX 275636 0769

Production: Lead & Silver No detailed return
Ownership: 1877-1884 WEST WHEAL MARY ANN MINING CO.
Management: Manager 1877 W.GEO.NETTLE
            Chief Agent 1875-1879 W.SKEAT; 1880-1881 J.RICH
            Secretary 1875-1884 W.GEO.NETTLE
Employment: Underground   Surface   Total
            1878   17   17
            1879   9   4   13
            1880   10   3   13
            1881   11   2   13
            1882   6   4   10
            1883   15   5   20
            1884   12   4   16

MARY ANNE   MENHENIOT   SX 288634 0770

Production: Lead & Silver   Ore(tons)   Lead(tons) Silver(ozs)   Value(£)
            1846   166.00   100.00   0.00   0.00
            1847   192.00   139.00   0.00   0.00
            1848   334.00   250.00   0.00   0.00
            1849   873.00   655.00   0.00   0.00
            1850   1186.00   890.30   0.00   0.00
            1851   1265.00   948.70   0.00   0.00
            1852   994.60   742.00   38600.00   0.00
            1853   1004.90   744.00   40160.00   0.00
            1854   1099.60   706.00   44470.00   0.00
            1855   1111.00   705.00   46530.00   0.00
            1856   1314.00   849.00   52638.00   0.00
            1857   1597.00   986.00   61625.00   0.00
            1858   1428.00   906.00   59796.00   0.00
            1859   1418.30   900.00   59400.00   0.00
            1860   1193.60   752.00   49632.00   0.00
            1861   1135.00   675.00   42761.00   0.00
            1862   910.70   594.00   44060.00   0.00
            1863   844.60   573.90   42402.00   0.00
            1864   752.30   496.50   35464.00   0.00
            1865   729.80   466.50   33120.00   0.00
            1866   830.30   531.20   38000.00   0.00
            1867   1048.60   786.00   56199.00   0.00
            1868   1139.80   854.20   61061.00   0.00
            1869   1005.90   754.40   53911.00   0.00
            1870   941.80   707.80   50610.00   0.00
            1871   1047.90   785.90   56199.00   0.00
            1872   1254.20   940.60   67210.00   0.00
            1873   759.70   569.20   40399.00   14653.60
            1874   538.00   403.50   28850.00   8597.40
            1875   6.10   4.50   0.00   93.60
            Comment 1846-1851 MARY ANN; 1868-1869 MARY ANN; 1875 MARY
            ANN
Ownership: 1895 H.R.LEWIS
            Comment 1871-1872 INC.TRELAWNEY; 1873 INC.TRELAWNEY
            SUSPENDED; 1874 STOPPED IN 1874; 1895 DRESSING FROM OLD
            BURROWS
Management: Manager 1859-1868 PETER CLYMO

                                  303

MARY ANNE   MENHENIOT   Continued

Chief Agent 1859 HY.HODGE; 1860–1867 HY.HODGE J.HARRIS JOHN
STEVENS; 1868–1874 W.SKEAT J.HARRIS JOHN STEVENS; 1895
B.BRYANT
Secretary 1859–1868 S.CLYMO (P); 1871–1874 W.C.NETTLE

MARY CONSOLS   BODMIN   SX 041653 0771

Production: Copper   Ore(tons) Metal(tons)   Value(£)
            1847   146.00   12.90   864.00
            1848   721.00   66.10   3702.30
            1849   656.00   50.00   3082.00
            1851   118.00   6.70   427.00
            Comment 1847–1849 (C); 1851 (C)MARY BODMIN

MARY GREAT CONSOLS   ST.NEOT   SX 187672 0772

Production: Copper   Ore(tons) Metal(tons)   Value(£)
            1854–1856 No detailed return
            1858–1862 No detailed return
            1864   No detailed return
            Comment 1854–1856 SEE MARY GREAT CONSOLS, DEVON; 1858–1862
            SEE MARY GREAT CONSOLS, DEVON; 1864 SEE MARY GREAT CONSOLS,
            DEVON
            Tin No detailed return
Ownership: 1912–1913 C.A.STOECKEL
            Comment 1863–1865 SUSPENDED; 1912 MARY:–OPENING ADIT NOT
            WORKING; 1913 MARY:CLOSED JUNE 1912
Management: Manager 1859–1862 THOS.RICHARDS; 1912–1913 W.A.PASCOE
            Chief Agent 1859–1862 W.THORPE
            Secretary 1859–1865 W.THORPE (P)

MARY,WEST   REDRUTH   SW 693442 0773

Production: Zinc   Ore(tons) Metal(tons)   Value(£)
            1878   7.00   0.00   10.50
            Copper   Ore(tons) Metal(tons)   Value(£)
            1878   6.00   0.10   8.50
            1880   3.00   0.30   18.50
            Arsenic Pyrite Ore(tons) Value(£)
            1878   5.00   3.00
Ownership: 1878–1879 HY.TELLAM; 1880–1881 HY.TELLAM & CO.
            Comment 1881 NOT WORKED
Management: Chief Agent 1878 SML.MICHELL; 1879–1881 HY.TELLAM
            Secretary 1878 HY.TELLAM
Employment: Underground   Surface   Total
            1878   3   3
            1880   3   3

304

MAUDE   CAMBORNE   SW 684437 0774

Production: Copper   Ore(tons) Metal(tons)   Value(£)
            1852-1853 No detailed return
            Comment 1852-1853 SEE EAST SETON

MAUDE,EAST   0775

Management: Chief Agent 1866-1868 JOS.WEBB

MAUDLIN   LANLIVERY   SX 084624 0776

Production: Copper   Ore(tons) Metal(tons)   Value(£)
            1867   3.00   0.40   27.80
            1869   21.00   1.10   60.50
            Comment 1867 (C); 1869 (C)
            Tin   Black(tons) Stuff(tons)   Tin(tons)   Value(£)
            1853   1.60   0.00   0.00   81.50
            1854   0.50   0.00   0.00   34.40
            1856   0.20   0.00   0.00   11.70
Ownership: 1907 CORNISH PROPRIETORY MINES LTD.
            Comment 1862-1867 MAUDLIN MINES; 1869-1872 MAUDLIN MINES
Management: Manager 1859-1867 WM.TREGAY; 1869-1872 WM.TREGAY
            Chief Agent 1864-1867 JOHN TREGAY; 1869-1871 JOHN TREGAY
            Secretary 1859-1860 W.P.CLEVERTON (P); 1861-1862 G.P.CARDOZA;
            1863-1867 G.H.CARDOZA; 1869-1872 G.H.CARDOZA; 1907
            H.C.JENKIN
Employment: Underground   Surface   Total
            1907   10   10

MAWLA   ST.AGNES   0777

Production: Copper No detailed return
            Tin No detailed return
Ownership: Comment 1860-1865 SEE STENCOOSE

MAXWELL   CROWAN   SW 641373 0778

Production: Copper No detailed return
Ownership: Comment 1865 SUSPENDED
Management: Manager 1859-1864 CLYMO
            Secretary 1859-1865 W.H.M.BLUES (P)

MEDLYN MOOR   WENDRON   SW 706335 0779

Production: Tin   Black(tons) Stuff(tons)   Tin(tons)   Value(£)
            1873   5.40   0.00   0.00   420.00
            1874   1.60   0.00   0.00   93.00
            1875   7.10   0.00   0.00   364.50
            1876   14.20   0.00   0.00   629.60
            1877   30.00   0.00   0.00   1232.60
            1878   14.80   0.00   0.00   500.20

                                  305

MEDLYN MOOR   WENDRON   Continued

```
            Tin          Black(tons)  Stuff(tons)   Tin(tons)    Value(£)
            1879            5.90          4.00         0.00        261.20
            1880            1.60          0.00         0.00         72.00
Ownership:  1876-1880 MEDLYN MOOR MINING CO.
Management: Manager 1873-1874 J.ROWE; 1876-1880 JOS.PRISK
            Chief Agent 1872 J.ROWE; 1875-1879 CHAS.ROWE
            Secretary 1875-1879 GREN.SHARP
Employment:              Underground      Surface         Total
            1878             24             17              41
            1879             16             17              33
            1880                             5               5
```


MELEDOR   ST.STEPHEN-IN-BRANNEL   0780

```
Production: Iron No detailed return
Ownership:  1906-1907 MELEDOR SYNDICATE
Management: Secretary 1906-1907 J.TREEBY BARRATT
Employment:              Underground      Surface         Total
            1907              6               7              13
```


MELLANEAR   ST.ERTH   SW 560361 0781

Production: Lead	Ore(tons)	Metal(tons)	Value(£)
1886	21.00	0.00	0.00
Zinc	Ore(tons)	Metal(tons)	Value(£)
1872	32.00	0.00	16.90
1878	83.70	0.00	90.00
1879	84.00	0.00	294.00
1882	153.00	68.80	428.00
1883	6.00	2.10	21.00
1886	285.00	0.00	562.00
1887	649.00	228.00	796.00
1888	131.00	0.00	249.00
1889	129.00	60.00	174.00
Copper	Ore(tons)	Metal(tons)	Value(£)
1865	323.00	13.90	1030.00
1866	381.00	17.00	1014.50
1867	182.00	5.90	341.50
1868	56.00	1.50	76.60
1870	182.00	21.40	1369.60
1871	996.00	56.90	3299.10
1872	1200.00	68.70	4622.90
1873	1645.00	91.50	5553.70
1877	3675.00	234.00	13442.60
1878	5714.80	373.50	18079.90
1879	6528.00	424.30	20612.00
1880	6777.30	420.70	23267.50
1881	6811.60	419.70	21437.60
1882	6741.00	455.00	26964.00
1883	6789.00	377.40	19401.00
1884	6928.00	0.00	16029.00
1885	6138.00	0.00	10731.00
1886	4019.00	0.00	7668.00

MELLANEAR   ST.ERTH   Continued

        Copper   Ore(tons) Metal(tons)   Value(£)
        1887   1939.00   0.00   4225.00
        1888   2857.00   0.00   8429.00
        1889   1009.00   0.00   1694.00
        Comment 1865–1868 (C); 1870–1873 (C)
        Tin   Black(tons) Stuff(tons) Tin(tons)   Value(£)
        1879   0.00   4.00   0.00   16.00
        1881   0.00   223.00   0.00   526.80
        1882   7.80   124.00   4.80   387.00
        1883   4.70   118.00   0.00   168.00
        1884   4.60   94.00   0.00   208.00
        1885   15.00   437.00   0.00   482.00
        1886   9.00   251.00   0.00   384.00
        1887   7.00   284.00   0.00   320.00
        1888   7.50   410.00   0.00   440.00
        1889   7.00   344.00   0.00   344.00
Ownership: 1876–1889 MELLANEAR COPPER MINING CO.LTD.
             Comment 1889 ABANDONED
Management: Manager 1863 RICH.STEPHENS; 1864 WM.PAULL; 1865–1868 W.GILL;
             1869 RICH.STEPHENS; 1870–1871 ED.ROGERSWELL; 1872 ED.ROGERS;
             1873–1874 J.MOYLE; 1875–1881 JOHN GILBERT
             Chief Agent 1864–1868 RICH.STEPHENS; 1872 J.MOYLE; 1877 JOHN
             WOOLCOCK; 1878–1879 JOHN WOOLCOCK WM.TOMS; 1880 JOHN WOOLCOCK
             SML.HARNS; 1881 SML.HARRIS WM.TOMS; 1882–1889 JOHN GILBERT
             Secretary 1863–1866 ALM.E.PAULL; 1869–1874 COMMITTEE; 1875
             JOHN TAYLOR & SONS; 1876 JOHN HAYLE; 1877 W.G.WILLIAMS;
             1878–1879 W.G.WILLIAMS (S); 1880–1881 WILLIAMS(S) &   .
             TAYLOR(P)
Employment: Underground   Surface   Total
        1878   139   146   285
        1879   137   161   298
        1880   130   163   293
        1881   134   166   300
        1882   133   159   292
        1883   130   146   276
        1884   109   129   238
        1885   97   103   200
        1886   80   80   160
        1887   65   61   126
        1888   84   62   146
        1889   48   40   88

MENADUE   LUXULYAN   SX 031594 0782

Production: Zinc   Ore(tons) Metal(tons)   Value(£)
        1870   14.00   0.00   48.30
        1871   1.00   0.00   3.80

MENANDARVA   LUXULYAN   0783

Production: Tin   Black(tons) Stuff(tons) Tin(tons)   Value(£)
        1876   33.60   0.00   0.00   1080.80
        Comment 1876 SEE ALSO MENANDARVA & BATHORNES

                                    307

MENANDARVA AND BATHORNES   LUXULYAN   0784

Production: Tin   Black(tons) Stuff(tons)   Tin(tons)   Value(£)
            1876   9.60   0.00   0.00   345.10
            Comment 1876 SEE ALSO MENANDARVA

MENGERN   WENDRON   SW 702301 0785

Production: Tin   Black(tons) Stuff(tons)   Tin(tons)   Value(£)
            1853   1.20   0.00   0.00   75.00
            1880—1884 No detailed return
            Comment 1853 MENGEARNE INC.TREGUNSTIC; 1880—1884 SEE
            COMBELLACK
Ownership: Comment 1880—1884 SEE COMBELLACK
Employment: Underground   Surface   Total
            1880   9   7   16
            1881   16   13   29
            1882   18   20   38
            1883   18   18   36
            1884   9   10   19
            Comment 1880—1884 INC.COMBELLACK

MERLYN MOOR   0786

Production: Tin   Black(tons) Stuff(tons)   Tin(tons)   Value(£)
            1873   3.30   0.00   0.00   260.60

MERTH   ST.AUSTELL   0787

Production: Tin   Black(tons) Stuff(tons)   Tin(tons)   Value(£)
            1907   14.50   0.00   0.00   1407.00
Ownership: 1906—1911 WHEAL MERTH LTD.; 1912—1913 CORNISH CONSOLS LTD.
            Comment 1909 SUSPENDED; 1910 SUSPENDED JULY 1910; 1911—1912
            SUSPENDED; 1913 ABANDONED FEB.1914
Management: Chief Agent 1906 J.L.TEALE; 1907—1910 A.B.CLIMAS
Employment: Underground   Surface   Total
            1906   14   16   30
            1907   20   29   49
            1908   30   42   72
            1909—1910   2   2
            1912   1   1
            1913   3   3

MESSER   LANIVET   SX 063640 0788

Production: Copper   Ore(tons) Metal(tons)   Value(£)
            1853   430.00   20.30   1960.20
            1854   317.00   22.30   2146.00
            1855   383.00   15.70   1501.30
            1859   190.00   10.70   964.80
            Comment 1853—1855 (C); 1859 (C)SEE ALSO TRETOIL
            Tin   Black(tons) Stuff(tons)   Tin(tons)   Value(£)
            1860—1861 No detailed return

308

MESSER   LANIVET   Continued

Comment 1860–1861 SEE TRETOIL
Ownership: Comment 1859–1861 SEE TRETOIL; 1863–1865 SUSPENDED
Management: Chief Agent 1862 W.WILLIAMS
Secretary 1862–1865 J.VIVIAN

METAL AND FLOW   BREAGE   SW 623295 0789

Production: Tin	Black(tons)	Stuff(tons)	Tin(tons)	Value(£)
1885	17.00	0.00	0.00	795.00
1886	52.00	0.00	0.00	2467.00
1887	45.40	0.00	0.00	2422.00
1888	43.10	0.00	0.00	2293.00
1889	47.00	0.00	0.00	2021.00
1890	48.60	0.00	0.00	2124.00
1891	47.00	0.00	0.00	2125.00
1892	51.60	0.00	0.00	2413.00
1893	54.40	0.00	0.00	2317.00
1894	20.40	0.00	0.00	718.00
1895	9.30	0.00	0.00	266.00
1896	27.90	0.00	0.00	825.00
1897	20.00	0.00	0.00	600.00
1898	42.50	0.00	0.00	1588.00
1900	15.20	0.00	0.00	982.00
1901	19.10	0.00	0.00	1038.00
1902	10.50	0.00	0.00	508.00
1903	7.20	0.00	0.00	365.00
1904	6.20	0.00	0.00	395.00
1905	17.50	0.00	0.00	1451.00
1906	2.50	0.00	0.00	239.00

Comment 1897 OPEN WORK; 1898 MINE & OPEN WORK FIGURES AGG.
Ownership: 1884–1885 WHEAL METAL & FLOW CO.; 1886–1890 METAL & FLOW CO.;
1891–1893 WHEAL METAL & FLOW MINING CO. C.B.; 1894–1899 WHEAL
METAL & FLOW MINING CO.; 1900–1901 WHEAL METAL & FLOW MINING
CO.LTD.; 1902–1906 BREAGE VALLEY MINING CO.
Comment 1894–1901 SURFACE WORK ONLY
Management: Chief Agent 1884–1888 WM.ARGALL; 1889–1899 STEP.P.CURTIS;
1900–1906 JAS.RICHARDS
Secretary 1891 C.B.PARRY; 1892 C.B.PARRY (S); 1893–1901
GEO.E.COLLINS (S)

Employment:	Underground	Surface	Total
1884		6	6
1885	7	48	55
1886	6	46	52
1887	8	46	54
1888	6	41	47
1889	8	40	48
1890	6	44	50
1891	6	36	42
1892	6	28	34
1893	9	33	42
1894		13	13
1895	4	17	21
1896	6	34	40
1898	5	20	25

METAL AND FLOW   BREAGE   Continued

	Underground	Surface	Total
1899		13	13
1900	2	27	29
1901		34	34
1902	4	17	21
1903	3	10	13
1904	3	22	25
1905	4	28	32

METAL,EAST   BREAGE   0790

Production: Tin No detailed return
Ownership: Comment 1859–1865 SEE NEW VOR

METAL,NEW   BREAGE   0791

Production: Tin No detailed return
Management: Chief Agent 1863–1864 CHAS.CARKEEK
            Secretary 1863–1864 ROBT.H.PIKE & SON

METAL,NORTH   BREAGE   SW 632306 0792

Production: Tin   Black(tons) Stuff(tons)   Tin(tons)   Value(£)
            1874   0.50   0.00   0.00   21.00
            1883   7.70   13.20   0.00   400.00
            1884   1.00   0.00   0.00   42.00
            Comment 1874 INC.HARRIET VALUE EST
Ownership: 1882–1883 NORTH METAL MINING CO.
            Comment 1872 INC.HARRIET; 1873 INC.HARRIET STOPPED IN 1873 OR
            1874; 1874 INC.HARRIET
Management: Manager 1872–1874 BEN.GUNDRY
            Chief Agent 1874 N.W.JAMES; 1882–1883 JOS.PRISK
Employment: Underground   Surface   Total
            1882   33   38   71
            1883   18   16   34
            1884   23   11   34

METAL,SITHNEY WHEAL   SITHNEY   SW 631300 0793

Production: Tin No detailed return
Management: Manager 1863–1864 THOS.GILL; 1865–1866 J.JULIAN
            Chief Agent 1862 WM.CHAPPEL WM.ROGERS; 1863–1864 F.FRANCIS;
            1865–1866 F.HARRIS
            Secretary 1862 JOHN BURGESS; 1863 ED.BURGESS; 1864–1866 JOHN
            BURGESS

METAL,WEST   BREAGE   SW 623295 0794

Production: Tin   Black(tons) Stuff(tons)   Tin(tons)   Value(£)
            1876   No detailed return

METAL,WEST   BREAGE   Continued

              Comment 1876 SEE NEW CARLEEN VOR
Ownership: Comment 1876 SEE NEW CARLEEN VOR

METAL,WEST   ST.JUST   SW 396312 0796

Production: Tin No detailed return
Ownership: Comment 1864-1865 SUSPENDED
Management: Manager 1860 JOHN BENNETT; 1861-1863 T.BENNETT
            Chief Agent 1860-1863 JAS.WHITE
            Secretary 1860-1865 J.N.R.MILLETT (P)

MICHELL   MARAZION   0797

Production: Lead & Silver   Ore(tons) Lead(tons) Silver(ozs) Value(£)
            1854   11.00   8.00   45.00   0.00
            Copper   Ore(tons) Metal(tons) Value(£)
            1848   157.00   8.20   404.40
            Comment 1848 (C)GREAT MICHELL CONSOLS
            Tin No detailed return
Ownership: Comment 1864-1866 SEE PROSPER UNITED

MID-CORNWALL MINES   ROCHE   0798

Production: Iron   Ore(tons)   Iron(%)   Value(£)
            1874   1655.00   0.00   1240.70
            Comment 1874 BH.CONSISTED OF 6 MINES

MILL POOL   ST.HILARY   SW 568301 0799

Production: Tin   Black(tons) Stuff(tons) Tin(tons)   Value(£)
            1852   7.20   0.00   0.00   0.00
            1853   6.40   0.00   0.00   389.40
            1855   6.60   0.00   0.00   437.80
            1856   90.30   0.00   0.00   6337.90
            1857   70.50   0.00   0.00   5335.60
            1858   70.20   0.00   0.00   4678.20
            1859   96.00   0.00   0.00   6697.00
            1860   73.30   0.00   0.00   5804.50
            1861   60.40   0.00   0.00   4453.10
            1862   41.90   0.00   0.00   2665.80
            1863   2.60   0.00   0.00   143.20
            Comment 1852 TIN ORE; 1862 TWO RETURNS AGGREGATED
Ownership: Comment 1862-1865 SUSPENDED
Management: Chief Agent 1859-1861 WM.OATS JNR. J.DANIELL; 1862-1865
            J.DANIELL
            Secretary 1859-1865 WM.PAINTER (P)

                              311

MILLETT   CROWAN   SW 629324 0800

Production: Lead   Ore(tons) Metal(tons)   Value(£)
            1870   5.50   3.90   0.00
            Copper No detailed return
Management: Manager 1869-1870 ROBT.RICHARDS; 1871 NICH.RICHARDS
            Secretary 1869-1871 THOS.ROACH

MILLS,NEW   LADOCK   0801

Production: Iron   Ore(tons)   Iron(%)   Value(£)
            1874   92.20   0.00   69.10
            1883   170.00   0.00   85.00
            Comment 1874 BH.; 1883 BH.
Ownership: 1874-1875 NEW MILLS IRON MINES CO.; 1876 PAWTON & NEW MILLS
            IRON MINES CO.; 1880-1882 PAWTON & NEW MILLS IRON MINES CO.
            Comment 1877-1879 SEE PAWTON; 1882 NOT WORKED
Management: Manager 1880-1881 J.R.HARVEY
            Chief Agent 1874-1875 ALF.HEWLETT; 1876 J.R.HARVEY; 1882
            J.R.HARVEY
Employment: Underground   Surface   Total
            1880   6   4   10

MINEAR DOWNS   ST.AUSTELL   0802

Production: Tin   Black(tons) Stuff(tons)   Tin(tons)   Value(£)
            1868   0.00   0.00   0.00   2921.10
            1869   58.70   0.00   0.00   4278.30
            1870   53.30   0.00   0.00   4005.20
            1871   117.10   0.00   0.00   9668.80
            1872   100.70   0.00   0.00   9057.10
            1873   76.70   0.00   0.00   5877.40
            1874   38.30   0.00   0.00   1995.70
            1875   34.70   0.00   0.00   2071.30
            1880   17.30   0.00   0.00   830.00
            1882   42.30   0.00   27.90   2516.00
            1883   71.00   0.00   0.00   3925.00
            1884   56.00   0.00   0.00   2675.00
            1885   34.00   0.00   0.00   1656.00
            1886   33.80   0.00   0.00   1884.00
            1887   20.50   0.00   0.00   1243.00
            1888   25.80   0.00   17.00   1626.00
            1889   26.50   0.00   0.00   1546.00
            1890   37.00   0.00   0.00   2083.00
            1891   No detailed return
            1895   7.00   0.00   0.00   273.00
            1896   2.00   0.00   0.00   57.00
            1900   16.00   0.00   0.00   1308.00
            1901   27.30   0.00   0.00   1931.00
            1902   24.60   0.00   0.00   1728.00
            1903   21.80   0.00   0.00   1673.00
            1904   6.90   0.00   0.00   462.00
            1905   5.20   0.00   0.00   402.00
            1906   6.50   0.00   0.00   639.00
            1907   21.00   0.00   0.00   2297.00

                                   312

Tin	Black(tons)	Stuff(tons)	Tin(tons)	Value(£)
1908	17.70	0.00	0.00	1463.00
1909	18.00	0.00	0.00	1480.00
1910	18.00	0.00	0.00	1688.00
1911	17.00	0.00	0.00	2046.00
1912	21.20	0.00	0.00	2943.00
1913	18.90	0.00	0.00	2487.00

Comment 1868 TIN STUFF; 1882—1890 OPEN WORKS; 1891 AT WORK;
1895—1896 OPEN WORKS; 1900—1913 OPEN WORK
Ownership: 1882—1891 J.LOVERING & CO.
             Comment 1882—1891 OPEN WORK
Management: Manager 1872 J.LOVERING
             Chief Agent 1873—1875 ROBT.MARTIN
             Secretary 1873—1875 J.LOVERING

MINERAL BOTTOM   PERRANZABULOE   SW 794515 0803

Production: Lead & Silver	Ore(tons)	Lead(tons)	Silver(ozs)	Value(£)
1870	95.00	71.20	0.00	0.00
1871	37.70	28.20	141.00	0.00

Management: Manager 1869—1871 RICH.NANCARROW
             Chief Agent 1864—1865 J.JULIFF; 1869—1871 ROBT.R.NANCARROW
             Secretary 1864—1865 RICH.CLOGG; 1869—1871 RICH.CLOGG

MINERAL COURT   ST.STEPHEN—IN—BRANNEL   0804

Production: Tin	Black(tons)	Stuff(tons)	Tin(tons)	Value(£)
1855	12.10	0.00	0.00	734.70
1856	1.30	0.00	0.00	87.50

MITCHELL   BREAGE   0805

Production: Tin	Black(tons)	Stuff(tons)	Tin(tons)	Value(£)
1862	11.80	0.00	0.00	716.90
1874	0.00	56.00	0.00	57.00
1875	0.00	24.00	0.00	0.00

Management: Manager 1874—1875 W.BUCKETT
             Secretary 1874—1875 W.BUCKETT

MITCHELL   LANIVET   0806

Production: Tin No detailed return
Ownership: Comment 1874—1875 SEE PROSPER(LANIVET); 1880—1881 SEE
             PROSPER(LANIVET)

MOLESWORTH   WADEBRIDGE   SW 978701 0807

Production: Copper	Ore(tons)	Metal(tons)	Value(£)
1886	5.00	0.00	30.00
1887	5.00	0.00	15.00

MOLESWORTH   WADEBRIDGE   Continued

Ownership: 1886 S.SERPELL
             Comment 1887 NOT WORKING
Management: Chief Agent 1886 S.SERPELL; 1887 H.PAINTER
Employment: Underground   Surface   Total
             1886   4   4   8

MOLESWORTH UNITED   BREAGE   0808

Ownership: 1880 F.E.BINGLEY; 1881-1882 MOLESWORTH UNITED MINING CO.
             Comment 1882 NOT WORKED
Management: Manager 1881 J.RICHARDS
             Chief Agent 1880 J.RICHARDS; 1882 J.RICHARDS
Employment: Underground   Surface   Total
             1880   4   4
             1881   6   6

MOLLINIS   BUGLE   SX 024591 0809

Production: Tin   Black(tons) Stuff(tons)   Tin(tons)   Value(£)
             1863-1864 No detailed return
             Comment 1863-1864 SEE GOONBARROW
Ownership: Comment 1864-1865 SEE GOONBARROW

MONTAGUE   SSCORRIER   SW 487381 0810

Production: Tin   Black(tons) Stuff(tons)   Tin(tons)   Value(£)
             1853   3.00   0.00   0.00   174.00

MORSHEAD   CALSTOCK   SX 435697 0811

Production: Copper No detailed return
Management: Chief Agent 1883 WM.B.COLLOM
Employment: Underground   Surface   Total
             1883   4   4

MORVAH CONSOLS   MORVAH   SW 407359 0812

Production: Tin   Black(tons) Stuff(tons)   Tin(tons)   Value(£)
             1874   5.90   0.00   0.00   328.50
             Comment 1874 VALUE EST
Management: Manager 1873 THOS.BENNETTS
             Secretary 1873 R.MARRACK

MORVAH HILL   MORVAH   SW 417357 0813

Production: Tin No detailed return
Ownership: Comment 1907 SEE BOSIGRAN CONSOLIDATED; 1908 SEE ROSEMERGY

                                     314

Production: Tin   Black(tons) Stuff(tons)   Tin(tons)   Value(£)
            1882   28.00   91.50   18.30   1652.00
            1883   20.00   67.00   0.00   1047.00
            1884   23.50   0.00   0.00   960.00
            1885   5.70   0.00   0.00   225.00
            Comment 1882-1883 TWO RETURNS AGGREGATED; 1884 WHITS
Ownership: 1880 WM.TREGAY & OTHERS; 1881-1885 MOUNT CARBIS MINING CO.;
            1888-1889 MOUNT CARBIS MINING CO.
            Comment 1889 SUSPENDED
Management: Chief Agent 1880-1885 WM.TREGAY; 1888-1889 LEN.TREGAY
Employment: Underground   Surface   Total
            1880   8   3   11
            1881   27   17   44
            1883   30   22   52
            1884   20   22   42
            1885   6   6
            1888   7   4   11

Production: Lead   Ore(tons) Metal(tons)   Value(£)
            1854   10.70   8.00   0.00
            1861   No detailed return
            Comment 1861 SEE TREBISKEN GREEN
            Zinc No detailed return
            Iron   Ore(tons)   Iron(%)   Value(£)
            1864   No detailed return
            1871   1350.00   0.00   1012.50
            1872   No detailed return
            1873   2242.30   0.00   1681.00
            1874-1877 No detailed return
            1880   379.80   0.00   189.90
            1905   240.00   48.00   60.00
            1907   50.00   45.00   20.00
            Comment 1864 SEE TREBISKEN; 1871 BH.; 1872 BH.SEE TREBISKEN;
            1873 BH.; 1874-1876 BH.SEE TREBISKEN; 1877 SEE TREBISKEN;
            1880 SP.INC.110 TONS BH (£55); 1905 HE.; 1907 HE.
Ownership: 1873-1874 AGRA BANKING CO.; 1875-1877 AGRA BANK & CO.;
            1878-1880 EDM.R.GRAY; 1881-1882 CORNISH STEEL & IRON ORE CO.;
            1883 CORNISH STEEL & IRON ORE CO.LTD.; 1899 JOHN HOLLOWAY;
            1902-1906 JOHN HOLLOWAY; 1907 EXECUTORS OF JOHN HOLLOWAY
            Comment 1875-1881 INC.TREBISKEN; 1882-1883 INC.TREBISKEN NOT
            WORKED; 1900-1901 SEE PERRAN(PERRANZABULOE); 1902
            INC.TREAMBLE; 1903 MOUNTS; 1904-1905 MOUNTS SUSPENDED;
            1906-1907 MOUNTS & OTHER MINES
Management: Manager 1873-1878 A.J.WHALLEY; 1880 PHIL.MICHELL
            Chief Agent 1860-1866 JOHN BALL; 1875-1876 PHIL.MICHELL;
            1878-1879 PHIL.MICHELL
            Secretary 1860-1863 T.WHITFORD (P); 1864-1865 EDW.CARTER;
            1866 MASTERMAN & CO.; 1877 PHIL.MICHELL; 1899 JOHN GILBERT
            (S)
Employment: Underground   Surface   Total
            1881   17   18   35
            1882   28   24   52

MOUNT NO.1   PERRANZABULOE   Continued

                    Underground   Surface   Total
          1903   5   1   6
          1905   3   3   6
          1906   10   2   12
          1907   6   45   51
          Comment 1881-1882 INC.TREBISKEN

MOUNT NO.2   PERRANZABULOE   SW 781565 0816

Production: Lead No detailed return
Ownership: Comment 1860-1868 SEE EAST BUDNICK

MOUNTS BAY   BREAGE   SW 579286 0817

Production: Copper No detailed return
Ownership: 1876 MOUNTS BAY MINING CO.
             Comment 1887-1888 SEE SYDNEY GODOLPHIN
Management: Chief Agent 1876 JOHN CURTIS

MOUNTS BAY CONSOLS   BREAGE   SW 579286 0818

Production: Copper   Ore(tons) Metal(tons)   Value(£)
             1882   110.00   11.00   659.00
             1883   59.00   5.30   352.00
             1884   2.00   0.00   2.00
             1885   No detailed return
             Comment 1885 SEE SHEPHERDS UNITED
             Tin   Black(tons) Stuff(tons)   Tin(tons)   Value(£)
             1881   0.00   7.50   0.00   12.80
             1882   1.70   27.80   1.00   93.00
             1883   27.00   0.00   0.00   1259.00
             1884   18.20   0.00   0.00   800.00
             Comment 1882 TWO RETURNS AGGREGATED
Ownership: 1880 MOUNTS BAY UNITED CO.; 1881 MOUNTS BAY CONSOLS MINING
             CO.LTD.; 1882-1886 MOUNTS BAY CONSOLS LTD.
             Comment 1880-1881 SYDNEY COVE; TREBARVAH; PEMBROKE COPPER
Management: Manager 1881 WM.ARGALL
             Chief Agent 1880 WM.ARGALL; 1882-1886 WM.ARGALL
Employment: Underground   Surface   Total
             1881   51   104   155
             1882   88   68   156
             1883   41   50   91
             1884   36   30   66
             1885   42   35   77
             1886   10   10   20

MOYLE   GWENNAP   SW 731407 0819

Production: Copper   Ore(tons) Metal(tons)   Value(£)
             1862   175.00   5.60   332.40
             1863   7.00   0.20   13.80

                                        316

MOYLE   GWENNAP   Continued

            Comment 1862-1863 (C)
            Tin   Black(tons) Stuff(tons)   Tin(tons)   Value(£)
            1861   12.80   0.00   0.00   843.30
            1862   21.70   0.00   0.00   1370.80
Ownership: Comment 1863-1865 SUSPENDED
Management: Manager 1860 JOS.MICHELL
            Chief Agent 1860 GEO.JOHNS; 1861-1862 RALPH P.GOLDSWORTHY
            GEO.JOHNS
            Secretary 1860-1862 RICH.GREENWOOD (P)

MULBERRY   LANIVET   SX 016659 0820

Production: Tin   Black(tons) Stuff(tons)   Tin(tons)   Value(£)
            1860   6.40   0.00   0.00   492.60
            1861   4.80   0.00   0.00   343.30
            1862   22.80   0.00   0.00   1513.80
            1863   32.60   0.00   0.00   2170.60
            1864   38.90   0.00   0.00   2041.70
            1865   48.20   0.00   0.00   2677.80
            1866   13.30   0.00   0.00   716.40
            1868   76.90   0.00   0.00   4155.10
            1869   94.70   0.00   0.00   6498.40
            1870   81.90   0.00   0.00   5888.00
            1871   84.90   0.00   0.00   6535.50
            1872   73.20   0.00   0.00   6188.70
            1873   71.60   0.00   0.00   5422.50
            1874   60.30   0.00   0.00   3276.80
            1875   40.90   0.00   0.00   2041.30
            1876   36.00   0.00   0.00   1538.20
            1877   No detailed return
            1878   26.40   0.00   0.00   918.70
            1880   40.30   0.00   0.00   1989.90
            1881   49.00   0.00   33.70   2761.90
            1882   59.80   0.00   39.00   3544.00
            1883   80.00   0.00   0.00   4000.00
            1884   29.60   0.00   0.00   1651.00
            1885   33.00   0.00   0.00   1532.00
            1886   29.10   0.00   0.00   1520.00
            1887   18.70   0.00   0.00   1088.00
            1888   26.80   0.00   17.60   1603.00
            1889   22.00   0.00   0.00   1147.00
            1890   35.00   0.00   0.00   1970.00
            1891   No detailed return
            1895   1.00   0.00   0.00   33.00
            1896   6.00   0.00   0.00   211.00
            1904   1.50   0.00   0.00   120.00
            1905   6.50   0.00   0.00   531.00
            1906   7.70   0.00   0.00   762.00
            1907   30.10   0.00   0.00   3021.00
            1908   21.00   0.00   0.00   1525.00
            1910   7.00   0.00   0.00   500.00
            1911   3.00   0.00   0.00   270.00
            1913   8.30   0.00   0.00   691.00
            Comment 1861 MULBERRY HILL; 1862 TWO RETURNS AGGREGATED;

                              317

MULBERRY   LANIVET   Continued

                1863-1864 MULBERRY HILL; 1868 MULBERRY HILL; 1870-1875
                MULBERRY HILL; 1878 MULBERRY HILL; 1880-1881 MULBERRY HILL;
                1883-1890 OPEN WORKS; 1891 AT WORK; 1895-1896 OPEN WORKS;
                1904-1908 OPEN WORKS; 1910-1911 OPEN WORKS; 1913 OPEN WORKS
Ownership: 1875-1877 DANBUR MARTIN & CO.; 1878-1879 MULBERRY AND PROSPER
                MINE CO.; 1880-1881 MULBERRY MINE CO.; 1882 MULBERRY TIN
                WORKS CO.LTD.; 1883 BAIN,FIELD,HITCHINGS & CO.; 1884-1885
                DAVID COCK; 1886-1891 BAIN,FIELD,HITCHINGS & CO.
                Comment 1868-1876 MULBERRY HILL; 1879 INC.PROSPER(LANIVET);
                1882-1891 OPEN WORKS
Management: Manager 1866-1877 W.L.MARTIN; 1878-1880 THOS.MARTIN; 1881
                DAVID COCK
                Chief Agent 1869-1873 THOS.HAMBLY; 1874 THOS.HAMBLY
                PHIL.RICH; 1875-1880 THOS.HAMBLY
                Secretary 1866-1868 WM.MARTIN & CO.; 1869-1872 MARTIN BROS.;
                1873-1874 W.L.MARTIN

MULBERRY,EAST   LANIVET   SX 019658 0821

Production: Tin   Black(tons) Stuff(tons)   Tin(tons)   Value(£)
            1873   1.10   0.00   0.00   92.80
            1874   4.00   0.00   0.00   180.00
            Comment 1874 VALUE EST

MUSIC,EAST   ST.AGNES   SW 703471 0822

Production: Copper No detailed return
Ownership: Comment 1860-1865 SUSPENDED
Management: Chief Agent 1859 AB.BENNETTS
            Secretary 1859 J.R.EARLE (P)

NANCEKUKE   PORTH TOWAN   SW 693470 0823

Production: Lead & Silver   Ore(tons) Lead(tons) Silver(ozs)   Value(£)
            1850   No detailed return
            1851   159.60   116.00   0.00   0.00
            1852   338.00   220.00   8300.00   0.00
            1853   250.00   169.00   5400.00   0.00
            1854   No detailed return
            Comment 1850 SEE TYWARNHAILE
            Tin   Black(tons) Stuff(tons)   Tin(tons)   Value(£)
            1853   40.00   0.00   0.00   887.00

NANCEMELLAN,SOUTH   GWITHIAN   0824

Production: Tin   Black(tons) Stuff(tons)   Tin(tons)   Value(£)
            1873   6.90   0.00   0.00   541.10

                                  318

NANCYGOLLEN   NANCYGOLLEN   SW 641323 0825

Production: Tin   Black(tons) Stuff(tons)   Tin(tons)   Value(£)
            1854   6.20   0.00   0.00   350.00
            1913   0.80   0.00   0.00   67.00
            Comment 1854 NANSEGOLLAN; 1913 INC.PENGELLY
Ownership: 1907 J.DEFRIES & SONS LTD.; 1912 HOOVER & MERTON; 1913
            PROSPIDNICK MINING CO.LTD.
            Comment 1907 INC.PENGELLY; 1912 TAKEN FROM A.J.ANGOVE
            18/6/1912; 1913 INC.PENGELLY
Employment: Underground   Surface   Total
            1912   6   6   12
            1913   2   2
            Comment 1912-1913 INC.PENGELLY

NANGILES   KEA   SW 764420 0826

Production: Zinc   Ore(tons) Metal(tons)   Value(£)
            1856   144.70   0.00   230.40
            1858   153.10   0.00   249.10
            1866   1.50   0.00   1.50
            Copper   Ore(tons) Metal(tons)   Value(£)
            1845-1848 No detailed return
            1863   156.00   10.60   862.00
            1864   397.00   23.80   1937.00
            1865   454.00   30.00   2424.00
            1866   200.00   11.40   763.80
            1867   237.00   14.10   980.20
            1868   117.00   5.70   366.00
            Comment 1845-1848 SEE ANDREW; 1863-1868 (C)
            Tin   Black(tons) Stuff(tons)   Tin(tons)   Value(£)
            1855   6.80   0.00   0.00   423.80
            1856   0.00   0.00   0.00   387.60
            1857   0.00   0.00   0.00   2.00
            1858   0.00   0.00   0.00   617.40
            1862   0.00   0.00   0.00   447.10
            1863   0.00   0.00   0.00   284.00
            1864   7.00   0.00   0.00   740.20
            1865   8.70   0.00   0.00   581.10
            1866   0.80   0.00   0.00   203.10
            1867   0.00   0.00   0.00   180.10
            1868   0.00   0.00   0.00   171.30
            1869   10.50   0.00   0.00   1241.00
            1870   16.50   0.00   0.00   1429.10
            1871   6.10   0.00   0.00   570.40
            1872   1.90   0.00   0.00   224.70
            1873   10.60   5.00   0.00   786.70
            1874   0.00   400.00   0.00   145.00
            1875   0.00   42.00   0.00   46.40
            1876   1.00   0.00   0.00   37.50
            1901   0.00   643.00   0.00   264.00
            1902   2.00   0.00   0.00   122.00
            1903   6.40   0.00   0.00   415.00
            1904   12.60   0.00   0.00   733.00
            1905   7.60   0.00   0.00   441.00
            Comment 1856-1858 TIN STUFF; 1862 TIN STUFF. TWO RETURNS

AGG.; 1863 TIN STUFF; 1864-1866 VALUE INC.TIN STUFF;
1867-1868 TIN STUFF; 1869-1872 VALUE INC.TIN STUFF

Arsenic	Ore(tons)	Metal(tons)	Value(£)
1869	3.50	0.00	513.50
1870	6.00	0.00	15.30
1871	4.00	0.00	4.50

Comment 1869 VALUE INCORRECT?; 1870-1871 CRUDE
Ownership: 1900-1901 W.J.TRYTHALL; 1902-1904 WM.BURROWS; 1905 DEITZSCH &
SHIFF; 1906 S.G.MINERALS SYNDICATE LTD.
Comment 1901 ABANDONED SEPT.1901
Management: Manager 1859-1866 JAS.ROWE; 1867-1868 JAS.ROWE JNR.;
1869-1874 JAS.ROWE; 1875 JOH.ROWE
Chief Agent 1859-1860 JOHN WASLEY; 1861-1864 ED.DOWER;
1865-1866 ED.DOWER J.RICHARDS; 1869-1874 JAS.ROWE JNR.;
1903-1904 WM.BURROWS; 1906 H.C.JENKIN
Secretary 1859-1860 JAS.HOLLOW (P); 1861-1871 BEN.MATTHEWS;
1872-1875 HY.G.SHARP

Employment:	Underground	Surface	Total
1901	8	3	11
1902	3	3	6
1903	5	3	8
1904	8	8	16
1905	4		4

NANJEATH   ST.STEPHEN-IN-BRANNEL   0827

Production: Iron No detailed return
Management: Chief Agent 1873 JAS.NICHOLSON

NANTALLON   LANIVET   SX 021670 0828

Production: Iron	Ore(tons)	Iron(%)	Value(£)
1859	1716.10	0.00	429.00
1860	408.80	0.00	121.00
1861	No detailed return		

Comment 1859-1861 BH.
Ownership: 1863-1865 EDW.CARTER & EXECUTORS OF T.WHITFORD
Management: Chief Agent 1863-1865 THOS.HAMBLEY

NANTURRAS   ST.HILARY   SW 551307 0829

Production: Tin	Black(tons)	Stuff(tons)	Tin(tons)	Value(£)
1904	5.00	0.00	0.00	350.00
1905	7.00	0.00	0.00	392.00

Ownership: 1903-1904 NANTURRAS MINING CO.
Comment 1904 OR PENBERTHY CROFTS; 1905 SEE PENBERTHY CROFTS
Management: Chief Agent 1903 J.MAYNE; 1904 T.PENBERTHY

Employment:	Underground	Surface	Total
1903	12	6	18
1904	21	17	38

NELSON   CAMBORNE   SW 640382 0830

Production: Copper   Ore(tons) Metal(tons)   Value(£)
            1862   17.00   1.20   94.70
            1863   6.00   0.40   25.50
            Comment 1862—1863 (C)
Ownership: Comment 1863—1865 SUSPENDED
Management: Manager 1859 GEO.R.ODGERS; 1860—1862 STEP.LEAN
            Chief Agent 1859 JOH.ANGROVE; 1861—1862 HUGH STEPHENS
            Secretary 1859—1860 C.WESCOMBE (P); 1861—1865 W.H.REYNOLDS
            (P)

NEPTUNE   PERRANUTHNOE   SW 541299 0831

Production: Tin No detailed return
Ownership: 1902 G.D.MCGREGOR

NEPTUNE,EAST   PERRANUTHNOE   SW 546301 0832

Management: Chief Agent 1867—1868 PETER FLOYD JAS.ROWE JNR.

NEPTUNE,OLD   PERRANUTHNOE   SW 546301 0833

Production: Copper   Ore(tons) Metal(tons)   Value(£)
            1863   30.00   1.40   105.00
            1864   10.00   0.60   45.50
            Comment 1863—1864 (C)
            Tin No detailed return
Ownership: Comment 1864—1865 SUSPENDED
Management: Manager 1862—1863 JOH.HARRIS
            Chief Agent 1862—1863 ED.HARVEY
            Secretary 1862—1865 O.WANDSEY

NEWQUAY   NEWQUAY   SW 810613 0834

Production: Lead   Ore(tons) Metal(tons)   Value(£)
            1845   73.00   43.00   0.00

NEWTON   ST.MELLION   SX 399699 0835

Production: Copper No detailed return
            Tin No detailed return
            Arsenic No detailed return
            Silver   Ore(tons) Metal(tons)   Value(£)
            1877   140.00   0.00   0.00
            1878   84.50   0.00   5806.70
            1879   17.30   0.00   181.60
            Comment 1878 ORE CONTAINED 26,800 OZ. SILVER
            Arsenic Pyrite Ore(tons)   Value(£)
            1874   938.80   450.00
            1875   3254.50   1627.30
            1876   1488.30   3062.00

                               321

NEWTON   ST.MELLION   Continued

Arsenic Pyrite   Ore(tons)   Value(£)
1877   809.70   1618.00
1879   32.20   22.50
1880   16.00   15.00
1897   50.00   50.00
Comment 1876-1877 ARSENICAL MUNDIC
Ownership: 1877-1879 WHEAL NEWTON LTD.; 1897 THOS.WHITE
              Comment 1874-1875 FORMERLY BARNARD; 1879 NOW CALLED SOUTH
              PRINCE OF WALES; 1880-1881 SEE FORTUNE(HARROWBARROW)
Management: Manager 1874-1879 HY.BENNETTS
              Chief Agent 1873 HY.BENNETTS; 1877 S.H.EMMENS; 1878
              WM.GRIFFIN
              Secretary 1874-1875 H.F.HAWKE; 1876-1877 G.EMMENS; 1878
              G.EMMENS S.BROOME
Employment: Underground   Surface   Total
              1878   73   12   85
              1879   32   3   35
              1880   19   6   25
              1881   30   4   34
              1897   3   3

NEWTON PITS   TREVAGUE   0836

Production: Manganese No detailed return
Ownership: 1900 NEWTON PITS CO.
Management: Secretary 1900 WM.BAWDEN

NEWTON SILVER MINE   ST.MELLION   SX 399699 0837

Production: Copper No detailed return
             Tin No detailed return
             Arsenic No detailed return
Ownership: 1879 NEWTON SILVER MINING CO.LTD.
Management: Manager 1879 H.BENNETT
             Secretary 1879 A.C.COX (S)

NINNIS DOWNS   ST.MEWAN   SW 980509 0838

Production: Tin   Black(tons) Stuff(tons)   Tin(tons)   Value(£)
             1884   0.50   0.00   0.00   22.00
             1887   1.10   0.00   0.00   70.00
             1888   No detailed return
             Comment 1888 NOT WORKING
Ownership: 1886 NINNIS DOWN CO.; 1887-1888 NINNIS DOWNS CO.
Management: Chief Agent 1883-1884 JACOB OLVER; 1886-1888 JACOB OLVER
Employment: Underground   Surface   Total
             1884   4   4   8
             1886   4   4
             1887   7   5   12

322

NORRIS   ST.CLEER   SX 248699 0839

Production: Tin   Black(tons) Stuff(tons)   Tin(tons)   Value(£)
            1863   12.20   0.00   0.00   639.90
            1864   36.20   0.00   0.00   2150.10
            1865   22.70   0.00   0.00   1325.30
            1866   21.90   0.00   0.00   1149.90
            1867   3.00   0.00   0.00   148.70
Management: Manager 1861–1862 JAS.NANCE
            Chief Agent 1860 J.ANDREWS JAS.NANCE; 1861–1866 WM.ANDREWS
            Secretary 1860–1866 J.FIELD JNR. (P)

NORTH DOWNS   SCORRIER   SW 705445 0841

Production: Copper   Ore(tons) Metal(tons)   Value(£)
            1845   306.00   24.00   1731.00
            1846   232.00   18.00   1284.40
            1847   183.00   14.50   1093.40
            1852   222.00   14.80   1333.50
            1853   471.00   36.10   3508.40
            1854   543.00   43.90   4582.10
            1855   188.00   10.80   1109.30
            1857   105.00   8.60   873.50
            1858   204.00   18.00   1688.50
            1859   536.00   44.10   4342.30
            1860   945.00   77.10   7145.50
            1861   1492.00   132.40   12179.20
            1862   1348.00   105.20   8859.30
            1863   1121.00   84.90   6783.40
            1864   656.00   43.90   3821.10
            1865   291.00   20.00   1575.30
            1866   186.00   12.00   789.50
            1867   326.00   28.60   2051.00
            1868   761.00   69.60   4776.90
            1869   726.00   70.40   4513.00
            1870   577.00   50.40   3039.00
            1871   463.00   34.10   2075.40
            Comment 1845–1847 (C); 1852–1855 (C); 1857–1871 (C)
            Tin   Black(tons) Stuff(tons)   Tin(tons)   Value(£)
            1857   2.30   0.00   0.00   180.70
            1858   1.00   0.00   0.00   40.60
            1860   5.20   0.00   0.00   352.70
            1868   1.10   0.00   0.00   57.60
            1869   1.90   0.00   0.00   85.40
            1870   0.70   0.00   0.00   46.80
Ownership: Comment 1912–1913 NORTH DOWN. SEE PEEVOR UNITED
Management: Manager 1859–1869 FRAN.PRYOR
            Chief Agent 1859–1860 JOHN PRINCE JNR.; 1861–1866 T.GRENFELL;
            1867–1868 J.GRENFELL; 1869 JOS.WILLIAMS
            Secretary 1859 FRAN.PRYOR (P); 1860 JOHN NICHOLAS (P);
            1861–1864 FRAN.PRYOR (P); 1865–1868 FRAN.PRYOR WM.WARD; 1869
            WM.WARD
Employment: Underground   Surface   Total
            1912–1913
            Comment 1912–1913 SEE PEEVOR UNITED

CM–M*   323

Production: Copper   Ore(tons) Metal(tons)   Value(£)
            1862   62.00   4.20   350.30
            1863   438.00   27.50   2170.90
            1864   924.00   55.20   4547.20
            1865   1080.00   62.00   4832.10
            1866   667.00   38.10   2437.80
            1867   1705.00   145.60   10427.00
            1868   2513.00   182.70   12271.80
            1869   1524.00   108.30   6702.50
            1870   196.00   14.00   847.60
            Comment 1862-1870 (C)
            Tin   Black(tons) Stuff(tons)   Tin(tons)   Value(£)
            1862   0.00   0.00   0.00   414.00
            1863   0.00   0.00   0.00   588.90
            1864   0.00   0.00   0.00   481.80
            1865   0.00   0.00   0.00   547.90
            1866   0.00   0.00   0.00   624.30
            1867   0.00   0.00   0.00   1660.10
            1868   0.00   0.00   0.00   1803.70
            1869   0.00   0.00   0.00   1956.10
            1870   0.00   0.00   0.00   538.00
            1892   0.20   0.00   0.00   11.00
            1893   No detailed return
            1895   0.00   10.00   0.00   5.00
            1896   No detailed return
            1898   10.00   0.00   0.00   450.00
            1899   No detailed return
            1901   0.70   0.00   0.00   42.00
            1902   2.10   0.00   0.00   137.00
            1903   0.00   7.00   0.00   102.00
            1904   0.00   20.00   0.00   6.00
            1905   0.00   38.00   0.00   168.00
            1906   1.00   0.00   0.00   80.00
            Comment 1862 TIN STUFF.TWO RETURNS AGG.; 1863-1870 TIN STUFF
            Arsenic   Ore(tons) Metal(tons)   Value(£)
            1898   16.00   0.00   165.00
            Arsenic Pyrite   Ore(tons)   Value(£)
            1897   248.00   170.00
            1900   33.00   66.00
            1901   55.00   111.00
            1902   21.00   21.00
Ownership: 1891-1893 WM.ROSEWARNE; 1895 JOHNS & SON; 1897 G.PENGELLY;
            1898-1907 ANGLO PENINSULA MINING & CHEMICAL CO.LTD
            Comment 1859 INC.TRESKERBY; 1860 INC.ROSE (SCORRIER)
Management: Manager 1861-1863 THOS.TRELEASE; 1866-1869 WM.RICH
            Chief Agent 1859 JOH.VIVIAN; 1860 JOS.VIVIAN MART.JENKIN;
            1861-1863 MART.JENKIN; 1864 JAS.CRAZE MART.JENKIN; 1865
            WM.RICH MART.JENKIN; 1866-1868 CORN.BAWDEN; 1869 W.ENNOR;
            1897-1907 WM.ROSEWARNE
            Secretary 1859-1860 RICH.GREENWOOD (P); 1861-1869 ED.KING
            (P)
Employment: Underground   Surface   Total
            1892   4   4
            1895   2   2
            1897   3   8   11

                                   324

NORTH DOWNS,GREAT   SCORRIER   Continued

	Underground	Surface	Total
1898	5	6	11
1899	4	4	8
1900	3	5	8
1901	3	6	9
1902	6	3	9
1903	5	3	8
1904	2	1	3
1905-1906	3	1	4

NORTHWOOD   ST.NEOT   SX 202699 0843

Production: Tin   Black(tons) Stuff(tons) Tin(tons)   Value(£)
            1875   2.10   0.00   0.00   107.50
Ownership: 1875 NORTH WOOD CO.
            Comment 1873 NORTHMOOR
Management: Manager 1872-1873 THOS.TREVILLION
            Chief Agent 1872 JAS.TREVILLION; 1874-1875 JAS.TREVILLION
            Secretary 1872-1873 RICH.CLOGG & SON

NUT   ST.ERTH   SW 574333 0844

Production: Tin   Black(tons) Stuff(tons) Tin(tons)   Value(£)
            1913   0.70   4.70   0.00   69.00
Ownership: 1913 WHEAL NUT LTD.
            Comment 1913 OPENING OLD SHAFTS.NO WORK UNDGRD.
Employment: Underground   Surface   Total
            1913   16   16

OATFIELDS   CROWAN   SW 632337 0845

Production: Copper No detailed return
            Tin No detailed return
Ownership: Comment 1864-1865 SEE CRENVER & ABRAHAM

OCEAN   ST.AGNES   SW 734528 0846

Production: Copper No detailed return
            Tin No detailed return
Ownership: Comment 1863-1865 SEE EAST POLBERRO

OKEL TOR   CALSTOCK   SX 444689 0847

Production: Lead   Ore(tons) Metal(tons)   Value(£)
            1857   1.20   0.90   0.00
            Copper   Ore(tons) Metal(tons)   Value(£)
            1859   323.00   12.20   1023.40
            1860   664.00   24.10   1925.30
            1861   1198.00   45.50   3632.20
            1862   538.00   18.60   1363.80

OKEL TOR   CALSTOCK   Continued

Copper	Ore(tons)	Metal(tons)	Value(£)
1864	905.00	37.00	2827.20
1865	1713.00	70.00	5158.10
1866	1786.00	80.20	5173.90
1867	1339.00	58.50	3927.90
1868	1635.00	77.10	4855.70
1869	1890.00	89.80	5208.70
1870	1170.00	47.80	2506.60
1871	53.00	1.90	102.00
1877	10.00	0.30	16.20
1880	No detailed return		
1883	66.00	3.80	218.00

Comment 1859-1862 (C); 1864-1871 (C); 1880 SEE COTEHELE
CONSOLS; 1883 INC.COTEHELE CONSOLS

Tin	Black(tons)	Stuff(tons)	Tin(tons)	Value(£)
1872	44.80	0.00	0.00	3726.80
1873	8.10	0.00	0.00	627.00
1882	12.20	0.00	7.90	732.00
1883	20.60	0.00	0.00	1093.00
1884	10.50	0.00	0.00	466.00
1885	18.00	0.00	0.00	712.00
1886	38.00	0.00	0.00	2040.00
1887	42.00	0.00	0.00	2550.00

Comment 1873 VALUE EST; 1883-1885 INC.COTEHELE CONSOLS

Arsenic	Ore(tons)	Metal(tons)	Value(£)
1855	151.00	0.00	0.00
1862	413.30	0.00	240.00
1863	210.60	0.00	126.70
1864	35.50	0.00	21.30
1866	35.00	0.00	21.00
1871	67.30	0.00	105.00
1872	27.00	0.00	84.00
1873	90.50	0.00	271.50
1876	218.00	0.00	1042.00
1878	317.00	0.00	2219.00
1879	479.00	0.00	3353.00
1880	690.80	0.00	6744.20
1882	761.30	0.00	3807.00
1883	892.00	0.00	4460.00
1884	551.50	0.00	4400.00
1885	430.00	0.00	3460.00
1886	248.00	0.00	1950.00
1887	119.00	0.00	1126.00

Comment 1855 ARSENICAL MUNDIC; 1862-1864 CRUDE; 1866 CRUDE;
1871 CRUDE; 1876 MOSTLY REFINED; 1882 FROM ARSENICAL PYRITES;
1883-1887 INC.COTEHELE

Arsenic Pyrite	Ore(tons)	Value(£)
1876	267.00	200.00
1877	400.00	215.50
1881	5400.00	4050.00
1882	4567.00	3425.00

Comment 1876-1877 ARSENICAL MUNDIC
Ownership: 1877-1880 HOR.N.LAY; 1881-1886 OKEL TOR CO.LTD.; 1887
CALSTOCK TIN & ARSENIC WORKS SYND.LTD.
Comment 1883-1886 INC.COTEHELE CONSOLS

OKEL TOR   CALSTOCK   Continued

Management: Manager 1859–1865 WM.B.COLLOM; 1869 JOHN EDWARDS; 1873
THOS.NIELL; 1874–1875 H.BENNETT; 1876–1877 JOHN BENNETT
Chief Agent 1859–1860 TRATTON; 1861–1863 J.EDWARDS; 1864–1865
J.METHERELL; 1866–1868 JOHN RODDA; 1870 J.GILLCHRIST; 1871
WM.B.COLLOM; 1872 J.BAILEY; 1878–1886 HY.BULFORD; 1887
H.T.BULFORD
Secretary 1860–1865 W.CHANNING (P); 1866–1870 WM.B.COLLOM;
1872 WM.B.COLLOM; 1874–1875 R.F.HAWKE; 1876–1877 H.A.LAY;
1878–1881 HOR.N.LAY

Employment:

	Underground	Surface	Total
1878	33	30	63
1879	51	49	100
1880	49	45	94
1881	67	73	140
1882	86	106	192
1883	64	108	172
1884	33	79	112
1885	27	42	69
1886	63	68	131
1887	62	80	142

Comment 1883–1886 INC.COTEHELE CONSOLS

ONSLOW CONSOLS,GREAT   ST.BREWARD   SX 096777 0848

Production: Copper

	Ore(tons)	Metal(tons)	Value(£)
1853	150.00	6.10	621.20
1855	608.00	19.20	1691.50
1856	173.00	5.30	393.70
1863	2.00	0.10	10.50

Comment 1853 (C); 1855–1856 (C); 1863 (C)
Arsenic No detailed return
Ownership: 1896–1898 THOS.PARKYN
Comment 1865 SUSPENDED; 1896–1898 ONSLOW CONSOLS
Management: Manager 1859 GEO.RICKARD; 1861–1864 GEO.RICKARD; 1869–1872
THOS.DUNN
Secretary 1861–1865 WIL.FORSTER (P); 1869–1872 C.WILLIAMS

OSBORNE   CROWAN   SW 586327 0849

Production: Tin

	Black(tons)	Stuff(tons)	Tin(tons)	Value(£)
1872	31.10	0.00	0.00	2608.80
1873	31.50	0.00	0.00	2314.70
1874	1.70	10.00	0.00	125.50

Ownership: Comment 1871 SEE CROWAN MINES; 1873 STOPPED IN 1873 OR 1874;
1874 STOPPED IN 1874; 1875 IN LIQUIDATION
Management: Manager 1872–1873 CARKEEK; 1874–1875 JOHN CARKEEK
Chief Agent 1872–1873 RICH.ROWE
Secretary 1872–1875 W.PAGE CARDOZA

OWEN VEAN   ST.HILARY   SW 543309 0850

Production: Tin   Black(tons) Stuff(tons)   Tin(tons)   Value(£)
            1883   5.90   0.00   0.00   309.00
            1884   65.80   0.00   0.00   2881.00
            1885   26.40   0.00   0.00   1144.00
            Comment 1883-1885 INC.TREGURTHA DOWNS
Ownership: 1882-1885 OWEN VEAN & TREGURTHA DOWNS MINING CO.
            Comment 1882-1885 INC.TREGURTHA DOWNS; 1909-1913 SEE HAMPTON
Management: Chief Agent 1882-1885 WM.DERRY
Employment: Underground   Surface   Total
            1882   41   41
            1883   40   35   75
            1884   90   58   148
            1885   63   66   129
            1909-1913
            Comment 1882-1885 INC.TREGURTHA DOWNS; 1909-1913 SEE HAMPTON

OWLES   ST.JUST   SW 366325 0851

Production: Copper   Ore(tons) Metal(tons)   Value(£)
            1878   183.00   16.60   169.90
            1879   162.80   10.20   345.50
            1882   134.00   8.40   536.00
            1888   37.00   0.00   148.00
            Tin   Black(tons) Stuff(tons)   Tin(tons)   Value(£)
            1853   191.00   0.00   0.00   13146.00
            1854   252.60   0.00   0.00   16136.00
            1855   242.40   0.00   0.00   16891.10
            1856   231.80   0.00   0.00   18576.40
            1857   233.60   0.00   0.00   20067.80
            1858   208.40   0.00   0.00   14306.70
            1859   207.70   0.00   0.00   16720.80
            1860   202.10   0.00   0.00   17223.90
            1861   190.80   0.00   0.00   14757.40
            1862   280.90   0.00   0.00   19907.70
            1863   226.90   0.00   0.00   16181.60
            1864   218.20   0.00   0.00   14828.00
            1865   204.50   0.00   0.00   11924.50
            1866   173.70   0.00   0.00   9147.20
            1867   60.80   0.00   0.00   3248.60
            1868   151.00   0.00   0.00   8735.10
            1869   287.10   0.00   0.00   20938.20
            1870   275.90   0.00   0.00   21674.20
            1871   236.00   0.00   0.00   19620.10
            1872   226.90   0.00   0.00   20248.20
            1873   260.00   0.00   0.00   20280.00
            1874   208.00   0.00   0.00   43520.00
            1875   290.50   0.00   0.00   14525.00
            1876   285.30   0.00   0.00   12850.00
            1877   289.00   0.00   0.00   11555.00
            1878   219.00   0.00   0.00   7725.30
            1879   205.30   0.00   0.00   8905.60
            1880   190.10   0.00   0.00   10722.50
            1881   277.50   0.00   189.00   14985.00
            1882   225.00   0.00   148.50   13500.00

Tin	Black(tons)	Stuff(tons)	Tin(tons)	Value(£)
1883	198.00	0.00	0.00	10493.00
1884	245.00	0.00	0.00	11086.00
1885	133.50	0.00	0.00	6650.00
1886	109.00	0.00	0.00	5995.00
1887	97.50	0.00	0.00	5850.00
1888	82.60	0.00	0.00	5785.00
1889	109.50	0.00	0.00	5777.00
1890	100.40	0.00	0.00	5750.00
1891	109.40	0.00	0.00	5450.00
1892	108.60	0.00	0.00	5966.00
1893	19.00	0.00	0.00	760.00

Comment 1862 TWO RETURNS AGGREGATED; 1873-1874 VALUE EST

Arsenic	Ore(tons)	Metal(tons)	Value(£)
1879	31.00	0.00	124.00
1880	14.60	0.00	61.80

Bismuth	Ore(tons)	Value(£)
1877	No detailed return	
1879	0.10	14.00

Comment 1877 NO ORE SOLD

Uranium	Ore(tons)	Value(£)
1877	No detailed return	
1878	0.20	0.00
1879	0.10	20.00

Comment 1877 NO ORE SOLD

Ownership: 1877-1878 R.BOYNS A.CHENHALLS & OTHERS; 1880-1881 WHEAL OWLES
ADVENTURERS; 1882-1886 JOHN WESTON & OTHERS; 1887
SPEN.J.WESTON & OTHERS; 1888-1890 THOS.BOLITHO & SONS &
OTHERS; 1891-1893 THOS.BOLITHO & SONS & OTHERS C.B.
Comment 1878-1881 INC.EDWARD; 1893 ABANDONED

Management: Manager 1859-1881 RICH.BOYNS
Chief Agent 1859 JOHN BOYNS JNR.; 1860 JOHN BOYNS JNR.JOHN
HATTO; 1861 JOHN HOLLOW WM.ROWE JAS.ROWE; 1862-1864 JOHN
HOLLOW WM.ROWE JOHN BOYNS; 1865-1868 JOHN HOLLOW JOHN BOYNS;
1869-1870 J.HOLLOW P.HARVEY W.ROWE J.BOYNS; 1871-1874
P.HARVEY W.ROWE J.BOYNS T.TREGEAR; 1875 P.HARVEY W.ROWE
J.BOYNS T.TREGEAR & WM.OATES,BENJ.ROWE; 1876-1877 JOHN BOYNS;
1878-1879 PHIL.HARVEY THOS.TREGEAR BEN.ROWE; 1880-1881
JAS.ROACH THOS.TREGEAR BEN.ROWE; 1882-1888 RICH.BOYNS;
1891-1893 RICH.BOYNS
Secretary 1859 JOHN BOYNS (P); 1860-1869 RICH.BOYNS (P);
1870-1876 RICH.BOYNS; 1877-1878 RICH.BOYNS & OTHERS;
1879-1880 WESTON&BOYNS & OTHERS; 1889-1891 RICH.BOYNS;
1892-1893 RICH.BOYNS (P)

Employment:	Underground	Surface	Total
1878	121	107	228
1879	106	82	188
1880	119	102	221
1881	118	112	230
1882	121	104	225
1883	108	95	203
1884	105	182	287
1885	70	60	130
1886	66	54	120
1887	60	50	110

OWLES   ST.JUST   Continued

                    Underground   Surface   Total
           1888   68   48   116
           1889   58   46   104
           1890   55   46   101
           1891   63   46   109
           1892   54   46   100
           1893   4   4

OXNAMS LANDS   ST.AGNES   0852

Production: Lead & Silver   Ore(tons) Lead(tons) Silver(ozs)   Value(£)
           1846   188.00   113.00   0.00   0.00
           1847   47.00   28.00   0.00   0.00
           1848   470.00   288.00   0.00   0.00
           1849   269.20   164.00   0.00   0.00
           1850   172.00   101.10   0.00   0.00
           1851   70.50   44.70   0.00   0.00
           1852   39.00   24.50   0.00   0.00
           1853   8.50   2.20   66.00   0.00
           1854   2.50   1.70   87.00   0.00
           1855   9.00   6.30   312.00   0.00
           Comment 1846-1850 OXNAMS

PADSTOW SILVER LEAD   PADSTOW   SW 896773 0854

Production: Lead & Silver   Ore(tons) Lead(tons) Silver(ozs)   Value(£)
           1876   50.00   37.50   177.00   0.00
           1877   24.00   18.00   90.00   360.00
           Comment 1876-1877 PADSTOW CONSOLS
           Copper No detailed return
Ownership: 1875-1876 PADSTOW MINING CO.; 1877 PADSTOW CONSOLS MINING
           CO.
           Comment 1877 PADSTOW CONSOLS
Management: Chief Agent 1875 HY.FRANCIS; 1876 T.ALDERSON BROWNE;
           1877-1010 CARVER F.H.CARVER

PALHARMON   TYWARDREATH   SX 084567 0855

Management: Chief Agent 1865 RICH.RICH
           Secretary 1865 R.FOSTER

PAR   LUXULYAN   SX 054558 0856

Production: Tin   Black(tons) Stuff(tons)   Tin(tons)   Value(£)
           1865   17.00   0.00   0.00   930.10
Ownership: 1907-1908 MINES AND COMMERCE LTD.
           Comment 1865 SUSPENDED; 1873 STOPPED IN 1873 & EARLY IN 1874
Management: Manager 1873 JOHN KESSELL
           Chief Agent 1864 WM.TREGAY J.BEARD; 1872 JOHN KESSELL;
           1907-1908 A.A.HUMPHRIES
           Secretary 1864 W.H.JENKIN; 1908 C.V.THOMAS (S)

Production: Copper No detailed return
Ownership: Comment 1859–1863 SEE FOWEY; 1864–1865 PAR AND ST.BLAZEY
           CONSOLS; 1905–1913 SEE TREGREHAN
Management: Chief Agent 1864–1865 U.BRYANT
           Secretary 1864–1865 WM.TAYLOR

PAR CONSOLS   ST.BLAZEY   SX 058530 0858

Production: Zinc   Ore(tons) Metal(tons)   Value(£)
            1858   11.20   0.00   38.40
            1860   5.00   0.00   3.90
            1867   299.80   0.00   606.70
            1868   209.10   0.00   350.70
            1869   25.00   0.00   47.40
            Copper   Ore(tons) Metal(tons)   Value(£)
            1845   5655.00   464.50   30881.30
            1846   6065.00   557.60   35144.20
            1847   6101.00   625.50   42953.30
            1848   8470.00   914.40   53253.90
            1849   7372.00   663.70   43102.50
            1850   7152.00   641.90   44090.30
            1851   6986.00   660.10   44353.20
            1852   5692.00   527.80   43818.20
            1853   4038.00   344.20   33899.00
            1854   3719.00   342.60   35343.60
            1855   4467.00   443.80   46788.30
            1856   4149.00   437.30   42796.40
            1857   4173.00   420.60   43643.60
            1858   3363.00   320.90   29862.30
            1859   3276.00   306.60   29273.50
            1860   3804.00   347.20   32587.50
            1861   3401.00   321.00   29423.30
            1862   3133.00   279.40   23600.90
            1863   2435.00   192.60   15092.50
            1864   1366.00   97.80   8313.50
            1865   843.00   67.80   5512.20
            1866   948.00   72.20   4716.00
            1867   1115.00   74.50   5238.70
            1868   1155.00   69.20   4637.90
            1869   417.00   21.60   1315.10
            Comment 1845–1869 (C)
            Tin   Black(tons) Stuff(tons)   Tin(tons)   Value(£)
            1852   18.10   0.00   0.00   0.00
            1855   200.80   0.00   0.00   13367.50
            1856   316.50   0.00   0.00   23694.20
            1857   309.90   0.00   0.00   24877.30
            1858   220.50   0.00   0.00   14719.60
            1859   371.40   0.00   0.00   26241.60
            1860   303.30   0.00   0.00   23765.60
            1861   335.90   0.00   0.00   24345.10
            1862   374.50   0.00   0.00   24895.40
            1863   307.10   0.00   0.00   20517.00
            1864   308.30   0.00   0.00   21039.40
            1865   278.40   0.00   0.00   15575.90

331

PAR CONSOLS   ST.BLAZEY   Continued

        Tin   Black(tons) Stuff(tons)   Tin(tons)   Value(£)
        1866   305.20   0.00   0.00   14727.90
        1867   108.60   0.00   0.00   5676.70
        1868   20.90   0.00   0.00   1042.60
        1869   6.30   0.00   0.00   413.10
        Comment 1852 TIN ORE; 1862 TWO RETURNS AGGREGATED
Management: Manager 1859 A.STEPHENS; 1860–1871 FRAN.PUCKEY
        Chief Agent 1859 FRAN.PUCKEY; 1861–1862 THOS.RICH
        JOS.HOSKING; 1863–1864 THOS.RICH JOS.HOSKING AB.STEPHENS;
        1865 THOS.RICH JOS.HOSKING; 1866 JOS.HOSKING; 1867–1868
        GEO.JOB; 1869–1871 JOHN PUCKEY
        Secretary 1859–1864 MAJOR DAVIS (P); 1865–1871
        WM.POLKINGHORNE

PAR CONSOLS,GREAT   ST.BLAZEY   0859

Production: Tin No detailed return
Ownership: 1880–1881 PAR GREAT CONSOLS CO.LTD.
        Comment 1880 TRGHR,BLWDO,BRN RYLTN,BRTY R,AVGGN; 1881 WHEAL
        ELIZABETH IS A PORTION OF SETT
Management: Secretary 1880–1881 ED.THOMAS (S)

PAR CONSOLS,NEW   ST.BLAZEY   0860

Ownership: 1895 NEW PAR CONSOLS LTD.
Management: Manager 1895 C.GREGORY

PAR CONSOLS,WEST   ST.BLAZEY   SX 073533 0861

Production: Copper   Ore(tons) Metal(tons)   Value(£)
        1858   155.00   18.80   1298.50
        1862   10.00   1.00   86.20
        1863   6.00   0.90   76.50
        1864   9.00   2.00   191.60
        Comment 1858 (C); 1862–1864 (C)
        Tin   Black(tons) Stuff(tons)   Tin(tons)   Value(£)
        1858   10.80   0.00   0.00   609.10
        1859   ‐44.20   0.00   0.00   3168.10
        1860   12.60   0.00   0.00   735.60
        1861   14.20   0.00   0.00   1012.80
        1862   4.10   0.00   0.00   229.00
        1863   9.70   0.00   0.00   479.60
        1864   5.70   0.00   0.00   263.40
        Comment 1858 1.7 ORE ($107.6) RET. E.DIST.; 1862 PART YEAR
        ONLY
Ownership: Comment 1864–1865 SUSPENDED; 1905–1913 SEE TREGREHAN
Management: Manager 1859–1860 JOHN WEBB
        Chief Agent 1859–1860 JOHN WEBB JNR.; 1861 JOHN WEBB
        H.G.WEBB; 1862 WM.WOOLCOCK H.G.WEBB; 1863 WM.WOOLCOCK
        Secretary 1859–1865 J.H.MURCHISON (P)

                                    332

PAR TIN   ST.BLAZEY   0862

Production: Tin   Black(tons) Stuff(tons)   Tin(tons)   Value(£)
            1884   0.30   0.00   0.00   13.00
            1885   No detailed return
Ownership: 1884-1886 PAR TIN MINING CO.; 1887-1888 PAR TIN MINING
            CO.LTD.
Management: Chief Agent 1884-1886 THOS.PARKYN; 1887 GEO.TREMAYNE
            Secretary 1888 FRAN.W.MICHELL
Employment: Underground   Surface   Total
            1884   15   2   17
            1885   8   12   20
            1886   8   4   12
            1887   20   20
            1888   13   16   29

PAR,SOUTH   ST.BLAZEY   0863

Production: Copper   Ore(tons) Metal(tons)   Value(£)
            1860-1861 No detailed return
            Comment 1860-1861 SEE EAST CRINNIS
            Tin No detailed return
Ownership: Comment 1861-1865 SEE EAST CRINNIS

PARBOLA   GWINEAR   SW 614363 0865

Production: Tin   Black(tons) Stuff(tons)   Tin(tons)   Value(£)
            1874   60.80   0.00   0.00   2745.00
            1875   49.60   0.00   0.00   2478.50
            1906   4.50   0.00   0.00   489.00
            1907   2.00   0.00   0.00   141.00
            1908   2.70   0.00   0.00   203.00
            1909   9.00   0.00   0.00   670.00
            1910   10.80   0.00   0.00   958.00
            Comment 1874 VALUE EST; 1875 NOW JENNINGS
Ownership: Comment 1875 NOW JENNINGS; 1877 STOPPED AUG.1877
Management: Manager 1872 J.TONKIN; 1873-1875 W.TREGONNING
            Chief Agent 1872-1875 ROBT.RICHARDS
            Secretary 1872-1875 JOHN R.DANIELL

PARBOLA,NORTH   GWINEAR   SW 611366 0866

Production: Tin No detailed return
Ownership: 1905 WM.MIDDLIN; 1906-1908 PARBOLA LTD.; 1909-1913
            A.BLACKBURN
            Comment 1906-1910 NORTH & SOUTH PARBOLA; 1911-1913 NORTH &
            SOUTH PARBOLA NOT WORKED
Management: Chief Agent 1906 BEN.NICHOLAS; 1909-1913 JOHN W.DAWE
Employment: Underground   Surface   Total
            1905   5   4   9
            1906   14   32   46
            1907   35   37   72
            1908   12   30   42
            1909   49   39   88

                                  333

PARBOLA,NORTH   GWINEAR   Continued

                    Underground   Surface   Total
          1910   22   4   26
          1911-1912   2   2
          Comment 1906-1912 NORTH & SOUTH PARBOLA

PARBOLA,SOUTH   GWINEAR   SW 614363 0867

Production: Tin No detailed return
Ownership: Comment 1906-1913 SEE NORTH PARBOLA
Employment: Underground   Surface   Total
          1906-1912
          Comment 1906-1912 SEE NORTH PARBOLA

PARK OF MINES   ST.ENODER   SW 911588 0868

Production: Tin   Black(tons) Stuff(tons)   Tin(tons)   Value(£)
          1873   89.20   0.00   0.00   6826.00
          1874   231.90   0.00   0.00   13744.80
          1875   159.60   0.00   0.00   8673.20
          1876   40.90   0.00   0.00   1761.60
          1877   2.20   0.00   0.00   87.50
          1878   32.20   0.00   0.00   1136.40
          1879   1.30   0.00   0.00   52.00
          1891   1.00   0.00   0.00   59.00
          1892   3.50   0.00   0.00   201.00
          1893   6.50   0.00   0.00   350.00
          1894   11.30   0.00   0.00   512.00
          1895   6.70   0.00   0.00   269.00
          1896   No detailed return
          Comment 1891-1896 NEW PARK OF MINES
Ownership: 1876 PARK OF MINES CONSOLIDATED CO.; 1877-1881 G.BRODERICK &
          PARKER; 1889-1890 PARK O'MINES CO.LTD.; 1891-1895 NEW PARK
          O'MINES CO. C.B.; 1901 ALB.GREGORY
          Comment 1889-1890 PARK O'MINES; 1891-1894 NEW PARK O'MINES;
          1895 NEW PARK O'MINES.SUSPENDED; 1901 PARKA OR NEW PARK OF
          MINE
Management: Manager 1872-1873 WM.EVANS; 1877-1881 JAS.EVANS
          Chief Agent 1872-1873 J.TAMBLYN; 1874-1875 WM.PASCOE
          J.TAMBLYN; 1876 WM.PASCOE; 1877 J.E.PARKER
          Secretary 1872-1881 S.H.F.COX; 1889-1890 FRAN.W.MICHELL;
          1891-1895 FRAN.W.MICHELL (P)
Employment: Underground   Surface   Total
          1878   12   20   32
          1879   20   20   40
          1880   8   19   27
          1881   26   17   43
          1882   36   17   53
          1889   4   1   5
          1890   5   5
          1891   8   8
          1892   6   6
          1893   7   7
          1894   13   7   20

                              334

PARK OF MINES   ST.ENODER   Continued

                    Underground   Surface   Total
            1895   8   6   14
            1901   4   3   7

PARK OF MINES,SOUTH   ST.ENODER   0869

Production: Tin No detailed return
Ownership: 1875-1876 G.BRODERICK & CO.; 1877 J.G.PARKER & OTHERS
Management: Manager 1875-1877 JAS.EVANS

PARK VENTON   MARAZION   0870

Production: Copper No detailed return
            Tin No detailed return
Ownership: Comment 1862-1865 SUSPENDED
Management: Manager 1860-1861 A.BENNETT
            Secretary 1860-1865 A.BENNETT (P)

PARK WYNN   ST.STEPHEN-IN-BRANNEL   SW 942530 0871

Production: Tin   Black(tons) Stuff(tons)   Tin(tons)   Value(£)
            1852   0.60   0.00   0.00   0.00
            Comment 1852 TIN ORE

PARKA CONSOLS   ST.ENODER   SW 911588 0872

Production: Tin   Black(tons) Stuff(tons)   Tin(tons)   Value(£)
            1881   4.00   0.00   2.50   215.50
            1882   0.40   0.00   0.30   23.00
            1883   23.20   0.00   0.00   1230.00
            Comment 1882-1883 PART OF TREVARREN UNITED
Ownership: 1880 PARKA MINES CONSOLS MINING CO.LTD.; 1881 PARKA MINE
            CONSOLS MINING CO.LTD.; 1882 TREVARREN UNITED MINING CO.
            Comment 1880-1881 PARKA MINE CONSOLS; 1882 PARKA MINES
            CONSOLS
Management: Manager 1881 W.BOER
            Chief Agent 1881 W.E.JOB
            Secretary 1880-1882 HY.BROWNE

PARKANCHY   SCORRIER   SW 723430 0873

Production: Tin No detailed return
            Tungsten   Ore(tons) Metal(tons)   Value(£)
            1911   17.00   0.00   800.00
            Wolfram No detailed return
Ownership: 1909-1911 PARKANCHY WOLFRAM LTD.
            Comment 1911 SUSPENDED IN LIQUIDATION
Employment: Underground   Surface   Total
            1909   6   31   37
            1911   31   29   60

                                    335

PATTY   ST.COLUMB MINOR   0874

Ownership: 1877-1881 G.BRODERICK & J.G.PARKER
Management: Manager 1877-1881 JAS.EVANS

PAUL DOWNS   LEEDSTOWN   SW 601331 0875

Production: Copper   Ore(tons) Metal(tons)   Value(£)
            1855   138.00   8.20   813.40
            Comment 1855 (C)
            Tin   Black(tons) Stuff(tons)   Tin(tons)   Value(£)
            1855   0.20   0.00   0.00   12.80

PAWTON   ST.BREOCK   SW 952701 0876

Production: Iron   Ore(tons)   Iron(%)   Value(£)
            1861   14400.00   0.00   4320.00
            1862   9876.00   0.00   2962.80
            1863   7618.00   0.00   5692.50
            1864   7848.90   0.00   2550.60
            1865   9626.00   0.00   2407.20
            1873   1000.00   0.00   1115.00
            1874   4372.00   0.00   2732.00
            1875   No detailed return
            Comment 1861-1864 BH.; 1865 BH.INC.TREMOOR; 1873-1875 BH.
Ownership: 1863-1865 FRED.LEVICK; 1873-1875 NATIVE IRON ORE CO.;
             1876-1882 PAWTON & NEW MILLS IRON MINES CO.
             Comment 1865 SUSPENDED; 1871 SEE PAWTON, DEVON; 1872 SEE
             LADOCK; 1877-1878 INC.NEW MILLS; 1879 INC.NEW MILLS NOT
             WORKED; 1880-1882 NOT WORKED
Management: Manager 1874 WALT.BAXTER
             Chief Agent 1863-1864 WM.VIVIAN; 1873-1874 W.H.HOSKING; 1875
             JOHN NETTING; 1876-1882 J.R.HARVEY
Employment: Underground   Surface   Total
            1878   12   2   14

PEDNANDREA UNITED   REDRUTH   SW 703420 0877

Production: Copper   Ore(tons) Metal(tons)   Value(£)
            1860   240.00   18.30   1589.20
            1861   180.00   13.50   1175.90
            1862   100.00   5.40   428.30
            1863   90.00   5.00   379.50
            1864   23.00   1.70   144.90
            1865   66.00   4.10   328.00
            1866   20.00   1.30   80.50
            1867   22.00   1.70   116.50
            1868   10.00   0.90   65.20
            1872   8.00   0.40   27.00
            1874   6.00   0.60   45.10
            1875   10.00   1.10   83.70
            1876   3.00   0.30   20.50
            1877   7.00   0.60   37.60
            1878   2.60   0.20   13.70

                                  336

Comment 1860–1868 (C)PEDNANDREA; 1872 (C)PEDNANDREA;
1874–1878 (C)PEDNANDREA

Tin	Black(tons)	Stuff(tons)	Tin(tons)	Value(£)
1854	43.00	0.00	0.00	108.40
1855	39.80	0.00	0.00	2574.00
1856	141.60	0.00	0.00	10528.50
1857	266.20	0.00	0.00	19182.30
1858	288.70	0.00	0.00	18175.70
1859	258.20	0.00	0.00	18825.40
1860	224.60	0.00	0.00	17558.70
1861	209.50	0.00	0.00	14976.70
1862	302.60	0.00	0.00	19720.90
1863	239.30	0.00	0.00	15641.30
1864	213.90	0.00	0.00	13187.00
1865	227.60	0.00	0.00	12713.70
1866	230.00	0.00	0.00	11339.60
1867	182.80	0.00	0.00	9707.80
1868	147.60	0.00	0.00	8296.00
1869	248.00	0.00	0.00	17767.00
1870	303.90	0.00	0.00	23067.80
1871	341.00	0.00	0.00	27910.00
1872	255.50	0.00	0.00	22889.10
1873	250.40	0.00	0.00	19651.00
1874	349.80	0.00	0.00	21220.00
1875	304.10	0.00	0.00	16147.50
1876	319.00	0.00	0.00	14391.80
1877	367.00	0.00	0.00	15400.00
1878	212.80	0.00	0.00	7573.10
1879	69.30	0.00	0.00	2627.20
1880	19.50	0.00	0.00	1658.60
1881	58.50	0.00	39.50	3455.40
1882	190.50	0.00	120.00	11146.00
1883	123.70	0.00	0.00	6552.00
1884	162.00	0.00	0.00	7386.00
1885	134.40	0.00	0.00	700.00
1886	150.70	0.00	0.00	8350.00
1887	153.00	0.00	0.00	9971.00
1888	153.90	0.00	0.00	10329.00
1889	159.60	0.00	0.00	8842.00
1890	139.00	0.00	0.00	7802.00
1891	76.10	0.00	0.00	4334.00
1893	0.00	120.00	0.00	120.00
1894	0.00	80.00	0.00	125.00
1897	0.00	49.00	0.00	15.00
1898	285.00	0.00	0.00	56.00
1899	2.20	0.00	0.00	157.00
1900	2.00	0.00	0.00	124.00

Comment 1854 PEDNANDREA SOLD IN STONE; 1855–1861 PEDNANDREA;
1862 PEDNANDREA. TWO RETURNS AGG.; 1863–1866 PEDNANDREA; 1879
PEDNANDREA & SPARNON CONSOLS; 1880 VALUE INC.TIN STUFF; 1882
PEDNANDREA AND SPARNON UNITED; 1885 VALUE VERY DOUBTFUL; 1898
BLACK TIN VERY DOUBTFULL

Arsenic	Ore(tons)	Metal(tons)	Value(£)
1857	48.40	0.00	96.80
1858	49.30	0.00	110.90

PEDNANDREA UNITED   REDRUTH   Continued

Arsenic	Ore(tons)	Metal(tons)	Value(£)
1863	53.30	0.00	53.30
1865	42.00	0.00	35.00
1866	75.00	0.00	83.00
1867	27.10	0.00	34.50
1868	55.30	0.00	163.90
1870	24.00	0.00	79.00
1873	36.80	0.00	105.60
1874	61.50	0.00	120.00
1875	74.00	0.00	148.00
1876	63.10	0.00	331.00
1877	98.00	0.00	272.30
1878	63.30	0.00	263.90
1879	19.10	0.00	76.40
1882	18.00	0.00	54.00
1885	10.00	0.00	50.00
1886	14.10	0.00	62.00
1887	30.00	0.00	164.00
1888	40.00	0.00	257.00
1889	35.00	0.00	229.00
1890	24.00	0.00	169.00
1891	42.00	0.00	299.00

Comment 1857–1858 PEDNANDREA; 1863 CRUDE; 1865 CRUDE; 1866
PEDNANDREA. CRUDE; 1867–1868 CRUDE; 1870 PEDNANDREA. CRUDE;
1873–1879 PEDNANDREA; 1885 SOOT; 1886 CRUDE
Ownership: 1876–1878 PEDNANDREA CONSOLIDATED MINE CO.; 1879 PEDNANDREA
MINE CO.; 1880–1881 PEDNANDREA UNITED MINE ADVENTURERS;
1882–1890 R.BAIN & OTHERS; 1893–1895 JOHN GOLDSWORTHY;
1897–1898 W.J.TRYTHALL; 1899–1902 JOHN RODDA & SONS
Comment 1873–1878 PEDNANDREA; 1879 PEDNANDREA AND SPARNON
CONSOLS; 1880 PEDNANDREA & SPARNON; 1893–1894 PEDNANDREA PART
OF; 1895 PEDNANDREA PART OF WORK STOPPED; 1897 PEDNANDREA;
1898 PEDNANDREA ABANDONED; 1899–1902 PEDNANDREA
Management: Manager 1859 JOHN CARPENTER; 1865–1878 WM.TREGAY
Chief Agent 1859 JAS.THOMAS; 1860 WM.TREGAY JAS.THOMAS;
1861–1863 WM.TREGAY JAS.THOMAS T.DELBRIDGE; 1864 WM.TREGAY
JAS.THOMAS; 1865–1871 JAS.THOMAS; 1872–1878 W.PRIDEAUX
J.POPE; 1879 WM.ROSEWARNE; 1880–1881 WM.ROSEWARNE JAS.THOMAS;
1889–1890 WM.RICH; 1893–1895 JOHN GOLDSWORTHY
Secretary 1859–1864 W.PAGE CARDOZA (P); 1865–1866 CARDOZA &
PEARCE; 1867–1868 CARDOZA & CHEGWIN; 1869–1875 W.PAGE
CARDOZA; 1879 R.BAIN; 1880–1881 A.F.TEAGUE (P); 1882–1890
R.S.TEAGUE

Employment:	Underground	Surface	Total
1878	96	172	268
1879		41	41
1880	26	31	57
1881	98	115	213
1882	138	128	266
1883	66	47	113
1884	83	55	138
1885	104	55	159
1886	126	46	172
1887	103	79	182
1888	99	89	188

338

	Underground	Surface	Total
1889	92	88	180
1890	92	78	170
1893	5		5
1894	6		6
1897	8		8
1898	5		5
1899	11		11
1900	6		6

PEEVOR   REDRUTH   SW 708442 0879

Production: Copper

Copper	Ore(tons)	Metal(tons)	Value(£)	
1882	5.10	0.20	16.00	
Tin	Black(tons)	Stuff(tons)	Tin(tons)	Value(£)
1872	1.80	0.00	0.00	711.50
1873	3.70	350.70	0.00	1082.60
1874	50.10	933.00	0.00	4066.40
1875	48.50	2315.90	0.00	5407.70
1876	116.90	2178.30	0.00	8065.50
1877	292.00	50.70	0.00	11533.20
1878	546.60	785.50	0.00	19893.90
1879	563.40	610.90	0.00	25142.00
1880	649.50	68.80	0.00	31677.90
1881	411.90	629.60	0.00	21871.60
1882	218.10	0.00	143.90	13365.00
1883	119.40	0.00	0.00	6325.00
1884	81.00	0.00	0.00	3555.00
1885	59.70	0.00	0.00	2631.00
1886	10.50	324.00	0.00	573.00
1887	4.50	157.00	0.00	250.00
1888	0.30	14.00	0.00	19.00
1912	8.00	0.00	0.00	917.00
1913	19.00	69.00	0.00	2120.00

Comment 1872 VALUE INC.TIN STUFF; 1874 VALUE EST; 1877 3
RETURNS AGGREGATED; 1886 TWO RETURNS AGGREGATED; 1912-1913
PEEVOR UNITED

Arsenic	Ore(tons)	Metal(tons)	Value(£)
1883	11.50	0.00	57.00

Ownership: 1877-1890 WHEAL PEEVOR MINING CO.; 1891-1893 WHEAL PEEVOR
MINING CO. C.B.; 1911-1913 PEEVOR UNITED MINES LTD.
Comment 1889-1892 SUSPENDED; 1893 ABANDONED
Management: Manager 1872 ED.ROGERS; 1873-1874 AB.T.JAMES; 1877
THOS.PRYOR; 1878 JOS.PRYOR; 1879 THOS.PRYOR
Chief Agent 1872-1874 WM.ARGENT; 1875-1879 WM.WHITE; 1880
WM.T.WHITE THOS.C.KING JOHN ANGOVE; 1881 THOS.C.KING
WM.T.WHITE; 1882-1884 WM.T.WHITE; 1911-1913 J.A.TEMBY
Secretary 1872-1881 THOS.PRYOR; 1885-1890 THOS.PRYOR;
1891-1893 THOS.PRYOR (P)

Employment:

	Underground	Surface	Total
1878	151	112	263
1879	158	118	276
1880	186	135	321
1881	183	139	319

PEEVOR   REDRUTH   Continued

	Underground	Surface	Total
1882	126	96	222
1883	89	75	164
1884	64	44	108
1885	70	37	107
1886	4	9	13
1887	3	5	8
1888		4	4
1890-1893		1	1
1912-1913			

Comment 1912-1913 SEE PEEVOR UNITED

PEEVOR UNITED   REDRUTH   SW 708442 0880

Production: Tin No detailed return
Ownership: 1912-1913 PEEVOR UNITED MINES LTD.
            Comment 1912-1913 INC.NORTH DOWNS & BLACK DOG
Management: Chief Agent 1912-1913 J.A.TEMBY
Employment: Underground Surface Total

	Underground	Surface	Total
1911	6	52	58
1912	16	29	45
1913	20	47	67

Comment 1912-1913 INC.PEEVOR,NORTH DOWNS AND BLACK DOG

PEEVOR,EAST   SCORRIER   0881

Production: Copper No detailed return
            Tin No detailed return
Ownership: 1880-1881 J.NORFOLK & OTHERS
Management: Chief Agent 1880-1881 THOS.PARKYN
Employment: Underground Surface Total

	Underground	Surface	Total
1881	2	1	3

PEEVOR,NEW   REDRUTH   0882

Production: Copper   Ore(tons) Metal(tons)   Value(£)

	Ore(tons)	Metal(tons)	Value(£)
1882	1.70	0.20	10.00

            Tin No detailed return
Ownership: 1880-1881 NEW WHEAL PEEVOR MINING CO.; 1882-1883 NEW PEEVOR
            MINING CO.
            Comment 1883 NOT WORKED
Management: Chief Agent 1880-1881 WM.T.WHITE THOS.C.KING JOHN ANGOVE;
            1882-1883 WM.T.WHITE
Employment: Underground Surface Total

	Underground	Surface	Total
1880	10	2	12
1881	6	1	7
1882	2		2

PEEVOR,WEST   REDRUTH   SW 705440 0883

Production: Tin   Black(tons) Stuff(tons)   Tin(tons)   Value(£)
            1879   0.00   135.20   0.00   78.00
            1880   12.10   0.00   0.00   486.80
            1881   0.00   807.90   0.00   1226.80
            1882   130.10   2193.00   0.00   6647.00
            1883   110.80   0.00   0.00   5923.00
            1884   17.80   0.00   0.00   745.00
            1885   1.00   60.00   0.00   33.00
            1886   2.00   50.00   0.00   120.00
            1887   0.50   36.80   0.00   24.00
            1888   0.70   20.00   0.00   35.00
            Comment 1882 TWO RETURNS AGGREGATED
            Arsenic   Ore(tons) Metal(tons)   Value(£)
            1884   20.00   0.00   100.00
Ownership: 1875-1876 WEST PEVOR MINING CO.; 1877 WEST PEEVOR MINING CO.;
            1878-1881 WEST WHEAL PEEVOR MINING CO.; 1882-1891 WEST PEEVOR
            MINING CO.; 1892-1893 WEST WHEAL PEEVOR MINING CO.
            Comment 1875-1877 WEST PEVOR; 1889-1892 SUSPENDED; 1893
            ABANDONED
Management: Chief Agent 1875-1876 AB.T.JAMES; 1877 THOS.PRYOR; 1878-1879
            THOS.PRYOR WM.S.WHITE; 1880-1881 THOS.PRYOR WM.T.WHITE;
            1882-1884 WM.T.WHITE
            Secretary 1877 THOS.PRYOR; 1885-1890 THOS.PRYOR; 1891-1893
            THOS.PRYOR (P)
Employment: Underground   Surface   Total
            1878   10   10
            1879   18   2   20
            1880   31   14   45
            1881   38   54   92
            1882   74   70   144
            1883   60   64   124
            1884   14   10   24
            1885   6   6
            1886   3   3
            1887   4   4
            1888   2   1   3
            1889-1892   1   1

PELYN WOOD   LANLIVERY   SX 097583 0885

Production: Copper No detailed return
            Tin No detailed return
Ownership: 1881-1882 PELYN WOOD MINING CO.; 1904-1905 W.H.PASCOE;
            1906-1907 PELYN MINES LTD.
            Comment 1862-1865 SUSPENDED; 1904-1907 PELYN
Management: Manager 1859-1861 JOS.RICHARDS
            Chief Agent 1859 WARMENGENTON; 1881 T.H.BENNETT
            Secretary 1861-1863 THOS.FULLER (P); 1882 WM.BATTYE;
            1906-1907 W.H.PASCOE
Employment: Underground   Surface   Total
            1881   9   9
            1882   12   12
            1906-1907   4   4

                                    341

PEMBROKE   PAR   SX 058526 0886

Production: Zinc   Ore(tons) Metal(tons)   Value(£)
            1859   69.00   0.00   155.50
            Comment 1859 INC.EAST CRINNIS
            Copper   Ore(tons) Metal(tons)   Value(£)
            1852   265.00   13.50   1103.10
            1853   1177.00   71.70   6831.50
            1854   1880.00   120.60   12010.80
            1855   1496.00   82.50   8002.30
            1856   1203.00   58.60   4923.00
            1857   905.00   47.10   4403.30
            1858   1058.00   57.80   4893.90
            1859   1149.00   60.10   5378.20
            1867   71.00   1.10   31.20
            1868   12.00   0.50   28.50
            1869   1.00   0.10   1.50
            1870   30.00   0.50   3.70
            Comment 1852-1859 (C) INC.EAST CRINNIS; 1867-1870 (C) OLD
            PEMBROKE
            Tin   Black(tons) Stuff(tons)   Tin(tons)   Value(£)
            1856   5.40   0.00   0.00   391.50
            1857   1.40   0.00   0.00   111.20
            Comment 1856-1857 INC.EAST CRINNIS

PEMBROKE COPPER   BREAGE   SW 627282 0887

Production: Copper No detailed return
Ownership: Comment 1880-1881 SEE MOUNTS BAY CONSOLS

PEMBROKE,NEW   ST.BLAZEY   SX 057530 0888

Production: Copper   Ore(tons) Metal(tons)   Value(£)
            1862   22.00   0.40   14.30
            1863   13.00   0.50   34.80
            1864   8.00   0.30   16.70
            1865   1.00   0.10   7.60
            1866   16.00   0.50   24.00
            1867   9.00   0.20   11.50
            1869   55.00   4.90   303.90
            1870   674.00   99.50   6093.60
            1871   445.00   45.90   2815.00
            1873   74.00   6.10   366.90
            1874   118.00   15.90   1070.80
            1875   442.00   54.10   4096.00
            1876   297.00   22.20   1427.00
            Comment 1862-1867 (C)PEMBROKE; 1869-1871 (C); 1873-1876 (C)
            Tin   Black(tons) Stuff(tons)   Tin(tons)   Value(£)
            1866   13.70   0.00   0.00   623.20
            1867   26.90   0.00   0.00   1417.50
            1868   35.10   0.00   0.00   1985.90
            1869   71.50   0.00   0.00   5097.00
            1870   80.20   0.00   0.00   5894.90
            1871   107.60   0.00   0.00   8576.50
            1872   127.10   0.00   0.00   11072.10

                                       342

PEMBROKE,NEW   ST.BLAZEY   Continued

	Black(tons)	Stuff(tons)	Tin(tons)	Value(£)
1873	138.00	0.00	0.00	10827.50
1874	144.60	0.00	0.00	8366.10
1875	112.50	0.00	0.00	6080.00
1876	58.70	0.00	0.00	2570.60
1877	9.70	0.00	0.00	387.70

Ownership: Comment 1876 ABANDONED OCT.1876
Management: Manager 1864-1877 FRAN.PUCKEY
             Chief Agent 1864-1872 JOHN PUCKEY; 1873-1876 CHAS.MERRETT
             Secretary 1864-1877 JOHN POLKINGHORNE

PEMPOLLS & CO. REGULUS   0890

Production: Copper   Ore(tons) Metal(tons) Value(£)
            1846   105.00   10.60   776.50
            1850   86.00   8.00   640.40
            Comment 1846 (C); 1850 (C)

PENBERTHY CROFTS   ST.HILARY   SW 555324 0891

Production: Lead   Ore(tons) Metal(tons) Value(£)
            1876   1.40   1.00   0.00
            1882   1.20   0.80   11.00
            1883   1.20   0.80   9.00
            Copper   Ore(tons) Metal(tons) Value(£)
            1875   32.00   1.80   138.40
            1882   1.00   0.10   3.00
            Comment 1875 (C)

	Black(tons)	Stuff(tons)	Tin(tons)	Value(£)
1874	27.20	0.00	0.00	1211.50
1875	36.20	0.00	0.00	1764.20
1876	12.70	0.00	0.00	709.50
1877	0.00	122.90	0.00	175.50
1879	3.80	0.00	0.00	137.70
1880	10.00	0.00	0.00	491.60
1881	26.30	0.00	17.10	1481.50
1882	28.70	0.00	18.90	1716.00
1883	4.20	15.00	0.00	225.00
1906	8.00	0.00	0.00	680.00
1907	5.00	0.00	0.00	411.00

            Comment 1874 VALUE EST; 1883 TWO RETURNS AGGREGATED
Ownership: 1877 PENBERTHY CROFTS MINING CO.; 1879-1884 PENBERTHY CROFTS
             MINING CO.; 1905-1907 NANTURRAS MINING CO.; 1908-1910 GURLYN
             CONSOLS TIN MINES LTD.
             Comment 1884 NOT WORKED; 1904 SEE NANTURRAS; 1905 FORMERLY
             NANTURRAS; 1909 SHUT AND FOR SALE MARCH 1910
Management: Manager 1873-1874 ALF.BRAY
             Chief Agent 1873-1877 PAUL ROACH; 1879-1884 JOHN CURTIS;
             1905-1910 T.PENBERTHY
             Secretary 1873-1877 THOS.FIELD; 1880 T.ABBOT; 1881 T.ABBOT
             (P); 1906-1907 READ & BRIGSTOCKE
Employment: Underground   Surface   Total
            1879   12   12

343

PENBERTHY CROFTS   ST.HILARY   Continued

                         Underground   Surface   Total
                1880   13   32   45
                1881   38   28   66
                1882   17   31   48
                1883   3   7   10
                1905   6   8   14
                1906   6   4   10
                1907   4   5   9
                1908   3   3

PENCORSE CONSOLS   ST.ENODER   SW 870558 0893

Production: Lead   Ore(tons) Metal(tons)   Value(£)
                1856   3.00   2.10   0.00
                1857   4.60   3.00   0.00
                1858   No detailed return
                1859   6.10   4.50   0.00
            Zinc   Ore(tons) Metal(tons)   Value(£)
                1854   500.00   0.00   0.00
                1855   770.00   0.00   0.00
                1856   928.20   0.00   2093.20
                1857   499.00   0.00   1643.80
                1858   373.00   0.00   1391.60
                1859   368.10   0.00   1223.30
            Copper No detailed return
Ownership: Comment 1860—1865 SUSPENDED
Management: Manager 1859—1867 H.B.GROSE
            Chief Agent 1859 H.B.GROSE
            Secretary 1859—1863 J.B.EALES (P)

PENCROM   LELANT   0894

Production: Tin No detailed return
Management: Manager 1869 JOHN SHARP
            Secretary 1869 W.STREET

PENDARVES AND ST.AUBYN   GWINEAR   SW 631378 0895

Production: Copper   Ore(tons) Metal(tons)   Value(£)
                1854   125.00   10.30   1066.00
            Comment 1854 (C)
            Tin   Black(tons) Stuff(tons)   Tin(tons)   Value(£)
                1856   0.00   0.00   0.00   128.30
            Comment 1856 TIN STUFF

PENDARVES CONSOLS   CAMBORNE   SW 656410 0896

Production: Copper   Ore(tons) Metal(tons)   Value(£)
                1851   258.00   15.30   971.70
            Comment 1851 (C)

                                      344

PENDARVES UNITED   CAMBORNE   SW 660392 0897

Production: Copper   Ore(tons) Metal(tons)   Value(£)
            1866   134.00   6.90   406.40
            1867   17.00   1.00   70.10
            1868   36.00   2.20   146.50
            1869   23.00   1.20   68.80
            1870   34.00   1.70   88.90
            1871   6.00   0.30   18.00
            1872   11.00   1.50   114.40
            1880   13.00   1.00   50.00
            1881   41.70   3.10   176.50
            Comment 1866-1872 (C)
            Tin   Black(tons) Stuff(tons)   Tin(tons)   Value(£)
            1867   456.70   0.00   0.00   25046.60
            1868   565.70   0.00   0.00   32363.90
            1869   467.00   0.00   0.00   33704.50
            1870   476.50   0.00   0.00   35985.50
            1871   381.50   0.00   0.00   30956.70
            1872   253.00   0.00   0.00   22445.50
            1873   48.50   0.00   0.00   3119.30
            1880   1.50   6.90   0.00   140.80
            1881   0.00   281.00   0.00   261.50
            Arsenic   Ore(tons) Metal(tons)   Value(£)
            1869   56.00   0.00   137.50
Ownership: 1880-1881 PENDARVES UNITED MINES CO.
            Comment 1865-1866 INC.CONDURROW & TRYPHENA PENDARVES;
            1867-1872 INC.CONDURROW & TRYPHENA PENDARVES & TOLCARNE &
            HARRIET; 1873 STOPPED IN 1873 OR 1874
Management: Manager 1865-1869 REG.T.GRYLLS; 1870 HY.BENNETTS; 1871
            J.PRISK; 1872-1873 ROBT.RICHARDS
            Chief Agent 1865 J.MOYLE J.DUNSTAN; 1866 HY.BENNETTS
            R.RICHARDS; 1867 JAS.PENBERTHY JOHN KING; 1868 WM.ROBINS
            J.DUNSTAN; 1869 R.RICHARDS T.ROBINS; 1870 W.PENBERTHY JOHN
            KING; 1871 ?FRANCIS WM.PENBERTHY; 1872-1873 CHAS.BISHOP
            WM.PENBERTHY; 1879-1881 JOHN BRENTON
            Secretary 1865-1870 REG.T.GRYLLS; 1871 GEO.LIGHTLEY (P);
            1872-1873 ALF.LANYON; 1880 SML.ABBOT; 1881 SML.ABBOT (P)
Employment: Underground   Surface   Total
            1880   30   12   42
            1881   14   8   22

PENDEEN CONSOLS   ST.JUST   SW 381359 0898

Production: Copper   Ore(tons) Metal(tons)   Value(£)
            1858   222.00   9.20   722.40
            1859   1091.00   44.80   3610.40
            1860   1494.00   67.80   5514.70
            1861   1653.00   78.40   6162.10
            1862   983.00   55.30   4107.40
            1863   570.00   26.30   1739.10
            1864   264.00   10.00   721.00
            1865   279.00   14.20   1023.10
            1866   213.00   12.60   821.30
            1867   60.00   4.30   293.30
            1868   17.00   1.80   97.00

                              345

PENDEEN CONSOLS   ST.JUST   Continued

	Copper	Ore(tons)	Metal(tons)	Value(£)
	1869	9.00	1.40	87.40
	1871	2.00	0.30	17.10

Comment 1858-1869 (C); 1871 (C)

	Tin	Black(tons)	Stuff(tons)	Tin(tons)	Value(£)
	1862	14.90	0.00	0.00	978.30
	1863	54.00	0.00	0.00	3540.10
	1864	57.80	0.00	0.00	3766.90
	1866	21.00	0.00	0.00	1058.20
	1867	1.00	0.00	0.00	41.30
	1870	5.30	0.00	0.00	337.70

Comment 1862 PART YEAR ONLY; 1864 TWO QUARTERS ONLY
Management: Manager 1859 JOHN CARTHEW; 1869 HY.EDDY
Chief Agent 1859 WM.EDDY; 1860-1861 WM.EDDY JAS.WARREN;
1862-1864 JAS.WARREN; 1865-1867 JAS.EDDY; 1868 HY.EDDY W.HALL
W.G.CANVILLE
Secretary 1859-1869 RICH.WHITE (P)

PENGELLY   NANCYGOLLEN   SW 641323 0899

Production: Tin   Black(tons) Stuff(tons)   Tin(tons)   Value(£)
            1913   No detailed return
            Comment 1913 SEE NANCYGOLLEN
Ownership: Comment 1907 SEE NANCYGOLLEN; 1912-1913 SEE NANCYGOLLEN
Employment: Underground   Surface   Total
            1912-1913
            Comment 1912-1913 SEE NANCYGOLLEN

PENGENNA   ST.KEW   SX 051790 0900

Production: Lead & Silver No detailed return
            Antimony   Ore(tons) Metal(tons)   Value(£)
            1861   15.00   0.00   0.00
            1862   No detailed return
            Comment 1862 A FEW CWTS RAISED AND SOLD
Management: Manager 1869-1871 JAS.POLGLAZE
            Chief Agent 1859-1865 E.HITCHINGS
            Secretary 1861-1865 NICH.ENNOR

PENHALDARVA   KENWYN   SW 789461 0901

Production: Lead & Silver   Ore(tons) Lead(tons) Silver(ozs) Value(£)
            1857   8.70   6.70   127.00   0.00
            1858   188.10   141.20   2820.00   0.00
            1859   72.50   54.30   1086.00   0.00
            1860   27.50   20.60   412.00   0.00
            1861   4.40   3.10   61.00   0.00
            Copper No detailed return
Ownership: Comment 1863-1865 SUSPENDED
Management: Manager 1859-1862 JAS.POPE
            Chief Agent 1859-1862 THOS.HODGE
            Secretary 1859-1865 GEO.PAINTER (P)

PENHALDARVA,SOUTH   KENWYN   SW 783454 0902

Production: Lead No detailed return
Ownership: Comment 1863-1865 SUSPENDED
Management: Manager 1859-1862 JAS.POPE
             Chief Agent 1859 THOS.HODGE; 1860-1862 JOHN POPE THOS.HODGE
             Secretary 1859-1865 GEO.PAINTER (P)

PENHALE   CUBERT   SW 764574 0903

Production: Iron   Ore(tons)   Iron(%)   Value(£)
            1858   6134.40   0.00   2754.00
            1859-1861 No detailed return
            1862   1000.00   0.00   700.00
            1863-1864 No detailed return
            Comment 1858 BH.& 81.4 TONS SP (£30.4); 1859 SEE GWINDRA;
            1862 BH.PENHALL; 1863-1864 PENHALL
Ownership: 1863 E.CARTER & CO
            Comment 1863 PENHALL. INC.HALWYN

PENHALE   CUBERT   SW 762580 0904

Production: Lead & Silver   Ore(tons)   Lead(tons) Silver(ozs)   Value(£)
            1849   50.00   35.00   0.00   0.00
            1850   124.00   86.70   0.00   0.00
            1851   65.00   42.00   0.00   0.00
            1852   7.00   5.00   50.00   0.00
            1853   230.00   154.00   2460.00   0.00
            1854   181.00   118.00   1830.00   0.00
            1855   125.00   82.00   1148.00   0.00
            1856   4.00   2.40   0.00   0.00
            1857   1.50   1.10   0.00   0.00
            1858-1859 No detailed return
            1867   37.80   28.30   0.00   0.00
            1869   173.10   130.00   650.00   0.00
            1870   266.00   199.50   997.00   0.00
            1871   121.50   90.40   450.00   0.00
            1872   No detailed return
            Comment 1867 PENHALE UNITED; 1869-1872 PENHALE UNITED
            Copper   Ore(tons) Metal(tons)   Value(£)
            1849   96.00   8.20   569.90
            Comment 1849 (C)
Ownership: Comment 1872 SEE PHOENIX(PERRANZABULOE)

PENHALE AND BARTON   ST.ENODER   SW 905578 0906

Production: Tin   Black(tons) Stuff(tons)   Tin(tons)   Value(£)
            1875   0.00   14.70   0.00   80.40
            1876   9.30   0.00   0.00   353.70
            1877   1.10   6.20   0.00   65.00
            1880   4.70   0.00   0.00   211.00
            1881   2.20   5.00   0.00   115.50
            1882   3.10   5.40   2.00   170.00
            Comment 1876 PENHALE & BURTON

CM-N   347

PENHALE AND BARTON   ST.ENODER   Continued

Ownership: 1876—1877 PENHALE MINING CO.; 1880 E.VICKERS & CO.; 1881—1886
             T.E.VICKERS
             Comment 1883—1886 NOT WORKED
Management: Manager 1880 JOHN BULLOCK
             Chief Agent 1874—1877 S.H.F.COX; 1880—1886 JAS.EVANS
             Secretary 1874 S.H.F.COX
Employment: Underground   Surface   Total
             1880   10   1   11
             1881   30   16   46
             1882   8   6   14

PENHALE AND LARKHOLES   ST.CLEER   SX 242689 0907

Production: Tin   Black(tons) Stuff(tons)   Tin(tons)   Value(£)
            1911   3.60   0.00   0.00   328.00
            Comment 1911 PENHALE. VALUE EST
Ownership: 1910—1912 DEVON & EAST CORNWALL DEVELOPMENT LTD.
            Comment 1910 STARTING MAY 1910; 1911 SUSPENDED NOV.1911; 1912
            NOT WORKED
Management: Chief Agent 1910—1912 WM.TALLOW
Employment: Underground   Surface   Total
            1910   12   12
            1911   6   19   25

PENHALE AND LOMAX   CUBERT   SW 762580 0908

Ownership: Comment 1870—1871 PENHALE & LOMAX UNITED; 1881 SEE PERRAN
            SILVER LEAD CONSOLS
Management: Manager 1869 RICH.PRYOR; 1870—1871 JOHN PRYOR
            Chief Agent 1869—1871 HY.BENNETTS
            Secretary 1869—1871 HY.L.PHILLIPS

PENHALE MOOR   ST.ENODER   SW 905578 0909

Production: Copper No detailed return
            Tin No detailed return
Ownership: Comment 1862—1865 SUSPENDED
Management: Manager 1860—1861 H.B.GROSE; 1864—1865 SML.BENNETTS
            Chief Agent 1863 SML.UREN
            Secretary 1860—1865 W.BRIGGS (P)

PENHALE WHEAL VOR   BREAGE   SW 624309 0910

Production: Tin   Black(tons) Stuff(tons)   Tin(tons)   Value(£)
            1866   23.70   0.00   0.00   1059.30
            1867   36.50   0.00   0.00   1962.50
            1868   22.00   0.00   0.00   1219.80
            1869   3.30   43.00   0.00   283.80
            1871   0.00   9.20   0.00   9.40
            1872   0.00   113.00   0.00   206.70
            1873   0.00   9.20   0.00   48.00

348

PENHALE WHEAL VOR   BREAGE   Continued

          Comment 1873 VALUE EST
Ownership: 1877 PENHALE TIN MINING CO.
Management: Manager 1866—1867 J.CHAPPEL; 1869—1877 W.H.MARTIN
          Chief Agent 1866—1867 R.MARTIN; 1868 W.H.MARTIN; 1869—1877
          JOHN RICHARDS
          Secretary 1866 J.HOLROYD; 1867—1877 WM.BATTYE

PENHALLOW MOOR   NEWLYN EAST   SW 826552 0914

Production: Tin No detailed return
          Iron No detailed return
Ownership: 1872—1873 CORNISH CONSOLIDATED IRON MINES CORP.LTD.; 1875
          CORNISH CONSOLIDATED CORPORATION
Management: Chief Agent 1872—1873 S.J.PITTAR; 1874—1875 JOHN PARKYN

PENHALLS   ST.AGNES   SW 723516 0915

Production: Copper   Ore(tons) Metal(tons)   Value(£)
          1870   11.00   1.30   80.00
          1871   13.80   0.80   0.00
          1879   10.50   1.00   60.90
          Comment 1870 (C)PENHALLES; 1871 (P)COPPER EST; 1879 (C)
          Tin   Black(tons) Stuff(tons) Tin(tons)   Value(£)
          1860   61.70   0.00   0.00   4974.60
          1861   44.20   0.00   0.00   3301.30
          1862   88.20   0.00   0.00   5718.40
          1863   49.20   0.00   0.00   3325.50
          1864   71.70   0.00   0.00   4599.10
          1865   91.60   0.00   0.00   5200.00
          1866   95.30   0.00   0.00   4663.40
          1867   137.50   0.00   0.00   7293.40
          1868   178.20   0.00   0.00   10052.10
          1869   221.90   0.00   0.00   16224.20
          1870   218.10   0.00   0.00   15706.70
          1871   211.00   0.00   0.00   16040.70
          1872   196.60   0.00   0.00   17318.50
          1873   204.30   0.00   0.00   15415.60
          1874   229.00   0.00   0.00   13193.90
          1875   189.90   0.00   0.00   9966.30
          1876   216.20   0.00   0.00   9560.60
          1877   234.00   0.00   0.00   9236.40
          1878   193.30   0.00   0.00   7007.20
          1879   197.20   0.00   0.00   7982.70
          1880   154.30   0.00   0.00   8198.00
          1881   91.10   0.00   62.60   5089.30
          1882   117.80   0.00   79.30   7061.00
          1883   91.00   0.00   0.00   4915.00
          1884   25.00   0.00   0.00   1153.00
          Comment 1862 TWO RETURNS AGGREGATED
Ownership: 1876—1886 PENHALLS MINING CO.
          Comment 1885—1886 NOT WORKED
Management: Manager 1859 MART.EDWARDS; 1860—1862 RICH.PRYOR; 1863—1865
          WM.HIGGINS; 1867 PETER VIAN; 1868—1881 SML.BENNETTS

                                     349

PENHALLS   ST.AGNES   Continued

Chief Agent 1859—1861 ART.GRIPE; 1862 WM.HIGGINS; 1864—1865
JAS.NANCE; 1866—1868 WM.HIGGINS; 1869—1875 WM.HIGGINS PETER
VIAN; 1876—1878 PETER VIAN; 1879 PETER VIAN ART.GRIPE;
1880—1881 RICH.HARRIS; 1882—1886 SML.BENNETTS
Secretary 1859—1874 JOS.NEWTON (P); 1876—1879 JAS.HICKEY;
1880—1881 WALT.PIKE

Employment:	Underground	Surface	Total
1878	103	80	183
1879	97	85	182
1880	76	64	140
1881	85	58	143
1882	78	52	130
1883	52	38	90
1884	37	40	77

PENHANGER   ST.IVE   SX 292675 0916

Production:	Lead & Silver	Ore(tons)	Lead(tons)	Silver(ozs)	Value(£)
	1857	1.10	0.70	0.00	0.00

Ownership: Comment 1861—1865 SUSPENDED
Management: Manager 1859—1860 ROBT.KNAPP
            Chief Agent 1859 JOSH.HUSBAND
            Secretary 1859—1860 W.GEO.NETTLE (P)

PENHELURICK,GREAT   REDRUTH   0917

Production: Copper No detailed return
            Tin No detailed return
Management: Manager 1861—1865 JOHN THOMAS
            Secretary 1861—1865 CHAS.FOX (P)

PENKEVILL   KENWYN   SW 748437 0918

Production: Copper   Ore(tons) Metal(tons)   Value(£)
            1855—1856 No detailed return
            Comment 1855—1856 PENKIVELL SEE CREEGBRAWSE
            Tin   Black(tons) Stuff(tons)   Tin(tons)   Value(£)
            1856—1873 No detailed return
            1875—1876 No detailed return
            Comment 1856—1873 SEE CREEGBRAWSE; 1875—1876 SEE CREEGBRAWSE
            Arsenic   Ore(tons) Metal(tons)   Value(£)
            1873   No detailed return
            1874   No detailed return
            Comment 1873—1874 SEE CREEGBRAWSE
Ownership: Comment 1859—1877 SEE CREEGBRAWSE

PENNANCE   GWENNAP   SW 713406 0919

Production:	Copper	Ore(tons)	Metal(tons)	Value(£)
	1866	40.00	2.30	135.00
	1868	71.00	3.60	213.10

350

PENNANCE   GWENNAP   Continued

Copper	Ore(tons)	Metal(tons)	Value(£)
1869	166.00	6.50	339.60
1870	25.00	2.70	166.10
1871	34.00	1.80	99.00
1872	105.00	9.80	752.40
1873	147.00	8.10	462.40

Comment 1866 (C); 1868—1872 (C)

Tin	Black(tons)	Stuff(tons)	Tin(tons)	Value(£)
1870	0.00	0.00	0.00	46.60
1871	0.00	0.00	0.00	111.50
1872	0.00	0.00	0.00	1.30

1880—1881 No detailed return
Comment 1870 PENZANCE TIN STUFF; 1871—1872 TIN STUFF;
1880—1881 SEE EAST BULLER

Ownership: Comment 1873 STOPPED IN 1873 OR 1874; 1874—1875 NEW PENNANCE
Management: Manager 1862—1871 JOHN RICHARDS; 1872—1875 S.TREDINNICK
             Chief Agent 1862—1871 JAS.HIGGINS; 1872 JOH.PAULL
             JAS.HIGGINS
             Secretary 1862—1868 WM.WILLIAMS; 1869—1875 SIR F.WILLIAMS

PENNANCE STAMPS   GWENNAP   0921

Production: Lead & Silver	Ore(tons)	Lead(tons)	Silver(ozs)	Value(£)
1856	14.00	5.00	80.00	0.00

PENROSE   SITHNEY   SW 634252 0922

Production: Lead & Silver	Ore(tons)	Lead(tons)	Silver(ozs)	Value(£)
1845	116.00	69.00	0.00	0.00
1846	11.00	7.00	0.00	0.00
1862	39.00	26.50	230.00	0.00
1863	59.80	0.00	0.00	0.00
1866	30.60	21.90	700.00	0.00

Management: Manager 1863—1868 J.HUNT
             Secretary 1863—1868 JAS.CLARKE

PENROSE,NEW   BREAGE   SW 600265 0923

Production: Copper No detailed return
             Tin No detailed return
Ownership: 1880—1881 RICH.THOMPSON & SON; 1882 NEW PENROSE TIN & COPPER
             MINING CO.LTD.
Management: Manager 1880—1881 JOHN CURTIS
             Chief Agent 1882 JOHN CURTIS
Employment:

	Underground	Surface	Total
1880	2		2
1881	4		4

PENSTRUTHAL CONSOLS   GWENNAP   SW 707398 0924

Production: Copper   Ore(tons) Metal(tons)   Value(£)
            1845   534.00   45.10   3344.80
            1846   165.00   13.00   926.80
            1847   No detailed return
            1875   22.00   2.90   237.90
            1876   156.00   9.80   603.90
            1877   156.10   12.80   759.90
            1878   89.10   5.40   254.70
            1879   27.00   1.50   75.70
            Comment 1845-1847 (C)PENSTRUTHAL; 1875-1876 (C)PENSTRUTHAL;
            1877 PENSTRUTHAL

            Tin   Black(tons) Stuff(tons)   Tin(tons)   Value(£)
            1873   4.20   143.50   0.00   666.40
            1874   36.90   0.00   0.00   2069.50
            1875   98.20   0.00   0.00   4914.00
            1876   82.60   0.00   0.00   3696.20
            1877   64.60   0.00   0.00   1959.80
            1878   46.50   0.00   0.00   1700.10
            1879   38.30   0.00   0.00   1521.30
            Comment 1873 PENSTRUTHAL; 1874 PENSTRUTHAL VALUE EST
Ownership: 1876 PENSTRUTHAL CONSOLS MINING CO.; 1877-1879 PENSTRUTHAL
            CONSOLS TIN & COPPER CO.LTD.
Management: Manager 1872 JOS.MICHELL; 1873-1877 WM.TEAGUE; 1878
            WM.THOMAS; 1879 WM.POLKINGHORNE
            Chief Agent 1872 JOHN KENDAL RICH.MORCORN W.G.WHITE;
            1873-1874 CHAS.BISHOP ED.MOYLE; 1875 WM.POLKINGHORNE
            WM.THOMAS WM.HANIBLE; 1876 WM.POLKINGHORNE WM.THOMAS;
            1877-1878 RICH.RENDLE WM.POLKINGHORNE; 1879 RICH.RENDLE
            WM.THOMAS
            Secretary 1872-1875 MAT.GREEN; 1877 MAT.GREEN; 1878-1879
            ED.ASHMEAD
Employment: Underground   Surface   Total
            1878   50   17   67
            1879   48   16   64

PENSTRUTHAL,NORTH   GWENNAP   SW 709318 0925

Production: Copper No detailed return
            Tin   Black(tons) Stuff(tons)   Tin(tons)   Value(£)
            1880   30.30   0.00   0.00   1537.80
            1881   23.10   0.00   15.90   1286.80
            1882   3.00   0.00   2.00   177.00
            1883   1.00   0.00   0.00   53.00
            1884   0.20   1.00   0.00   7.00
            1901   1.50   20.00   0.00   114.00
            1902   0.00   23.00   0.00   78.00
            1910   0.00   2.00   0.00   4.00
Ownership: 1880-1885 NORTH PENSTRUTHAL MINING CO.; 1900-1901
            PELMEAR,BARNETT & NICHOLLS; 1902 WM.BARNETT; 1908-1913
            MART.BOLITHO
            Comment 1911-1912 SUSPENDED AUG.1911: 1913 ABANDONED
Management: Manager 1880-1881 STEP.DAVEY
            Chief Agent 1881 WM.POLKINGHORNE; 1882-1883 STEP.DAVEY;
            1884-1885 WM.POLKINGHORNE
                                352

PENSTRUTHAL,NORTH   GWENNAP   Continued

Employment: Underground   Surface   Total
              1880   62   27   89
              1881   47   36   83
              1882   25   20   45
              1883   29   10   39
              1884   25   10   35
              1885   2   2
              1901   3   2   5
              1902   1   1
              1909-1910   4   4

PENSTRUTHAL,SOUTH   GWENNAP   SW 708393 0926

Production: Tin   Black(tons) Stuff(tons)   Tin(tons)   Value(£)
              1880   0.00   20.00   0.00   12.00
Ownership: 1880-1885 SOUTH PENSTRUTHAL MINING CO.
Management: Manager 1880-1881 STEP.DAVEY
            Chief Agent 1881 WM.POLKINGHORNE; 1882-1883 STEP.DAVEY; 1884
            THOS.DAVEY
            Secretary 1881 ED.ASHMEAD (S); 1885 RICH.RENDLE
Employment: Underground   Surface   Total
              1880   16   24   40
              1881   22   19   41
              1882   24   15   39
              1883   21   15   36
              1884   25   13   38
              1885   23   14   37

PENSTRUTHAL,WEST   REDRUTH   SW 703391 0927

Production: Copper No detailed return
Ownership: Comment 1861-1865 SEE SOUTH BULLER

PENTIRE   ST.ENDELLION   0928

Production: Lead   Ore(tons) Metal(tons)   Value(£)
              1846   34.00   20.00   0.00

PENTIRE GLAZE   ST.ENDELLION   SW 941799 0929

Production: Lead & Silver   Ore(tons) Lead(tons) Silver(ozs)   Value(£)
              1849   87.00   61.00   0.00   0.00
              1850   56.50   34.00   0.00   0.00
              1851   201.00   119.90   0.00   0.00
              1852   372.00   205.00   8200.00   0.00
              1853   273.00   194.00   8148.00   0.00
              1854   205.00   144.00   6048.00   0.00
              1855   133.00   90.00   3046.00   0.00
              1856   84.00   50.00   1820.00   0.00
              1857   No detailed return

353

PENTRASSOE   ST.EWE   0930

Production: Tin No detailed return
Ownership: Comment 1899 SEE COMMERCE
Employment: Underground   Surface   Total
            1899
            Comment 1899 SEE COMMERCE

PENTRUFF   ST.AUSTELL   SX 035554 0931

Production: Tin   Black(tons) Stuff(tons)   Tin(tons)   Value(£)
            1876   0.40   0.00   0.00   15.00
            1884   6.10   0.00   0.00   307.00
            1897-1898 No detailed return
            Comment 1897-1898 SEE BURNGULLOW
Ownership: 1884 MARTIN BROS.
Management: Secretary 1884 MARTIN BROS.
Employment: Underground   Surface   Total
            1884   11   6   17

PENWORTHA   PERRANZABULOE   SW 757526 0933

Production: Iron   Ore(tons)   Iron(%)   Value(£)
            1871   1550.00   0.00   1162.50
            Comment 1871 BH.

PENZANCE CONSOLS   SANCREED   SW 412290 0934

Production: Tin   Black(tons) Stuff(tons)   Tin(tons)   Value(£)
            1853   1.40   0.00   0.00   106.10
            1854   8.60   0.00   0.00   557.70

PERRAN   PERRANUTHNOE   SW 540298 0936

Production: Tin No detailed return
            Silver No detailed return
Ownership: 1903-1908 G.D.MCGREGOR & JAS.R.D.MCGREGOR
            Comment 1903-1905 LADDER WAY; 1908 SUSPENDED
Employment: Underground   Surface   Total
            1903   36   15   51
            1904   16   18   34
            1905   4   6   10
            1906   3   1   4
            1907   5   1   6
            1908   2   2

PERRAN NO.1   PERRANZABULOE   SW 751542 0935

Production: Lead   Ore(tons) Metal(tons)   Value(£)
            1863   0.50   0.30   0.00
            Zinc   Ore(tons) Metal(tons)   Value(£)
            1903   1.00   0.00   9.00

                              354

PERRAN NO.1   PERRANZABULOE   Continued

Comment 1903 FOR ZINC SEE BUDNIC CONSOLS

PERRAN NO.2   PERRANZABULOE   SW 779563 0937

Production: Lead No detailed return
            Copper No detailed return
            Tin No detailed return
Ownership: 1900—1901 JOHN HOLLOWAY
            Comment 1900—1901 PERRAN(MOUNT)
Management: Secretary 1900—1901 JOHN GILBERT (S)
Employment: Underground   Surface   Total
            1900   19   6   25
            1901   6   6   12

PERRAN CONSOLS   PERRANZABULOE   SW 767558 0938

Production: Tin   Black(tons) Stuff(tons)   Tin(tons)   Value(£)
            1869   26.30   0.00   0.00   1583.70
            1870   32.60   0.00   0.00   2137.40
            1871   50.80   0.00   0.00   3810.70
            1872   52.10   0.00   0.00   4568.70
            1873   33.80   0.00   0.00   2632.00
            1874   19.00   0.00   0.00   855.50
            1894   0.70   0.00   0.00   28.00
            1895   2.80   0.00   0.00   93.00
            1896   No detailed return
            Comment 1873—1874 VALUE EST; 1894—1896 PERRAN CONSOLS(WHEAL
            CREEG)
Ownership: 1894—1895 THOS.MICHELL
            Comment 1869—1871 FORMERLY VLOW; 1872 PERRAN GREAT CONSOLS
            FORMERLY VLOW; 1894—1895 CREEG SHAFT TRIBUTERS
Management: Manager 1869 WM.JOHNS; 1870—1872 JAS.NANCE; 1873 JAS.JULIFF
            Chief Agent 1869—1872 WM.JOHNS JNR.; 1873 JAS.NANCE
            Secretary 1869—1870 E.EMMERSON; 1871—1872 J.JOHNS (P)
            E.EMMERSON (S); 1873 J.JOHNS (P)
Employment: Underground   Surface   Total
            1894   1   1   2
            1895   4   1   5

PERRAN GREAT ST.GEORGE,WEST   PERRANZABULOE   SW 736528 0939

Production: Copper No detailed return
Ownership: Comment 1866—1868 FORMERLY PRUDENCE; 1869—1871 GREAT WEST
            ST.GEORGE
Management: Manager 1866 THOS.WASLEY; 1867—1871 WM.BAWDEN
            Chief Agent 1866 GEO.R.ODGERS
            Secretary 1866 C.THOMAS; 1867—1868 C.THOMAS J.SHERRIES

PERRAN IRON   PERRANZABULOE   SW 781565 0940

Production: Iron   Ore(tons)   Iron(%)   Value(£)
            1866   163.70   0.00   40.90
            Comment 1866 BH.PERRAN BAY
Ownership: 1863-1865 EDW.CARTER & EXECUTORS OF T.WHITFORD
            Comment 1864-1865 PERRAN BAY IRON

PERRAN IRON CO.,NEW   0941

Production: Manganese   Ore(tons) Metal(tons)   Value(£)
            1878   49.20   0.00   46.90
            Iron   Ore(tons)   Iron(%)   Value(£)
            1880   2728.00   0.00   1636.80
            Comment 1880 SP.PERRAN MINES

PERRAN MINES   PERRANZABULOE   0942

Production: Copper   Ore(tons) Metal(tons)   Value(£)
            1860   167.00   10.60   968.60
            1861   339.00   16.90   1411.20
            1862   200.00   9.50   715.00
            1863   92.00   5.80   429.00
            1865   10.00   0.40   31.70
            Comment 1860-1863 (C); 1865 (C)
            Tin   Black(tons) Stuff(tons)   Tin(tons)   Value(£)
            1861   11.50   0.00   0.00   751.60
            1862   8.10   0.00   0.00   457.90
            1863   4.90   0.00   0.00   299.90
            1864   4.90   0.00   0.00   310.20
            1865   2.70   0.00   0.00   134.40
            Comment 1862 TWO RETURNS AGGREGATED

PERRAN SILVER-LEAD CONS.   PERRANZABULOE   SW 761580 0943

Production: Lead & Silver   Ore(tons) Lead(tons) Silver(ozs)   Value(£)
            1883   14.00   8.40   280.00   133.00
Ownership: 1881-1884 PERRAN SILVER LEAD CONSOLS CO.LTD.
            Comment 1868 PERRAN SILVER LEAD; 1881 FORMERLY PENHALE AND
            LOMAX; 1884 NOT WORKED
Management: Manager 1868 M.WASLEY
            Chief Agent 1881-1882 RICH.PRYOR; 1883-1884 W.NINNIS
            Secretary 1881 RICH.PRYOR & SON
Employment: Underground   Surface   Total
            1881   13   32   45
            1882   33   17   50

PERRAN ST.GEORGE   PERRANZABULOE   SW 746535 0944

Production: Lead   Ore(tons) Metal(tons)   Value(£)
            1856   0.60   0.40   0.00
            1857   No detailed return

Zinc	Ore(tons)	Metal(tons)	Value(£)
1857	7.20	0.00	11.00

Copper	Ore(tons)	Metal(tons)	Value(£)
1845	1665.00	105.20	7322.80
1846	1782.00	126.70	8522.40
1847	2063.00	121.10	8030.40
1848	1803.00	112.40	6028.50
1849	828.00	78.20	5196.80
1850	1870.00	132.00	8754.80
1851	3753.00	230.80	14375.20
1852	3775.00	198.60	15385.80
1853	3820.00	175.40	16149.80
1854	2971.00	153.50	14464.90
1855	3322.00	159.10	14895.40
1856	2746.00	135.50	11594.00
1857	1106.00	42.10	3919.50
1858	307.00	12.50	1005.20
1859	314.00	17.00	1520.50

Comment 1845–1859 (C)

Tin	Black(tons)	Stuff(tons)	Tin(tons)	Value(£)
1855	0.10	0.00	0.00	1.80
1856	0.10	0.00	0.00	4.60
1858	1.20	0.00	0.00	61.10
1859	3.30	0.00	0.00	208.50
1860	6.60	0.00	0.00	410.60

Management: Manager 1859–1873 THOS.PILL
Chief Agent 1859 THOS.PILL; 1860 THOS.PILL JNR.
Secretary 1859 ?WAY SNR. (P); 1860–1862 M.T.HITCHINGS (P);
1864–1873 M.T.HITCHINGS (P)

PERRAN UNITED   PERRANZABULOE   SW 754539 0945

Production: Copper	Ore(tons)	Metal(tons)	Value(£)
1853	360.00	14.80	1126.50
1854	1753.00	62.30	5272.00
1855	2799.00	82.70	6609.10
1856	2785.00	87.50	6274.70
1857	1163.00	32.80	2752.30

Comment 1853 (C); 1854 (C)PERRAN INC.LEISURE; 1855
(C)INC.LEISURE; 1856–1857 (C)

PERRAN WHEAL VIRGIN   PERRANZABULOE   SW 772493 0946

Production: Lead & Silver	Ore(tons)	Lead(tons)	Silver(ozs)	Value(£)
1873	10.00	7.50	37.00	162.50
1874	3.60	2.70	22.00	0.00

Copper	Ore(tons)	Metal(tons)	Value(£)
1871	4.00	0.30	22.00

Comment 1871 (C)
Tin No detailed return
Ownership: Comment 1873 STOPPED IN 1873 & EARLY IN 1874
Management: Manager 1872–1873 GEO.TREMAYNE
Chief Agent 1870–1871 GEO.E.TREMAYNE; 1872–1873 CHAS.OATES

PERRAN WHEAL VIRGIN   PERRANZABULOE   Continued

Secretary 1870–1871 GEO.E.TREMAYNE; 1872–1873 THOS.PRYOR

PERRAN WHEAL VYVYAN   PERRANZABULOE   SW 751495 0947

Production: Lead No detailed return
Ownership: Comment 1873 STOPPED IN 1873 & EARLY IN 1874
Management: Manager 1869–1872 JAS.TREWARTHA; 1873 S.NANCARROW
Secretary 1869–1873 W.PAGE CARDOZA

PERRAN,GREAT   PERRANZABULOE   0948

Production:	Zinc	Ore(tons)	Metal(tons)	Value(£)
	1877	16.00	0.00	66.30

PERRAN,NEW   PERRANZABULOE   0949

Production:	Lead	Ore(tons)	Metal(tons)	Value(£)
	1878	0.80	0.50	7.90
	Zinc	Ore(tons)	Metal(tons)	Value(£)
	1878	134.30	0.00	256.60

PERRAN,WHEAL   PERRANZABULOE   SW 773547 0950

Production:	Tin	Black(tons)	Stuff(tons)	Tin(tons)	Value(£)
	1862	2.60	0.00	0.00	160.80
	1863	4.30	0.00	0.00	268.20

Comment 1862 WAS BUDNICK CONSOLS.PART YEAR

PERRANZABULOE IRON MINE   PERRANZABULOE   0951

Production: Iron No detailed return
Ownership: Comment 1859 TWO RETURNS AGGREGATED
Management: Manager 1859 JOHN BALL
Chief Agent 1859–1861 J.WEBB; 1862 HY.ROWE
Secretary 1859 T.WHITFORD J.WEBB(P); 1860–1861 J.WEBB (P);
1862 M.DAWKE

PERSEVERANCE   REDRUTH   SW 702411 0952

Production:	Tin	Black(tons)	Stuff(tons)	Tin(tons)	Value(£)
	1873	0.00	62.80	0.00	28.60
	1874	0.00	154.20	0.00	155.00
	1875	0.00	155.00	0.00	175.90
	1877	0.00	287.00	0.00	288.00
	1878	0.00	122.00	0.00	133.00

Comment 1874 TIN STUFF EST
Ownership: 1877 PERSEVERANCE MINING CO.LTD.
Comment 1873–1874 FORMERLY EAST UNY; 1879 NOW WORKED BY UNY
Management: Manager 1873–1879 WM.RICH

Secretary 1873–1876 ED.CRAWSHAY; 1877–1878 SML.ABBOT

Employment:	Underground	Surface	Total
1878	3		3

PHOENIX   LINKINHORNE   SX 267724 0953

Production: Copper

	Ore(tons)	Metal(tons)	Value(£)
1848	194.00	27.70	1643.40
1849	150.00	13.80	962.00
1850	720.00	110.20	8136.30
1851	962.00	147.00	10501.70
1852	1338.00	156.80	14002.60
1853	2068.00	217.00	21509.10
1854	3005.00	254.20	26699.90
1855	3543.00	234.80	23510.50
1856	4581.00	294.00	26330.60
1857	5170.00	310.00	30008.10
1858	4965.00	324.90	28728.40
1859	4260.00	303.40	28112.00
1860	4444.00	325.40	29066.20
1861	5337.00	429.00	37064.20
1862	4628.00	278.70	21218.40
1863	5656.00	288.50	20074.80
1864	5142.00	231.20	17606.10
1865	4662.00	191.00	13122.20
1866	2995.00	145.60	8668.70
1867	2182.00	128.50	8615.90
1868	2137.00	130.70	8519.70
1869	2444.00	176.80	11204.20
1870	1750.00	118.80	7069.60
1871	1309.00	90.50	5671.90
1872	1348.00	89.90	6986.50
1873	756.00	66.20	4315.80
1874	420.00	43.90	3034.60
1875	532.00	50.90	4029.20
1876	650.00	51.30	3630.60
1877	673.30	47.60	2883.80
1878	441.60	37.20	1911.50
1879	597.00	65.50	3291.70
1880	187.80	19.60	1128.50
1881	179.20	16.00	936.20
1882	228.80	19.10	1125.00
1883	89.20	9.70	555.00
1884	118.00	0.00	616.00
1885	61.00	0.00	353.00
1886	14.00	0.00	50.00
1888	102.00	0.00	598.00
1889	52.00	0.00	82.00
1896	96.00	0.00	147.00

Comment 1848–1865 (C); 1866–1870 (C)PHOENIX MINES; 1871–1876
(C); 1877–1886 PHOENIX & WEST PHOENIX UNITED; 1888–1889
PHOENIX UNITED; 1896 PHOENIX & WEST PHOENIX UNITED

Tin	Black(tons)	Stuff(tons)	Tin(tons)	Value(£)
1853	9.50	0.00	0.00	606.50

PHOENIX   LINKINHORNE   Continued

Tin	Black(tons)	Stuff(tons)	Tin(tons)	Value(£)
1854	5.50	0.00	0.00	324.50
1855	11.80	0.00	0.00	663.00
1856	9.50	0.00	0.00	644.00
1857	4.80	0.00	0.00	314.40
1858	4.80	0.00	0.00	287.30
1859	2.60	0.00	0.00	132.10
1860	4.50	0.00	0.00	323.80
1861	13.10	0.00	0.00	839.00
1862	21.50	0.00	0.00	1261.70
1863	28.40	0.00	0.00	1825.10
1864	131.80	0.00	0.00	7317.60
1865	178.20	0.00	0.00	8658.20
1866	275.70	0.00	0.00	12510.20
1867	358.00	0.00	0.00	14736.50
1868	349.30	0.00	0.00	18762.40
1869	421.20	0.00	0.00	28495.90
1870	405.00	0.00	0.00	28891.20
1871	406.10	0.00	0.00	21540.70
1872	364.90	0.00	0.00	29340.70
1873	343.20	0.00	0.00	26754.00
1874	312.40	0.00	0.00	16012.60
1875	490.90	0.00	0.00	23645.10
1876	693.30	0.00	0.00	28491.00
1877	494.50	0.00	0.00	18996.30
1878	678.60	0.00	0.00	23875.10
1879	564.60	0.00	0.00	22520.20
1880	416.00	0.00	0.00	20771.10
1881	504.90	0.00	347.10	27483.30
1882	586.00	0.00	389.70	34187.00
1883	549.10	0.00	0.00	28804.00
1884	552.10	0.00	0.00	24780.00
1885	596.00	0.00	0.00	27864.00
1886	611.00	0.00	0.00	33666.00
1887	786.00	0.00	0.00	48071.00
1888	733.30	0.00	0.00	44725.00
1889	775.30	0.00	0.00	39914.00
1890	648.00	0.00	0.00	32924.00
1891	483.90	0.00	0.00	23715.00
1892	603.70	0.00	0.00	32166.00
1893	576.70	0.00	0.00	28727.00
1894	458.40	0.00	0.00	18320.00
1895	278.70	0.00	0.00	10702.00
1896	261.00	0.00	0.00	9578.00
1897	265.20	0.00	0.00	10153.00
1898	90.60	0.00	0.00	3368.00
1907	10.20	0.00	0.00	722.00
1908	16.20	0.00	0.00	1019.00
1909	26.30	0.00	0.00	1821.00
1910	24.20	0.00	0.00	2038.00
1911	14.00	0.00	0.00	1517.00
1912	2.20	0.00	0.00	238.00
1913	6.60	0.00	0.00	663.00

Comment 1857-1861 PHOENIX MINES; 1862 TWO RETURNS AGGREGATED;
1873 VALUE EST; 1876-1880 PHOENIX & WEST PHOENIX UNITED;

1881-1889 INC.WEST PHOENIX; 1890-1898 PHOENIX UNITED;
1907-1913 PHOENIX UNITED

Ownership: 1877-1887 PHOENIX & WEST PHOENIX UNITED MINING CO.; 1888-1890
PHOENIX UNITED MINING CO.; 1891-1892 PHOENIX & W.PHOENIX
UNITED MINING CO.CB; 1893-1898 PHOENIX & W.PHOENIX UNITED
MINING CO.; 1907-1910 COSMOPOLITAN PROPRIETORY LTD.;
1911-1913 PHOENIX MINES (CORNWALL) LTD.
Comment 1876-1887 PHOENIX AND WEST PHOENIX UNITED; 1888-1890
PHOENIX UNITED; 1891-1897 PHOENIX & WEST PHOENIX UNITED; 1898
IN LIQUIDATION

Management: Manager 1859-1860 JAS.SECCOMBE JNR.; 1866 THOS.RICH; 1867
WM.HOSKING; 1868-1869 JOS.HOSKING; 1872 JOS.HOSKING;
1873-1877 WM.HOSKING; 1878-1880 JOHN TRUSCOTT; 1881
J.TRUSCOTT J.PRISK
Chief Agent 1859-1860 JOHN HORE; 1861-1862 JOHN HORE HY.VIAL;
1863-1865 HY.VIAL MARK WHITFORD; 1866 WM.HOSKING; 1867
JOS.HOSKING CHAS.WEST; 1868-1869 WM.HOSKING CHAS.WEST; 1872
CHAS.WEST JOS.HOSKING JNR.; 1873-1875 CHAS.WEST HY.HOSKING;
1876-1877 HY.HOSKING THOS.HOSKING CHAS.WEST; 1878
HY.HARVEY,JOS.HOSKING,J.L.WILLIAMS; 1879-1881 HY.HARVEY
JOS.HOSKING ED.BURN; 1882-1886 JOHN TRUSCOTT; 1887
WM.POLKINGHORNE; 1889-1898 JOHN WILLIAMS; 1907-1908
J.R.ARCHIBALD; 1909-1910 W.NANCE WILLIAMS; 1911-1913
WM.THOMAS
Secretary 1859 CROFT (P); 1861 SML.UREN; 1862 S.NETTLE;
1864-1865 M.ALLEN; 1866 CHAS.WEST; 1867-1869 WM.POLKINGHORNE;
1872-1879 WM.POLKINGHORNE; 1880 WM.POLKINGHORNE (S); 1881
C.W.POLKINGHORNE(S); 1888-1891 WM.POLKINGHORNE; 1892-1898
WM.POLKINGHORNE (P)

Employment:

	Underground	Surface	Total
1878	230	301	531
1879	210	243	453
1880	214	240	454
1881	209	220	429
1882	241	250	491
1883	197	267	464
1884	195	224	419
1885	213	229	442
1886	229	249	478
1887	260	280	540
1888	299	288	587
1889	284	298	582
1890	250	281	531
1891	210	250	460
1892	255	229	484
1893	263	279	542
1894	183	242	425
1895	100	140	240
1896	100	150	250
1897	82	143	225
1898		44	44
1907	40	68	108
1908	43	78	121
1909	63	87	150
1910	56	97	153

PHOENIX   LINKINHORNE   Continued

	Underground	Surface	Total
1911	61	95	156
1912	45	5	50
1913	65	49	114

Comment 1883-1887 INC.WEST PHOENIX; 1891-1898 INC.WEST
PHOENIX

PHOENIX   PERRANZABULOE   SW 765581 0954

Production: Lead & Silver

Year	Ore(tons)	Lead(tons)	Silver(ozs)	Value(£)
1872	20.80	15.60	78.00	0.00
1873	33.50	25.00	125.00	551.30
1874	12.70	9.50	47.00	183.10
1875	12.80	9.80	50.00	178.00
1876	0.80	12.00	60.00	11.00

Comment 1872-1875 PHOENIX SILVER LEAD; 1876 PHOENIX SILVER
LEAD. LEAD FIGURE WRONG?
Ownership: 1875-1877 PHOENIX SILVER LEAD MINING CO.
Comment 1872 EMBRACES PENHALE(CUBERT) & EAST GOLDEN;
1874-1875 PHOENIX SILVER LEAD; 1876 IN LIQUIDATION; 1877
STOPPED AUG.1877
Management: Chief Agent 1873 WM.PARRY; 1874-1875 RICH.PRYOR & SON;
1876-1877 RICH.PRYOR
Secretary 1873 RICH.PRYOR; 1876-1877 HY.L.PHILLIPS

PHOENIX DUNSLEY   LINKINHORNE   SX 268718 0955

Production: Tin No detailed return
Management: Manager 1871 T.SYMONS; 1873-1875 WM.SKEWIS
Chief Agent 1872 J.J.SIMMONS
Secretary 1871-1875 WM.SKEWIS & M.BAWDEN

PHOENIX,EAST   LINKINHORNE   SX 274722 0956

Production: Copper

Year	Ore(tons)	Metal(tons)	Value(£)
1865	31.00	1.20	82.70

Comment 1865 (C)

Tin

Year	Black(tons)	Stuff(tons)	Tin(tons)	Value(£)
1865	2.10	0.00	0.00	106.30
1866	7.50	0.00	0.00	299.50
1868	3.40	0.00	0.00	181.00
1869	6.60	0.00	0.00	382.90
1870	7.70	0.00	0.00	472.70
1871	1.90	0.00	0.00	97.60
1873	2.20	0.00	0.00	147.50

Management: Manager 1863-1873 JAS.SECCOMBE
Chief Agent 1861-1862 JAS.SECCOMBE FRAN.RENALS; 1863-1873
FRAN.RENALS
Secretary 1861-1873 JAS.SECCOMBE

PHOENIX,NEW   ALTARNUN   SX 235794 0957

Production: Zinc No detailed return
            Manganese   Ore(tons) Metal(tons)   Value(£)
            1887   387.00   0.00   1053.00
            1890   83.00   0.00   83.00
            Tin   Black(tons) Stuff(tons)   Tin(tons)   Value(£)
            1884   2.90   0.00   0.00   133.00
            1885   12.00   0.00   0.00   552.00
            1886   18.80   0.00   0.00   1069.00
            1887   15.00   0.00   0.00   1014.00
            1888   No detailed return
            Comment 1888 NOT WORKING
Ownership: 1884 J.BAILEY; 1885—1887 J.BAILEY & OTHERS; 1890—1891 NEW
            PHOENIX MINING CO.; 1899 NEW PHOENIX MINING CO.
            Comment 1887 IN LIQUIDATION; 1891 SUSPENDED
Management: Chief Agent 1884 JOHN VERCOE; 1890 BICK.ANTHONY
            Secretary 1885—1887 J.BAILEY
Employment: Underground   Surface   Total
            1884   2   3   5
            1885   44   28   72
            1886   26   20   46
            1890   30   2   32

PHOENIX,NORTH   LINKINHORNE   SX 273734 0958

Production: Copper   Ore(tons) Metal(tons)   Value(£)
            1863   7.00   0.40   31.50
            Comment 1863 (C)
            Tin No detailed return
Ownership: 1908—1913 NORTH PHOENIX SYNDICATE LTD.
            Comment 1910—1912 SUSPENDED; 1913 SUSPENDED LIQUIDATED
Management: Manager 1859—1870 JOHN MARTIN
            Chief Agent 1908—1910 W.NANCE WILLIAMS
            Secretary 1859—1870 JOHN W.DINGLE (P)
Employment: Underground   Surface   Total
            1908   2   2
            1909   3   3

PHOENIX,SOUTH   LINKINHORNE   SX 262716 0959

Production Copper No detailed return
            Manganese No detailed return
            Tin   Black(tons) Stuff(tons)   Tin(tons)   Value(£)
            1873   7.00   0.00   0.00   451.20
            1874   5.60   0.00   0.00   249.20
            1875   3.30   0.00   0.00   155.70
            1876   2.20   0.00   0.00   95.00
            1877   14.40   0.00   0.00   664.30
            1878   19.40   0.00   0.00   609.20
            1879   9.20   0.00   0.00   323.00
            1880   6.40   0.00   0.00   321.00
            1882   6.30   0.00   4.10   343.00
            1883   17.30   0.00   0.00   889.00
            1884   26.80   0.00   0.00   1384.00

363

PHOENIX,SOUTH   LINKINHORNE   Continued

```
         Tin       Black(tons)  Stuff(tons)   Tin(tons)    Value(£)
         1888        10.70        0.00         0.00         538.00
         1889        44.00        0.00         0.00        2125.00
         1890        61.40        0.00         0.00        2797.00
         1891       128.90        0.00         0.00        5825.00
         1892        67.70        0.00         0.00        3121.00
         1893      No detailed return
         Comment 1883 INC.CARADON
```
Ownership: 1877 CHAS.PEARSON; 1878-1890 HY.HOUSEMAN; 1891-1893 SOUTH
            PHOENIX MINING CO. C.B.; 1906-1910 CORNISH CONSOLIDATED TIN
            MINES LTD.; 1911 CORNISH CONSOLS LTD.
            Comment 1864-1865 SUSPENDED; 1875-1876 IN LIQUIDATION;
            1884-1887 NOT WORKED; 1888 STARTED FEB.1888; 1892 SUSPENDED;
            1893 ABANDONED; 1909 SUSPENDED; 1910 IN LIQUIDATION; 1911
            ABANDONED
Management: Manager 1859-1860 JOS.BARTH; 1875-1881 JAS.KELLY
            Chief Agent 1859-1860 JOS.BARTH; 1861-1863 JAS.BARKELL;
            1871-1874 JAS.KELLY; 1906-1911 J.DANIELL
            Secretary 1859 J.TRATHAN (P); 1860 JAS.BARKELL (P); 1861-1863
            J.J.TRATHAN (P); 1871-1877 CHAS.PEARSON; 1878-1879
            HY.HOUSEMAN; 1880-1881 RICH.CROSS; 1882-1884 RICH.CLOGG;
            1885-1886 W.J.LAVINGTON; 1887-1891 WM.H.RULE; 1892-1893
            WM.H.RULE (P)
Employment: Underground   Surface   Total
            1878   12   11   23
            1879   12   10   22
            1880   16   12   28
            1881   16   11   27
            1882   21   17   38
            1883   24   18   42
            1885   1   1
            1888   48   31   79
            1889   46   41   87
            1890   65   57   122
            1891   66   68   134
            1892   69   62   131
            1906   5   15   20
            1907   30   36   66
            1908   41   33   74
            1909   4   4
            1911   2   2

PHOENIX,WEST   LINKINHORNE   SX 256721 0960

Production: Copper   Ore(tons) Metal(tons)   Value(£)
            1877-1886 No detailed return
            1896   No detailed return
            Comment 1877-1886 SEE PHOENIX (LINKINHORNE); 1896 SEE PHOENIX
            (LINKINHORNE)
            Tin   Black(tons) Stuff(tons)   Tin(tons)   Value(£)
            1872   19.80   0.00   0.00   1590.80
            1873   69.60   0.00   0.00   5425.00
            1874   85.20   0.00   0.00   4452.60
            1875   133.20   0.00   0.00   6699.00
```
```

PHOENIX,WEST   LINKINHORNE   Continued

        Tin   Black(tons) Stuff(tons)   Tin(tons)   Value(£)
        1876—1889 No detailed return
        Comment 1873 VALUE EST; 1876—1889 SEE PHOENIX
Ownership: Comment 1876—1887 SEE PHOENIX(LINKINHORNE); 1891—1898 SEE
        PHOENIX(LINKINHORNE)
Management: Manager 1870—1871 JOS.HOSKING; 1873—1875 JOS.HOSKING
        Chief Agent 1870—1871 CHAS.WEST WM.HOSKING; 1873—1875
        CHAS.WEST WM.HOSKING
        Secretary 1870—1871 WM.POLKINGHORNE; 1873—1875
        WM.POLKINGHORNE
Employment: Underground   Surface   Total
        1883—1887
        1891—1898
        Comment 1883—1887 SEE PHOENIX(LINKINHORNE); 1891—1898 SEE
        PHOENIX(LINKINHORNE)

PHOENIX,WEST   LINKINHORNE   SX 256721 0961

Production: Tin No detailed return
Ownership: 1880—1886 WEST PHOENIX ADVENTURERS
Management: Manager 1880—1881 RICH.CROSS
        Chief Agent 1882 JOHN HOLMAN
        Secretary 1880—1881 JOHN HOLMAN W.GLUYAS; 1883—1886
        RICH.CLOGG
Employment: Underground   Surface   Total
        1880   12   14   26
        1881   10   13   23
        1882   16   8   24
        1883   22   6   28
        1884   18   9   27
        1885   10   7   17
        1886   10   5   15

PINK   ST.DAY   SW 724427 0962

Production: Copper   Ore(tons) Metal(tons)   Value(£)
        1848   183.00   14.60   811.30
        1849   354.00   24.10   1562.50
        1850   228.00   15.50   1040.30
        Comment 1848—1850 (C)
        Tin   Black(tons) Stuff(tons)   Tin(tons)   Value(£)
        1855   0.10   0.00   0.00   4.80
Ownership: Comment 1864—1868 SEE CLINTON

PLUSHEYS   LINKINHORNE   SX 303726 0963

Production: Lead No detailed return
Ownership: 1883—1886 PLUSHEYS MINING CO.; 1887—1888 PLUSHEYS MINING
        CO.LTD.
Management: Chief Agent 1883—1887 THOS.TRELEASE
        Secretary 1888 GEO.ROBSON

                                365

PLUSHEYS   LINKINHORNE   Continued

Employment: Underground   Surface   Total
              1883   6   5   11
              1884   6   6
              1885   9   7   16
              1886   10   16   26
              1887   3   3
              1888   6   4   10

POLBERRO   ST.AGNES   SW 715515 0964

Production: Copper   Ore(tons) Metal(tons)   Value(£)
              1849   35.00   5.90   416.50
              1850   296.00   17.10   1116.00
              1851   578.00   27.60   1666.10
              1854   143.00   7.70   747.70
            Comment 1849-1851 (C); 1854 (C)
            Tin   Black(tons) Stuff(tons)   Tin(tons)   Value(£)
              1852   222.30   0.00   0.00   0.00
              1853   282.50   0.00   0.00   18998.60
              1854   263.40   0.00   0.00   18000.00
              1855   283.20   0.00   0.00   18368.00
              1856   262.00   0.00   0.00   19466.40
              1857   237.10   0.00   0.00   18643.30
              1858   267.40   0.00   0.00   17058.70
              1859   258.70   0.00   0.00   19255.90
              1860   217.40   0.00   0.00   17410.10
              1861   213.00   0.00   0.00   15629.40
              1862   191.40   0.00   0.00   12867.40
              1863   221.70   0.00   0.00   14947.20
              1864   163.80   0.00   0.00   10484.10
              1865   168.10   0.00   0.00   9280.70
              1866   146.70   0.00   0.00   7205.40
              1867   104.50   0.00   0.00   5556.70
              1868   116.00   0.00   0.00   6559.70
              1869   122.70   0.00   0.00   8661.40
              1870   115.20   0.00   0.00   8448.00
              1871   104.40   0.00   0.00   8345.30
              1872   84.40   0.00   0.00   7345.70
              1873   64.80   0.00   0.00   4802.20
              1874   52.40   0.00   0.00   2900.40
              1875   19.10   250.00   0.00   1076.80
              1876   16.50   0.00   0.00   706.80
              1877   17.10   0.00   0.00   651.00
              1878   20.40   0.00   0.00   720.10
              1879   17.00   0.00   0.00   586.10
              1880   9.10   0.00   0.00   438.30
              1881   12.00   0.00   7.50   613.00
              1882   19.80   0.00   12.90   1188.00
              1883   14.90   0.00   0.00   797.00
              1884   6.70   0.00   0.00   318.00
              1885   6.30   0.00   0.00   296.00
              1886   12.10   0.00   0.00   739.00
              1887   5.30   0.00   0.00   335.00
              1888   7.10   0.00   0.00   491.00

                                366

Tin	Black(tons)	Stuff(tons)	Tin(tons)	Value(£)
1889	4.50	0.00	0.00	227.00
1890	6.90	0.00	0.00	362.00
1891	2.40	1.10	0.00	124.00
1892	7.00	0.00	0.00	94.00
1893	2.00	3.00	0.00	98.00
1894	4.10	0.00	0.00	171.00
1895	3.00	0.00	0.00	112.00
1896	1.50	0.00	0.00	48.00

Comment 1852 TIN ORE; 1855 POLBERRO CONSOLS; 1856 POLBERRO
MINES; 1862 TWO RETURNS AGGREGATED; 1876 NEW POLBERROW; 1877
POLBERROW; 1891 TWO RETURNS AGGREGATED
Ownership: 1877-1879 JOHN TAYLOR & SONS & OTHERS; 1880-1881 JOHN TAYLOR
& SONS; 1882 JOHN TAYLOR & SONS & OTHERS; 1883-1890 POLBERRO
MINING CO.; 1891-1894 POLBERRO MINING CO. C.B.; 1895-1902
POLBERRO MINING CO.
Comment 1859-1868 POLBERRO CONSOLS; 1875-1876 POLBERROW NEW
ADVENTURE; 1892-1895 UNITED WITH TREVAUNANCE; 1896-1897
UNITED WITH TREVAUNANCE SUSPENDED; 1898 TREVAUNANCE
SUSPENDED; 1899-1902 TREVAUNANCE
Management: Manager 1859-1863 JOHN HANCOCK; 1864-1870 HANCOCK TAYLOR &
SONS; 1871-1872 TAYLOR & SONS HANCOCK; 1873-1878 JOHN
HANCOCK
Chief Agent 1859-1862 NICH.DUNSTON; 1863 NICH.DUNSTON
J.B.BENNETTS; 1864-1865 NICH.DUNSTON TAYLOR & SON; 1866-1872
NICH.DUNSTON; 1873 JOHN TAYLOR & SONS RICH.PRYOR; 1874-1875
JOHN TAYLOR & SONS; 1876 THOS.NINNIS; 1879-1882
THOS.M.NINNIS; 1884-1890 WM.VIVIAN; 1891 G.COULTER HANCOCK
(P); 1892-1898 G.COULTER HANCOCK (P) CHAS.THOMAS; 1899-1900
G.COULTER HANCOCK (P); 1901-1902 G.COULTER HANCOCK (P)
CHAS.THOMAS
Secretary 1859-1861 MICH.MORCOM (P); 1862 J.B.BENNETTS; 1863
JOHN TAYLOR & SONS; 1864-1870 J.B.BENNETTS; 1871-1872
J.H.BENNETTS; 1873-1875 J.P.BENNETTS; 1876 TAYLOR & SONS &
OTHERS; 1877-1878 JOHN TAYLOR & SONS; 1879 JOHN TAYLOR; 1883
THOS.M.NINNIS; 1891-1902 F.J.HARVEY (S)

Employment:	Underground	Surface	Total
1878	6	16	22
1879	7	10	17
1880	10	7	17
1881	18	5	23
1882	16	10	26
1883	14	10	24
1884	14	2	16
1885	10	5	15
1886	11	17	28
1887	17	10	27
1888	17	8	25
1889	25	7	32
1890	23	5	28
1891		2	2
1892	16	6	22
1893	18	9	27
1894	23	9	32
1895	20	13	33

POLBERRO,EAST   ST.AGNES   SW 702501 0965

Production: Copper No detailed return
            Tin No detailed return
Ownership: Comment 1863-1865 INC.BETSY,BLUE HILLS & OCEAN; 1873 STOPPED
            IN 1873 & EARLY IN 1874
Management: Manager 1863 JAS.CROSS; 1872-1873 WM.JOHNS
            Chief Agent 1864-1865 ART.GRIPE
            Secretary 1863-1865 JOHN CLAY; 1872-1873 S.A.POPE

POLBERRO,WEST   ST.AGNES   SW 701598 0966

Production: Copper   Ore(tons) Metal(tons)   Value(£)
            1856   205.00   9.10   739.70
            Comment 1856 (C)
            Tin   Black(tons) Stuff(tons) Tin(tons)   Value(£)
            1855   1.50   0.00   0.00   80.50
            1856   6.10   0.00   0.00   393.40
            1857   0.10   0.00   0.00   8.10

POLBREEN   ST.AGNES   SW 718503 0967

Production: Copper   Ore(tons) Metal(tons)   Value(£)
            1873   6.00   0.80   46.50
            Comment 1873 (C)
            Tin   Black(tons) Stuff(tons) Tin(tons)   Value(£)
            1858   12.60   0.00   0.00   858.90
            1859   25.90   0.00   0.00   1964.50
            1860   47.90   0.00   0.00   3829.80
            1861   51.70   0.00   0.00   3828.80
            1862   60.70   0.00   0.00   4152.00
            1863   52.20   0.00   0.00   3538.20
            1864   28.60   0.00   0.00   1843.70
            1865   26.40   0.00   0.00   1459.90
            1866   45.00   0.00   0.00   2256.40
            1867   23.10   0.00   0.00   1231.80
            1868   1.30   0.00   0.00   67.60
            1871   30.70   0.00   0.00   2429.70
            1872   48.40   0.00   0.00   4289.90
            1873   30.40   0.00   0.00   2386.80
            1874   5.70   0.00   0.00   343.00
Ownership: Comment 1873 STOPPED IN 1873 & EARLY IN 1874
Management: Manager 1859-1867 JOHN HANCOCK; 1872-1873 JOHN NANCARROW
            Chief Agent 1863 J.B.BENNETTS; 1868 JOHN HANCOCK; 1869-1871
            JOHN NANCARROW; 1872 J.BRYANT HY.DALE
            Secretary 1861 MICH.MORCOM; 1862 J.B.BENNETTS; 1863 JOHN
            TAYLOR & SONS; 1864 J.B.BENNETTS; 1865-1867 JOHN CLAY;
            1869-1873 Y.CHRISTIAN

POLBREEN,NEW   ST.AGNES   SW 717504 0968

Production: Tin No detailed return
Ownership: 1898-1901 NEW POLBREEN MINING CO.; 1903-1904 NEW POLBREEN
            MINING CO.

POLBREEN,NEW   ST.AGNES   Continued

Comment 1903-1904 SUSPENDED
Management: Chief Agent 1898-1901 JOEL HOOPER; 1903-1904 JOEL HOOPER
Secretary 1898-1901 F.J.HARVEY; 1903-1904 F.J.HARVEY
Employment: Underground   Surface   Total
    1899   12   12   24
    1900   16   11   27
    1901   18   12   30

POLBREEN,WEST   ST.AGNES   SW 714501 0969

Production: Tin   Black(tons) Stuff(tons) Tin(tons)   Value(£)
    1873   3.40   0.00   0.00   272.00
    1888   1.80   0.00   0.00   109.00
    1889   1.50   0.00   0.00   89.00
Ownership: 1880-1881 WEST POLBREEN MINE CO.; 1882-1889 WEST POLBREEN
MINING CO.
Comment 1873 STOPPED IN 1873 & EARLY IN 1874
Management: Manager 1870-1873 JOHN NANCARROW; 1881)
Chief Agent 1872-1873 J.BRYANT; 1880-1889 WM.VIVIAN
Secretary 1863 JOHN CARTHEW; 1864-1866 W.DAVIES; 1871-1873
Y.CHRISTIAN; 1881 J.COULTER HANCOCK (P
Employment: Underground   Surface   Total
    1881   6   6
    1882   8   8
    1883   13   1   14
    1884   11   2   13
    1885   10   6   16
    1886   11   6   17
    1887   16   5   21
    1888-1889   12   8   20

POLCREBO   CROWAN   SW 648331 0970

Production: Tin   Black(tons) Stuff(tons) Tin(tons)   Value(£)
    1883   12.00   424.00   0.00   499.00
    1884   11.50   407.50   0.00   519.00
    1885   30.00   304.30   0.00   964.00
    1886   17.00   470.00   0.00   852.00
    1887   25.00   672.00   0.00   1265.00
    1888   14.00   510.30   0.00   724.00
    1889   2.00   137.00   0.00   97.00
    1890   0.00   90.00   0.00   66.00
Ownership: 1882-1890 POLCREBO TIN MINING CO.
Comment 1873 STOPPED IN 1873 & EARLY IN 1874
Management: Chief Agent 1872-1873 BEN.GUNDRY; 1879 WM.CHAPPEL; 1882-1890
W.H.MARTIN
Secretary 1872-1873 JAS.WILLIAMS
Employment: Underground   Surface   Total
    1882   9   9   18
    1883   27   8   35
    1884   35   13   48
    1885   32   10   42
    1886   26   9   35

369

	Underground	Surface	Total
1887	28	8	36
1888	33	7	40
1889	19	5	24
1890	15	5	20

POLDICE   ST.DAY   SW 740427 0971

Production: Zinc   Ore(tons) Metal(tons)   Value(£)

Zinc	Ore(tons)	Metal(tons)	Value(£)
1872	7.00	0.00	4.00
1879	5.70	0.00	20.00

Copper	Ore(tons)	Metal(tons)	Value(£)
1845	2485.00	153.20	10889.30
1846	1985.00	124.40	8563.80
1847	1873.00	121.10	8547.60
1848	1127.00	81.50	4859.60
1849	791.00	55.00	3699.80
1850	922.00	65.30	4564.10
1851	703.00	48.30	3182.90
1852	568.00	38.80	3258.30
1868	899.00	58.00	3582.90
1869	1462.00	95.70	5760.90
1870	545.00	33.30	1836.60
1871	439.00	27.20	1601.10
1872	162.00	9.60	717.70
1873	126.00	7.30	462.00
1874	2.00	0.10	6.50
1875	5.00	0.20	14.20
1876	27.00	1.20	60.20
1877	21.00	1.00	42.50
1878	19.90	1.00	42.40
1879	1.30	0.10	2.00

Comment 1845–1852 (C); 1868–1876 (C); 1877 THE POLDICE

Tin	Black(tons)	Stuff(tons)	Tin(tons)	Value(£)
1867	41.90	0.00	0.00	1976.90
1868	112.80	0.00	0.00	5563.40
1869	157.00	0.00	0.00	10378.30
1870	275.50	0.00	0.00	19591.10
1871	307.50	0.00	0.00	23352.60
1872	212.70	0.00	0.00	17608.00
1873	176.20	0.00	0.00	13651.60
1874	20.60	0.00	0.00	1111.90
1875	24.00	15.90	0.00	1309.00
1876	3.10	17.80	0.00	235.60
1877	4.80	7.60	0.00	198.30
1878	6.30	0.00	0.00	201.50
1879	0.00	100.00	0.00	126.00
1888–1889	No detailed return			
1890	2.40	5117.00	0.00	1690.00
1891	3.00	3367.00	0.00	710.00
1892	0.00	66.00	0.00	77.00
1898	0.00	42.00	0.00	17.00
1899	0.00	74.00	0.00	44.00
1900	No detailed return			

Tin	Black(tons)	Stuff(tons)	Tin(tons)	Value(£)
1905	No detailed return			
1906	0.00	511.00	0.00	416.00
1907	0.00	512.00	0.00	343.00
1908	0.00	695.00	0.00	461.00
1909	0.00	411.00	0.00	329.00
1910	0.00	381.00	0.00	53.00
1912	0.00	223.00	0.00	409.00
1913	0.00	933.00	0.00	675.00

Comment 1874 THE POLDICE; 1876-1877 THE POLDICE; 1888-1889
SEE ST.DAY MANOR; 1890-1891 TWO RETURNS AGGREGATED; 1905 SEE
CUSGARNE; 1906 SEE ALSO CUSGARNE; 1909 TWO RETURNS
AGGREGATED

Arsenic	Ore(tons)	Metal(tons)	Value(£)
1867	38.90	0.00	87.40
1870	50.00	0.00	105.60
1872	105.00	0.00	200.00
1873	68.00	0.00	264.10
1889	8.00	0.00	42.00
1890	12.00	0.00	71.00
1891	7.00	0.00	47.00
1893	7.00	0.00	80.00
1895	11.00	0.00	94.00
1896	3.00	0.00	44.00

Comment 1867 CRUDE; 1870 CRUDE; 1893 INC.UNITY WOOD, MAID &
GORLAND. WASTE
Pyrites No detailed return
Ochre No detailed return

Arsenic Pyrite	Ore(tons)	Value(£)
1878	21.00	9.40
1879	86.70	42.00
1893	124.00	14.00

Comment 1893 INC.UNITY WOOD,MAID & GORLAND.WASTE
Ownership: 1868-1869 POLDICE COMPANY; 1877 F.M.WILLIAMS; 1878-1879
POLDICE CO.; 1880 WEST POLDICE UNITED CO.; 1888-1890
THOS.BEHENNA & OTHERS; 1891 T.VERGER; 1892-1894 W.J.TRYTHALL;
1895-1896 T.H.LETCHER; 1898-1900 W.J.TRYTHALL; 1905-1908
T.H.LETCHER; 1909-1913 J.T.LETCHER
Comment 1869-1871 FORMERLY ST.DAY UNITED; 1873 STOPPED IN
1873 & EARLY IN 1874; 1880 POLDICE UNITED; 1888 TRIBUTERS
INC.GORLAND UNITY & MAID; 1889 ABANDONED INC.UNITY & MAID;
1890 INC.UNITY & MAID; 1891-1893 INC.UNITY MAID & GOOUGUMPAS;
1894-1896 INC.UNITY MAID & GOOUGUMPAS OCCASIONAL WORK; 1900
IN LIQUIDATION; 1905-1908 T.H.LETCHER LORDS AGENT; 1909-1913
J.T.LETCHER LORDS AGENT
Management: Manager 1869-1871 T.COCK; 1873 JOS.COCK
Chief Agent 1868 T.GILBERT; 1869 JOS.RALPH; 1870-1871
THOS.BRAY JOHN GERRANS JOS.BLIGHT; 1874 JOS.COCK; 1875
S.TREDINNICK; 1876-1878 HY.MICHELL; 1879-1880 JOS.BLIGHT;
1888-1890 T.H.LETCHER
Secretary 1870-1872 SIR F.WILLIAMS; 1873-1876 F.M.WILLIAMS;
1877 HY.MICHELL; 1878 F.M.WILLIAMS; 1891-1894 T.H.LETCHER

Employment:	Underground	Surface	Total
1878	3	6	9
1879	3	4	7

	Underground	Surface	Total
1888	21	20	41
1889	7	21	28
1890	18	29	47
1891	4	1	5
1892	3		3
1893	4		4
1894	2		2
1895		9	9
1896		11	11
1898–1900	4		4
1905		7	7
1906		8	8
1908		16	16
1909		7	7
1910		11	11
1911	4	8	12
1912	2	13	15
1913	2	14	16

Comment 1888 INC.GORLAND,UNITY NO.2 & MAID; 1889–1890
INC.UNITY NO.2 & MAID; 1891–1896 INC.UNITY NO.2 MAID &
GOOUGUMPAS

POLDICE AND EAST MAID   ST.DAY   SW 740427 0972

Production: Tin   Black(tons) Stuff(tons)   Tin(tons)   Value(£)
          1891   0.00   10.00   0.00   35.00

POLDICE (BEHENNA)   ST.DAY   SW 740427 0973

Production: Tin No detailed return
Ownership: 1911–1913 THOS.BEHENNA & JOH.BEHENNA
            Comment 1911–1912 SUSPENDED; 1913 NOT WORKED

POLDICE HALVANERS   ST.DAY   SW 740427 0974

Production: Tin   Black(tons) Stuff(tons)   Tin(tons)   Value(£)
          1892   3.10   1874.00   0.00   336.00

POLDICE (H.T.TRYTHALL)   ST.DAY   SW 740427 0975

Production: Tin   Black(tons) Stuff(tons)   Tin(tons)   Value(£)
          1892   0.00   89.00   0.00   16.00

POLDICE (LETCHER)   ST.DAY   SW 740427 0976

Production: Tin   Black(tons) Stuff(tons)   Tin(tons)   Value(£)
          1895   0.50   1498.00   0.00   131.00
          1896   1.00   1355.00   0.00   142.00
          Comment 1905–1906 SEE CUSGARNE

POLDICE (R.MARTIN & SON)   ST.DAY   SW 740427 0977

Production: Tin   Black(tons) Stuff(tons)   Tin(tons)   Value(£)
            1893   0.00   87.00   0.00   83.00
            1894   0.00   70.00   0.00   56.00
            1895   0.00   126.00   0.00   75.00
            1896   0.00   35.00   0.00   22.00
            1897   0.00   4.00   0.00   1.00

POLDICE (W.J.TRYTHALL)   ST.DAY   SW 740427 0978

Production: Tin   Black(tons) Stuff(tons)   Tin(tons)   Value(£)
            1893   0.00   159.00   0.00   50.00
            1894   0.00   160.60   0.00   53.00
            1895-1896 No detailed return

POLDICE (WATERS & TEAGUE)   ST.DAY   SW 740427 0979

Production: Tin   Black(tons) Stuff(tons)   Tin(tons)   Value(£)
            1892   0.00   24.00   0.00   17.00
            1893   0.00   95.00   0.00   63.00

POLDICE,OLD (PART OF)   ST.DAY   SW 740427 0980

Production: Copper   Ore(tons) Metal(tons)   Value(£)
            1875   199.00   19.80   1556.80
            1876   336.00   29.40   2053.60
            1877   213.30   15.20   898.90
            1878   0.80   0.10   1.70
            1881   0.70   0.10   2.50
            1884   48.00   0.00   143.00
            1909   9.00   0.00   40.00
            Comment 1875-1876 (C); 1909 OLD POLDICE
            Tin   Black(tons) Stuff(tons)   Tin(tons)   Value(£)
            1899   0.00   11.00   0.00   16.00
            1900   0.00   75.00   0.00   31.00
            1910   0.00   41.00   0.00   16.00
            1911   0.00   267.00   0.00   322.00
            Comment 1899-1900 OLD POLDICE; 1910-1911 OLD POLDICE
Ownership: 1892-1894 MARTIN,RICHARD & SON; 1895-1898 RICH.MARTIN & SON;
            1899 T.H.LETCHER; 1909-1910 W.C.B.SYNDICATE LTD.
            Comment 1898-1899 T.H.LETCHER LORDS AGENT; 1909 OLD POLDICE;
            1910 OLD POLDICE CLOSED SINCE MAR.1910
Management: Chief Agent 1898 T.H.LETCHER
            Secretary 1909 WM.J.COCK
Employment: Underground   Surface   Total
            1892-1894   2   2
            1895   4   4
            1896-1897   1   1
            1899   1   1
            1909   5   5
            1910   1   1
            Comment 1909 OLD POLDICE

                                 373

POLDICE, PART OF   ST.DAY   SW 740427 0981

Production: Copper No detailed return
           Tin No detailed return
Ownership: 1905-1910 W.J.TRYTHALL
           Comment 1910 NOT WORKED
Employment: Underground   Surface   Total

	Underground	Surface	Total
1905	4		4
1906	14	1	15
1907	12	2	14
1908	10	2	12
1909	5		5

POLDICE,WEST   GWENNAP   SW 735435 0982

Production: Lead

	Ore(tons)	Metal(tons)	Value(£)
1881	0.30	0.20	3.50

Copper

	Ore(tons)	Metal(tons)	Value(£)
1874	3.00	0.20	16.80
1882	18.50	5.20	79.00
1883	27.30	2.20	105.00

Comment 1874 (C); 1882-1883 WEST POLDICE UNITED

Tin

	Black(tons)	Stuff(tons)	Tin(tons)	Value(£)
1873	2.80	0.00	0.00	193.30
1874	38.70	64.40	0.00	2424.00
1875	13.90	1381.60	0.00	3904.50
1876	45.10	1783.10	0.00	4597.30
1877	158.40	0.00	0.00	6328.80
1878	6.70	261.60	0.00	548.30
1879	0.00	367.90	0.00	280.00
1880	0.00	73.50	0.00	45.00
1881	56.10	0.00	37.00	3173.80
1882	143.20	0.00	79.00	7608.00
1883	110.60	0.00	0.00	5305.00
1884	48.80	0.00	0.00	1844.00
1889	No detailed return			

Comment 1881-1884 WEST POLDICE UNITED; 1889 SEE TOLGULLOW
UNITED (POLDICE)

Arsenic

	Ore(tons)	Metal(tons)	Value(£)
1882	97.00	0.00	465.00
1883	101.10	0.00	602.00
1884	87.40	0.00	551.00

Comment 1882-1883 WEST POLDICE UNITED; 1884 WEST POLDICE
UNITED.SOOT

Pyrites No detailed return

Arsenic Pyrite

	Ore(tons)	Value(£)
1878	22.50	10.90
1879	40.40	20.00
1880	48.30	34.50

Ownership: 1877-1879 WEST POLDICE MINING CO.; 1880-1884 WEST POLDICE
           UNITED MINING CO.; 1905 WHEAL MERTH LTD.
           Comment 1873 STOPPED IN 1873 & EARLY IN 1874; 1880-1884 WEST
           POLDICE UNITED; 1905 OLD WEST POLDICE
Management: Manager 1873-1874 JOS.COCK; 1881 WM.TEAGUE
           Chief Agent 1875-1879 HY.MICHELL; 1880 WM.TEAGUE; 1881 JOHN
           PENHALL; 1882-1884 WM.TEAGUE; 1905 JOHN DANIELL

POLDICE,WEST   GWENNAP   Continued

        Secretary 1873–1876 F.M.WILLIAMS; 1877 HY.MICHELL; 1881
        WM.TEAGUE (S); 1905 WM.THOMAS
Employment: Underground   Surface   Total
        1878   10   54   64
        1879   7   7
        1880   29   62   91
        1881   100   125   225
        1882   165   176   341
        1883   129   123   252
        1884   71   68   139

POLDORY   ST.DAY   SW 757417 0984

Production: Tin   Black(tons) Stuff(tons)   Tin(tons)   Value(£)
        1907   13.10   0.00   0.00   1094.00
        1908   5.90   0.00   0.00   361.00
        1909   11.30   0.00   0.00   827.00
        1910   15.00   0.00   0.00   1100.00
        1911   18.60   6.10   0.00   912.00
        1912   0.00   20.00   0.00   24.00
        1913   3.80   0.00   0.00   231.00
        Comment 1907–1911 INC.SUNNY CORNER
Ownership: 1906–1910 R.H.MOORE; 1911–1912 SUNNY CORNER & POLDORY
        SYNDICATE LTD.; 1913 J.T.ODGERS, RICH ODGERS, J.J.DUNSTAN
        Comment 1906–1912 INC.SUNNY CORNER
Employment: Underground   Surface   Total
        1906   10   9   19
        1907   15   21   36
        1908   24   20   44
        1909   14   9   23
        1910   15   12   27
        1911–1912   9   10   19
        1913   2   2
        Comment 1906–1912 INC.SUNNY CORNER

POLGEAR   WENDRON   SW 694367 0986

Production: Iron No detailed return
Ownership: Comment 1863–1865 SUSPENDED
Management: Manager 1860–1862 WM.PASCOE
        Secretary 1860–1862 JOHN CADY (P)

POLGOOTH TIN MINE   ST.AUSTELL   SX 001504 0987

Production: Tin   Black(tons) Stuff(tons)   Tin(tons)   Value(£)
        1886   54.50   0.00   0.00   3000.00
        1887   No detailed return
Ownership: 1886 JOS.FELL; 1887–1889 POLGOOTH TIN MINE LTD.; 1890–1895
        MRS.G.BROWNE
        Comment 1889 IN LIQUIDATION; 1891–1895 POLGOOTH
Management: Chief Agent 1886 JAS.TONKIN; 1887–1888 SID.HODGKINSON;
        1889–1893 JAS.TONKIN; 1894–1895 THOS.ELLERY

                              375

POLGOOTH TIN MINE   ST.AUSTELL   Continued

                    Secretary 1891–1892 C.S.BROWNE
Employment: Underground   Surface   Total
              1886   36   16   52
              1887   44   17   61
              1890   6   2   8

POLGOOTH UNITED   ST.AUSTELL   SX 001504 0988

Production: Tin   Black(tons) Stuff(tons)   Tin(tons)   Value(£)
              1874   3.90   0.00   0.00   180.50
              1875   24.10   0.00   0.00   1260.50
              1876   30.00   0.00   0.00   1446.50
              1877   0.00   5.00   0.00   30.50
              1882   6.50   0.00   4.30   390.00
            Comment 1874 VALUE EST; 1882 GREAT POLGOOTH UNITED
Ownership: 1877 JAS.HADDOCK; 1880–1882 GREAT POLGOOTH UNITED MINING
            CO.LTD.
            Comment 1881 INCS.TREGREHAN, BELOWDO, BRYNN ROYALTON, BURTHY
            ROW, ALVIGGAN & BURNGULLOW; 1882 GREAT POLGOOTH UNITED
Management: Chief Agent 1873–1875 THOS.WASLEY; 1876–1877 JOHN WASLEY;
            1880 T.R.PARKYN; 1881 WM.RICHARDS; 1882 SILAS PASCOE
            Secretary 1873–1877 W.BROWNE
Employment: Underground   Surface   Total
              1881   20   18   38
              1882   11   28   39

POLGOOTH,GREAT   ST.AUSTELL   SX 001504 0989

Production: Tin   Black(tons) Stuff(tons)   Tin(tons)   Value(£)
              1852   355.50   0.00   0.00   0.00
              1853   260.50   0.00   0.00   17745.00
              1854   96.60   0.00   0.00   6273.80
              1855   283.80   0.00   0.00   18254.60
              1856   209.40   0.00   0.00   14137.00
              1857   66.80   0.00   0.00   5200.80
              1858   51.70   0.00   0.00   3388.50
              1859   30.10   0.00   0.00   2149.80
              1860   44.00   0.00   0.00   3361.90
              1861   70.30   0.00   0.00   5143.50
              1862   73.10   0.00   0.00   4788.00
              1863   42.40   0.00   0.00   2724.80
              1864   40.80   0.00   0.00   2569.60
              1865   12.30   0.00   0.00   654.60
              1866   48.70   0.00   0.00   2473.80
              1869   25.60   0.00   0.00   1786.70
              1870   27.70   0.00   0.00   2033.90
              1871   49.90   0.00   0.00   3938.00
              1872   16.80   0.00   0.00   1512.60
              1873   55.10   0.00   0.00   4168.30
              1874   22.20   0.00   0.00   995.50
              1875   14.00   0.00   0.00   721.60
              1876   15.00   0.00   0.00   645.50
              1877   7.00   0.00   0.00   280.00

POLGOOTH,GREAT   ST.AUSTELL   Continued

Tin	Black(tons)	Stuff(tons)	Tin(tons)	Value(£)
1878	7.00	0.00	0.00	246.90
1879	5.00	0.00	0.00	175.00
1880	11.00	0.00	0.00	550.00
1881	4.70	0.00	3.10	248.40
1882	4.40	0.00	2.90	220.00
1883	3.50	0.00	0.00	179.00
1884	1.50	0.00	0.00	68.00
1885	1.10	0.00	0.00	45.00
1886	1.20	0.00	0.00	61.00
1887	1.00	0.00	0.00	60.00
1888	3.00	0.00	0.00	180.00
1889	2.70	0.00	0.00	140.00
1890	11.20	0.00	0.00	516.00
1891	8.30	0.00	0.00	445.00
1892	7.00	0.00	0.00	388.00
1893	5.00	0.00	0.00	261.00
1894	5.40	0.00	0.00	209.00

Comment 1852 TIN ORE; 1862 TWO RETURNS AGGREGATED; 1864 3
QUARTERS; 1874 VALUE EST; 1881-1889 POLGOOTH; 1890 AGGREGATED
TO INC.OLD MINE; 1891 INC.OLD MINE
Ownership: 1877-1878 W.WEST AND CO.; 1880-1891 RICH.HANCOCK
            Comment 1880-1891 POLGOOTH
Management: Manager 1862-1872 RICH.HANCOCK
            Chief Agent 1861 RICH.HANCOCK; 1873-1890 RICH.HANCOCK
            Secretary 1860-1872 WM.BROWNE (P); 1873-1876 WM.BROWNE; 1879
            RICH.HANCOCK

Employment:		Underground	Surface	Total
	1878	2	7	9
	1879	1	6	7
	1880		5	5
	1881		6	6
	1882-1883		3	3
	1884		2	2
	1885		4	4
	1886		3	3
	1887		2	2
	1888		4	4
	1889		3	3
	1890		2	2
	1891	8	3	11
	1892	5	4	9
	1893	5	2	7
	1894	4	2	6
	1895		1	1

Comment 1891 G.BROWNE & RICH.HANCOCK AMALG.

POLGOOTH,NEW   ST.AUSTELL   SX 001504 0990

Production: Tin	Black(tons)	Stuff(tons)	Tin(tons)	Value(£)
1881	0.20	0.00	0.10	9.00

Ownership: 1881 JAS.A.MEREDITH
Management: Chief Agent 1881 JAS.A.MEREDITH

POLGOOTH,NEW   ST.AUSTELL   Continued

Employment: Underground   Surface   Total
                1881   4   4

POLGOOTH,NORTH GREAT   ST.MEWEN   SX 001504 0991

Production: Tin No detailed return
Ownership: Comment 1880 ELIZABETH IS A PORTION OF SETT
Management: Secretary 1880 JOHN EDWARDS
Employment: Underground   Surface   Total
                1880   16   16

POLGOOTH,OLD   ST.AUSTELL   SX 001504 0992

Production: Tin   Black(tons) Stuff(tons)   Tin(tons)   Value(£)
                1883–1885 No detailed return
                1890–1891 No detailed return
                Comment 1883–1885 SEE TREGONTREES; 1890–1891 SEE
                POLGOOTH,GREAT?
Ownership: Comment 1883 SEE TREGONTREES
Employment: Underground   Surface   Total
                1883–1885
                Comment 1883–1885 SEE TREGONTREES

POLGOOTH,SOUTH   ST.AUSTELL   SW 989499 0993

Production: Tin   Black(tons) Stuff(tons)   Tin(tons)   Value(£)
                1880   4.60   0.00   0.00   211.50
                1881   26.60   0.00   17.30   1551.80
                1882   6.40   0.00   4.50   371.00
                1896   2.60   0.00   0.00   92.00
                1897   16.00   0.00   0.00   644.00
                1898   10.00   0.00   0.00   424.00
                1899   5.00   0.00   0.00   348.00
                Comment 1896–1899 INC.TRELOWETH
Ownership: 1879–1886 SOUTH POLGOOTH MINING CO.; 1890 J.NORFOLK;
                1891–1893 J.NORFOLK(P); 1894–1895 J.NORFOLK; 1902 C.VIVIAN
                LADDS; 1905–1908 C.VIVIAN LADDS
                Comment 1886 NOT WORKED; 1892 SUSPENDED; 1894–1895 NOT
                WORKING; 1896–1900 SEE TRELOWETH; 1902 SOUTH POLGOOTH
                TRELOWETH; 1903–1904 SEE TRELOWETH; 1905–1908 SUSPENDED
Management: Manager 1881 THOS.PARKYN
                Chief Agent 1879–1880 T.R.PARKYN JNR.; 1881 WM.KENDALL;
                1890–1895 WM.JOHNS
                Secretary 1881–1886 JOHN DYER (S)
Employment: Underground   Surface   Total
                1879   12   12
                1880   13   30   43
                1881   24   31   55
                1882   12   13   25
                1883   2   2   4
                1884   2   2
                1885   1   1

                                378

POLGOOTH,SOUTH   ST.AUSTELL   Continued

	Underground	Surface	Total
1890	4	3	7
1891—1892		1	1
1893	5		5
1894		1	1
1896—1901			
Comment 1896—1901 SEE TRELOWETH

POLGOOTH,WEST   ST.AUSTELL   SW 981501 0994

Production: Tin   Black(tons) Stuff(tons)   Tin(tons)   Value(£)
            1852   1.20   0.00   0.00   0.00
            Comment 1852 INC.HEWAS TIN ORE

POLGRAIN   WENDRON   0995

Production: Tin   Black(tons) Stuff(tons)   Tin(tons)   Value(£)
            1880   No detailed return
            Comment 1880 SEE GARLIDNA

POLHARMAN   TYWARDREATH   SX 084567 0996

Production: Copper No detailed return
            Comment 1865—1867 SEE POLHARMAN, DEVON
            Tin No detailed return
Management: Chief Agent 1865—1866 PHIL.RICH ROBT.HOOPER; 1867 PHIL.RICH
            Secretary 1865—1867 R.FOSTER

POLHIGEY MOOR   WENDRON   SW 705353 0997

Production: Tin   Black(tons) Stuff(tons)   Tin(tons)   Value(£)
            1862   3.60   0.00   0.00   214.50
            1863   10.90   0.00   0.00   659.70
            1864   8.60   0.00   0.00   433.90
            1873   9.40   0.00   0.00   664.80
            Comment 1862 PART YEAR ONLY; 1873 NEW POLHIGEY MOOR
Ownership: Comment 1865 SUSPENDED; 1873 STOPPED IN 1873 & EARLY IN 1874
Management: Manager 1860—1861 WM.POLKINGHORNE; 1862—1864 SML.BENNETTS
            Chief Agent 1859—1860 R.MARTIN; 1861 SML.BENNETTS; 1872—1873
            JOHN TONKIN
            Secretary 1859—1865 WM.POLKINGHORNE (P); 1872—1873 JOHN
            TONKIN

POLLARD   ST.CLEER   SX 245700 0999

Production: Copper No detailed return
Management: Manager 1859—1860 JAS.NANCE
            Chief Agent 1859—1860 JAS.NANCE; 1861—1865 W.C.COCK
            Secretary 1859 E.A.CROUCH (P); 1860 F.TAYLOR (P); 1861—1865
            JOHN TAYLOR

CM-O   379

Production:

Zinc	Ore(tons)	Metal(tons)	Value(£)
1897	15.00	0.00	42.00
1899	42.00	11.50	62.00
1900	20.00	4.50	28.00

Copper	Ore(tons)	Metal(tons)	Value(£)
1856	100.00	15.30	1635.70
1858	52.00	5.20	483.70
1859	272.00	37.80	3788.30
1860	484.00	46.70	4298.70
1861	856.00	65.70	5679.30
1862	1323.00	81.00	6393.00
1863	990.00	80.70	6267.00
1864	743.00	55.80	4718.40
1865	488.00	35.60	2795.40
1866	481.00	40.50	2771.70
1867	354.00	24.10	1668.00
1898	8.00	0.00	14.00
1899	224.00	0.00	940.00
1900	6.00	0.00	27.00

Comment 1856 (C); 1858-1867 (C)

Tin	Black(tons)	Stuff(tons)	Tin(tons)	Value(£)
1866	0.00	0.00	0.00	21.00
1867	0.20	0.00	0.00	11.30

Comment 1866 TIN STUFF

Arsenic Pyrite	Ore(tons)	Value(£)
1897	34.00	38.00
1898	30.00	27.00

Ownership: 1897-1908 POLMEAR SYNDICATE
Comment 1907-1908 SUSPENDED
Management: Manager 1865-1866 ROBT.MUSTON
Chief Agent 1860-1864 WM.ROWE; 1866 J.BEARD; 1897-1900 P.A.LE FEUVRE; 1901-1904 J.F.C.ABELSPIES; 1906-1908 RICH.H.WILLIAMS
Secretary 1859-1862 JOHN DALLY (P); 1863 COMMITTEE; 1864-1866 FRAN.BARRATT

Employment:

	Underground	Surface	Total
1897	7	4	11
1898	27	20	47
1899	17	32	49
1900	1	10	11
1901	18	10	28
1902	14	7	21
1903	17	5	22
1904	19	4	23
1905		6	6
1906	9	2	11
1907	3	3	6

POLMEAR,EAST   ST.AUSTELL   1001

Production:

Copper	Ore(tons)	Metal(tons)	Value(£)
1862	30.00	2.50	214.50
1864	27.00	2.60	215.30

Comment 1862 (C); 1864 (C)

POLMEAR,SOUTH   ST.AUSTELL   SX 035515 1002

Production: Lead & Silver   Ore(tons) Lead(tons) Silver(ozs)   Value(£)
            1880   13.30   7.00   150.00   58.60
            Zinc   Ore(tons) Metal(tons)   Value(£)
            1872   97.30   0.00   98.40
            1873   124.00   0.00   99.20
            1876   202.00   0.00   291.20
            1877   86.80   0.00   390.30
            Copper   Ore(tons) Metal(tons)   Value(£)
            1869   25.00   0.90   45.60
            1876   9.00   0.80   57.80
            Comment 1869 (C); 1876 (C)
Ownership: 1877-1880 FRAN.BARRATT; 1881 MRS.FRAN.BARRATT; 1883-1884
            MRS.FRAN.BARRATT; 1892-1893 SOUTH POLMEAR MINE ADVENTURERS
            C.B.
            Comment 1892-1893 SUSPENDED
Management: Manager 1876-1880 FRAN.BARRATT
            Chief Agent 1880-1881 J.NICHOLLS; 1882 WALT.J.NICHOLLS;
            1892-1893 THOS.ELLERY
            Secretary 1876-1880 FRAN.BARRATT; 1881 WALT.J.NICHOLLS;
            1883-1884 WALT.J.NICHOLLS; 1892-1893 JOS.ROGERS (P)
Employment: Underground   Surface   Total
            1878-1879   6   6
            1880   10   10
            1881-1883   4   4
            1884   2   2

POLMEAR,WEST   ST.AUSTELL   SX 029518 1003

Production: Copper No detailed return
            Tin No detailed return
Ownership: Comment 1859-1860 SEE CARVATH UNITED; 1861-1862 SEE ALSO
            CARVATH UNITED; 1863-1865 SUSPENDED. SEE ALSO CARVATH UNITED
Management: Manager 1862-1863 WM.BODY
            Chief Agent 1861 WM.BODY; 1862-1863 RICH.HANCOCK
            WM.POLKINGHORNE; 1865 J.BEARD
            Secretary 1861 H.W.HIGMAN (P); 1862-1863 C.E.TREFFRY;
            1864-1865 WM.POLKINGHORNE

POLPEOR   LELANT   SW 510362 1005

Production: Tin   Black(tons) Stuff(tons)   Tin(tons)   Value(£)
            1874   215.50   0.00   0.00   11952.80
            1875   177.10   0.00   0.00   8645.80
            Comment 1873 SEE KITTY(LELANT); 1874 KITTY(LELANT):TWO
            RETURNS AGG; 1875 NOW SISTERS
Ownership: 1875 POLPEOR CO.
            Comment 1873-1874 SEE KITTY(LELANT); 1875 SEE ALSO SISTERS;
            1876-1881 SEE SISTERS
Management: Chief Agent 1875 WM.ROSEWARNE

POLROSE   BREAGE   SW 613310 1006

Production: Tin   Black(tons) Stuff(tons)   Tin(tons)   Value(£)
            1873   27.60   0.00   0.00   1966.10
            1874   61.30   0.00   0.00   3257.60
            1875   48.30   0.00   0.00   1002.90
            1876   61.00   0.00   0.00   2654.00
            1877   60.90   0.00   0.00   2194.80
            1878   37.20   0.00   0.00   1327.30
            1883   0.10   3.00   0.00   3.00
Ownership: 1877-1884 POLROSE MINING CO.
Management: Manager 1872 SML.REED; 1874-1878 CHAS.ROACH; 1880
            WM.BENNETTS; 1881 W.H.WATSON
            Chief Agent 1872 CHAS.BOTTRAL; 1873 CHAS.ROACH CHAS.BOTTRAL;
            1874-1878 CHAS.BOTTRAL; 1879 WM.BENNETTS CHAS.ROACH; 1880
            CHAS.ROACH W.H.WATSON; 1881 WM.BENNETTS
            Secretary 1872 JOHN R.DANIELL; 1873-1879 THOS.ROACH;
            1880-1881 JOHN WATSON; 1882-1884 W.H.WATSON
Employment: Underground   Surface   Total
            1878   38   23   61
            1879   11   2   13
            1880   16   8   24
            1881   24   9   33
            1882   14   9   23
            1883   12   9   21
            1884   12   8   20

POLTRAVORGIE   PORT ISAAC   SX 013795 1007

Production: Antimony No detailed return
Ownership: 1912-1913 S.A.TAMBLYN
            Comment 1912 REOPENED JUNE 1912; 1913 NOT WORKED

POLYEAR   ST.MEWAN   SW 994508 1008

Production: Tin   Black(tons) Stuff(tons)   Tin(tons)   Value(£)
            1882   3.00   0.00   2.20   183.00
            1885   5.10   0.00   0.00   255.00
            1886   35.90   0.00   0.00   2630.00
            1887   8.20   0.00   0.00   589.00
            1888   7.50   0.00   0.00   573.00
            1889   6.20   0.00   0.00   366.00
            1890   81.00   0.00   0.00   464.00
            1891   18.00   0.00   0.00   977.00
            1892   3.10   0.00   0.00   171.00
            1893   1.40   0.00   0.00   70.00
            Comment 1882 POLGEAR CONSOLS; 1886 POLGEAR; 1890 BLACK TIN
            DOUBTFUL
Ownership: 1881-1883 JOHN EDWARDS; 1885-1886 POLGEAR ADVENTURERS;
            1887-1893 POLYEAR TIN MINING CO.LTD.
            Comment 1881 POLGEAR CONSOLS; 1882 POLYARE CONSOLS; 1883
            POLYARE CONSOLS. NOT WORKED; 1885-1886 POLGEAR CONSOLS
Management: Chief Agent 1881-1883 JOHN EDWARDS; 1885-1886 JOHN EDWARDS;
            1888-1893 JOHN ROWE
            Secretary 1887-1891 R.B.FASTNEDGE; 1892-1893 R.B.FASTNEDGE

Employment: (S)

	Underground	Surface	Total
1881	10	10	20
1882	10	11	21
1885	27	15	42
1886	14	21	35
1887	23	12	35
1888	15	13	28
1889	16	13	29
1890	30	22	52
1891	34	22	56
1892	20	14	34
1893	3		3

POLZEATH   PADSTOW   SW 938791 1009

Production: Lead

	Ore(tons)	Metal(tons)	Value(£)
1855	5.00	3.20	0.00
1856	1.20	0.80	0.00

Comment 1856 POLZEATH CONSOLS

POOL HALVANS   CAMBORNE   1010

Production: Copper

	Ore(tons)	Metal(tons)	Value(£)
1880	805.00	30.00	1415.00

POOL,EAST   ILLOGAN   SW 674415 1011

Production: Copper

	Ore(tons)	Metal(tons)	Value(£)
1845	929.00	73.30	5430.20
1846	633.00	43.80	3009.40
1847	608.00	34.10	2188.70
1848	1088.00	64.80	3550.30
1849	1136.00	66.90	4138.40
1850	1422.00	70.10	4313.90
1851	2050.00	106.80	6502.50
1852	1740.00	86.50	6845.10
1853	2157.00	98.60	9068.10
1854	2034.00	99.20	9455.30
1855	2287.00	134.70	13590.50
1856	2821.00	178.70	16584.40
1857	2535.00	139.00	13607.90
1858	1300.00	66.80	5840.70
1859	1280.00	58.70	5223.50
1860	1896.00	93.50	7997.00
1861	2435.00	120.30	10208.70
1862	2944.00	149.40	11596.80
1863	3264.00	155.80	11355.60
1864	2805.00	126.20	9817.50
1865	2083.00	93.50	6646.00
1866	1808.00	97.30	6233.00
1867	1582.00	76.70	4997.90

Copper	Ore(tons)	Metal(tons)	Value(£)
1868	1753.00	79.80	4764.60
1869	1641.00	85.00	4858.40
1870	1854.00	90.70	4927.70
1871	2430.00	125.60	7140.60
1872	2520.00	138.40	10992.40
1873	1933.00	95.90	5525.40
1874	1920.00	95.50	5342.50
1875	1823.00	85.60	5780.40
1876	1806.00	90.90	5725.60
1877	1527.00	90.60	5265.80
1878	2564.60	151.20	6983.90
1879	2362.30	112.50	4416.90
1880	1350.10	75.60	3190.50
1881	903.90	48.80	2435.70
1882	1093.30	54.60	3300.00
1883	453.00	26.50	1451.00
1884	376.00	0.00	952.00
1885	229.00	0.00	778.00
1888	15.00	0.00	90.00
1889	26.00	0.00	106.00
1890	312.00	0.00	800.00
1891	451.00	0.00	1551.00
1892	15.00	0.00	91.00
1893	729.00	0.00	1556.00
1894	730.00	0.00	1092.00
1895	373.00	0.00	340.00
1897	64.00	0.00	125.00
1898	16.00	0.00	40.00
1899	22.00	0.00	106.00
1901	10.00	0.00	48.00
1902	31.00	0.00	122.00
1903	27.00	0.00	172.00
1904	27.00	0.00	163.00
1905	70.00	0.00	106.00
1906	31.00	0.00	341.00
1907	68.00	0.00	346.00
1908	23.00	0.00	167.00
1909	25.00	0.00	130.00
1910	12.00	0.00	59.00
1911	16.00	0.00	97.00
1912	23.00	0.00	162.00
1913	21.00	0.00	100.00

Comment 1845-1876 (C); 1882 TWO RETURNS.ORE 381 HALVANS 712;
1883 TWO RETURNS.ORE 48 HALVANS 405; 1884-1885 HALVANS;
1897-1899 EAST POOL & AGAR UNITED; 1901-1913 EAST POOL & AGAR
UNITED

Tin	Black(tons)	Stuff(tons)	Tin(tons)	Value(£)
1852	20.80	0.00	0.00	0.00
1853	22.00	0.00	0.00	1566.60
1854	49.60	0.00	0.00	2852.10
1855	38.40	0.00	0.00	2099.00
1856	49.70	0.00	0.00	3066.80
1857	51.50	0.00	0.00	3383.90
1858	36.40	0.00	0.00	2061.00

Tin	Black(tons)	Stuff(tons)	Tin(tons)	Value(£)
1859	44.10	0.00	0.00	2729.60
1860	58.30	0.00	0.00	3854.50
1861	49.80	0.00	0.00	2885.30
1862	122.10	0.00	0.00	6623.00
1863	134.90	0.00	0.00	7646.50
1864	136.90	0.00	0.00	7510.00
1865	257.10	0.00	0.00	13363.40
1866	379.70	0.00	0.00	16360.40
1867	321.40	0.00	0.00	14618.40
1868	364.70	0.00	0.00	17269.60
1869	325.40	0.00	0.00	19185.20
1870	351.90	0.00	0.00	21704.20
1871	110.00	0.00	0.00	7046.80
1872	90.70	0.00	0.00	6726.60
1873	384.10	0.00	0.00	24038.10
1874	419.90	0.00	0.00	19392.70
1875	571.30	0.00	0.00	24553.30
1876	364.10	6664.00	0.00	25141.10
1877	392.00	4941.00	0.00	22969.80
1878	489.80	5904.00	0.00	25884.80
1879	864.50	2919.30	0.00	39490.90
1880	1268.20	821.10	0.00	60479.30
1881	1317.70	2130.00	0.00	67682.20
1882	1544.30	327.80	1019.20	88366.00
1883	1429.00	1026.00	0.00	69986.00
1884	1587.00	746.00	0.00	70161.00
1885	1518.80	271.00	0.00	71518.00
1886	1446.00	172.00	0.00	79564.00
1887	1236.00	103.00	0.00	77458.00
1888	1179.00	0.00	0.00	71478.00
1889	1105.00	0.00	0.00	56060.00
1890	995.00	0.00	0.00	50987.00
1891	1070.80	0.00	0.00	51963.00
1892	1004.00	0.00	0.00	49115.00
1893	938.30	0.00	0.00	41657.00
1894	940.00	0.00	0.00	33930.00
1895	877.00	0.00	0.00	30155.00
1896	529.30	0.00	0.00	14086.00
1897	309.00	0.00	0.00	8300.00
1898	770.00	0.00	0.00	29787.00
1899	816.00	0.00	0.00	54459.00
1900	739.00	0.00	0.00	54314.00
1901	699.00	0.00	0.00	40774.00
1902	663.00	0.00	0.00	39725.00
1903	706.00	0.00	0.00	45845.00
1904	661.00	0.00	0.00	45249.00
1905	585.50	0.00	0.00	47235.00
1906	644.50	0.00	0.00	64469.00
1907	636.00	0.00	0.00	67041.00
1908	711.00	0.00	0.00	56005.00
1909	586.00	0.00	0.00	46236.00
1910	507.00	0.00	0.00	45616.00
1911	479.00	0.00	0.00	53872.00
1912	501.00	0.00	0.00	63940.00

Tin	Black(tons)	Stuff(tons)	Tin(tons)	Value(£)
1913	528.50	0.00	0.00	62176.00

Comment 1852 TIN ORE; 1862 TWO RETURNS AGGREGATED; 1882–1887
EAST POOL & EAST POOL HALVANS AGGREGATED; 1897–1913 INC.AGAR

Cobalt	Ore(tons)	Metal(tons)	Value(£)
1871	3.00	0.00	120.00
1872	1.00	0.00	20.00
1873	0.30	0.00	12.00
1876	No detailed return		

Comment 1876 SEE BISMUTH

Arsenic	Ore(tons)	Metal(tons)	Value(£)
1854	93.10	0.00	188.10
1855	84.00	0.00	0.00
1856	No detailed return		
1857	108.80	0.00	216.90
1858	27.30	0.00	54.60
1859	86.40	0.00	172.60
1860	64.40	0.00	116.00
1861	67.70	0.00	109.50
1862	143.00	0.00	152.50
1863	49.30	0.00	49.30
1864	133.00	0.00	118.00
1865	180.50	0.00	121.30
1866	213.00	0.00	235.60
1867	244.10	0.00	539.50
1868	201.20	0.00	768.10
1869	241.00	0.00	922.60
1870	347.80	0.00	1003.10
1871	133.30	0.00	300.00
1872	385.00	0.00	698.00
1873	341.30	0.00	771.80
1874	514.00	0.00	2441.50
1875	477.60	0.00	2265.80
1876	646.00	0.00	3906.50
1877	622.00	0.00	2731.00
1878	376.00	0.00	1635.60
1879	545.00	0.00	2754.00
1880	No detailed return		
1881	383.00	0.00	2373.40
1882	358.70	0.00	1979.00
1883	359.00	0.00	2337.00
1884	655.00	0.00	4167.00
1885	568.00	0.00	3638.00
1886	411.00	0.00	2491.00
1887	507.00	0.00	3085.00
1888	474.00	0.00	3000.00
1889	334.00	0.00	2219.00
1890	612.00	0.00	4278.00
1891	640.00	0.00	5160.00
1892	422.00	0.00	2902.00
1893	529.00	0.00	4322.00
1894	620.00	0.00	5690.00
1895	508.00	0.00	5038.00
1896	653.00	0.00	8135.00
1897	440.00	0.00	5510.00

Arsenic	Ore(tons)	Metal(tons)	Value(£)
1898	493.00	0.00	4991.00
1899	584.00	0.00	6368.00
1900	465.00	0.00	6592.00
1901	542.00	0.00	5220.00
1902	407.00	0.00	2319.00
1903	326.00	0.00	1954.00
1904	325.00	0.00	1828.00
1905	418.00	0.00	2432.00
1906	778.00	0.00	10032.00
1907	646.00	0.00	15364.00
1908	614.00	0.00	7186.00
1909	520.00	0.00	4076.00
1910	512.00	0.00	3543.00
1911	401.00	0.00	1946.00
1912	445.00	0.00	3879.00
1913	250.00	0.00	1964.00

Comment 1855 WHITE ARSENIC; 1859–1871 CRUDE; 1884–1885 SOOT;
1886 CRUDE; 1897–1913 INC.AGAR

Tungsten	Ore(tons)	Metal(tons)	Value(£)
1861	6.50	0.00	29.00
1862	16.20	0.00	63.00
1863	13.00	0.00	67.40
1864	2.30	0.00	14.00
1865	2.00	0.00	11.00
1867	10.50	0.00	62.00
1868	9.20	0.00	67.30
1869	25.40	0.00	323.60
1870	40.80	0.00	534.70
1871	20.00	0.00	228.00
1872	75.30	0.00	847.00
1873	33.00	0.00	422.80
1874	32.80	0.00	545.00
1875	46.10	0.00	382.30
1876	23.50	0.00	172.50
1877	15.00	0.00	150.00
1878	10.00	0.00	100.00
1879	10.90	0.00	103.50
1880	1.00	0.00	8.50
1881	49.30	0.00	495.40
1882	54.90	0.00	717.00
1883	111.00	0.00	1443.00
1884	63.00	0.00	1089.00
1885	374.00	0.00	4799.00
1886	138.00	0.00	2206.00
1887	51.00	0.00	1264.00
1888	60.00	0.00	1625.00
1889	0.50	0.00	8.00
1890	104.00	0.00	1848.00
1891	137.30	0.00	3337.00
1892	125.00	0.00	3000.00
1893	22.00	0.00	420.00
1894	No detailed return		
1895	No detailed return		
1896	21.00	0.00	984.00

Tungsten	Ore(tons)	Metal(tons)	Value(£)
1897	101.00	0.00	1616.00
1898	74.00	0.00	3552.00
1899	22.00	0.00	1141.00
1900	7.00	0.00	257.00
1901	3.00	0.00	98.00
1902	9.00	0.00	273.00
1903	7.00	0.00	216.00
1904	13.00	0.00	1031.00
1905	49.00	0.00	2216.00
1906	120.00	0.00	7523.00
1907	73.00	0.00	8733.00
1908	57.50	0.00	4550.00
1909	60.00	0.00	5172.00
1910	75.00	0.00	8972.00
1911	48.00	0.00	5374.00
1912	43.00	0.00	4327.00
1913	45.50	0.00	5036.00

Comment 1862 150 TONS RAISED BUT NOT SOLD; 1878 VALUE EST.;
1886 INC.97 TONS LEAVINGS; 1898-1913 INC.AGAR
Wolfram No detailed return

Arsenic Pyrite	Ore(tons)	Value(£)
1893	1.00	3.00

Bismuth	Ore(tons)	Value(£)
1872	2.00	0.00
1873	1.20	68.70
1874	0.00	38.40
1876	0.40	0.00
1877	0.40	15.20

Comment 1876 INCLUDES COBALT

Uranium	Ore(tons)	Value(£)
1876	No detailed return	
1877	0.10	11.80
1878	0.30	0.00
1879	0.10	21.30

Comment 1876 PRODUCTION BUT NO SALES
Ownership: 1877-1890 EAST POOL MINING CO.; 1891-1893 EAST POOL MINING
CO. C.B.; 1894-1896 EAST POOL MINING CO.; 1897-1911 EAST POOL
& AGAR UNITED MINING CO.; 1912-1913 EAST POOL & AGAR LTD.
Comment 1897-1913 EAST POOL & AGAR UNITED
Management: Manager 1859 WM.S.GARBY; 1861-1873 WM.S.GARBY; 1876-1880 JOHN
MAYNARD; 1881 CHAS.BISHOP
Chief Agent 1861-1864 JOHN MAYNARD NICH.TAMBLYN; 1865-1867
JOHN MAYNARD A.JAMES; 1868 JOHN MAYNARD A.HOSKING; 1869 JOHN
MAYNARD JAS.HOSKING; 1870 JOHN MAYNARD JOHN HOSKING;
1871-1874 JOH.MAYNARD JOH.HOSKING J.M.OPIE (S); 1875
JOH.MAYNARD CHAS.BISHOP J.M.OPIE (S); 1876-1879 CHAS.BISHOP
J.M.OPIE (S); 1880-1881 JOHN PENHALL STEP.CURTIS J.MARTIN;
1882-1902 CHAS.BISHOP; 1903-1904 WM.RICHARDS; 1905-1908
JOS.TAMBLYN; 1909-1913 JOS.JENNINGS
Secretary 1859-1860 COMMITTEE (P); 1861-1866 W.PRISK (P);
1867-1870 COMMITTEE; 1871-1874 COMMITTEE (P); 1875-1880 THE
COMMITTEE; 1881 COMMITTEE (P); 1888-1890 J.H.MAYNE; 1891-1902
J.H.MAYNE (P)

Employment:

	Underground	Surface	Total
1878	192	117	309
1879	238	99	337
1880	253	105	358
1881	` 280	108	388
1882	307	111	418
1883	276	120	396
1884	262	438	700
1885	265	426	691
1886	321	401	722
1887	278	429	707
1888	270	415	685
1889	289	431	720
1890	343	378	721
1891	323	373	696
1892	338	357	695
1893	289	370	659
1894	264	349	613
1895	235	329	564
1896	106	260	366
1897	147	278	425
1898	236	340	576
1899	231	346	577
1900	250	301	551
1901	209	338	547
1902	233	280	513
1903	266	276	542
1904	295	281	576
1905	323	276	599
1906	319	322	641
1907	338	311	649
1908	333	314	647
1909	303	273	576
1910	286	261	547
1911	249	260	509
1912	247	258	505
1913	305	246	551

Comment 1897-1913 INC.AGAR

POOL,NORTH   REDRUTH   SW 675422 1012

Production:

Zinc	Ore(tons)	Metal(tons)	Value(£)
1857	23.30	0.00	11.60
Copper	Ore(tons)	Metal(tons)	Value(£)
1845	217.00	10.80	788.50
1846	291.00	12.70	783.70
1847	3096.00	211.40	14660.30
1848	5576.00	370.40	21262.50
1849	6447.00	433.10	27717.10
1850	7048.00	390.20	25617.90
1851	5914.00	334.00	21078.50
1852	5196.00	292.20	23280.60
1853	4729.00	283.70	27847.90
1854	3499.00	189.60	18791.80

Copper	Ore(tons)	Metal(tons)	Value(£)
1855	2205.00	112.80	11166.10
1856	1325.00	77.20	7128.80
1857	1062.00	58.00	5645.20
1858	403.00	21.00	1840.70
1859	557.00	28.80	2598.80
1866	16.00	0.90	59.20
1867	18.00	1.60	113.80
1868	51.00	4.10	255.80
1869	40.00	2.90	181.50
1870	80.00	6.10	347.00
1871	13.00	0.90	56.90

Comment 1845–1859 (C); 1866–1871 (C)

Tin	Black(tons)	Stuff(tons)	Tin(tons)	Value(£)
1856	1.70	0.00	0.00	102.90
1858	2.60	0.00	0.00	147.80

Arsenic	Ore(tons)	Metal(tons)	Value(£)
1857	17.20	0.00	34.30

Ownership: Comment 1876 NEW NORTH POOL
Management: Manager 1862–1865 RICH.TREDINNICK; 1869–1872 WM.C.VIVIAN
             F.CLYMO; 1873 JOS.VIVIAN & SON; 1874–1876 NICH.CLYMO
             Chief Agent 1866–1868 WM.C.VIVIAN
             Secretary 1869–1873 C.THOMAS; 1874–1876 WM.C.VIVIAN

PORKELLIS UNITED   WENDRON   SW 693328 1013

Production: Tin	Black(tons)	Stuff(tons)	Tin(tons)	Value(£)
1852	190.60	0.00	0.00	0.00
1853	26.40	0.00	0.00	1611.60
1854	118.50	0.00	0.00	7799.70
1855	205.10	0.00	0.00	13291.60
1856	236.00	0.00	0.00	15682.10
1857	271.10	0.00	0.00	18770.80
1858	221.10	0.00	0.00	12893.60
1859	9.40	0.00	0.00	620.30

Comment 1852 TIN ORE; 1856 PORKELLIS; 1858 PORKELLIS
Ownership: Comment 1860 SUSPENDED

PORKELLIS,NEW   WENDRON   SW 693328 1014

Production: Tin	Black(tons)	Stuff(tons)	Tin(tons)	Value(£)
1859	15.30	0.00	0.00	1189.10

PORTHELLY,NORTH   ST.MINVER   SW 927766 1015

Production: Lead & Silver	Ore(tons)	Lead(tons)	Silver(ozs)	Value(£)
1863	7.50	5.70	0.00	0.00

Comment 1863 NORTH PORTHILLY
Management: Chief Agent 1860 E.HITCHINGS; 1861–1866 SAMP.DYER
             Secretary 1860 NICH.ENNOR (P); 1861–1866 GEO.RICKARD (P)

PORTHLEDDEN   ST.JUST   SW 353319 1016

Production: Tin   Black(tons) Stuff(tons)   Tin(tons)   Value(£)
            1908   12.60   0.00   0.00   942.00
            1909   9.10   0.00   0.00   730.00
            1910   7.80   0.00   0.00   631.00
            1911   10.20   0.00   0.00   1088.00
            1912   20.40   0.00   0.00   2544.00
            1913   19.50   0.00   0.00   2260.00
Ownership: 1908-1913 FRAN.OATES
Employment: Underground   Surface   Total
            1908   4   14   18
            1909-1910   14   14   28
            1911   7   12   19
            1912   8   14   22
            1913   19   7   26

PORTHLEVEN   BREAGE   1017

Production: Lead   Ore(tons) Metal(tons)   Value(£)
            1845   80.00   48.00   0.00
            1846   82.00   49.00   0.00

PORTHTOWAN TIN & ARSENIC WORKS   PORTHTOWAN   9999

Production: Arsenic   Ore(tons) Metal(tons)   Value(£)
            1894   No detailed return
            Comment 1894 SEE BISSOE POOL

PRAAH   1018

Production: Tin   Black(tons) Stuff(tons)   Tin(tons)   Value(£)
            1896   1.00   0.00   0.00   24.00

PRAED CONSOLS   LELANT   SW 518360 1019

Production: Tin No detailed return
Ownership: Comment 1861-1865 SUSPENDED
Management: Manager 1860 JOHN STEVENS
            Secretary 1860-1863 THOS.FIELD JNR. (P)

PRIDEAUX   LUXULYAN   SX 071554 1020

Production: Iron   Ore(tons)   Iron(%)   Value(£)
            1862   1200.00   0.00   650.00
            1863   No detailed return
            1864   1000.00   0.00   0.00
            Comment 1862-1864 BH.
Ownership: 1863-1868 ANDREWS & CO.
Management: Chief Agent 1863-1868 JOHN PEARD

PRIDEAUX WOOD   LUXULYAN   SX 072557 1021

Production: Copper   Ore(tons) Metal(tons)   Value(£)
            1852   495.00   23.70   1810.80
            1853   407.00   15.60   1318.90
            1854   177.00   7.30   645.80
            1861   166.00   7.70   614.90
            1862   113.00   4.80   341.80
            1863   118.00   4.80   338.30
            Comment 1852-1854 (C); 1861-1863 (C)
            Tin   Black(tons) Stuff(tons)   Tin(tons)   Value(£)
            1855   0.80   0.00   0.00   58.50
            1857   18.50   0.00   0.00   1442.00
            1858   7.80   0.00   0.00   478.30
            1859   17.80   0.00   0.00   1236.20
            1860   5.90   0.00   0.00   451.90
            1861   4.90   0.00   0.00   323.60
            1862   2.90   0.00   0.00   177.70
            1863   6.50   0.00   0.00   408.00
            1904   1.30   0.00   0.00   50.00
            1905   0.00   820.00   0.00   182.00
            1906   6.50   0.00   0.00   600.00
            1907   11.00   0.00   0.00   920.00
            1908   3.70   0.00   0.00   219.00
            1909   3.00   0.00   0.00   50.00
            1911   4.00   0.00   0.00   420.00
            1912   20.90   0.00   0.00   2336.00
            1913   15.60   0.00   0.00   1420.00
            Comment 1904-1909 PRIDEAUX; 1911-1913 PRIDEAUX
Ownership: 1904-1910 E.A.JONES & CO.; 1911-1912 PRIDEAUX MINE CO.; 1913
            CENTRAL CONSOLIDATED TIN EXPLOR.CO.
            Comment 1863-1865 SUSPENDED; 1904-1909 PRIDEAUX; 1910
            PRIDEAUX ABANDONED; 1911 PRIDEAUX REOPENED AUG.1911;
            1912-1913 PRIDEAUX
Management: Manager 1860-1862 FRAN.PUCKEY
            Chief Agent 1859 RICH.JNR.; 1860-1862 JOHN PUCKEY; 1913
            H.C.MCDONAN
            Secretary 1859-1863 MAJOR DAVIS (P)
Employment: Underground   Surface   Total
            1904   10   10   20
            1905   5   5   10
            1906   6   15   21
            1907   8   10   18
            1908   6   4   10
            1909   6   5   11
            1911   6   9   15
            1912   11   20   31
            1913   10   12   22

PRINCE ALBERT CONSOLS   PERRANZABULOE   SW 797537 1022

Production: Copper No detailed return
            Tin   Black(tons) Stuff(tons)   Tin(tons)   Value(£)
            1852   5.00   0.00   0.00   0.00
            1853   3.00   0.00   0.00   194.10
            1854   1.00   0.00   0.00   60.60

                                        392

PRINCE ALBERT CONSOLS   PERRANZABULOE   Continued

          Comment 1852 TIN ORE; 1854 PRINCE ALBERT
Ownership: Comment 1863–1864 PRINCE ALBERT; 1865 PRINCE ALBERT
          SUSPENDED
Management: Chief Agent 1863–1864 RICH.DAVIES

PRINCE OF WALES   ST.MELLION   SX 401705 1023

Production: Lead & Silver   Ore(tons) Lead(tons) Silver(ozs)   Value(£)
            1871   17.50   13.10   65.00   0.00
            1873   5.60   4.20   21.00   120.30
            Copper   Ore(tons) Metal(tons) Value(£)
            1865   9.00   0.60   45.40
            1866   122.00   13.20   949.90
            1867   1173.00   107.80   8445.40
            1868   1581.00   138.10   10371.90
            1869   1215.00   106.00   7269.40
            1870   1049.00   87.40   5339.80
            1871   788.00   63.70   4095.30
            1872   871.00   65.50   5201.60
            1873   798.00   54.30   3309.80
            1874   751.00   50.80   3306.60
            1875   406.00   23.00   1699.60
            1876   167.00   9.00   567.40
            1877   337.00   19.70   1049.10
            1878   37.00   0.90   23.30
            1881   72.00   5.40   311.40
            1882   352.60   18.90   1047.00
            1883   421.00   16.00   1060.00
            1884   476.00   0.00   1005.00
            1885   44.00   0.00   202.00
            1886   48.00   0.00   216.00
            1887   49.00   0.00   171.00
            1888   234.00   0.00   1601.00
            1889   53.00   0.00   210.00
            1890   56.00   0.00   284.00
            1891   44.00   0.00   198.00
            1894   9.00   0.00   33.00
            1901   45.00   0.00   477.00
            1902   21.00   0.00   195.00
            1903   16.00   0.00   175.00
            1904   76.00   0.00   497.00
            1905   5.00   0.00   77.00
            Comment 1865–1876 (C); 1884 TWO RETURNS ORE.331 BURNT 145
            Tin   Black(tons) Stuff(tons) Tin(tons)   Value(£)
            1871   4.70   0.00   0.00   391.30
            1874   3.90   0.00   0.00   190.20
            1880   0.80   0.00   0.00   27.00
            1881   14.00   0.00   9.30   750.00
            1882   23.40   0.00   14.50   1374.00
            1883   26.40   0.00   0.00   1317.00
            1884   30.50   0.00   0.00   1444.00
            1885   23.20   0.00   0.00   1073.00
            1886   28.00   0.00   0.00   1576.00
            1887   33.60   0.00   0.00   2087.00

                                    393

Tin	Black(tons)	Stuff(tons)	Tin(tons)	Value(£)
1888	20.80	0.00	0.00	1278.00
1889	27.60	0.00	0.00	1483.00
1890	60.30	0.00	0.00	3200.00
1891	46.90	0.00	0.00	2258.00
1892	26.00	0.00	0.00	1188.00
1893	93.40	0.00	0.00	4475.00
1894	18.90	0.00	0.00	754.00
1895-1896 No detailed return				
1900	32.40	0.00	0.00	2379.00
1901	209.90	0.00	0.00	13832.00
1902	143.50	0.00	0.00	9244.00
1903	42.20	0.00	0.00	3124.00
1904	72.20	0.00	0.00	4455.00
1905	47.90	0.00	0.00	2810.00
1907	4.80	0.00	0.00	391.00
1908	6.90	0.00	0.00	577.00
1911	43.50	0.00	0.00	4856.00
1912	10.00	0.00	0.00	919.00
1913	29.20	0.00	0.00	2955.00

Silver	Ore(tons)	Metal(tons)	Value(£)
1874	2.50	0.00	20.00
1878	1.80	0.00	11.40

Arsenic Pyrite	Ore(tons)	Value(£)
1877	35.00	27.00
1882	16.80	12.00
1883	5.00	6.00
1885	14.00	12.00
1888	31.00	40.00
1889	21.00	31.00
1890	34.00	26.00
1891	382.00	266.00
1892	95.00	65.00
1893	13.00	7.00
1894	485.00	322.00
1895	1243.00	668.00
1896	2444.00	2094.00
1897	1941.00	1306.00
1898	512.00	348.00
1899	377.00	206.00
1906	325.00	390.00
1907	207.00	214.00
1910	123.00	58.00

Comment 1877 ARSENICAL MUNDIC
Ownership: 1877-1879 PRINCE OF WALES MINING CO.; 1880-1887 JOHN WATSON & OTHERS; 1888-1890 C.B.PARRY; 1891 PRINCE OF WALES MINING CO. C.B.; 1892-1893 PRINCE OF WALES MINING CO.LTD.; 1894 EAST CORNWALL TIN SYNDICATE; 1895-1899 EAST CORNWALL TIN SYNDICATE LTD.; 1900-1910 CALSTOCK TIN & COPPER CO.LTD.; 1911-1912 PRINCE OF WALES MINING CO.; 1913 PRINCE OF WALES MINE LTD. Comment 1879 SEE PRINCE OF WALES, DEVON
Management: Manager 1865-1866 JOHN HITCHINGS (S); 1867 W.C.COCK; 1868-1875 JOHN GIFFORD; 1880 JOHN ANDREWS Chief Agent 1864-1865 H.SPRAGUE; 1866 JOHN GIFFORD WM.GIFFORD; 1868 JOHN HITCHINGS (S) W.C.COCK; 1869 H.E.CROKER

(P) WM.GIFFORD; 1870–1871 H.E.CROKER (P) F.PHILLIPS;
1872–1873 T.PHILLIPS; 1874 JOHN PRYOR; 1876 H.E.CROKER;
1877–1879 JOHN ANDREWS; 1880–1899 STEP.ROBERTS; 1902
C.FRED.THOMAS; 1903–1905 R.GILNICKI; 1906 C.W.SECCOMBE;
1907–1910 W.L.MOLE; 1911–1912 JAS.A.MICHELL; 1913 E.S.KING &
CO.
Secretary 1864 H.E.CROKER; 1865–1868 H.E.CROKER (P);
1869–1871 JEHU.HITCHINGS (S); 1875–1877 JOHN PRYOR; 1878–1881
H.E.CROKER; 1891–1893 C.B.PARRY (S); 1894–1895 JAS.H.CROFTS
(S); 1896–1899 F.R.GRANT (S)

Employment:	Underground	Surface	Total
1878	6		6
1879	4		4
1880	12	14	26
1881	35	38	73
1882	48	35	83
1883	54	40	94
1884	40	30	70
1885	35	25	60
1886	41	30	71
1887	52	32	84
1888	49	37	86
1889	67	55	122
1890	83	60	143
1891	47	36	83
1892	44	26	70
1893	76	49	125
1894	18	14	32
1895	18	6	24
1896	31	6	37
1897	27	2	29
1898	6	1	7
1899	6		6
1900	38	32	70
1901	132	78	210
1902	95	56	151
1903	80	47	127
1904	53	55	108
1905	7	4	11
1906	18	6	24
1907	20	20	40
1908	43	24	67
1909	7	8	15
1910	7	6	13
1911	94	62	156
1912	9	36	45
1913	96	57	153

PRINCE OF WALES,SOUTH   ST.MELLION   1024

Production: Copper No detailed return
Ownership: Comment 1879 SEE NEWTON
Management: Manager 1869 WM.KNOTT
            Secretary 1869 J.J.HAMBLY

PRINCE OF WALES,WEST   CALSTOCK   SX 389707 1025

Production: Copper No detailed return
            Tin   Black(tons) Stuff(tons)   Tin(tons)   Value(£)
            1900   2.40   0.00   0.00   67.00
            1903   0.90   0.00   0.00   67.00
            1905   1.80   0.00   0.00   155.00
            1906   2.00   0.00   0.00   178.00
            1908   2.00   0.00   0.00   119.00
            1910   0.70   0.00   0.00   59.00
            1911   1.00   0.00   0.00   96.00
            1913   0.80   0.00   0.00   84.00
            Arsenic   Ore(tons) Metal(tons)   Value(£)
            1908   3.00   0.00   15.00
Ownership: 1900—1912 WEST PRINCE OF WALES TIN SYNDICATE; 1913 WEST
            PRINCE CO.LTD.
            Comment 1909—1911 SUSPENDED; 1912 NOT WORKED
Management: Manager 1869—1871 JEHU.HITCHINGS (S)
            Chief Agent 1869—1871 JOHN GIFFORD
            Secretary 1869—1871 H.E.CROKER (P); 1900—1912 FRED.WRIGHT
            (S)
Employment: Underground   Surface   Total
            1900   2   1   3
            1901   8   3   11
            1902   1   1
            1903   2   2
            1905   2   2   4
            1906   4   2   6
            1907   8   3   11
            1908   10   7   17
            1910   1   1
            1911   2   2   4
            1913   6   2   8

PRINCE ROYAL   PERRANZABULOE   SW 748522 1026

Production: Lead   Ore(tons) Metal(tons)   Value(£)
            1874   0.70   0.40   12.40
            Tin   Black(tons) Stuff(tons)   Tin(tons)   Value(£)
            1887   1.00   0.00   0.00   59.00
            1888   3.50   0.00   0.00   194.00
Ownership: 1875 DUIGAN & WALSALL; 1876 PRINCESS ROYAL CO.; 1877 DUIGAN &
            WALSALL; 1885—1886 PRINCE ROYAL ADVENTURERS; 1887—1889 PRINCE
            ROYAL MINE CO.
            Comment 1877 IDLE; 1889 SUSPENDED
Management: Chief Agent 1874—1877 HY.BENNETTS; 1885—1888 SML.BENNETTS;
            1889 W.K.MICHELL
            Secretary 1873—1874 W.W.STABLES; 1876—1877 J.H.COLLINS (S)
Employment: Underground   Surface   Total
            1885   8   8
            1886   9   3   12
            1887   12   5   17
            1889   6   4   10

PRINCE VICTOR   LAUNCESTON   1027

Production: Lead No detailed return
Management: Manager 1875-1876 JAS.EVANS
            Secretary 1875-1876 BRODERICK & PARKER

PRINCESS OF WALES   CALLINGTON   SX 378705 1028

Production: Copper No detailed return
Ownership: Comment 1873 STOPPED IN 1873
Management: Manager 1869-1871 T.FOOT
            Chief Agent 1867 GEO.RICKARD; 1868 WM.GIFFORD; 1869
            THOS.VOSPER (P) GEO.RICKARD; 1870-1873 GEO.RICKARD
            Secretary 1868 WM.WARD; 1869-1871 WM.WARD (S)

PROSPER   1029

Production: Copper   Ore(tons) Metal(tons)   Value(£)
            1862   6.00   0.50   41.70
            1863   4.00   0.20   13.10
            1871   90.00   4.10   212.30
            1872   194.00   2.80   26.40
            Comment 1862-1863 (C)WEST CENTRAL DISTRICT; 1871-1872 (C)WEST
            CENTRAL DISTRICT
            Tin No detailed return

PROSPER   BREAGE   SW 592272 1030

Production: Copper No detailed return
            Tin No detailed return
Ownership: Comment 1865 SUSPENDED; 1871 NO PLACE GIVEN
Management: Manager 1859-1860 HUGH STEPHENS; 1863-1864 HUGH STEPHENS
            Chief Agent 1859-1861 ED.WILLIAMS; 1862 ED.WILLIAMS
            F.BLEWETT; 1863-1864 SML.STEPHENS
            Secretary 1859-1865 HUGH STEPHENS (P)

PROSPER   LANIVET   SX 030642 1031

Production: Tin   Black(tons) Stuff(tons) Tin(tons)   Value(£)
            1861   6.30   0.00   0.00   441.50
            1862   16.80   0.00   0.00   1085.20
            1863   9.40   0.00   0.00   586.40
            1864   3.10   0.00   0.00   192.10
            1865   0.80   0.00   0.00   40.20
            1870   37.30   0.00   0.00   2601.00
            1871   39.50   0.00   0.00   3035.70
            1872   5.50   0.00   0.00   436.60
            1874   11.30   0.00   0.00   621.90
            1878   16.60   0.00   0.00   550.70
            1880   17.60   0.00   0.00   894.80
            1881   17.40   0.00   0.00   942.20
            1882   20.60   0.00   0.00   1170.00
            1883   52.00   0.00   0.00   2833.00

                                  397

Tin	Black(tons)	Stuff(tons)	Tin(tons)	Value(£)
1884	34.30	0.00	0.00	1593.00
1885	14.00	0.00	0.00	682.00
1886	19.40	0.00	0.00	1102.00
1887	15.10	0.00	0.00	944.00
1888	22.30	0.00	14.70	1554.00
1889	22.60	0.00	0.00	1269.00
1890	20.70	0.00	0.00	1238.00
1891	No detailed return			
1895	6.00	0.00	0.00	232.00
1896	7.00	0.00	0.00	200.00
1897	5.20	0.00	0.00	194.00
1898	2.00	0.00	0.00	95.00
1899	3.60	0.00	0.00	262.00
1900	8.90	0.00	0.00	704.00
1901	5.50	0.00	0.00	477.00
1902	7.90	0.00	0.00	560.00
1903	8.60	0.00	0.00	690.00
1904	5.50	0.00	0.00	400.00
1905	7.50	0.00	0.00	700.00
1906	7.00	0.00	0.00	736.00
1907	7.50	0.00	0.00	835.00
1908	7.50	0.00	0.00	585.00
1909	8.20	0.00	0.00	625.00
1910	7.50	0.00	0.00	675.00
1911	6.50	0.00	0.00	700.00
1912	6.50	0.00	0.00	800.00
1913	9.70	0.00	0.00	1230.00

Comment 1861 WEST CENTRAL DISTRICT; 1862 TWO RETURNS
AGGREGATED; 1863–1865 WEST CENTRAL DISTRICT; 1870–1872 WEST
CENTRAL DISTRICT; 1878 PROSPER 1 MITCHELL; 1880–1881 PROSPER
MITCHELL; 1882–1890 OPEN WORK; 1891 AT WORK; 1895–1913 OPEN
WORKS

Arsenic	Ore(tons)	Metal(tons)	Value(£)
1870	148.60	0.00	381.60
1871	60.00	0.00	67.50

Comment 1870–1871 CRUDE

Ownership: 1878 MULBERRY & PROSPER MINE CO.; 1880–1881 PROSPER &
MITCHELL MINING CO.; 1882–1883 WHEAL PROSPER MINING CO.LTD.;
1884–1891 JOHN LEYS
Comment 1874–1875 INC.MITCHELL; 1879 SEE MULBERRY; 1880–1881
INC.MITCHELL; 1882–1890 OPEN WORK
Management: Manager 1878 THOS.MARTIN; 1881 THOS.HAMBLY
Chief Agent 1874–1875 THOS.HAMBLY; 1880 THOS.MARTIN; 1881
T.R.COCK
Secretary 1874–1875 W.L.MARTIN; 1881 THOS.MARTIN

PROSPER AND MICHELL UTD.   MARAZION   SW 534318 1033

Production: Tin	Black(tons)	Stuff(tons)	Tin(tons)	Value(£)
1862	3.70	0.00	0.00	251.50
1863	19.80	0.00	0.00	1301.60
1864	20.10	0.00	0.00	1274.40
1865	22.60	0.00	0.00	1237.10

PROSPER AND MICHELL UTD.   MARAZION   Continued

```
        Tin          Black(tons) Stuff(tons)   Tin(tons)    Value(£)
        1866            7.50        0.00          0.00       397.60
        Comment 1862 PART YEAR ONLY; 1864-1866 PROSPER & MITCHELL
        UNITED
Ownership:  Comment 1864-1866 PROBABLY PART OF PROSPER UNITED
```

PROSPER UNITED   MARAZION   SW 534318 1034

Production: Lead & Silver   Ore(tons) Lead(tons) Silver(ozs)   Value(£)

Year	Ore(tons)	Lead(tons)	Silver(ozs)	Value(£)
1863	10.60	7.00	0.00	0.00
1864	10.90	7.10	100.00	0.00
1865	22.90	14.70	200.00	0.00
1866	23.50	17.40	0.00	0.00
1867	9.20	6.90	0.00	0.00

Comment 1867 FOR AG SEE GREAT RETALLACK

Copper   Ore(tons) Metal(tons)   Value(£)

Year	Ore(tons)	Metal(tons)	Value(£)
1845	5993.00	406.20	29144.00
1846	3589.00	252.30	17509.50
1847	2257.00	207.00	14348.00
1848	1372.00	72.80	4185.80
1849	115.00	5.30	294.50
1850	99.00	5.30	323.80
1862	606.00	35.90	2890.20
1863	2143.00	108.40	7947.50
1864	3616.00	169.90	13276.20
1865	4384.00	210.70	15310.70
1866	4523.00	234.70	14400.30
1867	5108.00	233.50	14624.30
1868	4055.00	156.30	9033.80
1869	629.00	35.10	2059.70
1870	52.00	2.50	134.20
1872	15.00	0.70	56.20

Comment 1845-1850 (C) PROSPER; 1862-1870 (C); 1872 (C)

Tin   Black(tons) Stuff(tons)   Tin(tons)   Value(£)

Year	Black(tons)	Stuff(tons)	Tin(tons)	Value(£)
1862	20.90	0.00	0.00	1399.80
1863	129.60	0.00	0.00	8564.10
1864	118.00	0.00	0.00	7192.40
1865	86.70	0.00	0.00	4525.50
1866	162.00	0.00	0.00	7686.90
1867	191.40	0.00	0.00	9831.20
1868	183.60	0.00	0.00	9977.50
1869	86.80	0.00	0.00	5859.30

Comment 1862 PART YEAR ONLY

Arsenic   Ore(tons) Metal(tons)   Value(£)

Year	Ore(tons)	Metal(tons)	Value(£)
1863	76.50	0.00	65.70
1865	153.50	0.00	113.50
1866	196.90	0.00	216.90
1867	247.40	0.00	340.00
1868	290.60	0.00	686.30
1869	98.40	0.00	246.40

```
            Comment 1863 CRUDE; 1865-1868 CRUDE
Ownership:  Comment 1864-1866 PROSPER & MICHELL UNITED
Management: Manager 1860-1861 THOS.RICHARDS
            Chief Agent 1860-1861 W.H.MARTIN; 1862-1865 STEP.LEAN
```

W.H.MARTIN; 1866 J.NICHOLLS W.H.MARTIN; 1867 J.NICHOLLS
W.HALL W.G.CANVILLE; 1868 J.NICHOLLS
Secretary 1860–1864 JOHN HOSKING (P); 1865–1867 C.WESCOMBE;
1868 R.R.MICHELL

PROSPER,GREAT   ST.DENNIS   SW 594577 1035

Production: Tin	Black(tons)	Stuff(tons)	Tin(tons)	Value(£)
1888	0.50	0.00	0.30	25.00
1889	1.10	0.00	0.00	58.00
1890	0.60	0.00	0.00	28.00
1891	No detailed return			

Comment 1888–1890 OPEN WORK; 1891 AT WORK
Ownership: 1888–1891 GREAT PROSPER CHINA CLAY & STONE WORKS
             Comment 1888–1891 OPEN WORKS

PROSPER,WEST   1036

Production: Tin	Black(tons)	Stuff(tons)	Tin(tons)	Value(£)
1865	4.00	0.00	0.00	213.80
1866	3.00	0.00	0.00	152.40

PROSPIDNICK   SITHNEY   SW 642317 1037

Production: Tin	Black(tons)	Stuff(tons)	Tin(tons)	Value(£)
1861	4.90	0.00	0.00	286.30
1862	20.40	0.00	0.00	1253.70

Ownership: Comment 1863–1865 SUSPENDED
Management: Chief Agent 1860 PHIL.ROGERS; 1861–1862 RICH.KENDALL
             Secretary 1860–1865 JOHN WATSON

PROSPIDNICK,NEW   SITHNEY   1038

Production: Tin	Black(tons)	Stuff(tons)	Tin(tons)	Value(£)
1863	13.30	0.00	0.00	841.80

Ownership: Comment 1863–1865 SUSPENDED
Management: Chief Agent 1862 W.H.BISHOP
             Secretary 1862 JOHN WATSON

PROVIDENCE   GWINEAR   SW 595348 1039

Production: Copper	Ore(tons)	Metal(tons)	Value(£)
1845	2615.00	205.70	14749.00
1846	1105.00	92.80	6296.00
1847	876.00	73.90	5201.00

Comment 1845–1847 (C)

Production: Tin No detailed return
Ownership: 1912-1913 WENDRON CORNWALL TIN MINES LTD.
            Comment 1912 PROSPECTING; 1913 NOT WORKED
Employment: Underground   Surface   Total
            1912
            Comment 1912 SEE WEST VOR

PROVIDENCE MINES   LELANT   SW 523384 1041

Production: Copper   Ore(tons) Metal(tons)   Value(£)
            1845   433.00   31.30   2325.70
            1846   222.00   11.50   683.60
            1851   141.00   6.50   362.50
            1853   244.00   14.60   1406.10
            1854   346.00   17.20   1679.40
            1855   168.00   7.10   675.30
            1862   10.00   0.90   0.00
            1864   16.00   1.50   0.00
            1866   5.00   0.50   0.00
            1873   10.00   1.20   0.00
            1876   30.00   1.40   0.00
            1878   3.40   0.10   0.00
            Comment 1845-1846 (C); 1851 (C); 1853-1855 (C); 1862 (C);
            1864 (C); 1866 (C); 1873 (C); 1876 (C)
            Tin   Black(tons) Stuff(tons)   Tin(tons)   Value(£)
            1853   95.50   0.00   0.00   6443.40
            1854   154.90   0.00   0.00   10347.20
            1855   275.90   0.00   0.00   18252.90
            1856   242.10   0.00   0.00   18143.90
            1857   289.40   0.00   0.00   23276.40
            1858   332.50   0.00   0.00   21141.50
            1859   353.50   0.00   0.00   25667.30
            1860   321.00   0.00   0.00   25483.70
            1861   271.30   0.00   0.00   19741.20
            1862   421.00   0.00   0.00   27748.60
            1863   338.90   0.00   0.00   22488.10
            1864   385.10   0.00   0.00   24150.90
            1865   386.40   0.00   0.00   21497.60
            1866   406.40   0.00   0.00   19357.20
            1867   389.70   0.00   0.00   20128.30
            1868   348.40   0.00   0.00   19200.30
            1869   349.30   0.00   0.00   24146.50
            1870   326.30   0.00   0.00   23513.60
            1871   266.70   0.00   0.00   20858.40
            1872   263.80   0.00   0.00   22541.00
            1873   269.50   0.00   0.00   19878.70
            1874   187.80   0.00   0.00   11002.40
            1875   159.40   0.00   0.00   7970.40
            1876   131.40   0.00   0.00   5567.60
            1877   90.10   0.00   0.00   3370.00
            1878   13.90   0.00   0.00   488.60
            1885   3.00   0.00   0.00   100.00
            1886   5.50   0.00   0.00   200.00
            1887   6.00   0.00   0.00   390.00

Tin	Black(tons)	Stuff(tons)	Tin(tons)	Value(£)
1888	3.00	0.00	0.00	180.00
1908	11.10	0.00	0.00	891.00
1909	5.00	0.00	0.00	470.00
1910	35.40	0.00	0.00	2630.00

Comment 1862 TWO RETURNS AGGREGATED; 1875 THE PROVIDENCE;
1878 IDLE SINCE JUNE 1878; 1887-1888 PROVIDENCE UNITED

Ownership: 1876-1878 PROVIDENCE MINES CO.; 1884-1887 PROVIDENCE UNITED
MINING CO.; 1888 J.J.NICHOLLS; 1906-1907 TASMASION
EXPLORATION CO.LTD.; 1908-1913 PROVIDENCE TIN MINES LTD.
Comment 1877 STOPPED SEP.1877; 1884-1886 PROVIDENCE UNITED;
1887-1888 PROVIDENCE; 1906-1909 PROVIDENCE; 1910 PROVIDENCE
SUSPENDED SEP.1910; 1911-1913 PROVIDENCE NOT WORKED

Management: Manager 1859 ARUN.ANTHONY; 1860-1877 WM.HOLLOW
Chief Agent 1859 WM.HOLLOW JNR.; 1860 THOS.ANTHONY; 1861-1864
PHIL.ROGERS WM.DUNSTON; 1865-1868 PHIL.ROGERS J.WHITE; 1869
PHIL.ROGERS R.MARTIN; 1870-1871 PHIL.ROGERS BEN.MARTIN
P.ROGERS JNR.; 1872 PHIL.ROGERS SML.ROGERS BEN.MARTIN; 1873
PHIL.ROGERS; 1874-1876 SML.ROGERS; 1878 ED.TRYTHALL;
1906-1908 F.ROUSE PAUL; 1911-1913 JOH.PENBERTHY
Secretary 1859-1862 SML.HIGGS (P); 1863-1872 SML.HIGGS & SON;
1873-1877 ED.TRYTHALL; 1884-1888 JAS.NICHOLLS

Employment:	Underground	Surface	Total
1878		20	20
1884	4	2	6
1885	10		10
1886	9		9
1887	11		11
1906	8		8
1907	70	90	160
1908	20	30	50
1909	20	31	51
1910	28	41	69

PROVIDENCE,EAST   LELANT   SW 528387 1042

Production: Tin	Black(tons)	Stuff(tons)	Tin(tons)	Value(£)
1863	10.20	0.00	0.00	625.20
1864	17.50	0.00	0.00	1078.20
1865	27.00	0.00	0.00	1506.30
1866	19.10	0.00	0.00	479.20
1867	12.60	0.00	0.00	644.20
1870	9.00	0.00	0.00	690.30
1871	1.60	0.00	0.00	119.10

Comment 1864 9 MONTHS ONLY
Ownership: Comment 1901 SEE HAWKS POINT
Management: Manager 1859 WM.HOLLOW; 1869 J.NANCARROW
Chief Agent 1859 THOS.UREN; 1860 THOS.UREN THOS.RICHARDS;
1861-1864 THOS.UREN; 1865 J.NANCARROW; 1866-1868 J.NANCARROW
W.WHITE; 1869 W.WHITE
Secretary 1859-1862 JAS.HOLLOW (P); 1863-1869 THOS.HOLLOW
(P)

PROVIDENCE,NORTH   ST.IVES   SW 420403 1043

Production: Copper No detailed return
            Tin No detailed return
Management: Manager 1859 WM.BISHOP
            Chief Agent 1860 RICH.HALE
            Secretary 1860 WM.BISHOP (P)

PROVIDENCE,SOUTH   LELANT   SW 523381 1044

Production: Tin   Black(tons) Stuff(tons)   Tin(tons)   Value(£)
            1855   1.90   0.00   0.00   135.40
            1856   4.10   0.00   0.00   275.60
            1857   0.40   0.00   0.00   30.80
            1875   12.00   0.00   0.00   600.00
            1876   20.00   15.00   0.00   880.50
            1877   33.80   0.00   0.00   1171.50
            1878   4.00   0.00   0.00   141.10
            1879   1.00   0.00   0.00   38.00
            1880   1.10   0.00   0.00   54.60
            1881   0.80   0.00   0.20   37.00
            1882   0.70   0.00   0.40   36.00
Ownership: 1876 S.BARDLEY JNR. & CO.; 1877-1881 SML.BEWLEY & CO.;
            1882-1883 SML.BEWLEY; 1906-1909 WORVAS DOWNS MINING CO.
            Comment 1877 LATE GIEW SUSPENDED; 1878 IDLE SINCE JUNE 1878;
            1879-1880 ABOVE ADIT & HEAPS ONLY; 1909 ABANDONED JAN.1910
Management: Manager 1859 HUGH STEPHENS; 1869-1871 JAS.EVANS; 1873-1874
            JAS.CRAZE; 1876 WM.BRABYN
            Chief Agent 1859 H.MARTIN; 1869-1871 THOS.UREN; 1873-1875
            WM.BRABYN; 1877-1878 JAS.W.CRAZE WM.PASCOE; 1879-1883
            WM.PASCOE; 1906-1909 COR.MILLETT
            Secretary 1859 HUGH STEPHENS; 1869-1871 CHRIS.STEPHENS;
            1873-1874 WM.PASCOE; 1875 ED.PASCOE
Employment: Underground   Surface   Total
            1878   10   10
            1879   8   8
            1880-1881   2   2   4
            1882   3   3
            1906   2   8   10
            1907   10   5   15
            1908   1   1
            1909   3   2   5

PROVIDENCE,WEST   ST.ERTH   SW 582345 1046

Production: Copper   Ore(tons) Metal(tons)   Value(£)
            1851   194.00   28.80   2070.60
            1852   146.00   20.70   1868.90
            1853   158.00   17.90   1967.10
            1854   234.00   25.20   2616.60
            1855   360.00   29.70   3052.50
            1856   184.00   16.50   1605.40
            1857   190.00   15.50   1574.60
            1862   10.00   0.60   51.00
            Comment 1851-1857 (C); 1862 (C)

                              403

PROVIDENCE,WEST   ST.ERTH   Continued

Tin   Black(tons) Stuff(tons)   Tin(tons)   Value(£)
1852   40.20   0.00   0.00   0.00
1853   80.00   0.00   0.00   4309.40
1854   33.50   0.00   0.00   2883.00
1855   143.20   0.00   0.00   9835.10
1856   100.60   0.00   0.00   7855.30
1857   102.60   0.00   0.00   8567.00
1858   51.70   0.00   0.00   3480.70
1859   66.50   0.00   0.00   4901.20
1860   39.70   0.00   0.00   3164.40
1861   30.00   0.00   0.00   2221.20
1862   33.80   0.00   0.00   2656.10
Comment 1852 TIN ORE
Ownership: Comment 1862-1865 SUSPENDED
Management: Manager 1859-1861 JOHN THOMAS
             Chief Agent 1859-1861 ED.THOMAS
             Secretary 1859-1862 THOS.W.ROBINSON (P)

PROVIDENCE,WEST   TOWEDNACK   SW 494385 1047

Production: Copper No detailed return
Tin   Black(tons) Stuff(tons)   Tin(tons)   Value(£)
1881   17.50   0.00   11.50   939.10
1882   29.90   0.00   16.70   1733.00
1883   18.50   0.00   0.00   980.00
Comment 1881-1883 WEST PROVIDENCE (BUZZA)
Ownership: 1880-1883 WEST PROVIDENCE MINING CO.
Management: Chief Agent 1880-1881 WM.BUGLEHOLE
             Secretary 1882-1883 EDW.BOND
Employment: Underground   Surface   Total
1881   72   48   120
1882   30   21   51
1883   4   4

PRUDENCE   PERRANZABULOE   SW 736528 1048

Production: Copper   Ore(tons) Metal(tons)   Value(£)
1845   513.00   24.10   1477.20
1846   No detailed return
1847   142.00   6.30   369.10
1848   114.00   5.90   297.10
1849   164.00   8.10   423.70
1862   49.00   2.10   156.20
1863   78.00   3.10   206.20
1864   47.00   2.00   153.30
1865   92.00   4.30   295.80
1883   0.30   0.10   9.00
Comment 1845-1849 (C); 1862-1865 (C); 1883 PRECIPITATE
Tin   Black(tons) Stuff(tons)   Tin(tons)   Value(£)
1856   0.20   0.00   0.00   10.50
1874   0.40   0.00   0.00   14.20
1879   0.60   0.00   0.00   23.60
Ownership: 1880-1881 WHEAL PRUDENCE MINING CO.; 1883-1884 SML.HARRIS

404

Comment 1865 SUSPENDED; 1866-1868 SEE WEST PERRAN GREAT
ST.GEORGE
Management: Manager 1861-1862 JOS.VIVIAN; 1864 W.H.THOMAS R.PRYOR; 1881
WALT.PIKE
Chief Agent 1863 STEP.THOMAS; 1864 SML.TRURAN; 1880
WALT.PIKE; 1883-1884 SML.HARRIS
Secretary 1861-1862 THOMAS (P); 1863-1865 C.MILLET THOMAS;
1880 WALT.PIKE

Employment:	Underground	Surface	Total
1880	3		3
1883		1	1

PRUDENCE,SOUTH   ST.AGNES   SW 738536 1049

Production: Tin	Black(tons)	Stuff(tons)	Tin(tons)	Value(£)
1882	18.00	0.00	11.70	1060.00
1884	23.70	0.00	0.00	1026.00

PRUSSIA   REDRUTH   SW 707447 1050

Production: Copper	Ore(tons)	Metal(tons)	Value(£)
1883	13.00	0.70	43.00

Comment 1883 PRUSSIA & CARDREW UNITED

Tin	Black(tons)	Stuff(tons)	Tin(tons)	Value(£)
1873	0.00	505.70	0.00	797.50
1874	0.00	432.10	0.00	275.50
1875	0.00	5139.10	0.00	10240.50
1876	0.00	405.10	0.00	739.10
1877	30.00	613.20	0.00	2354.30
1878	130.70	0.00	0.00	4633.50
1879	0.00	1045.00	0.00	910.90
1880	0.00	1388.80	0.00	903.40
1881	0.00	1703.20	0.00	1125.10
1882	23.40	1626.00	0.00	1170.00
1883	3.50	504.00	0.00	185.00

Comment 1874 VALUE EST; 1879-1883 INC.CARDREW
Ownership: 1873-1876 LEAN,JOSE & CO.; 1877-1881 WHEAL PRUSSIA MINING
CO.; 1882-1883 WHEAL PRUSSIA ADVENTURERS
Comment 1873-1874 FORMERLY EAST TRELEIGH WOOD; 1879-1880
PRUSSIA & CARDREW UNITED; 1882-1883 PRUSSIA & CARDREW UNITED
Management: Manager 1873-1878 WM.TREGAY; 1879-1881 JOS.PRYOR
Chief Agent 1874-1877 R.SMITHERN; 1878 JOHN POPE; 1882-1883
J.PRYOR
Secretary 1877-1878 WM.TREGAY; 1879 SML.ABBOT; 1880
R.F.TEAGUE; 1881 R.S.TEAGUE (P)

Employment:	Underground	Surface	Total
1878	55	30	85
1879	45	17	62
1880	51	27	78
1881	65	27	92
1882	67	30	97
1883		4	4

Comment 1880-1883 INC.CARDREW

QUEEN   CALSTOCK   SX 394701 1051

Production: Lead & Silver   Ore(tons) Lead(tons) Silver(ozs)   Value(£)
            1870   5.10   3.70   0.00   0.00
            Copper No detailed return
            Tin No detailed return
            Silver   Ore(tons) Metal(tons)   Value(£)
            1871   5.10   0.00   421.60
            Arsenic Pyrite   Ore(tons)   Value(£)
            1888   1212.00   909.00
            1889   1361.00   1020.00
            1890   438.00   328.00
            1891   140.00   105.00
Ownership: 1888-1891 COOMBE LTD.
            Comment 1881 SEE LANGFORD; 1890 SUSPENDED
Management: Manager 1870-1871 WM.KNOTT
            Chief Agent 1870-1871 WM.DOBLE; 1888 J.HAMMERSLEY; 1889-1891
            C.W.PHILLIPS
            Secretary 1870-1871 J.T.BARNARD
Employment: Underground   Surface   Total
            1888   41   15   56
            1889   36   12   48
            1890   17   17
            1891   4   4

QUEENS   ROCHE   1053

Production: Iron   Ore(tons)   Iron(%)   Value(£)
            1876   50.00   0.00   37.50
            Comment 1876 BH.
Ownership: 1874 QUEENS IRON MINE CO.; 1875-1877 QUEENS IRON MINE
            CO.LTD.
Management: Chief Agent 1874-1877 JOHN KESSELL

RACER   ST.IVES   SW 489393 1054

Production: Tin   Black(tons) Stuff(tons) Tin(tons)   Value(£)
            1907   0.00   700.00   0.00   300.00
            1908   0.00   632.00   0.00   360.00
Ownership: 1906-1908 WHEAL RACER LTD.
Employment: Underground   Surface   Total
            1906   9   5   14
            1907   9   15   24
            1908   8   7   15

RAMOTH   PERRANZABULOE   SW 762548 1055

Production: Tin   Black(tons) Stuff(tons) Tin(tons)   Value(£)
            1879   2.70   0.00   0.00   104.10
            1882   0.40   0.00   0.30   21.00
            1902   0.60   0.00   0.00   39.00
Ownership: 1901 JAS.MICHELL; 1902 JAS.MICHELL & ZIMAN
            Comment 1901 INC.NORTH LEISURE; 1902 INC.NORTH LEISURE
            SUSPENDED

RAMOTH   PERRANZABULOE   Continued

Management: Secretary 1901-1902 JAS.T.ELLERY
Employment: Underground   Surface   Total
            1901   7   7
            1902   4   4
            Comment 1901-1902 INC.NORTH LEISURE

RANSOM   TOWEDNACK   SW 499395 1056

Production: Tin   Black(tons) Stuff(tons)   Tin(tons)   Value(£)
            1863-1877 No detailed return
            Comment 1863-1876 SEE ROSEWALL HILL; 1877 SEE GOOLE PELLAS
            Arsenic   Ore(tons) Metal(tons)   Value(£)
            1869   No detailed return
            Comment 1869 SEE ROSEWALL HILL
Ownership: Comment 1860-1875 SEE ROSEWALL HILL; 1876 SEE GOOLE PELLAS

RAPSON   WENDRON   1057

Ownership: 1912 WENDRON CORNWALL TIN MINES LTD.; 1913 NIGERIAN FINANCE
            SYNDICATE LTD.
            Comment 1912-1913 PROSPECTING
Management: Secretary 1913 J.S.ALLEN (S)
Employment: Underground   Surface   Total
            1912
            1913   6   6   12
            Comment 1912 SEE WEST VOR

RASHLEIGH   LUXULYAN   SX 164575 1058

Production: Iron No detailed return
Ownership: 1871-1872 C.E.TREFFRY & CO.; 1873-1874 ROWE & CO.
Management: Chief Agent 1871-1874 WM.FLOYD

RED RIVER TIN BURNINGS   9998

Production: Arsenic   Ore(tons) Metal(tons)   Value(£)
            1875   69.10   0.00   333.50
            1876   71.10   0.00   173.00
            1880   42.60   0.00   119.30
            Comment 1876 RED RIVER; 1880 RED RIVER

REDMOOR   STOKE CLIMSLAND   SX 356711 1059

Production: Lead & Silver Ore(tons) Lead(tons) Silver(ozs)   Value(£)
            1858   91.00   51.00   3570.00   0.00
            1859   135.90   72.00   4320.00   0.00
            1860   39.50   21.00   1260.00   0.00
            1861   No detailed return
            1883   4.00   3.00   20.00   55.00
            1884   0.30   0.00   0.00   3.00

                                     407

Comment 1884 FOR AG SEE ZZ SUNDRIES

Copper	Ore(tons)	Metal(tons)	Value(£)
1862	25.00	1.30	103.10
1873	6.00	0.40	25.20
1878	7.20	4.50	370.00
1883	11.90	0.50	12.00
1884	4.00	0.00	4.00

Comment 1862 (C); 1873 (C); 1878 (C)PRECIPITATE

Tin	Black(tons)	Stuff(tons)	Tin(tons)	Value(£)
1863	0.00	0.00	0.00	1022.60
1869	9.90	0.00	0.00	658.30
1870	29.40	0.00	0.00	2184.90
1871	19.10	0.00	0.00	1480.40
1872	4.40	0.00	0.00	404.00
1873	1.70	0.00	0.00	98.40
1874–1877 No detailed return				
1878	2.00	0.00	0.00	70.50
1883	22.80	0.00	0.00	1170.00
1907	9.60	0.00	0.00	816.00
1908	21.20	0.00	0.00	1473.00
1910	4.00	0.00	0.00	400.00
1911	50.50	0.00	0.00	6441.00
1912	53.50	0.00	0.00	7247.00
1913	31.50	0.00	0.00	4300.00

Comment 1863 RADMOOR; 1874–1875 SEE EMMENS UNITED; 1910–1913
OPEN WORK
Iron No detailed return

Arsenic	Ore(tons)	Metal(tons)	Value(£)
1869	13.00	0.00	31.30
1870	21.10	0.00	47.50
1871	23.20	0.00	29.00
1883	84.10	0.00	533.00
1907	24.00	0.00	600.00
1908	114.00	0.00	864.00

Comment 1870–1871 RETURNED AS SOOT

Tungsten	Ore(tons)	Metal(tons)	Value(£)
1907	2.40	0.00	240.00

Silver	Ore(tons)	Metal(tons)	Value(£)
1878	7.70	0.00	173.70

Comment 1878 ARGENTIFEROUS CU PRECIPITATE

Arsenic Pyrite	Ore(tons)	Value(£)
1878	5.70	3.40

Ownership: 1877–1878 EMMENS & CO.LTD.; 1879–1883 H.BENNETT & R.P.COATH;
1884–1886 NEW REDMOOR MINING CO.; 1907 CALLINGTON DEVELOPMENT
SYNDICATE LTD.; 1908–1910 C.D.PHILLIPS N.T.
Comment 1874 SEE EMMENS UNITED; 1881 NOT WORKED; 1885–1886
NOT WORKED; 1888–1893 SEE CALLINGTON UNITED; 1910 NOT WORKED
Management: Manager 1859–1868 THOS.TAYLOR; 1869–1870 F.BENNETT; 1877
G.EMMENS
Chief Agent 1859 THOS.TAYLOR; 1870 WM.GIFFORD; 1871–1873 JOHN
GIFFORD; 1878–1881 H.BENNETT S.B.BARNETT; 1882–1886
H.BENNETT
Secretary 1859 E.A.CROUCH (P); 1860–1864 H.E.CROKER (P);
1865–1870 H.E.CROKER & WATSON(S); 1871–1873 WATSON(S) &
CROKER(P); 1878–1881 G.EMMENS

REDMOOR   STOKE CLIMSLAND   Continued

Employment: Underground   Surface   Total
            1878   5   5
            1882   55   27   82
            1883   11   2   13
            1884   2   2
            1888-1893
            1907   14   21   35
            1908   26   39   65
            Comment 1888-1893 SEE CALLINGTON UNITED

REDRUTH CONSOLS   REDRUTH   SW 694426 1060

Production: Copper   Ore(tons) Metal(tons)   Value(£)
            1846   71.00   7.00   511.10
            1847   93.00   7.70   559.50
            1848   212.00   10.40   567.00
            Comment 1846-1848 (C)

REETH   BREAGE   SW 590307 1032

Production: Tin   Black(tons) Stuff(tons)   Tin(tons)   Value(£)
            1906   3.30   0.00   0.00   364.00
            1907   10.50   0.00   0.00   1014.00
            1908   8.50   0.00   0.00   611.00
            1909   5.90   0.00   0.00   395.00
            1910   3.80   0.00   0.00   294.00
Ownership: 1905-1913 WHEAL REETH MINING & EXPLORATION SYN.LTD
            Comment 1907-1910 SEE GODOLPHIN; 1911-1912 SEE ALSO
            GODOLPHIN. WORKING DUMPS; 1913 TURNING OVER DUMPS
Management: Secretary 1905-1913 ED.MICHELL & SONS
Employment: Underground   Surface   Total
            1906   11   7   18
            1907
            1908-1910   8   8
            1911-1912   10   10
            1913   9   9
            Comment 1907-1912 SEE ALSO GODOLPHIN

REETH   LELANT   SW 590307 1061

Production: Tin   Black(tons) Stuff(tons)   Tin(tons)   Value(£)
            1853   33.20   0.00   0.00   1958.20
            1855   137.40   0.00   0.00   8422.10
            1856   99.50   0.00   0.00   6760.10
            1857   47.30   0.00   0.00   3447.60
            1858   54.40   0.00   0.00   3292.70
            1859   46.30   0.00   0.00   3113.00
            1860   95.90   0.00   0.00   7200.00
            1861   107.20   0.00   0.00   8089.50
            1862   172.30   0.00   0.00   11251.90
            1863   178.00   0.00   0.00   11702.30
            1864   184.50   0.00   0.00   11374.30

Tin	Black(tons)	Stuff(tons)	Tin(tons)	Value(£)
1865	178.00	0.00	0.00	9402.00
1866	146.30	0.00	0.00	6933.20
1867	31.40	0.00	0.00	1447.60

Comment 1862 TWO RETURNS AGGREGATED
Ownership: Comment 1863-1864 OR REETH CONSOLS
Management: Manager 1859 SML.HIGGINS; 1860-1862 HIGGINS & STEVENS; 1863
                HIGGINS & STEVENS; 1864-1865 J.HIGGINS
                Chief Agent 1859 THOS.TREVENNEN; 1860-1863 THOS.TREVENNEN
                JOHN WHITBURN; 1864-1865 THOS.UREN JOHN WHITBURN
                Secretary 1859-1860 RODD,PEARCE,HIGGS; 1861 SML.HIGGS; 1862
                SML.HIGGS & S.H.RODD; 1863-1865 SML.HIGGS & E.H.RODD

REETH CONSOLS   LELANT   1062

Production: Tin	Black(tons)	Stuff(tons)	Tin(tons)	Value(£)
1855	223.60	0.00	0.00	13654.80
1856	264.40	0.00	0.00	17288.90
1857	181.30	0.00	0.00	13635.10
1858	163.50	0.00	0.00	9502.80
1859	55.10	0.00	0.00	3813.40

Ownership: Comment 1863-1864 SEE REETH

REGENT   CHARLESTOWN   SX 050523 1063

Production: Copper No detailed return
Ownership: Comment 1859-1865 SEE WEST CRINNIS

RELEATH,EAST   WENDRON   1064

Production: Copper No detailed return
            Tin No detailed return
Ownership: Comment 1863-1865 SUSPENDED
Management: Chief Agent 1860-1862 T.VINCENT
                Secretary 1860 T.VINCENT (P)

RELISTIAN CONSOLS   GWINEAR   SW 604368 1065

Production: Copper	Ore(tons)	Metal(tons)	Value(£)	
1877	2.60	0.10	11.10	
Tin	Black(tons)	Stuff(tons)	Tin(tons)	Value(£)
1875	0.10	2.70	0.00	2.50
1876	0.10	1.00	0.00	8.70
1877	5.00	1.90	0.00	197.70
Arsenic	Ore(tons)	Metal(tons)	Value(£)	
1877	2.80	0.00	10.00	
1878	1.20	0.00	6.10	

Ownership: 1875-1877 RELISTIAN CONSOLS CO.; 1907 J.DEFRIES & SONS LTD.
                Comment 1907 RELISTIAN
Management: Chief Agent 1875-1877 JOHN CURTIS
                Secretary 1876-1877 W.H.WATSON

RELISTIAN,EAST   GWINEAR   SW 607368 1066

Production: Copper   Ore(tons) Metal(tons)   Value(£)
            1847   93.00   6.50   454.30
            Comment 1847 (C)

RELUBBUS   ST.HILARY   SW 572318 1067

Production: Tin No detailed return
Ownership: 1913 AFRIC SYNDICATE LTD.
Employment: Underground   Surface   Total
            1913   4   1   5

REPERRY   LANIVET   SX 044634 1068

Production: Tin   Black(tons) Stuff(tons)   Tin(tons)   Value(£)
            1864   0.10   0.00   0.00   2.00
            1870   3.00   0.00   0.00   202.90
Management: Manager 1869-1871 THOS.PARKYN
            Secretary 1869-1871 G.H.POULTON

RESPRYN   ST.WINNOW   SX 098634 1069

Production: Copper No detailed return
Ownership: Comment 1861-1865 SUSPENDED
Management: Manager 1859 WM.TREGAY
            Chief Agent 1860 J.VARESE

RESTINNIS   ST.BLAZEY   SX 055554 1070

Production: Iron   Ore(tons)   Iron(%)   Value(£)
            1873   515.00   0.00   386.70
            1874   372.00   0.00   279.00
            1875   1060.00   0.00   498.00
            1876   1108.00   0.00   664.80
            1877   1400.00   0.00   840.00
            Comment 1873-1877 BH.INC.TREFFRY
Ownership: 1873-1874 RESTINNIS & TREFFRY IRON MINES CO.; 1875-1876
            ST.BLAZEY MINERAL CO.LTD.
            Comment 1873-1876 INC.TREFFRY
Management: Manager 1873-1876 THOS.FLOYD

RESTORMEL ROYAL   LOSTWITHIEL   SX 098614 1071

Production: Iron   Ore(tons)   Iron(%)   Value(£)
            1858   12956.20   0.00   4539.70
            1859   6915.60   0.00   2593.60
            1860   2980.20   0.00   1117.60
            1861   5518.90   0.00   2069.60
            1862   2611.70   0.00   1305.80
            1863   2294.00   0.00   1006.10
            1864   8357.70   0.00   3911.40

CM-P   411

RESTORMEL ROYAL   LOSTWITHIEL   Continued

Iron	Ore(tons)	Iron(%)	Value(£)
1865	6170.10	0.00	2622.50
1866	4190.30	0.00	2516.30
1867	1005.20	0.00	351.80
1868	608.30	0.00	212.90
1869	1089.30	0.00	587.40
1870	No detailed return		
1871	3954.90	0.00	2337.60
1872	7521.80	0.00	3095.10
1873	4609.50	0.00	2587.80
1874	1087.00	0.00	951.70
1875	3510.00	0.00	1408.00
1876	4120.00	0.00	2472.00
1877	1056.00	0.00	475.00
1878	344.50	0.00	206.70
1879	No detailed return		
1880	1724.80	0.00	1120.60
1881	2010.00	0.00	1256.20
1882	848.00	44.00	550.00
1883	200.00	0.00	100.00
1910	688.00	42.00	250.00
1911	312.00	42.00	90.00

Comment 1858-1883 BH.; 1910-1911 BH.

Ownership: 1863-1874 JOHN TAYLOR & SONS; 1875-1883 RESTORMEL IRON MINING
CO.LTD.; 1908-1910 RESTORMEL ROYAL IRON MINES; 1911-1913 NEW
RESTORMEL IRON MINING CO.LTD.
Comment 1859-1862 ROYAL RESTORMEL IRON MINES; 1877-1881
RESTORMEL & RESTORMEL ROYAL; 1882-1883 RESTORMEL; 1908-1912
RESTORMEL; 1913 RESTORMEL CLOSED DOWN

Management: Manager 1859-1861 JOHN TAYLOR; 1862 JOHN TAYLOR & SONS;
1875-1876 CHAS.D.TAYLOR; 1877-1883 GEO.H.TAYLOR
Chief Agent 1859-1873 WM.COOK; 1874 WM.COOK CHAS.D.TAYLOR;
1875-1876 JOHN WOOLCOCK; 1877-1881 JAS.WILLIAMS; 1910
F.I.LESLIE; 1911-1913 JOHN T.TONKIN
Secretary 1859 JAS.BENNETTS (P); 1860-1862 JOHN BENNETTS (P);
1908-1910 H.O.JARRETT

Employment:

	Underground	Surface	Total
1878-1879	2		2
1880	30	6	36
1881-1883	2		2
1908	18	6	24
1909	30	11	41
1910	23	15	38
1911	12	5	17
1912	10	2	12

RESTRONGNET   ST.DAY   SW 808385 1072

Production: Tin	Black(tons)	Stuff(tons)	Tin(tons)	Value(£)
1876	8.30	0.00	0.00	209.20

RESTRONGNET STREAM   FEOCK POINT   SW 808385 1073

Production: Tin No detailed return
Ownership: Comment 1874 STOPPED IN 1874
Management: Manager 1873–1874 CHAS.D.TAYLOR
Secretary 1873–1874 JOHN TAYLOR & SONS

RESURGY   1074

Production:	Iron	Ore(tons)	Iron(%)	Value(£)
	1872	No detailed return		

Comment 1872 BH.SEE KNIGHTOR

RETALLACK,GREAT   PERRANZABULOE   SW 792558 1075

Production:	Lead & Silver	Ore(tons)	Lead(tons)	Silver(ozs)	Value(£)
	1863	1.90	1.30	0.00	0.00
	1867	86.60	64.90	550.00	0.00
	1868	41.90	31.40	500.00	0.00
	1869	44.00	33.00	300.00	0.00
	1870	21.50	16.10	0.00	0.00
	1871	2.40	1.80	8.00	0.00

Comment 1867 AG SOMETIMES INCS NEW CHIV.,; 1868 NORTH
CHIVENOR, CRENVER & ABRAHAM, FALMOUTH & SPERRIES,; 1869
JANE,WEST JANE, PROSPER UNITED,NORTH RETALLACK; 1870 &
TRESAVEAN FOR 1867–9

	Zinc	Ore(tons)	Metal(tons)	Value(£)
	1858	150.00	0.00	350.50
	1859	250.00	0.00	700.00
	1860	2371.00	0.00	5670.20
	1861	5561.00	0.00	11335.90
	1862	589.90	0.00	916.50
	1863	708.70	0.00	1694.20
	1866	274.80	0.00	575.20
	1867	16.90	0.00	26.00
	1874	50.00	0.00	150.00
	1875	420.00	0.00	1091.80
	1876	680.00	0.00	1972.70
	1877	280.00	0.00	675.00
	1878	179.80	0.00	407.50
	1879	22.20	0.00	77.50
	1880	84.50	0.00	235.60

	Iron	Ore(tons)	Iron(%)	Value(£)
	1858	4017.00	0.00	1807.70
	1859	6609.10	0.00	1957.50
	1860	200.00	0.00	40.00
	1861	No detailed return		

Comment 1858–1861 BH.
Ownership: 1876–1882 GREAT RETALLACK MINING CO.
Management: Manager 1859–1866 W.H.REYNOLDS; 1872–1881 JOHN HARRIS
Chief Agent 1859–1866 W.H.MIDDLETON; 1867–1871 GEO.R.ODGERS
JOHN HARRIS; 1882 JOHN HARRIS
Secretary 1859–1877 JOHN WATSON (P); 1878–1881 JOHN WATSON
(S)

RETALLACK,GREAT   PERRANZABULOE   Continued

Employment: Underground   Surface   Total
           1878   8   12   20
           1879   4   1   5
           1880   8   11   19
           1881   4   1   5

RETALLACK,NORTH   PERRANZABULOE   SW 787562 1076

Production: Lead & Silver   Ore(tons) Lead(tons) Silver(ozs)   Value(£)
           1869   4.90   3.60   0.00   0.00
           1870   No detailed return
           Comment 1869 FOR AG SEE GREAT RETALLACK
Management: Chief Agent 1866—1871 GEO.ROGERS
           Secretary 1867—1868 JOHN HARRIS; 1869—1871 JOHN WATSON

RETANNA HILL   WENDRON   SW 716329 1078

Production: Tin   Black(tons) Stuff(tons)   Tin(tons)   Value(£)
           1853   0.20   0.00   0.00   21.70
           1864   0.00   0.00   0.00   71.00
           Comment 1853 RETANNA; 1864 TIN STUFF

RETEW   ST.ENODER   SW 920571 1079

Production: Iron   Ore(tons)   Iron(%)   Value(£)
           1880   248.00   0.00   148.80
           1881   1400.00   0.00   840.00
           Comment 1880 BH.; 1881 BH.INC COLDVREATH & TREWERTHA & RH
Ownership: 1880—1881 WEST OF ENGLAND IRON ORE CO.; 1882 WEST OF ENGLAND
           IRON ORE CO.LTD.
Management: Chief Agent 1880 DAVID COCK; 1881 JAS.WILLIAMS DAVID COCK;
           1882 DAVID COCK
Employment: Underground   Surface   Total
           1880   6   2   8
           1881   2   6   8
           1882   4   1   5

RETIRE   WITHIEL   SX 006645 1080

Production: Iron   Ore(tons)   Iron(%)   Value(£)
           1858   2988.30   0.00   1887.80
           1859   No detailed return
           1860   600.00   0.00   240.00
           1861   2566.00   0.00   1026.40
           1862   2118.00   0.00   1258.00
           1863   1977.00   0.00   623.10
           1864   2146.50   0.00   644.00
           1865   3143.90   0.00   943.20
           1866   2640.00   0.00   792.00
           1867   2641.20   0.00   792.30
           1869   588.00   0.00   340.50

                                    414

RETIRE   WITHIEL   Continued

        Iron   Ore(tons)   Iron(%)   Value(£)
        1870   350.00   0.00   87.50
        Comment 1858-1867 BH.; 1869-1870 BH.
Ownership: 1863-1866 JOS.MORCOM; 1867-1872 MORCOM & WEST
             Comment 1873 SEE WITHIEL
Management: Chief Agent 1863-1872 WM.BUTTON

RICHARDS FRIENDSHIP   ST.HILARY   1052

Production: Copper   Ore(tons) Metal(tons)   Value(£)
             1854-1855 No detailed return
             Comment 1854-1855 SEE RICHARDS FRIENDSHIP, DEVON

RIVER TAMAR   CALSTOCK   SX 417723 1081

Production: Copper No detailed return
             Tin   Black(tons) Stuff(tons)   Tin(tons)   Value(£)
             1870   0.00   0.00   0.00   17.50
             Comment 1870 TIN STUFF
Ownership: Comment 1859-1860 RIVER TAMAR COPPER COMPANY; 1863-1865
             SUSPENDED
Management: Chief Agent 1859-1862 J.COCK
             Secretary 1859-1865 T.NICHOLLS (P)

ROBARTES   ILLOGAN   SW 680437 1082

Management: Manager 1866-1868 JAS.CHAMPION
             Chief Agent 1872-1873 JOHN GOLDSWORTHY
             Secretary 1866-1868 JAS.ANGOVE; 1872-1873 F.R.WILSON (P)

ROBINS   ST.NEOT   SX 184684 1083

Production: Tin   Black(tons) Stuff(tons)   Tin(tons)   Value(£)
             1852   4.40   0.00   0.00   0.00
             1853   10.20   0.00   0.00   511.20
             1854   9.00   0.00   0.00   497.10
             1855   2.80   0.00   0.00   146.40
             Comment 1852 TIN ORE

ROCHE CONSOLS   ROCHE   SW 972625 1084

Production: Tin No detailed return
Management: Manager 1870-1872 THOS.PARKYN

ROCK   ST.AGNES   SW 725810 1085

Production: Tin No detailed return
Ownership: Comment 1863 SEE SOUTH WHEAL KITTY; 1864-1871 SEE WEST KITTY;
             1872 SEE ST.AGNES CONSOLS

ROCK HILL   BUGLE   SX 014582 1086

Production: Tin   Black(tons) Stuff(tons)   Tin(tons)   Value(£)
            1870   6.20   0.00   0.00   464.50
            1871   2.50   0.00   0.00   196.20
            1872   12.10   0.00   0.00   1108.30
            1906   3.10   0.00   0.00   327.00
            1907   12.30   0.00   0.00   1252.00
            Comment 1870–1872 ROCK HILL UNITED; 1906–1907 ROCK HILL
            UNITED
Ownership: Comment 1860–1868 ROCK MILL
Management: Chief Agent 1860–1871 H.COWLEY
            Secretary 1869–1871 JAS.POLGLAZE

ROCKS   BUGLE   SX 014582 1087

Production: Tin   Black(tons) Stuff(tons)   Tin(tons)   Value(£)
            1852   56.00   0.00   0.00   0.00
            1853   99.80   0.00   0.00   7097.20
            1855   62.60   0.00   0.00   4863.30
            1856   7.20   0.00   0.00   363.10
            1857   4.30   0.00   0.00   262.70
            1873   20.00   0.00   0.00   1619.40
            1874   0.00   100.00   0.00   145.30
            1882   5.00   6.60   0.00   618.00
            1883   17.00   0.00   0.00   976.00
            Comment 1852 INC.TREVERBYN TIN ORE; 1853 INC.TREVERBYN; 1855
            INC.TREVERBYN. TWO RETURNS AGG.; 1856–1857 ROCK; 1873
            INC.GOONBARROW; 1874 INC.GOONBARROW; 1883 INC.CARNESMERRY
            Iron No detailed return
Ownership: 1880–1883 ROCKS TIN MINING CO.LTD.
            Comment 1872–1873 INC.GOONBARROW; 1880–1883 ROCKS TIN
Management: Manager 1880–1881 DAVID COCK
            Chief Agent 1872–1873 WOOD HORE; 1882–1883 DAVID COCK
            Secretary 1872–1873 J.W.COOM
Employment: Underground   Surface   Total
            1880   50   50
            1881   98   1   99
            1882   45   67   112
            1883   21   28   49

ROCKS,EAST   LUXULYAN   1088

Production: Iron No detailed return
Ownership: 1873 EAST ROCKS IRON CO.
            Comment 1872 SEE CANNA; 1873 SEE ALSO CANNA; 1874 SEE CANNA
Management: Manager 1873 J.W.COOM

RODD,GREAT   NORTH HILL   SX 286751 1089

Production: Tin No detailed return
Ownership: 1877 C.H.MAUNDER & OTHERS; 1878 GREAT WHEAL RODD SILVER LEAD
            CO.
            Comment 1878 IN LIQUIDATION. SEE ALSO GREAT RODD, DEVON;

                                    416

RODD,GREAT   NORTH HILL   Continued

1879-1881 SEE GREAT RODD,DEVON
Management: Chief Agent 1877 THOS.F.HOSKING
Employment: Underground   Surface   Total
            1878   4   4

RODNEY   MARAZION   SW 531314 1090

Production: Copper   Ore(tons) Metal(tons)   Value(£)
            1845   132.00   7.00   465.30
            1846   407.00   24.20   1548.60
            1847   724.00   40.40   2691.00
            1848   302.00   15.00   813.00
            Comment 1845-1848 (C)
Ownership: 1907 CORNISH TIN SYNDICATE LTD.
            Comment 1912-1913 SEE HAMPTON
Employment: Underground   Surface   Total
            1912-1913
            Comment 1912-1913 SEE HAMPTON

ROSE   PERRANZABULOE   1092

Production: Lead & Silver   Ore(tons) Lead(tons) Silver(ozs) Value(£)
            1845   57.00   34.00   0.00   0.00
            1846   375.00   224.00   0.00   0.00
            1847   378.00   227.00   0.00   0.00
            1848   399.00   239.00   0.00   0.00
            1849   107.00   70.00   0.00   0.00
            1850-1854 No detailed return
            1861   2.50   1.60   26.00   0.00
            Iron   Ore(tons)   Iron(%)   Value(£)
            1865   70.00   0.00   17.50
            Comment 1865 BH.

ROSE   SCORRIER   SW 718449 1093

Production: Copper   Ore(tons) Metal(tons)   Value(£)
            1862   79.00   4.40   341.40
            1863   571.00   41.90   4323.70
            1864   2730.00   173.40   14648.00
            1865   3717.00   223.00   17361.80
            1866   2357.00   149.30   9747.10
            1867   1725.00   134.50   9468.50
            1868   1184.00   79.80   5086.00
            1869   248.00   16.90   1038.40
            1870   206.00   10.90   618.90
            1873   71.00   4.20   258.70
            Comment 1862 (C); 1863 (C) TWO RETURNS AGGREGATED; 1864-1870
            (C); 1873 (C)
            Tin   Black(tons) Stuff(tons) Tin(tons)   Value(£)
            1862   0.00   0.00   0.00   24.60
            1865   0.00   0.00   0.00   27.30
            1866   0.00   0.00   0.00   136.80

                                     417

Tin	Black(tons)	Stuff(tons)	Tin(tons)	Value(£)
1867	0.00	0.00	0.00	210.20
1868	0.00	0.00	0.00	59.40
1869	0.00	252.70	0.00	288.60
1870	0.00	0.00	0.00	47.10
1872	0.00	377.30	0.00	112.70
1873	0.00	3.20	0.00	231.40
1874	0.00	42.00	0.00	19.00
1875	0.00	650.10	0.00	358.60

Comment 1862 TIN STUFF. PART YEAR ONLY; 1865-1866 TIN STUFF;
1867 TIN STUFF . RETURN FROM 1868 STATS.; 1870 TIN STUFF;
1873-1875 ROSE UNITED

Iron	Ore(tons)	Iron(%)	Value(£)
1876	65.00	0.00	48.70

Comment 1876 SP. OLD ROSE
Ownership: Comment 1860 SEE GREAT NORTH DOWNS; 1866-1868 INC.EAST DOWNS;
             1872-1873 ROSE UNITED
Management: Manager 1862 SAMP.WATERS; 1864-1868 GEO.TREMAYNE; 1872-1873
             GEO.TREMAYNE
             Chief Agent 1863 GEO.TREMAYNE; 1867-1868 SML.G.TRURAN
             Secretary 1863-1868 HY.MICHELL; 1872-1873 HY.MICHELL

ROSE   SITHNEY   SW 638247 1094

Production: Lead No detailed return
Ownership: 1879 WALT.THOMPSON
            Comment 1863-1864 SUSPENDED; 1879 OLD ROSE; 1880-1881 OLD
            ROSE SEE LOMAX
Management: Manager 1859-1862 THOS.MARTIN
            Chief Agent 1860-1864 WM.ROWE; 1865 WM.ROWE SML.TRAVERN; 1879
            WM.ARGOLL
            Secretary 1859 J.SYMONS (P); 1860-1865 THOS.W.ROBINSON (P);
            1879 H.THOMPSON

Employment:	Underground	Surface	Total
1879	2		2
1880	3		3

ROSE AND CHIVERTON   NEWLYN EAST   SW 818536 1095

Production: Lead & Silver	Ore(tons)	Lead(tons)	Silver(ozs)	Value(£)
1870	30.00	21.00	0.00	0.00
1871	7.50	5.60	28.00	0.00
1872	No detailed return			

Ownership: Comment 1866-1868 EAST ROSE & CHIVERTON; 1869-1870 FORMERLY
            EAST ROSE; 1872 SEE EAST ROSE
Management: Chief Agent 1866-1871 JAS.EVANS
            Secretary 1869-1871 GEO.STILL

ROSE DOWN,WEST   LINKINHORNE   SX 273713 1096

Production: Copper	Ore(tons)	Metal(tons)	Value(£)
1872	10.00	0.60	49.20

Comment 1872 (C)
Management: Manager 1860-1866 JAS.SECCOMBE; 1867-1871 JOHN TRUSCOTT
            Chief Agent 1861-1871 JOHN HORE
            Secretary 1860 JOHN HARDING (P); 1861-1867 JAS.SECCOMBE (P);
            1868-1871 WM.THORNE

ROSE,EAST   NEWLYN EAST   SW 837554 1097

Production: Lead & Silver   Ore(tons)   Lead(tons)   Silver(ozs)   Value(£)
            1845   7883.00   4729.00   0.00   0.00
            1846   5191.00   3114.00   0.00   0.00
            1847   6424.00   3854.00   0.00   0.00
            1848   5333.00   3191.00   0.00   0.00
            1849   4758.70   2856.00   0.00   0.00
            1850   4206.00   2524.10   0.00   0.00
            1851   3192.90   2234.00   0.00   0.00
            1852   2381.10   1607.00   48000.00   0.00
            1853   1357.10   925.00   27499.00   0.00
            1854   1215.00   828.30   24621.00   0.00
            1855   2343.10   1510.30   46760.00   0.00
            1856   2691.00   1776.00   53280.00   0.00
            1857   1199.20   791.30   23739.00   0.00
            1858   726.00   416.50   10400.00   0.00
            1859   728.00   386.00   13090.00   0.00
            1860   607.30   322.00   10948.00   0.00
            1861   147.00   66.00   2376.00   0.00
            1865   44.00   31.00   1221.00   0.00
            1866   147.70   102.90   4015.00   0.00
            1872   6.00   4.50   22.00   0.00
            1882   40.00   30.00   0.00   0.00
            1883   155.40   114.70   0.00   1548.00
            1884   85.00   0.00   0.00   595.00
            1885   50.00   0.00   0.00   456.00
            1886   58.00   0.00   0.00   0.00
            Comment 1856 RECENTLY STOPPED WORKING
            Zinc   Ore(tons)   Metal(tons)   Value(£)
            1855   No detailed return
            1857   31.00   0.00   4.70
            1859   19.00   0.00   8.60
            1882   70.00   31.50   199.00
            1883   161.20   56.30   167.00
            Comment 1855 SEE GREAT BADDERN
            Copper   Ore(tons)   Metal(tons)   Value(£)
            1850   67.00   9.50   734.20
            1851   93.00   11.10   789.50
            Comment 1850-1851 (C)
Ownership: 1880 EAST WHEAL ROSE MINING CO.; 1881-1886 EAST WHEAL ROSE
            MINING CO.LTD.
            Comment 1860-1865 SUSPENDED; 1866-1870 SEE ROSE & CHIVERTON;
            1872 FORMERLY ROSE & CHIVERTON
Management: Manager 1859 JAS.EVANS; 1873-1876 THOS.DOIDGE; 1880-1881
            THOS.DOIDGE
            Chief Agent 1859 JOHN MIDDLETON; 1866 SML.G.TRURAN JAS.EVANS;
            1867-1872 JAS.EVANS; 1881 WM.SKEWIS; 1882-1886 THOS.DOIDGE

Secretary 1859 HY.BORROW (P); 1867-1870 HY.WHITWORTH;
1871-1876 HY.WHITWORTH (P); 1880-1881 HY.BROWNE (P)

Employment:	Underground	Surface	Total
1881	22	81	103
1882	71	68	139
1883	60	56	116
1884	32	27	59
1885	31	31	62
1886	37	25	62

ROSE,NORTH   NEWLYN EAST   SW 838560 1098

Production: Lead & Silver	Ore(tons)	Lead(tons)	Silver(ozs)	Value(£)
1847	84.00	50.00	0.00	0.00
1848	80.00	49.00	0.00	0.00
1849	75.70	46.00	0.00	0.00
1850	119.00	72.00	0.00	0.00
1851	354.00	214.10	0.00	0.00
1852	321.30	211.00	5900.00	0.00
1853	230.60	175.00	5400.00	0.00
1854	No detailed return			
1855	77.00	50.00	1309.00	0.00
1856-1857	No detailed return			

ROSE,OLD   PORTHLEVEN   1100

Production: Iron No detailed return
Ownership: 1863-1865 EDW.CARTER
Management: Chief Agent 1863-1865 THOS.PARKYN

ROSEARROCK   PORT ISAAC   1101

Ownership: 1912 J.B.MAY
Comment 1912 OPENING ADIT

Employment:	Underground	Surface	Total
1912	1		1

ROSECLIFF   NEWQUAY   SW 819623 1102

Ownership: Comment 1866 INC.TOLCARNE
Management: Chief Agent 1866 J.PHILLIPS

ROSEMERGY   MORVAH   SW 420363 1103

Production: Tin No detailed return
Ownership: 1908 ROSEMERGY & MORVAH SYNDICATE LTD.
Comment 1907 SEE BOSIGRAN CONSOLIDATED; 1908 INC.MORVAH HILL
Management: Manager 1908 J.KIMBER

ROSEVALE   ZENNOR   SW 458379 1104

Production: Tin   Black(tons) Stuff(tons)   Tin(tons)   Value(£)
            1908   0.20   0.00   0.00   14.00
            1909   0.40   0.00   0.00   36.00
            1910   1.80   0.00   0.00   165.00
            1911   0.80   0.00   0.00   85.00
            1913   0.40   0.00   0.00   54.00
Ownership: 1906-1912 W.NANKERVIS; 1913 RAYFIELD CORNWALL SYND.LTD.
            Comment 1912-1913 RESTARTED OCT.1912
Management: Chief Agent 1913 W.NANKERVIS
Employment: Underground   Surface   Total
            1906   6   6
            1909   4   4
            1910   6   6
            1911   4   1   5
            1912   14   2   16
            1913   20   16   36

ROSEWALL HILL   TOWEDNACK   SW 497392 1105

Production: Tin   Black(tons) Stuff(tons)   Tin(tons)   Value(£)
            1863   116.90   0.00   0.00   7617.00
            1864   72.10   0.00   0.00   4510.30
            1865   65.80   0.00   0.00   3602.70
            1866   112.30   0.00   0.00   5379.50
            1867   97.90   0.00   0.00   4884.20
            1868   134.90   0.00   0.00   7375.50
            1869   103.20   0.00   0.00   7160.60
            1870   107.70   0.00   0.00   7905.80
            1871   91.20   0.00   0.00   7946.20
            1872   95.60   0.00   0.00   8343.90
            1873   140.50   0.00   0.00   11130.50
            1874   131.00   0.00   0.00   7487.80
            1875   98.10   0.00   0.00   5009.80
            1876   60.70   0.00   0.00   2580.00
            1877   No detailed return
            Comment 1863 ROSEWARNE HILL & RANSOM; 1864 ROSEWALL HILL &
            RANSOM UNITED; 1865-1866 INC.RANSOM; 1867-1874 ROSEWALL HILL
            & RANSOME UNITED; 1875 INC.RANSOM; 1876 INC.RANSOM NOW GOOLE
            PELLAS; 1877 INC.RANSOM SEE GOOLE PELLAS
            Arsenic   Ore(tons) Metal(tons)   Value(£)
            1869   11.50   0.00   17.20
            Comment 1869 ROSEWALL HILL & RANSOM UNITED
Ownership: Comment 1859 ROSEWALL HILL & ROSEWARNE UNITED; 1860-1875
            ROSEWALL HILL & RANSOM UNITED; 1876 SEE GOOLE PELLAS
Management: Manager 1859-1868 THOS.TREWEEKE; 1869-1871 JOSH.DANIELL;
            1872-1874 WM.BUGLEHOLE
            Chief Agent 1859 PAUL ROACH; 1860 JOHN THOMAS; 1861-1866
            ED.THOMAS; 1867-1868 FRANK TREWEEKE SML.UREN; 1869 SML.UREN;
            1870-1871 WM.BUGLEHOLE; 1872-1874 JOHN WHITE; 1875
            THOS.BUGLEHOLE
            Secretary 1859-1864 THOS.TREWEEKE JNR(P); 1865-1874
            THOS.W.ROBINSON; 1875 SML.ABBOT

ROSEWARNE   GWINEAR   SW 598372 1106

Production: Zinc   Ore(tons) Metal(tons) Value(£)
    1856   10.00   0.00   15.00
    Copper No detailed return
    Tin   Black(tons) Stuff(tons) Tin(tons) Value(£)
    1853   4.00   0.00   0.00   196.90
    1855   0.50   0.00   0.00   28.40
    1860   125.20   0.00   0.00 9605.20
    Comment 1860 INC.HERLAND
Ownership: Comment 1859–1864 INC.HERLAND; 1865 INC.HERLAND SUSPENDED
Management: Manager 1859–1864 HUGH STEPHENS
    Chief Agent 1859 HY.WOOLCOCK; 1860–1864 ED.BLEWETT
    Secretary 1859–1865 HUGH STEPHENS (P)

ROSEWARNE CONSOLS   GWINEAR   SW 621361 1107

Production: Copper   Ore(tons) Metal(tons) Value(£)
    1858   201.00   8.80   734.80
    1861   73.00   6.10   552.30
    1862   476.00   56.70   4864.90
    1863   523.00   43.00   3383.20
    1864   697.00   73.00   6261.60
    1865   425.00   44.40   3676.90
    1866   278.00   21.50   1426.10
    1867   143.00   9.60   674.20
    1868   111.00   7.40   481.50
    1869   16.00   1.70   109.90
    Comment 1858 (C); 1861–1869 (C)
    Tin   Black(tons) Stuff(tons) Tin(tons) Value(£)
    1860   42.00   0.00   0.00 3551.70
    1861   11.20   0.00   0.00   896.50
    1862   2.90   0.00   0.00   245.30
    Comment 1862 VALUE INC.TIN STUFF
Management: Manager 1859–1864 JAS.RICHARDS; 1865–1871 J.NANCARROW
    Chief Agent 1859–1861 JAS.RICHARDS; 1862–1864 JAS.BERRYMAN;
    1865–1871 R.KNUCKEY
    Secretary 1859–1862 JAS.HOLLOW (P); 1863–1864 THOS.HOLLOW;
    1867–1871 J.CHAPMAN

ROSEWARNE DOWNS   GWINEAR   9997

Production: Arsenic   Ore(tons) Metal(tons) Value(£)
    1894   No detailed return
    Comment 1894 SEE BISSOE POOL

ROSEWARNE TIN STREAM   GWINEAR   9995

Production: Arsenic   Ore(tons) Metal(tons) Value(£)
    1886   15.00   0.00   34.00
    1888   26.00   0.00   78.00
    Comment 1886 CRUDE; 1888 ROSEWARNE MILL

ROSEWARNE UNITED   GWINEAR   SW 619367 1108

Production: Zinc   Ore(tons) Metal(tons)   Value(£)
            1857   44.30   0.00   66.60
            Copper   Ore(tons) Metal(tons)   Value(£)
            1854   731.00   54.00   5561.70
            1855   2527.00   218.80   23169.20
            1856   2451.00   189.10   18126.30
            1857   1514.00   114.40   11924.80
            1858   915.00   70.30   6597.40
            1859   1103.00   101.80   9875.20
            1860   1222.00   105.50   10037.40
            1861   879.00   81.10   7532.50
            1862   516.00   43.80   3697.60
            1863   269.00   20.00   1596.60
            1864   473.00   33.30   2825.70
            1865   930.00   87.30   7204.30
            1866   974.00   85.90   5875.20
            1867   757.00   58.10   4184.50
            1868   125.00   7.10   462.40
            Comment 1854-1868 (C)
            Tin   Black(tons) Stuff(tons)   Tin(tons)   Value(£)
            1858   0.00   0.00   0.00   86.40
            1860   0.00   0.00   0.00   175.40
            1861   0.00   0.00   0.00   432.30
            1862   0.00   0.00   0.00   1078.70
            1863   0.00   0.00   0.00   279.80
            1874   32.50   32.50   0.00   1476.90
            Comment 1858 TIN STUFF; 1860-1861 TIN STUFF; 1862 TIN STUFF.
            TWO RETURNS AGG.; 1863 TIN STUFF
Ownership: 1912-1913 W.MIDDLIN
            Comment 1873 SUSPENDED; 1913 NOT WORKED
Management: Manager 1859-1864 THOS.RICHARDS; 1865 THOS.RICHARDS & SON;
            1872 JOHN JAMES
            Chief Agent 1859-1864 HY.WOOLCOCK; 1865 J.TREMBY; 1872 JOHN
            PASCOE; 1873 EDW.HOSKING JOHN PASCOE; 1874-1875 EDW.HOSKING
            Secretary 1859-1862 WM.HUTHNANCE (P); 1863-1865
            THOS.RICHARDS; 1872 WALT.PIKE; 1873 JOHN WATSON
Employment: Underground   Surface   Total
            1912   1   1   2

ROSEWARNE UNITED,WEST   GWINEAR   SW 607368 1109

Production: Lead   Ore(tons) Metal(tons)   Value(£)
            1859   0.40   0.10   0.00
            Comment 1859 WEST ROSEWARNE
            Copper No detailed return
Management: Manager 1859-1862 WM.RICHARDS
            Chief Agent 1859-1864 RICH.REYNOLDS
            Secretary 1859-1862 WM.RICHARDS (P)

ROSEWARNE,EAST   GWINEAR   SW 631378 1110

Production: Copper   Ore(tons) Metal(tons)   Value(£)
            1857   197.00   17.70   1829.00

                                      423

ROSEWARNE, EAST   GWINEAR   Continued

Copper	Ore(tons)	Metal(tons)	Value(£)
1858	120.00	8.40	797.20
1859	112.00	8.00	760.20
1860	458.00	42.20	4003.80
1861	474.00	39.50	3642.50
1862	506.00	44.60	3889.70
1863	731.00	73.40	5978.70
1864	889.00	86.70	7615.90
1865	1080.00	110.40	9093.70
1866	825.00	74.80	5362.20
1867	713.00	54.60	3890.60
1868	1001.00	67.30	4437.80
1869	811.00	54.30	3346.90
1870	318.00	19.70	1174.70
1873	4.00	0.20	15.60

Comment 1857–1870 (C); 1873 (C)

Tin	Black(tons)	Stuff(tons)	Tin(tons)	Value(£)
1859	0.00	0.00	0.00	36.40
1860	0.00	0.00	0.00	7.20
1866	0.00	0.00	0.00	18.60
1870	0.00	0.00	0.00	14.70
1873	25.60	0.00	0.00	1996.00

Comment 1859–1860 TIN STUFF; 1866 TIN STUFF; 1870 TIN STUFF;
1873 NEW EAST ROSEWARNE VALUE EST
Management: Manager 1859–1863 JOHN DELBRIDGE
Chief Agent 1859–1865 JOHN JAMES; 1866–1869 CHAS.GLASSON
Secretary 1859–1866 ED.KING (P); 1867–1869 ED.KING R.H.PIKE &
SON

ROSEWARNE, NEW   GWINEAR   SW 614368 1111

Production: Copper

	Ore(tons)	Metal(tons)	Value(£)
1863	114.00	13.60	1151.50
1864	309.00	30.90	2767.00
1865	501.00	37.80	3067.50
1866	138.00	10.60	801.40
1873	144.00	17.50	1268.60
1874	46.00	4.40	289.40
1875	89.00	6.30	463.20
1876	144.00	13.90	961.40
1877	3.00	0.10	8.40

Comment 1863–1866 (C); 1873–1876 (C)

Tin	Black(tons)	Stuff(tons)	Tin(tons)	Value(£)
1863	0.00	0.00	0.00	441.80
1864	0.00	0.00	0.00	2204.70
1865	0.00	0.00	0.00	1957.60
1866	0.00	0.00	0.00	630.10
1872	33.10	0.00	0.00	2928.70
1873	73.30	0.00	0.00	5713.00
1874	41.70	0.00	0.00	2496.20
1875	11.00	18.90	0.00	630.50
1876	0.50	20.50	0.00	46.20

Comment 1863–1866 TIN STUFF; 1873 VALUE EST

ROSEWARNE,NEW   GWINEAR   Continued

         Arsenic   Ore(tons) Metal(tons)   Value(£)
         1873   35.50   0.00   71.00
         1874   27.90   0.00   73.20
         1875   15.00   0.00   89.00
Management: Manager 1870—1871 GEO.R.ODGERS; 1872 JOHN JAMES
            Chief Agent 1870—1872 JOS.RULE; 1873—1874 EDW.HOSKING
            JOS.RULE; 1875 EDW.HOSKING WM.BENNETTS JOS.RULE; 1876—1877
            WM.BENNETTS JOS.RULE
            Secretary 1870—1872 WALT.PIKE; 1873—1877 JOHN WATSON

ROSEWARNE,NEW WEST   GWINEAR   SW 607368 1112

Production: Copper No detailed return
         Tin   Black(tons) Stuff(tons)   Tin(tons)   Value(£)
         1873   34.90   0.00   0.00   2737.90
         1874   9.90   0.00   0.00   611.90
         Arsenic   Ore(tons) Metal(tons)   Value(£)
         1874   15.70   0.00   39.10
Management: Manager 1872—1874 JOHN CURTIS; 1875 J.P.CURTIS
            Chief Agent 1872—1875 EZ.JOHNS
            Secretary 1872—1875 T.ROBINSON

ROSEWARNE,NORTH   GWINEAR   SW 609372 1113

Production: Lead & Silver No detailed return
         Copper   Ore(tons) Metal(tons)   Value(£)
         1873   5.90   8.60   619.40
         Comment 1873 (C)
         Tin   Black(tons) Stuff(tons)   Tin(tons)   Value(£)
         1872   0.00   0.00   0.00   52.20
         1873   0.00   0.10   0.00   8.30
         1874   0.00   2.30   0.00   7.50
         Comment 1872 TIN STUFF
Ownership: Comment 1865 SUSPENDED; 1873 SUSPENDED
Management: Manager 1870—1871 GEO.R.ODGERS; 1872 JOHN JAMES; 1874—1875
            EDW.HOSKING
            Chief Agent 1860—1864 JOHN TYACKE; 1872 JOHN SWEET; 1873
            EDW.HOSKING; 1874—1875 WM.BENNETT
            Secretary 1860—1865 A.BINNS (P); 1870—1872 WALT.PIKE;
            1873—1875 JOHN WATSON

ROSKEAR   CAMBORNE   SW 656410 1114

Production: Copper   Ore(tons) Metal(tons)   Value(£)
         1864   11.00   0.60   47.00
         1866   12.00   0.70   50.70
         Comment 1864 (C); 1866 (C)

Production: Tin   Black(tons) Stuff(tons)   Tin(tons)   Value(£)
            1878   0.00   2.50   0.00   2.60

ROSKEAR, NORTH   CAMBORNE   SW 656413 1116

Production: Zinc   Ore(tons) Metal(tons)   Value(£)
            1856   6.00   0.00   9.80
            Copper   Ore(tons) Metal(tons)   Value(£)
            1845   6430.00   543.60   40955.50
            1846   5052.00   396.20   29207.60
            1847   5773.00   459.60   33480.30
            1848   5489.00   433.60   26072.40
            1849   5660.00   420.20   26756.20
            1850   5077.00   396.10   27950.30
            1851   4548.00   376.30   25844.70
            1852   3202.00   247.40   20466.50
            1853   3050.00   223.70   21608.70
            1854   2216.00   163.50   17042.50
            1855   2154.00   149.50   15453.30
            1856   2234.00   146.40   13795.40
            1857   2247.00   148.20   14971.80
            1858   2314.00   162.00   14624.30
            1859   1587.00   107.50   10111.70
            1860   1811.00   143.70   13840.60
            1861   1369.00   111.80   10358.50
            1862   1205.00   92.30   7904.70
            1863   1319.00   104.60   8364.30
            1864   1035.00   89.90   7758.20
            1865   983.00   78.10   6334.70
            1866   610.00   39.90   2832.20
            1867   330.00   24.20   1716.40
            1868   319.00   24.90   1576.50
            1869   209.00   16.80   1072.40
            1870   33.00   2.80   164.40
            1871   36.00   3.30   220.70
            1872   46.00   4.00   323.90
            1873   13.00   1.10   70.00
            Comment 1845-1873 (C)
            Tin   Black(tons) Stuff(tons)   Tin(tons)   Value(£)
            1855   16.20   0.00   0.00   1022.40
            1856   27.10   0.00   0.00   1865.60
            1857   33.50   0.00   0.00   2438.20
            1858   28.50   0.00   0.00   1688.10
            1859   36.90   0.00   0.00   2487.50
            1860   49.00   0.00   0.00   3705.50
            1861   104.50   0.00   0.00   7461.10
            1862   151.90   0.00   0.00   9812.20
            1863   111.80   0.00   0.00   7293.30
            1864   103.60   0.00   0.00   6263.30
            1865   190.40   0.00   0.00   10580.40
            1866   142.00   0.00   0.00   6996.90
            1867   43.20   0.00   0.00   2153.30
            1868   48.90   0.00   0.00   2705.10
            1869   110.20   0.00   0.00   7789.60

ROSKEAR,NORTH   CAMBORNE   Continued

        Tin   Black(tons) Stuff(tons)   Tin(tons)   Value(£)
        1870   88.10   0.00   0.00   6410.00
        1871   114.60   0.00   0.00   9123.80
        1872   87.20   0.00   0.00   7772.60
        1873   66.70   0.00   0.00   5407.80
        1874   30.90   0.00   0.00   1875.40
        Comment 1862 TWO RETURNS AGGREGATED; 1874 VALUE EST
        Arsenic   Ore(tons) Metal(tons)   Value(£)
        1867   153.60   0.00   230.40
        1868   162.90   0.00   513.10
        1869   181.90   0.00   461.60
        1870   214.80   0.00   739.80
        1871   190.30   0.00   355.30
        1872   224.30   0.00   400.00
        1873   110.80   0.00   341.80
        1874   99.00   0.00   247.50
        Comment 1867-1868 CRUDE; 1869 TWO RETURNS AGGREGATED;
        1870-1871 CRUDE
Ownership: Comment 1862-1865 ALSO ROSKEAR,NOWETH; 1873 STOPPED IN 1873 &
        EARLY IN 1874
Management: Manager 1859-1864 JOS.VIVIAN; 1865-1868 JOS.VIVIAN & SON;
        1870-1873 JAS.HOSKING
        Chief Agent 1859 JAS.DUNKIN; 1861 JAS.DUNKIN; 1862-1865
        J.HOSKING RICH.ANGOVE; 1866 FRAN.HOSKING RICH.ANGOVE;
        1867-1868 RICH.ANGOVE; 1869 RICH.ANGOVE RICH.GOLDSWORTHY;
        1870-1873 RICH.ANGOVE; 1874 JAS.HOSKING
        Secretary 1859 WM.DARKE (P); 1860-1862 THOS.FIELD JNR. (P);
        1863-1869 THOS.W.FIELD; 1870-1873 THOS.PRYOR

ROSKEAR,SOUTH   CAMBORNE   SW 656410 1117

Production: Zinc   Ore(tons) Metal(tons)   Value(£)
        1878   32.40   0.00   64.70
        Copper   Ore(tons) Metal(tons)   Value(£)
        1845   1464.00   113.90   8738.00
        1846   1376.00   95.80   6952.30
        1847   756.00   58.10   4210.70
        1848   1095.00   88.00   5312.90
        1849   877.00   56.20   3588.10
        1850   370.00   20.20   1346.10
        1874   62.00   3.00   182.20
        1875   78.00   3.90   277.40
        1876   111.00   7.60   534.60
        1877   66.00   4.20   257.30
        1878   41.90   3.30   172.00
        1879   18.80   1.10   55.40
        1881   8.10   0.50   29.60
        Comment 1845-1850 (C); 1874 (C)TWO RETURNS AGGREGATED;
        1875-1876 (C)
        Tin   Black(tons) Stuff(tons)   Tin(tons)   Value(£)
        1858   1.30   0.00   0.00   75.60
        1874   18.10   0.00   0.00   975.00
        1875   39.30   0.00   0.00   2028.00
        1876   63.30   0.00   0.00   2709.00

                                   427

ROSKEAR,SOUTH   CAMBORNE   Continued

Tin	Black(tons)	Stuff(tons)	Tin(tons)	Value(£)
1877	23.30	0.00	0.00	920.80
1878	29.30	0.00	0.00	1032.70
1879	0.00	150.00	0.00	165.00
1881	0.00	10.00	0.00	2.10
1882	2.50	150.00	1.40	35.00
Arsenic	Ore(tons)	Metal(tons)	Value(£)	
1874	25.00	0.00	25.00	
1875	201.00	0.00	201.00	
1876	257.80	0.00	1283.80	
1877	57.10	0.00	256.50	
1878	94.20	0.00	422.70	

Comment 1876 SOOT
Ownership: 1872–1873 COST BOOK COMPANY; 1875–1878 SOUTH ROSKEAR CO.;
             1879 SOUTH ROSKEAR COPPER MINING CO.; 1880–1882 SOUTH ROSKEAR
             TIN & COPPER MINING CO.
             Comment 1873 STOPPED IN 1873 & EARLY IN 1874; 1880–1882 NOT
             WORKED
Management: Manager 1873 WM.SKEWIS ·
             Chief Agent 1874 MOSES BOWDEN; 1875 JAS.HOSKING; 1876
             WM.SKEWIS; 1877–1878 JAS.HOSKING; 1879 JAS.HOSKING JOHN
             BRENTON; 1880–1882 JAS.HOSKING
Employment: Underground   Surface   Total
1878	65	36	101
1879	41	35	76
1880–1881		2	2
1882		9	9

ROSKEAR,WEST   CAMBORNE   SW 624402 1118

Production: | Lead & Silver | Ore(tons) | Lead(tons) | Silver(ozs) | Value(£) |
|--------|-----------|------------|-------------|----------|
| 1874 | 2.00 | 1.20 | 0.00 | 36.00 |
| 1875 | 2.80 | 2.00 | 0.00 | 0.00 |
| 1876 | 6.20 | 4.60 | 0.00 | 0.00 |
| 1877 | 20.80 | 15.20 | 60.00 | 325.70 |
| 1878 | 10.60 | 7.70 | 30.00 | 205.00 |
| 1879 | 5.20 | 4.00 | 15.00 | 78.90 |
| Zinc | Ore(tons) | Metal(tons) | Value(£) | |
| 1876 | 39.00 | 0.00 | 120.00 | |
| 1877 | 57.00 | 0.00 | 228.00 | |
| 1878 | 174.80 | 0.00 | 407.40 | |
| 1879 | 70.40 | 0.00 | 153.20 | |
| 1882 | 6.00 | 2.50 | 12.00 | |
| Copper | Ore(tons) | Metal(tons) | Value(£) | |
| 1873 | 20.00 | 1.10 | 61.00 | |
| 1874 | 151.00 | 8.20 | 506.20 | |
| 1875 | 72.00 | 4.60 | 345.30 | |
| 1876 | 95.00 | 6.10 | 426.20 | |
| 1877 | 69.10 | 2.90 | 176.60 | |
| 1878 | 55.70 | 2.30 | 97.10 | |
| 1879 | 13.90 | 0.50 | 19.20 | |

Comment 1873–1876 (C)
Tin	Black(tons)	Stuff(tons)	Tin(tons)	Value(£)
1874	0.00	85.00	0.00	65.00

428

ROSKEAR,WEST   CAMBORNE   Continued

Tin	Black(tons)	Stuff(tons)	Tin(tons)	Value(£)
1875	0.00	69.30	0.00	0.00
1877	0.00	20.40	0.00	54.00
1879	0.00	5.10	0.00	11.20

Ownership: 1877–1884 WEST ROSKEAR MINING CO.
     Comment 1879 IN LIQUIDATION; 1880 NOT WORKED
Management: Manager 1877–1880 HUGH STEPHENS
     Chief Agent 1874 JAS.EVANS HUGH STEPHENS; 1875–1876 HUGH
     STEPHENS; 1879 WM.BENNETTS; 1881–1883 J.NICHOLLS; 1884
     CHRIS.LOBB
     Secretary 1875–1878 JAS.EVANS

Employment:	Underground	Surface	Total
1878	31	42	73
1880		1	1
1881	27	16	43
1882	21	9	30
1883	17	9	26
1884		1	1

ROSKROW UNITED   PERRANAWORTHAL   SW 761378 1119

Production: Lead No detailed return
     Copper No detailed return
Ownership: 1912–1913 TRESIDER & DOIDGE
     Comment 1865 FORMERLY SOUTH TRESAVEAN; 1912 REOPENING
     DEC.1912; 1913 EXPLORING ROSKROW
Management: Manager 1869–1872 D.TREBILCOCK
     Chief Agent 1865–1867 JOHN WHITBURN

ROSTIDYON & DEEP LEVEL   ST.WENN   1120

Production: Manganese	Ore(tons)	Metal(tons)	Value(£)
1874	60.00	0.00	270.00

     Iron No detailed return
Ownership: 1874–1876 DEMELZA HAEM.ORE CO.
     Comment 1874 ROSTIDGON & DEEP LEVEL; 1875 LITTLE SKEWS &
     TRELIVER; 1876 LITTLE SKEWS & TRELIVER & GLEBELAND; 1876
     GLEBELAND & LITTLE SKEWS
Management: Chief Agent 1874–1876 JOHN CARR

ROWE   HELSTON   1121

Production: Zinc	Ore(tons)	Metal(tons)	Value(£)
1873	5.00	0.00	12.50

ROYALTON,GREAT   ROCHE   SW 978618 1123

Production: Tin	Black(tons)	Stuff(tons)	Tin(tons)	Value(£)
1868	19.00	0.00	0.00	1128.20
1869	39.90	0.00	0.00	2987.40
1870	15.30	0.00	0.00	1176.30

ROYALTON,GREAT   ROCHE   Continued

Tin	Black(tons)	Stuff(tons)	Tin(tons)	Value(£)
1871	13.10	0.00	0.00	994.00
1899	0.20	0.00	0.00	9.00
1900	0.50	0.00	0.00	40.00
1901	0.00	422.00	0.00	216.00
1908	0.00	1034.00	0.00	585.00

Comment 1868-1870 ROYALTON
Ownership: 1899 W.J.PARKYN; 1900-1901 WRIGHT,WILLIAMS & CO.; 1906-1910
R.J.TURNER
Comment 1868-1869 ROYALTON; 1872 SEE TREWORLACK; 1881
ROYALTON; 1901 ABANDONED MARCH 1901; 1910 NOT WORKED
Management: Manager 1869 H.REYNOLDS; 1872 THOS.PARKYN
Chief Agent 1868 WM.KITTO; 1869 THOS.PARKYN D.TONKIN;
1870-1871 THOS.PARKYN; 1906-1910 J.T.HARVEY
Secretary 1869 JAS.H.CROFTS WM.WARD; 1870-1871 JAS.H.CROFTS;
1881 C.J.POUCHEY

Employment:	Underground	Surface	Total
1899	10	11	21
1900	24	27	51
1901	10	7	17
1906	8	12	20
1907	4	6	10
1908	14	18	32
1909	2		2

RUBY   LUDGVAN   1124

Production: Tin No detailed return
Management: Manager 1872-1873 J.RICHARDS

RUBY   ST.AUSTELL   SX 035557 1125

Production: Iron	Ore(tons)	Iron(%)	Value(£)
1862	2000.00	0.00	1250.00
1863	No detailed return		
1864	7925.10	0.00	3962.60
1865	6545.90	0.00	1966.80
1866	7951.80	0.00	2385.30
1867	1772.00	0.00	531.60
1868	5937.20	0.00	1781.20
1869	No detailed return		
1870	5406.90	0.00	1622.10
1871	9154.20	0.00	2746.30
1872	5359.00	0.00	4824.00
1873	6649.00	0.00	7313.90
1874	6280.00	0.00	4710.50
1876	4521.00	0.00	1356.30
1877	616.00	0.00	369.60

Comment 1862-1863 BH.; 1864-1869 BH.INC.KNIGHTOR; 1870-1871
RH.INC.KNIGHTOR; 1872-1874 BH.INC.TRETHURGY; 1876 BH.DITTO.
PART RAISED IN 1875; 1877 BH.
Ownership: 1863-1871 FRANCIS BARRATT & CO.; 1873 ALF.ALLOTT & J.W.DAY;
1874-1877 ALF.ALLOTT; 1893 JOHN HOLLOWAY

RUBY   ST.AUSTELL   Continued

        Comment 1864-1871 INC.KNIGHTOR; 1872-1877 INC.TRETHURGY; 1894
        ABANDONED
Management: Chief Agent 1863 FRANCIS BARRATT; 1864-1871 FRANCIS BARRATT
        JNR; 1872-1873 J.W.DAY; 1874-1876 ED.FAULCONER; 1877 J.BEARD;
        1893 JOHN ROWE
        Secretary 1893 T.M.GEORGE (S)
Employment: Underground   Surface   Total
        1894   5   1   6

RUBY   WENDRON   SW 701322 1126

Production: Copper   Ore(tons) Metal(tons)   Value(£)
        1847   136.00   12.50   914.20
        Comment 1847 (C)
        Tin No detailed return
Ownership: 1913 MEDLYN TIN LANDS LTD.
        Comment 1913 SEE GARLIDNA
Employment: Underground   Surface   Total
        1913   2   23   25

RUTHERN   ST.COLUMB MAJOR   1128

Production: Tin   Black(tons) Stuff(tons) Tin(tons)   Value(£)
        1909   0.40   0.00   0.00   30.00
        1910   2.00   0.00   0.00   200.00
Ownership: 1909-1913 RUTHERN TIN MINE CO.
        Comment 1911-1912 WORKING DUMPS; 1913 WORKING DUMPS CLOSED
        FEB.1913
Management: Chief Agent 1909-1913 GEO.BELLAMY
Employment: Underground   Surface   Total
        1909   8   8
        1910   4   4
        1911   5   5
        1912   3   3
        1913   2   2   4

RUTHERS   ST.COLUMB MAJOR   SW 923601 1129

Production: Manganese   Ore(tons) Metal(tons)   Value(£)
        1874   24.60   0.00   108.00
        1875   25.00   0.00   100.00
        1880   722.60   0.00   903.10
        1881   300.00   0.00   375.00
        Comment 1880 RUTHOS
        Tin   Black(tons) Stuff(tons) Tin(tons)   Value(£)
        1889   3.00   2.00   0.00   150.00
        1890   3.00   0.00   0.00   150.00
        1891   0.10   0.00   0.00   8.00
        Comment 1889 TWO RETURNS AGGREGATED
Ownership: 1875 CORNISH CONS CORP.; 1878-1879 NEW PERRAN MINERALS CO.;
        1880-1881 WEST OF ENGLAND MINERALS CO.; 1885-1886 RUTHERS
        MINING CO.; 1889-1891 JOHN HOOPER AND OTHERS

Comment 1886 NOT WORKED; 1891 IN LIQUIDATION
Management: Manager 1878-1879 JAS.HENDERSON
           Chief Agent 1874-1875 JOHN PARKYN; 1880-1881 G.H.NORTH;
           1889-1891 JOHN HOOPER
           Secretary 1885-1886 G.PEVERALL
Employment: Underground   Surface   Total
           1878
           1880-1881
           1889   4   3   7
           1890-1891   4   6   10
           Comment 1878 SEE RUTHERS(IRON); 1880-1881 SEE RUTHERS(IRON)

RUTHERS(IRON)   ST.COLUMB MAJOR   SW 923601 1130

Production: Iron   Ore(tons)   Iron(%)   Value(£)
           1872   1242.00   0.00   745.00
           1880   134.50   0.00   94.10
           1881   900.00   0.00   675.00
           Comment 1872 BH.; 1880-1881 RH.
Ownership: 1872-1873 CORNISH CONS.IRON MINES CORP.LTD.; 1875-1876
           CORNISH CONS.IRON MINES CORP.LTD.; 1877 GREAT PERRAN MINE
           ADVENRS.; 1878 NEW PERRAN MINERAL CO.; 1879 W.R.ROEBUCK AND
           OTHERS; 1880-1881 CORNISH STEEL IRON ORE CO.; 1882-1883
           NEWQUAY MINING CO.LTD.
           Comment 1875 RUTHERN; 1881 HM INSP.GIVES NEWQUAY MIN.CO.;
           1882-1883 RUTHOS
Management: Manager 1873 ART.PETO; 1875-1878 JOHN PARKIN; 1879
           JAS.HENDERSON; 1880-1881 J.WHITAKER BUSHE
           Chief Agent 1872 S.J.PITTAR; 1877-1878 JAS.HENDERSON;
           1879-1883 JOHN H.JAMES
Employment: Underground   Surface   Total
           1878   1   1
           1880   24   13   37
           1881   22   8   30
           1882   2   2
           1883   1   1
           Comment 1878 COULD BE RUTHERS; 1880-1881 COULD BE RUTHERS

SALLY   ILLOGAN   SW 679469 1131

Production: Zinc No detailed return
           Tin No detailed return
Management: Chief Agent 1861-1866 JOHN PRINCE
           Secretary 1861-1862 JOHN PRINCE (P); 1863-1866 JOHN PRINCE
           JNR.

SAMPSON,ROYAL   ENDELLION   SX 028819 1133

Production: Iron No detailed return
           Silver No detailed return
Ownership: 1863-1871 ROYAL SAMPSON MINING CO.
Management: Manager 1863-1867 JOHN THOMAS

SAMPSON,ROYAL   ENDELLION   Continued

        Chief Agent 1863-1871 NICH.THOMAS
        Secretary 1863-1865 COMMITTEE

SAVATH   LUXULIAN   SX 020613 1134

Production: Tin No detailed return
            Iron No detailed return
Management: Manager 1873 B.J.BAWDEN
            Chief Agent 1872-1873 R.G.BAWDEN; 1874 B.J.BAWDEN

SCORRIER CONSOLS   ST.AGNES   SW 724456 1136

Production: Copper No detailed return
            Tin   Black(tons) Stuff(tons)   Tin(tons)   Value(£)
            1874   6.90   0.00   0.00   338.70
            1875   1.90   0.00   0.00   107.20
Ownership: Comment 1862-1864 SUSPENDED
Management: Manager 1865 JOHN WATERS
            Chief Agent 1861 JOSH.DANIELL
            Secretary 1861-1864 TIM.PAINTER (P); 1865 HY.MICHELL

SETON   CAMBORNE   SW 656415 1137

Production: Zinc   Ore(tons) Metal(tons)   Value(£)
            1857   7.40   0.00   12.00
            1858   10.30   0.00   20.60
            1859   3.00   0.00   6.00
            1861   3.00   0.00   6.00
            1876   6.10   0.00   8.10
            Comment 1859 ORE & VALUE EST. 1861 ORE EST
            Copper   Ore(tons) Metal(tons)   Value(£)
            1845   2035.00   148.70   11352.30
            1846   4590.00   408.70   30160.20
            1847   5402.00   451.50   33012.80
            1848   5483.00   433.70   26496.80
            1849   4953.00   333.40   21577.30
            1850   6009.00   404.20   27554.60
            1851   5471.00   372.70   24708.40
            1852   5286.00   314.50   25347.70
            1853   4802.00   295.90   29593.20
            1854   4945.00   262.90   25645.20
            1855   3754.00   196.90   19442.30
            1856   3150.00   173.70   15616.20
            1857   2527.00   146.00   14497.00
            1858   2383.00   137.80   12035.00
            1859   2336.00   141.50   13077.20
            1860   2077.00   108.00   9649.30
            1861   2378.00   127.50   11197.20
            1862   3290.00   212.50   17272.00
            1863   4035.00   264.90   20533.20
            1864   5003.00   292.70   24206.10
            1865   6118.00   362.50   27774.20

                              433

Copper	Ore(tons)	Metal(tons)	Value(£)
1866	6191.00	378.40	24391.90
1867	5057.00	301.30	20020.00
1868	4622.00	256.50	15847.00
1869	3366.00	186.80	10752.00
1870	1913.00	107.20	5923.40
1871	1595.00	102.60	6223.40
1872	891.00	52.40	4079.40
1873	418.00	28.20	1783.70
1874	165.00	11.00	679.20
1875	37.00	1.40	99.70
1876	13.00	0.70	41.40
1877	10.00	0.40	38.00

Comment 1845-1876 (C)

Tin	Black(tons)	Stuff(tons)	Tin(tons)	Value(£)
1855	4.10	0.00	0.00	273.30
1856	21.20	0.00	0.00	1495.70
1857	13.80	0.00	0.00	999.00
1858	24.80	0.00	0.00	1569.60
1859	25.80	0.00	0.00	1791.20
1860	38.40	0.00	0.00	2941.70
1861	62.10	0.00	0.00	4512.50
1862	47.30	0.00	0.00	2962.80
1863	33.50	0.00	0.00	2109.30
1864	42.10	0.00	0.00	3162.10
1865	52.00	0.00	0.00	2834.30
1866	39.80	0.00	0.00	1908.40
1867	38.20	0.00	0.00	2020.70
1868	62.30	0.00	0.00	3574.50
1869	74.00	0.00	0.00	5322.80
1870	102.70	0.00	0.00	7705.50
1871	109.00	0.00	0.00	9070.60
1872	70.00	0.00	0.00	6176.90
1873	92.50	0.00	0.00	7891.10
1874	68.50	0.00	0.00	3864.70
1875	4.10	0.00	0.00	206.50
1876	0.00	26.40	0.00	13.10

Comment 1862 TWO RETURNS AGGREGATED; 1874 SEATON

Arsenic	Ore(tons)	Metal(tons)	Value(£)
1857	54.80	0.00	109.90
1858	56.60	0.00	99.70
1859	58.80	0.00	113.70
1860	71.80	0.00	138.00
1861	155.60	0.00	275.60
1862	57.00	0.00	100.00
1863	110.80	0.00	110.80
1864	136.00	0.00	101.00
1865	94.00	0.00	83.00
1866	99.00	0.00	137.70
1867	65.50	0.00	157.20
1868	133.50	0.00	436.20
1869	50.00	0.00	112.30
1870	119.60	0.00	366.20
1871	265.90	0.00	401.50
1872	135.10	0.00	225.00

SETON   CAMBORNE   Continued

        Arsenic   Ore(tons) Metal(tons)   Value(£)
        1873   125.80   0.00   382.80
        1874   88.60   0.00   310.00
        1875   114.30   0.00   511.60
        Comment 1859-1870 CRUDE
        Fluorspar   Ore(tons)   Value(£)
        1871   10.50   5.80
Ownership: 1876 WHEAL SETON MINING CO.
             Comment 1873 STOPPED IN 1873 & EARLY IN 1874; 1874-1875 NOW
             PART OF WEST SETON
Management: Manager 1859-1860 HY.SKEWIS; 1861-1871 ROBT.WILLIAMS;
             1872-1876 WM.TEAGUE & SON
             Chief Agent 1859-1860 ROBT.WILLIAMS; 1861-1864 WM.ROWE;
             1865-1869 WM.ROWE JOHN POPE JNR; 1870-1871 WM.ROWE; 1872
             CHRIS.LOBB R.RUNDAL; 1873-1876 CHRIS.LOBB
             Secretary 1859-1871 T.H.TILLY (P); 1872-1876 WM.TEAGUE

SETON,EAST   ILLOGAN   SW 661418 1138

Production: Copper   Ore(tons) Metal(tons)   Value(£)
            1847   250.00   18.60   1359.60
            1848   113.00   8.60   500.80
            1852   195.00   12.60   1019.10
            1853   148.00   8.00   775.60
            1870   192.00   17.10   1061.90
            1871   187.00   15.60   960.30
            1872   264.00   17.50   1237.40
            1873   227.00   15.30   932.10
            1874   137.00   8.20   485.60
            Comment 1847-1848 (C); 1852-1853 (C)INC.MAUDE; 1870-1874 (C)
            Tin   Black(tons) Stuff(tons)   Tin(tons)   Value(£)
            1871   0.00   60.00   0.00   63.10
            1873   0.00   139.00   0.00   21.50
            1874   0.00   50.00   0.00   27.80
            1875   0.00   17.00   0.00   0.00
Ownership: Comment 1872 SEE ALSO EMILY HENRIETTA; 1873 SEE ALSO EMILY
            HENRIETTA STOPPED IN 1873 & EARLY IN 1874
Management Manager 1860-1869 JOS.VIVIAN; 1870-1873 WM.PASCOE; 1875-1876
            THOS.PRYOR
            Chief Agent 1860-1869 WM.THOMAS JNR; 1870-1873 HY.ARTHUR;
            1874 W.R.REYNOLDS
            Secretary 1860-1866 ALM.E.PAULL (P); 1867-1869 JOHN WATSON;
            1870-1871 FRAN.PRYOR; 1872-1873 THOS.PRYOR

SETON,NEW   CAMBORNE   SW 646410 1139

Production: Copper   Ore(tons) Metal(tons)   Value(£)
            1863   5.00   0.40   30.40
            Comment 1863 (C)
Ownership: Comment 1873 STOPPED IN 1873 & EARLY IN 1874
Management: Manager 1865-1872 MAL.BATH; 1873-1874 MAL.BATH JOH.THOMAS;
            1875 MAL.BATH
            Chief Agent 1859-1864 MAL.BATH

                                435

SETON,NEW   CAMBORNE   Continued

            Secretary 1859 R.MATTHEWS (P); 1860-1871 BEN.MATTHEWS (P);
            1872-1875 OLI.MATTHEWS

SETON,SOUTH   CAMBORNE   SW 637406 1140

Production: Copper No detailed return
Management: Manager 1859-1864 JOS.HIGGINS; 1867-1871 MAL.BATH
            Chief Agent 1865-1871 JOH.THOMAS
            Secretary 1859-1863 JOHN TIPPETT (P); 1864-1871 BEN.MATTHEWS

SETON,VIOLET   CAMBORNE   SW 632413 1141

Production: Lead   Ore(tons) Metal(tons)   Value(£)
            1882   24.30   18.20   474.00
            1883   23.50   17.20   409.00
            1885   8.00   0.00   122.00
            Zinc   Ore(tons) Metal(tons)   Value(£)
            1882   144.60   57.60   181.00
            1883   59.70   20.60   56.00
            1885   68.00   0.00   59.00
            Copper   Ore(tons) Metal(tons)   Value(£)
            1881   430.40   29.30   1532.50
            1882   244.70   19.50   822.00
            1883   147.80   6.00   346.00
            1884   106.00   0.00   299.00
            1885   135.00   0.00   316.00
            1886   69.00   0.00   86.00
            1888   138.00   0.00   564.00
Ownership: 1878-1887 VIOLET SETON MINING CO.; 1888-1893 ?FLUYDER
            Comment 1889-1890 SUSPENDED; 1891 PUMPING RESUMED IN
            DECEMBER; 1892-1893 ABANDONED
Management: Manager 1877-1881 JOHN NICHOLLS
            Chief Agent 1879 WM.TONKIN WM.ARGALL; 1880-1881 SAMP.HAMMER;
            1882-1883 JOHN NICHOLLS; 1884-1893 SAMP.HAMMER
            Secretary 1877 ?FLUYDER
Employment: Underground   Surface   Total
            1878   6   2   8
            1879   20   23   43
            1880   41   28   69
            1881   52   73   125
            1882   51   60   111
            1883   38   28   66
            1884   43   24   67
            1885   33   24   57
            1886   20   19   39
            1887   19   16   35
            1888   19   22   41
            1891   30   30
            1892   20   11   31

                                     436

Production: Copper

	Ore(tons)	Metal(tons)	Value(£)
1848	236.00	19.00	1095.50
1849	739.00	59.00	3981.20
1850	639.00	44.60	3007.90
1851	532.00	36.90	2459.40
1852	885.00	76.10	6622.90
1853	1938.00	109.00	10098.10
1854	2307.00	153.00	15487.70
1855	3064.00	206.10	21205.70
1856	4026.00	285.00	27042.50
1857	4851.00	356.40	36296.60
1858	5482.00	410.20	37881.70
1859	6839.00	526.70	49894.30
1860	7155.00	518.90	48015.20
1861	6957.00	492.10	44488.00
1862	6284.00	438.10	36219.70
1863	6346.00	427.80	33341.90
1864	5412.00	369.30	31189.50
1865	6117.00	439.90	34748.00
1866	5889.00	433.20	29648.60
1867	6375.00	491.50	34258.30
1868	6703.00	512.10	34017.40
1869	6511.00	509.90	31755.10
1870	4825.00	397.60	23497.00
1871	3809.00	309.70	19159.10
1872	2800.10	219.90	17497.90
1873	2789.00	212.80	13774.50
1874	2085.00	152.90	10027.00
1875	1423.00	99.10	7507.80
1876	3643.00	271.50	18415.50
1877	3314.70	241.00	13611.20
1878	2200.10	169.90	8262.70
1879	1384.90	110.70	5251.40
1880	500.30	40.20	2242.00
1881	542.30	49.80	2645.40
1882	376.60	29.20	1848.00
1883	398.20	31.20	1748.00
1884	290.00	0.00	1076.00
1885	165.00	0.00	481.00
1886	70.00	0.00	197.00
1887	48.00	0.00	110.00
1888	195.00	0.00	823.00
1889	57.00	0.00	101.00
1890	21.00	0.00	85.00

Comment 1848-1849 (C)WEST SEATON; 1850 (C); 1851 (C)WEST
SEATON; 1852 (C); 1853-1855 (C)WEST SEATON; 1856 (C); 1857
(C)WEST SEATON; 1858-1871 (C); 1872 TWO RETURNS AGGREGATED;
1873-1876 (C)

Tin	Black(tons)	Stuff(tons)	Tin(tons)	Value(£)
1856	0.00	0.00	0.00	7.00
1857	0.00	0.00	0.00	109.10
1860	3.50	0.00	0.00	262.80
1861	1.90	0.00	0.00	125.70
1862	28.10	0.00	0.00	1574.80
1863	7.50	0.00	0.00	471.80

Tin	Black(tons)	Stuff(tons)	Tin(tons)	Value(£)
1864	21.90	0.00	0.00	1365.30
1865	19.60	0.00	0.00	1038.20
1866	40.80	0.00	0.00	1934.60
1867	16.30	0.00	0.00	834.90
1868	52.70	0.00	0.00	2935.10
1869	62.40	103.20	0.00	4466.70
1870	96.10	0.00	0.00	7250.90
1871	134.50	0.00	0.00	10818.50
1872	116.80	146.10	0.00	9945.80
1873	137.00	0.00	0.00	10230.70
1874	177.50	0.00	0.00	9874.90
1875	62.90	0.00	0.00	3270.00
1876	171.30	0.00	0.00	7050.50
1877	233.10	0.00	0.00	9499.70
1878	239.60	0.00	0.00	8362.00
1879	299.30	0.00	0.00	12099.70
1880	188.50	219.00	0.00	10207.10
1881	169.90	0.00	100.80	9474.40
1882	153.90	0.00	103.80	9456.00
1883	241.30	0.00	0.00	12824.00
1884	203.50	0.00	0.00	9238.00
1885	189.50	0.00	0.00	8936.00
1886	242.20	0.00	0.00	13523.00
1887	208.60	0.00	0.00	13637.00
1888	219.00	0.00	0.00	14820.00
1889	201.60	0.00	0.00	11078.00
1890	167.60	0.00	0.00	8956.00

Comment 1856-1857 TIN STUFF; 1862 TWO RETURNS AGGREGATED;
1870 VALUE INC.TIN STUFF

Arsenic	Ore(tons)	Metal(tons)	Value(£)
1862	32.70	0.00	43.10
1864	18.50	0.00	14.00
1865	37.50	0.00	31.40
1866	85.50	0.00	106.90
1867	74.30	0.00	191.30
1869	151.90	0.00	418.00
1870	105.40	0.00	275.30
1871	82.60	0.00	125.20
1872	92.70	0.00	162.50
1873	171.90	0.00	494.20
1874	186.90	0.00	487.10
1875	148.70	0.00	484.80
1876	160.20	0.00	800.80
1877	203.70	0.00	561.70
1878	200.10	0.00	773.30
1879	223.10	0.00	983.30
1880	157.60	0.00	936.80
1881	242.20	0.00	1158.20
1882	235.60	0.00	1301.00
1883	263.30	0.00	1463.00
1884	226.50	0.00	1401.00
1885	159.00	0.00	800.00
1886	232.00	0.00	913.00
1887	190.00	0.00	1105.00

Arsenic	Ore(tons)	Metal(tons)	Value(£)
1888	237.00	0.00	1361.00
1889	219.00	0.00	1347.00
1890	130.00	0.00	837.00

Comment 1862 CRUDE; 1864–1867 CRUDE; 1870–1871 CRUDE; 1876 SOOT; 1884–1885 SOOT; 1886 CRUDE

Arsenic Pyrite	Ore(tons)	Value(£)
1880	5.80	8.90

Ownership: 1877–1881 WEST WHEAL SETON ADVENTURERS; 1882–1886 WEST WHEAL SETON MINING CO.; 1887–1891 WEST WHEAL SETON ADVENTURERS
Comment 1874–1875 INC.SETON; 1890 NOW ABANDONED; 1891 ABANDONED

Management: Manager 1869 MAL.BATH; 1872–1873 MAL.BATH; 1874–1876 JOH.THOMAS; 1878–1881 W.R.RUTTER
Chief Agent 1859–1860 CHAS.THOMAS; 1861–1866 CHAS.THOMAS MAL.BATH JOHN JENNINGS; 1867–1868 MAL.BATH JOHN JENNINGS; 1869 JOHN JENNINGS; 1870 JOHN JENNINGS MAL.BATH; 1871 MAL.BATH JOHN JAMES; 1872–1873 JOHN JENNINGS WM.PASCOE; 1874–1877 MAL.BATH; 1878 MAL.BATH STEP.TERRILL; 1879–1881 MAL.BATH; 1882–1884 CHAS.THOMAS; 1885–1887 W.R.RUTTER; 1888–1890 JOHN JAMES
Secretary 1859–1870 BEN.MATTHEWS (P); 1871 BEN.MATTHEWS; 1872–1873 OLI.MATTHEWS; 1874–1876 THOS.PRYOR; 1877–1881 THOS.PRYOR (P)

Employment:	Underground	Surface	Total
1878	158	214	372
1879	146	178	324
1880	119	157	276
1881	118	143	261
1882	110	164	274
1883	110	156	266
1884	90	136	226
1885	100	131	231
1886	95	131	226
1887	110	134	244
1888	130	151	281
1889	129	140	269
1890	124	137	261

SEYMOUR   CHACEWATER   SW 738452 1143

Production: Zinc

	Ore(tons)	Metal(tons)	Value(£)
1877	20.00	0.00	85.50
1878	5.00	0.00	12.50

Copper

	Ore(tons)	Metal(tons)	Value(£)
1877	12.00	0.80	54.50
1878	1.50	0.10	5.50

Ownership: 1877–1878 NICH.ROBERTS
Management: Chief Agent 1877–1878 NICH.ROBERTS
Secretary 1877–1879 NICH.ROBERTS

Employment:	Underground	Surface	Total
1878	1	4	5

SEYMOUR,EAST   CHACEWATER   1144

Production: Zinc   Ore(tons) Metal(tons)   Value(£)
            1876   48.00   0.00   254.00

SHARP TOR,WEST   LINKINHORNE   SX 260732 1145

Production: Copper   Ore(tons) Metal(tons)   Value(£)
            1865   103.00   14.00   1225.10
            1866   68.00   7.90   520.60
            1867   11.00   0.80   58.30
            Comment 1865-1867 (C)
            Tin   Black(tons) Stuff(tons)   Tin(tons)   Value(£)
            1867   2.60   0.00   0.00   73.60
Ownership: 1907 J.DEFRIES & SONS LTD.
Management: Manager 1859 THOS.MORRIS
            Chief Agent 1859-1865 WM.RICHARDS
            Secretary 1859-1860 THOS.MORRIS (P); 1861-1865 WM.RICHARDS
            (P)

SHEBA CONSOLS,GREAT   STOKE CLIMSLAND   SX 372737 1146

Production: Copper   Ore(tons) Metal(tons)   Value(£)
            1854   786.00   28.90   2476.20
            1855   1146.00   33.10   2588.00
            1856   779.00   21.70   1318.70
            1857   950.00   27.90   2144.80
            1858   337.00   8.20   477.00
            Comment 1854 (C)GREAT SHEBA; 1855-1857 (C); 1858 (C)GREAT
            SHEBA
Ownership: Comment 1864-1865 SEE WEST MARTHA
Management: Manager 1859-1865 HY.RICKARD
            Chief Agent 1859 HY.RICKARD
            Secretary 1859-1861 ROBT.SARGEANT; 1862-1863 FRAN.PRYOR;
            1864-1865 HY.RICKARD

SHELTON   BUGLE   SX 007571 1147

Production: Tin   Black(tons) Stuff(tons)   Tin(tons)   Value(£)
            1870   78.80   0.00   0.00   5424.20
            1871   93.70   0.00   0.00   6811.20
            1872   21.70   0.00   0.00   1596.00
            Comment 1870-1872 SHELTON CLAY AND TIN COMPANY
Ownership: Comment 1869-1872 SHILTON. SEE BUNNIE

SHEPHERDS ROSE UNITED   PERRANZABULOE   1149

Production: Lead No detailed return
Ownership: 1880-1881 SHEPHERDS WHEAL ROSE CO.
            Comment 1880 FORMERLY NORTH CHIVERTON,ALBERT & EAST ANNA
Management: Manager 1880-1881 J.OWEN
            Chief Agent 1880 STEP.RICHARDS; 1881 STEP.RICHARDS H.MINERS

SHEPHERDS ROSE UNITED   PERRANZABULOE   Continued

Employment: Underground   Surface   Total
                1881   13   13

SHEPHERDS UNITED   BREAGE   1150

Production: Copper   Ore(tons) Metal(tons)   Value(£)
             1885   5.00   0.00   3.00
             Comment 1885 MOUNTS BAY SECTION.UNDRESSED ORE
             Tin   Black(tons) Stuff(tons)   Tin(tons)   Value(£)
             1885   39.00   0.00   0.00   1812.00

SHEPHERDS,NORTH   NEWLYN EAST   SW 811547 1151

Production: Lead & Silver No detailed return
Management: Chief Agent 1865—1866 HY.BENNETTS

SHEPHERDS,OLD   NEWLYN EAST   SW 817541 1152

Production: Lead & Silver   Ore(tons)   Lead(tons) Silver(ozs)   Value(£)
             1883   133.40   98.70   3603.00   1255.00
             1884   171.00   0.00   0.00   1297.00
             1885   145.00   0.00   0.00   1215.00
             1886   122.00   0.00   0.00   0.00
             Comment 1884 FOR AG SEE ZZ SUNDRIES
             Zinc   Ore(tons)   Metal(tons)   Value(£)
             1883   21.60   7.30   74.00
             1884   41.00   0.00   128.00
             1885   98.00   0.00   289.00
             1886   14.00   0.00   45.00
Ownership: 1880 OLD SHEPHERDS MINING CO.; 1881—1886 OLD SHEPHERDS MINING
            CO.LTD.
            Comment 1860—1865 SHEPHERDS; 1880 SEE ALSO H.M.INSPECTOR'S
            LISTS
Management: Manager 1880 JOHN B.REYNOLDS; 1881 R.NANCARROW & THOMAS
            Chief Agent 1860—1865 THOS.RICHARDS; 1880 RICH.NANCARROW
            THOS.DOIDGE; 1881 JAS.NANCARROW; 1882—1884 RICH.NANCARROW;
            1885—1886 RICH.NANCARROW JAS.NANCARROW
            Secretary 1860—1865 THOS.RICHARDS (P); 1880—1881 HY.BROWNE
            (P)
Employment: Underground   Surface   Total
             1880   3   3
             1881   49   62   111
             1882   57   68   125
             1883   60   31   91
             1884   63   42   105
             1885   46   27   73
             1886   48   21   69

                                   441

SHEPHERDS,WEST   PERRANZABULOE   SW 758519 1153

Production: Lead No detailed return
Ownership: 1880–1881 WEST SHEPHERDS MINING CO.LTD.; 1882–1884 WEST
            SHEPHERDS MINING CO.
            Comment 1883 NOT WORKED; 1884 WEST GREAT SHEPHERDS
Management: Manager 1880–1881 CHAS.CRAZE
            Chief Agent 1881–1884 JAS.M.CRAZE
            Secretary 1880 JOH.THOMAS; 1881 JOH.THOMAS (P)
Employment: Underground   Surface .   Total
            1880   12   11   23
            1881   12   10   22
            1882   28   10   38
            1884   14   13   27

SICILY   BROADOAK   SX 136651 1155

Production: Lead & Silver No detailed return
Management: Chief Agent 1861–1865 J.SYMONS
            Secretary 1861–1862 W.G.PAINTER (P); 1863–1865 M.G.PAINTER

SILVER & LANTEGLOS   ST.TEATH   1156

Production: Lead No detailed return
Ownership: 1883–1885 WHEAL SILVER & LANTEGLOS MINING CO.
Management: Chief Agent 1883–1885 WM.BENNETTS
Employment: Underground   Surface   Total
            1883   6   6   12
            1884   7   6   13
            1885   8   4   12

SILVER HILL   CALSTOCK   SX 377696 1157

Production: Lead & Silver No detailed return
            Copper No detailed return
            Tin No detailed return
Ownership: 1880–1884 SILVER HILL MINING CO.
            Comment 1883 NOT WORKED
Management: Chief Agent 1881–1884 GEO.RICKARD
Employment: Underground   Surface   Total
            1881   13   4   17
            1882   12   4   16
            1884   3   1   4

SILVER VALLEY   CALSTOCK   SX 385702 1158

Production: Lead   Ore(tons) Metal(tons)   Value(£)
            1852   8.70   6.50   0.00
            1853–1854 No detailed return
            Comment 1852–1854 INC.BROTHERS

                                      442

Production: Lead & Silver No detailed return
Ownership: Comment 1865 SUSPENDED
Management: Manager 1860-1864 JOH.ANGOVE
            Secretary 1860-1865 THOS.W.ROBINSON (P)

SILVER VALLEY   LISKEARD   SX 255716 1160

Production: Tin No detailed return
Ownership: 1907-1912 CHAS.F.ROSEBY; 1913 WEST OF ENGLAND TIN CORP.LTD.
           Comment 1910 NOT WORKED; 1913 DISCONTINUED JUNE 1913
Management: Manager 1913 E.S.KING
            Chief Agent 1907-1912 R.KNOTT
Employment: Underground   Surface   Total
            1908   8   8
            1909   2   2
            1911   4   4   8
            1912   6   10   16
            1913   8   4   12

SILVER VEIN   LOSTWITHIEL   SX 122596 1161

Production: Lead & Silver   Ore(tons) Lead(tons) Silver(ozs)   Value(£)
            1861   3.50   2.60   40.00   0.00
            Copper   Ore(tons) Metal(tons) Value(£)
            1877   0.00   1.80   72.00
            1878   1.30   0.80   102.00
            Comment 1877-1878 PRECIPITATE
            Silver   Ore(tons) Metal(tons)   Value(£)
            1877   0.30   0.00   0.00
Ownership: 1877 RICH.HINGSTON
Management: Manager 1860 J.SQUIRE; 1864-1866 JAS.SECCOMBE
            Chief Agent 1860 F.SQUIRE; 1861-1862 EDW.BURN F.SQUIRE; 1863
            EDW.BURN; 1865-1866 H.BURN; 1877-1881 PETER TEMBY
            Secretary 1860-1863 W.MANSEL (P); 1864-1866 JAS.SECCOMBE;
            1877-1881 RICH.HINGSTON

SILVERWELL   ST.AGNES   SW 750487 1163

Production: Lead & Silver No detailed return
            Zinc No detailed return
            Tin   Black(tons) Stuff(tons)   Tin(tons)   Value(£)
            1899   No detailed return
            Comment 1899 SILVERWEEL. SEE EAST DOWNS
Ownership: 1898-1901 C.D.TEAGUE; 1903-1904 C.D.TEAGUE; 1905 CORNISH
           MINING TRUST LTD.
           Comment 1903-1904 SUSPENDED
Management: Manager 1905 W.SLADE OLVER
Employment: Underground   Surface   Total
            1899   13   13
            1900   11   10   21
            1901   2   2   4
            1905   10   9   19

Production: Arsenic   Ore(tons) Metal(tons)   Value(£)
            1891   27.00   0.00   215.00

SISTERS   LELANT   SW 509363 1164

Production: Copper   Ore(tons) Metal(tons)   Value(£)
            1845   577.00   37.20   2595.30
            1846   910.00   68.00   4375.30
            1847   1134.00   100.80   7341.90
            1848   365.00   29.20   1729.10
            Comment 1845-1848 (C)
            Tin   Black(tons) Stuff(tons)   Tin(tons)   Value(£)
            1875   36.10   0.00   0.00   1728.00
            1876   263.90   0.00   0.00   10778.10
            1877   399.80   0.00   0.00   15534.70
            1878   600.10   0.00   0.00   21167.60
            1879   569.50   0.00   0.00   21411.90
            1880   440.20   0.00   0.00   21831.70
            1881   369.70   0.00   249.60   19422.60
            1882   365.50   0.00   241.20   21180.00
            1883   343.70   0.00   0.00   17686.00
            1884   359.60   0.00   0.00   16070.00
            1885   300.90   0.00   0.00   13847.00
            1886   282.00   0.00   0.00   15287.00
            1887   240.60   0.00   0.00   14810.00
            1888   226.20   0.00   0.00   14217.00
            1889   208.90   0.00   0.00   11174.00
            1890   129.80   0.00   0.00   7249.00
            1891   34.00   0.00   0.00   1580.00
            1892   29.80   0.00   0.00   1448.00
            1893   29.70   0.00   0.00   1250.00
            1894   32.00   0.00   0.00   1091.00
            1895   27.20   0.00   0.00   827.00
            1896   25.20   0.00   0.00   709.00
            1897   19.90   0.00   0.00   558.00
            1898   14.00   0.00   0.00   427.00
            1899   14.70   0.00   0.00   830.00
            1901   5.20   0.00   0.00   261.00
            1908   2.20   0.00   0.00   158.00
            Comment 1875 WAS POLPEOR.RETURN FROM OCT.75; 1877 SEE
            MARY(LELANT), MARGARET & TRENCROM
Ownership: 1876-1881 WHEAL SISTERS ADVENTURERS; 1882-1889 WHEAL SISTERS
            MINING CO.; 1890 WHEAL SISTERS MINING SYNDICATE; 1891-1893
            WHEAL SISTERS MINE SYNDICATE(P); 1894-1900 WHEAL SISTERS MINE
            SYNDICATE; 1906-1910 CORNISH CONSOLIDATED TIN MINES LTD.;
            1911 CORNISH CONSOLS LTD.
            Comment 1875-1876 INC.KITTY, POLPEOR, MARY & MARGARET;
            1877-1881 INC.KITTY, POLPEOR, MARY, MARGARET & TRENCROM; 1909
            SUSPENDED; 1910 IN LIQUIDATION; 1911 ABANDONED
Management: Manager 1875-1881 WM.ROSEWARNE
            Chief Agent 1875 SIMON THOMAS MAT.CURNOW; 1876 SIMON THOMAS
            MAT.CURNOW JNR; 1878-1881 MAT.CURNOW SIMON THOMAS N.RICHARDS;
            1882-1889 WM.ROSEWARNE; 1890-1891 F.W.THOMAS; 1892-1900 SIMON
            THOMAS; 1906 J.L.TEALE; 1907-1910 A.B.CLIMAS

SISTERS   LELANT   Continued

Secretary 1875 THOS.W.FIELD; 1876-1881 THOS.W.FIELD (S);
1891-1900 F.W.THOMAS (P)

Employment:	Underground	Surface	Total
1878	307	177	484
1879	280	240	520
1880	246	211	457
1881	234	211	445
1882	217	203	420
1883	216	207	423
1884	209	196	405
1885	205	191	396
1886	206	202	408
1887	185	192	377
1888	184	185	369
1889	184	187	371
1890	10	30	40
1891	12	35	47
1892	10	25	35
1893	9	18	27
1894	8	21	29
1895	5	26	31
1896		24	24
1897		22	22
1898		18	18
1899		17	17
1907	50	83	133
1908	49	97	146
1909		3	3
1911		4	4

SITHNEY   SITHNEY   SW 635293 1165

Production: Tin No detailed return
Ownership: Comment 1859-1864 SITHNEY & CARNMEAL UNITED MINES; 1865
            SITHNEY & CARNMEAL UNITED MINES SUSPENDED
Management: Manager 1859-1864 W.M.MARTIN
            Chief Agent 1859-1864 WM.CHAPPEL
            Secretary 1859-1865 FRED.HILL (P)

SKEWS,LITTLE   ST.WENN   SW 974654 1166

Production: Iron No detailed return
Ownership: Comment 1875-1876 SEE ROSTIDYON & DEEP LEVEL

SLADE   ST.IVE   SX 298702 1167

Production: Copper No detailed return
Ownership: Comment 1859-1860 SEE GREAT CARADON

SLIMEFORD   CALSTOCK   SX 437698 1168

Production: Tin   Black(tons) Stuff(tons)   Tin(tons)   Value(£)
            1871   5.90   0.00   0.00   361.50
            1872   10.00   0.00   0.00   763.10

SPARNON   REDRUTH   SW 704417 1169

Production: Copper   Ore(tons) Metal(tons)   Value(£)
            1865   22.00   0.60   40.30
            Comment 1865 (C)
            Tin   Black(tons) Stuff(tons)   Tin(tons)   Value(£)
            1867   1.50   0.00   0.00   75.00
            1879   No detailed return
            1882   No detailed return
            Comment 1879 SEE PEDNANDREA UNITED; 1882 SEE PEDNANDREA
            UNITED
Ownership: Comment 1879-1880 SEE PEDNANDREA UNITED
Management: Manager 1865-1873 W.TREGAY
            Chief Agent 1865-1868 EL.CHEGWIN; 1869 EL.CHEGWIN JAS.THOMAS;
            1870-1871 JAS.THOMAS
            Secretary 1865-1873 W.PAGE CARDOZA

SPEARNE CONSOLS   ST.JUST   SW 371338 1170

Production: Tin   Black(tons) Stuff(tons)   Tin(tons)   Value(£)
            1853   14.50   0.00   0.00   910.00
            1855   47.30   0.00   0.00   3000.60
            1856   35.40   0.00   0.00   2473.40
            1857   29.10   0.00   0.00   2202.20
            1858   30.40   0.00   0.00   1875.00
            1859   28.60   0.00   0.00   1940.80
            1860   36.90   0.00   0.00   2750.10
            1861   32.20   0.00   0.00   2373.50
            1862   69.10   0.00   0.00   4365.60
            1863   77.10   0.00   0.00   4983.50
            1864   36.30   0.00   0.00   2224.90
            1865   40.60   0.00   0.00   2134.10
            1866   23.10   0.00   0.00   1019.10
            1873   33.90   0.00   0.00   2550.40
            1874   No detailed return
            Comment 1860 SPEARN CONSOLS; 1862 TWO RETURNS AGGREGATED;
            1873 SPEARN CONSOLS; 1874 SEE SPEARNE MOOR
Ownership: 1876-1877 SPEARNE MOOR MINING CO.
            Comment 1877 INC.SPEARNE MOOR SUSPENDED SEPT.1877
Management: Manager 1859-1861 JOHN CARTHEW; 1862-1864 WM.TREMBATH;
            1869-1873 JOHN WALLIS; 1874-1877 JAS.BENNETTS
            Chief Agent 1859-1861 WM.TREMBATH; 1865 THOS.BENNETTS;
            1867-1868 JOHN WALLIS; 1874-1877 CHAS.ELLIS
            Secretary 1859-1861 RICH.PEARCE (P); 1862-1865 JAS.COULSON;
            1867-1873 RICH.WHITE; 1874-1877 ED.TRYTHALL

Production: Copper   Ore(tons) Metal(tons)   Value(£)
            1862   7.00   1.30   112.20
            1866   3.00   1.20   72.10
            1872   10.00   1.10   94.80
            1874   7.00   1.70   121.50
            Comment 1862 (C); 1866 (C); 1872 (C); 1874 (C)

Tin	Black(tons)	Stuff(tons)	Tin(tons)	Value(£)
1854	45.50	0.00	0.00	2962.40
1855	61.70	0.00	0.00	3959.10
1856	44.70	0.00	0.00	3200.60
1857	35.70	0.00	0.00	2725.00
1858	51.60	0.00	0.00	3219.20
1859	44.00	0.00	0.00	3215.70
1860	81.80	0.00	0.00	6393.10
1861	85.90	0.00	0.00	6188.20
1862	92.60	0.00	0.00	5962.80
1863	77.10	0.00	0.00	4877.30
1864	75.20	0.00	0.00	4546.40
1865	91.60	0.00	0.00	5863.80
1866	89.50	0.00	0.00	4214.00
1867	84.90	0.00	0.00	4316.40
1868	82.00	0.00	0.00	4338.20
1869	75.40	0.00	0.00	5025.50
1870	80.30	0.00	0.00	5575.10
1871	78.20	0.00	0.00	5724.40
1872	80.70	0.00	0.00	6318.00
1873	73.50	0.00	0.00	4954.40
1874	94.90	0.00	0.00	4797.10
1875	78.50	0.00	0.00	3611.30
1876	47.60	0.00	0.00	1844.20
1877	7.30	0.00	0.00	235.40
1878	1.00	0.00	0.00	35.50
1879	0.90	0.00	0.00	29.80

            Comment 1858-1861 SPEARN MOOR; 1862 TWO RETURNS AGGREGATED;
            1864-1869 SPEARN MOOR; 1873 SPEARN MOOR; 1874 UNITED WITH
            SPEARNE CONSOLS; 1875-1877 SPEARN MOOR; 1878 SPEARN MOOR MINE
            CLOSED; 1879 SPEARN MOOR
            Arsenic   Ore(tons) Metal(tons)   Value(£)
            1875   7.40   0.00   35.00
            Comment 1875 SPEARN MOOR
Ownership: 1876-1878 SPEARNE MOOR MINING CO.; 1879 SPEARN MOOR CO.
            Comment 1877 SEE SPEARNE CONSOLS
Management: Manager 1859 THOS.ANTHONY; 1860-1877 JAS.BENNETTS; 1879
            JAS.BENNETTS
            Chief Agent 1859 CHAS.ELLIS; 1860 CHAS.ELLIS RICH.JAMES;
            1861-1876 CHAS.ELLIS; 1878 ED.TRYTHALL
            Secretary 1859-1870 SML.HIGGS (P); 1871-1872 SML.HIGGS & SON;
            1873-1876 ED.TRYTHALL
Employment: Underground   Surface   Total
            1878   3   3

SPEED,SOUTH   ST.IVES   SW 523381 1172

Production: Copper   Ore(tons) Metal(tons)   Value(£)
            1851   105.00   5.20   319.30
            Comment 1851 (C)
            Tin   Black(tons) Stuff(tons)   Tin(tons)   Value(£)
            1852   3.40   0.00   0.00   0.00
            1854   34.70   0.00   0.00   2241.00
            1855   15.90   0.00   0.00   982.00
            Comment 1852 TIN ORE; 1854 TWO RETURNS AGGREGATED

SPEEDWELL   CUBERT   1173

Production: Iron   Ore(tons)   Iron(%)   Value(£)
            1858   5466.80   0.00   1573.80
            1859-1861 No detailed return
            Comment 1858-1861 BH.

SPEEDWELL   ST.HILARY   SW 559284 1174

Production: Copper   Ore(tons) Metal(tons)   Value(£)
            1852   436.00   29.50   2525.00
            1853   333.00   23.50   2369.30
            1854   108.00   6.50   662.50
            Comment 1852-1854 (C)
            Tin   Black(tons) Stuff(tons)   Tin(tons)   Value(£)
            1853   0.40   0.00   0.00   21.50
            1872   0.00   0.00   0.00   6.10
            Comment 1872 TIN STUFF

SPERRIES   ZENNOR   SW 469382 1176

Production: Zinc   Ore(tons) Metal(tons)   Value(£)
            1872   18.00   0.00   36.00
            Tin   Black(tons) Stuff(tons)   Tin(tons)   Value(£)
            1855   0.90   0.00   0.00   51.70

SQUIRE   ST.ERTH   SW 557342 1177

Production: Copper   Ore(tons) Metal(tons)   Value(£)
            1852   58.00   6.70   573.90
            1853   82.00   8.50   909.70
            Comment 1852-1853 (C)

ST.AGNES CONSOLS   ST.AGNES   SW 723515 1178

Production: Copper   Ore(tons) Metal(tons)   Value(£)
            1846   527.00   24.80   1491.50
            1847   No detailed return
            1876   48.00   4.50   343.70
            1877   21.00   2.70   147.50
            Comment 1846-1847 (C); 1876 (C); 1877 NEW ST. AGNES

ST.AGNES CONSOLS   ST.AGNES   Continued

        Tin   Black(tons) Stuff(tons)   Tin(tons)   Value(£)
        1872   2.90   0.00   0.00   232.50
        1873   11.00   0.00   0.00   866.30
        1874   10.60   50.00   0.00   429.70
        1875   1.20   0.00   0.00   57.10
        1876   3.40   0.00   0.00   129.50
        1877   1.30   0.00   0.00   48.50
        1911   15.00   0.00   0.00   1713.00
        Comment 1875 1 ORE (\52) RET. AS W.DIST.; 1877 NEW ST.AGNES;
        1911 ST.AGNES CONSOLIDATED
Ownership: 1877 NEW ST.AGNES MINING CO.
        Comment 1872 FORMERLY ROCK; 1875 IN LIQUIDATION; 1877 NEW
        ST.AGNES
Management: Manager 1872-1876 JOS.VIVIAN
        Chief Agent 1872-1877 WM.VIVIAN
        Secretary 1872 JAS.EVANS; 1873-1876 REYNOLDS & CO.; 1877
        REYNOLDS

ST.ANDREW   GWITHIAN   SW 596403 1179

Production: Copper   Ore(tons) Metal(tons)   Value(£)
        1845   370.00   14.90   826.30
        1846-1847 No detailed return
        Comment 1845-1847 (C)
        Tin No detailed return

ST.AUBYN   ST.DAY   SW 711419 1180

Production: Tin   Black(tons) Stuff(tons)   Tin(tons)   Value(£)
        1901   0.00   308.00   0.00   23.00
        Comment 1901 ST.AUBYN & OTHERS
        Fluorspar   Ore(tons)   Value(£)
        1892   19.00   10.00
        1893   111.00   45.00
        1894   94.00   23.00
        Comment 1892 OPENWORKS; 1893 OPENWORKS, SUNDRY WORKERS; 1894
        OPENWORKS, SUNDRY WORKERS. INC.WEST DAMSEL
Ownership: 1909-1913 MARC.RUTHENBURG
        Comment 1910-1913 NOT WORKED
Employment: Underground   Surface   Total
        1909   2   2

ST.AUBYN AND GRYLLS   BREAGE   SW 568290 1181

Production: Copper   Ore(tons) Metal(tons)   Value(£)
        1850   80.00   8.50   578.80
        1851   110.00   9.40   639.80
        1852   75.00   6.20   564.80
        1853   180.00   14.80   1386.90
        1854   116.00   6.80   689.70
        1856   93.00   5.40   464.70
        1857   151.00   9.20   880.50

                                     449

ST.AUBYN AND GRYLLS   BREAGE   Continued

```
Copper          Ore(tons) Metal(tons)    Value(£)
1858             216.00      18.00        1682.40
1859              94.00       7.40         738.30
1863               4.00       0.20          16.30
```
Comment 1850—1854 (C); 1856—1859 (C); 1863 (C)ST.AUBYN'S &
GRYLLS
```
Tin             Black(tons) Stuff(tons)   Tin(tons)   Value(£)
1854               3.50       0.00          0.00       224.90
1855              30.40       0.00          0.00      1935.70
1856              45.60       0.00          0.00      3205.80
1857              80.40       0.00          0.00      6160.70
1858              99.60       0.00          0.00      6396.50
1859             121.40       0.00          0.00      8555.40
1860             142.10       0.00          0.00     11109.80
1861             121.30       0.00          0.00      8688.90
1862              66.90       0.00          0.00      4366.70
1863               0.00       0.00          0.00       464.30
```
Comment 1858—1859 ST.AUBYNS & GRYLLS; 1862 VALUE INC.TIN
STUFF
Ownership: 1912—1913 ATLAS BORING CO.LTD.
             Comment 1868—1871 SEE GREAT WESTERN MINES; 1912 OPENING
             SHAFT; 1913 CLEANING ADIT NOT WORKED
Management: Manager 1859—1862 JOHN CURTIS
             Chief Agent 1859—1862 JOHN DAVEY
             Secretary 1859—1862 WM.VAWDREY (P)

ST.AUBYN UNITED   ST.DAY   SW 714442 1182

Production: Copper   Ore(tons) Metal(tons)   Value(£)
            1871   4.00   0.60   37.30
            1872   2.00   0.30   23.90
            1873   221.00   20.80   1286.90
            1874   410.00   33.60   2263.90
            1875   280.00   21.80   1674.90
            1876   375.00   29.20   1975.40
            1877   54.00   4.30   284.80
            1879   15.00   1.50   83.50
            1880   11.40   0.90   63.70
            1881   6.70   0.50   29.20
            Comment 1871—1876 (C)
            Tin   Black(tons) Stuff(tons)   Tin(tons)   Value(£)
            1871   0.00   0.00   0.00   2714.10
            1872   0.00   0.00   0.00   2607.60
            1873   0.00   3116.80   0.00   2066.20
            1874   25.00   226.00   0.00   1427.00
            1875   0.00   546.80   10.20   169.40
            1876   0.00   379.00   7.50   95.20
```
            Comment 1874 TIN STUFF EST
Ownership:  1877—1884 ST.AUBYN UNITED MINES CO.
Management: Manager 1869—1876 JOHN MICHELL; 1877 WM.MICHELL; 1878—1879
            JOHN MICHELL; 1881 JOHN JENNINGS
            Chief Agent 1877 JOHN MICHELL; 1880 F.W.DABB JOHN JENNINGS;
            1882—1884 JOHN JENNINGS
            Secretary 1869—1880 F.W.DABB; 1881 F.W.DABB (P)
```

ST.AUBYN UNITED   ST.DAY   Continued

Employment: Underground   Surface   Total
              1878   18   5   23
              1879   25   5   30
              1880   22   8   30
              1881   25   8   33
              1882   23   9   32
              1883   21   8   29
              1884   6   7   13

ST.AUSTELL   ST.AUSTELL   1183

Production: Iron   Ore(tons)   Iron(%)   Value(£)
              1858   742.00   0.00   222.60
              1859   816.50   0.00   244.60
              1860   1452.50   0.00   435.70
              1861   214.00   0.00   53.50
              1865   191.80   0.00   57.60
              1866   379.00   0.00   113.70
            Comment 1858-1861 BH.; 1865-1866 BH.
Ownership: 1863-1865 JOS.MORCOM
Management: Chief Agent 1864-1865 WM.BUTTON

ST.AUSTELL CONSOLS   ST.STEPHEN-IN-BRANNEL   SW 967512 1184

Production: Lead   Ore(tons) Metal(tons)   Value(£)
              1860   2.40   1.80   0.00
              1861   No detailed return
            Zinc No detailed return
            Copper   Ore(tons) Metal(tons)   Value(£)
              1846   75.00   5.00   318.80
              1862   17.00   0.90   65.40
              1863   11.00   0.60   42.30
            Comment 1846 (C); 1862-1863 (C)
            Tin   Black(tons) Stuff(tons)   Tin(tons)   Value(£)
              1854   6.00   0.00   0.00   322.40
              1855   19.70   0.00   0.00   1244.80
              1856   94.20   0.00   0.00   6881.30
              1857   132.20   0.00   0.00   10021.20
              1858   129.30   0.00   0.00   8152.60
              1859   159.90   0.00   0.00   11485.10
              1860   110.90   0.00   0.00   8701.80
              1861   111.60   0.00   0.00   7996.30
              1862   84.80   0.00   0.00   5586.50
              1875   23.00   0.00   0.00   1177.90
              1876   51.20   0.00   0.00   2323.90
              1877   87.50   0.00   0.00   3480.20
              1878   62.40   0.00   0.00   2168.70
              1879   1.50   0.00   0.00   60.30
            Comment 1854-1856 ST.AUSTLE CONSOLS
            Nickel   Ore(tons) Metal(tons)   Value(£)
              1854   75.00   0.00   0.00
              1855   39.00   0.00   0.00
              1857   1.00   0.00   50.60

CM-Q*   451

Nickel	Ore(tons)	Metal(tons)	Value(£)
1858	1.50	0.00	78.60
1859	2.70	0.00	121.60
1860	3.30	0.00	134.40
1861	0.80	0.00	23.50

Comment 1854—1855 INC. COBALT

Arsenic	Ore(tons)	Metal(tons)	Value(£)
1859	1.80	0.00	2.10

Comment 1859 CRUDE

Uranium	Ore(tons)	Value(£)
1858	1.20	21.70
1859	0.10	10.50
1863	0.20	23.00

Ownership: 1875—1877 ST.AUSTELL CONSOLS ADVENTURERS; 1878—1881
ST.AUSTELL CONSOLS MINE CO.; 1899—1901 RICH.H.WILLIAMS
Comment 1862—1865 SUSPENDED
Management: Manager 1859—1861 RICH.WILLIAMS; 1876—1881 JOHN NICHOLLS
Chief Agent 1859—1864 JOHN TRUSCOTT; 1875 JOHN NICHOLLS;
1876—1881 SAMP.HAMMER
Secretary 1859—1862 JOHN WILLIAMS (P); 1863 JOHN H.WILLIAMS

Employment:	Underground	Surface	Total
1878	58	56	114
1879	18	14	32
1899	2	2	4

ST.BLAZEY   ST.BLAZEY   SX 066553 1185

Production: Tin No detailed return
Ownership: 1875—1881 NEW ST.BLAZEY CO.
Comment 1864—1865 SEE PAR
Management: Chief Agent 1875—1881 PHIL.RICH

ST.BLAZEY CONSOLS   ST.BLAZEY   SX 066553 1186

Production: Tin	Black(tons)	Stuff(tons)	Tin(tons)	Value(£)
1852	9.70	0.00	0.00	0.00
1855	4.50	0.00	0.00	136.70

Comment 1852 TIN ORE
Ownership: Comment 1873 STOPPED IN 1873 & EARLY IN 1874
Management: Manager 1873 PHIL.RICH
Secretary 1872 PHIL.RICH

ST.BLAZEY,NEW   ST.BLAZEY   SX 066553 1187

Production: Tin	Black(tons)	Stuff(tons)	Tin(tons)	Value(£)
1874	2.10	0.00	0.00	95.50
1875	2.50	0.00	0.00	105.00

Ownership: 1876—1879 NEW ST.BLAZEY CO.
Management: Chief Agent 1874—1879 PHIL.RICH

ST.BREWARD CONSOLS   ST.BREWARD   SX 096777 1188

Production: Copper No detailed return
Ownership: 1876 ST.BREWARD MINING CO.; 1877-1878 ST.BREWARD MINING
            CO.LTD.
            Comment 1876-1878 ST.BREWARD GREAT ONSLOW
Management: Chief Agent 1872-1873 THOS.DUNN; 1876-1878 THOS.DUNN
Employment: Underground   Surface   Total
            1878   1   1

ST.COLOMB   ST.COLOMB   1190

Production: Iron   Ore(tons)   Iron(%)   Value(£)
            1858   96.10   0.00   39.50
            1872   290.00   0.00   217.00
            Comment 1858 BH.; 1872 BH.

ST.DAY HALVANERS   ST.DAY   9998

Production: Arsenic   Ore(tons) Metal(tons)   Value(£)
            1893   7.00   0.00   80.00
            Comment 1893 POLDICE,UNITY WOOD,MAID & GORLAND.WASTE

ST.DAY MANOR   ST.DAY   1191

Production: Tin   Black(tons) Stuff(tons)   Tin(tons)   Value(£)
            1888   28.10   2131.50   0.00   1653.00
            1889   18.50   2588.00   0.00   878.00
            Comment 1888-1889 INC.POLDICE,GORLAND,UNITY NO.1 & MAID
            Fluorspar   Ore(tons)   Value(£)
            1888   10.00   5.00
            1889   No detailed return
            Comment 1888 VARIOUS OWNERS; 1889 VARIOUS OWNERS

ST.DAY UNITED   ST.DAY   SW 735425 1192

Production: Copper   Ore(tons) Metal(tons)   Value(£)
            1852   260.00   16.80   1551.30
            1853   1236.00   82.60   8831.10
            1854   2103.00   128.30   13049.80
            1855   3175.00   230.80   24008.80
            1856   2954.00   199.20   18905.80
            1857   2165.00   142.60   14638.90
            1858   1841.00   102.20   9178.80
            1859   2542.00   137.40   12698.50
            1860   2205.00   110.70   9599.60
            1861   1678.00   101.90   9167.40
            1862   733.00   32.80   2435.50
            1863   483.00   20.10   1372.20
            1864   392.00   18.10   1453.50
            1865   360.00   20.60   1562.20
            1866   231.00   13.50   885.30
            1867   69.00   4.10   256.70

                                  453

Comment 1852-1853 (C)GREAT DAY UNITED; 1854-1867 (C)

Tin	Black(tons)	Stuff(tons)	Tin(tons)	Value(£)
1855	135.60	0.00	0.00	7463.90
1856	140.30	0.00	0.00	8173.10
1857	154.20	0.00	0.00	9929.00
1858	172.30	0.00	0.00	9702.10
1859	165.90	0.00	0.00	11099.40
1860	180.60	0.00	0.00	13849.90
1861	289.10	0.00	0.00	18904.30
1862	436.60	0.00	0.00	24768.70
1863	475.20	0.00	0.00	27355.10
1864	480.40	0.00	0.00	25238.50
1865	462.40	0.00	0.00	23046.50
1866	257.90	0.00	0.00	10922.30
1867	39.50	0.00	0.00	1796.40
1894	0.00	113.00	0.00	32.00
1895	No detailed return			

Comment 1862 TWO RETURNS AGGREGATED; 1894-1895 ST.DAY UNITED
(WATERS & TEAGUE)

Fluorspar	Ore(tons)	Value(£)
1860	48.90	48.80

Ownership: 1892-1894 WATERS & TEAGUE(P)
              Comment 1869-1871 SEE POLDICE; 1892-1893 PAYNTERS TO BILLINGS
              SHAFT; 1894 HOLMANS TO BILLINGS SHAFT
Management: Manager 1859-1863 FRAN.PRYOR
              Chief Agent 1859 JOHN GILBERT; 1860-1862 JOHN RALPH; 1863
              EL.RALPH; 1864-1865 JOS.COCK JOHN GILBERT
              Secretary 1859-1860 ED.KING & F.PRYOR(P); 1861-1862
              FRAN.PRYOR (P); 1863-1865 ED.KING

Employment:	Underground	Surface	Total
1892-1894	2		2

ST.DENNIS CONSOLS   ST.DENNIS   SW 594577 1193

Production: Tin	Black(tons)	Stuff(tons)	Tin(tons)	Value(£)
1860	56.70	0.00	0.00	4029.00
1861	13.00	0.00	0.00	982.20

Ownership: Comment 1860-1865 SUSPENDED
Management: Manager 1859 ?RICHARD
              Chief Agent 1874 THOS.PARKYN
              Secretary 1859-1865 J.TAYLOR & SONS (P)

ST.DENNIS CROWN   ST.DENNIS   SW 595575 1194

Production: Tin	Black(tons)	Stuff(tons)	Tin(tons)	Value(£)
1901	0.40	0.00	0.00	29.00
1903	12.00	0.00	0.00	1017.00
1904	7.00	0.00	0.00	549.00
1905	2.50	0.00	0.00	203.00

Ownership: 1901-1905 ST.DENNIS CROWN CHINA CLAY CO.LTD.; 1906 ST.DENNIS
              CROWN MINING CO.LTD.
              Comment 1906 ABANDONED MARCH 1907
Management: Chief Agent 1901-1906 A.E.CLOW

ST.DENNIS CROWN   ST.DENNIS   Continued

Employment:

	Underground	Surface	Total
1901	4	9	13
1902	24	27	51
1903	33	24	57
1904	35	28	63
1905	10	18	28
1906	8	17	25

ST.ERTH ALLUVIALS   ST.ERTH   SW 656318 1195

Production: Tin

	Black(tons)	Stuff(tons)	Tin(tons)	Value(£)
1913	57.00	0.00	0.00	6515.00

Comment 1913 OPEN WORK

ST.GEORGE   PERRANPORTH   SW 749515 1196

Production: Lead & Silver No detailed return
Ownership: 1879-1882 J.BRODERICK & J.G.PARKER
           Comment 1879-1880 NOT WORKED
Management: Manager 1875-1878 J.G.PARKER
           Chief Agent 1875-1878 JAS.EVANS
           Secretary 1875-1878 J.BRODERICK; 1881-1882 J.G.PARKER
Employment:

	Underground	Surface	Total
1878	8	2	10
1881	10	1	11
1882		13	13

ST.GEORGE,GREAT   PERRANPORTH   SW 746535 1197

Production: Copper

	Ore(tons)	Metal(tons)	Value(£)
1865	10.00	0.30	25.20

Comment 1865 (C)

Tin

	Black(tons)	Stuff(tons)	Tin(tons)	Value(£)
1875	0.50	0.00	0.00	25.10
1886	6.30	0.00	0.00	224.00
1887	7.50	0.00	0.00	337.00
1888	15.10	0.00	0.00	834.00
1889	18.90	0.00	0.00	865.00
1890	12.80	0.00	0.00	617.00
1891	8.10	0.00	0.00	396.00
1892	4.50	0.00	0.00	95.00
1893	2.50	0.00	0.00	120.00
1894	1.90	0.00	0.00	43.00
1899	6.20	0.00	0.00	267.00

Comment 1875 INC.DROSKIN; 1886 WESTERN PART; 1887-1889
INC.DROSKIN; 1890-1894 INC.DROSKIN W.PORTION; 1899
INC.DROSKIN

Tungsten

	Ore(tons)	Metal(tons)	Value(£)
1899	1.50	0.00	33.00

Comment 1899 INC.DROSKIN
Ownership: 1891-1893 THOS.MICHELL & OTHERS; 1899 NOBELS EXPLOSIVES
           CO.LTD.

455

ST.GEORGE,GREAT   PERRANPORTH   Continued

Comment 1887 GT.ST.GEORGE & DROSKIN UTD.; 1891-1893
GT.ST.GEORGE & DROSKIN UTD.; 1894 ABANDONED; 1899 GREAT
ST.GEORGE & DROSKYN
Management: Chief Agent 1891-1893 THOS.MICHELL; 1899 JOS.TURNER
Secretary 1891 THOS.MICHELL

Employment:	Underground	Surface	Total
1891	3	3	6
1892		3	3
1893		1	1
1899	9	1	10

Comment 1891-1893 INC.DROSKIN; 1899 INC.DROSKIN

ST.GEORGE,WEST   PERRANZABULOE   SW 736528 1198

Production: Copper No detailed return

Tin	Black(tons)	Stuff(tons)	Tin(tons)	Value(£)
1867	152.00	6.90	0.00	436.30
1868	65.00	3.10	0.00	169.10

Comment 1867 (C); 1868 (C)WEST GREAT ST.GEORGE

ST.ISSEY   ST.ISSEY   SW 955717 1199

Production: Lead No detailed return
Copper No detailed return
Management: Manager 1860-1872 JOS.HARRIS
Chief Agent 1870-1872 MART.RICKARD
Secretary 1860-1872 JOS.HARRIS (P)

ST.IVES CONSOLS   ST.IVES   SW 506397 1200

Production: Copper	Ore(tons)	Metal(tons)	Value(£)
1856	27.00	6.40	649.10
1859	27.00	5.50	567.60
1863	14.00	1.70	145.90
1870	12.00	1.90	127.20
1873	37.00	3.40	247.20
1874	7.00	1.40	95.40
1875	3.00	0.50	42.10

Comment 1856 (C); 1859 (C); 1863 (C); 1870 (C); 1873-1875
(C)

Tin	Black(tons)	Stuff(tons)	Tin(tons)	Value(£)
1853	61.00	0.00	0.00	3506.80
1854	201.00	0.00	0.00	11500.00
1855	208.90	0.00	0.00	12460.30
1856	212.60	0.00	0.00	13852.20
1857	206.90	0.00	0.00	14732.40
1858	246.90	0.00	0.00	14810.30
1859	340.60	0.00	0.00	24010.40
1860	293.70	0.00	0.00	22870.20
1861	249.90	0.00	0.00	17343.20
1862	331.10	0.00	0.00	20480.40
1863	279.80	0.00	0.00	17534.60

Tin	Black(tons)	Stuff(tons)	Tin(tons)	Value(£)
1864	251.50	0.00	0.00	14906.90
1865	207.90	0.00	0.00	10700.50
1866	245.20	0.00	0.00	11576.80
1867	256.40	0.00	0.00	13493.90
1868	255.80	0.00	0.00	13737.40
1869	253.00	0.00	0.00	17190.80
1870	181.90	0.00	0.00	12796.40
1871	176.00	0.00	0.00	13187.70
1872	157.60	0.00	0.00	13109.30
1873	184.40	0.00	0.00	13827.20
1874	138.30	0.00	0.00	7499.50
1875	152.90	0.00	0.00	7952.00
1876	72.70	0.00	0.00	2889.00
1877	67.20	0.00	0.00	2521.00
1878	66.60	0.00	0.00	2132.60
1879	62.00	0.00	0.00	2252.90
1880	56.70	0.00	0.00	2618.50
1881	52.60	0.00	34.00	2755.00
1882	57.80	0.00	37.50	3468.00
1883	54.30	0.00	0.00	2878.00
1884	54.80	0.00	0.00	2424.00
1885	49.30	0.00	0.00	2264.00
1886	40.10	0.00	0.00	2126.00
1887	24.00	0.00	0.00	1360.00
1888	9.90	0.00	0.00	596.00
1889	5.70	0.00	0.00	233.00
1890	5.10	0.00	0.00	234.00
1891	4.70	0.00	0.00	210.00
1892	4.00	0.00	0.00	158.00
1896	3.00	0.00	0.00	65.00
1911	3.00	0.00	0.00	332.00
1912	20.00	0.00	0.00	2344.00
1913	18.00	0.00	0.00	1962.00

Comment 1862 TWO RETURNS AGGREGATED; 1896 OPEN WORK

Ownership: 1877–1892 GEO.TREWEEKE; 1909–1913 ST.IVES CONSOLIDATED MINES LTD.
Comment 1877–1881 ABOVE ADIT & HEAPS ONLY

Management: Manager 1859 ED.BAWDEN; 1860–1864 J.NANCARROW; 1865–1871
RICH.MARTIN; 1872–1874 JOHN GILBERT; 1879–1881 THOS.MICHELL
Chief Agent 1859 THOS.MICHELL; 1860–1864 RICH.MARTIN;
1865–1868 JOSH.DANIELL; 1869–1871 MART.GEORGE H.POLLARD
JAS.EVANS; 1872–1874 RICH.MARTIN MART.GEORGE; 1875–1878
THOS.MICHELL; 1882–1888 THOS.MICHELL; 1910–1912
H.P.ROBERTSON; 1913 F.C.CANN
Secretary 1859 WM.DARKE (P); 1860–1874 PHIL.APLIN (P);
1875–1881 GEO.TREWEEKE; 1889–1890 GEO.TREWEEKE

Employment:	Underground	Surface	Total
1878	3	63	66
1879	9	60	69
1880	11	56	67
1881	20	47	67
1882	24	52	76
1883	31	43	74
1884	27	36	63

ST.IVES CONSOLS   ST.IVES   Continued

	Underground	Surface	Total
1885	19	34	53
1886	20	32	52
1887	14	21	35
1888	12	18	30
1889		5	5
1890		4	4
1891		7	7
1892		6	6
1909	20	11	31
1910	54	61	115
1911	36	67	103
1912	61	62	123
1913	27	59	86

ST.IVES WHEAL ALLEN   ST.IVES   SW 499399 1201

Production: Copper

	Ore(tons)	Metal(tons)	Value(£)
1868	2.00	0.50	35.50

Comment 1868 (C)

Tin

	Black(tons)	Stuff(tons)	Tin(tons)	Value(£)
1862	19.50	0.00	0.00	1202.20
1863	23.20	0.00	0.00	1460.30
1864	20.90	0.00	0.00	1226.60
1865	32.30	0.00	0.00	1671.80
1866	15.70	0.00	0.00	729.70
1867	13.00	0.00	0.00	535.50
1868	0.80	0.00	0.00	45.40

Comment 1862 TWO RETURNS AGGREGATED
Ownership: Comment 1859 ST.IVES WHEAL ELLEN
Management: Manager 1863-1866 J.NANCARROW
Chief Agent 1859 HY.TAYLOR; 1860 R.GEO.T.TAYLOR HY.TAYLOR;
1861 RICH.TAYLOR HY.TAYLOR; 1862 RICH.GEORGE HY.TAYLOR;
1863-1865 A.S.BRYANT JNR; 1866 J.DANIELL
Secretary 1859-1862 THOS.RICHARDS (P); 1863-1866
THOS.W.ROBINSON

ST.IVES,WEST   ST.IVES   SW 482404 1202

Production: Tin No detailed return
Management: Chief Agent 1866-1871 THOS.UREN
Secretary 1867-1868 J.R.REYNOLDS; 1869-1871 CHRIS.STEPHENS

ST.JUST AMALGAMATED   ST.JUST   SW 352316 1203

Production: Tin

	Black(tons)	Stuff(tons)	Tin(tons)	Value(£)
1869	180.50	0.00	0.00	12619.40
1870	162.80	0.00	0.00	11837.30
1871	149.30	0.00	0.00	11531.50
1872	141.70	0.00	0.00	12035.20
1873	152.50	0.00	0.00	5660.80
1874	151.60	0.00	0.00	9009.10

ST.JUST AMALGAMATED   ST.JUST   Continued

	Tin Black(tons)	Stuff(tons)	Tin(tons)	Value(£)
1875	101.40	0.00	0.00	4872.50
1876	8.60	0.00	0.00	360.80
1877	25.10	0.00	0.00	986.60

Ownership: 1877-1878 ST.JUST AMALGAMATED MINING CO.
              Comment 1874 CLOSED 10TH JULY 1874; 1875 SUSPENDED
Management: Manager 1868-1873 RICH.PRYOR; 1874 RICH.PRYOR & SON;
              1876-1878 JAS.BENNETTS
              Chief Agent 1868 J.WHITE; 1869 THOS.GUNDRY NICH.BARTLE;
              1870-1872 THOS.RICHARDS NICH.BARTLE; 1873-1874 THOS.RICHARDS
              WM.BAWDEN; 1875 JAS.BENNETTS CHAS.CLEMENS; 1876-1878
              CHAS.CLEMENS
              Secretary 1869-1874 HY.L.PHILLIPS; 1875-1878 RICH.BOYNS
Employment: Underground   Surface   Total
              1878   26   16   42

ST.JUST CONSOLS   ST.JUST   SW 358298 1204

Production: Copper No detailed return

	Tin Black(tons)	Stuff(tons)	Tin(tons)	Value(£)
1864	1.60	0.00	0.00	116.80
1867	No detailed return			

              Comment 1864 3 MONTHS ONLY; 1867 SEE CAPE CORNWALL
Management: Manager 1869 W.WILLIAMS
              Chief Agent 1860 JOHN CARTHEW
              Secretary 1860 JOHN CARTHEW (P); 1869 W.AUGWIN

ST.JUST UNITED   ST.JUST   SW 352316 1205

Production: Copper No detailed return

	Tin Black(tons)	Stuff(tons)	Tin(tons)	Value(£)
1862	28.00	0.00	0.00	1852.70
1863	158.70	0.00	0.00	11198.80
1864	223.30	0.00	0.00	14268.20
1865	206.20	0.00	0.00	11390.00
1878	27.50	0.00	0.00	979.50
1879	20.10	0.00	0.00	806.00
1880	36.60	0.00	0.00	1937.10
1881	83.90	0.00	55.60	4802.60
1882	188.90	0.00	122.90	10960.00
1883	218.80	0.00	0.00	11888.00
1884	270.00	0.00	0.00	12900.00
1885	157.50	0.00	0.00	7768.00
1886	98.30	0.00	0.00	5757.00
1887	67.00	0.00	0.00	4394.00
1888	50.00	0.00	0.00	3555.00
1893	0.50	0.00	0.00	28.00
1894	8.00	0.00	0.00	320.00
1895-1896	No detailed return			
1897	3.00	0.00	0.00	100.00
1898	No detailed return			
1899	0.60	0.00	0.00	40.00
1900	8.00	0.00	0.00	600.00

Tin	Black(tons)	Stuff(tons)	Tin(tons)	Value(£)
1901	12.30	0.00	0.00	900.00
1902	15.00	0.00	0.00	1050.00
1903	15.00	0.00	0.00	1000.00
1904	8.00	0.00	0.00	560.00
1905	6.00	0.00	0.00	553.00
1906	6.50	0.00	0.00	600.00
1907	8.70	0.00	0.00	947.00
1908	13.40	0.00	0.00	1037.00
1909	16.80	0.00	0.00	1350.00

Comment 1862 PART YEAR ONLY; 1893 JOHN MERRIFIELD & OTHERS;
1894-1895 ST.JUST UNITED CONSOLS; 1899-1909 ST.JUST UNITED
CONSOLS

Ownership: 1879 ST.JUST AMALGAMATED MINING CO.; 1880-1888 ST.JUST UNITED
MINING CO.; 1893 JOHN MERRIFIELD & OTHERS; 1894-1909
JAS.CHENHALLS
Comment 1867-1868 ST.JUST UNITED LIM; 1893 TRIBUTERS STOPPED
WORK IN MAY; 1894-1896 ST.JUST UNITED CONSOLS BELLAN PART;
1897 ST.JUST UNITED CONSOLS BELLAN PART SUSPENDED OCTOBER;
1898 ST.JUST UNITED CONSOLS BELLAN PART SUSPENDED OCTOBER
1897; 1899-1902 ST.JUST UNITED CONSOLS BELLAN PART; 1903-1909
ST.JUST UNITED CONSOLS

Management: Manager 1864-1865 RICH.PRYOR; 1879-1881 JAS.BENNETTS
Chief Agent 1861-1863 JOHN CARTHEW; 1864-1865 W.R.RUTTER
R.WEARNE WM.WHITE; 1867 R.WEARNE RALPH P.GOLDSWORTHY; 1868
T.TREGLOWN; 1879-1881 CHAS.CLEMENS; 1887-1888 JAS.BENNETTS
Secretary 1861-1863 JOHN CARTHEW (P); 1864-1865 WM.AUGWIN;
1867-1868 RICH.PRYOR; 1879 RICH.BOYNS; 1882-1886 RICH.BOYNS;
1893 JAS.CHENHALLS

Employment:

	Underground	Surface	Total
1879	28	31	59
1880	74	59	133
1881	96	50	146
1882	123	50	173
1883	131	59	190
1884	145	70	215
1885	108	48	156
1886	98	29	127
1887	70	30	100
1888	39	16	55
1893	21	3	24
1894	14	5	19
1895	20	6	26
1896	17	5	22
1897	1	1	2
1899	14	4	18
1900	10	4	14
1901	11	18	29
1902	10	14	24
1903	12	6	18
1904	4	7	11
1905	2	6	8
1906	4	8	12
1907	6	6	12
1908	10	8	18

ST.JUST UNITED   ST.JUST   Continued

                    Underground   Surface   Total
          1909   7   8   15
          Comment 1893 TWO RETURNS AGGREGATED

ST.JUST UNITED,EAST   ST.JUST   SW 359308 1206

Production: Tin   Black(tons) Stuff(tons)   Tin(tons)   Value(£)
          1867   19.70   0.00   0.00   969.20
Ownership: Comment 1865 SEE BOSORNE
Management: Manager 1866 RICH.PRYOR
           Chief Agent 1866 R.WEARNE WM.WHITE
           Secretary 1866 WM.AUGWIN

ST.MICHAEL PENKEVILL   1207

Production: Tin   Black(tons) Stuff(tons)   Tin(tons)   Value(£)
          1853   4.80   0.00   0.00   244.70
          1855   30.90   0.00   0.00   1739.80
          Comment 1855 ST.MICHAEL PENVIL

ST.NEOT   ST.NEOT   SX 184646 1208

Production: Tin No detailed return
Ownership: 1907-1911 ST.NEOT MINING SYNDICATE LTD.
           Comment 1911 ABANDONED SEPT.1911 IN LIQUIDATION
Management: Manager 1869-1872 JOHN GLUYAS
           Secretary 1869-1872 RICH.CLOGG; 1909-1911 B.BRYANT
Employment: Underground   Surface   Total
          1907   9   2   11
          1908   15   1   16
          1909   7   5   12
          1910   10   3   13

ST.STEPHEN   LAUNCESTON   1209

Production: Manganese   Ore(tons) Metal(tons)   Value(£)
          1867-1868 No detailed return
          1873   17.00   0.00   114.80
          1874   10.00   0.00   35.50
          Comment 1867-1868 SEE CHILLATON (DEVON)

ST.STEPHENS   ST.STEPHEN-IN-BRANNEL   1210

Production: Tin   Black(tons) Stuff(tons)   Tin(tons)   Value(£)
          1873   3.70   0.00   0.00   320.30
          1874   16.90   0.00   0.00   983.50
          Plumbago No detailed return
Ownership: 1906 KINGSWAY SYNDICATE LTD.
           Comment 1873 ST.STEPHENS TIN
Management: Manager 1873-1875 JOHN NICHOLLS

                              461

ST.STEPHENS   ST.STEPHEN-IN-BRANNEL   Continued

           Secretary 1873-1875 WM.WARD; 1906 M.MONROE (S)
Employment: Underground   Surface   Total
           1906   3   3   6

ST.STEPHENS   ST.AUSTELL   1211

Production: Iron   Ore(tons)   Iron(%)   Value(£)
            1874   400.00   0.00   320.00
            Comment 1874 BH.
Ownership: 1874-1876 ST.STEPHENS IRON ORE CO.
Management: Manager 1874-1876 G.B.SANDEMAN

ST.VINCENT   ALTERNUN   SX 207795 1213

Production: Iron No detailed return
Ownership: 1880-1881 ST.VINCENT IRON ORE CO.
Management: Chief Agent 1880-1881 JOHN DINGLE

STAMPING MILLS   ST.AGNES   1214

Production: Tin   Black(tons) Stuff(tons) Tin(tons)   Value(£)
            1859   10.10   0.00   0.00   594.70
            1860   9.70   0.00   0.00   652.50
            1861   0.80   0.00   0.00   61.80

STANLEY   1215

Production: Tin   Black(tons) Stuff(tons) Tin(tons)   Value(£)
            1852   1.00   0.00   0.00   0.00
            Comment 1852 TIN ORE

STANNACH WHITE   1216

Production: Tin   Black(tons) Stuff(tons) Tin(tons)   Value(£)
            1880   1.00   0.00   0.00   57.80

STENCOOSE   ST.AGNES   SW 709458 1217

Production: Copper No detailed return
            Tin No detailed return
Ownership: 1912-1913 HY.GRIPE
            Comment 1860-1861 STENCOOSE & MAWLA UNITED; 1862-1865
            STENCOOSE & MAWLA UNITED SUSPENDED; 1912-1913 REOPENING JULY
            1912
Management: Chief Agent 1860-1861 NICH.REED
            Secretary 1860-1865 MOSES BAWDEN (P)
Employment: Underground   Surface   Total
            1912   4   4
            1913   2   2

                                    462

Production: Tin No detailed return
Management: Chief Agent 1863-1865 RICH.H.WILLIAMS
            Secretary 1863-1865 RICH.H.WILLIAMS

STRAY PARK   CAMBORNE   SW 654398 1219

Production: Copper   Ore(tons) Metal(tons)   Value(£)
            1857   184.00   5.70   435.40
            1860   653.00   46.50   4223.40
            1861   622.00   40.70   3530.00
            1862   445.00   26.40   2094.90
            1863   320.00   17.20   1265.50
            1864   85.00   3.50   278.50
            1866   11.00   0.60   34.90
            1867   8.00   0.40   30.00
            Comment 1857 (C); 1860-1864 (C); 1866-1867 (C)
            Tin   Black(tons) Stuff(tons)   Tin(tons)   Value(£)
            1853   8.50   0.00   0.00   533.60
            1855   27.80   0.00   0.00   1763.60
            1856   14.30   0.00   0.00   1023.60
            1857   8.20   0.00   0.00   726.20
            1861   38.60   0.00   0.00   2819.20
            1862   98.50   0.00   0.00   6704.30
            1863   59.60   0.00   0.00   4038.80
            1864   66.50   0.00   0.00   4261.30
            1865   87.80   0.00   0.00   4996.80
            1866   56.80   0.00   0.00   2971.30
            1867   10.60   0.00   0.00   559.10
            1868   15.60   0.00   0.00   850.00
            1869   44.80   0.00   0.00   3170.50
            1870   15.40   0.00   0.00   1081.90
            Comment 1857 INC.CAMBORNE VEAN; 1862 TWO RETURNS AGGREGATED
Ownership: Comment 1870 SEE DOLCOATH; 1871-1872 INCORPORATED WITH
            DOLCOATH; 1873 AMALGAMATED WITH NORTH DOLCOATH; 1875 UNITED
            WITH DOLCOATH
Management: Manager 1859-1861 CHAS.THOMAS; 1862-1866 CHAS.THOMAS & SON;
            1867-1869 JOH.THOMAS; 1873-1874 JOH.THOMAS
            Chief Agent 1859-1860 RICH.PRYOR; 1861-1866 EDM.ROGERS
            WM.SKEWIS; 1867 WM.SKEWIS; 1868-1869 W.C.COCK
            Secretary 1859-1869 ELIAS DUNSTERVILLE; 1873-1874 DOLCOATH
            COMMITTEE

STRAY PARK,WEST   CAMBORNE   SW 647393 1220

Production: Copper   Ore(tons) Metal(tons)   Value(£)
            1854   96.00   7.20   739.50
            1855   127.00   9.40   972.10
            1856   164.00   12.00   1138.20
            1857   310.00   23.50   2406.50
            1858   306.00   20.90   1889.90
            1859   381.00   29.40   2758.20
            1860   377.00   27.50   2626.90
            1861   485.00   37.80   3537.30

STRAY PARK,WEST   CAMBORNE   Continued

Copper   Ore(tons) Metal(tons)   Value(£)
1862   461.00   36.10   3060.20
1863   438.00   34.90   2761.70
1864   209.00   16.90   1469.80
1865   152.00   12.80   1044.70
1866   93.00   7.90   616.20
1868   5.00   0.20   11.50
1875   2.00   0.10   6.00
Comment 1854-1866 (C); 1868 (C); 1875 (C)
Tin   Black(tons) Stuff(tons)   Tin(tons)   Value(£)
1855   2.20   0.00   0.00   188.00
1856   0.00   0.00   0.00   146.20
1857   0.00   0.00   0.00   27.80
1858   0.00   0.00   0.00   124.40
1859   0.00   0.00   0.00   288.00
1860   0.00   0.00   0.00   399.70
1861   0.00   0.00   0.00   276.30
1862   0.00   0.00   0.00   217.50
1863   0.00   0.00   0.00   332.70
1864   0.00   0.00   0.00   227.50
1865   0.00   0.00   0.00   243.20
1866   0.00   0.00   0.00   58.50
Comment 1856-1861 TIN STUFF; 1862 TIN STUFF. TWO RETURNS
AGGREGATED; 1863-1866 TIN STUFF
Management: Manager 1859 JOS.PURAN; 1860-1861 JOS.VIVIAN; 1862-1865
J.VIVIAN W.THOMAS
Chief Agent 1859-1860 HY.BERRYMAN; 1861 WM.THOMAS
HY.BERRYMAN; 1862-1865 NICH.THOMAS
Secretary 1859-1863 ROBT.H.PIKE (P); 1864-1865 ROBT.H.PIKE &
SON

SUNNY CORNER   ST.DAY   1221

Production: Tin   Black(tons) Stuff(tons)   Tin(tons)   Value(£)
            1907-1911 No detailed return
            Comment 1907-1911 SEE POLDORY
Ownership: Comment 1906-1912 SEE POLDORY (ST.DAY)
Employment: Underground   Surface   Total
            1906-1912
            Comment 1906-1912 SEE POLDORY (ST.DAY)

SWANPOOL   BUDOCK   SW 801312 1222

Production: Lead & Silver   Ore(tons) Lead(tons) Silver(ozs)   Value(£)
            1854   370.00   226.00   5650.00   0.00
            1855   476.00   276.00   6624.00   0.00
            1856   320.20   162.50   4875.00   0.00
            1857   1561.00   790.00   654.00   0.00
            1858   2400.00   480.00   9600.00   0.00
            1859   706.10   141.20   2824.00   0.00
            1860   189.40   37.80   756.00   0.00
Ownership: Comment 1862-1865 SUSPENDED
Management: Manager 1859 PHILLIPS & DARLINGTON

464

Chief Agent 1859-1861 JOHN KITTO
Secretary 1859 J.A.PHILLIPS (P)

SYDNEY COVE   BREAGE   SW 577282 1223

Production: Copper No detailed return
Ownership: Comment 1880-1881 SEE MOUNTS BAY CONSOLS

SYDNEY GODOLPHIN   BREAGE   SW 579286 1224

Production:	Tin	Black(tons)	Stuff(tons)	Tin(tons)	Value(£)
	1853	15.20	0.00	0.00	1061.00
	1886	21.70	0.00	0.00	1107.00
	1887	10.90	0.00	0.00	723.00
	1888	34.80	0.00	0.00	2078.00
	1889	8.00	0.00	0.00	383.00
	1906	6.10	0.00	0.00	525.00
	1907	3.00	0.00	0.00	277.00
	1908	0.40	0.00	0.00	19.00

Comment 1906-1908 SIDNEY GODOLPHIN

	Arsenic	Ore(tons)	Metal(tons)	Value(£)
	1889	2.00	0.00	4.00

Ownership: 1887-1889 SYDNEY GODOLPHIN MINE CO.; 1905-1907 PRAA
SYNDICATE; 1908-1912 GRYLLS MINES LTD.; 1913 PRAA SYNDICATE
Comment 1887-1888 FORMERLY MOUNTS BAY MINE; 1905-1907 SIDNEY
GODOLPHIN PENGERSICH; 1910-1911 NOT WORKED
Management: Chief Agent 1887-1888 WM.ARGALL & SON; 1905-1912
FRAN.BARRATT
Secretary 1913 J.TREEBY BARRATT

Employment:		Underground	Surface	Total
	1887	30	25	55
	1888	40	42	82
	1889	6	14	20
	1905	2	11	13
	1906	4	15	19
	1907	6	14	20
	1908		2	2
	1911-1913		1	1

TALLACK   ST.AGNES   SW 707483 1225

Ownership: 1878 RICHARDS,POWER & CO.
Management: Chief Agent 1878 WM.VIVIAN

Employment:		Underground	Surface	Total
	1878	2		2

TAMAR RIVER   CALSTOCK   SX 417723 1226

Production:	Arsenic No detailed return		
	Arsenic Pyrite	Ore(tons)	Value(£)
	1897	130.00	173.00

TAMAR RIVER   CALSTOCK   Continued

          Arsenic Pyrite   Ore(tons)   Value(£)
          1898   104.00   305.00
Ownership: 1897-1898 BROGAN & MALLOCK
             Comment 1898 ABANDONED
Management: Chief Agent 1897-1898 I.JAMES
Employment: Underground   Surface   Total
          1897   12   8   20
          1898   6   4   10

TAMAR SLAG   1227

Production: Copper   Ore(tons) Metal(tons)   Value(£)
            1852   90.00   5.40   419.00
            Comment 1852 (C)

TAMAR WORKS   1228

Production: Zinc   Ore(tons) Metal(tons)   Value(£)
            1872   10.00   0.00   20.00

TAMAR,CONSOLIDATED   CALSTOCK   1229

Production: Tin   Black(tons) Stuff(tons) Tin(tons)   Value(£)
            1912   0.00   369.00   0.00   222.00
            1913   2.60   0.00   0.00   278.00
            Arsenic   Ore(tons) Metal(tons)   Value(£)
            1912   40.10   0.00   635.00
            Arsenic Pyrite Ore(tons)   Value(£)
            1910   500.00   500.00
            1911   1150.00   1260.00
            1912   1778.00   1524.00
            1913   35.00   29.00
Ownership: 1909-1910 H.DAVIS; 1911-1913 CONSOLIDATED TAMAR MINES LTD.
Management: Chief Agent 1909-1913 JOS.CARTER
Employment: Underground   Surface   Total
          1909   8   6   14
          1910   42   31   73
          1911   26   56   82
          1912   16   10   26
          1913   12   8   20

TAMAR,NEW SOUTH   LANDULPH   SX 433629 1230

Production: Lead & Silver No detailed return
Ownership: Comment 1864-1865 SUSPENDED
Management: Chief Agent 1861-1863 THOS.TREVILLION WM.BENNETTS
             Secretary 1861 WM.WYMOND (P); 1862-1865 RICH.CLOGG

TAMPING QUARRY   1231

Production: Tin   Black(tons) Stuff(tons)   Tin(tons)   Value(£)
            1874   0.00   120.00   0.00   90.00

TAMWORTH   ST.NEOT   SX 204670 1232

Ownership: 1907 ASSOCIATED TAMWORTH MINES LTD.
            Comment 1908 SEE KILLHAM
Management: Secretary 1907 DAVID BLOTTON (S)

TEHIDY   ILLOGAN   SW 682417 1233

Production: Copper   Ore(tons) Metal(tons)   Value(£)
            1855   131.00   11.90   1240.50
            1856   305.00   25.20   2375.20
            1857   339.00   31.20   3182.80
            1858   246.00   18.00   1729.20
            1859   91.00   5.40   508.70
            Comment 1855-1859 (C)
            Tin   Black(tons) Stuff(tons)   Tin(tons)   Value(£)
            1858   0.60   0.00   0.00   2.10
            1859   0.00   0.00   0.00   1.50
            1860   0.00   0.00   0.00   0.80
            Comment 1859-1860 TIN STUFF
Ownership: Comment 1865 SUSPENDED
Management: Manager 1859-1860 JOHN POPE; 1862-1864 JOHN POPE
            Chief Agent 1859-1861 JOHN POPE
            Secretary 1859-1860 GEO.SWAN (P); 1861-1865 JOHN POPE (P)

TEHIDY TIN STREAMS   ILLOGAN   SW 685420 9997

Production: Arsenic Pyrite Ore(tons)   Value(£)
            1884   10.00   10.00   .

TERRACE HILL   LOSTWITHIEL   1234

Production: Lead & Silver No detailed return
Management: Chief Agent 1861 F.SQUIRE
            Secretary 1861 F.SQUIRE (P)

TERRAS   ST.STEPHEN-IN-BRANNEL   SW 932528 1236

Production: Tin   Black(tons) Stuff(tons)   Tin(tons)   Value(£)
            1873   1.50   0.00   0.00   90.60
            1884   5.00   0.00   0.00   226.00
            1887   7.00   0.00   0.00   420.00
            1888   16.20   0.00   0.00   975.00
            1889   10.70   0.00   0.00   506.00
            1890   1.70   0.00   0.00   90.00
            Comment 1884-1890 NEW TERRAS
Ownership: 1882-1886 NEW TERRAS MINING CO.LTD.; 1887-1890 NEW TERRAS TIN

MINING CO.LTD.; 1891–1892 JOHN H.H.JAMES; 1893 F.L.FRICKER
Comment 1873 STOPPED IN 1873 & IN EARLY 1874; 1874 STOPPED;
1882–1889 NEW TERRAS; 1890 NEW TERRAS. IN LIQUIDATION.
SUSPENDED; 1891–1892 NEW TERRAS.SUSPENDED
Management: Chief Agent 1869 J.W.B.DAINTY; 1870–1871 MART.RICKARD;
1873–1874 RICH.LARCHIN; 1882–1884 JOHN H.JAMES; 1885–1886
R.EADE
Secretary 1870–1874 W.F.PEARCE; 1887 A.G.ONSLOW; 1888–1889
JOHN N.LAMB

Employment:

	Underground	Surface	Total
1882	9	2	11
1883	6	20	26
1884	11	14	25
1885	5	20	25
1886	6	14	20
1887	14	24	38
1888	24	40	64
1889–1890	21	22	43
1891–1892		1	1
1893	2	1	3

TERRAS,SOUTH   ST.STEPHEN—IN—BRANNEL   SW 935524 1237

Production: Tin

	Black(tons)	Stuff(tons)	Tin(tons)	Value(£)
1873	2.30	0.00	0.00	175.00
1881	4.20	0.00	2.80	235.00

Comment 1873 VALUE EST

Iron

	Ore(tons)	Iron(%)	Value(£)
1873	680.00	0.00	510.00
1874	450.00	0.00	412.50
1876	500.00	0.00	375.00
1877	No detailed return		
1880	3000.00	0.00	1800.00
1881	793.00	0.00	475.80

Comment 1873–1874 BH.; 1876 MO.; 1877 NO ORE SOLD; 1880
BH.471 TONS ONLY WERE SOLD; 1881 BH.
Ochre No.detailed return
Ownership: 1877–1878 SOUTH TERRAS MINING CO.LTD.; 1879 SOUTH TERRAS
MINING CO.; 1880–1888 SOUTH TERRAS TIN MINING CO.LTD.; 1913
SOCIETE INDUSTRIELLE DU RADIUM LTD.
Comment 1913 NOT WORKED
Management: Chief Agent 1874–1875 JOHN H.JAMES; 1877 JOHN H.JAMES;
1879–1883 JOHN H.JAMES; 1885–1888 JOHN FRAZER
Secretary 1874–1875 RICH.LARCHIN; 1878 JOHN H.JAMES

Employment:

	Underground	Surface	Total
1878	8	7	15
1879	6	5	11
1880	24	14	38
1881	13	10	23
1882	11	3	14
1883	8	3	11
1884	6	8	14
1885–1886	10	3	13
1887	6	2	8

TERRAS,SOUTH   ST.STEPHEN—IN—BRANNEL   Continued

	Underground	Surface	Total
1888		11	11

THOMAS   PORT ISAAC   1238

Ownership: 1913 J.B.MAY
Comment 1913 OPENING ADIT NOT WORKED 1913
Employment:

	Underground	Surface	Total
1913	1		1

TIN ERA   ST.CLEER   SX 209747 1239

Production: Tin

	Black(tons)	Stuff(tons)	Tin(tons)	Value(£)
1887	2.00	0.00	0.00	120.00
1888	No detailed return			

Comment 1887 FORMERLY GOODEVERE
Ownership: 1888—1891 GOODEVERE MINING CO.LTD.
Comment 1887 SEE GOODEVERE; 1891 ABANDONED
Management: Chief Agent 1888—1891 R.KNOTT
Secretary 1888 R.CUMMING; 1889—1891 R.CUMMING (S)
Employment:

	Underground	Surface	Total
1888	6		6
1889	7	1	8
1890	7		7

TIN HILL   ST.STEPHEN   SW 957543 1240

Production: Tin

	Black(tons)	Stuff(tons)	Tin(tons)	Value(£)
1869	1.10	0.00	0.00	68.00
1870	1.10	0.00	0.00	83.00
1872	0.10	0.00	0.00	6.50
1881	1.80	0.00	1.10	94.50

Ownership: 1881—1882 TIN HILL MINING CO.LTD.
Management: Chief Agent 1881 GEO.H.EUSTICE
Secretary 1882 GEO.H.EUSTICE
Employment:

	Underground	Surface	Total
1881	17	20	37

TIN VALLEY   ST.ENODER   1241

Ownership: 1912—1913 ATLAS BORING CO.LTD.
Comment 1912 OPENING SHAFT DEC.1912; 1913 NOT WORKED
Employment:

	Underground	Surface	Total
1912	8	3	11

TIN VALLEY   ST.NEOT   SX 187672 1242

Production: Tin

	Black(tons)	Stuff(tons)	Tin(tons)	Value(£)
1870	1.90	0.00	0.00	119.70

TIN VALLEY   ST.NEOT   Continued

             Arsenic   Ore(tons) Metal(tons)   Value(£)
             1870   11.50   0.00   32.00
             Comment 1870 CRUDE
Ownership: 1869-1871 TIN VALLEY CO.
Management: Manager 1869-1871 RICH.SOUTHEY
             Chief Agent 1872-1875 J.COCKING
             Secretary 1872-1875 THOS.WOODWARD

TINCROFT   ILLOGAN   SW 670409 1243

Production: Lead   Ore(tons) Metal(tons)   Value(£)
            1859   22.00   15.00   0.00
            Copper   Ore(tons) Metal(tons)   Value(£)
            1845   5644.00   412.30   30627.50
            1846   5550.00   364.60   24897.20
            1847   5473.00   333.10   22371.60
            1848   4252.00   257.50   13968.00
            1849   4765.00   281.70   17051.70
            1850   7012.00   398.20   25390.30
            1851   7827.00   474.00   29706.30
            1852   9559.00   515.40   39867.40
            1853   8765.00   370.20   33103.10
            1854   3791.00   207.40   18097.10
            1855   4480.00   187.70   17242.30
            1856   3331.00   163.40   13895.80
            1857   2374.00   124.10   11637.50
            1858   2256.00   125.80   11033.00
            1859   1820.00   86.90   7513.50
            1860   1718.00   71.10   5541.20
            1861   2118.00   88.40   6940.80
            1862   1993.00   85.90   6163.80
            1863   714.00   34.80   2473.90
            1864   440.00   22.90   1865.30
            1865   468.00   24.90   1882.80
            1866   303.00   15.70   934.40
            1867   76.00   4.60   320.00
            1868   249.00   16.40   1064.00
            1869   122.00   7.80   470.20
            1870   40.00   2.40   129.00
            1872   122.00   6.10   473.10
            1876   104.00   5.60   352.20
            1878   116.30   4.00   212.20
            1882   47.50   14.50   262.00
            1883   223.00   14.50   840.00
            1884   159.00   0.00   415.00
            1885   111.00   0.00   258.00
            1886   174.00   0.00   437.00
            1887   186.00   0.00   412.00
            1888   792.00   0.00   4646.00
            1889   923.00   0.00   2742.00
            1890   477.00   0.00   1664.00
            1891   401.00   0.00   1582.00
            1892   246.00   0.00   583.00
            1893   505.00   0.00   1339.00

                              470

Copper	Ore(tons)	Metal(tons)	Value(£)
1894	634.00	0.00	1672.00
1895	658.00	0.00	1745.00
1896–1911 No detailed return			
1913	No detailed return		

Comment 1845–1870 (C); 1872 (C); 1876 (C); 1892–1893 INC.CARN
BREA; 1896–1911 SEE CARN BREA; 1913 SEE CARN BREA

Tin	Black(tons)	Stuff(tons)	Tin(tons)	Value(£)
1852	159.00	0.00	0.00	0.00
1853	169.80	0.00	0.00	11114.20
1854	146.20	0.00	0.00	8973.70
1855	225.70	0.00	0.00	13853.30
1856	150.70	0.00	0.00	10401.20
1857	178.00	0.00	0.00	13514.40
1858	159.30	0.00	0.00	9797.30
1859	254.40	0.00	0.00	16782.70
1860	220.80	0.00	0.00	16017.40
1861	242.10	0.00	0.00	16843.30
1862	361.00	0.00	0.00	23690.60
1863	359.60	0.00	0.00	24105.50
1864	471.40	0.00	0.00	29771.70
1865	359.40	0.00	0.00	20152.40
1866	498.50	0.00	0.00	24848.50
1867	44.30	0.00	0.00	2324.60
1868	268.80	0.00	0.00	14919.40
1869	392.10	0.00	0.00	28016.60
1870	787.60	0.00	0.00	59134.50
1871	820.80	0.00	0.00	65948.50
1872	829.40	0.00	0.00	74206.30
1873	693.00	0.00	0.00	53574.60
1874	685.90	0.00	0.00	38350.00
1875	730.30	0.00	0.00	35923.90
1876	794.40	0.00	0.00	33936.60
1877	795.90	0.00	0.00	31887.10
1878	754.10	0.00	0.00	25896.10
1879	510.50	0.00	0.00	20418.00
1880	617.80	0.00	0.00	30987.30
1881	547.50	0.00	364.00	29397.00
1882	415.00	0.00	259.00	22941.00
1883	455.00	0.00	0.00	23500.00
1884	456.70	0.00	0.00	19706.00
1885	504.70	0.00	0.00	22848.00
1886	543.50	0.00	0.00	29604.00
1887	574.10	0.00	0.00	36188.00
1888	476.70	0.00	0.00	30233.00
1889	694.70	0.00	0.00	37199.00
1890	893.90	0.00	0.00	47619.00
1891	1031.10	0.00	0.00	50894.00
1892	878.30	0.00	0.00	41219.00
1893	904.60	0.00	0.00	40847.00
1894	864.60	0.00	0.00	31680.00
1895	789.10	0.00	0.00	25578.00
1896–1913 No detailed return				

Comment 1852 TIN ORE; 1862 TWO RETURNS AGGREGATED; 1896–1913
SEE CARN BREA

Arsenic	Ore(tons)	Metal(tons)	Value(£)
1857	86.00	0.00	176.20
1858	53.60	0.00	110.30
1859	58.80	0.00	120.30
1860	51.30	0.00	93.10
1861	70.00	0.00	112.50
1865	100.00	0.00	75.00
1875	11.20	0.00	25.50
1877	71.30	0.00	230.40
1880	46.20	0.00	0.00
1882	69.00	0.00	366.00
1883	68.00	0.00	375.00
1884	69.50	0.00	382.00
1885	47.70	0.00	262.00
1886	28.10	0.00	147.00
1887	102.00	0.00	561.00
1888	83.00	0.00	506.00
1889	202.00	0.00	1271.00
1890	227.00	0.00	1499.00
1891	268.00	0.00	2017.00
1892	288.00	0.00	2029.00
1893	360.00	0.00	2816.00
1894	404.00	0.00	3694.00
1895	553.00	0.00	5314.00

1896—1913 No detailed return
Comment 1859—1861 CRUDE; 1865 CRUDE; 1884—1885 SOOT; 1886
CRUDE; 1896—1913 SEE CARN BREA

Tungsten	Ore(tons)	Metal(tons)	Value(£)

1896—1898 No detailed return
1901   No detailed return
1903—1913 No detailed return
Comment 1896—1898 SEE CARN BREA; 1901 SEE CARN BREA;
1903—1913 SEE CARN BREA

Ownership: 1876—1890 TINCROFT MINING CO.; 1891 TINCROFT MINING CO.CB;
1892—1895 TINCROFT MINE ADVENTURERS CB
Comment 1895 INC.COOKS KITCHEN; 1896—1913 SEE CARN BREA

Management: Manager 1859—1865 WM.TEAGUE; 1866 W.TEAGUE J.JEWELL;
1867—1871 WM.TEAGUE; 1872—1881 WM.TEAGUE & SON
Chief Agent 1859—1861 JAS.ANDREWS; 1862—1865 JAS.ANDREW
WM.TEAGUE JNR; 1866 JAS.ANDREWS WM.TEAGUE JNR ANT.BRAY;
1867—1868 JAS.ANDREWS WM.TEAGUE JNR; 1869—1871 JAS.ANDREWS
SML.MARTIN WM.TEAGUE JNR; 1872—1880 JAS.ANDREWS WM.ANDREWS
SML.MARTIN; 1881 JOHN HAMMILL SML.MARTIN; 1882—1890
WM.TEAGUE; 1892—1895 WM.TEAGUE
Secretary 1859—1862 HIR.WILLIAMS (P); 1863—1881 WM.TEAGUE;
1887—1891 F.W.DABB; 1892—1895 F.W.DABB (P)

Employment:	Underground	Surface	Total
1878	173	317	490
1879	227	263	490
1880	188	272	460
1881	182	256	438
1882	241	255	496
1883	236	263	499
1884	198	248	446
1885	190	271	461

TINCROFT   ILLOGAN   Continued

	Underground	Surface	Total
1886	181	283	464
1887	191	276	407
1888	222	289	511
1889	235	298	533
1890	291	314	605
1891	297	325	622
1892	291	314	605
1893	285	327	612
1894	273	316	589
1895	291	352	643
1896-1913			

Comment 1896-1913 SEE CARN BREA

TINCROFT CONSOLS,OLD   TOWEDNACK   SW 500361 1245

Production: Tin

	Black(tons)	Stuff(tons)	Tin(tons)	Value(£)
1875	0.00	15.00	0.00	18.10
1876	4.50	0.00	0.00	187.20
1877	0.00	270.40	0.00	102.70
1878	0.00	199.40	0.00	100.80
1879	0.00	34.20	0.00	28.00

Comment 1876 TIN CROFT CONSOLS; 1877-1879 NEW TINCROFT UNITED
Ownership: 1876-1881 JOHN B.REYNOLDS & CO.
            Comment 1875 OLD TINCROFT; 1907 SEE BALDHU (LUDGVAN)
Management: Manager 1875-1876 JAS.POPE; 1877-1878 JOS.S.VIVIAN
            Chief Agent 1875-1876 STEP.POPE; 1877-1878 WM.VIVIAN; 1879
            WM.VIVIAN JOS.S.VIVIAN; 1880-1881 WM.VIVIAN
            Secretary 1875 JOHN B.REYNOLDS; 1877-1878 JOHN B.REYNOLDS
Employment:

	Underground	Surface	Total
1878-1879	4		4
1907			

Comment 1907 SEE BALDHU (LUDGVAN)

TINDENE   BREAGE   SW 572314 1247

Production: Tin

	Black(tons)	Stuff(tons)	Tin(tons)	Value(£)
1887	5.00	0.00	0.00	340.00
1888	17.10	0.00	0.00	1026.00
1889	39.30	0.00	0.00	2312.00
1890	47.00	0.00	0.00	2866.00
1891	48.70	0.00	0.00	2878.00
1892	51.80	0.00	0.00	3011.00
1893	7.30	0.00	0.00	316.00

Arsenic

	Ore(tons)	Metal(tons)	Value(£)
1889	3.00	0.00	7.00
1891	3.00	0.00	10.00
1892	4.00	0.00	7.00
1893	1.00	0.00	4.00

Ownership: 1882 TINDENE MINING CO.; 1886-1889 TINDENE ADVENTURERS; 1890
            TINDENE CO.LTD.; 1891-1893 TINDENE MINE CO.LTD.
            Comment 1892 SUSPENDED; 1893 ABANDONED

TINDENE   BREAGE   Continued

Management: Chief Agent 1882 JOHN POPE; 1886—1891 WM.STEPHENS; 1892—1893
            JOHN CURTIS
            Secretary 1887 JOHN CARTER; 1888—1890 JOHN CURTIS; 1891—1893
            ED.ASHMEAD (S)
Employment: Underground   Surface   Total
            1882   12   3   15
            1886   3   6   9
            1887   28   19   47
            1888   32   24   56
            1889   37   26   63
            1890   50   23   73
            1891   53   28   81
            1892   39   21   60
            1893   3   3   6

TINDENE,SOUTH   BREAGE   SW 571305 1248

Production: Tin   Black(tons) Stuff(tons) Tin(tons)   Value(£)
            1891   0.00   23.00   0.00   11.00
Ownership: 1890—1891 SOUTH TINDENE CO.
            Comment 1891 SUSPENDED
Management: Chief Agent 1890—1891 JOHN CURTIS
Employment: Underground   Surface   Total
            1890—1891   2   2

TING TANG CONSOLS   ST.DAY   SW 730410 1249

Production: Copper   Ore(tons) Metal(tons)   Value(£)
            1845   100.00   5.90   408.50
            1846   414.00   24.70   1643.70
            1847   327.00   19.10   1329.80
            Comment 1845—1847 (C)

TOKENBURY   ST.IVE   SX 291708 1250

Management: Secretary 1859 E.A.CROUCH (P); 1860—1861 WM.TAYLOR (P)

TOLCARNE   CAMBORNE   SW 655381 1252

Production: Copper   Ore(tons) Metal(tons)   Value(£)
            1860   528.00   28.10   2504.30
            1861   476.00   30.90   2713.70
            1862   602.00   35.90   2837.10
            1863   1065.00   56.00   4126.10
            1864   1021.00   50.20   5901.70
            1865   791.00   40.50   2998.20
            1866   733.00   41.30   2692.70
            1867   381.00   19.30   1266.80
            1868   7.00   0.30   20.80
            1871   18.00   0.80   42.60
            1872   6.00   0.20   14.10

474

TOLCARNE   CAMBORNE   Continued

Comment 1860—1868 (C); 1871—1872 (C)
Tin   Black(tons) Stuff(tons)   Tin(tons)   Value(£)
1861   0.00   0.00   0.00   780.00
1862   2.00   0.00   0.00   1447.80
1865   0.00   0.00   0.00   162.00
Comment 1861 TIN STUFF; 1862 " VALUE INC.TIN STUFF.3 RTNS.
Ownership: Comment 1866 SEE ROSECLIFF; 1867—1872 SEE PENDARVES UNITED
Management: Manager 1859—1863 JOS.JEWELL; 1864—1865 TAYLOR & SON JEWELL
             Chief Agent 1859—1865 ALEX.CHEGWIN
             Secretary 1859—1862 J.TAYLOR & SON (P); 1863 TAYLOR
             J.P.BENNETTS; 1864—1865 J.P.BENNETTS

TOLCARNE,NORTH   CAMBORNE   SW 660392 1253

Production: Copper   Ore(tons) Metal(tons)   Value(£)
1873   5.00   0.30   17.10
Comment 1873 (C)

TOLCARNE,SOUTH   CAMBORNE   SW 656386 1254

Production: Copper   Ore(tons) Metal(tons)   Value(£)
1872   22.00   1.00   63.90
1873   20.00   0.90   42.00
1874   5.00   0.20   13.00
1875   41.00   4.50   331.30
1876   27.00   1.40   83.80
1877   36.00   1.30   101.70
1878   37.30   2.10   120.20
1879   39.50   3.30   145.40
1883   40.90   3.50   258.00
1884   543.00   0.00   2515.00
1885   374.00   0.00   1185.00
1886   No detailed return
Comment 1872—1876 (C); 1886 SEE WEST CONDURROW
Tin   Black(tons) Stuff(tons)   Tin(tons)   Value(£)
1876   8.70   17.00   0.00   3979.00
1881   0.00   0.00   0.00   389.20
1883   127.50   0.00   0.00   7212.00
1884   64.80   0.00   0.00   3166.00
1885   12.00   168.50   0.00   523.00
Comment 1881 TIN STUFF
Ownership: 1876—1885 SOUTH TOLCARNE MINING CO.
Management: Manager 1873—1875 JOS.VIVIAN & SON; 1876—1879 WM.RICH
             Chief Agent 1872 JOS.VIVIAN & SON JAS.PAULL; 1873—1875
             JAS.PAULL; 1876—1877 WM.HAMBLY; 1878—1879 JAS.KNOTWELL;
             1880—1881 THOS.ANGOVE JAS.KNOTWELL; 1882—1884 THOS.ANGOVE;
             1885 JOHN JENNINGS
             Secretary 1874 HICKEY TAYLOR & SONS; 1875—1881 JAS.HICKEY
Employment: Underground   Surface   Total
1878   12   4   16
1879   10   6   16
1880   12   10   22
1881   20   14   34

TOLCARNE,SOUTH   CAMBORNE   Continued

              Underground   Surface   Total
      1882   50   72   122
      1883   81   107   188
      1884   58   16   74
      1885   35   13   48

TOLCARNE,WEST   CROWAN   SW 641373 1255

Production: Copper No detailed return
Ownership: Comment 1864-1865 SUSPENDED
Management: Manager 1862-1863 FRAN.PRYOR
            Chief Agent 1860-1861 FRAN.PRYOR JOHN JENKIN; 1862-1863 JOHN
            JENKIN
            Secretary 1860-1863 W.J.DUNSFORD (P)

TOLDIS   ST.COLUMB MAJOR   SW 923598 1256

Production: Iron No detailed return
            Ochre No detailed return
Ownership: 1899-1903 INDIAN QUEENS MINING & COLOUR CO.LTD.
            Comment 1903 SUSPENDED
Management: Manager 1873 J.S.COLLARD
            Chief Agent 1873 BORLASE
Employment: Underground   Surface   Total
      1899   3   7   10
      1900   2   1   3
      1901   5   5
      1902   8   8

TOLDISH   ST.ENODER   SW 911558 1269

Production: Tin No detailed return
            Arsenic Pyrite No detailed return
Ownership: 1909-1912 TOLDISH TIN MINE CO.; 1913 FRADDON MINING SYNDICATE
            LTD.
            Comment 1910 SUSPENDED OCT.1910; 1911 REOPENED NOV.1911;
            1912-1913 NOT WORKED
Employment: Underground   Surface   Total
      1909   3   3
      1910   4   4   8
      1912   2   2

TOLGARRICK   ST.STEPHEN   SW 935524 1257

Production: Arsenic   Ore(tons) Metal(tons)   Value(£)
      1884   9.50   0.00   34.00
            Comment 1884 SOOT
            Plumbago No detailed return
Ownership: Comment 1911-1913 SEE NEW CROW HILL

TOLGULLOW UNITED   GWENNAP   SW 735435 1258

Production: Copper   Ore(tons) Metal(tons)   Value(£)
            1885   3.00   0.00   5.00
            1886   8.00   0.00   30.00
            1887   6.00   0.00   18.00
            1888   3.00   0.00   17.00
            Tin   Black(tons) Stuff(tons)   Tin(tons)   Value(£)
            1885   70.10   0.00   0.00   3392.00
            1886   162.20   0.00   0.00   8999.00
            1887   171.70   0.00   0.00   10481.00
            1888   105.00   0.00   0.00   6845.00
            1889   26.70   0.00   0.00   1030.00
            1890   0.00   91.50   0.00   169.00
            1891   0.00   91.00   0.00   139.00
            1892   0.00   116.00   0.00   197.00
            1893   0.00   126.00   0.00   130.00
            1894   0.00   78.00   0.00   84.00
            1895   0.00   78.00   0.00   52.00
            1896   0.00   69.00   0.00   135.00
            1897   0.00   60.00   0.00   48.00
            1898   0.00   104.00   0.00   52.00
            1899   0.00   100.00   0.00   49.00
            1900   0.00   198.00   0.00   173.00
            1901   0.00   214.00   0.00   174.00
            1902   0.00   102.00   0.00   78.00
            1903   0.00   57.00   0.00   55.00
            1904   0.00   52.00   0.00   38.00
            1905   0.00   37.00   0.00   33.00
            1906   0.00   40.00   0.00   71.00
            1907   0.00   30.00   0.00   70.00
            1908   0.00   25.00   0.00   46.00
            1909   1.00   0.00   0.00   69.00
            1910   0.00   9.00   0.00   8.00
            Comment 1890-1892 TOLGULLOW; 1893 TOLGULLOW M.RICHARDS &
            OTHERS
            Arsenic   Ore(tons) Metal(tons)   Value(£)
            1885   39.70   0.00   227.00
            1886   70.10   0.00   399.00
            1887   47.00   0.00   318.00
            1888   29.00   0.00   175.00
            1889   28.00   0.00   73.00
            Comment 1885 SOOT; 1886 CRUDE
            Arsenic Pyrite Ore(tons) Value(£)
            1885   112.00   85.00
            1886   40.00   25.00
Ownership: 1885-1886 TOLGULLOW UNITED ADVENTURERS; 1887-1889 TOLGULLOW
            UNITED MINES LTD.; 1892-1896 MAT.RICHARDS & OTHERS; 1897
            WM.JEFFREY & OTHERS; 1898 WM.JEFFREY & RICH.LIGHT; 1899-1900
            RICH.LIGHT & SONS; 1901-1904 WM.JEFFREY,RICH.LIGHT & SONS;
            1905-1911 RICH.LIGHT & SONS
            Comment 1888 TOLGULLOW MINE; 1889 TOLGULLOW NOW ABANDONED;
            1892-1910 T.H.LETCHER LORDS AGENT; 1911 T.H.LETCHER LORDS
            AGENT ABANDONED JUNE 1911
Management: Chief Agent 1892-1911 T.H.LETCHER
            Secretary 1885-1886 HY.MICHELL; 1887-1888 EDG.TAYLOR

                                 477

TOLGULLOW UNITED   GWENNAP   Continued

Employment: Underground   Surface   Total
            1885   79   53   132
            1886   145   224   369
            1887   145   70   215
            1888   75   49   124
            1889   26   29   55
            1892–1896   4   4
            1897   1   1
            1898   4   4
            1899   2   2
            1900   2   2   4
            1901   4   4
            1902–1909   2   2
            1910   1   1

TOLGULLOW UNITED MINES   GWENNAP   SW 735435 1259

Production: Copper No detailed return
            Tin   Black(tons) Stuff(tons)   Tin(tons)   Value(£)
            1889   9.50   446.00   0.00   476.00
            Comment 1889 TOLGULLOW WEST POLDICE PORTION
            Arsenic   Ore(tons) Metal(tons)   Value(£)
            1889   15.00   0.00   37.00
            1892   8.00   0.00   50.00
            Comment 1889 WEST POLDICE SECTION; 1892 TOLGULLOW.POLDICE
            ADVENTURERS
Ownership: 1889–1891 TINNERS & HALVANERS
            Comment 1890–1891 ABANDONED
Management: Secretary 1889–1891 T.S.LETCHER
Employment: Underground   Surface   Total
            1889   10   15   25
            1890   8   8
            1891   3   3

TOLGUS   REDRUTH   SW 686431 1260

Production: Tin   Black(tons) Stuff(tons)   Tin(tons)   Value(£)
            1908   0.00   54.00   0.00   44.00

TOLGUS CONSOLS   REDRUTH   SW 669421 1261

Production: Copper No detailed return
Ownership: 1877–1880 TOLGUS CONSOLS CO.LTD.
Management: Manager 1877 NICH.CLYMO; 1878–1880 WM.C.VIVIAN
            Secretary 1877–1878 WM.C.VIVIAN
Employment: Underground   Surface   Total
            1878   9   4   13
            1879   6   2   8

TOLGUS UNITED,OLD   REDRUTH   SW 689435 1262

Production: Zinc   Ore(tons) Metal(tons)   Value(£)
            1859   0.00   0.00   17.00
            Copper   Ore(tons) Metal(tons)   Value(£)
            1862   18.00   0.80   59.90
            Comment 1862 (C)
Ownership: Comment 1865 SUSPENDED
Management: Manager 1859-1860 GEO.REYNOLDS; 1861-1864 WM.PASCOE
            Chief Agent 1859-1864 WM.GILBERT
            Secretary 1859-1860 WM.CHARLES (P); 1861 W.H.PASCOE (P);
            1862-1865 WM.CHARLES

TOLGUS,EAST   REDRUTH   SW 694426 1263

Production: Zinc   Ore(tons) Metal(tons)   Value(£)
            1862   11.10   0.00   11.10
            Copper   Ore(tons) Metal(tons)   Value(£)
            1856   124.00   7.00   650.00
            1857   230.00   14.30   1346.10
            1858   675.00   41.70   3866.90
            1859   644.00   36.80   3444.20
            1860   153.00   7.60   678.60
            1861   123.00   6.80   586.40
            1862   29.00   1.40   106.60
            1863   7.00   0.40   31.50
            Tin   Black(tons) Stuff(tons)   Tin(tons)   Value(£)
            1857   0.00   0.00   0.00   94.40
            1858   0.00   0.00   0.00   310.60
            1859   0.00   0.00   0.00   504.10
            1860   31.40   0.00   0.00   2590.30
            1861   17.90   0.00   0.00   1367.30
            1862   8.20   0.00   0.00   564.20
            1863   0.80   0.00   0.00   51.50
            1865   0.00   0.00   0.00   43.70
            Comment 1857-1859 TIN STUFF; 1862 TWO RETURNS AGGREGATED;
            1865 TIN STUFF
Ownership: Comment 1865 SUSPENDED
Management: Manager 1859-1865 JOS.JEWELL
            Chief Agent 1861-1863 ANT.BRAY; 1864 JOHN TAYLOR & SONS; 1865
            ANT.BRAY.
            Secretary 1859-1861 J.P.BENNETTS (P); 1862-1863 TAYLOR & SONS
            BENNETTS; 1864-1865 J.P.BENNETTS

TOLGUS,GREAT NORTH   REDRUTH   SW 688440 1264

Production: Copper   Ore(tons) Metal(tons)   Value(£)
            1862   5.00   0.20   12.60
            Comment 1862 (C)
Ownership: Comment 1861-1865 SUSPENDED
Management: Manager 1860 T.DALE
            Secretary 1860 SPARGO & COWLING (P); 1861-1865 H.COWLING

                                       479

TOLGUS,GREAT SOUTH   REDRUTH   SW 687422 1265

Production: Zinc

	Ore(tons)	Metal(tons)	Value(£)
1859	0.00	0.00	6.70

Copper

	Ore(tons)	Metal(tons)	Value(£)
1855	618.00	37.80	3817.80
1856	1052.00	72.50	6945.90
1857	3036.00	213.70	21182.90
1858	3265.00	239.60	22054.50
1859	3346.00	249.20	23571.40
1860	3297.00	250.40	23472.00
1861	1681.00	141.80	13076.80
1862	928.00	88.40	7566.10
1863	784.00	79.70	6458.90
1864	524.00	48.20	4220.00
1865	969.00	66.10	5255.60
1866	939.00	57.50	3667.80
1867	562.00	40.90	2911.40
1868	735.00	43.30	2793.50
1869	20.00	1.10	70.50
1871	5.00	0.20	10.50

Comment 1855–1869 (C); 1871 (C)

Tin

	Black(tons)	Stuff(tons)	Tin(tons)	Value(£)
1858	0.00	0.00	0.00	4.50
1859	0.00	0.00	0.00	144.20
1860	0.00	0.00	0.00	404.10
1861	0.00	0.00	0.00	724.80
1862	0.00	0.00	0.00	1476.00
1863	0.00	0.00	0.00	1199.60
1864	0.00	0.00	0.00	1977.00
1865	0.00	0.00	0.00	1930.00
1866	9.40	0.00	0.00	755.80
1867	14.60	0.00	0.00	766.80
1868	4.50	0.00	0.00	239.80
1869	66.00	0.00	0.00	4625.70
1870	35.30	0.00	0.00	2617.00

Comment 1858–1861 TIN STUFF; 1862 TIN STUFF. TWO RETURNS
AGGREGATED; 1863–1865 TIN STUFF; 1866 VALUE INC.TIN STUFF;
1869–1870 VALUE INC.TIN STUFF

Arsenic

	Ore(tons)	Metal(tons)	Value(£)
1867	5.80	0.00	12.40
1869	5.40	0.00	11.50
1870	3.50	0.00	9.10

Comment 1867 CRUDE; 1870 CRUDE

Management: Manager 1859–1868 JOHN DAWE; 1869 JOHN RODDA
Chief Agent 1859–1861 SIMON KNEEBONE; 1862–1869 JOHN DAWE JNR
SIMON KNEEBONE
Secretary 1859–1861 RICH.LYLE (P); 1862–1869 GEO.LIGHTLEY

TOLGUS,NORTH   REDRUTH   SW 693443 1266

Production: Tin No detailed return
Ownership: Comment 1872–1874 SEE TRELEIGH WOOD UNITED

TOLGUS,SOUTH   REDRUTH   SW 685426 1267

Production: Lead   Ore(tons) Metal(tons)   Value(£)
            1864   0.50   0.30   0.00
            Zinc   Ore(tons) Metal(tons)   Value(£)
            1865   8.00   0.00   4.00
            Copper   Ore(tons) Metal(tons)   Value(£)
            1847   262.00   20.20   1446.70
            1848   367.00   26.30   1516.10
            1849   962.00   83.30   5671.40
            1850   1803.00   173.70   12535.90
            1851   2427.00   208.80   14323.40
            1852   2521.00   210.40   18605.80
            1853   2522.00   161.70   15847.20
            1854   1361.00   114.10   11870.90
            1855   1157.00   102.30   10842.50
            1856   1867.00   156.60   14991.60
            1857   1975.00   168.60   17777.20
            1858   2780.00   222.20   20182.90
            1859   2721.00   224.10   21497.60
            1860   2516.00   204.70   19594.50
            1861   2281.00   172.30   15844.80
            1862   2480.00   175.70   14742.40
            1863   2379.00   163.60   12854.40
            1864   2310.00   154.70   13158.50
            1865   1205.00   67.50   5271.30
            1866   620.00   36.70   2649.60
            1867   82.00   3.00   180.70
            Comment 1847-1867 (C)
            Tin   Black(tons) Stuff(tons)   Tin(tons)   Value(£)
            1855   2.10   0.00   0.00   8.20
            1856   0.00   0.00   0.00   7.70
            1858   0.00   0.00   0.00   19.60
            1862   0.00   0.00   0.00   32.30
            1863   0.00   0.00   0.00   100.20
            1864   0.00   0.00   0.00   121.10
            1865   0.00   0.00   0.00   400.80
            1866   0.00   0.00   0.00   341.00
            1867   1.20   0.00   0.00   69.90
            Comment 1855 FIGURES DUBIOUS; 1856 TIN STUFF; 1858 TIN STUFF;
            1862-1866 TIN STUFF; 1867 VALUE INC.TIN STUFF
Ownership: 1906-1908 T.H.LETCHER; 1909-1910 J.T.LETCHER
            Comment 1906-1908 T.H.LETCHER LORDS AGENT; 1909-1910
            J.T.LETCHER LORDS AGENT
Management: Manager 1864-1865 JOHN TAYLOR & SONS
            Chief Agent 1859-1865 JOS.JEWELL J.WILLIAMS
            Secretary 1859-1865 JOHN HAYE (P)
Employment: Underground   Surface   Total
            1906   3   3
            1908   6   6

TOLGUS,WEST   REDRUTH   SW 679427 1268

Production: Zinc   Ore(tons) Metal(tons)   Value(£)
            1864   12.50   0.00   13.10
            1865   9.80   0.00   10.30

                              481

TOLGUS,WEST   REDRUTH   Continued

Zinc	Ore(tons)	Metal(tons)	Value(£)
1871	20.60	0.00	20.60
1875	35.10	0.00	38.60
1882	25.70	11.50	71.00
Copper	Ore(tons)	Metal(tons)	Value(£)
1861	161.00	9.70	837.20
1862	140.00	7.80	623.40
1863	190.00	13.10	1009.80
1864	581.00	43.00	3644.60
1865	918.00	58.50	4620.90
1866	1660.00	100.60	6382.40
1867	2059.00	123.40	8377.80
1868	2118.00	134.90	8732.40
1869	1343.00	93.80	5643.00
1870	2688.00	192.20	10935.70
1871	2250.00	144.60	8842.70
1872	2759.00	192.80	15344.60
1873	2679.00	194.00	11903.20
1874	3366.00	302.80	20701.90
1875	3532.00	329.40	25458.40
1876	3623.00	303.70	20912.70
1877	3995.20	397.70	24241.00
1878	3752.00	411.20	22334.90
1879	3044.10	312.00	16267.00
1880	2984.20	292.80	16300.20
1881	1320.70	104.50	6664.00
1882	1025.40	116.60	7820.00
1883	970.50	100.70	6032.00

Comment 1861-1876 (C)

Tin	Black(tons)	Stuff(tons)	Tin(tons)	Value(£)
1882	0.10	9.00	0.10	3.00

Ownership: 1877-1881 JOHN TAYLOR & SONS; 1882-1883 WEST WHEAL TOLGUS MINING CO.
Management: Manager 1862-1872 JOHN TAYLOR & SONS; 1873-1878 JOHN HANCOCK; 1879-1881 JOHN GILBERT
Chief Agent 1862-1868 JOS.JEWELL WM.GRIBBLE; 1869 JOS.JEWELL JNR WM.GRIBBLE R.HANCOCK; 1870-1871 JAS.VISICK WM.GRIBBLE RICH.HANCOCK; 1872 JAS.VISICK WM.GRIBBLE JOHN HANCOCK; 1873 JAS.VISICK WM.GRIBBLE; 1874-1876 WM.GRIBBLE JAS.VIGUR RICH.PENROSE; 1877-1881 WM.GRIBBLE JAS.VIGUR; 1882-1883 JOHN GILBERT
Secretary 1862-1874 JOHN HAYE (P); 1875-1876 J.TAYLOR & SONS & OTHERS; 1877-1881 JOHN HAYE (P)

Employment:	Underground	Surface	Total
1878	123	104	227
1879	127	109	236
1880	120	108	228
1881	34	43	77
1882	43	49	92
1883	56	31	87

TOLVADDEN   MARAZION   SW 532304 1270

Production: Lead   Ore(tons) Metal(tons)   Value(£)
            1863   3.00   2.10   0.00
            Zinc   Ore(tons) Metal(tons)   Value(£)
            1865   8.00   0.00   16.00
            Copper   Ore(tons) Metal(tons)   Value(£)
            1857   472.00   33.50   3157.30
            1858   1532.00   124.00   11130.30
            1859   2442.00   151.60   13855.10
            1860   2110.00   117.80   10545.80
            1861   1353.00   74.30   6464.90
            1862   1019.00   61.30   4989.90
            1863   652.00   39.00   2971.50
            1864   789.00   45.00   3718.70
            1865   300.00   16.10   1261.60
            1866   73.00   4.30   337.30
            Comment 1857-1866 (C)
            Tin   Block(tons) Stuff(tons) Tin(tons)   Value(£)
            1864   0.00   0.00   0.00   893.40
            1865   0.00   0.00   0.00   1459.20
            Comment 1864-1865 TIN STUFF
Management: Manager 1859 AB.BENNETT
            Chief Agent 1859-1864 FRAN.GUNDRY; 1865 FRAN.GUNDRY
            JAS.THOMAS
            Secretary 1859-1865 AB.BENNETT (P)

TOLVADDEN,EAST   MARAZION   1271

Production: Copper No detailed return
Management: Manager 1869-1872 AB.BENNETT
            Secretary 1869-1872 AB.BENNETT

TOLVADDEN,NORTH   ST.HILARY   1272

Production: Copper No detailed return
Ownership: Comment 1864-1865 SUSPENDED
Management: Manager 1859-1861 PETER FLOYD
            Chief Agent 1862-1863 FRAN.GUNDRY
            Secretary 1862-1863 AB.BENNETT; 1865 AB.BENNETT

TOLVADDEN,WEST   PAUL   SW 470279 1273

Production: Copper   Ore(tons) Metal(tons)   Value(£)
            1862   28.00   1.30   100.70
            1863   9.00   0.30   22.50
            Comment 1862-1863 (C)
            Tin No detailed return
Ownership: Comment 1862-1865 SUSPENDED
Management: Manager 1859-1860 JAS.THOMAS
            Chief Agent 1859 CHAS.THOMAS; 1861 CHAS.THOMAS
            Secretary 1859-1860 JOHN DICKINSON (P); 1861-1865 J.H.DINGLE
            (P)

CM-R*   483

TONKIN   CALLINGTON   SX 369705 1274

Production: Copper   Ore(tons) Metal(tons) Value(£)
     1870–1873 No detailed return
     Comment 1870–1873 SEE FLORENCE(CALLINGTON)
Ownership: Comment 1870–1873 SEE FLORENCE(CALLINGTON); 1874 SEE EMMENS
     UNITED; 1875 SEE FLORENCE(CALLINGTON)

TORDOWN   CALLINGTON   1275

Production: Manganese No detailed return
Management: Chief Agent 1880 WM.LITTLEJOHN
     Secretary 1880 JOS.ROBINSON

TOWAN   PORTH TOWAN   SW 696481 1276

Production: Copper   Ore(tons) Metal(tons) Value(£)
     1863   134.00   4.50   273.60
     1864   25.00   0.70   41.20
     1865   25.00   0.60   31.20
     Comment 1863–1864 (C); 1865 (C)GREAT TOWAN
     Tin No detailed return

TOWAN,NEW   PORTH TOWAN   SW 699472 1277

Management: Chief Agent 1866–1873 RICH.PRYOR; 1874–1876 RICH.PRYOR & SON
     Secretary 1869–1876 HY.L.PHILLIPS

TOWAN,SOUTH   PORTH TOWAN   SW 696475 1278

Production: Copper No detailed return
Management: Manager 1869–1871 WM.TREGAY
     Secretary 1869–1871 G.H.CARDOZA

TOWAN,WEST   PORTH TOWAN   SW 683473 1279

Production: Zinc   Ore(tons) Metal(tons) Value(£)
     1857   1.40   0.00   1.40
     Copper No detailed return
     Tin   Black(tons) Stuff(tons) Tin(tons) Value(£)
     1852   49.00   0.00   0.00   0.00
     1853   114.00   0.00   0.00   7886.40
     1854   100.60   0.00   0.00   7117.60
     1855   88.40   0.00   0.00   5920.90
     1856   76.10   0.00   0.00   5813.70
     1857   84.40   0.00   0.00   6799.80
     1858   37.90   0.00   0.00   2474.00
     1881   1.50   0.00   1.00   75.00
     1885   3.00   0.00   0.00   120.00
     Comment 1852 TIN ORE
Ownership: 1881–1886 JOHN R.CHIDLEY
     Comment 1861–1865 SUSPENDED; 1882 NOT WORKED; 1886 NOT

TOWAN,WEST   PORTH TOWAN   Continued

            WORKED
Management: Chief Agent 1860 T.W.HUNT
            Secretary 1860 JOHN LITTLE (P); 1881-1886 HY.DALE
Employment: Underground   Surface   Total
            1881   19   22   41
            1883   4   4
            1884   10   6   16
            1885   4   2   6

TOWER CONSOLS   ROCHE   SW 988597 1280

Production: Iron No detailed return
Ownership: Comment 1873 SEE MAGNETIC
Management: Chief Agent 1872 DAVID COCK; 1874 DAVID COCK

TOY TOR   HELSTON   2000

Production: Tin No detailed return
Ownership: Comment 1869-1870 SEE TOY TOR, DEVON

TRANNACK   SITHNEY   SW 661296 1281

Production: Zinc   Ore(tons) Metal(tons)   Value(£)
            1864   3.50   0.00   3.50
            Copper   Ore(tons) Metal(tons)   Value(£)
            1851   59.00   6.50   457.00
            1852   196.00   16.40   1359.20
            1862   35.00   1.20   91.90
            1863   23.00   0.80   50.90
            1864   82.00   4.60   369.90
            1865   33.00   1.50   113.50
            1866   49.00   2.00   107.30
            1868   8.00   0.30   25.80
            Comment 1851-1852 (C)INC.BOSCENCE; 1862-1866 (C); 1868 (C)
            Tin   Black(tons) Stuff(tons) Tin(tons)   Value(£)
            1855   25.90   0.00   0.00   1711.70
            1863   0.00   0.00   0.00   10.90
            Comment 1855 TRANACH INC.BOSCENCE; 1863 TIN STUFF
Management: Chief Agent 1862 WM.TRURAN; 1863-1866 W.CLIFT; 1869-1871
            WM.BAWDEN
            Secretary 1862-1866 JOHN REED; 1869-1871 JOHN REED

TREAMBLE   PERRANZABULOE   SW 786559 1282

Production: Lead & Silver Ore(tons) Lead(tons) Silver(ozs) Value(£)
            1862   0.30   0.20   4.00   0.00
            1873   20.00   15.00   75.00   240.00
            1874   11.30   8.00   48.00   0.00
            1878   0.80   0.50   0.00   7.70
            Zinc   Ore(tons) Metal(tons)   Value(£)
            1882   7.00   2.80   14.00

                                    485

Iron	Ore(tons)	Iron(%)	Value(£)
1859	2391.00	0.00	597.00
1860	1312.20	0.00	328.10
1864	222.40	0.00	55.60
1873	426.00	0.00	319.50
1874	9438.00	0.00	7550.40
1875	958.20	0.00	766.40
1876	No detailed return		
1881	160.00	0.00	96.00
1882	589.00	45.00	300.00
1892	691.00	43.00	156.00
1893	124.00	42.00	22.00
1894	955.00	42.00	167.00

Comment 1859-1860 BH.; 1864 BH.TRESAMBLE; 1873-1876 BH.;
1881-1882 BH.; 1892 BH.; 1893 BH.OPEN WORK; 1894 BH.
Ownership: 1863-1865 EDW.CARTER & EXECUTORS OF T.WHITFORD; 1872-1875
CORNISH CONSOLIDATED FE MINES CORP.LTD; 1876 LEWIS BIDDER &
NOBLE; 1877 GREAT PERRAN MINES ADVENTURERS; 1878 NEW PERRAN
MINERAL CO.LTD.; 1879 W.R.ROWEBUCK & OTHERS; 1880-1881
CORNISH STEEL IRON ORE CO.; 1882-1883 NEWQUAY MINING CO.LTD.;
1892 EDW.CARTER P
Comment 1863-1865 TRAMBLE; 1880-1881 TWO RETURNS AGGREGATED;
1902 SEE MOUNT NO 1
Management: Manager 1873 ART.PETO; 1874 WM.PARKYN; 1875-1876 JOHN PARKYN;
1878 JAS.HENDERSON
Chief Agent 1863-1865 JOHN BALL; 1872 R.PALMER; 1877
JAS.HENDERSON; 1879 JOHN H.JAMES; 1880-1881 JOHN H.JAMES
J.WHITAKER BUSHE; 1882-1883 JOHN H.JAMES
Secretary 1880 W.R.ROEBUCK; 1881 NEWQUAY MINE CO.LTD.
Employment:

	Underground	Surface	Total
1878	10	11	21
1879		3	3
1881	30	14	44
1882	26	18	44
1883		1	1
1892		4	4

TREASURY,WEST   CROWAN   SW 598345 1283

Production: Copper

	Ore(tons)	Metal(tons)	Value(£)
1845	267.00	17.80	1274.40
1846	483.00	35.10	2444.50
1847	318.00	21.80	1511.90
1848	944.00	78.60	4648.10
1849	1362.00	101.90	6897.40
1850	1442.00	129.80	9186.20
1851	857.00	73.80	5104.70
1852	1595.00	124.80	10831.90
1853	1317.00	82.40	8258.80
1854	913.00	57.90	5843.50

Comment 1845-1854 (C)

TREBARTHA LEMARNE   NORTH HILL   SX 255776 1284

Production: Tin   Black(tons) Stuff(tons)   Tin(tons)   Value(£)
            1884   11.10   0.00   0.00   453.00
            1885   0.70   42.00   0.00   24.00
            1886   5.10   0.00   0.00   257.00
            1887   1.50   0.00   0.00   83.00
            1888   1.60   0.00   0.00   100.00
            Comment 1885 WHITS
            Arsenic   Ore(tons) Metal(tons)   Value(£)
            1885   75.00   0.00   152.00
            1886   99.00   0.00   495.00
            1887   27.00   0.00   156.00
            Comment 1886 CRUDE
            Tungsten   Ore(tons) Metal(tons)   Value(£)
            1886   11.00   0.00   273.00
            1887   1.00   0.00   24.00
            1888   2.20   0.00   54.00
            1889   No detailed return
            Comment 1886-1889 TUNGSTATE OF SODA
            Wolfram No detailed return
            Arsenic Pyrite   Ore(tons)   Value(£)
            1888   18.00   14.00
Ownership: 1881-1888 TREBARTHA LEMARNE MINING CO.
Management: Chief Agent 1881 E.NICHOLLS; 1884-1886 E.W.TEMBY; 1887-1888
            HY.BENNETTS
            Secretary 1882-1883 E.NICHOLLS
Employment: Underground   Surface   Total
            1881   6   6
            1883   10   30   40
            1884   11   10   21
            1885   12   23   35
            1886   5   12   17
            1887   12   10   22
            1888   4   4   8

TREBARVAH   PERRANUTHNOE   SW 543293 1285

Production: Copper   Ore(tons) Metal(tons)   Value(£)
            1852   474.00   31.20   2635.60
            1853   466.00   27.80   2758.30
            1854   260.00   20.90   2167.80
            1855   237.00   17.80   1834.30
            1856   256.00   22.40   2164.90
            1857   328.00   23.70   2373.80
            1858   309.00   22.20   2083.50
            1859   525.00   36.00   3341.20
            1861   402.00   27.30   2560.90
            1873   91.00   7.50   473.80
            1874   98.00   7.00   404.90
            Comment 1852-1853 (C); 1854 (C)TREVARVAH; 1855-1859 (C); 1861
            (C); 1873-1874 (C)
            Tin   Black(tons) Stuff(tons)   Tin(tons)   Value(£)
            1853   5.00   0.00   0.00   247.80
            1855   0.10   0.00   0.00   0.70
            1873   0.00   178.70   0.00   164.70

                                        487

TREBARVAH   PERRANUTHNOE   Continued

Tin	Black(tons)	Stuff(tons)	Tin(tons)	Value(£)
1874	0.00	18.00	0.00	27.80

Comment 1874 VALUE EST

Iron	Ore(tons)	Iron(%)	Value(£)
1873	888.60	0.00	711.70
1874	839.00	0.00	629.00
1875	No detailed return

Comment 1873—1875 BH.
Ownership: 1873—1875 TREBARVAH MINING CO.
            Comment 1861—1865 SUSPENDED; 1880—1881 SEE MOUNTS BAY
            CONSOLS
Management: Manager 1859 OSBORNE; 1872—1873 WM.HOLLOW
            Chief Agent 1860 FRAN.HOSKING; 1872—1873 THOS.LAITY
            Secretary 1859—1860 R.R.MITCHELL (P); 1872—1876 S.H.F.COX

TREBEIGH CONSOLS   ST.IVE   SX 302697 1286

Production: Lead No detailed return

Silver	Ore(tons)	Metal(tons)	Value(£)
1877	0.50	0.00	0.00

Ownership: 1876—1881 TREBEIGH CONSOLS CO.
            Comment 1879—1881 NOT WORKED
Management: Chief Agent 1876—1878 JOHN GIFFORD
            Secretary 1878 T.B.LAWS

Employment:	Underground	Surface	Total
1878	8	4	12

TREBELL CONSOLS   LANIVET   SX 055633 1287

Production: Tin	Black(tons)	Stuff(tons)	Tin(tons)	Value(£)
1852	1.10	0.00	0.00	0.00
1853	2.60	0.00	0.00	179.30
1854	12.10	0.00	0.00	776.10

Comment 1852 TIN ORE; 1854 TREBELE

TREBISKEN   PERRANZABULOE   SW 779563 1289

Production: Iron	Ore(tons)	Iron(%)	Value(£)
1858	10771.50	0.00	2693.00
1859	4338.20	0.00	1084.30
1860	2640.90	0.00	406.30
1861	1860.40	0.00	465.10
1864	2876.00	0.00	719.00
1865	6193.80	0.00	1548.40
1866	1234.60	0.00	308.70
1871	1435.50	0.00	970.90
1872	8293.00	0.00	4975.80
1873	2464.00	0.00	1848.00
1874	2545.00	0.00	2036.00
1875	374.00	0.00	299.20
1876	516.70	0.00	359.60
1877	282.80	0.00	169.10

Comment 1858 BH.TREVISKEN; 1859–1861 BH.; 1864 BH.INC.MOUNT
           NO.1; 1865–1866 BH.; 1871 BH.; 1872 BH.INC.MOUNT NO.1; 1873
           BH.; 1874 BH.INC.TREBISKEN GREEN & MOUNT.NO.1; 1875–1877
           BH.INC.MOUNT NO.1
Ownership: 1863–1865 EDW.CARTER & EXECUTORS OF T.WHITFORD; 1871 CORNISH
           CONSOLIDATED FE MINES CORP.LTD; 1872–1874 AGRA BANKING CO.
           Comment 1871 INC.DUCHY & PERU; 1875–1883 SEE MOUNT NO.1
Management: Chief Agent 1863–1865 JOHN BALL; 1871 R.PALMER; 1872–1874
           A.J.WHALLEY
Employment: Underground   Surface   Total
           1881–1882
           Comment 1881–1882 SEE MOUNT NO.1

TREBISKEN GREEN   PERRANZABULOE   SW 785570 1290

Production: Lead & Silver   Ore(tons)   Lead(tons) Silver(ozs)   Value(£)
           1861   7.40   5.10   80.00   0.00
           Comment 1861 INC.MOUNT NO.1
           Tin   Black(tons) Stuff(tons)   Tin(tons)   Value(£)
           1861   3.90   0.00   0.00   134.30
           Iron   Ore(tons)   Iron(%)   Value(£)
           1874   No detailed return
           Comment 1874 GREEN SEE TREBISKEN

TREBULLETT   LEZANT   SX 328788 1291

Production: Antimony   Ore(tons) Metal(tons)   Value(£)
           1891   10.00   4.30   250.00
           Comment 1891 VALUE INCS. DUMFRIES
Ownership: 1890 W.ROOKE; 1891–1892 THE ENGLISH ANTIMONY CO.LTD.
           Comment 1891–1892 SUSPENDED
Management: Chief Agent 1890–1892 J.JOLL
           Secretary 1891 W.C.ELBOROUGH (S); 1892 E.FERRINGS (S)
Employment: Underground   Surface   Total
           1890   15   3   18
           1891   10   8   18

TREBURGETT,EAST   ST.TEATH   1292

Production: Lead & Silver No detailed return
Ownership: 1884–1885 EAST TREBURGETT MINING CO.
Management: Chief Agent 1884–1885 WM.BENNETTS
Employment: Underground   Surface   Total
           1884   3   3
           1885   4   2   6

TREBURGETT,OLD   ST.TEATH   SX 057797 1293

Production: Lead & Silver   Ore(tons)   Lead(tons) Silver(ozs)   Value(£)
           1871   29.50   22.10   110.00   0.00
           1872   140.00   105.00   525.00   0.00

TREBURGETT,OLD   ST.TEATH   Continued

Lead & Silver	Ore(tons)	Lead(tons)	Silver(ozs)	Value(£)
1873	342.40	256.50	1282.00	9677.80
1874	508.30	381.20	2285.00	15145.10
1875	426.70	327.50	2000.00	11386.70
1876	444.80	333.60	2035.00	11223.80
1877	234.30	175.50	1059.00	5620.50
1878	37.30	27.00	167.00	407.00
1879	14.50	11.00	66.00	160.00

Zinc	Ore(tons)	Metal(tons)	Value(£)
1871	40.60	0.00	136.40
1877	3.30	0.00	4.60

Comment 1877 OLD TREBURGET

Copper	Ore(tons)	Metal(tons)	Value(£)
1878	0.10	0.10	1.00

Iron	Ore(tons)	Iron(%)	Value(£)
1875	62.00	0.00	49.60

Comment 1875 SP.& SOME BROWN HAEM.
Ownership: 1875-1879 OLD TREBURGETT CO.LTD.; 1880 CHAS.BENNETT
             WM.HARRIS
             Comment 1877 SOLD IN SEPT.1877; 1878-1879 ABANDONED JUNE
             1878; 1881 TREBURGETT
Management: Manager 1873-1878 WM.HANCOCK; 1881 WM.HARRIS
             Chief Agent 1872 WM.HANCOCK; 1875 WM.F.BRYANT; 1876-1879
             W.T.BRYANT; 1880 WM.HARRIS
             Secretary 1872-1877 F.R.WILSON (S); 1878 JOHN TUCKER; 1881
             CHAS.BENNETT

Employment:	Underground	Surface	Total
1878	16	20	36
1880	4		4

TREBURLAND   ALTARNUN   SX 237794 1294

Production: Tin	Black(tons)	Stuff(tons)	Tin(tons)	Value(£)
1881	0.80	0.00	0.50	40.50

             Comment 1881 TREBURLAND LAUNCESTON
             Wolfram No detailed return
Ownership: 1881 J.B.BROOKS & OTHERS; 1900 MINING ESTATES SYNDICATE LTD.;
             1913 TREBURLAND WOLFRAM & TIN CO.LTD.
             Comment 1901-1902 SEE ANNIE(ALTARNUN)
Management: Manager 1913 E.SKEWIS
             Chief Agent 1881 J.DAINTY; 1900 WM.WADGE

Employment:	Underground	Surface	Total
1881		4	4
1900	7		7
1913	21	11	32

TREDINNEY CONSOLS   ST.BURYAN   1295

Production: Tin No detailed return
Management: Chief Agent 1862-1868 W.H.RICHARDS

490

TREDINNICK   ST.ERNEY   SX 360597 1296

Production: Lead   Ore(tons) Metal(tons)   Value(£)
            1876   14.30   10.00   0.00

TREEN DOWNS   ZENNOR   SW 436383 1297

Production: Tin   Black(tons) Stuff(tons)   Tin(tons)   Value(£)
            1906   0.10   0.00   0.00   8.00
Ownership: 1906-1907 DAN.M.THOMAS & PARTY
Employment: Underground   Surface   Total
            1906-1907   4   4

TREFFRY   LUXULYAN   SX 056554 1298

Production: Iron   Ore(tons)   Iron(%)   Value(£)
            1862   2000.00   0.00   600.00
            1863   No detailed return
            1872   1217.00   0.00   730.20
            1873-1877 No detailed return
            Comment 1862 HE. TREFFRY CONSOLS; 1863 TREFFRY CONSOLS; 1872
            BH.; 1873-1877 BH.SEE RESTINNIS
Ownership: 1872 SML.MOSS
            Comment 1873-1876 SEE RESTINNIS

TREFREW   1300

Production: Lead   Ore(tons) Metal(tons)   Value(£)
            1852   49.00   37.50   0.00
            1853   No detailed return
            1854   22.00   17.50   0.00
            Comment 1852 TWO RETURNS AGGREGATED

TREFULACH   ST.ENODER   SW 902561 1301

Production: Tin No detailed return
Ownership: Comment 1862-1865 SUSPENDED
Management: Chief Agent 1860 JAS.POPE; 1861 JAS.POPE THOS.HODGE; 1862
            THOS.HODGE
            Secretary 1861-1865 M.G.PAINTER (P)

TREFUSIS   REDRUTH   SW 707410 1302

Production: Copper   Ore(tons) Metal(tons)   Value(£)
            1853   113.00   9.10   929.40
            1854   196.00   13.10   1351.50
            1862   13.00   0.70   57.40
            Comment 1853-1854 (C); 1862 (C)
            Tin   Black(tons) Stuff(tons)   Tin(tons)   Value(£)
            1853   3.00   0.00   0.00   108.60
            1854   17.80   0.00   0.00   1206.10
            1855   51.70   0.00   0.00   3082.40

TREFUSIS   REDRUTH   Continued

Tin	Black(tons)	Stuff(tons)	Tin(tons)	Value(£)
1856	19.20	0.00	0.00	1267.60
1857	20.50	0.00	0.00	1523.20
1858	25.00	0.00	0.00	1556.30
1859	24.30	0.00	0.00	1611.60
1860	27.30	0.00	0.00	2066.20
1861	23.00	0.00	0.00	1585.80
1862	17.50	0.00	0.00	1109.50
1863	4.50	0.00	0.00	278.80
1864	0.00	0.00	0.00	13.60

Comment 1862 TWO RETURNS AGGREGATED; 1864 TIN STUFF
Ownership: Comment 1860–1865 SUSPENDED
Management: Manager 1859 Z.CARKEEK
            Chief Agent 1859 JOHN POPE JNR; 1860–1865 J.TREGONNING
            Secretary 1859 THOS.RICHARDS (P)

TREFUSIS,EAST   REDRUTH   SW 708414 1303

Production: Copper   Ore(tons) Metal(tons)   Value(£)

1862	3.00	0.20	15.70

            Comment 1862 (C)
Ownership: Comment 1863–1865 SUSPENDED
Management: Manager 1859–1862 THOS.RICHARDS
            Chief Agent 1859–1860 JAS.POPE; 1861–1862 J.S.HOSKING
            Secretary 1859–1865 THOS.RICHARDS (P)

TREGARDOCK   ST.TEATH   SX 041839 1305

Production: Lead & Silver   Ore(tons) Lead(tons) Silver(ozs)   Value(£)

1853	32.00	14.00	686.00	0.00
1854	18.90	8.10	0.00	0.00
1856–1857 No detailed return				
1860	9.00	6.50	0.00	0.00

            Comment 1860 FOR AG SEE WEST CHIVERTON
            Copper No detailed return
            Iron No detailed return
Ownership: 1875–1877 TREGARDOCK MINE CO.
Management: Manager 1869–1875 JOHN SPARGO
            Chief Agent 1859–1867 GOLDSWORTHY; 1868 JOHN SPARGO;
            1875–1877 SML.SPARGO
            Secretary 1861–1867 W.E.CUMMINS (P); 1869–1875 JOHN PETERS

TREGAWNE   ST.BREOCK   1306

Production: Iron   Ore(tons)   Iron(%)   Value(£)

1859	130.50	0.00	32.60
1860–1861 No detailed return			

            Comment 1859 BH.& SOME SPATHOSE
Ownership: 1863–1865 EDW.CARTER & EXECUTORS OF T.WHITFORD
Management: Chief Agent 1863–1865 THOS.HAMBLY

492

TREGEAGLE   ST.NEOT   SX 175687 1307

Production: Tin   Black(tons) Stuff(tons)   Tin(tons)   Value(£)
            1870   11.70   0.00   0.00   785.90
            1871   34.50   0.00   0.00   2382.20
            1873   18.50   0.00   0.00   0.00
            1874   7.80   0.00   0.00   705.00
            1884   0.40   0.00   0.00   12.00
            1885   1.20   0.00   0.00   47.00
            1887   0.90   0.00   0.00   70.00
            1888   11.10   0.00   0.00   587.00
            1889   7.60   0.00   0.00   345.00
            1890   18.60   0.00   0.00   833.00
            1891   7.70   0.00   0.00   360.00
            1909   4.20   0.00   0.00   297.00
            1910   2.00   0.00   0.00   140.00
            1912   2.00   0.00   0.00   200.00
            Comment 1874 VALUE EST
Ownership: 1882-1892 JOS.TAYLOR & CO.; 1907-1913 TREGEAGLE MINE LTD.
            Comment 1891-1892 SUSPENDED; 1910 CLOSED MAY 1910; 1911-1913
            REOPENED JAN.1912
Management: Manager 1872 J.BENNETT; 1873-1874 JOS.TAYLOR; 1875-1876
            P.TEMBY; 1911-1913 D.W.STEWART
            Chief Agent 1872-1874 JOHN GLUYAS; 1882-1892 ED.PASCOE
            Secretary 1872 J.GLUBB; 1873-1874 HY.TAYLOR; 1875-1876
            JOS.TAYLOR & CO.
Employment: Underground   Surface   Total
            1882-1883   4   1   5
            1884-1885   5   4   9
            1886   5   7   12
            1887   8   5   13
            1888   31   21   52
            1889   28   24   52
            1890   30   29   59
            1891   1   1   2
            1908   10   14   24
            1909   12   11   23
            1910   9   12   21
            1912   13   28   41
            1913   1   1

TREGEARE   EGLOSKERRY   1308

Production: Manganese No detailed return
            Arsenic   Ore(tons) Metal(tons)   Value(£)
            1890   45.00   0.00   165.00
            Comment 1890 TREGEARNE
Ownership: 1890 J.E.BOCKEMER; 1891 TREGEARE MANGANESE MINING CO.
            Comment 1891 IN LIQUIDATION
Management: Chief Agent 1890 FRED.ASHWELL
            Secretary 1891 HIBBERT,BULL & CO.
Employment: Underground   Surface   Total
            1890   16   16

Production: Tin No detailed return
Ownership: 1906 S.J.HANCOCK
Employment: Underground   Surface   Total
            1906   10   10

TREGEMBO   ST.HILARY   SW 571319 1310

Production: Copper   Ore(tons) Metal(tons)   Value(£)
            1884   6.00   0.00   45.00
            Tin   Black(tons) Stuff(tons) Tin(tons)   Value(£)
            1880   0.00   35.00   0.00   0.00
            1883   31.00   0.00   0.00   1650.00
            1884   11.30   0.00   0.00   544.00
            1887   41.10   0.00   0.00   2746.00
            1888   46.80   0.00   0.00   3124.00
            1889   12.50   0.00   0.00   694.00
            Comment 1880 UNSOLD
            Arsenic   Ore(tons) Metal(tons)   Value(£)
            1888   5.00   0.00   5.00
Ownership: 1880—1886 HARVEY & CO.; 1887—1889 TREGEMBO MINING CO.
            Comment 1885 NOT WORKED; 1889 NOW ABANDONED
Management: Manager 1881 ED.CHEGWIN
            Chief Agent 1880 ED.CHEGWIN; 1882—1886 ED.CHEGWIN; 1887—1888
            THOS.HODGE
            Secretary 1880 ED.CHEGWIN; 1881 W.J.RAWLINGS (P)
Employment: Underground   Surface   Total
            1880   6   1   7
            1881   18   6   24
            1882   20   12   32
            1883   50   47   97
            1884   6   4   10
            1886   24   44   68
            1887   40   30   70
            1888—1889   63   41   104

TREGEMBO,EAST   CROWAN   SW 586327 1311

Production: Tin   Black(tons) Stuff(tons) Tin(tons)   Value(£)
            1883   7.00   0.00   0.00   371.00
            1884   3.70   0.00   0.00   179.00
            1885   4.10   0.00   0.00   180.00
            1886   12.00   0.00   0.00   720.00
            1887   13.40   0.00   0.00   862.00
            1888   12.70   0.00   0.00   834.00
            1889   16.40   0.00   0.00   919.00
            1890   22.10   0.00   0.00   1269.00
            1891   14.50   0.00   0.00   769.00
            1892   17.10   0.00   0.00   955.00
            1893   12.40   0.00   0.00   601.00
            1894   10.00   0.00   0.00   375.00
Ownership: 1882—1890 GREN.SHARP & OTHERS; 1891—1893 GREN.SHARP & OTHERS
            CB; 1894 GREN.SHARP & OTHERS
Management: Chief Agent 1882—1886 MARK R.CHEGWIN; 1887—1894 JOS.PRISK

```
         Secretary 1891-1894 JOHN BIDDER (S)
Employment:            Underground    Surface       Total
         1882              5                           5
         1883              8            5             13
         1884             12            5             17
         1885             26            5             31
         1886             32            8             40
         1887             38            8             46
         1888             30            8             38
         1889             28            7             35
         1890             30           14             44
         1891             27           12             39
         1892             24            9             33
         1893             20            5             25
         1894             16            6             22
```

TREGILLIA   ST.HILARY   1312

Production:	Tin	Black(tons)	Stuff(tons)	Tin(tons)	Value(£)
	1855	0.50	0.00	0.00	33.20

TREGILLIA STREAM WORKS   ST.HILARY   1313

Production: Tin No detailed return
Management: Chief Agent 1861-1865 RICH.GRUBB
            Secretary 1861-1865 THOS.W.ROBINSON (P)

TREGONETHA   ST.WENN   SW 952627 1314

Production:	Iron	Ore(tons)	Iron(%)	Value(£)
.	1873	288.00	0.00	216.00
	1874	413.00	0.00	268.40
	1875	600.00	0.00	495.00
	1876	242.00	0.00	145.20

            Comment 1873-1876 BH.
Ownership: 1872-1874 ST.WENN IRON MINES CO.; 1875 THOS.FENWICK; 1876
            JOHN TAYLOR & CO.
Management: Chief Agent 1872 SML.COCK J.W.COOM; 1873 J.W.COOM; 1874-1875
            S.R.COCK; 1876 PETER TEMBY

TREGONTREES   ST.AUSTELL   1315

Production:	Tin	Black(tons)	Stuff(tons)	Tin(tons)	Value(£)
	1883	22.00	0.00	0.00	1166.00
	1884	22.10	0.00	0.00	1069.00
	1885	27.30	0.00	0.00	1349.00

            Comment 1883-1885 AND OLD POLGOOTH
Ownership: 1883-1885 TREGONTREES & OLD POLGOOTH CONSOLS MINING CO.
            Comment 1883-1884 TREGONTREES & OLD POLGOOTH CONSOLS; 1885
            TREGONTREES & OLD POLGOOTH CONSOLS IN LIQUIDATION
Management: Chief Agent 1883 SILAS PASCOE

TREGONTREES   ST.AUSTELL   Continued

Secretary 1884 R.B.FASTNEDGE
Employment: Underground   Surface   Total
        1883   16   25   41
        1884-1885   12   28   40
    Comment 1883-1885 INC.OLD POLGOOTH

TREGORDON   WADEBRIDGE   SX 001736 1316

Production: Lead   Ore(tons) Metal(tons)   Value(£)
        1849   28.80   20.00   0.00
        1850   20.00   14.00   0.00
        1851   15.00   7.00   0.00
        1852   9.00   6.00   0.00
        1854   No detailed return
    Comment 1852 MINE STOPPED 5TH MAY 1852

TREGOSS   ROCHE   SW 977617 1317

Production: Tin   Black(tons) Stuff(tons)   Tin(tons)   Value(£)
        1873   1.80   0.00   0.00   125.50
        1874   0.90   0.00   0.00   50.00
    Comment 1873 TREGROSS; 1874 TREGROSS VALUE EST
Ownership: Comment 1872 FORMERLY ARCHIE
Management: Chief Agent 1872-1874 THOS.PARKYN
            Secretary 1872-1873 JAS.H.CROFTS

TREGREHAN CONSOLS   ST.BLAZEY   SX 045540 1318

Production: Tin   Black(tons) Stuff(tons)   Tin(tons)   Value(£)
        1889   2.10   0.00   0.00   115.00
        1890   2.20   0.00   0.00   115.00
        1891   2.20   0.00   0.00   114.00
        1892   0.70   0.00   0.00   32.00
        1893   3.10   0.00   0.00   136.00
        1909   2.70   0.00   0.00   196.00
        1911   3.20   0.00   0.00   309.00
        Wolfram No detailed return
Ownership: 1889-1894 TREGREHAN CONSOLS MINING CO.LTD.; 1905 J.H.COLLINS
            & SONS; 1906-1907 ARGOLIS COPPER MINING SYNDICATE LTD.;
            1908-1913 TREGREHAN MINING SYNDICATE LTD.
            Comment 1880 TREGREHAR. SEE PAR GREAT CONSOLS; 1881
            TREGREHAR. SEE GREAT POLGOOTH; 1905-1910 PAR CONSOLS, WEST,
            PAR & TREGREHAN CONSOLS; 1911-1912 PAR CONSOLS, WEST, PAR &
            TREGREHAN CONSOLS SUSPENDED OCT.1911; 1913 PAR CONSOLS, WEST,
            PAR & TREGREHAN CONSOLS NOT WORKED
Management: Manager 1892-1894 F.W.MICHELL (P)
            Chief Agent 1892-1894 JAS.PENROSE; 1906-1907 J.H.COLLINS &
            SONS
            Secretary 1889-1891 F.W.MITCHELL; 1892-1894 C.GREGORY
Employment: Underground   Surface   Total
        1889   6   4   10
        1890   13   12   25

496

TREGREHAN CONSOLS   ST.BLAZEY   Continued

	Underground	Surface	Total
1891	14	19	33
1892	17	21	38
1893	19	13	32
1894	3		3
1905	8		8
1906-1907	6		6
1908	15	20	35
1909	8	3	11
1910	6		6
1911	12	13	25
1912		1	1

TREGREHAN,SOUTH   ST.AUSTELL   1320

Production: Tin No detailed return
Ownership: 1913 SOUTH TREGREHAN SYNDICATE
            Comment 1913 IN LIQUIDATION
Management: Chief Agent 1913 WM.WEDLAKE
Employment: Underground   Surface   Total
            1913   15   1   16

TREGUDDER   PADSTOW   1322

Ownership: 1875 TREGUDDER MINING CO.

TREGULLOW CONSOLS   ST.AGNES   SW 718462 1323

Production: Copper No detailed return
            Tin No detailed return
Ownership: Comment 1861-1865 SUSPENDED
Management: Chief Agent 1860 JOHN DALE
            Secretary 1860 THOS.SPARGO (P)

TREGUNE CONSOLS,GREAT   ALTARNUN   SX 224797 1324

Production: Copper   Ore(tons) Metal(tons)   Value(£)
            1862   24.00   2.60   221.30
            1864   7.00   0.80   65.60
            1865   21.00   1.90   161.20
            Comment 1862 (C)GREAT TREGUNE; 1864-1865 (C)GREAT TREGUNE
Management: Manager 1859-1860 JOHN SPARGO
            Chief Agent 1859-1860 JOHN ROWE; 1861-1865 WM.RICHARDS JOHN
            ROWE
            Secretary 1859-1860 F.S.THOMAS (P); 1861-1865 CHAS.PEARSON
            (P)

TREGUNSTIC   WENDRON   SW 706304 1325

Production: Tin   Black(tons) Stuff(tons)   Tin(tons)   Value(£)
            1853   No detailed return
            Comment 1853 SEE MENGERN

TREGURTHA DOWNS   ST.HILARY   SW 539311 1326

Production: Tin   Black(tons) Stuff(tons)   Tin(tons)   Value(£)
            1883-1885 No detailed return
            1886   31.50   20.00   0.00   1740.00
            1887   114.00   76.00   0.00   6740.00
            1888   194.00   0.00   0.00   13144.00
            1889   166.00   0.00   0.00   9443.00
            1890   190.00   0.00   0.00   11231.00
            1891   No detailed return
            1892   113.00   0.00   0.00   6444.00
            1893   144.10   0.00   0.00   7714.00
            1894   132.00   0.00   0.00   5716.00
            1895   31.40   0.00   0.00   1272.00
            1899   8.00   0.00   0.00   538.00
            1900   13.50   0.00   0.00   1056.00
            Comment 1883-1885 SEE OWEN VEAN; 1886-1887 TWO RETURNS
            AGGREGATED; 1891 SEE HELENA
            Arsenic   Ore(tons) Metal(tons)   Value(£)
            1888   8.00   0.00   32.00
            1893   6.00   0.00   25.00
            1894   8.00   0.00   35.00
            1895   9.00   0.00   35.00
Ownership: 1886 TREGURTHA DOWNS CO.; 1887-1890 TREGURTHA DOWNS MINING
            CO.LTD.; 1892-1895 WM.WIGHTON P; 1899-1902 TREGURTHA DOWNS
            CONSOLIDATED MINES LTD
            Comment 1882-1885 SEE OWEN VEAN; 1891 SEE HELENA; 1892 LATE
            HELENA; 1901-1902 IN LIQUIDATION; 1909-1913 SEE HAMPTON
Management: Chief Agent 1886-1887 HY.PRICE; 1888-1890 JOHN PRISK; 1892
            H.G.RICHARDS; 1893-1895 WM.ROSEWARNE
            Secretary 1887 CHAS.F.ROSEBY
Employment: Underground   Surface   Total
            1882-1885
            1886   70   46   116
            1887   112   63   175
            1888   137   77   214
            1889   146   84   230
            1890   180   76   256
            1892   161   76   237
            1893   105   66   171
            1894   138   35   173
            1895   104   62   166
            1899   25   17   42
            1900   26   36   62
            1901   11   11
            1909-1913
            Comment 1882-1885 SEE OWEN VEAN; 1909-1913 SEE HAMPTON

                                      498

TREGWOLLANS   ST.STEPHEN-IN-BRANNEL   1327

Production: Iron   Ore(tons)   Iron(%)   Value(£)
            1858   995.20   0.00   298.60
            1859   501.20   0.00   150.40
            1860   1257.20   0.00   327.40
            1861   No detailed return
            Comment 1858-1860 BH.

TREHANE   MENHENIOT   SX 287638 1328

Production: Lead & Silver   Ore(tons)   Lead(tons) Silver(ozs)   Value(£)
            1847   312.00   206.00   0.00   0.00
            1848   422.00   279.00   0.00   0.00
            1849   459.60   290.00   0.00   0.00
            1850   430.00   254.10   0.00   0.00
            1851   414.00   285.80   0.00   0.00
            1852   506.00   375.00   22000.00   0.00
            1853   473.00   311.00   14574.00   0.00
            1854   570.40   335.00   14580.00   0.00
            1855   459.00   326.00   14996.00   0.00
            1856   313.00   222.00   12876.00   0.00
            1857   172.00   89.00   623.00   0.00
            1858-1861 No detailed return

TREHILL   STOKE CLIMSLAND   SX 372737 1329

Production: Copper   Ore(tons) Metal(tons)   Value(£)
            1860   245.00   7.80   543.00
            1861   210.00   5.40   343.00
            1862   85.00   2.30   147.40
            Comment 1860-1862 (C)
Management: Chief Agent 1860 GEO.ROWE; 1861-1863 HY.RICKARD
            Secretary 1860 W.S.TROTTER (P); 1861-1863 E.S.CODD (P)

TRELAWNEY   MENHENIOT   SX 287635 1330

Production: Lead & Silver   Ore(tons)   Lead(tons) Silver(ozs)   Value(£)
            1845   280.00   168.00   0.00   0.00
            1846   529.00   316.00   0.00   0.00
            1847   883.00   640.00   0.00   0.00
            1848   413.00   298.00   0.00   0.00
            1849   1296.50   934.00   0.00   0.00
            1850   1496.00   1131.40   0.00   0.00
            1851   1112.40   840.30   0.00   0.00
            1852   971.90   721.00   33906.00   0.00
            1853   979.00   718.00   29797.00   0.00
            1854   782.80   569.00   26174.00   0.00
            1855   836.90   592.00   26235.00   0.00
            1856   1029.00   761.00   34093.00   0.00
            1857   1108.00   831.00   39075.00   0.00
            1858   1307.00   980.00   49000.00   0.00
            1859   1332.70   999.00   49950.00   0.00
            1860   1251.10   940.00   47000.00   0.00

                                        499

Lead & Silver	Ore(tons)	Lead(tons)	Silver(ozs)	Value(£)
1861	798.00	495.00	31150.00	0.00
1862	1005.00	668.00	42080.00	0.00
1863	1073.80	730.20	45990.00	0.00
1864	800.00	534.00	38034.00	0.00
1865	832.70	532.70	37770.00	0.00
1866	879.30	562.50	30464.00	0.00
1867	1272.20	954.00	51516.00	0.00
1868	756.10	567.00	30618.00	0.00
1869	874.60	666.00	35964.00	0.00
1870	389.70	292.20	15768.00	0.00
1871	363.40	272.50	14688.00	0.00

1883-1884 No detailed return
Comment 1883-1884 SEE HONY

Arsenic	Ore(tons)	Metal(tons)	Value(£)
1900	167.00	0.00	1673.00
1901	224.00	0.00	1788.00
1902	235.00	0.00	1400.00

Arsenic Pyrite	Ore(tons)	Value(£)
1898	803.00	1405.00
1899	2621.00	3931.00
1900	1725.00	2588.00
1901	1577.00	3155.00
1902	502.00	376.00

Ownership: 1896-1902 ANGLO PENINSULA MINING & CHEMICAL CO.LTD
Comment 1871-1873 SEE MARY ANNE; 1880-1884 TRELAWNY SEE HONY;
1902 ABANDONED 1902
Management: Manager 1859 RICH.JENKIN; 1860 WM.JENKIN; 1865-1866 JOHN
PRYOR; 1867-1869 WM.JOHNS
Chief Agent 1859 THOS.GRENFELL; 1860 WM.BRYANT THOS.GRENFELL;
1861-1863 FRAN.PRYOR RICH.PRYOR THOS.GRENFELL; 1864
H.LAVINGTON JOHN PRYOR THOS.GRENFELL; 1865-1868
THOS.GRENFELL; 1869-1870 THOS.GRENFELL JOHN PRYOR; 1897-1902
HY.BENNETTS
Secretary 1859-1860 JOHN PHILLIPS (P); 1861-1862 THOS.PRYOR
(P); 1863 DUNSFORD & RANKEN; 1864 WM.JOHNS; 1865-1868
H.LAVINGTON; 1869-1870 LAVINGTON(S)PRYOR(P)

Employment:	Underground	Surface	Total
1880-1884			
1897	34	21	55
1898	33	16	49
1899	54	31	85
1900	69	38	107
1901	73	45	118
1902	72	42	114

Comment 1880-1884 SEE HONY

TRELAWNEY,NEW   ST.IVE   SX 311683 1331

Production: Arsenic Pyrite	Ore(tons)	Value(£)
1895	51.00	51.00

Comment 1895 FROM OLD HEAPS.VAL.EST.
Management: Chief Agent 1869-1870 JOHN TRUSCOTT
Secretary 1869-1870 WM.THORNE

TRELAWNEY,NORTH   ST.IVE   SX 297656 1332

Production: Lead & Silver Ore(tons) Lead(tons) Silver(ozs) Value(£)

	Ore(tons)	Lead(tons)	Silver(ozs)	Value(£)
1852	13.00	9.50	260.00	0.00
1853	5.00	3.00	60.00	0.00
1854	35.90	20.50	290.00	0.00
1855	108.20	74.50	1124.00	0.00
1856	21.00	10.50	185.00	0.00
1857	33.50	25.00	275.00	0.00
1858	No detailed return			
1859	47.00	35.00	1820.00	0.00
1860	No detailed return			
1861	63.50	0.00	0.00	0.00

Ownership: Comment 1864–1865 SUSPENDED
Management: Manager 1859–1863 HY.HODGE
     Secretary 1859–1865 MART.RICKARD (P)

TRELAWNY   1333

Management: Chief Agent 1866 E.H.DINGLE

TRELEATHER,NORTH   PADSTOW   SW 912768 1334

Production: Lead & Silver No detailed return
     Copper No detailed return
Management: Manager 1860–1872 JOS.HARRIS
     Chief Agent 1860–1872 N.FAULL
     Secretary 1860–1872 G.OXLEY (P)

TRELEIGH CONSOLS,NEW   REDRUTH   SW 696439 1336

Production: Zinc   Ore(tons) Metal(tons) Value(£)

	Ore(tons)	Metal(tons)	Value(£)
1869	7.30	0.00	5.40

Comment 1869 NEW TRELEIGH

Copper   Ore(tons) Metal(tons) Value(£)

	Ore(tons)	Metal(tons)	Value(£)
1845	1637.00	123.00	9269.80
1846	2151.00	169.50	12190.70
1847	2308.00	217.90	16213.80
1848	1266.00	112.80	6956.20
1849	1254.00	102.60	6872.90
1850	1355.00	89.40	6105.20
1851	1137.00	81.80	5451.40
1852	980.00	72.10	6198.60
1853	343.00	19.20	1750.00
1854	119.00	6.70	647.70
1855	197.00	6.20	538.90
1859	207.00	10.80	926.60
1860	203.00	13.00	1160.60
1861	555.00	33.40	2985.30
1862	177.00	9.00	693.80
1863	485.00	22.10	1617.40
1864	523.00	21.00	1650.90
1865	224.00	6.50	398.50
1866	212.00	11.70	762.10

TRELEIGH CONSOLS,NEW   REDRUTH   Continued

             Copper   Ore(tons) Metal(tons)   Value(£)
             1867   95.00   4.80   332.20
             1868   23.00   1.40   104.60
             1869   192.00   10.30   602.10
             1870   39.00   1.30   52.30
             Comment 1845-1855 (C)TRELEIGH CONSOLS; 1859-1865 (C)NEW
             TRELEIGH; 1866 (C)NEW TRELEIGH AND TRELEIGH CONSOLS AGG.;
             1867-1870 (C)NEW TRELEIGH
             Tin No detailed return
Management: Manager 1859-1871 FRAN.PRYOR
             Chief Agent 1859-1863 JOHN PRINCE JNR; 1864-1865 SML.MICHELL;
             1866-1868 SML.MICHELL J.MICHELL; 1869-1871 SML.MICHELL
             Secretary 1859-1863 FRAN.PRYOR (P); 1864-1871 W.NICHOLSON
             (P)

TRELEIGH WOOD   REDRUTH   SW 699433 1337

Production: Copper   Ore(tons) Metal(tons)   Value(£)
             1875   13.00   1.20   94.20
             1876   10.00   0.80   55.20
             1877   14.60   1.00   53.00
             Comment 1875-1876 (C)
             Tin   Black(tons) Stuff(tons)   Tin(tons)   Value(£)
             1873   64.10   0.00   0.00   4291.00
             1874   116.20   0.00   0.00   5910.60
             1875   95.70   0.00   0.00   4329.70
             1876   50.10   0.00   0.00   1904.30
             1877   131.30   0.00   0.00   4843.70
             1878   75.70   0.00   0.00   2670.00
             1879   0.00   0.00   0.00   20.50
             Comment 1879 TIN STUFF
Ownership: Comment 1876 TRELEIGH; 1880-1881 NOT WORKED
Management: Manager 1875-1876 EDW.HOSKING
             Chief Agent 1872-1874 EDW.HOSKING J.HAWIS JOHN WILLIAMS; 1875
             S.GOLDSWORTHY JOHN WILLIAMS; 1876 S.GOLDSWORTHY; 1877-1881
             WM.GOLDSWORTHY
             Secretary 1872-1875 JOHN WATSON; 1876-1881 T.B.LAWS
Employment: Underground   Surface   Total
             1878   58   50   108

TRELEIGH WOOD UNITED   REDRUTH   SW 693443 1338

Production: Tin   Black(tons) Stuff(tons)   Tin(tons)   Value(£)
             1873   0.00   35.90   0.00   23.90
             1874   0.00   42.30   0.00   23.60
Ownership: Comment 1872-1874 FORMERLY NORTH TOLGUS; 1875 SUSPENDED
Management: Chief Agent 1872-1874 THOS.PRYOR JEFFREY BROWNE; 1875
             THOS.PRYOR
             Secretary 1872-1875 F.R.WILSON

TRELEIGH WOOD,EAST   REDRUTH   SW 707447 1339

Production: Tin No detailed return
Ownership: 1872 LEAN JOSE & CO.; 1875 LEAN JOSE & CO.
            Comment 1873-1874 SEE PRUSSIA
Management: Manager 1872 LEN.TREGAY; 1875 LEN.TREGAY

TRELEIGH WOOD,NORTH   REDRUTH   1340

Production: Zinc   Ore(tons) Metal(tons)   Value(£)
            1876   21.00   0.00   95.00
            Comment 1876 NORTH TRELEIGH
            Copper   Ore(tons) Metal(tons)   Value(£)
            1876   2.00   0.20   18.00
            Comment 1876 (C)
            Tin No detailed return
Management: Manager 1872-1874 WM.TREGAY

TRELIVER   ST.COLUMB MAJOR   SW 921606 1341

Production: Iron   Ore(tons)   Iron(%)   Value(£)
            1858   977.50   0.00   219.90
            1859   326.50   0.00   73.50
            1860   37.00   0.00   14.80
            1861   37.00   0.00   9.20
            Comment 1858-1861 BH.
Ownership: 1863-1865 EDW.CARTER & EXECUTORS OF T.WHITFORD
            Comment 1875-1876 TRELIVER LEVEL SEE ROSTIDYON
Management: Chief Agent 1863-1865 WM.BUCKTHOUGHT

TRELOGAN   ST.COLUMB MINOR   SW 823611 1342

Production: Lead   Ore(tons) Metal(tons)   Value(£)
            1852   11.50   8.20   0.00
            1853-1854 No detailed return

TRELOW   ST.ISSEY   SW 921695 1343

Production: Lead & Silver No detailed return
Ownership: 1876-1877 TRELOW MINING CO.
Management: Chief Agent 1876-1877 T.ALDERSON BROWNE

TRELOWETH   LELANT   SW 989499 1288

Production: Copper   Ore(tons) Metal(tons)   Value(£)
            1854   182.00   11.90   1204.70
            1855   385.00   18.10   1765.20
            1856   376.00   22.40   2083.00
            1857   274.00   19.90   2023.50
            1858   391.00   23.70   2120.80
            1859   586.00   30.00   2671.30
            1860   1030.00   59.70   5447.20

                                  503

TRELOWETH   LELANT   Continued

```
              Copper          Ore(tons) Metal(tons)    Value(£)
              1861             704.00      61.10        5572.00
              1862             548.00      43.20        3547.20
              1863             640.00      43.30        3348.90
              1864             717.00      47.40        4016.30
              1865             440.00      22.90        1834.30
              1866              10.00       0.60          47.20
              Comment 1854-1866 (C)
Ownership:    Comment 1865 SUSPENDED
Management:   Manager 1859-1864 THOS.RICHARDS
              Chief Agent 1859 M.JAMES; 1860-1864 WM.ROSEWARNE
              Secretary 1859-1865 THOS.RICHARDS (P)
```


TRELOWETH   ST.MEWAN   1344

```
Production:   Tin          Black(tons) Stuff(tons)   Tin(tons)   Value(£)
             1896-1899 No detailed return
             1900              6.00       0.00        0.00        443.00
             1901              4.00       0.00        0.00        280.00
             Comment 1896-1897 SEE SOUTH POLGOOTH; 1898-1899 TRELOWETH
             MEWAN SEE SOUTH POLGOOTH
Ownership:   1896-1900 J.NORFOLK; 1901 J.NORFOLK & SONS TRUSTEES;
             1903-1904 C.VIVIAN LADDS
             Comment 1896 PART OF SOUTH POLGOOTH; 1897-1901 TRELOWETH
             SHAFT SOUTH POLGOOTH; 1902 SEE SOUTH POLGOOTH; 1903 OR SOUTH
             POLGOOTH; 1904 OR SOUTH POLGOOTH SUSPENDED; 1905 SEE SOUTH
             POLGOOTH
Management:  Chief Agent 1896-1900 WM.JOHNS
Employment:               Underground     Surface      Total
             1896             4              3           7
             1897             5              4           9
             1898             5              3           8
             1899             8              4          12
             1900             8              3          11
             1901             8              4          12
             Comment 1896-1901 INC.SOUTH POLGOOTH
```


TRELYON CONSOLS   ST.IVES   SW 518388 1345

```
Production:  Copper          Ore(tons) Metal(tons)    Value(£)
            1851             100.00       9.10         635.50
            Comment 1851 (C)
            Tin          Black(tons) Stuff(tons)   Tin(tons)   Value(£)
            1853             75.20       0.00        0.00        4569.90
            1854             46.90       0.00        0.00        3284.60
            1855             30.70       0.00        0.00        1964.80
            1856             31.40       0.00        0.00        2183.30
            1857             29.80       0.00        0.00        2247.00
            1858             36.80       0.00        0.00        2317.10
            1859             53.40       0.00        0.00        3884.30
            1860             73.00       0.00        0.00        5780.50
            1861             48.40       0.00        0.00        3544.80
            1862             72.90       0.00        0.00        4624.20
```

504

TRELYON CONSOLS   ST.IVES   Continued

Tin	Black(tons)	Stuff(tons)	Tin(tons)	Value(£)
1863	68.70	0.00	0.00	4614.90
1864	58.50	0.00	0.00	3698.90
1865	68.50	0.00	0.00	3813.60
1866	75.10	0.00	0.00	3630.70
1867	87.30	0.00	0.00	4613.10
1868	80.40	0.00	0.00	4450.90
1869	59.00	0.00	0.00	4110.40
1870	51.60	0.00	0.00	3772.40
1871	55.50	0.00	0.00	4367.60
1872	50.10	0.00	0.00	4306.80
1873	47.10	0.00	0.00	3575.60
1874	13.60	0.00	0.00	729.20

Comment 1862 TWO RETURNS AGGREGATED
Ownership: Comment 1863-1872 LATE VENTURE
Management: Manager 1859-1869 RICH.JAMES; 1870-1873 ED.POOLEY
              Chief Agent 1859-1869 ED.POOLEY
              Secretary 1859-1861 GEO.HIGGS (P); 1862-1872 SML.HIGGS & SON;
              1873 SML.HIGGS

TREMADOC   ST.NEOT   1346

Production:
Tin	Black(tons)	Stuff(tons)	Tin(tons)	Value(£)
1873	3.80	0.00	0.00	112.10
1875	3.00	0.00	0.00	139.50

Ownership: Comment 1872 TREMADOC BRIDGE
Management: Manager 1872 WM.BAWDEN
              Chief Agent 1873-1876 NICH.VIVIAN
              Secretary 1872-1875 W.T.HANCOCK; 1876 HERB.J.MARSHALL

TREMAYNE   GWINEAR   SW 591348 1347

Production:
Copper	Ore(tons)	Metal(tons)	Value(£)
1848	864.00	58.30	3130.30
1849	1646.00	85.20	5030.90
1850	834.00	60.40	4026.90
1851	525.00	28.50	1724.10
1852	756.00	42.80	3434.90
1853	289.00	14.80	1557.30
1856	276.00	15.40	1355.40
1863	3.00	0.40	27.30
1865	8.00	1.00	83.40
1868	55.00	3.30	213.60

Comment 1848-1853 (C); 1856 (C); 1863 (C); 1865 (C); 1868
(C)

Tin	Black(tons)	Stuff(tons)	Tin(tons)	Value(£)
1852	63.40	0.00	0.00	0.00
1853	44.00	0.00	0.00	3023.70
1855	159.00	0.00	0.00	10912.80
1856	118.30	0.00	0.00	9082.00
1857	85.30	0.00	0.00	7134.80
1858	78.30	0.00	0.00	5314.60
1859	100.70	0.00	0.00	7436.30

TREMAYNE   GWINEAR   Continued

Tin	Black(tons)	Stuff(tons)	Tin(tons)	Value(£)
1860	98.10	0.00	0.00	7942.80
1861	76.80	0.00	0.00	5790.40
1862	182.10	0.00	0.00	12603.00
1863	191.30	0.00	0.00	13537.60
1864	143.90	0.00	0.00	9543.50
1865	142.90	0.00	0.00	8279.40
1866	35.80	0.00	0.00	1979.50
1867	8.80	0.00	0.00	631.70
1868	0.00	0.00	0.00	274.10
1869	0.00	0.00	0.00	183.90
1870	0.00	0.00	0.00	192.70
1871	0.00	0.00	0.00	122.00
1872	0.00	0.00	0.00	107.20

Comment 1852 TIN ORE; 1862 TWO RETURNS AGGREGATED; 1866 TWO
RETURNS AGGREGATED; 1867 VALUE INC.TIN STUFF
Management: Manager 1859-1866 R.WILLIAMS
Chief Agent 1859-1866 JOHN WILLIAMS
Secretary 1859 R.R.MICHELL (P); 1860-1862 THOS.FIELD JNR (P);
1863-1866 THOS.W.FIELD

TREMAYNE,WEST   GWINEAR   SW 565338 1348

Production: Copper

	Ore(tons)	Metal(tons)	Value(£)
1869	49.00	3.00	181.80
1870	18.00	1.30	76.50

Comment 1869-1870 (C)

Tin	Black(tons)	Stuff(tons)	Tin(tons)	Value(£)
1870	0.00	0.00	0.00	38.20
1872	0.00	0.00	0.00	110.20

Comment 1870 TIN STUFF; 1872 TIN STUFF
Management: Manager 1868 STEP.ROBERTS

TREMENHERE   WENDRON   SW 678296 1349

Production: Copper

	Ore(tons)	Metal(tons)	Value(£)
1873	18.00	2.50	163.70
1874	14.00	1.60	114.40
1875	19.00	1.40	105.80

Comment 1873-1875 (C)

Tin	Black(tons)	Stuff(tons)	Tin(tons)	Value(£)
1858-1867 No detailed return				
1871	3.30	0.00	0.00	241.60
1872	2.30	0.00	0.00	171.70
1873	17.20	0.00	0.00	1082.20
1874	21.50	0.00	0.00	1073.80
1875	10.90	0.00	0.00	492.10
1876	3.70	0.00	0.00	104.90

Comment 1858-1867 SEE TREVENEN
Ownership: Comment 1859-1866 SEE TREVENEN; 1872-1873 INC.TREWORLIS
Management: Chief Agent 1868 J.M.HENTY; 1869-1871 RICH.QUENTRALL;
1872-1874 RICH.QUENTRALL WM.DUNSTAN
Secretary 1869-1871 HY.ROGERS; 1872 J.J.ROGERS; 1873-1874

          HY.ROGERS

TREMOOR   LANIVET   SX 021648 1350

Production: Iron   Ore(tons)   Iron(%)   Value(£)
            1859   347.00   0.00   86.70
            1860   245.00   0.00   61.30
            1861   375.50   0.00   93.90
            1865   No detailed return
            1876   399.80   0.00   239.40
            1877   322.60   0.00   193.60
            Comment 1859—1861 BH.& SOME SPATHOSE; 1865 BH.SEE PAWTON;
            1876—1877 BH.& SOME SPATHOSE
Ownership: 1863—1865 EDW.CARTER & CO.; 1874 TREMORE IRON ORE CO.; 1877
            WM.WARD & SONS
            Comment 1874 TREMORE; 1877 TREEMOOR
Management: Chief Agent 1863—1865 THOS.HAMBLY; 1874 SML.HOCKADAY; 1877
            THOS.HAMBLY

TRENANCE   MULLION   SW 674173 1351

Production: Copper No detailed return
Ownership: Comment 1860—1865 SUSPENDED

TRENCROM   LELANT   SW 515368 1352

Production: Copper   Ore(tons) Metal(tons)   Value(£)
            1862   2.00   0.60   51.90
            Comment 1862 (C)TRENEROM
            Tin   Black(tons) Stuff(tons)   Tin(tons)   Value(£)
            1861   48.80   0.00   0.00   3120.40
            1862   91.40   0.00   0.00   5435.10
            1863   93.20   0.00   0.00   5577.10
            1864   118.70   0.00   0.00   6612.50
            1865   129.30   0.00   0.00   6536.30
            1866   130.90   0.00   0.00   5726.40
            1867   65.80   0.00   0.00   3110.40
            1872   2.60   0.00   0.00   202.30
            1873   25.10   0.00   0.00   1628.30
            1874   107.70   0.00   0.00   5321.70
            1875   123.50   0.00   0.00   5457.90
            1877   No detailed return
            1913   5.90   0.00   0.00   703.00
            Comment 1862 TWO RETURNS AGGREGATED; 1875 96 ORE,£4299 RET.
            IN W.C.DIST.; 1877 SEE MARY (LELANT) & ALSO SISTERS; 1913
            TRENCROM HILL
Ownership: 1911—1913 M.MILTON & H.MILTON
            Comment 1877—1881 SEE MARY (LELANT) & SEE SISTERS; 1911—1913
            TRENCROM HILL
Management: Manager 1859 THOS.RICHARDS; 1872—1874 RICH.JAMES
            Chief Agent 1859 RICH.HOLLOW; 1860—1861 FRANK BENNETTS
            RICH.HOLLOW; 1862 WM.ARTHUR RICH.HOLLOW; 1863—1864 WM.ARTHUR

CM-S   507

TRENCROM   LELANT   Continued

HY.WOOLCOCK; 1865 WM.ARTHUR WM.ROSEWARNE; 1870–1871
RICH.JAMES; 1872–1874 NICH.RICHARDS; 1875 NICH.RICHARDS
MAT.CURNOW; 1911–1913 A.GERRY
Secretary 1859–1861 R.R.MICHELL (P); 1862–1865 THOS.RICHARDS;
1870 R.R.MICHELL; 1871 THOS.W.FIELD; 1872–1875 FIELD &
MICHELL

Employment:	Underground	Surface	Total
1911	4	9	13
1912	4	10	14
1913	16	5	21

TRENITHICK   WENDRON   SW 673297 1353

Production: Tin No detailed return
Ownership: Comment 1859 SEE TREWORLIS; 1860–1861 TREVITHICK SEE
TREWORLIS; 1863–1868 SEE TREWORLIS

TRENOW CONSOLS   PERRANUTHNOE   SW 533297 1354

Production: Copper	Ore(tons)	Metal(tons)	Value(£)
1845	2306.00	257.10	19565.70
1846	1786.00	166.00	11845.20
1847	433.00	32.70	2412.20
1855	619.00	38.60	3907.00
1856	275.00	16.30	1437.40

Comment 1845–1847 (C); 1855–1856 (C)

Tin	Black(tons)	Stuff(tons)	Tin(tons)	Value(£)
1857	0.00	0.00	0.00	1413.00
1912	No detailed return			

Comment 1857 TIN STUFF; 1912 TRENOWE
Ownership: 1912 JAS.VERRANT
Comment 1912 SINKING OCT.1912 ABANDONED JAN.1913

TRENOWETH   ST.AUSTELL   1356

Production: Tin	Black(tons)	Stuff(tons)	Tin(tons)	Value(£)
1899	0.00	27.00	0.00	25.00

Comment 1899 OPEN WORK
Ownership: 1899–1901 E.A.TREGILGAS

Employment:	Underground	Surface	Total
1899	2	2	4
1900	5		5
1901		4	4

TRENOWETH,EAST   TEWINGTON   1357

Production: Tin	Black(tons)	Stuff(tons)	Tin(tons)	Value(£)
1877	11.30	0.00	0.00	453.80
1878	29.70	0.00	0.00	1116.90
1879	89.60	0.00	0.00	3337.70
1880	11.10	0.00	0.00	425.40

TRENOWETH,EAST   TEWINGTON   Continued

Tin	Black(tons)	Stuff(tons)	Tin(tons)	Value(£)
1882	0.90	0.00	0.60	58.00
1886	5.80	0.00	0.00	272.00
1887	4.60	0.00	0.00	225.00
1888	4.60	0.00	0.00	239.00
1889	2.90	0.00	0.00	123.00
1890	25.30	0.00	0.00	1508.00
1891	No detailed return			

TRENUTE   LEZANT   SX 329790 1358

Production: Antimony No detailed return
Ownership: 1909-1913 LEO VAN DE WATER
           Comment 1910-1913 NOT WORKED
Management: Chief Agent 1909-1913 J.JOLL

TRENWITH   ST.IVES   SW 573402 1359

Production: Copper

	Ore(tons)	Metal(tons)	Value(£)
1845	203.00	21.20	1535.20
1846	79.00	6.90	485.80
1847	No detailed return		
1855	87.00	5.60	558.10
1856	89.00	9.30	880.70
1863	2.00	0.30	23.20

Comment 1845-1846 (C); 1855-1856 (C); 1863 (C)

Tin	Black(tons)	Stuff(tons)	Tin(tons)	Value(£)
1856	17.70	0.00	0.00	1062.80
1857	1.20	0.00	0.00	83.30
1912	3.50	0.00	0.00	436.00
1913	5.20	0.00	0.00	591.00

Plumbago No detailed return

Uranium	Ore(tons)	Value(£)
1911	67.00	0.00
1912	42.00	0.00
1913	95.00	0.00

Ownership: 1908 ST.IVES CONSOLIDATED MINES LTD.; 1909 BRITISH RADIUM
           CORPORATION; 1910-1911 ST.IVES CONSOLIDATED MINES LTD.;
           1912-1913 BRITISH RADIUM CORPORATION LTD.
Management: Chief Agent 1910-1912 H.P.ROBERTSON
Employment:

	Underground	Surface	Total
1909	35	30	65
1910	46	32	78
1911	53	18	71
1912	45	17	62
1913	26	9	35

TREORE   PORT GAVERNE   SX 020800 1360

Production: Antimony No detailed return
Ownership: 1885 TREORE MINING CO.
Management: Chief Agent 1885 FRANK YOUNG

509

TRERANK   ROCHE   SW 980600 1361

Production: Iron   Ore(tons)   Iron(%)   Value(£)
            1866   124.00   0.00   37.20
            1867   540.00   0.00   175.20
            1868   508.70   0.00   152.50
            1869   No detailed return
            Comment 1866-1869 BH.
Ownership: 1865-1868 WM.BROWNE
Management: Chief Agent 1865-1868 WM.HOOPER

TREREW   CRANTOCK   SW 814581 1362

Production: Lead No detailed return
Ownership: Comment 1863-1865 SUSPENDED
Management: Manager 1869 T.HANCOCK; 1870-1872 T.HANCOCK J.GILBERT
            Chief Agent 1860-1862 JOHN MIDDLETON
            Secretary 1860-1865 JOHN MIDDLETON (P); 1869-1872
            J.P.BENNETTS

TRESAVEAN   GWENNAP   SW 721394 1363

Production: Lead   Ore(tons) Metal(tons)   Value(£)
            1856   4.00   2.60   0.00
            1869   0.60   0.40   0.00
            1870   No detailed return
            Comment 1867-1869 FOR SILVER SEE GREAT RETALLACK; 1869-1870
            INC TRETHARUP; 1870 INC.TRETHARUP
            Copper   Ore(tons) Metal(tons)   Value(£)
            1845   6683.00   375.90   26194.20
            1846   5855.00   329.50   22007.30
            1847   4588.00   291.20   20271.70
            1848   3723.00   227.60   12910.90
            1849   3334.00   184.40   11484.20
            1850   2680.00   128.00   8042.10
            1851   1850.00   86.10   5100.90
            1852   2565.00   100.10   7669.00
            1853   3212.00   129.50   11461.50
            1854   3990.00   156.70   14402.40
            1855   3169.00   125.60   11774.50
            1856   2577.00   93.00   7599.30
            1857   1946.00   71.80   6284.70
            1858   865.00   30.30   2425.50
            1859   227.00   7.70   624.70
            1860   443.00   16.00   1298.10
            1861   489.00   16.40   1309.30
            1862   298.00   9.40   659.40
            1863   101.00   3.20   218.20
            1864   125.00   4.20   315.20
            1865   67.00   2.70   185.70
            1866   121.00   5.30   331.10
            1867   143.00   5.60   364.30
            1868   134.00   6.60   412.80
            1869   136.00   7.80   447.50
            1870   33.00   1.80   95.80

                               510

Copper	Ore(tons)	Metal(tons)	Value(£)
1873	5.00	0.30	14.40
1877	1.20	0.10	5.30

Comment 1845-1870 (C); 1873 (C)

Tin	Black(tons)	Stuff(tons)	Tin(tons)	Value(£)
1855	12.20	0.00	0.00	730.20
1856	4.30	0.00	0.00	274.80
1857	3.40	0.00	0.00	142.80
1858	2.00	0.00	0.00	53.50
1860	2.60	0.00	0.00	196.70
1861	3.60	0.00	0.00	159.30
1862	8.80	0.00	0.00	472.70
1863	3.80	0.00	0.00	190.90
1864	0.40	0.00	0.00	22.30
1869	0.00	0.00	0.00	82.40
1871	0.00	943.10	0.00	1288.00
1872	8.10	0.00	0.00	688.90
1874	0.00	388.40	0.00	592.20
1875	0.00	180.90	0.00	181.60
1876	0.00	31.60	0.00	13.40
1877	0.00	17.50	0.00	16.60
1883	25.50	0.00	0.00	1300.00
1884	12.40	0.00	0.00	522.00
1885	13.20	0.00	0.00	608.00
1886	4.10	0.00	0.00	223.00
1901	0.00	81.00	0.00	64.00
1907	0.00	3.00	0.00	4.00
1908	0.10	0.00	0.00	7.00
1910	0.00	3.80	0.00	3.00

Comment 1860-1861 INC.TRETHARUP; 1862 INC.TRETHARUP. TWO
RETURNS AGG.; 1863 INC.TRETHARUP. TWO RETURNS AGG..VALUE
INC.TIN STUFF; 1864 INC.TRETHARUP; 1869 INC.TRETHARUP TIN
STUFF; 1871-1872 INC.TRETHARUP

Ownership: 1880-1886 TRESAVEAN MINING CO.LTD.; 1900-1901 TRESAVEAN
DEVELOPING SYNDICATE; 1906 THOS.BLANDFORD; 1907-1913
TRESAVEAN MINES LTD.
Comment 1859-1872 INC.TRETHARUP; 1873 INC.TRETHARUP STOPPED
IN 1873 & IN 1874; 1886 NOT WORKED; 1900 TRESAVEAN
UNITED.INC.BREWER,COMFORD,TRETHELLAN & TREVISKEY; 1901
TRESAVEAN UNITED.INC.BREWER,COMFORD,TRETHALLAN &
TREVISKEY.STOPPED DEC.1901; 1910-1913 UNWATERING

Management: Manager 1869 JOS.ODGERS; 1873-1875 JOHN BLIGHT; 1881
JOH.JAMES
Chief Agent 1859-1861 JOHN POPE JOS.ODGERS; 1862 JOHN POPE
JOS.ODGERS JOHN DAVEY; 1863-1868 JAS.POPE JOS.ODGERS; 1870
JOS.ODGERS JOHN BLIGHT; 1871 MOYLE JOHN BLIGHT; 1872
THOS.EDWARDS JOHN BLIGHT; 1873 THOS.EDWARDS; 1874-1875
E.H.JAMES; 1876-1877 J.MICHELL; 1882-1886 JOS.PRISK;
1909-1910 C.BRACKENBURY; 1911-1912 A.LANGLEY
Secretary 1859-1871 EDM.MICHELL (P); 1872-1873 HY.ROGERS;
1874-1875 E.MICHELL; 1876-1877 E.MICHELL I.MICHELL; 1880-1881
ED.HARVEY (S); 1900-1901 WM.HAMBLY (S)

Employment: Underground   Surface   Total
        1881   16   58   74
        1882   74   106   180

TRESAVEAN   GWENNAP   Continued

	Underground	Surface	Total
1883	32	48	80
1884	22	22	44
1885	20	19	39
1901	4		4
1907	19	21	40
1908	16	50	66
1909	28	42	70
1910	56	59	115
1911	80	40	120
1912	76	63	139
1913	106	84	190

TRESAVEAN BARRIER   GWENNAP   1364

Production: Copper   Ore(tons) Metal(tons)   Value(£)
            1845   874.00   83.10   6055.00
            1846   430.00   30.00   2046.20
            1847   545.00   44.70   3193.00
            1848   277.00   18.90   1068.40
            Comment 1845-1848 (C)

TRESAVEAN,NORTH   GWENNAP   SW 718395 1366

Production: Copper   Ore(tons) Metal(tons)   Value(£)
            1879-1885 No detailed return
            Comment 1879-1885 SEE COMFORD
            Tin   Black(tons) Stuff(tons)   Tin(tons)   Value(£)
            1880-1884 No detailed return
            Comment 1880-1884 SEE COMFORD; 1893 SEE COMFORD
Ownership: Comment 1879-1886 SEE COMFORD
Employment: Underground   Surface   Total
            1880-1885
            Comment 1880-1885 SEE COMFORD

TRESAVEAN,SOUTH   GWENNAP   SW 761378 1367

Production: Copper No detailed return
Ownership: Comment 1863-1864 SUSPENDED; 1865 SEE ROSKROW UNITED
Management: Chief Agent 1860-1861 S.WHITBURN; 1862 T.ROGERS
            Secretary 1860-1862 C.R.WEBB (P)

TRESAVEAN,WEST   GWENNAP   SW 715397 1368

Production: Tin No detailed return
Ownership: 1877-1879 WEST TRESAVEAN MINING CO.
Management: Chief Agent 1877-1879 THOS.PARKYN

TRESELLYN   ALTARNUN   SX 188788 1369

Production: Tin   Black(tons) Stuff(tons)   Tin(tons)   Value(£)
            1887   0.30   0.00   0.00   15.00
            1888   0.70   0.00   0.00   47.00
            1889   0.10   0.00   0.00   8.00
            1892   0.10   0.00   0.00   3.00
            1893   2.00   0.00   0.00   106.00
            1894   0.80   0.00   0.00   26.00
            Comment 1887-1889 TRESILLAN; 1892-1894 TRESILLAN
Ownership: 1877-1878 TRESELLYN TIN MINING CO.LTD.; 1886 TRESILLAN TIN
            CO.; 1887-1891 WILD REED & MAYNE; 1892-1894 REED & MAYNE;
            1912-1913 THOS.BRENTON
            Comment 1859-1861 TRESLYON; 1862-1867 TRELYON; 1869-1871
            TREVELLYN; 1878 IN LIQUIDATION; 1886-1889 TRESILLAN;
            1890-1891 TRESILLAN SUSPENDED; 1892-1893 TRESILLIAN; 1894
            TRESILLIAN NOT WORKING SINCE JUNE; 1912 TRESELLIAN
            PROSPECTING NOV.1912; 1913 TRESELLIAN SUSPENDED JAN.1914
Management: Manager 1859-1867 JOHN SPARGO; 1869-1872 M.W.BAWDEN;
            1873-1878 JOS.HODGE
            Chief Agent 1859-1860 JOHN SPARGO; 1886-1894 SML.MAYNE
            Secretary 1859-1867 JOHN SPARGO (P); 1869-1872 ROBT.BELL;
            1873-1878 B.CLARKE (S)
Employment: Underground   Surface   Total
            1878   3   3
            1886   10   10   20
            1887   8   6   14
            1888   8   3   11
            1889   3   3
            1891   1   1
            1892   1   2   3
            1893   4   6   10
            1894   4   2   6
            1912   5   5

TRESELYAN & SCADDICK CONSOLS   ALTARNUN   SX 188788 1370

Production: Copper No detailed return
            Tin No detailed return
Ownership: Comment 1860-1867 TRESELYAN & SCADDICK CONSOLS
Management: Chief Agent 1860-1867 JOHN SPARGO
            Secretary 1861-1867 G.J.SOPER (P)

TRESIBBLE   LUXULYAN   SX 017594 1371

Production: Iron   Ore(tons)   Iron(%)   Value(£)
            1859   1832.80   0.00   458.20
            1860-1861 No detailed return
            Comment 1859 BH.& SOME SPATHOSE

TRESKERBY   SCORRIER   SW 717437 1372

Production: Copper   Ore(tons) Metal(tons)   Value(£)
            1880   38.00   2.20   125.70

                             513

TRESKERBY   SCORRIER   Continued

Tin   Black(tons) Stuff(tons)   Tin(tons)   Value(£)
1889   9.00   250.00   0.00   461.00
Ownership: Comment 1859 SEE NORTH DOWNS
Management: Chief Agent 1861—1865 JOHN DAVEY
             Secretary 1861—1865 T.H.TILLY (P)

TRESKERBY,EAST   KEA   SW 728460 1373

Production: Copper   Ore(tons) Metal(tons)   Value(£)
            1863   8.00   0.50   43.80
            1864   10.00   1.10   91.70
            1865   18.00   1.20   95.30
            Comment 1863—1865 (C)
            Tin   Black(tons) Stuff(tons)   Tin(tons)   Value(£)
            1864   0.00   0.00   0.00   641.70
Ownership: Comment 1865 SUSPENDED
Management: Chief Agent 1860—1864 JOHN NANCARROW
             Secretary 1860 H.SIMS (P); 1861—1864 W.G.SPARKE (P); 1865
             HY.MICHELL

TRESKERBY,NEW   ST.AGNES   1374

Production: Copper No detailed return
Ownership: Comment 1862—1865 SUSPENDED
Management: Manager 1860—1861 W.TREGONNING
             Secretary 1860—1865 W.PAGE CARDOZA (P)

TRESKERBY,NORTH   ST.AGNES   SW 720451 1375

Production: Copper   Ore(tons) Metal(tons)   Value(£)
            1858   157.00   8.90   832.60
            1860   856.00   60.80   5472.80
            1861   1758.00   118.40   10632.80
            1862   2323.00   146.60   11831.90
            1863   2326.00   138.50   10606.50
            1864   1974.00   110.00   9322.30
            1865   2058.00   131.00   10318.70
            1866   1821.00   130.70   8570.60
            1867   1774.00   125.10   8768.00
            1868   1109.00   74.60   4881.10
            1869   965.00   73.40   4607.60
            1870   1518.00   119.30   7024.70
            1871   728.00   55.10   3379.30
            1872   156.00   11.20   872.40
            1874   21.00   1.50   120.70
            1875   41.00   2.90   207.00
            1876   59.00   4.40   305.50
            1877   143.40   11.60   670.40
            1878   132.90   8.00   444.00
            1879   194.60   10.50   577.60
            1880   5.00   0.30   15.50
            1883   1.50   0.10   4.00

                                    514

TRESKERBY,NORTH   ST.AGNES   Continued

Comment 1858 (C); 1860–1872 (C); 1874–1876 (C)
Tin	Black(tons)	Stuff(tons)	Tin(tons)	Value(£)
1860	0.00	0.00	0.00	117.60
1861	0.00	0.00	0.00	317.20
1862	0.00	0.00	0.00	1000.50
1863	0.00	0.00	0.00	1748.30
1864	0.00	0.00	0.00	718.90
1865	0.00	0.00	0.00	463.70
1866	7.60	0.00	0.00	343.80
1867	8.70	0.00	0.00	450.60
1868	12.50	0.00	0.00	692.80
1869	10.60	0.00	0.00	755.90
1870	9.70	0.00	0.00	691.10
1871	6.50	0.00	0.00	486.40
1872	7.20	0.00	0.00	576.10
1873	6.20	0.00	0.00	523.50
1874	5.50	0.00	0.00	239.60
1875	11.20	0.00	0.00	434.80
1876	15.00	2.50	0.00	505.50
1877	2.20	0.00	0.00	84.60
1878	0.00	16.30	0.00	3.40
1882	0.10	8.00	0.10	2.00
1884	0.10	10.00	0.00	2.00
1890	0.00	63.00	0.00	26.00

Comment 1860–1861 TIN STUFF; 1862 TIN STUFF. TWO RETURNS
AGGREGATED; 1863–1865 TIN STUFF
Arsenic	Ore(tons)	Metal(tons)	Value(£)
1876	13.30	0.00	45.50
1877	3.00	0.00	12.00

Arsenic Pyrite	Ore(tons)	Value(£)
1876	16.90	12.80
1877	8.40	17.50
1878	32.00	48.00
1879	35.50	17.10
1882	3.00	2.00

Comment 1876–1877 ARSENICAL MUNDIC; 1877 VAL.INC.3.1 TONS
IRON PYRITES
Ownership: 1876–1886 NORTH TRESKERBY MINING CO.; 1888–1890 J.W.VICKERS;
1891 EXECUTORS OF J.W.VICKERS
Comment 1886 NOT WORKED; 1890 NOW ABANDONED; 1891 ABANDONED
Management: Manager 1859 JOHN VIVIAN; 1860–1862 FRAN.PRYOR; 1863–1873
RICH.PRYOR; 1874–1876 RICH.PRYOR & SON; 1878–1879
MART.GEORGE
Chief Agent 1859 GEO.THOMAS; 1860–1862 RICH.KITTO; 1863–1864
JOHN TREGONNING; 1865–1866 JOHN TREGONNING THOS.JENKINS;
1867–1868 JOHN TREGONNING THOS.JENKIN; 1869–1873 THOS.JENKIN;
1877 JOHN NANCARROW; 1880 J.H.HAMBLY; 1881 J.H.HAMBLY
W.GOLDSWORTHY; 1882–1886 RICH.PRYOR; 1888–1891 WM.RICH
Secretary 1859 C.R.WEBB (P); 1860–1864 BEN.MATTHEWS (P); 1865
B.MATTHEWS J.EDWARDS; 1866–1869 BEN.MATTHEWS; 1870–1877
HY.L.PHILLIPS; 1878–1881 HY.WHITWORTH
Employment:	Underground	Surface	Total
1878	45	23	68
1879	48	19	67
1880	2	1	3

CM-S*   515

TRESKERBY,NORTH   ST.AGNES   Continued

                    Underground   Surface   Total
          1882   6   6
          1883   2   2
          1884   35   24   59
          1885   42   32   74
          1888   8   2   10
          1889   7   2   9
          1890   7   7

TRETHARUP   GWENNAP   SW 717392 1377

Production: Lead   Ore(tons) Metal(tons)   Value(£)
            1869-1870 No detailed return
            Comment 1869-1870 SEE TRESAVEAN
            Copper No detailed return
            Tin   Black(tons) Stuff(tons)   Tin(tons)   Value(£)
            1860-1864 No detailed return
            1869   No detailed return
            1871-1872 No detailed return
            Comment 1860-1864 SEE TRESAVEAN; 1869 SEE TRESAVEAN;
            1871-1872 SEE TRESAVEAN
Ownership: Comment 1859-1873 SEE TRESAVEAN

TRETHELLAN   GWENNAP   SW 717390 1378

Production: Copper   Ore(tons) Metal(tons)   Value(£)
            1845   2612.00   143.30   9978.90
            1846   1517.00   77.50   5118.10
            1847   1304.00   65.30   4313.80
            1848   1015.00   51.60   2749.70
            1849   1287.00   54.90   3132.40
            1850   1045.00   48.20   3022.30
            1851   523.00   25.60   1535.90
            1852   459.00   18.70   1392.70
            1853   517.00   17.10   1528.90
            1854   442.00   15.10   1351.90
            1855   344.00   9.80   809.90
            1856   258.00   7.30   533.10
            Comment 1845-1856 (C)
Ownership: Comment 1860-1865 SUSPENDED; 1900-1901 SEE TRESAVEAN UNITED
Management: Manager 1859 WM.RICHARDS
            Chief Agent 1859 JOHN DYER
            Secretary 1859 WM.RICHARDS (P)

TRETHELLAN,WEST   GWENNAP   SW 714386 1379

Production: Copper   Ore(tons) Metal(tons)   Value(£)
            1845   295.00   15.40   1067.70
            1846   137.00   6.70   431.30
            1847   435.00   20.20   1316.30
            1849   239.00   12.90   826.10
            1850   182.00   7.80   476.90

                                      516

TRETHELLAN,WEST   GWENNAP   Continued

      Comment 1845-1847 (C); 1849-1850 (C)

TRETHIN   ADVENT   SX 094843 1380

Production: Lead & Silver No detailed return
Ownership: 1876-1877 TRETHIN MINING CO.
Management: Chief Agent 1876-1877 RAYN.ST.STEPHEN

TRETHOSA   ST.STEPHEN-IN-BRANNEL   SW 946554 1381

Production: Iron No detailed return
Ownership: 1864 WM.BROWNE
Management: Chief Agent 1864 STEP.COLLINS

TRETHURGY   ST.AUSTELL   SX 035554 1382

Production: Iron   Ore(tons)   Iron(%)   Value(£)
           1872-1874 No detailed return
           1876   No detailed return
           Comment 1872-1874 BH.SEE RUBY; 1876 BH.SEE RUBY
Ownership: Comment 1872-1877 SEE RUBY

TRETOIL   LANIVET   SX 067641 1383

Production:	Ore(tons)	Metal(tons)	Value(£)	
Zinc				
1858	156.00	0.00	468.60	
Copper	Ore(tons)	Metal(tons)	Value(£)	
1845	658.00	51.50	3499.80	
1846	569.00	38.70	2344.40	
1847	328.00	20.50	1326.20	
1858	365.00	19.20	1705.50	
1859	100.00	5.80	441.50	
1860	405.00	21.10	1835.70	

Comment 1845-1847 (C); 1858 (C); 1859 (C)INC.MESSER; 1860
(C)

Tin	Black(tons)	Stuff(tons)	Tin(tons)	Value(£)
1858	78.00	0.00	0.00	5858.00
1859	3.80	0.00	0.00	237.10
1860	3.20	0.00	0.00	256.30
1861	15.70	0.00	0.00	1175.90
1874	12.30	0.00	0.00	625.00
1875	13.60	0.00	0.00	709.60
1882	0.70	0.00	0.50	42.00

Comment 1860-1861 INC.MESSER; 1874 VALUE EST

Iron	Ore(tons)	Iron(%)	Value(£)
1858	183.50	0.00	197.30
1859	70.00	0.00	78.70
1860-1861 No detailed return			
1873	1600.00	0.00	1200.00
1874	275.00	0.00	197.00

           Comment 1858-1861 BH.; 1873-1874 BH.

TRETOIL   LANIVET   Continued

Ownership: 1875-1876 TRETOIL TIN & IRON CO.LTD.; 1882-1885
            J.POSTLETHWAITE & CO.
            Comment 1859-1861 TRETOIL & MESSER UNITED; 1862-1865
            SUSPENDED; 1882-1885 OPEN WORK
Management: Manager 1859-1861 RICH.RICH; 1872-1875 T.HOOPER
            Chief Agent 1859-1861 THOS.GEORGE; 1864-1865 THOS.GEORGE;
            1874 W.FLETCHER PAGAN; 1883-1885 W.WILLIAMS
            Secretary 1859-1865 WM.CHARLES (P); 1872 W.FLETCHER PAGAN;
            1873-1874 JOHN FLETCHER PAGAN
Employment: Underground   Surface   Total
            1883   4   4
            1884   4   4
            1885   2   2

TREVANION   BUGLE   1384

Production: Tin   Black(tons) Stuff(tons)   Tin(tons)   Value(£)
            1872   No detailed return
            1875-1876 No detailed return
            Comment 1872 SEE TREVERBYN; 1875-1876 SEE TREVERBYN

TREVARRACK   LELANT   SW 524372 1387

Production: Tin No detailed return
Ownership: Comment 1869-1871 TREVARRACK UNITED
Management: Manager 1869-1871 THOS.UREN; 1872-1875 JAS.POPE
            Chief Agent 1872-1874 JOHN PEARCE; 1875 JOHN PEARCE
            STEP.ANDREWS
            Secretary 1869-1874 CHRIS.STEPHENS; 1875 CHRIS.STEPHENS JOHN
            B.REYNOLDS

TREVARREN UNITED MINES   ST.AUSTELL   1388

Production: Tin   Black(tons) Stuff(tons)   Tin(tons)   Value(£)
            1882-1883 No detailed return
            Comment 1882 INC.GOVER,INDIAN QUEENS & PARKA; 1883 INC.GOVER
            & PARKA.
Ownership: 1883-1884 TREVARREN UNITED MINING CO.
Management: Secretary 1883-1884 HY.BROWNE
Employment: Underground   Surface   Total
            1883   32   22   54

TREVAUNANCE   ST.AGNES   SW 711508 1389

Production: Copper   Ore(tons) Metal(tons)   Value(£)
            1884   10.00   0.00   70.00
            1885   48.00   0.00   279.00
            1886   200.00   0.00   959.00
            1887   31.00   0.00   225.00
            1888   248.00   0.00   1471.00
            1889   15.00   0.00   52.00

Comment 1884–1889 TREVAUNANCE UNITED

Tin	Black(tons)	Stuff(tons)	Tin(tons)	Value(£)
1855	24.00	0.00	0.00	1581.20
1856	29.30	0.00	0.00	2269.70
1857	27.70	0.00	0.00	2274.20
1858	32.00	0.00	0.00	2054.20
1859	27.00	0.00	0.00	1956.10
1860	36.00	0.00	0.00	2835.80
1861	26.70	0.00	0.00	1919.30
1862	27.80	0.00	0.00	1799.60
1863	25.20	0.00	0.00	1719.90
1864	15.80	0.00	0.00	1036.80
1865	29.10	0.00	0.00	1592.60
1866	10.00	0.00	0.00	466.80
1867	30.50	0.00	0.00	1625.30
1868	12.20	0.00	0.00	666.60
1869	19.90	0.00	0.00	1422.10
1870	20.40	0.00	0.00	1570.30
1871	34.00	0.00	0.00	2709.10
1872	30.70	0.00	0.00	2650.60
1873	24.20	0.00	0.00	1786.60
1874	18.90	0.00	0.00	855.00
1875	8.00	0.00	0.00	400.10
1876	9.00	0.00	0.00	400.50
1877	12.00	0.00	0.00	456.50
1878	7.20	0.00	0.00	261.20
1879	8.10	0.00	0.00	324.00
1880	10.50	0.00	0.00	530.70
1881	9.60	0.00	5.90	556.80
1882	4.00	0.00	2.80	234.00
1883	6.60	0.00	0.00	351.00
1884	10.00	0.00	0.00	452.00
1906	6.00	0.00	0.00	603.00
1907	5.00	0.00	0.00	568.00
1908	7.90	0.00	0.00	654.00
1909	3.70	0.00	0.00	312.00
1910	4.40	0.00	0.00	464.00
1911	14.00	0.00	0.00	1681.00
1912	24.50	0.00	0.00	3040.00
1913	6.70	0.00	0.00	704.00

Comment 1855 TREVANANCE; 1862 TWO RETURNS AGGREGATED;
1883–1884 TREVAUNANCE UNITED
Ownership: 1882–1891 TREVAUNANCE UNITED ADVENTURERS; 1906–1911
A.S.B.SAWLE; 1912–1913 TREVAUNANCE MINE CO.LTD.
Comment 1887–1890 TREVAUNANCE UNITED; 1891 TREVANANCE UNITED
SUSPENDED; 1892–1902 SEE POLBERRO CONSOLS
Management: Manager 1859–1860 RICH.NEWTON; 1879–1881 JOHN SMITH
Chief Agent 1859–1861 RICH.NEWTON; 1862–1863 JOHN GOYNE;
1864–1866 JAS.NANCE; 1867–1873 JOHN BLIGHT; 1874 JOSH.NEWTON;
1875–1879 G.COULTER HANCOCK; 1882–1890 WM.VIVIAN
Secretary 1859–1874 JOSH.NEWTON (P); 1875–1881 F.GILB ENNYS

Employment:	Underground	Surface	Total
1878	15		15
1879	13		13
1880	9	7	16

	Underground	Surface	Total
1881	6		6
1882	9		9
1883	15	2	17
1884	22	17	39
1885	23	9	32
1886	24	6	30
1887	23	7	30
1888	20	7	27
1889	23	6	29
1890	16	4	20
1906	5	2	7
1907	4	2	6
1908-1909	4	1	5
1910	12	6	18
1911	12	8	20
1912-1913	14	15	29

TREVEDDOE,GREAT   WARLEGGAN   SX 151695 1390

Production: Copper

	Ore(tons)	Metal(tons)	Value(£)
1862	17.00	0.80	57.50
1901	154.00	0.00	586.00
1902	158.00	0.00	370.00
1903	1081.00	0.00	3151.00
1904	570.00	0.00	978.00
1906	17.00	0.00	141.00
1908	110.00	0.00	564.00
1909	166.00	0.00	708.00
1910	46.00	0.00	137.00
1911	34.00	0.00	153.00

Comment 1862 (C); 1901-1904 TREVODDOE; 1906 TREVODDOE;
1908-1911 TREVODDOE

Tin	Black(tons)	Stuff(tons)	Tin(tons)	Value(£)
1860	44.50	0.00	0.00	2676.40
1861	13.10	0.00	0.00	735.50
1862	1.00	0.00	0.00	52.90
1863	17.70	0.00	0.00	893.50
1864	15.30	0.00	0.00	749.10
1865	11.80	0.00	0.00	517.40
1866	19.40	0.00	0.00	752.30
1867	8.60	0.00	0.00	379.60
1868	8.30	0.00	0.00	374.90
1869	9.30	0.00	0.00	484.80
1870	15.50	0.00	0.00	787.90
1871	20.70	0.00	0.00	1231.30
1887	9.00	0.00	0.00	405.00
1888	12.40	0.00	8.10	574.00
1890	26.00	0.00	0.00	1500.00
1891	No detailed return			
1898	20.20	0.00	0.00	708.00
1899	23.00	0.00	0.00	1567.00
1900	50.00	0.00	0.00	3490.00
1901	68.70	0.00	0.00	3313.00

Tin	Black(tons)	Stuff(tons)	Tin(tons)	Value(£)
1902	53.20	0.00	0.00	3404.00
1903	39.80	0.00	0.00	2594.00
1904	25.50	0.00	0.00	870.00
1905	47.00	0.00	0.00	3442.00
1906	53.00	0.00	0.00	4596.00
1907	67.00	0.00	0.00	5916.00
1908	58.00	0.00	0.00	3508.00
1909	55.00	0.00	0.00	3201.00
1910	47.00	0.00	0.00	3108.00
1911	13.50	‧0.00	0.00	1332.00
1912	34.50	0.00	0.00	4150.00
1913	37.00	0.00	0.00	3786.00

Comment 1862 INC.CABELLA PART YEAR ONLY; 1864 INC.CABELLA;
1868–1871 INC.CABELLA; 1887 FORMERLY WHISPER OPEN WORK; 1888
TREVEDDOE OPEN WORK; 1890 TREVEDDOE OPEN WORK; 1891 TREVEDDOE
OPEN WORK AT WORK; 1898 TREVEDDOE OPEN WORK AT WORK; 1899
TREVEDDOE OPEN WORK AT WORK TWO RETURNS AGG.; 1900–1913
TREVEDDOE

Ownership: 1887 TREVEDDOE MINING CO.; 1890–1891 TREVEDDOE MINING CO.;
1899 NICHOLLS & CO.; 1900–1913 TREVEDDOE MINING CO.LTD.
Comment 1864–1871 INC.CABELLA; 1872 SEE WHISPER; 1887
TREVEDDOE FORMERLY WHISPER OPEN WORK; 1890–1891 TREVEDDOE
FORMERLY WHISPER OPEN WORK; 1901–1913 TREVEDDOE

Management: Manager 1869–1871 TIM.ROWSE
Chief Agent 1861 JAS.POLGLAZE T.SIMMONS; 1862–1868
JAS.POLGLAZE; 1899 JOHN MOSS; 1900–1904 ROBT.SACK; 1905–1913
W.H.ADAMS
Secretary 1861–1868 JOHN WATSON; 1869–1871 R.LARKIN

Employment:	Underground	Surface	Total
1899	9	8	17
1900	16	23	39
1901	32	26	58
1902	32	28	60
1903	36	22	58
1904	20	12	32
1905	22	19	41
1906	30	27	57
1907	31	24	55
1908	29	21	50
1909	30	26	56
1910	25	21	46
1911	12	20	32
1912	16	10	26
1913	14	12	26

TREVEGA   TOWEDNACK   SW 482405 1391

Production: Tin	Black(tons)	Stuff(tons)	Tin(tons)	Value(£)
1907	1.60	0.00	0.00	111.00
1908	4.00	0.00	0.00	320.00
1909	6.70	0.00	0.00	550.00
1910	9.00	0.00	0.00	864.00
1911	9.00	0.00	0.00	990.00

TREVEGA   TOWEDNACK   Continued

Tin	Black(tons)	Stuff(tons)	Tin(tons)	Value(£)
1912	5.70	0.00	0.00	735.00
1913	2.00	0.00	0.00	180.00

Ownership: 1907-1908 TREVEGA MINE LTD.; 1909-1913 STURTEVANT ENGINEERING
            CO.LTD.
            Comment 1908 IN LIQUIDATION
Employment: Underground   Surface   Total
               1907   18   12   30
               1908   5   4   9
               1909   7   5   12
               1910   12   8   20
               1911   8   4   12
               1912   5   6   11
               1913   7   5   12

TREVEGEAN,GREAT   ST.JUST   SW 366289 1392

Production: Tin	Black(tons)	Stuff(tons)	Tin(tons)	Value(£)
1873	3.50	0.00	0.00	262.70
1874	8.80	0.00	0.00	393.00
1875	6.10	0.00	0.00	339.80
1876	0.60	0.00	0.00	22.00

            Comment 1873-1876 GREAT TREVAGEAN
Management: Manager 1872-1873 PETER EDDY
            Chief Agent 1874 RICH.NICHOLAS; 1875-1876 PETER EDDY
            Secretary 1872-1876 W.CHENHALLS

TREVELGUE   PORTH   1393

Production: Lead No detailed return
Ownership: 1889 G.TANGYE
Management: Secretary 1889 G.TANGYE
Employment: Underground   Surface   Total
               1889   4   4

TREVELL,SOUTH   1394

Production: Tin	Black(tons)	Stuff(tons)	Tin(tons)	Value(£)
1864	0.10	0.00	0.00	8.90

TREVELLYAN   PERRANUTHNOE   SW 554305 1395

Production: Copper No detailed return

Tin	Black(tons)	Stuff(tons)	Tin(tons)	Value(£)
1852	35.60	0.00	0.00	0.00
1853	52.60	0.00	0.00	3418.40
1854	35.80	0.00	0.00	2356.70
1855	45.80	0.00	0.00	2875.30
1856	48.00	0.00	0.00	3332.20
1857	35.80	0.00	0.00	2637.00
1858	15.70	0.00	0.00	1040.80

TREVELLYAN   PERRANUTHNOE   Continued

            Comment 1852 TREVELYAN TIN ORE; 1853—1858 TREVELYAN
Ownership: Comment 1862—1865 SUSPENDED
Management: Manager 1859 GEO.R.ODGERS
            Chief Agent 1859 JOHN OSBORNE; 1860—1861 RICH.KENDALL
            Secretary 1859—1861 JOHN WATSON (P)

TREVELLYAN,EAST   PERRANUTHNOE   SW 556305 1396

Production: Copper No detailed return
            Tin No detailed return
Ownership: Comment 1862—1865 SUSPENDED
Management: Manager 1859 PETER FLOYD
            Chief Agent 1859 PETER FLOYD
            Secretary 1859 J.D.BRUNTON (P); 1860—1861 ED.KING (P);
            1862—1863 R.R.MICHELL

TREVELLYAN,WEST   PERRANUTHNOE   SW 552307 1397

Production: Copper   Ore(tons) Metal(tons)   Value(£)
            1860   348.00   29.60   2717.00
            1861   225.00   21.00   1858.50
            1862   216.00   17.90   1487.70
            1863   112.00   9.80   773.80
            1864   76.00   5.40   458.10
            Comment 1860—1864 (C)TREVELYAN
            Tin   Black(tons) Stuff(tons)   Tin(tons)   Value(£)
            1859   0.00   0.00   0.00   108.20
            1860   0.00   0.00   0.00   17.80
            Comment 1859—1860 TIN STUFF
Ownership: Comment 1864—1865 SUSPENDED
Management: Manager 1860 GEO.R.ODGERS
            Chief Agent 1860—1863 JOHN OSBORNE
            Secretary 1860 JOHN WATSON (P); 1861 GEO.R.ODGERS (P);
            1862—1865 JOHN WATSON

TREVEN   CRANTOCK   1398

Production: Lead   Ore(tons) Metal(tons)   Value(£)
            1872   1.00   0.70   0.00
            Copper   Ore(tons) Metal(tons)   Value(£)
            1873   15.00   1.20   88.10
            Comment 1873 (C)
            Iron No detailed return
Management: Chief Agent 1872 JOHN HANCOCK

TREVEN   ST.ERTH   SW 573342 1399

Production: Tin No detailed return
Management: Manager 1869—1871 R.STEVENS
            Chief Agent 1872 SML.J.REED
            Secretary 1869—1872 R.STEVENS

                                        523

TREVENA   BREAGE   SW 608288 1400

Production: Tin   Black(tons) Stuff(tons)   Tin(tons)   Value(£)
            1852   5.40   0.00   0.00   0.00
            Comment 1852 TIN ORE

TREVENEN   WENDRON   SW 680297 1401

Production: Copper   Ore(tons) Metal(tons)   Value(£)
            1867   2.00   0.20   14.70
            Comment 1867 (C)TREVENNEN
            Tin   Black(tons) Stuff(tons)   Tin(tons)   Value(£)
            1854   36.10   0.00   0.00   2328.80
            1855   24.30   0.00   0.00   1616.00
            1856   5.60   0.00   0.00   300.90
            1858   0.00   0.00   0.00   546.70
            1859   5.10   0.00   0.00   357.60
            1860   31.80   0.00   0.00   2522.20
            1861   46.50   0.00   0.00   3668.30
            1862   44.50   0.00   0.00   3046.60
            1863   92.30   0.00   0.00   6429.00
            1864   107.40   0.00   0.00   7022.20
            1865   67.50   0.00   0.00   3912.60
            1866   80.70   0.00   0.00   4125.40
            1867   57.60   0.00   0.00   3159.40
            Comment 1858 INC.TREMENHERE TIN STUFF; 1859-1861
            INC.TREMENHERE; 1862 INC.TREMENHERE. TWO RETURNS AGG.;
            1863-1867 INC.TREMENHERE
Ownership: Comment 1859-1866 INC.TREMENHERE
Management: Manager 1859 JOHN TRURAN; 1860-1862 J.WEBB; 1863-1866
            W.MEDLAND
            Chief Agent 1859-1860 JOHN TIPPETT; 1861-1864 W.TIPPETT;
            1865-1866 C.GEORGE
            Secretary 1859-1866 C.WESCOMB (P)

TREVENNA   ST.NEOT   SX 181685 1402

Production: Copper No detailed return
            Tin No detailed return
Ownership: 1866-1872 TREVENNA TIN & COPPER MINING CO.
Management: Manager 1866-1868 W.H.WILCOCK; 1869-1872 JOHN GLUYAS

TREVENNA,SOUTH   ST.NEOT   SX 184646 1403

Management: Chief Agent 1866-1868 J.JENNINGS

TREVENNEN,EAST   WENDRON   1404

Production: Tin   Black(tons) Stuff(tons)   Tin(tons)   Value(£)
            1860   0.00   0.00   0.00   39.80
            1861   3.60   0.00   0.00   255.40
            1862   1.20   0.00   0.00   84.10
            1874   2.80   0.00   0.00   119.40

                                   524

TREVENNEN,EAST   WENDRON   Continued

        Tin   Black(tons) Stuff(tons)   Tin(tons)   Value(£)
        1877   4.00   0.00   0.00   160.10
        Comment 1860 TIN STUFF
Ownership: Comment 1859-1862 INC.WEST LOVELL; 1863-1865 INC.WEST LOVELL
            SUSPENDED
Management: Manager 1859-1862 WM.TRURAN
            Secretary 1859 T.P.TYACKE (P); 1860-1865 I.G.PLOMER (P)

TREVENNEN,NEW   WENDRON   SW 684298 1405

Production: Copper   Ore(tons) Metal(tons)   Value(£)
        1863   3.00   0.30   21.70
        Comment 1863 (C)NEW TREVENNING
        Tin   Black(tons) Stuff(tons)   Tin(tons)   Value(£)
        1862   2.20   0.00   0.00   136.40
        1863   1.30   0.00   0.00   81.50
        Comment 1862 NEW TREVENEN.TWO RETURNS AGG.
Ownership: Comment 1872-1874 SEE NEW TRUMPET & LOVELL UNITED
Management: Manager 1869-1871 JAS.EVANS
            Chief Agent 1862-1866 JOS.PHILLIPS; 1867-1871 J.JAMES
            Secretary 1862-1868 WM.CARNE; 1869-1871 C.CARNE

TREVERBYN   BUGLE   SX 033564 1406

Production: Tin   Black(tons) Stuff(tons)   Tin(tons)   Value(£)
        1852-1853 No detailed return
        1855   No detailed return
        1872   0.90   0.00   0.00   81.30
        1875   0.10   0.00   0.00   1.60
        1876   1.40   0.00   0.00   62.30
        Comment 1852-1853 SEE ROCKS; 1855 SEE ROCKS; 1872
        INC.TREVANION; 1875-1876 INC.TREVANION
        Iron   Ore(tons)   Iron(%)   Value(£)
        1865   204.50   0.00   56.20
        1872-1875 No detailed return
        Comment 1865 BH.; 1872-1875 BH.SEE KNIGHTOR
Ownership: 1864-1866 MARTIN & NICHOLLS; 1871-1873 J.DAY & CO.
            Comment 1873-1875 SEE ALSO LADOCK; 1876 SEE LADOCK
Management: Chief Agent 1864-1866 CAPT.DENNIS; 1871-1873 J.DAY

TREVERO   1407

Production: Copper   Ore(tons) Metal(tons)   Value(£)
        1869   7.00   0.60   39.20
        Comment 1869 (C)

TREVETHOE   UNY LELANT   SW 528357 1408

Production: Tin   Black(tons) Stuff(tons)   Tin(tons)   Value(£)
        1903   1.70   0.00   0.00   99.00
        1905   5.80   0.00   0.00   450.00

TREVETHOE   UNY LELANT   Continued

Ownership: 1903-1904 TREVETHOE CO.LTD.; 1905 WHEAL MERTH LTD.
            Comment 1905 IN LIQUIDATION
Management: Chief Agent 1903-1904 WM.THOMAS; 1905 JOHN DANIELL
            Secretary 1905 WM.THOMAS
Employment: Underground   Surface   Total
            1903   22   11   33
            1904   37   13   50
            1905   4   4

TREVINCE CONSOLS   GWENNAP   1409

Production: Copper   Ore(tons) Metal(tons)   Value(£)
            1880   25.70   1.60   17.70
            1881   25.70   0.50   17.70
Ownership: 1880-1881 TREVINCE CONSOLS MINING CO.
Management: Chief Agent 1880-1881 JOHN MAYNE
Employment: Underground   Surface   Total
            1880   13   13
            1881   3   3

TREVINNICK   ST.KEW   SX 007784 1410

Production: Lead & Silver   Ore(tons) Lead(tons) Silver(ozs)   Value(£)
            1892   15.00   0.00   0.00   105.00
            1893-1894 No detailed return
            Tin No detailed return
            Antimony   Ore(tons) Metal(tons)   Value(£)
            1892   6.00   2.50   98.00
            1893   No detailed return
Ownership: 1890 TREVINNICK MINING SYNDICATE; 1891-1894 TREVINNICK MINING
            SYNDICATE LTD.; 1913 STANNARIES MINING & SMELTING CO.
            Comment 1892-1894 SUSPENDED; 1913 TREVANNICK
Management: Chief Agent 1876-1878 F.C.MAITLAND; 1890-1891 RICH.RICH
            Secretary 1876-1878 W.H.MAITLAND; 1891 W.R.LOCKING; 1892-1894
            W.R.LOCKING (S)
Employment: Underground   Surface   Total
            1891   7   1   8
            1892   6   3   9
            1913   6   2   8

TREVISKEY   GWENNAP   SW 727396 1411

Production: Copper   Ore(tons) Metal(tons)   Value(£)
            1845   767.00   82.10   6436.20
            1846   885.00   89.70   6623.20
            1847   2162.00   198.60   14678.50
            1848   1909.00   156.90   9386.20
            1849   2421.00   232.30   15849.50
            1850   2884.00   285.80   20628.50
            1851   2459.00   214.10   14724.10
            1852   1938.00   139.90   11657.20
            1853   1226.00   79.10   8089.20

TREVISKEY   GWENNAP   Continued

Copper   Ore(tons) Metal(tons)   Value(£)
1854   756.00   50.30   5151.80
1855   380.00   21.90   2179.70
Comment 1845-1855 (C)
Tin   Black(tons) Stuff(tons)   Tin(tons)   Value(£)
1855   0.40   0.00   0.00   21.70
1856   0.50   0.00   0.00   32.20
Ownership: Comment 1861-1862 TREVISKEY UNITED; 1863-1865 TREVISKEY
            UNITED SUSPENDED; 1900-1901 SEE TRESAVEAN UNITED
Management: Chief Agent 1861 MARK TREWARTHA
            Secretary 1861-1862 D.TREBILCOCK (P)

TREVISSA   ST.ENODER   SW 870561 1412

Production: Lead   Ore(tons) Metal(tons)   Value(£)
            1860   4.40   3.20   0.00
            1861   No detailed return
            Comment 1860 FOR AG SEE CALSTOCK CONSOLS
            Zinc   Ore(tons) Metal(tons)   Value(£)
            1860   639.70   0.00   1606.20
            Copper No detailed return

TREVOOLE   CROWAN   SW 640373 1413

Production: Copper   Ore(tons) Metal(tons)   Value(£)
            1857   428.00   22.80   2138.30
            1858   812.00   44.40   3788.40
            1859   1207.00   57.60   5062.60
            1860   640.00   25.30   2080.10
            1861   658.00   24.70   1934.00
            Comment 1857-1861 (C)
            Tin   Black(tons) Stuff(tons)   Tin(tons)   Value(£)
            1857   0.00   0.00   0.00   38.60
            1858   0.00   0.00   0.00   83.20
            1859   0.00   0.00   0.00   12.00
            1860   0.00   0.00   0.00   51.30
            1862   0.00   0.00   0.00   37.50
            Comment 1857-1860 TIN STUFF; 1862 TIN STUFF
Ownership: Comment 1861-1865 SUSPENDED
Management: Manager 1859-1860 HUGH STEPHENS
            Chief Agent 1859 ED.BLEWETT; 1860 JOHN LEAN
            Secretary 1859-1860 HUGH STEPHENS (P)

TREVORVAS   HELSTON   1414

Production: Tin   Black(tons) Stuff(tons)   Tin(tons)   Value(£)
            1912   0.00   46.00   0.00   31.00
Ownership: 1912-1913 WM.HARRY,THOS.JAMES & WM.KING
            Comment 1913 NOT WORKED 1913
Employment: Underground   Surface   Total
            1912   3   3

527

TREWAVAS   BREAGE   SW 600625 1415

Production: Copper   Ore(tons) Metal(tons)   Value(£)
            1845   977.00   73.10   5452.50
            1846   397.00   33.40   2330.70
            1847   No detailed return
            Comment 1845-1847 (C)

TREWEATHA   MENHENIOT   SX 291654 1416

Production: Lead & Silver   Ore(tons) Lead(tons) Silver(ozs)   Value(£)
            1853   78.00   56.00   4866.00   0.00
            1854   331.00   230.00   19320.00   0.00
            1855   316.00   195.00   13884.00   0.00
            1856   351.00   206.50   5598.00   0.00
            1857   336.00   215.00   14190.00   0.00
            1858   215.00   138.00   8694.00   0.00
            1859   133.40   84.50   5383.00   0.00
            1860   154.00   9.00   585.00   0.00
            1861   10.40   7.00   112.00   0.00
            1862   38.70   27.00   432.00   0.00
            1863   105.20   80.10   1044.00   0.00
            1864   88.50   66.10   858.00   0.00
            1865   35.70   22.40   726.00   0.00
            1866   141.70   89.50   2980.00   0.00
            1867   313.50   234.70   9938.00   0.00
            1868   183.80   137.50   5754.00   0.00
            1869   339.80   254.80   10704.00   0.00
            1870   356.00   267.00   11214.00   0.00
            1871   686.60   514.90   21630.00   0.00
            1872   154.70   120.70   7464.00   0.00
            Comment 1859 TWO RETURNS AGGREGATED; 1872 SUSPENDED AUG.1872
Ownership: Comment 1859-1862 TREWATHA
Management: Manager 1859 THOS.RICHARDS; 1860 FRAN.PRYOR; 1868 T.FOOT;
            1869-1871 T.FOOT,T.HORSWILL
            Chief Agent 1859 WM.ROWE; 1860-1864 JOHN SCOBLE; 1865-1867
            JOHN SCOBLE T.FOOT; 1868-1871 JOHN SCOBLE; 1872 T.FOOT
            Secretary 1859 THOS.RICHARDS (P); 1860-1863 JAS.WOOLFERSTAN
            (P); 1865-1866 J.HORSWILL; 1867-1868 THOS.HORSWILL; 1869-1871
            WARD,LITTLEWOOD

TREWEETHA   ST.ENODER   SW 934537 1417

Production: Iron   Ore(tons)   Iron(%)   Value(£)
            1881   No detailed return
            Comment 1881 SEE RETEW

TREWELLARD   ST.JUST   1418

Production: Tin No detailed return
Management: Manager 1859 HY.WILLIAMS
            Chief Agent 1859 WM.TREGIRE

                              528

Production: Lead   Ore(tons) Metal(tons)   Value(£)
            1852   1.70   1.30   0.00

TREWEY DOWNS   ZENNOR   SW 461368 1421

Production: Tin   Black(tons) Stuff(tons)   Tin(tons)   Value(£)
            1906   0.20   0.00   0.00   15.00
            1909   19.00   0.00   0.00   1490.00
Ownership: 1906 TREWEY DOWNS MINING CO.LTD.; 1907-1910 TREWEY CONSOLS
            CO.LTD.; 1911-1913 CORNISH MINES LTD.
            Comment 1910-1911 NOT WORKED; 1913 ABANDONED
Management: Secretary 1906-1909 J.B.CORNISH (S)
Employment: Underground   Surface   Total
            1906   10   2   12
            1907   12   14   26
            1908   10   30   40
            1909   26   11   37

TREWHEELA   ST.ENODER   SW 911572 1422

Production: Iron   Ore(tons)   Iron(%)   Value(£)
            1859   265.80   0.00   66.40
            1860   265.80   0.00   66.40
            1861   95.00   0.00   23.70
            Comment 1859 BH.& SOME SPATHOSE; 1860-1861 BH.INC.BENNALLACK

TREWINT CONSOLS   ALTARNUN   SX 212800 1424

Production: Tin   Black(tons) Stuff(tons)   Tin(tons)   Value(£)
            1887   0.60   0.00   0.00   32.00
            1888   4.50   0.00   0.00   113.00
            1889   2.20   0.00   0.00   78.00
Ownership: 1882-1883 WM.HARRIS & OTHERS; 1886 MESSRS.DELLBRIDGE;
            1887-1889 TREWINT CONSOLS MINING CO.; 1901 JOHN S.SAWVEY
            Comment 1882 TREWENT DOWNS; 1883 NOT WORKED; 1886 TREWINT
            DOWN; 1889 SUSPENDED; 1901 TREWINT DOWN; 1906-1910 TREWINT.
            SEE HORSE BURROW (ALTARNUN)
Management: Chief Agent 1882-1883 FRAN.SPRY; 1886-1889 FRAN.SPRY
Employment: Underground   Surface   Total
            1882   3   1   4
            1886   6   6
            1887   6   6   12
            1888   6   6
            1889   5   3   8
            1901   2   2   4
            1907
            Comment 1907 SEE HORSE BURROW (ALTARNUN)

TREWITTEN   ST.CLEER   SX 229717 1425

Production: Tin No detailed return
Ownership: 1882-1883 TREWITTEN MINING CO.LTD.
Management: Chief Agent 1882-1883 CHAS.HOLMAN
Employment: Underground   Surface   Total
            1882   12   7   19
            1883   14   27   41

TREWOLLACK WHEAL ROSE,GT.   ST.COLUMB MINOR   SW 849616 1426

Production: Lead & Silver No detailed return
Management: Chief Agent 1865-1866 J.F.WILCOX SNR

TREWOON CONSOLS   1428

Production: Tin   Black(tons) Stuff(tons)   Tin(tons)   Value(£)
            1870   0.20   0.00   0.00   12.40

TREWOON UNITED   1429

Production: Tin   Black(tons) Stuff(tons)   Tin(tons)   Value(£)
            1869   5.40   0.00   0.00   396.70

TREWORLACK   ST.COLUMB MINOR   SW 849616 1430

Production: Lead No detailed return
            Tin No detailed return
Ownership: Comment 1872 FORMERLY GREAT ROYALTON; 1873 STOPPED IN 1873 &
            EARLY IN 1874; 1876-1881 TREWOLLOCK
Management: Manager 1869-1871 HY.RICHARDS
            Chief Agent 1872-1873 JOHN HANCOCK; 1876-1881 JAS.EVANS
            Secretary 1869-1873 U.P.HARRIS; 1876-1881 G.BRODERICK

TREWORLIS   WENDRON   SW 672291 1431

Production: Copper   Ore(tons) Metal(tons)   Value(£)
            1861   138.00   7.50   683.80
            1862   407.00   22.60   1813.80
            1863   121.00   4.30   267.20
            1864   23.00   1.20   106.40
            1868   15.00   0.90   81.70
            Comment 1861-1864 (C); 1868 (C)
            Tin   Black(tons) Stuff(tons) Tin(tons)   Value(£)
            1860   9.20   0.00   0.00   636.20
            1861   24.50   0.00   0.00   1736.30
            1862   24.60   0.00   0.00   1611.50
            1863   17.40   0.00   0.00   1118.30
            1864   16.50   0.00   0.00   1072.80
            1866   30.60   0.00   0.00   1368.90
            1867   29.00   0.00   0.00   1447.00
            1868   52.80   0.00   0.00   2752.90

                                 530

TREWORLIS   WENDRON   Continued

Tin	Black(tons)	Stuff(tons)	Tin(tons)	Value(£)
1869	39.10	0.00	0.00	2569.00
1870	48.40	0.00	0.00	3362.50
1871	29.50	0.00	0.00	2132.20
1872	5.80	0.00	0.00	439.90
1874	25.70	0.00	0.00	1373.80
1882	1.00	40.00	0.70	45.00

Comment 1860 TREWORLAS; 1862 TWO RETURNS AGGREGATED; 1882 HALVANS
Ownership: Comment 1859-1861 INC.TRENITHICK; 1863-1868 INC.TRENITHICK; 1872-1873 SEE TREMENHERE
Management: Manager 1859-1864 JOHN BURGAN; 1870-1871 RICH.QUENTRALL
Chief Agent 1859-1861 J.DUNSTAN; 1862-1864 B.DUNSTAN; 1865-1867 R.DUNSTAN; 1868-1869 WM.DUNSTAN
Secretary 1859-1871 HY.ROGERS (P)

TRISTRAM   ST.BLAZEY   1432

Production:	Tin	Black(tons)	Stuff(tons)	Tin(tons)	Value(£)
	1856	21.10	0.00	0.00	1315.20

TROON   ST.ENODER   SW 891576 1433

Production: Tin No detailed return
Management: Chief Agent 1873-1877 THOS.WASLEY
Secretary 1873-1877 HORN & KELLY

TRUGO   ST.COLUMB MAJOR   SW 894606 1434

Production:	Copper	Ore(tons)	Metal(tons)	Value(£)
	1881	43.00	2.80	144.70

Ownership: 1880-1881 TRUGO COPPER MINING CO.
Management: Manager 1880-1881 JAS.TAMBLYN
Chief Agent 1876-1877 S.H.F.COX; 1880-1881 E.CARTER
Secretary 1876-1877 S.H.F.COX; 1880-1881 GEO.CARTER & OTHERS

Employment:		Underground	Surface	Total
	1880	10	4	14
	1881	10	7	17

TRUMPET CONSOLS   WENDRON   SW 674302 1435

Production:	Copper	Ore(tons)	Metal(tons)	Value(£)
	1862	3.00	0.50	41.30
	1878	0.80	0.30	16.70
	1883	21.50	3.70	268.00
	1884	153.00	0.00	1039.00
	1885	303.00	0.00	1121.00
	1886	67.00	0.00	300.00
	1888	15.00	0.00	76.00

Comment 1862 (C); 1878 (C); 1883-1886 NEW TRUMPET CONSOLS; 1888 NEW TRUMPET CONSOLS

CM-T   531

Tin	Black(tons)	Stuff(tons)	Tin(tons)	Value(£)
1854	196.10	0.00	0.00	13875.20
1855	214.30	0.00	0.00	14740.70
1856	167.90	0.00	0.00	13142.50
1857	142.30	0.00	0.00	11798.60
1858	101.20	0.00	0.00	6720.70
1859	93.90	0.00	0.00	7286.50
1860	109.10	0.00	0.00	9072.90
1861	129.00	0.00	0.00	9805.50
1862	86.70	0.00	0.00	6101.80
1863	114.90	0.00	0.00	7953.70
1864	120.20	0.00	0.00	7725.40
1865	134.40	0.00	0.00	7447.50
1866	223.30	0.00	0.00	10872.50
1867	284.90	0.00	0.00	15049.90
1868	263.60	0.00	0.00	14378.90
1869	248.10	0.00	0.00	17495.70
1870	273.50	0.00	0.00	19984.40
1871	298.80	0.00	0.00	23386.70
1872	253.90	0.00	0.00	21834.40
1873	211.70	0.00	0.00	15291.70
1874	154.30	0.00	0.00	8208.10
1875	165.90	0.00	0.00	8036.20
1876	139.60	0.00	0.00	5776.50
1877	100.90	0.00	0.00	3919.60
1878	52.10	0.00	0.00	1726.50
1879	5.70	0.00	0.00	226.00
1880	4.60	0.00	0.00	204.50
1881	0.90	0.00	0.50	35.30
1882	1.70	0.00	1.20	95.00
1883	1.20	0.00	0.00	60.00
1884	1.10	0.00	0.00	43.00
1885	0.90	0.00	0.00	46.00
1886	4.50	0.00	0.00	234.00
1887	3.10	0.00	0.00	180.00
1888	0.90	0.00	0.00	55.00

Comment 1862 TWO RETURNS AGGREGATED; 1881-1888 NEW TRUMPET CONSOLS

Ownership: 1876-1881 TRUMPET CONSOLS ADVENTURERS; 1882-1886 NEW TRUMPET CONSOLS MINING CO.; 1887-1888 NEW TRUMPET CONSOLS MINING CO. LTD.
Comment 1875 DREAM SECTION ABANDONED; 1881 TRUMPET & NEW TRUMPET CONSOLS; 1882-1888 NEW TRUMPET CONSOLS

Management: Manager 1859-1869 R.QUENTRALL; 1871-1881 RICH.QUENTRALL
Chief Agent 1862-1874 R.GLUYAS; 1875-1881 JOHN GLUYAS;
1882-1883 RICH.QUENTRALL & SON; 1884-1888 THOS.QUENTRALL
Secretary 1859-1862 R.R.MICHELL (P); 1863-1876 HY.ROGERS;
1877-1881 HY.ROGERS (S)

Employment:
	Underground	Surface	Total
1878	3	37	40
1879		18	18
1880		7	7
1881	10	5	15
1882	25	6	31
1883	33	5	38

TRUMPET CONSOLS   WENDRON   Continued

	Underground	Surface	Total
1884-1885	30	7	37
1886	17	5	22
1887	9	2	11
1888	8	3	11

TRUMPET UNITED   WENDRON   1436

Production: Tin   Black(tons) Stuff(tons)   Tin(tons)   Value(£)
            1864   7.40   0.00   0.00   471.20
            1865   3.30   0.00   0.00   178.00
Ownership: Comment 1865 SUSPENDED
Management: Chief Agent 1862-1864 GEO.R.ODGERS
            Secretary 1862-1865 JOHN WATSON (P)

TRUMPET & LOVELL UNITED,NEW   WENDRON   1437

Production: Tin   Black(tons) Stuff(tons)   Tin(tons)   Value(£)
            1875   2.80   0.00   0.00   137.20
            1876   4.00   0.00   0.00   160.20
            Comment 1875-1876 LOVELL UNITED, NOW CALLED NEW TRUMPET
Ownership: Comment 1872-1874 INC.NEW TREVENNEN
Management: Manager 1872-1876 RICH.QUENTRALL
            Chief Agent 1872-1876 THOS.JAMES
            Secretary 1872-1876 HY.ROGERS

TRUMPET,EAST   WENDRON   SW 679304 1438

Production: Tin   Black(tons) Stuff(tons)   Tin(tons)   Value(£)
            1868   8.90   0.00   0.00   495.90
            1869   11.80   0.00   0.00   847.70
            1870   4.50   0.00   0.00   343.30
            1871   3.60   0.00   0.00   287.00
            1872   0.60   0.00   0.00   45.60
Management: Manager 1866-1871 RICH.QUENTRALL
            Chief Agent 1866-1871 W.ODGERS
            Secretary 1866-1871 HY.ROGERS

TRUNGLE   GWINEAR   1439

Production: Lead & Silver No detailed return
            Zinc No detailed return
            Copper No detailed return
Ownership: Comment 1863-1865 SUSPENDED
Management: Chief Agent 1860-1862 THOS.RICHARDS W.TRELEASE
            Secretary 1860-1865 THOS.RICHARDS (P)

TRUNGLE   GWINEAR   SW 593375 1440

Production: Lead & Silver   Ore(tons)   Lead(tons) Silver(ozs)   Value(£)
            1860   0.60   0.50   0.00   0.00
            1861   0.90   0.60   10.00   0.00
            Comment 1860 FOR AG SEE GREAT ALFRED
            Zinc   Ore(tons) Metal(tons)   Value(£)
            1861   45.40   0.00   95.30
            Copper No detailed return
Ownership: Comment 1865 SUSPENDED
Management: Chief Agent 1862–1864 RICH.REYNOLDS
            Secretary 1862–1865 THOS.RICHARDS (P)

TRURO   PERRANZABULOE   SW 761529 1441

Production: Lead No detailed return
Management: Chief Agent 1862–1866 H.MIDDLETON

TRUSCOTT   LAUNCESTON   SX 305857 1442

Production: Manganese No detailed return
            Iron   Ore(tons)   Iron(%)   Value(£)
            1876   400.00   0.00   240.00
            1877   140.00   0.00   84.00
            Comment 1876 BH.MANGENTIFEROUS ORE; 1877 BH.
Ownership: 1876–1877 W.NEWTON & W.JONES
Management: Manager 1876–1877 W.NEWTON

TRUTHALL   1443

Production: Tin   Black(tons) Stuff(tons)   Tin(tons)   Value(£)
            1856   0.00   0.00   0.00   52.90
            Comment 1856 TIN STUFF

TRYPHENA   CAMBORNE   SW 654389 1444

Production: Copper   Ore(tons) Metal(tons)   Value(£)
            1847   27.00   6.50   496.10
            1848   45.00   9.40   651.10
            1862   2.00   0.20   14.00
            Comment 1847–1848 (C); 1862 TRYPHENA PENDARVES(C)
            Tin   Black(tons) Stuff(tons)   Tin(tons)   Value(£)
            1855   25.80   0.00   0.00   1799.00
            1856   5.20   0.00   0.00   379.90
            Comment 1855 TRYPHINA
Ownership: Comment 1863–1864 TRYPHENA PENDARVES. SUSPENDED; 1865–1872
            SEE PENDARVES UNITED
Management: Manager 1860–1862 RICH.PRYOR
            Chief Agent 1860–1864 JOS.RULE
            Secretary 1860–1862 ELIAS DUNSTERVILLE (P); 1863–1864 ELIAS
            DUNSTERVILLE (P)

TUNDRA   ZENNOR   1446

Production: Tin No detailed return
Ownership: 1913 ROBT.DE RUSTAFJAELL
            Comment 1913 NOT WORKED

TUNNEL MINE   CALLINGTON   1447

Production: Copper No detailed return
            Tin No detailed return
            Silver No detailed return
Ownership: 1885-1887 TUNNEL MINING CO.
Management: Chief Agent 1885-1887 GEO.RICKARD
Employment: Underground   Surface   Total
            1885   19   23   42
            1886   10   3   13
            1887   1   1

TURNAVORE   ST.AGNES   SW 717513 1448

Production: Tin   Black(tons) Stuff(tons)   Tin(tons)   Value(£)
            1856   2.10   0.00   0.00   118.10
            1857   0.90   0.00   0.00   57.00
            1858   0.90   0.00   0.00   41.70

TURNAVORE STAMPS   ST.AGNES   SW 717513 1449

Production: Tin   Black(tons) Stuff(tons)   Tin(tons)   Value(£)
            1858   7.70   0.00   0.00   364.50

TWELVE HEADS   GWENNAP   9995

Production: Arsenic Pyrite   Ore(tons)   Value(£)
            1894   No detailed return
            Comment 1894 SEE KILLICOR

TYRINGHAM CONSOLS   TOWEDNACK   SW 494385 1450

Production: Tin No detailed return
Ownership: Comment 1863-1865 SUSPENDED; 1880-1881 SEE BOSSOW
Management: Chief Agent 1860-1862 THOS.MICHELL
            Secretary 1860-1865 THOS.RICHARDS (P)

TYWARNHAILE   ST.AGNES   SW 702472 1451

Production: Lead   Ore(tons) Metal(tons)   Value(£)
            1850   39.50   28.50   0.00
            1852   57.00   41.00   0.00
            1853   No detailed return
            Comment 1850 INC.NANCEKUKE

                                  535

Copper	Ore(tons)	Metal(tons)	Value(£)
1848	1788.00	110.60	6089.70
1849	4721.00	293.20	18718.00
1850	6487.00	379.50	24638.40
1851	5649.00	321.20	20304.50
1852	2092.00	105.90	7476.20
1853	375.00	18.70	1766.40
1854	200.00	12.70	1267.00
1855	222.00	12.20	1159.60
1856	192.00	10.40	856.80
1857	253.00	13.10	1174.00
1858	225.00	11.20	992.60
1859	292.00	16.20	1427.20
1860	684.00	31.70	2656.60
1861	2351.00	102.20	8519.30
1862	2977.00	123.50	9147.50
1863	1868.00	99.70	7633.10
1864	1965.00	85.40	6514.30
1865	29.00	1.00	63.10
1867	49.00	2.40	153.50
1868	105.00	5.20	328.90
1869	51.00	2.60	156.00
1877	40.00	1.30	47.00
1903	99.00	0.00	247.00
1907	227.00	0.00	869.00
1908	315.00	0.00	315.00

Comment 1848-1852 (C); 1853 (C)TYWARNHAYLE; 1854 (C); 1855-1858 (C)TYWARNHAYLE; 1859 (C); 1860 (C)GREAT TYWARNHAILE; 1861-1865 (C); 1867-1869 (C); 1877 TYWARNHAYLE
Ownership: 1901-1902 BAINBRIDGE SEYMOUR & CO.; 1903-1904 TYWARNHAILE MINE LTD.; 1905-1906 TYWARNHAILE SYNDICATE LTD.; 1907-1908 TYWARNHAILE MINING CO.LTD.
Comment 1864-1865 SUSPENDED; 1904 SUSPENDED; 1908 ABANDONED
Management: Manager 1863 JOHN NICHOLS
Chief Agent 1859-1861 JOHN EDWARDS R.HAMPTON; 1862 JOHN HAMPTON JOHN NICHOLLS JOHN DAWE; 1863 JOHN DAWE; 1901-1904 WM.ROBERTS; 1905-1908 W.R.THOMAS
Secretary 1859 P.P.INSKIP (P); 1860-1861 ROBT.H.PIKE (P); 1862-1863 C.WESCOMBE (P)

Employment:	Underground	Surface	Total
1902	6	24	30
1903	14	28	42
1905		25	25
1906	23	33	56
1907	38	40	78
1908	22	27	49

TYWARNHAILE,EAST   SCORRIER   SW 707476 1452

Production: Copper No detailed return
Tin No detailed return
Ownership: 1875-1876 EAST TYWARNHAILE MINING CO.
Management: Chief Agent 1876 J.CHAMPION

UNION   REDRUTH   SW 671417 1453

Production: Copper   Ore(tons) Metal(tons)   Value(£)
            1862   13.00   0.60   55.20
            1863   11.00   0.60   42.20
            1866   53.00   4.90   363.60
            Comment 1862-1863 (C); 1866 (C)
            Tin   Black(tons) Stuff(tons)   Tin(tons)   Value(£)
            1852   1.00   0.00   0.00   0.00
            1853   14.10   0.00   0.00   1039.90
            1854   2.00   0.00   0.00   164.20
            1856   34.10   0.00   0.00   2504.30
            1857   6.40   0.00   0.00   631.70
            1858   0.00   0.00   0.00   816.40
            1859   0.00   0.00   0.00   2315.00
            1860   0.00   0.00   0.00   622.50
            1861   5.60   0.00   0.00   377.80
            1862   14.80   0.00   0.00   968.00
            1863   5.20   0.00   0.00   332.00
            1864   0.90   0.00   0.00   58.00
            1866   0.00   0.00   0.00   9.20
            1882   0.20   12.00   0.00   10.00
            Comment 1852 TIN ORE; 1857 VALUE INC.TIN STUFF; 1858-1860 TIN
            STUFF; 1862 TWO RETURNS AGGREGATED; 1866 TIN STUFF
            Arsenic   Ore(tons) Metal(tons)   Value(£)
            1866   5.80   0.00   5.80
            Comment 1866 CRUDE
Ownership: 1881-1884 WHEAL UNION MINING CO.
            Comment 1884 NOT WORKED
Management: Chief Agent 1860 THOS.GLANVILLE; 1861-1866 THOS.GLANVILLE
            M.TAYLOR; 1879-1881 WM.TEAGUE JNR; 1882-1884 JOS.PRISK
            Secretary 1859-1861 RICH.LYLE (P); 1862-1866 GEO.LIGHTLEY;
            1881 JOHN THOMAS
Employment: Underground   Surface   Total
            1881   10   8   18
            1882   16   9   25

UNITED HILLS   PORTH TOWAN   SW 697474 1454

Production: Copper   Ore(tons) Metal(tons)   Value(£)
            1845   3348.00   201.20   14154.00
            1846   3431.00   206.40   13759.60
            1847   3388.00   209.10   14414.70
            Comment 1845-1847 (C)

UNITED MINES   GWENNAP   SW 745412 1455

Production: Zinc   Ore(tons) Metal(tons)   Value(£)
            1856   90.20   0.00   203.50
            1857   96.60   0.00   182.50
            1860   83.80   0.00   106.60
            Copper   Ore(tons) Metal(tons)   Value(£)
            1845   14374.00   1012.20   74908.10
            1846   11956.00   802.60   56471.10
            1847   12888.00   864.60   61026.80

                                  537

Copper	Ore(tons)	Metal(tons)	Value(£)
1848	11373.00	724.60	41796.80
1849	12228.00	736.50	47249.10
1850	9881.00	638.50	43197.50
1851	8181.00	585.00	39001.90
1852	10233.00	646.20	55462.50
1853	11764.00	656.70	62598.50
1854	11827.00	603.10	58668.50
1855	10300.00	564.00	55882.80
1856	9400.00	552.20	50772.90
1857	9175.00	507.80	49383.50
1858	9311.00	489.40	42635.90
1859	9550.00	479.50	43069.50
1860	8800.00	423.80	36058.70
1861	4273.00	165.70	13116.60

Comment 1845-1861 (C)

Tin	Black(tons)	Stuff(tons)	Tin(tons)	Value(£)
1853	5.00	0.00	0.00	300.40
1855	26.80	0.00	0.00	1569.60
1856	16.10	0.00	0.00	974.20
1857	7.90	0.00	0.00	523.40
1858	9.90	0.00	0.00	569.80
1859	8.80	0.00	0.00	581.50
1860	16.30	0.00	0.00	1188.20
1861	16.60	0.00	0.00	1096.30
1862	44.30	0.00	0.00	2563.20
1880	0.00	39.00	0.00	47.10
1881	0.00	119.00	0.00	81.50

Arsenic	Ore(tons)	Metal(tons)	Value(£)
1860	4.00	0.00	7.00

Comment 1860 CRUDE
Ownership: 1880-1881 UNITED MINES MINING CO.; 1898 W.J.TRYTHALL
             Comment 1862-1868 SEE CLIFFORD AMALGAMATED
Management: Manager 1859-1860 JOHN DAVEY; 1861 JOHN RICHARDS
             Chief Agent 1859-1861 RICH.GREY; 1880-1881 WM.KELLOW & SONS
             Secretary 1859 COMMITTEE; 1860-1861 WILLIAMS & SONS (P)

UNITY   REDRUTH   1456

Production: Tin No detailed return
Ownership: 1893 ED.N.TRELOAR(P)
             Comment 1893 TRIBUTER

UNITY CONSOLS   GWINEAR   SW 608355 1457

Production: Lead	Ore(tons)	Metal(tons)	Value(£)
1856	0.30	0.20	0.00

1857-1859 No detailed return

Copper	Ore(tons)	Metal(tons)	Value(£)
1851	306.00	16.70	1035.00
1852	272.00	17.90	1414.40
1854	114.00	5.50	532.00
1856	159.00	10.30	934.10

Copper	Ore(tons)	Metal(tons)	Value(£)
1857	445.00	25.60	2509.60
1858	506.00	31.60	2795.80
1859	214.00	12.80	1235.70
1860	363.00	18.40	1624.10
1861	255.00	18.30	1595.60
1862	109.00	6.30	501.10
1863	8.00	0.80	62.20
1864	25.00	1.60	129.20
1865	53.00	3.80	311.50

Comment 1851-1852 (C); 1854 (C); 1856-1865 (C)

Tin	Black(tons)	Stuff(tons)	Tin(tons)	Value(£)
1853	38.60	0.00	0.00	2095.70
1854	51.40	0.00	0.00	0.00
1855	83.20	0.00	0.00	4130.70
1857	3.70	0.00	0.00	236.90
1858	0.00	0.00	0.00	89.50
1859	0.00	0.00	0.00	27.50

Arsenic	Ore(tons)	Metal(tons)	Value(£)
1854	103.60	0.00	0.00
1855	106.00	0.00	0.00
1858	3.80	0.00	6.00

Comment 1855 WHITE ARSENIC

Management: Manager 1859-1866 W.H.REYNOLDS
Chief Agent 1859-1866 ABEL.PEARCE
Secretary 1859-1866 JOHN WATSON (P)

UNITY NO.1   ST.DAY   SW 736430 1458

Production:
Tin	Black(tons)	Stuff(tons)	Tin(tons)	Value(£)
1852	11.30	0.00	0.00	0.00
1854	22.00	0.00	0.00	1283.30
1856	6.10	0.00	0.00	382.20

1888-1893 No detailed return
Comment 1852 TIN ORE; 1888-1889 SEE ST.DAY MANOR; 1890-1893
SEE GORLAND
Ownership: Comment 1888-1891 SEE GORLAND
Employment   Underground   Surface   Total
1888-1891
Comment 1888-1891 SEE GORLAND

UNITY NO.2   ST.DAY   SW 736430 1459

Production: Tin No detailed return
Ownership: Comment 1888-1896 SEE POLDICE
Employment: Underground   Surface   Total
1888-1896
Comment 1888-1896 SEE POLDICE

Production: Lead   Ore(tons) Metal(tons)   Value(£)
            1876   1.40   1.10   7.40
            Copper   Ore(tons) Metal(tons)   Value(£)
            1875   30.00   2.00   143.60
            1876   52.00   3.90   268.50
            1877   75.20   5.50   333.80
            1878   2.70   0.10   5.40
            Comment 1875-1876 (C)
            Tin   Black(tons) Stuff(tons)   Tin(tons)   Value(£)
            1872   10.10   0.00   0.00   907.90
            1873   47.30   0.00   0.00   3331.40
            1874   90.00   0.00   0.00   2414.90
            1875   92.60   0.00   0.00   4858.50
            1876   128.10   0.00   0.00   5362.00
            1877   71.30   0.00   0.00   2845.50
            1878   2.40   0.00   0.00   85.70
            1904   0.00   19.00   0.00   43.00
            Arsenic   Ore(tons) Metal(tons)   Value(£)
            1875   11.00   0.00   28.00
            1876   69.30   0.00   374.00
            1877   43.10   0.00   215.50
            1893   No detailed return
            Comment 1893 SEE POLDICE
            Wolfram No detailed return
            Arsenic Pyrite Ore(tons)   Value(£)
            1878   21.00   9.00
            1893   No detailed return
            1900   No detailed return
            Comment 1893 SEE POLDICE; 1900 SEE CREEGBRAWSE
Ownership: 1876-1881 WHEAL UNITY WOOD CO.; 1897 WM.MOYLE & JOS.JEFFREY;
            1903-1905 JAS.COLLINS
            Comment 1898-1900 SEE CREEGBRAWSE (BEHENNA)
Management: Manager 1872-1881 GEO.E.TREMAYNE
            Chief Agent 1871 GEO.E.TREMAYNE; 1872-1881 WM.MAYNE
            Secretary 1871-1876 HY.MICHELL; 1877-1881 HY.MICHELL (S)
Employment: Underground   Surface   Total
            1878   13   13
            1898-1900
            1904-1905   2   2
            Comment 1898-1900 SEE CREEGBRAWSE (BEHENNA)

Production: Copper   Ore(tons) Metal(tons)   Value(£)
            1854   66.00   7.00   757.40
            Comment 1854 (C)
            Tin   Black(tons) Stuff(tons)   Tin(tons)   Value(£)
            1899   0.00   63.00   0.00   97.00
            1900   0.00   34.00   0.00   80.00
            1901   0.00   34.00   0.00   27.00
            1902   0.00   20.00   0.00   11.00
            1903   0.00   1.00   0.00   3.00
            1904   0.00   61.00   0.00   17.00
Ownership: 1899-1900 WM.MOYLE JOS.JEFFREY; 1901-1902 JOS.JEFFREY

UNITY WOOD,EAST   KENWYN   Continued

        THOS.BEHENNA; 1903—1904 THOS.BEHENNA
Management: Chief Agent 1899—1904 CHAS.PENGELLY
Employment: Underground   Surface   Total
            1899—1900   2   2
            1902   2   2
            1903   1   1
            1904   2   2

UNITY WOOD,WEST   GWENNAP   SW 733433 1462

Production: Tin   Black(tons) Stuff(tons)   Tin(tons)   Value(£)
            1902   0.00   20.00   0.00   22.00
            Arsenic No detailed return
Ownership: 1898—1899 JOHN SANDOUR; 1901—1902 JOHN SANDOUR JOS.JEFFREY
Management: Chief Agent 1899 CHAS.PENGELLY
Employment: Underground   Surface   Total
            1899   2   2
            1902   2   2

UNITY,NORTH   GWINEAR   SW 607358 1463

Production: Copper   Ore(tons) Metal(tons)   Value(£)
            1856   161.00   8.70   770.10
            Comment 1856 (C)

UNY   REDRUTH   SW 694409 1464

Production: Copper   Ore(tons) Metal(tons)   Value(£)
            1854   110.00   5.70   541.50
            1855   418.00   20.80   1945.20
            1856   145.00   6.10   500.30
            1857   178.00   7.00   653.60
            1858   137.00   7.90   648.70
            1859   258.00   12.10   1086.40
            1860   228.00   8.70   745.50
            1862   261.00   20.50   1712.40
            1863   353.00   27.70   2156.10
            1864   292.00   19.30   1621.70
            1865   105.00   6.20   488.10
            1866   9.00   0.70   50.30
            1877   8.00   0.50   19.20
            1884   58.00   0.00   353.00
            1885   7.00   0.00   28.00
            1888   3.00   0.00   18.00
            1890   8.00   0.00   20.00
            Comment 1854—1860 (C); 1862—1866 (C)
            Tin   Black(tons) Stuff(tons)   Tin(tons)   Value(£)
            1853   13.20   0.00   0.00   884.70
            1854   75.80   0.00   0.00   5698.50
            1855   131.50   0.00   0.00   8683.90
            1856   88.80   0.00   0.00   6506.00
            1857   79.30   0.00   0.00   6258.80

                                     541

Tin	Black(tons)	Stuff(tons)	Tin(tons)	Value(£)
1858	117.60	0.00	0.00	7521.90
1859	96.20	0.00	0.00	7069.40
1860	103.60	0.00	0.00	8341.80
1861	137.00	0.00	0.00	10133.50
1862	262.00	0.00	0.00	16795.30
1863	230.30	0.00	0.00	14779.50
1864	242.00	0.00	0.00	14606.00
1865	303.60	0.00	0.00	16090.10
1866	199.30	0.00	0.00	9289.50
1867	220.30	0.00	0.00	11243.10
1868	273.70	0.00	0.00	14897.20
1869	240.70	0.00	0.00	16748.80
1870	245.50	0.00	0.00	18201.70
1871	293.70	0.00	0.00	23602.20
1872	212.30	0.00	0.00	18492.80
1873	252.30	0.00	0.00	19973.80
1874	320.10	0.00	0.00	17827.10
1875	314.90	0.00	0.00	15703.50
1876	339.70	0.00	0.00	15168.60
1877	358.90	0.00	0.00	14500.70
1878	423.80	0.00	0.00	15025.20
1879	290.30	0.00	0.00	11489.00
1880	225.70	0.00	0.00	11990.30
1881	193.00	0.00	130.30	10826.70
1882	233.00	0.00	154.40	15378.00
1883	248.00	0.00	0.00	12885.00
1884	243.00	0.00	0.00	10653.00
1885	65.20	0.00	0.00	2725.00
1887	16.00	0.00	0.00	1125.00
1888	78.20	0.00	0.00	4952.00
1889	132.40	0.00	0.00	7423.00
1890	141.30	0.00	0.00	7982.00
1891	177.50	0.00	0.00	9648.00
1892	167.90	0.00	0.00	6976.00
1893	26.00	0.00	0.00	1310.00

Ownership: 1877-1890 WHEAL UNY MINING CO.; 1891 WHEAL UNY MINING
CO.(CB); 1892-1893 WHEAL UNY CO.LTD.
Comment 1879 SEE PERSEVERANCE; 1893 ABANDONED
Management: Manager 1859-1863 SML.COADE; 1864-1868 JOHN DAWE; 1869-1879
WM.RICH; 1881 WM.HAMBLY
Chief Agent 1859 WILLIAMS; 1861-1863 M.ROGERS; 1864-1871
SML.COADE MAT.ROGERS; 1872-1879 SML.COADE JNR MAT.ROGERS;
1880 HY.EDDY WM.PROFIT JNR MAT.ROGERS; 1881 WM.PROFIT JNR
JAS.WHITE; 1882-1886 WM.HAMBLY; 1887-1888 C.F.BISHOP;
1889-1892 JAS.WHITE
Secretary 1859-1862 ROBT.H.PIKE (P); 1863-1872 ROBT.H.PIKE &
SON; 1873 H.CRAWSHAW; 1874-1877 ROBT.H.PIKE & SON; 1878-1879
WALT.PIKE JAS.HICKEY; 1880 WALT.PIKE; 1881 ROBT.H.PIKE & SON;
1891 C.B.PARRY (S); 1892-1893 ED.ASHMEAD (S)

Employment:	Underground	Surface	Total
1878	98	183	281
1879	115	166	281
1880	125	163	288
1881	121	165	286

	Underground	Surface	Total
1882	109	172	281
1883	93	143	236
1884	84	140	224
1885	75	104	179
1886	4	4	8
1887	35	63	98
1888	108	89	197
1889	109	53	162
1890	114	120	234
1891	142	133	275
1892	126	138	264
1893	32	35	67

UNY,EAST   REDRUTH   SW 702411 1466

Production: Copper

	Ore(tons)	Metal(tons)	Value(£)	
1881	10.20	0.80	35.40	
1882	116.90	12.30	701.00	
1883	156.80	14.10	712.00	
1884	70.00	0.00	217.00	

Tin

	Black(tons)	Stuff(tons)	Tin(tons)	Value(£)
1853	2.00	0.00	0.00	119.40
1854	12.00	0.00	0.00	0.00
1873	0.00	14.00	0.00	28.50
1881	0.00	188.40	0.00	0.00
1882	25.30	785.20	16.30	1266.00
1883	0.50	20.20	0.00	26.00
1884	17.30	0.00	0.00	444.00

Comment 1873 TIN STUFF EST; 1853–1854 EAST UNY CONSOLS
Ownership: 1880–1881 EAST WHEAL UNY MINING CO.; 1882–1887 PRISK & WEBBER
Comment 1873–1874 SEE PERSEVERANCE; 1885–1887 NOT WORKED
Management: Manager 1880–1881 S.HY.EDDY
Chief Agent 1880 J.K.HARVEY; 1881 J.K.HARVEY WM.HOOPER; 1882–1887 WM.HOOPER
Secretary 1872 ROBT.H.PIKE & SON; 1881 RICH.BOYNS (P)
Employment:

	Underground	Surface	Total
1881	30	13	43
1882	36	16	52
1883	47	16	63
1884	26	10	36

URANIUM MINES   ST.STEPHEN–IN–BRANNEL   SW 935524 1467

Production: Plumbago No detailed return
Arsenic Pyrite No detailed return

Uranium	Ore(tons)	Value(£)
1890	22.00	2200.00
1891	31.00	620.00
1892	37.00	740.00
1893	25.00	500.00
1894	19.00	815.00

Uranium	Ore(tons)	Value(£)
1895	40.00	2071.00
1896	35.00	1500.00
1897	30.00	1367.00
1898	26.00	1185.00
1899	7.50	275.00
1900	41.00	1517.00
1901	79.00	2923.00
1902	52.00	2028.00
1903	6.00	234.00
1905	103.00	0.00
1906	11.00	0.00
1907	No detailed return	
1908	71.00	7550.00
1909	6.40	0.00
1910	13.00	0.00

Comment 1905–1906 FROM WASTE HEAPS

Ownership: 1889–1890 THE URANIUM MINING CO.; 1891–1898 URANIUM MINES LTD.; 1899–1904 MINERALS RESEARCH SYNDICATE LTD.; 1905 BETHMANN & KUPFER; 1906–1912 BRITISH METALLIFEROUS MINES LTD.
Comment 1891–1898 IN LIQUIDATION; 1903–1905 SUSPENDED; 1911–1912 NOT WORKED

Management: Chief Agent 1889–1891 W.R.THOMAS; 1893–1898 JOHN BRENTON; 1899–1904 E.J.TURNER; 1905 T.H.NORTH
Secretary 1891–1898 A.J.LEEZE (S); 1906–1908 A.J.LEESE (S); 1909–1912 T.H.NORTH

Employment:

	Underground	Surface	Total
1889	22	11	33
1890	19	19	38
1891	2	3	5
1892	4	2	6
1893	7	1	8
1894	10	1	11
1895	8	2	10
1896	6	2	8
1897	7	4	11
1898	7	3	10
1899	6	2	8
1900	14	11	25
1901	17	15	32
1902	16	10	26
1903	4	2	6
1904		1	1
1905		4	4
1906	17	9	26
1907	21	11	32
1908	22	14	36
1909	31	17	48
1910	26	19	45
1911		1	1

VADDEN,NEW   PERRANUTHNOE   SW 550292 1468

Production: Copper No detailed return
            Tin   Black(tons) Stuff(tons)   Tin(tons)   Value(£)
            1862   0.00   0.00   0.00   653.00
            Comment 1862 TIN STUFF. TWO RETURNS AGGREGATED
Ownership: Comment 1864–1865 SUSPENDED
Management: Manager 1859 PETER FLOYD; 1860–1863 PETER FLOYD & SON
            Chief Agent 1859 PETER FLOYD
            Secretary 1859–1860 J.D.BRUNTON (P); 1861–1865 J.H.DINGLE
            (P)

VENTON   ST.IVE   SX 293663 1469

Production: Lead   Ore(tons) Metal(tons)   Value(£)
            1852   4.50   3.30   0.00
            1853   No detailed return

VENTONWYN   ST.STEPHEN–IN–BRANNEL   SW 962504 1470

Production: Tin   Black(tons) Stuff(tons)   Tin(tons)   Value(£)
            1899   0.60   0.00   0.00   40.00
            1902   8.90   0.00   0.00   631.00
            1903   41.60   0.00   0.00   2956.00
            1904   29.30   0.00   0.00   2107.00
            1905   37.00   0.00   0.00   3001.00
            1906   26.70   0.00   0.00   2564.00
            1907   3.70   0.00   0.00   230.00
            1908   13.60   0.00   0.00   824.00
            1909   6.00   0.00   0.00   321.00
            1912–1913 No detailed return
            Comment 1912–1913 SEE GREAT DOWGAS
Ownership: 1885 JOHN PETER; 1886 VENTONWYN TIN MINING CO.; 1892–1893
            R.COTTON(P); 1894–1899 R.COTTON; 1900 ROBT.SACH; 1901–1905
            JOHN MORRIS & CO.; 1906 VENTONWYN MINING SYNDICATE; 1907–1908
            JEWELL JOHN HOCKING
            Comment 1909–1913 SEE GREAT DOWGAS
Management: Chief Agent 1892–1900 JAS.COLLINS; 1901–1903 ROBT.SACH;
            1904–1906 W.H.ADAMS
            Secretary 1885–1886 JOHN PETER
Employment: Underground   Surface   Total
            1885   5   5
            1886   4   4
            1892   6   6
            1893   8   14   22
            1894   10   6   16
            1896   4   4
            1897   4   1   5
            1898–1899   6   3   9
            1900   4   2   6
            1901   12   8   20
            1903   26   17   43
            1904   22   18   40
            1905   29   22   51
            1906   23   20   43

                                    545

VENTONWYN   ST.STEPHEN—IN—BRANNEL   Continued

                    Underground   Surface   Total
        1907   8   8
        1908   5   6   11
        1909   6   7   13
        1910
        1912—1913
        Comment 1909 SEE ALSO GREAT DOWGAS; 1910 SEE GREAT DOWGAS;
        1912—1913 SEE GREAT DOWGAS

VENTURE   ST.IVES   SW 520390 1471

Production: Tin No detailed return
Ownership: Comment 1863—1872 SEE TRELYON CONSOLS

VERRANT   GOLDSITHNEY   SW 546306 1472

Production: Tin   Black(tons) Stuff(tons)   Tin(tons)   Value(£)
        1905   2.80   0.00   0.00   245.00
        1906   9.70   0.00   0.00   980.00
        1909   0.00   18.00   0.00   18.00
Ownership: 1905 G.D.MCGREGOR; 1906 CORNWALL MINES SYNDICATE LTD.;
        1907—1908 G.D.MCGREGOR
        Comment 1907—1908 INC.CAROLINE; 1909—1913 SEE GOLDSITHNEY
Employment: Underground   Surface   Total
        1905   17   12   29
        1906   34   21   55
        1907   8   6   14
        Comment 1907 INC.CAROLINE

VINCENT   ALTARNUN   SX 209795 1473

Production: Tin   Black(tons) Stuff(tons)   Tin(tons)   Value(£)
        1872   14.00   0.00   0.00   1059.70
        1873   19.70   0.00   0.00   2407.20
        1874   15.30   0.00   0.00   768.10
        1875   3.20   0.00   0.00   138.80
        1881   10.00   0.00   6.60   550.00
Ownership: 1880 WHEAL VINCENT TIN MINING CO.LTD.; 1881—1883 VINCENT TIN
        MINING CO.LTD.; 1913 PREMIER EXPLORATION CO.
        Comment 1882 NOT WORKED IN LIQUIDATION; 1883 IN LIQUIDATION
Management: Manager 1872 JOHN GIFFORD; 1873—1874 T.ODGERS
        Chief Agent 1872 RICH.THOMAS; 1875 CHAS.J.SIMS; 1881 J.SPRY
        Secretary 1872—1875 E.NICHOLLS; 1880 W.B.LORD; 1881
        CEC.QUENTIN (S)
Employment: Underground   Surface   Total
        1880   11   21   32
        1881   20   15   35
        1913   10   25   35

VIRGIN   ST.DAY   SW 747420 1475

Production: Copper   Ore(tons) Metal(tons)   Value(£)
            1845   655.00   41.70   2964.30
            1846   946.00   72.10   4956.80
            1847   464.00   30.70   2216.30
            Comment 1845-1847 (C)

VIRGIN,CHIVERTON WHEAL   MITHIAN   1476

Production: Lead No detailed return
Management: Manager 1869-1871 GEO.TREMAYNE
            Chief Agent 1869-1871 CHAS.OATES
            Secretary 1869-1870 FRAN.PRYOR; 1871 THOS.PRYOR

VIRGIN,WEST   ST.DAY   SW 743419 1477

Ownership: Comment 1865 SUSPENDED
Management: Chief Agent 1861-1864 W.PRYOR
            Secretary 1861-1865 FRAN.PRYOR (P)

VLOW   PERRANZABULOE   SW 767558 1478

Production: Tin   Black(tons) Stuff(tons)   Tin(tons)   Value(£)
            1864   23.50   0.00   0.00   1152.40
            1865   62.60   0.00   0.00   2950.80
            1866   26.20   0.00   0.00   1065.00
            1911   2.80   0.00   0.00   206.00
            1912   2.40   0.00   0.00   299.00
            1913   1.40   0.00   0.00   177.00
            Comment 1864 OR NORTH BUDNICK
Ownership: 1911-1913 F.MARSHALL
            Comment 1866 SUSPENDED; 1869-1872 SEE PERRAN CONSOLS;
            1911-1913 TRIBUTERS
Management: Manager 1864 JOHN TONKIN; 1865 R.M.KITTO
            Chief Agent 1864-1865 WM.JOHNS
            Secretary 1864-1866 RICH.COWLING
Employment: Underground   Surface   Total
            1911   4   4
            1912   5   5
            1913   2   2   4

VOR & LEEDS UNITED,WEST   LEEDSTOWN   1479

Production: Tin No detailed return
Ownership: 1881 WEST VOR & LEEDS UNITED MINING CO.LTD.
Management: Chief Agent 1881 STEP.HARRIS
            Secretary 1881 J.H.SYMONDS (S)
Employment: Underground   Surface   Total
            1881   4   4

                              547

VOR,EAST   SITHNEY   SW 646305 1481

Production: Tin   Black(tons) Stuff(tons)   Tin(tons)   Value(£)
            1856   31.20   0.00   0.00   1909.20
            1857   0.70   0.00   0.00   45.30
Management: Chief Agent 1864—1865 JAS.POLLARD
            Secretary 1864—1865 W.WATSON

VOR,GREAT   BREAGE   SW 621302 1482

Production: Copper No detailed return
            Tin   Black(tons) Stuff(tons)   Tin(tons)   Value(£)
            1853   26.00   0.00   0.00   1938.40
            1854   187.40   0.00   0.00   13090.00
            1855   313.10   0.00   0.00   19192.50
            1856   424.70   0.00   0.00   29681.50
            1857   696.60   0.00   0.00   53584.20
            1858   573.90   0.00   0.00   36662.20
            1859   381.10   0.00   0.00   30195.20
            1860   233.10   0.00   0.00   19082.70
            1861   214.70   0.00   0.00   15876.30
            1862   338.40   0.00   0.00   22917.00
            1863   348.40   0.00   0.00   24655.80
            1864   576.20   0.00   0.00   38280.20
            1865   847.80   0.00   0.00   49420.20
            1866   844.90   0.00   0.00   43798.50
            1867   658.50   0.00   0.00   36867.90
            1868   661.00   0.00   0.00   38166.70
            1869   576.90   0.00   0.00   41879.70
            1870   362.70   0.00   0.00   27884.90
            1871   320.10   0.00   0.00   26574.70
            1872   297.60   0.00   0.00   26465.10
            1873   305.00   0.00   0.00   23645.40
            1874   83.50   0.00   0.00   4429.00
            1875   40.60   0.00   0.00   1824.90
            1876   47.40   0.00   0.00   1749.80
            1877   13.10   0.00   0.00   494.70
            1883   0.10   4.00   0.00   2.00
            1884   0.10   3.00   0.00   2.00
            1885   No detailed return
            1907   9.10   0.00   0.00   892.00
            1908   13.10   0.00   0.00   961.00
            1909   32.70   0.00   0.00   2455.00
            1910   9.50   0.00   0.00   747.00
            Comment 1854 GREAT VOR UNITED; 1862 TWO RETURNS AGGREGATED;
            1867 GREAT VOR UNITED; 1883—1885 NEW GREAT VOR; 1907—1910
            VOR
            Arsenic   Ore(tons) Metal(tons)   Value(£)
            1876   41.00   0.00   24.00
Ownership: 1876—1877 GREAT WHEAL VOR UNITED MINES CO.; 1881—1885 NEW
            GREAT WHEAL VOR TIN MINING CO.; 1906—1913 WHEAL VOR LTD.
            Comment 1869—1877 GREAT VOR UNITED; 1881—1885 NEW GREAT VOR;
            1906—1909 VOR; 1910—1913 VOR SUSPENDED MARCH 1910
Management: Manager 1859—1863 THOS.GILL; 1864—1867 T.JULYAN; 1869—1877
            STEP.HARRIS; 1881 H.COWLING
            Chief Agent 1859—1860 ROBT.RICHARDS; 1861 F.FRANCIS

                                  548

VOR,GREAT   BREAGE   Continued

STEP.HARRIS; 1862-1864 F.FRANCIS; 1865 WALT.HARRIS; 1866-1868
STEP.HARRIS; 1869 G.M.HENTY J.JAMES J.LANYON; 1870-1871
J.HARRIS J.JAMES J.LANYON; 1872 J.JAMES J.LANYON; 1882-1885
JOS.PRISK; 1906-1912 WM.THOMAS
Secretary 1859-1870 G.T.NOAKES (P); 1871-1872 J.J.TRURAN (S);
1873-1875 ARGALL(P) & TRURAN(S); 1876-1877 J.J.TRURAN (S);
1913 F.W.THOMAS (S)

Employment:	Underground	Surface	Total
1881	5		5
1882	4	15	19
1883	6		6
1884	6	2	8
1885	8	14	22
1906	16	49	65
1907	27	65	92
1908	47	56	103
1909	37	53	90
1910	13	36	49

VOR,GREAT EAST   SITHNEY   SW 635310 1483

Production: Tin No detailed return
Management: Manager 1881 H.COWLING

VOR,NEW   BREAGE   SW 633306 1484

Production: Tin No detailed return
Ownership: Comment 1888-1891 NEW VOR & WHEAL WALLIS CO.; 1859-1861
            INC.EAST METAL; 1862-1865 INC.EAST METAL SUSPENDED; 1888 NEW
            VOR & WHEAL WALLIS; 1889-1891 NEW VOR & WHEAL WALLIS.
            SUSPENDED
Management: Manager 1859-1863 JOS.VIVIAN
            Chief Agent 1859-1861 NICH.THOMAS; 1888-1891 E.R.RIDINGTON
            Secretary 1859-1861 G.T.OLDFIELD (P)

Employment:	Underground	Surface	Total
1888		7	7

VOR,NORTH   BREAGE   SW 627314 1485

Production: Tin No detailed return
Ownership: Comment 1863-1865 SUSPENDED
Management: Manager 1859-1861 JAS.WHITE
            Chief Agent 1859-1862 RICH.WHITE
            Secretary 1859-1865 J.N.R.MILLETT (P)

VOR,PENHALE WHEAL   BREAGE   SW 624309 1486

Production: Tin No detailed return
Management: Manager 1866-1867 WM.BENNETTS
            Chief Agent 1866-1867 W.H.MARTIN
            Secretary 1866-1867 WM.CHAPPEL

VOR,WEST   BREAGE   SW 614298 1487

Production: Tin   Black(tons) Stuff(tons)   Tin(tons)   Value(£)
            1878   8.00   0.00   0.00   1.50
Ownership: 1878-1881 WEST VOR UNITED MINING CO.LTD.; 1912 WENDRON
            CORNWALL TIN MINES LTD.; 1913 CORNISH PROPERTIES DEVELOPMENT
            CO.LTD.
            Comment 1912 PROSPECTING SEPT.1912; 1913 PROSPECTING
Management: Manager 1878-1880 STEP.HARRIS; 1881 WM.WILLIAMS
            Chief Agent 1865-1866 J.SOUTHEY
            Secretary 1865-1866 T.MILLS; 1878-1881 THOS.HUNTER (S)
Employment: Underground   Surface   Total
            1878   7   7
            1879   5   5
            1880   4   4
            1912   6   6   12
            1913   7   6   13
            Comment 1912 INC.PROVIDENCE,LOVELL & RAPSON

VYVYAN   CONSTANTINE   SW 734294 1488

Production: Copper   Ore(tons) Metal(tons)   Value(£)
            1845   378.00   22.20   1462.90
            1846   300.00   18.50   1108.10
            1847   234.00   13.10   813.80
            1848   92.00   8.30   461.90
            1849   237.00   15.10   887.70
            1850   242.00   16.00   1061.10
            1852   178.00   10.30   777.40
            1853   124.00   7.30   650.70
            1855   163.00   12.20   1240.50
            1856   147.00   10.10   915.60
            1862   35.00   2.00   144.70
            1863   62.00   4.10   270.80
            1864   63.00   3.70   297.10
            Comment 1845-1847 (C)VIVIAN; 1848-1850 (C); 1852-1853 (C);
            1855-1856 (C); 1862-1864 (C)
            Tin   Black(tons) Stuff(tons)   Tin(tons)   Value(£)
            1855   4.10   0.00   0.00   253.40
            1856   16.50   0.00   0.00   1130.40
            1857   14.00   0.00   0.00   990.70
            1858   6.60   0.00   0.00   423.90
            1859   11.40   0.00   0.00   636.10
            1862   5.60   0.00   0.00   345.40
            1863   26.50   0.00   0.00   1632.70
            1864   6.90   0.00   0.00   410.20
            Comment 1862 PART YEAR ONLY
Ownership: Comment 1865 SUSPENDED
Management: Manager 1859-1864 JAS.HAMPTON
            Chief Agent 1859-1860 JAS.HAMPTON; 1861-1864 J.NICHOLLS
            Secretary 1859-1861 T.WESTLAKE; 1862-1865 C.WESCOMBE

WALLIS   BREAGE   SW 632306 1480

Ownership: Comment 1888-1891 SEE NEW VOR

WARD   1489

Production: Tin   Black(tons) Stuff(tons)   Tin(tons)   Value(£)
            1875   0.60   0.00   0.00   16.50
            Comment 1875 OR NORTH WITHY BROOK

WELLINGTON MINES   ST.HILARY   SW 561295 1490

Production: Copper   Ore(tons) Metal(tons)   Value(£)
            1847   194.00   21.80   1564.60
            1848   289.00   39.70   2351.40
            1849   672.00   62.30   4253.60
            1850   1034.00   90.30   6177.30
            1851   430.00   30.00   1946.70
            Comment 1847-1851 (C)
            Tin No detailed return
Ownership: Comment 1860-1861 SUSPENDED; 1868-1871 WELLINGTON SEE GREAT
             WESTERN MINES
Management: Secretary 1859 R.R.MICHELL (P)

WENDRON CONSOLS   WENDRON   SW 689320 1491

Production: Copper   Ore(tons) Metal(tons)   Value(£)
            1865   24.00   5.10   415.20
            1866   24.00   3.40   278.00
            Comment 1865-1866 (C)
            Tin   Black(tons) Stuff(tons)   Tin(tons)   Value(£)
            1854   10.00   0.00   0.00   653.80
            1856   130.30   0.00   0.00   8575.50
            1857   269.90   0.00   0.00   20394.00
            1858   201.80   0.00   0.00   12875.30
            1859   233.90   0.00   0.00   18017.50
            1860   223.60   0.00   0.00   17969.80
            1861   255.20   0.00   0.00   18491.30
            1862   320.00   0.00   0.00   20913.50
            1863   206.60   0.00   0.00   13558.90
            1864   117.00   0.00   0.00   7094.50
            1865   63.20   0.00   0.00   3366.40
            1866   26.30   0.00   0.00   1424.20
            Comment 1862 TWO RETURNS AGGREGATED
Management: Manager 1859-1864 R.J.TAYLOR
            Chief Agent 1859-1860 RICH.JENKIN; 1861 ED.JENKIN; 1862-1864
            WM.JOHNS; 1865 WM.JOHNS JOHN WATERS
            Secretary 1859-1865 FRED.HILL (P)

WENDRON CONSOLS,NEW   WENDRON   SW 681317 1492

Management: Chief Agent 1862 W.GLUYAS; 1863-1865 R.GLUYAS
            Secretary 1862-1865 FRED.HILL

                                   551

WENDRON CONSOLS,WEST   WENDRON   1493

Production: Tin No detailed return
Management: Manager 1860-1862 RICH.KENDALL
            Chief Agent 1860 GEO.R.ODGERS; 1861-1862 W.HOSKING
            Secretary 1860-1862 T.E.WATSON (P)

WENDRON UNITED   WENDRON   SW 697342 1494

Production: Tin   Black(tons) Stuff(tons)   Tin(tons)   Value(£)
            1853   3.20   0.00   0.00   221.90
            1854   1.90   0.00   0.00   118.40
            1864   12.30   0.00   0.00   751.70
            1870   0.00   0.00   0.00   717.80
            1871   0.00   0.00   0.00   465.70
            1872   0.00   0.00   0.00   284.80
            1873   0.00   110.00   0.00   219.90
            1875   1.50   0.00   0.00   157.00
            1876   0.00   0.00   0.00   4.90
            Comment 1873 TIN STUFF EST; 1875 VALUE INC.TIN STUFF; 1876
            TIN STUFF
Ownership: Comment 1863-1865 SUSPENDED
Management: Manager 1859-1862 WM.PASCOE
            Chief Agent 1859-1860 JOHN BOLITHO; 1861-1862 R.CADY; 1875
            JOS.TREGONNING
            Secretary 1859-1862 JOHN CADY (P); 1863-1865 ROBT.H.PIKE &
            SON

WENDRON,SOUTH   WENDRON   SW 705301 1495

Production: Tin   Black(tons) Stuff(tons)   Tin(tons)   Value(£)
            1875   2.10   0.00   0.00   93.50
            1876   1.90   0.00   0.00   79.70
            1877   2.30   0.00   0.00   93.60
            1879   1.30   0.00   0.00   50.00
            1880   2.00   0.00   0.00   90.00
Ownership: 1876-1881 SOUTH WENDRON MINING CO.
Management: Manager 1875-1878 JOS.PRISK; 1879-1880 JOHN H.JAMES
            Chief Agent 1880-1881 SML.JAMES
            Secretary 1875 H.MIDDLETON; 1876-1878 JOS.PRISK
Employment: Underground   Surface   Total
            1878   4   1   5
            1879   12   8   20
            1880   3   2   5

WENTWORTH CONSOLS   PERRANZABULOE   SW 787514 1496

Production: Lead No detailed return
Ownership: Comment 1865 SUSPENDED
Management: Manager 1864 J.JULIFF
            Chief Agent 1864 J.HARRIS
            Secretary 1864-1865 RICH.CLOGG

WEST DOWN   LANIVET   SX 027636 1497

Production: Tin   Black(tons) Stuff(tons)   Tin(tons)   Value(£)
           1852   2.50   0.00   0.00   0.00
           Comment 1852 WEST DOWNS. TIN ORE
           Iron   Ore(tons)   Iron(%)   Value(£)
           1859   1092.20   0.00   273.00
           1860   703.00   0.00   281.20
           1861   No detailed return
           1873   No detailed return
           1874   103.50   0.00   51.70
           Comment 1859-1861 BH.& SOME SPATHOSE; 1873 SEE LANIVET; 1874
           BH.& SOME SPATHOSE
Ownership: Comment 1873-1874 SEE LANIVET

WEST DOWN END   EGLOSKERBY   SX 253855 1498

Production: Manganese   Ore(tons) Metal(tons)   Value(£)
           1875   No detailed return
           1877-1878 No detailed return
           1879   45.00   0.00   60.00
           1880   70.00   0.00   122.50
           1881   No detailed return
           Comment 1875 SEE LIDCOTT; 1877 SEE LIDCOTT; 1878 SEE LIDCOTT;
           1879 SEE ALSO LIDCOTT; 1881 WEST END DOWN. SEE LIDCOTT
Ownership: 1891 E.G.B.LETHBRIDGE
           Comment 1875-1882 WEST DOWN. SEE LIDCOTT; 1891 WEST DOWN
           ENDS. SUSPENDED

WEST OF ENGLAND   1503

Production: Copper   Ore(tons) Metal(tons)   Value(£)
           1875   22.00   1.50   132.00
           Comment 1875 (C)

WHEATLEY BLUE   1504

Production: Lead & Silver   Ore(tons)   Lead(tons) Silver(ozs)   Value(£)
           1852   126.00   90.00   1200.00   0.00
           1853   No detailed return

WHISPER   WARLEGGAN   SX 152696 1505

Production: Tin   Black(tons) Stuff(tons)   Tin(tons)   Value(£)
           1872   20.40   0.00   0.00   1401.90
           1873   4.30   0.00   0.00   296.70
           1885   12.50   0.00   0.00   562.00
           1886   11.00   0.00   0.00   350.00
           1887   No detailed return
           Comment 1885-1886 OPEN WORKS; 1887 SEE TREVEDDOE,GREAT
Ownership: 1885 ROBT.LARCHIN; 1886 WHEAL WHISPER TIN & COPPER CO.
           Comment 1872 INC.GREAT TREVEDDOE & CABILLA; 1873 STOPPED IN
           1873 & EARLY IN 1874; 1885-1886 OPEN WORK; 1887 SEE GREAT

TREVEDDOE; 1890–1891 SEE GREAT TREVEDDOE
Management: Manager 1872–1873 TIM.ROWSE
            Secretary 1872–1873 A.LARCHIN

WHITE ALICE   PENARTH   SW 697347 1506

Production: Tin No detailed return
Ownership: 1913 WHITE ALICE TIN MINE LTD.
             Comment 1913 PROSPECTING
Management: Manager 1913 A.S.MURDOCK
Employment: Underground   Surface   Total
             1913   24   5   29

WHITLEIGH   TAMERTON FOLIOTT, DEVON   SX 483598 1507

Production: Lead & Silver   Ore(tons) Lead(tons) Silver(ozs)   Value(£)
             1854   26.00   11.50   790.00   0.00

WILHELMINA   ALTARNUN   SX 209794 1508

Production: Tin   Black(tons) Stuff(tons)   Tin(tons)   Value(£)
             1885   10.00   0.00   0.00   450.00
             1886   40.30   0.00   0.00   2289.00
             1887   44.00   0.00   0.00   2689.00
             1888   No detailed return
             1889   3.70   0.00   0.00   174.00
             Comment 1888 NOT WORKING
Ownership: 1884–1885 J.BAILEY & OTHERS; 1886–1889 SUSAN BAILEY
             Comment 1887–1888 IN LIQUIDATION; 1889 NOW ABANDONED
Management: Chief Agent 1884 JOHN VERCOE; 1889 FRAN.SPRY
             Secretary 1885–1888 J.BAILEY
Employment: Underground   Surface   Total
             1884   6   2   8
             1885   24   22   46
             1886   33   45   78
             1889   30   14   44

WILLIAM   LUXULYAN   SX 038625 1510

Production: Tin   Black(tons) Stuff(tons)   Tin(tons)   Value(£)
             1863   0.70   0.00   0.00   46.70
             1864   2.80   0.00   0.00   200.10
             1865   5.10   0.00   0.00   326.40
             1866   2.60   0.00   0.00   156.90
Ownership: 1902 E.A.TREGILGAS; 1906–1907 TIN PROSPECTOR SYNDICATE LTD.
             Comment 1906–1907 JOHN HAYE OWNER
Employment: Underground   Surface   Total
             1902   4   4
             1906   6   4   10

WILLIAMS   LATCHLEY   SX 408738 1511

Production: Copper   Ore(tons) Metal(tons)   Value(£)
            1848   170.00   13.90   753.70
            Comment 1848 (C)

WILLIAMS ORE   1512

Production: Copper   Ore(tons) Metal(tons)   Value(£)
            1874   59.00   4.10   271.20
            Comment 1874 (C)3 RETURNS AGGREGATED

WITHIEL   WITHIEL   SX 006645 1513

Production: Iron   Ore(tons)   Iron(%)   Value(£)
            1872   850.00   0.00   510.00
            1873   100.50   0.00   75.00
            Comment 1872-1873 BH.
Ownership: 1872-1874 T.RICHARDSON & CO.
            Comment 1873 FORMERLY RETIRE
Management: Manager 1872-1874 WM.RICHARDS

WITHY BROOK,NORTH   1514

Production: Tin   Black(tons) Stuff(tons)   Tin(tons)   Value(£)
            1875   No detailed return
            Comment 1875 SEE WARD

WOOD MINE   ST.DAY   1515

Production: Copper   Ore(tons) Metal(tons)   Value(£)
            1867   5.00   0.20   12.40
            Comment 1867 (C)
            Tin   Black(tons) Stuff(tons)   Tin(tons)   Value(£)
            1855   0.90   0.00   0.00   42.00
            1856   0.60   0.00   0.00   43.10
            1868   1.10   0.00   0.00   56.50
            1903   0.00   89.00   0.00   20.00
            1904   0.00   109.00   0.00   46.00
            Comment 1903 INC.DAMSEL; 1904 INC.DAMSEL & BULLEN
Ownership: 1903-1904 T.H.LETCHER
            Comment 1903-1904 T.H.LETCHER LORDS AGENT
Employment: Underground   Surface   Total
            1903   1   1   2

WOOD MINE,EAST   1516

Production: Tin   Black(tons) Stuff(tons)   Tin(tons)   Value(£)
            1856   0.50   0.00   0.00   29.90

                                    555

WOODCLOSE   ST.MEWAN   SW 983503 1517

Production: Tin   Black(tons) Stuff(tons)   Tin(tons)   Value(£)
            1885   4.60   0.00   0.00   207.00
Ownership: 1882-1883 HANNAM & LYELL; 1885 HANNAM & LYELL
Management: Chief Agent 1882-1883 JOHN EDWARDS; 1885 JOHN EDWARDS
Employment: Underground   Surface   Total
            1884   43   32   75
            1885   20   18   38

WOODLEY   LANIVET   SX 027636 1518

Production: Iron   Ore(tons)   Iron(%)   Value(£)
            1861   70.70   0.00   17.70
            Comment 1861 HE.

WORK,EAST GREAT   BREAGE   SW 609316 1519

Production: Tin No detailed return
Management: Manager 1867 JOS.VIVIAN & SON
            Secretary 1867 C.THOMAS

WORK,GREAT   BREAGE   SW 596307 1520

Production: Lead   Ore(tons) Metal(tons)   Value(£)
            1861   0.50   0.30   0.00
            Copper   Ore(tons) Metal(tons)   Value(£)
            1847   46.00   6.70   427.80
            1848   143.00   23.40   1438.00
            1849   95.00   9.60   650.00
            1859   65.00   6.10   556.10
            1862   27.00   2.60   216.30
            1863   32.00   3.30   251.90
            1865   22.00   2.20   175.40
            1866   25.00   1.90   133.70
            1867   35.00   2.70   184.00
            1868   10.00   1.00   65.00
            Comment 1847-1849 (C); 1859 (C); 1862-1863 (C); 1865-1868
            (C)
            Tin   Black(tons) Stuff(tons)   Tin(tons)   Value(£)
            1853   116.00   0.00   0.00   8094.40
            1854   194.20   0.00   0.00   13866.00
            1855   156.60   0.00   0.00   11024.40
            1856   187.80   0.00   0.00   15229.10
            1857   194.90   0.00   0.00   16822.10
            1858   204.90   0.00   0.00   14475.30
            1859   211.70   0.00   0.00   17627.50
            1860   195.30   0.00   0.00   16831.00
            1861   190.80   0.00   0.00   14862.20
            1862   328.80   0.00   0.00   23886.80
            1863   303.20   0.00   0.00   22263.80
            1864   328.60   0.00   0.00   23024.60
            1865   284.30   0.00   0.00   17467.70
            1866   359.10   0.00   0.00   19255.90

Tin	Black(tons)	Stuff(tons)	Tin(tons)	Value(£)
1867	302.40	0.00	0.00	17658.60
1868	249.30	0.00	0.00	15087.50
1869	240.40	0.00	0.00	18444.50
1870	289.90	0.00	0.00	23339.70
1871	241.90	0.00	0.00	20883.50
1872	205.30	0.00	0.00	19174.50
1873	168.50	0.00	0.00	13905.80
1874	56.00	0.00	0.00	3389.40
1875	73.50	0.00	0.00	3722.50
1876	80.30	0.00	0.00	3568.10
1877	55.00	0.00	0.00	2311.70
1878	63.60	0.00	0.00	2309.30
1879	1.50	0.00	0.00	60.00
1881	44.00	0.00	29.00	2693.00
1882	51.00	0.00	36.50	3263.00
1883	102.70	0.00	0.00	6108.00
1884	109.50	0.00	0.00	5643.00
1885	44.20	0.00	0.00	2351.00
1888	7.60	0.00	0.00	482.00
1889	48.40	0.00	0.00	2830.00
1890	66.80	0.00	0.00	3839.00
1891	50.70	0.00	0.00	2800.00
1892	31.20	0.00	0.00	1807.00
1893	51.60	0.00	0.00	2702.00
1894	14.10	0.00	0.00	606.00
1895-1896 No detailed return				
1897	44.60	0.00	0.00	1909.00
1898	50.00	0.00	0.00	2433.00
1899	42.40	0.00	0.00	3393.00
1900	36.60	0.00	0.00	3397.00
1901	24.50	0.00	0.00	1848.00
1904	6.80	0.00	0.00	554.00
1905	0.00	53.00	0.00	438.00
1912	5.70	0.00	0.00	489.00

Comment 1862 TWO RETURNS AGGREGATED; 1868 VALUE INC.TIN
STUFF; 1873 GREAT WORK CONSOLS; 1874 GREAT WORK CONSOLS VALUE
EST; 1888-1889 GREAT WORK CONSOLIDATED; 1890-1891 GREAT WORK
CONSOLS; 1892-1896 OLD GREAT WORK; 1904-1905 GREAT WORK ETC

Arsenic	Ore(tons)	Metal(tons)	Value(£)
1888	2.00	0.00	5.00

Comment 1888 GREAT WORK CONSOLS
Ownership: 1876-1881 GREAT WORK MINING CO.; 1882-1885 WM.TEAGUE; 1888
GREAT WORK CONSOLIDATED MINES CO.; 1889-1890 J.ATTWOOD; 1891
GREAT WORK MINE SYNDICATE LTD.; 1892-1894 OLD GREAT WORK MINE
SYNDICATE LTD.; 1897-1898 GREAT WORK SYNDICATE; 1899 GREAT
WORK SYNDICATE LTD.; 1900-1901 GREAT WORK TIN MINING
DEVEL.SYND.LTD.; 1904-1905 ROSEWARNE & CO.; 1906 SOUTH WEST
CORNWALL MINES LTD.
Comment 1869-1872 GREAT WORK CONSOLS; 1873 GREAT WORK CONSOLS
STOPPED IN 1873 & 1874; 1874-1877 GREAT WORK CONSOLS;
1888-1890 GREAT WORK CONSOLS; 1892-1893 OLD GREAT WORK; 1894
OLD GREAT WORK SUSPENDED; 1901 ABANDONED AUG.1901; 1904-1905
INC.GODOLPHIN STAMPS ONLY; 1907-1912 SEE GODOLPHIN
Management: Manager 1859-1868 N.TREDINNICK; 1869-1871 S.TREDINNICK; 1872

WORK,GREAT   BREAGE   Continued

CHAS.CRAZE; 1873–1874 P.E.WILLIAMS; 1875 G.WILLIAMS;
1876–1881 WM.TEAGUE
Chief Agent 1859–1861 J.JOHNS; 1862–1871 J.JOHNS
THOS.EDWARDS; 1872 P.WILLIAMS; 1874 ED.MOYLE; 1876–1881
G.WILLIAMS; 1882–1885 WM.TEAGUE; 1891–1894 JOS.PRISK;
1897–1901 JOS.PRISK
Secretary 1859–1862 JAS.CLARKE (P); 1863–1871 COMMITTEE;
1872–1873 WM.TYACKE; 1874–1879 WM.TEAGUE; 1888 GEO.H.EUSTICE;
1889–1890 J.ATTWOOD; 1891–1894 R.B.FASTNEDGE (S)

Employment:	Underground	Surface	Total
1878	7	9	16
1879	7	2	9
1880	32	17	49
1881	43	65	108
1882	50	56	106
1883	66	57	123
1884	50	61	111
1885	13	53	66
1888		44	44
1889	6	63	69
1890	46	32	78
1891	12	6	18
1892	21	44	65
1893	44	45	89
1894	28	15	43
1897	55	21	76
1898	48	24	72
1899	52	31	83
1900	56	30	86
1901	57	2	59
1904		9	9
1905		7	7
1907–1912			

Comment 1904–1905 INC.GODOLPHIN.STAMPS ONLY; 1907–1912 SEE
GODOLPHIN

WORK,NEW GREAT   GERMOE   SW 588302 1521

Production: Tin	Black(tons)	Stuff(tons)	Tin(tons)	Value(£)
1892	0.00	94.00	0.00	111.00
1893	9.60	14.00	0.00	517.00
1894	12.80	0.00	0.00	502.00
1895	4.00	0.00	0.00	135.00
1896	No detailed return			

Ownership: 1890 NEW GREAT WORK TIN MINE CO.; 1891–1893 NEW GREAT WORK
TIN MINE CO.LTD.; 1894–1895 NEW GREAT WORK TIN MINE CO.
Comment 1894–1895 ABOVE ADIT ONLY; 1906–1910 SEE LADY
GWENDOLIN
Management: Chief Agent 1890–1891 SIMON THOMAS; 1892–1895 WM.STEPHENS
Secretary 1891–1895 C.V.THOMAS (S)

Employment:	Underground	Surface	Total
1890	12	8	20
1891	29	56	85
1892	8	3	11

WORK,NEW GREAT   GERMOE   Continued

                    Underground   Surface   Total
            1893   24   17   41
            1894   8   7   15

WORK,NORTH GREAT   BREAGE   SW 582317 1522

Production: Copper   Ore(tons) Metal(tons)   Value(£)
            1862   4.00   0.20   14.30
            Comment 1862 (C)
            Tin No detailed return
Management: Manager 1859-1861 JOS.VIVIAN; 1862-1864 JOHN POPE; 1865
            JOS.VIVIAN
            Chief Agent 1860-1861 JOHN POPE
            Secretary 1859 JOS.VIVIAN (P); 1860-1865 JOS.VIVIAN JNR (P)

WORK,SOUTH GREAT   ST.HILARY   SW 562301 1523

Production: Tin   Black(tons) Stuff(tons)   Tin(tons)   Value(£)
            1872   4.10   0.00   0.00   323.40
            1873   8.80   0.00   0.00   604.50
            1874   38.50   0.00   0.00   1493.00
            1875   37.00   0.00   0.00   1699.50
            1876   8.80   0.00   0.00   361.70
Management: Manager 1873-1876 SML.J.REED
            Chief Agent 1869-1871 E.CHEGWIN; 1872 SML.J.REED
            Secretary 1869-1871 WM.WATSON; 1872-1876 GREN.SHARP

WORK,WEST GREAT   BREAGE   SW 577312 1524

Production: Copper   Ore(tons) Metal(tons)   Value(£)
            1867   20.00   1.70   114.00
            1868   67.00   4.60   272.50
            1869   34.00   2.70   159.60
            1870   40.00   4.30   270.30
            1872   6.00   0.60   48.60
            Comment 1867-1870 (C); 1872 (C)
            Tin   Black(tons) Stuff(tons)   Tin(tons)   Value(£)
            1862   9.20   0.00   0.00   577.20
            1863   25.40   0.00   0.00   1595.20
            1864   7.60   0.00   0.00   492.20
            1865   43.80   0.00   0.00   2563.20
            1866   41.90   0.00   0.00   2105.00
            1867   89.00   0.00   0.00   4885.30
            1868   25.90   0.00   0.00   1410.90
            1869   91.50   0.00   0.00   6633.80
            1870   41.80   0.00   0.00   3031.80
            1871   36.90   0.00   0.00   2912.70
            1872   28.90   0.00   0.00   2545.10
            1873   21.10   0.00   0.00   1656.60
            1874   14.20   0.00   0.00   739.00
            1875   6.20   0.00   0.00   309.80
            1876   5.60   0.00   0.00   236.50

                                    559

WORK,WEST GREAT   BREAGE   Continued

Tin	Black(tons)	Stuff(tons)	Tin(tons)	Value(£)
1905	8.60	0.00	0.00	661.00
1907	27.00	0.00	0.00	3165.00
1911	0.00	1.60	0.00	10.00
1912	0.00	5.10	0.00	10.00

Comment 1862 TWO RETURNS AGG.
Management: Manager 1860–1861 CHAS.CARKEEK; 1862–1876 SML.J.REED
Chief Agent 1860–1861 JOS.REED SML.REED; 1865–1866
THOS.CORFIELD; 1867–1868 STEP.CHEGWIN; 1872–1876 E.CHEGWIN
Secretary 1860–1861 WM.PAINTER (P); 1862–1866 ALM.E.PAULL
(P); 1867–1876 WM.WATSON

WORTHY,GREAT   CROWAN   SW 641323 1525

Production: Tin No detailed return
Ownership: 1882–1885 GREAT WHEAL WORTHY MINING CO.
Management: Chief Agent 1882–1885 JOS.PRISK
Employment: Underground   Surface   Total
| 1882 | 3 | | 3 |
| 1883 | 4 | | 4 |

WORVAS DOWNS   LELANT   SW 511383 1526

Production: Tin No detailed return
Ownership: 1904 CORN.MILLETT; 1905–1909 WORVAS DOWNS MINING CO.;
1911–1913 JOHN PENBERTHY
Comment 1864–1865 SUSPENDED; 1905–1908 INC.BALNOON CONSOLS;
1909 INC.BALNOON CONSOLS ABAND.JAN.1910; 1913 CLOSED DOWN
DURING 1913
Management: Chief Agent 1860 THOS.RICHARDS JAS.HOLLOW; 1861–1863
RICH.HARRY; 1906–1909 CORN.MILLETT; 1911–1913 I.PENBERTHY
Secretary 1860 RICH.HARRY (P); 1861–1863 JAS.HOLLOW (P)
Employment: Underground   Surface   Total

1904	7	10	17
1905	13	16	29
1906	35	30	65
1907	43	27	70
1908	2	3	5
1909	3	2	5
1911	5		5
1912–1913	2		2

Comment 1905–1909 INC.BALNOON CONSOLS

WREY CONSOLS   ST.IVE   SX 297665 1527

Production: Lead & Silver

	Ore(tons)	Lead(tons)	Silver(ozs)	Value(£)
1853	8.00	6.00	120.00	0.00
1854	214.00	167.00	4175.00	0.00
1855	894.90	602.00	15050.00	0.00
1856	1064.00	760.00	23560.00	0.00
1857	896.00	583.00	17490.00	0.00
1858	658.00	459.00	13700.00	0.00

WREY CONSOLS   ST.IVE   Continued

        Lead & Silver   Ore(tons) Lead(tons) Silver(ozs)   Value(£)
        1859   613.40   427.70   12832.00   0.00
        1860   650.50   448.50   13455.00   0.00
        1861   394.00   263.00   5920.00   0.00
        1862   43.00   28.10   616.00   0.00
        1863-1866 No detailed return
        Comment 1863-1866 WREY SEE LUDCOTT
Ownership: Comment 1859 WREY; 1862-1865 WREY SEE LUDCOTT
Management: Manager 1859-1861 PETER CLYMO
             Chief Agent 1859 WM.HANCOCK; 1860-1861 WM.HANCOCK MARK
             WHITFORD
             Secretary 1859-1861 PETER CLYMO (P)

WREY,NORTH   ST.IVE   SX 302679 1528

Production: Lead & Silver   Ore(tons) Lead(tons) Silver(ozs)   Value(£)
            1858   10.00   3.00   72.00   0.00
            1859-1861 No detailed return
Ownership: Comment 1864-1865 SUSPENDED
Management: Chief Agent 1860-1863 THOS.KEMP
             Secretary 1861-1863 J.B.BELCOMBE (P)

YANKEE BOY   ST.JUST   SW 322325 1529

Production: Tin   Black(tons) Stuff(tons) Tin(tons)   Value(£)
            1896   0.00   40.00   0.00   25.00
Management: Manager 1896 THOS.TREGEAR
Employment: Underground   Surface   Total
            1896   4   2   6

ZION   CALSTOCK   SX 431696 1530

Production: Copper   Ore(tons) Metal(tons)   Value(£)
            1855   271.00   14.40   1374.30
            1856   499.00   29.00   2638.10
            1857   273.00   12.70   1221.60
            Comment 1855-1857 (C)

ZZ SUNDRIES   1531

Production: Lead & Silver   Ore(tons) Lead(tons) Silver(ozs)   Value(£)
            1850   13.00   10.00   0.00   0.00
            1884   0.00   0.00   5000.00   0.00
            1896   7.00   5.00   0.00   50.00
            1905   248.00   134.00   7936.00   2074.00
            1906   200.00   112.00   7000.00   2337.00
            1907   190.00   106.00   7220.00   2405.00
            1908   87.00   47.00   3219.00   674.00
            Comment 1850 MINES UNDER 10 TONS; 1884 AG OUTPUT OF CORNISH
            MINES; 1896 FROM QUARRIES. QUARRIES ACT; 1905-1908 FROM
            QUARRIES

                              561

Barytes	Ore(tons)	Value(£)
1876	688.50	0.00
1881	5.00	2.50

Comment 1876 DEVON AND CORNWALL SUNDRIES

Zinc	Ore(tons)	Metal(tons)	Value(£)
1860	52.00	0.00	145.60
1868	60.00	0.00	210.00
1870	33.00	0.00	113.90
1871	22.00	0.00	82.50
1874	75.00	0.00	160.00
1877	61.00	0.00	211.00
1879	39.00	0.00	136.50
1880	330.00	0.00	1095.30

Comment 1860 PADSTOW SHIPPED AT; 1870 TWO RETURNS AGGREGATED;
1880 TWO RETURNS AGGREGATED

Copper	Ore(tons)	Metal(tons)	Value(£)
1845	706.00	48.90	3391.60
1846	1021.00	66.10	4181.10
1847	1420.00	96.80	6432.60
1848	1276.00	92.40	5128.20
1849	1310.00	88.10	5559.80
1850	1386.00	93.60	6228.80
1851	1772.00	112.50	7124.70
1852	2487.00	141.50	11258.90
1853	2831.00	145.50	13471.80
1854	3225.00	154.70	14407.00
1883	9.00	0.40	22.00

Comment 1845-1854 (C)
Manganese No detailed return

Tin	Black(tons)	Stuff(tons)	Tin(tons)	Value(£)
1874	0.00	150.00	0.00	125.00
1883	16.60	0.00	0.00	913.00
1884	34.00	0.00	0.00	1526.00

Comment 1874 WASTE HEAPS OF ABANDONED MINES

Iron	Ore(tons)	Iron(%)	Value(£)
1866	1700.00	0.00	510.00
1870	1000.00	0.00	250.00
1871	1250.00	0.00	937.00

Comment 1866 SHIPPED FROM PADSTOW; 1870-1871 SUNDRY SMALLER
QUANTITIES

Arsenic	Ore(tons)	Metal(tons)	Value(£)
1878	147.80	0.00	291.00
1899	284.00	0.00	2840.00

Comment 1878 TIN STREAM BURNING HOUSES; 1899 EST. SUNDRY
SMALL LOTS OF MIXED ORES.